D1755504

GARLIC

The Science and Therapeutic Application of *Allium sativum* L. and Related Species

Second Edition

GARLIC

The Science and Therapeutic Application of *Allium sativum* L. and Related Species

Second Edition

Edited by

HEINRICH P. KOCH, PH.D., M.PHARM.
Professor of Pharmaceutical Chemistry and Biopharmaceutics
University of Vienna
Vienna, Austria

LARRY D. LAWSON, PH.D.
Research Scientist
Nature's Way Products, Inc.
Murdock Madaus Schwabe Group
Springville, Utah

Williams & Wilkins
A WAVERLY COMPANY

BALTIMORE • PHILADELPHIA • LONDON • PARIS • BANGKOK
BUENOS AIRES • HONG KONG • MUNICH • SYDNEY • TOKYO • WROCLAW

Editor: David C. Retford
Managing Editors: Kathleen Courtney Millet and Leah Hayes
Production Coordinator: Linda Carlson
Copy Editor: John G. Guardiano
Designer: Wilma Rosenberger
Illustration Planner: Lorraine Wrzosek
Composition: Mario Fernández
Printer: Victor Graphics

Copyright © 1996
Williams & WIlkins
351 West Camden Street
Baltimore, Maryland 21201-2436 USA

Rose Tree Corporate Center
1400 North Providence Road
Building II, Suite 5025
Media, Pennsylvania 19063-2043 USA

All rights reserved. This book is protected by copyright. No part of this book may be reproduced in any form or by any means, including photocopying, or utilized by any information storage and retrieval system without written permission from the copyright owner.

Accurate indications, adverse reactions, and dosage schedules for drugs are provided in this book, but it is possible that they may change. The reader is urged to review the package information data of the manufacturers of the medications mentioned.

Printed in the United States of America

First Edition in German, © 1988, Urban & Schwarzenberg

This is a translation of HP Koch, G Hahn: *Knoblauch*, Copyright Urban & Schwarzenberg München-Wien-Baltimore 1988.

Library of Congress Cataloging in Publication Data

Koch, Heinrich P.
   [Knoblauch. English]
   Garlic: the science and therapeutic application of Allium sativum L. and related species / Heinrich P. Koch, Larry D. Lawson.—2nd ed.
     p. cm.
   Rev. and translated ed. of: Knoblauch / Heinrich P. Koch. München; Baltimore: Urban & Schwarzenberg, © 1988.
   Includes bibliographical references and index.
   ISBN 0-683-18147-5
   1. Garlic—Therapeutic use. I. Lawson, Larry D. II. Title.
   [DNLM: 1. Garlic. 2. Medicine, Herbal. QV 766 K76K 1996a]
RM666.G15K6313 1996
615'.324324—dc20
DNLM/DLC
for Library of Congress   95-25682
                                                                                                                CIP

The Publishers have made every effort to trace the copyright holders for borrowed material. If they have inadvertently overlooked any, they will be pleased to make the necessary arrangements at the first opportunity.

                                                                96 97 98 99 00
                                                 1 2 3 4 5 6 7 8 9 10

Reprints of chapter(s) may be purchased from Williams & Wilkins in quantities of 100 or more. Call Isabella Wise, Special Sales Department, (800) 358-3583.

Foreword

Garlic, the "spice of life," is unique among members of the plant kingdom. Its use as a spice is well documented in ancient Sumerian stone recipe tablets, in quotations from the Bible, the Talmud, and the writings of the Prophet Mohammed. For example, in promoting the consumption of garlic, the 3rd-century Babylonian Talmud observes that garlic "satiates and warms [the body] and makes the face shine and increases semen and kills 'lice' [i.e., parasites] in the intestines, and some say it instills love and eliminates jealousy." Garlic's usage in medicinal remedies is recorded in the earliest known medical writings, such as *Papyrus Ebers* scrolls. Garlic is hardly bashful, announcing its presence with its characteristic "garlicky" aroma—an adjective widely used in the field of olfactory sciences for similar-smelling compounds. Those ingesting garlic-laced dishes are endowed with a distinctive breath, recognized, for example, by Shakespeare ("eat no onions nor garlic, for we are to utter sweet breath"), and recently chemically characterized. Today, the culinary virtues of garlic are enthusiastically celebrated each year in garlic festivals around the world.

For those who believe that for all things in Nature there is a reason, explaining the presence of the strong-smelling compounds in garlic, which undoubtedly cost considerable biochemical energy to create, presents a challenge. The odoriferous garlic compounds contain the elements sulfur and selenium, the former widely used by nature where a smell serves a defensive or attractive/communicative purpose, as in the case of the skunk or the urine of certain mammals. Perhaps garlic's sulfur and selenium compounds, being repugnant to some predators such as garlic-devouring worms or other denizens of the soil where garlic fights for survival, gives the plant an evolutionary edge, which is then passed from generation to generation. The sulfur and selenium aroma compounds are fragile, requiring Nature ingeniously to package all of the ingredients necessary for their creation in the garlic clove in stable form, necessitating only cutting or crushing for release on demand. On consuming garlic, do we also gain a health-protecting edge? From epidemiological studies involving the consumption of fresh garlic, the answer is a resounding YES! Clinical studies on health benefits from the use of various garlic extracts or commercial products are encouraging but, because of the many variables involved, less definitive.

Where can we turn for detailed information on our favorite spice, garlic? Until recently it was necessary to search out in a technical library review articles in various magazines or journals, or tersely written chapters in scientific monographs. Fortunately for all of us, the authors have now brought together for the first time in a book, a lucid, comprehensive history of the botany, pharmacology, and chemistry of garlic in a form clearly understandable as well as entertaining to readers of various backgrounds, along with extensive references to the original scientific and historical literature. Those who appreciate garlic and want to learn more about its colorful history and the science that defines its culinary virtues and medicinal activity, as well as those who professionally research the "stinking rose," owe the authors, all well known for their individual contributions to the growing area of garlic research, a hearty round of thanks for this concise volume. Enjoy!

—*Eric Block, Ph.D.*
Professor of Chemistry
State University of New York at Albany
Albany, New York

Preface *to the Second Edition*

The extraordinarily great success of the first edition of this book, which was published in German seven years ago, prompted us to prepare an English translation in order to make this text also accessible to those people who are not familiar with the German language. During the translation work, however, we became aware of the need for an updating of the scientific content of the manuscript, since the field of garlic research has expanded almost exponentially in the course of the past few years (Koch, 1992). This work soon outgrew the physical capacity of the senior author, and so he had to look for coauthors who were able and willing to take over part of this tremendous work.

Fortunately, three experts, all of them top scientists and long involved in garlic research, could be found for this job. The coauthor of the first edition, Dr. Hahn of Cologne, of course, spontaneously agreed to revise his own two introductory chapters dealing with the historical and botanical aspects of the garlic plant. Dr. Larry Lawson of Utah is now responsible for the completely rewritten third chapter on the composition and chemistry of garlic and garlic products; Dr. Siegers of Lübeck for the part on methods of analysis of commercial garlic preparations (Chapter 4); and Dr. Reuter of Cologne for the therapeutic effects (Chapter 5). Due to the temporary disablement of Professor Siegers, much of his chapter was written by Dr. Pentz from Siegers' research group. The revision and updating of the last three chapters (6–8) was done by the senior author himself, who also added some smaller parts to the preceding chapters and is responsible for the overall coordination of the whole text. Well-appreciated also are the many suggestions and editing help of Dr. Lawson in these revisions as well as the elimination of "germanicisms" from the updated text written by the German authors.

Garlic research is a highly active special discipline of phytomedicine, the science of the medicinally used plants. During the past eight years (since the first edition appeared) more innovative development was achieved in this field than in the decades before. We have attempted to cover both the newer and the older literature as completely as possible, with an editorial deadline of July 1995. When talking about garlic one must necessarily also "look over the fence" a little bit. Thus we have, here and there, included many cross-references to other *Allium* species, mainly onion, although our references for these plants are far from complete.

The reference list now contains 2580 entries (2240 on garlic), as compared to 1275 in the first edition. In compiling the reference list, it was supposed that the most frequently spoken "World Languages" (English, French, German, Spanish, Italian) are common to the majority of scientists, and so the titles of papers published in these languages are listed in the original, while English translations are presented for titles in all other languages, with the original language noted in parentheses. To assist the reader in obtaining access to the references, the respective abstract citations (Chemical Abstracts, Chemisches Zentralblatt, International Pharmaceutical Abstracts) have been added wherever possible.

A special category is the patent literature, where innovations are found earlier than in full papers; therefore, patents have been cited as far as possible from retrieval in Chemical Abstracts and, in part, from Chemisches Zentralblatt. A computer search in Chemical Abstracts on patents with the keywords "garlic" and compounds from it like "allicin," "diallyl disulfide," and the like revealed 274 citations for the years 1950 through May 1994, of which 53% (65% since 1990) are Japanese patents. However, we selected for the reference list only those patents that deal either with processing or with therapeutical and/or pharmaceutical applications of garlic.

The following is a summary of the most important changes since the first edition. In the historical and botanical sections (Chapters 1 and 2) only minor updating was necessary. The most remarkable novelty might be the successful interspecific hybridization between *A. sativum* and *A. cepa*, between *A. cepa* and *A. schoenoprasum*, and between *A. cepa* and *A. fistulosum,* and the creation of transgenic garlic plants, the practical consequences of which are not fully conceivable as yet.

Much progress has been made in the quantitation of garlic-derived organosulfur compounds, including the discovery of several new ones (Chapter 3). These advances were the result of the application of modern high performance liquid chromatography (HPLC) methodology to garlic analysis. Several new thiosulfinates were discovered, all containing the *trans*-1-propenyl group. The second most abundant sulfur compound in garlic, γ-glutamyl-*S*-*trans*-1-propenylcysteine, was discovered only recently, even though this type of compound had been sought in the early 1960s. An old error was revealed by recent demonstrations that garlic, unlike onion, contains no *S*-propyl compounds. HPLC has made possible the rapid analysis of alliin and the γ-glutamylcysteines, and it has permitted the analysis of allicin, other thiosulfinates, ajoenes, and allyl pentato octasulfides, compounds which had previously eluded analysis by gas chromatography (GC). Analytical methods for these compounds have allowed the study of variations that occur between garlic plants, with respect to strain type, soil conditions, climate, country of origin, and so on, as well as variations that occur as a result of garlic processing. Lastly, and most recently, it was discovered that the necessary separation of alliin from alliinase in garlic cloves is achieved by the location of these compounds in different cells rather than in different compartments of all cells, as is the case with onions.

Chapter 4 has been completely reorganized, especially with regard to the analytical methods used for determining the main sulfur compounds of garlic and garlic preparations. Since the first edition of this book, new instrumental methodology has been applied toward investigating garlic's constituents to a far-reaching extent. Therefore, an update of the methodological approaches was deemed appropriate. Today, a large number of interesting compounds, supposed to be active and beneficial for human health, can be specifically determined with modern procedures, and quality control of health products made from garlic is based on these methods.

Concerning the clinical applications of garlic in recent years, special interest has been given to its antiarteriosclerotic effects. Thus, while all sections of Chapter 5 were updated, the section dealing with specific influences on the heart and circulatory system and serum lipid–lowering was particularly expanded, and clinical results were summarized in comprehensive tables. Special attention was also given to the anticancer effects of garlic and its constituents, as well as to their effects on free radicals (antioxidant effects). Finally, a new section on processed, deodorized garlic extracts has been included.

Remarkable progress has been made in the area of biopharmaceutics in the broadest sense. A first breakthrough occurred with the pharmacokinetics of the most important organosulfur constituents of the *Allium* species. The results show that alliin, allicin, the sulfides, vinyldithiins, ajoenes, and *S*-allylcysteine are quickly absorbed from the gastrointestinal tract, followed by rapid and extensive metabolism in vivo, so that practically none of the original compounds survive intact in the body. The influence of garlic or its constituents on the drug metabolizing enzyme system (cytochrome P-450, etc.) is dose-dependent and extends from slight potentiation to strong depression of the respective enzyme activities.

In toxicology, as a whole, no fundamental new findings were found during the past few years, although some spectacular cases of incidental intoxication due to unskilled application of garlic chopped in oil were reported. An additional subsection has been added dealing with possible ecological hazards due to environmental contamination of garlic crops, possibly causing potential danger to human and/or animal consumers.

Last, but not least, it is a pleasure for us to express here our gratitude to all coworkers and secretaries, to the translator of the first edition, Dr. Sigrid Klein of Utah, and to all of the others who gave us their invaluable assistance in the retrieval of the vast literature, as well as in the finalization of the revised edition of "our book." We thank, too, the sponsors who actually made the appearance of this edition possible, and, finally, the publisher.

Heinrich P. Koch
Larry D. Lawson
July 1995

Preface *to the First Edition*

"We remember the fish, which we did eat in Egypt freely; the cucumbers, and the melons, and the leeks, and the onions, and the garlick: But now our soul is dried away: there is nothing at all, beside this manna, before our eyes."
—Numbers 11:5–6

Onions, leek, and garlic are without doubt among the oldest crops of mankind which have been used as medicinal plants, spices, and food. It is still unknown when man encountered garlic as a wild plant for the first time and where this doubtlessly interesting event occurred. It certainly was not the plant which has been cultivated since ancient times and which is now available in so many hybrid variations. We do not know the original form of garlic, which is botanically classified in the genus *Allium,* which comprises more than 600 species. But it did exist and it motivated populations of ancient cultures to use this plant and its bulb to improve the taste of their hunted prey and gathered fruits. Then, when people established settlements, they cultivated their most treasured seasoning plants close to their dwellings. In the meantime, they had discovered how to use garlic as a remedy for various ailments. Thus, the cultivated species which were grown in southern Europe and in the countries of Asia were developed through thousands of years, and are still included in many gardens.

The attitude among people toward garlic, however, was often divided. It certainly evolved during the development of urban settlements. Not everyone appreciates the "strongly aromatic bulb" and the dishes prepared with it. This rejection is not unique to our day; even Luidprant, the Bishop of Cremona, who served as the ambassador of Otto The Great at the court of Byzantium, took offense at the garlic odor of the Byzantine emperor. Often we are affected in the same way. If we have consumed garlic, we do not perceive that our fellow man has also "enjoyed" this food. However, if he has eaten it, we turn up our noses. Large parts of populations of southern European countries and Asia are proponents of garlic, whereas in central and northern Europe garlic is more or less shunned. Perhaps the reason for this is twofold; not only does the original species emerge from the Southeast, but its medicinal effectiveness is better known and appreciated there than in the North. The inhabitants of these areas with a warmer climate may have been dependent upon such a disinfectant spice. In the meantime, many people, even in the northern regions, have recognized the value of garlic. In studying the literature published during the last decades, it can be seen that intensive studies have not been limited to chemists of natural substances. Physicians also experimented with garlic and what was considered at the beginning of the century to be controversial medicine is today accepted by many practitioners. We now know that garlic is not only a spice but also a medicine of high value.

Medical research on garlic as a remedy for high blood pressure and arteriosclerosis has increased for several years. Therefore, it may be useful to summarize the currently known facts about garlic and its diverse applications in a monograph. The number of "senior citizens," the older people of our society, is steadily increasing, putting more and more emphasis on diseases of old age. In addition, the population has become more and more health conscious, a very gratifying aspect. The trend toward self-medication, especially with plant medicines, cannot be overlooked. Many of the remedies have a long tradition in folk medicine. Yet, here a deficit of information exists, not only with the customer, who becomes confused by the media, but also the pharmacist and the practicing physician who wish to advise their patients appropriately. Therefore, encouraged by numerous inquiries by the general population, the pharmaceutical industry, and decision makers in medical professions and health administration, we have attempted to evaluate the information accumulated over the years and to make it available to interested parties in a scientific summary, yet in a readable form and perhaps even entertaining at times.

At this point, we thank the publishing house and all promoters of this project for their cooperation and support. It was not easily accomplished, but we only realized this as the work progressed. The choice and arrangement of subjects is subjective and may not totally satisfy every reader. Deficiencies, however, can be remedied; therefore, we gratefully accept all critical comments and additional suggestions. They will possibly be considered in a new edition. We also thank our coworkers: Miss Iris Biedermann of Vienna and Mrs. Anneliese Mayer of Cologne, who have been of great assistance in the evaluation of the botanical literature. All those mentioned have contributed to the expeditious completion of this book through their selfless and untiring collaboration in collecting and evaluating the extensive specialty literature.

—Heinrich P. Koch and Gottfried Hahn
Vienna and Cologne, March 1988

Contributors

GOTTFRIED HAHN, PH.D.
Botanist, Biologist and Historian
Overath, Germany

HEINRICH P. KOCH, PH.D., M.PHARM.
Professor of Pharmaceutical Chemistry and Biopharmaceutics
University of Vienna
Vienna, Austria

LARRY D, LAWSON, PH.D.
Research Scientist
Nature's Way Products, Inc.
Murdock Madaus Schwabe Group
Springville, Utah

REINHARD PENTZ, DR.RER.NAT
Senior Lecturer
University of Lübeck
Lübeck, Germany

HANS D. REUTER, PH.D.
Professor of Phytomedicine
University of Cologne
Cologne, Germany

CLAUS-PETER SIEGERS, M.D.
Professor of Toxicology
University of Lübeck
Lübeck, Germany

Acknowledgment

The authors wish to thank the organizations listed below for their support of this publication; however, to assure the absence of influence on the contents of the book, no portion of it was reviewed by the sponsors (other than the person of the second editor) prior to publication.

Murdock Madaus Schwabe/Nature's Way Products, Inc.
10 Mountain Springs Parkway
Springville, Utah 84663

Lichtwer Pharma GmbH.
Berlin, D13435
POB 260326
Germany

Pure-Gar Inc.
10029 South Tacoma Way
Tacoma, Washington 98498

Global Marketing Assoc.
3450 3rd Street, Suite 3A
San Francisco, California 94124

American Botanical Council
P.O. Box 201660
Austin, Texas 78720

Contents

Foreword by Eric Block v
Preface to the Second Edition vii
Preface to the First Edition ix

1 History, Folk Medicine, and Legendary Uses of Garlic 1
GOTTFRIED HAHN

- 1.1 DERIVATION OF THE NAME 1
- 1.2 APPLICATION BY ANCIENT POPULATIONS 2
- 1.3 THE IMPORTANCE OF GARLIC 11
- 1.4 GARLIC IN FOLK MEDICINE AND AS A LEGENDARY AND MAGICAL PLANT 19

2 Botanical Characterization and Cultivation of Garlic 25
GOTTFRIED HAHN

- 2.1 A BRIEF OVERVIEW OF THE GENUS *ALLIUM* 25
- 2.2 DESCRIPTION OF *ALLIUM SATIVUM* 28
- 2.3 CULTIVATION OF GARLIC IN GARDENS AND AS AN AGRICULTURAL CROP 34

3 The Composition and Chemistry of Garlic Cloves and Processed Garlic 37
LARRY D. LAWSON

- 3.1 THE GENERAL COMPOSITION OF GARLIC 37
- 3.2 THE SULFUR COMPOUNDS OF GARLIC 38
- 3.2.1 Discovery of the Sulfur Compounds of Garlic: Allyl Sulfides, the First Clue 38
- 3.2.2 Total Known Sulfur Compounds in Garlic Cloves 41
- 3.2.3 Glutamylcysteines: Important Storage Compounds ... 41
- 3.2.4 Cysteine Sulfoxides: Precursors of the Thiosulfinates . 46
- 3.2.5 Alliinase: Lysis of the Cysteine Sulfoxides 48
- 3.2.6 Allicin and Other Thiosulfinates: Odorous But Medicinal . 53
- 3.2.7 S-Alkylcysteines 69
- 3.2.8 Allithiamine 70
- 3.2.9 Scordinin: A Plant Growth Hormone 71
- 3.2.10 Biogenesis of Garlic's Sulfur Compounds 72
- 3.2.11 The Function of Garlic's Sulfur Compounds 74
- 3.3 NONSULFUR COMPOUNDS IN GARLIC 76
- 3.3.1 Carbohydrates: Sugars, Fructans, Pectins 76
- 3.3.2 Enzymes in Garlic 77
- 3.3.3 Protein, Free Amino Acids, and Dipeptides 79
- 3.3.4 Lipids: The True Oil in Garlic 81
- 3.3.5 Sterols and Hydrocarbons: Unsaponifiable Lipids ... 84
- 3.3.6 Total Lipophilic Compounds 84
- 3.3.7 Steroid and Triterpenoid Glycosides (Saponins) 85
- 3.3.8 Vitamins and Minerals 86
- 3.3.9 Organoselenium Compounds 88

3.3.10	Flavonoid Pigments and Phenols	89
3.3.11	Adenosine, Guanosine, and Nucleic Acids	90
3.3.12	Miscellaneous Compounds: Phytic Acid, Plant Hormones, Lectins	91
3.3.13	Machado's Garlicin—A Mystery	91
3.4	GARLIC PROCESSING	92
3.4.1	Garlic Powder Products	93
3.4.2	Oil of Steam-Distilled Garlic ("Essential Oil")	99
3.4.3	Oil of Oil-Macerated Garlic	99
3.4.4	Ether-Extracted Oil	102
3.4.5	Extract of Garlic Aged in Dilute Alcohol	103
3.4.6	Garlic Aged ("Picked") in Dilute Acetic Acid	106
3.4.7	Total Thioallyl Compounds	107

4 Garlic Preparations: *Methods for Qualitative and Quantitative Assessment of Their Ingredients* ..109
RHEINHARD PENTZ AND CLAUS-PETER SIEGERS

4.1	INTRODUCTION	109
4.2	THE MAIN INGREDIENT OF DIFFERENT GARLIC PREPARATIONS	110
4.2.1	Dry Garlic Powders	110
4.2.2	Dry Extracts with Conservation of Alliin	111
4.2.3	Garlic Juice	112
4.2.4	Steam-Distilled Garlic Oil	112
4.2.5	Oil-Macerated Garlic Products	113
4.2.6	Aged Alcoholic Garlic Extract	113
4.2.7	Comparison and Value of Different Garlic Preparations	114
4.2.8	Odorless Garlic Preparations	114
4.3	ANALYTICAL DETERMINATION OF CONSTITUENTS IN GARLIC AND GARLIC PREPARATIONS	116
4.3.1	Alliin and Related Cysteine Sulfoxides	116
4.3.2	Allicin and Other Thiosulfinates	118
4.3.3	Polysulfides	122
4.3.4	Vinyldithiins	126
4.3.5	Ajoenes	127
4.3.6	Glutamyl Peptides	128
4.3.7	Elemental Sulfur	129
4.3.8	Hydrolysis to Thiols	130
4.4	OTHER GARLIC COMPONENTS	131
4.5	PROPOSALS FOR STANDARDIZATION OF GARLIC PREPARATIONS	132
4.5.1	Garlic Powder Preparations	132
4.5.2	Garlic Oil Preparations	133
4.6	SUMMARY	134

5 Therapeutic Effects and Applications of Garlic and Its Preparations 135
HANS D. REUTER, HEINRICH P. KOCH, AND LARRY D. LAWSON

5.1	EFFECTS ON THE HEART AND CIRCULATORY SYSTEM	135
5.1.1	Cholesterol and Lipid-Lowering Affects	136
5.1.2	Effects on Blood Pressure, Vascular Resistance, and Heart Function	148
5.1.3	Effects on Blood Coagulation, Fibrinolysis, and Blood Flow	154
5.1.4	Antithrombotic (Platelet Antiaggregatory) Effects	156
5.2	ANTIBIOTIC EFFECTS	162
5.2.1	Antibacterial Effects	164
5.2.2	Antifungal Effects	168
5.2.3	Antiprotozoal Effects	172
5.2.4	Antiviral Effects	172
5.2.5	Antiparasitic Effects	173
5.2.6	Insecticidal and Repellent Effects	174
5.2.7	Machado's Garlicin	175
5.3	ANTICANCER EFFECTS	176
5.3.1	Anticancer Effects: Epidemiological Studies	176

5.3.2	Anticancer Effects: Animal and IN Vitro Studies	178
5.3.3	Anticancer Effects: Active Compounds	186
5.4	ANTIOXIDANT EFFECTS	187
5.4.1	Antioxidant Effects: In Vitro Studies	187
5.4.2	Antioxidant Effects: In Vivo Studies	190
5.4.3	Antioxidant Effects: Active Compounds	190
5.5	IMMUNOMODULATORY EFFECTS	191
5.6	ANTIINFLAMMATORY EFFECTS	192
5.7	HYPOGLYCEMIC EFFECTS	193
5.8	HORMONE-LIKE EFFECTS	196
5.9	ENHANCEMENT OF THIAMINE ABSORPTION	197
5.10	EFFECTS ON ORGANIC AND METABOLIC DISTURBANCES	198
5.10.1	Effects on Enzyme Activities	199
5.10.2	Antihepatotoxic Effects	201
5.10.3	Dyspepsia and Indigestion	202
5.10.4	Respiratory Diseases	204
5.10.5	Antidote for Heavy Metal Poisoning and Other Toxins	205
5.10.6	Other Effects and Applications	206
5.11	GARLIC IN HOMEOPATHY	207
5.12	DOSING OF GARLIC AND ITS PREPARATIONS	208
5.13	DEODORIZED GARLIC EXTRACTS	209
5.13.1	Aged Garlic Extract	209
5.13.2	Odorless Garlic Extract	210
5.14	OTHER ACTIVE COMPOUNDS IN GARLIC	211

6 Biopharmaceutics of Garlic's Effective Compounds ... 213
Heinrich P. Koch

6.1	PHARMACOKINETICS OF THE SULFUR-CONTAINING ACTIVE COMPONENTS OF GARLIC	213
6.2	PHARMACOKINETICS OF THE VINYLDITHIINS	218
6.3	PHARMACOKINETICS OF ALLITHIAMINE	219
6.4	PHARMACOKINETICS OF MACHADO'S GARLICIN	220
6.5	DOES GARLIC ACT AS AN "ABSORPTION ENHANCER?"	220

7 Toxicology and Undesirable Effects of Garlic ... 221
Heinrich P. Koch

7.1	ACUTE, SUBACUTE, AND CHRONIC TOXICITY OF GARLIC	222
7.2	TREATMENT OF ACUTE TOXICITY	224
7.3	TOXICOLOGY OF MACHADO'S GARLICIN	224
7.4	TOXICOLOGY OF SCORDININ	224
7.5	TOXICOLOGY OF A COAGULATION INHIBITOR FROM GARLIC	224
7.6	LOCAL IRRITATION	225
7.7	ALLERGIES TO GARLIC	226
7.8	INTERACTION WITH OTHER DRUGS	227
7.9	ECOLOGICAL HAZARDS	227

8 Non-Medical Applications of Garlic and Curiosities ... 229
Heinrich P. Koch

References ... *235*
Index ... *321*

1
History, Folk Medicine, and Legendary Uses of Garlic

GOTTFRIED HAHN

1.1 DERIVATION OF THE NAME

Botanists call it *Allium sativum* L. The origin of the genus name remains unknown. A linguistic connection is often made with the Latin word *olere*, "to smell," because of its strong odor. On the other hand, it is possibly derived from the Greek word άλλεσδαι (hallestai), "to leap out." This expression describes the fast growth of the secondary bulbs, the so-called "cloves," which seem to leap out all at once from the primary bulb. Since the Roman poet Plautus (250–184 BC) knew the term *Allium* for garlic, it is thought that a derivation through *alum* (comfrey) from the old Aryan *áluh* or *álukám* is possible because of its use there as a spice. These Indo-European substantives are translated into "bulb," "edible root." Less likely is the explanation that *Allium* is derived from *alare* or *halare*. Both words mean "to breathe," "to exhale," referring to the odor. Also, the Celtic word *all*, meaning "warm," "burning," has been considered as a derivation (Pictet, 1877).

Older scientific designations are *Allium domesticum* and *Scordium*. An interesting popular designation in German is *Bauern Theriak*, and in Latin *Theriaca rusticorum*, which translate into "farmer's theriac" (Höfler, 1908; Schneider, 1974; Marzell, 1967). In ancient pharmaceutical practice, *theriac* was considered as an exceptionally useful and rather costly medicine, which often could only be prepared under the supervision of governmental or municipal officers. King Mithridates of Pontus (132–63 BC) has been credited with the discovery of theriac. Theriac was included in the *Pharmacopoeia Germanica* of 1882. If this precious medicine can be considered the same as garlic, as depicted by the term farmer's theriac, then *Allium sativum* was of equal renown among country folk as the true famous theriac. Here, it was used for many ailments. The saying "garlic is the cure-all of the peasants" originated with the Bourbons in France. This was meant to be contemptuous, yet it carried a deep truth. In addition to versions in the French language, which includes the term "farmer's theriac" (*thériaque des paysans* or *triacle des villains*), other languages have adopted similar terms. For example:

boeretheriak	(Dutch)
countryman's treacle	(English)
böndernis theragelse	(Danish)
bondens triakelse	(Swedish)

The German word *Knoblauch* evolved from the Old High German word *chlofalauh* or *chlobilouh* and the Old Saxon *clofloc* through several shiftings of sound. In Middle High German this odoriferous plant was called *klufloc*, and in the neighboring Dutch, *cluflooc* or *cloflooc*. All these names are a combination of the words *lauch* and *clobo* or *klobo* (Grimm, 1876). *Lauch* is found in the Old Nordic languages as *Laukar*. It was already inscribed on small metal plates in runic characters as discovered on an artifact in the southern Swedish province of Schonen. The wearer of the amulet was supposedly guaranteed good health. *Klobo*, *clobo*, or *clufu* mean split stock or, in the Old Anglo-Saxon, clove. The English word "clove," which means young bulb, originated therefrom, and we still refer to the garlic clove. *Knoblauch* actually means split leek or cloved leek. Also the Polish name *czosnek*, the Russian *chesnok*, and the Czechoslovakian *cesnek*, often with the additive *kuchynský* (kitchen), have the same meaning.

In the German dialects there are many variations:

Knufflauw	Westphalia
Knuflauk	Waldeck
Knoflak, Knuflak	Area of Göttingen
Knopfloch	Lübeck
Knoploch, Knub(e)loch, Knofloch	Various parts of the Rhineland
Knewelauch	Braunschweig
Gnuuwluch	Naumburg
Know(e)loch	Palatinate
Know(e)lich, Knoblich	Formerly in Bohemia
Knobluch, Knuwlet	Silesia
Knoufl	Carinthia
Knoflach	Tyrol
Knobl(e)t	Vorarlberg
Chnoblech, Chnoblecht, Chnoblich	Switzerland

Names such as *Judenkost* and *Joddelook* in the area of the Lower Rhine and the neighboring Netherlands, *Juddezeh* in the area of Coblenz, and *Judnavanilli* in some areas of the Palatinate, show that the Jewish population especially appreciated garlic and very likely knew its favorable effects. Garlic, however, was not favored by the rest of the population, as the name *Stinkertzwiebel* (stinking bulb) implies.

In the areas of Bentheim and Osnabrück, Germany, the country folk still call garlic *Süntjanslaof* (St. John's leek), perhaps because it blooms around the 24th of June. This name, however, is also used for other leeks. Names such as "wind clove," "old age clove," or "stomach clove" point to garlic's widespread application as food or medicine.

The Italians, who are very fond of garlic as a seasoning, call it *aglio*. In Swiss-Roman Graubünden (Switzerland) it is known as *Agl*, and in Swiss-Italian Tessin, as *ai*. In France it is referred to as *ail* or *ail commun*, and occasionally as *Perdrix (Chapon) de Gascogne*, and in Spain as *ajo*, a word that is often used in strong expressions.

In Denmark, garlic is *hvidlok*, and in Sweden it is *vitlök*, both words containing the term "leek." In China, the symbol for garlic (pronounced *dasuan*) means "big leek." In England it is simply "garlic," and "garlicky" means garlic-like (and *alliaceous* is often used synonymously). In the Old Anglo-Saxon it was called *garleac*. This term originates from gar = ger = spear and leek (a general name for *Allium* plants), hence "spear-leek," referring to the spear-like shape of the leaves (Candolle, 1884).

In the 13th century, when the use of family names was introduced in central Europe, the cultivation of garlic had been so well established that it contributed to name-building. Brechenmacher (1929) wrote about a woman named Mechthild Cnoblochin, who in 1286 lived in Esslingen, Germany. In Goethe's ancestry there was in 1291 a "Heynemannus dictus Cnobeloch." Today the name occurs in several variations, such as Knobloch, Knoblich, Knoblach, Knoblau, and others.

The name was even used for the designation of settlements, such as the village of Knoblauch near Magdeburg, Germany, and other examples in East Havelland and East Prussia. In Hessia, some villages have their common garlic plot, and similar references have also been observed in other countries. Frequently, the name was connected with the development of a coat of arms. Hausbuchmeister, a fine artist who worked during the period 1475–1500 in the area of the Middle Rhine and who obtained his name from a voluminous parchment scroll, the *Wolfegger Hausbuch*, created a copper etching in which a youth with a garlic coat of arms is depicted (Figure 1.1) (Hess, 1994; Behling, 1957).

1.2 APPLICATION BY ANCIENT POPULATIONS

Botanical researchers consider the regions of Central Asia as the place of garlic's origin (Hyams, 1971; Hehn, 1894). People who lived by hunting or who moved with their herds through vast areas of land like nomads must have felt the need for seasoned food very early in their evolution. Probably first in

Figure 1.1. Garlic coat of arms, held by a young man. *Hausbuchmeister,* around 1475 (9.6 × 8.0 cm).

line was salt, but plant products with leeks and garlic among them soon followed. *Allium sativum* L. appears to have been in use by humans since the Neolithic Age. True, it was not equally valued everywhere. There were always people who rejected it, as we observe today. The strong taste and odor are far less valued in the northern countries than in the southern ones. Numerous scientific and popular articles dealing somewhat thoroughly with garlic's history are available (Feng et al., 1990; Funke, 1977; Mayer, 1951; Reinhard, 1984, 1986; Strübing, 1967; Drobnik, 1938).

Written accounts by the Sumerians dating back to 2600–2100 BC are among the oldest reports. Since *Allium sativum* was already known at that time as a cultured plant, its use as a spice, medicine, and condiment would precede that era. It was probably known by the advanced civilizations of the Indus Valley. It is likely that from here it was introduced to the Chinese. A legend from India relates that when the Original Ocean (Urmeer) was searched by gods and demons for its inestimable wealth, many precious things came to light, including, among others, the most desirable nectar:

> An intense fight arose between the gods and the demons for the partaking of the nectar. In order to avoid further disagreements, *Vishnu,* the creator of the uni-

verse, appeared and started to distribute the nectar among the *Devas* [gods]. Cunningly, one of the demons sneaked into the group and also obtained a mouthful of nectar. Sun and moon had seen the crime and reported it immediately to *Vishnu*. Before the demon could swallow the nectar, *Vishnu* had the demon beheaded. The nectar ran out of its throat onto the ground, and exactly at this spot grew a plant—garlic. It contained many valuable characteristics of the nectar. But, since it had gone through the mouth of a demon, it had also developed *"tamsik,"* which means sinful aphrodisiac properties.

The most important textbooks of Sanskrit medicine from the Brahmanic Era, known as *Ayurveda* (Science of Life), were probably codified about 500 AD. Their content is much older, however, especially the three large collections (*Samhita*) of the legendary Old-Indian physicians Charaka, Susruta, and Vagbhata. They knew garlic, called *mahushuea,* as a tonic and as a remedy for skin diseases, loss of appetite, dyspepsia, cough, anorexia, rheumatic conditions, abdominal diseases, spleen enlargement, and hemorrhoids—obviously almost a panacea (Weiss, 1983).

The British Lieutenant Hamilton Bower, on a journey to East Turkestan in 1890, discovered there a manuscript which was written about 400 AD. Buddhistic monks were probably the authors, using much older sources for their book. Hoernle translated and published this manuscript in Calcutta in 1897. Here, the origin of garlic is ascribed to the drops of blood that fell onto the earth when Janar-Dana decapitated Asura, the ruler of the Old-Indian gods. According to the *Bower Manuscript,* garlic was used as an additive to an aphrodisiac drink. Garlic is even praised in a hymn, which Aschoff translated as *Knoblauchlied* (Garlic Song) (Hoernle, 1897; Aschoff, 1900):

> On the salubrious mountain, where the salubrious plants grow, live the *Munis,* men with enlightened minds. They check the taste, the characteristics, the forms, the strengths and names of all salubrious plants.

Garlic is designated as a panacea in a manuscript found in 1889 in the Ruins of Mingat near Kuchar in Central Asia. This manuscript, found in a chamber under a Buddhist stupa, consisted of 56 sheets made of birch bark. Therefore, it was called the *Birch Bark Manuscript.* The sheets show double-sided writing in Old Sanskrit.

Through the knowledge of Indian medicine, garlic, often called *uccatá,* gained access into Tibet and China. Whether China obtained it through India, or whether *Allium sativum* reached the "Middle Kingdom" (i.e., the Chinese provinces) through direct access cannot be determined. The famous *Materia Medica* of the Chinese, known as *Pen Ts'ao Kang Mu,* is still relatively young. It was written during the years 1552–1578 AD by Li Shi Chen, an official of the government, and goes back to much older prescription collections. Bretschneider determined the names of the plants (Bretschneider, 1882). Among the strongly scented genera, which were also used as kitchen seasonings, were *A. sativum, A. odorum,* and *A. fistulosum.* A work called *Rhaya* is supposedly much older. Allegedly, it was written by Chou Kung in 1100 BC (probably it is much younger). Garlic is listed here as *Resolventium. Allium* species were still obtainable in every Chinese pharmacy around 1880–1890. The Chinese were also familiar with the cultivation of other *Allium* species. A Chinese book on agriculture, called *Ts'i min yao shu* (Important Rules for People to Earn their Livelihood Peacefully), is dated to the 5th century AD. It gives suggestions for the cultivation of *A. sativum, A. fistulosum, A. odorum,* and other *Allium* species. Garlic was not only valued as drug, but also as seasoning, and not only the bulbs, but the entire plant was also used. Wherever Chinese people settled, we find not only their culture, but their medicines as well. Thus, a Chinese drug store in Malaysia, Indonesia, or elsewhere, will always carry the drugs customary in China. Hooper published a paper in which the names of the Chinese drugs and their local terms are listed (Hooper, 1938). The author, a botanist, became acquainted with these drugs in Malaysia. Several *Allium* species were found in this collection, among others *A. bakeri* Regel (*Chiu chiu pai, Kew pak*) and *A. odorum* L. (*Chiu hsiu tzü, Kow chow chu*). After both names is the addition "and other species," indicating that *A. sativum* was certainly included (Li, 1969).

Most of the Chinese medicinal plants gained entry to Japan by way of Korea. We do not know whether *Allium sativum* was found in ancient Japanese medicine formulations at the time of the physicians Abe Manao and Idzumo Hirosada in the 9th century (Thoms, 1926). Japan patterned its agriculture after that of China, especially for spices and seasonings. The Japanese used *Allium sativum* as a remedy for colds and as an aphrodisiac. Apparently, it was sufficient simply to hang up garlic bulbs in the house to prevent coughs and hoarseness. The Japanese explored the medicinal application of the "aromatic bulbs" in their clinics. Thus, a "garlic sanatorium" was established in which *Allium sativum* was used extensively against numerous diseases. One of the therapeutic possibilities is the spraying of patients with a garlic solution in a shower room specifically designed for this purpose.

Turning now to the Mediterranean region, we find that garlic played an important role in Egypt. In addition to inscriptions, graphic depictions, and findings in sepulchers, there are papyri which give information on medicinal and seasoning plants (Loret, 1904). The famous *Papyrus Ebers,* which was translated into English by the Danish scientist B. Ebbell (1937), is apparently based on older Egyptian writings and knowledge. It was found in 1872 near the region of old Uêset, the "Thebes with one hundred gates," during excavations led by the German Egyptologist Georg Ebers (Ebers, 1875; Joachim, 1890). At first he was not aware of what a treasure had fallen into his hands. While the document was painstakingly deciphered, it became apparent that it was an old Egyptian medicinal manual with more than 800 different formulations. It is cause for speculation that even at the time of the pyramid construction, Egyptian medicine did not stand isolated but dealt with the same problems and used the same drug substances as did the physicians in Babylon, Ur, and Uruk. It can hardly be determined whether the discoverer of the medicinal knowledge lived at the Nile, the Euphrates, or the Tigris, but it can be safely said that garlic did not grow wild in Egypt and had to be imported from the areas of the Euphrates and Tigris. Later, however, it was cultivated by the Egyptians. No less than 22 drug formulations in the famous book contain garlic. Garlic was also found in sepulchers. Schweinfurth (1887) reported such findings in the sepulchers of Assasif near Thebes and of Dra-Abu-Negga. Moreover, remains of garlic bulbs were found in the burial chambers of Tutankhamen, who was buried in 1352 BC.

Egyptians appreciated cleanliness very much, at least the upper classes. The well-situated Egyptians not only washed themselves several times a day, but also attended to their body for three days each month with emetics and laxatives as well as with enemas in order to free it from all uncleanliness. For this purpose, onions and garlic were used, among others. We have a report from the Greek historian Herodotus (about 490–420 BC) who traveled through Egypt around 450 BC, and who was rather concerned about the laborers at the pyramids. A report of Herodotus in *History of Egypt* contains the following interesting description (Herodotus, 1963):

> On the pyramid, there is an inscription in Egyptian hieroglyphs. It reports the amount of radishes, onions and garlic which were consumed by the construction workers of the pyramid. I remember very clearly that the interpreter who read this inscription to me mentioned that an amount of 1600 silver talents were spent for this purpose.

If this figure is compared with today's currency, the impressive sum of 10 million dollars was spent over 20 years for 360,000 workers—and this just for radishes, onions, and garlic! The scientists of the Cairo Museum question this number, with good reason, and it is conceivable that the interpreter exaggerated slightly. Nevertheless, the report by Herodotus shows that garlic, onions, and radishes were considered to be absolutely necessary to keep the workers in good health.

Botany and pharmacognosy were highly developed in Egypt. Some plant families were already grouped together, and were differentiated as palms, water-lilies (Nymphaceae), grains, climbers, and onion plants.

The ancient Egyptians also connected religious concepts with plants, and such relationships still exist today in folklore. In the fusion of religion and daily life it was not so much a question of the times as it was a matter of priestly interpretation that lifted the veneration of medicinal plants into the mystical sphere. If it was the gods who had sent diseases, then it was only natural that healing be attributed to them as well. Thus, garlic and onions were held in high regard in the eyes of the people; they were considered to be consecrated and sacred. Pliny the Elder (23–79 AD) reported that the Egyptians took an oath by appealing to garlic and onion.

ALLIUM CEPASQUE INTER DEOS IN JURE JURANDO HABET AEGYPTUS.

(While taking an oath, garlic and onion were presented to the gods of Egypt.)

The Roman satirist Juvenal (60–127 AD) decisively said that in this way the Egyptians were probably growing their gods in kitchen gardens (Juvenal, 1858).

Ghosts and demons were considered responsible for the outbreak of diseases. However, the Egyptians also took natural causes into account, which they designated as "worms." Today, we would refer to germs. The knowledge about such origins presumably comes from the physicians of Babylon and the land ashore of the rivers Euphrates and Tigris. Real worms as causes for disease were known in all warm countries. If the worm could not be traced during the course of a disease, it was concluded that it was so small that it could not be seen by the human eye. Thus, the Babylonian physicians had reached a level of hypothetical thinking equal to that of the physicians during the European medieval era prior to the discovery of the microscope. Since intestinal worms could be successfully treated with garlic, this drug was also used for other infectious diseases. A Babylonian cuneiform tablet from 3000 BC includes a prescription of an obscure tonic which contained a considerable amount of garlic. Garlic held a prominent position among the common medicines of the Euphrates-Tigris area. R. Campbell Thompson (1923, 1924) intensively studied these drug mixtures while he worked as an assistant at the British Museum in London, and he rendered to us many suggestions. According to Oefele (1902), a cuneiform tablet from Niffer contained the following prescription:

> Reed seed and dates, mix and drink with palm wine. Calf milk and bitter principles, drink with palm wine. Cypress and bitter principles, drink with palm wine. Excrements and bitter principles, drink with palm wine. Oak leaves and bitter principles, drink with palm wine. Papyrus, wine and bitter principles, drink with palm wine. Tentacle of the worm *"ipkhu"* and bitter principle, drink with palm wine. Garlic and bitter principles, drink with palm wine. Drink the root of the tree *susu* and bitter principles with oil and palm wine.

On another tablet (No. 46226) owned by the British Museum, an unknown writer tells us of a medicinal herb garden of the Babylonian King Mardukpaliddin (722–710 BC). Here, 64 plants are mentioned, and the list begins with onion and garlic.

The knowledge of the healing power of garlic that we observe with the Babylonians and Assyrians presumably originates from the high culture of the Sumerians. The library of the Assyrian King Assurbanipal in Nineveh (669–627 BC) yielded a number of Sumerian texts concerning the exorcising of diseases, and these also refer to dates, onions, and leek species.

The ancient Iranian people also consumed garlic in considerable amounts. It was reported that the Persian king's court ate more than 26 kilograms of garlic and more than 13 kilograms of onions per day. In a very late edition of the renowned *Zend Avesta*, the holy book of Zoroaster, a physician named Thrita is mentioned, whom the gods awarded with 10,000 medicinal plants (Jastrow, 1914). Among them were *Allium* species, Aloe, Benzoe, Cannabis, and pomegranates. The Minoan cultures also knew of garlic. The Cretian palaces were the economic centers where books were kept on incoming and outgoing merchandise. Unfortunately, they were written on degradable materials, but it can still be proven that among the numerous spices, the leek species held an important position.

The Israelites acquired garlic from the Egyptians (Woenig, 1886). This rapidly growing group of people lived for a long time under Egyptian supremacy. We do not know whether they knew *Allium sativum* prior to that time, but it is very possible, since Abraham's family (or Jewish tribes) came from the Euphrates-Tigris area. While in Egypt, however, it became part of their steady diet. Dieffenbach mentioned the following plants that were cultivated by the Israelites (Dieffenbach, 1902):

> Wheat, barley, grapes, figs, pomegranates, olive tree, lentils, beans, spelt, millet, pumpkin, melons, leek species, onions, caraway, dill, mustard, flax, dates, and apples.

Garlic was called *Schûm, Shûm,* or *Schûmin* (Woenig, 1886). The lamentation of the people is understandable, when upon the *Exodus* this valuable nutritive, seasoning, and medicinal plant was no longer available:

> We remember the fish, which we did eat in Egypt freely; the cucumbers, and the melons, and the leeks, and the onions, and the garlick. (Numbers 11:5)

The Talmud (ancient Rabbinic writings), which contains elements of various time periods, differentiates between dietetically and medicinally used plants. *Allium sativum* belonged to both. As reported in the *Midrash* (Interpretation of the Law), King Solomon (965–926 BC) prepared his meals according to medicinal-dietetic principles. Medicinal science rose to an appreciable climax under Jesus Sirach (author of a collection of proverbs of the Old Testament from the 2nd century BC). At this time, garlic and onion were associated with foods that also had cultural importance (Ridley, 1900). Often it cannot be determined whether or not the species *Allium sativum* is involved in the Biblical interpretation, since there are over 60 different kinds of leek plants known in the Mediterranean area. Garlic was used medicinally for indigestion, as a diuretic and spasmolytic, and also as an aphrodisiac (Maurizio, 1928). Raw (uncooked) garlic with bread was well-liked, a use which is still common among Jews, though living now in other countries. Garlic was also fed to animals, e.g., to dogs and chickens. The use of garlic among the Jews is still very important (Ratner, 1916). Often this was not only a characteristic of lifestyle among Jewish ethnic groups, but reason enough for derogatory remarks by the non-Jewish population, who detested garlic, as can be found in antiquity (Wagner, 1793).

The Arabs presumably knew both wild and cultivated forms of *Allium sativum*. They used all parts of the plant for treatment of worms, snake bites, vermin control, skin rashes, dentistry, treatment of eye diseases, menstrual abnormalities, and in veterinary medicine. Often, the Jewish physician, who was in Arabic service, spread the medicinal qualities of garlic, especially during the early medieval era. Thus, around 950 AD, the Jewish physician Chasdai Ben Isaak Ibn Shaprut worked at the court of caliph Abd ar-Rahman III of Cordoba. He obtained a richly illustrated volume of the works of Dioscurides (see below) from Byzantium and established an exact pharmaceutical science, based on his knowledge of medicinal literature and his own experience. His notes were also incorporated into the outstanding work of the great Arabian botanist Ibn Al

Baithar (1197–1248 AD), which contained treatises of more than 1400 medicinal plants, including garlic (Ibn al Baithar, 1840).

The Greeks became acquainted with garlic through Egypt and the Orient (Baumann, 1982; Demitsch, 1889). They called it *Skorodon*, and the garlic dealer was named *Skorodopolos*. A broth made of garlic and salt, called *Skorodalmae*, was favored by the people of ancient Greece, but was detested by the aristocracy. It was also believed to be unpleasant to the gods, as no one who smelled of garlic was allowed into the temples of various deities. Like the Libyan Aphrodite, so also the mother of the gods excluded the garlic eater from the cult.

> When the clever philosopher and despiser of the gods, Stilpo (about 330 BC), had partaken of garlic and retreated into a shrine to sleep, a goddess appeared to him in his dream and said: "You are a philosopher and are not afraid to transgress the law?" He answered: "Give me something else to eat, and I will refrain from eating garlic."

Homer (8th century BC) knew of the effectiveness of garlic and combined it with mystic conceptions. He describes the visit of Odysseus with the sorceress Circe in the *Odyssey*. This graceful woman transfers the companions of the King of Ithaca into swine with a magic drink. Only Odysseus himself is able to avoid this fate. However, he would have been lost if the god Hermes (Roman Mercury) had not come to his aid. He gave him the magic plant *Moly* (μῶλυ) with the words:

> Nevertheless, she [Circe] will not be able to transform you. The virtue of this medicinal plant will prevent her.

Homer then describes the plant further: "Its root was black, milk-white its blossom, *Moly* is it called by the gods" (*Odyssey* X, 302 ff.), and the Roman poet Ovid (43 BC–17 AD) reports this story too (*Metamorphoses* XIV, 291) (Homer, 1971; Berendes, 1891, 1907; Koch, 1995):

PACIFER HUIC FLOREM CYLENNIUS DEDERAT ALBUM, MOLY VOCANT SUPERI, NIGRA RADICE TENETUR.

> (The peace god of Cylene gave it white blossoms and black roots, the great ones call it Moly.)

There has been much discussion about this idea. Presumably it is an *Allium* species, a garlic-like plant from the mountains. The Greek natural scientist and philosopher Theophrastos (371–287 BC), a scholar of Plato and Aristotle, investigated this legendary plant (Sprengel, 1822). He found a leek species in the meadows of the Kyllene Mountains which was held sacred by the people of that area and which grew in the holy realms of the god Hermes. We do not know whether it is *Allium moly* L., which we value in our gardens because of its golden yellow blossoms, or *Allium magicum* L., described by *Linnaeus* much later. Since both species have red or yellowish-red blossoms, while Homer talked of only white blossoms, it is possible that *Moly* is to be identified as the true garlic or because of its black root, as *Allium nigrum* L. In any case, it is a well-known and valued leek plant (Koch, 1995).

The region Megaris near the Saronic Gulf was famous and notorious: the inhabitants of this area apparently cultivated garlic and also consumed it in considerable amounts. Garlic is found in numerous Greek kitchen recipes as well as in medicines. Frequently it was used together with onions, bay leaves, and myrrh as an ointment. In Hellenistic medicine it was used for inhalations. The Greek physician Hippocrates (about 460–370 BC) recommended *Allium sativum* as a diuretic, laxative, and emmenagogue. It was also used for pneumonia and externally for putrid wounds. Many allusions to garlic are contained in the *Corpus Hippocraticum*, which was not, as the name might suggest, written by Hippocrates himself, but by his students around 400 BC. In these writings *A. sativum*, *A. porrum*, and *A. cepa* were counted as dietetic laxatives, and onions and garlic were added to emmenagogues (Hippokrates, 1897; Dierbach, 1824; Kapferer & Sticker, 1934).

Another botanical text originating from Herodot was revised by Kanngiesser (1911);

among the 66 plants mentioned are garlic and onions. The Greek comedy poet Aristophanes (445–385 BC) considered garlic as a symbol of physical strength. Perhaps this allusion goes back to an old tradition of Greek athletes, who ate garlic before the contests in Olympia. In one of his comedies, Aristophanes wrote: "A garlic breakfast will make you hotter for fighting." This comment by one of the characters was directed at a butcher. Perhaps this was the origin of our famous salami? He also suggested that the disloyal wife chew garlic in the morning to prove her innocence to her husband returning home from watch (in *The Wives at the Feast of Demeter Thesmophores*). A pupil of Theophrast, the Greek physician Erasistratos of Julio (305–257 BC), used garlic with sulfur and vinegar for the external treatment of eczema. The famous Pythagoras (580–500 BC) considered garlic to be the king of spices—an opinion, however, with which many Greeks did not agree.

The ancient Romans became aware of garlic through the Greeks (Fraas, 1845; Lenz, 1859). There, too, opinions about the odorous bulb were divided. Many liked it, but others could not endure the scent. The Romans called garlic bulbs *alliinae*. The garlic dealers who peddled their wares from house to house were named *alliarii*. In a comedy by the Latin poet Plautus (254–184 BC), the insult *"Man, go to the devil, you smell like garlic!"* is hurled at a commoner. And the famous Roman scholar Marcus Terentius Varro (116–27 BC) says in one of his writings:

> Our fathers and great-grandfathers were rather courageous men, though their words had a crude odor of garlic and onions.

We can possibly assume that, as the wealth of the Roman population increased, the use of garlic decreased. The Roman emperor Marcus Aurelius (121–180 AD) did not like garlic. When he traveled through Palestine, he was not only offended by the noise in the cities, but also by the odor of garlic. Marcus Porcius Cato (234–149 BC) mentioned *Allium* in his work *De re rustica*, among 120 agricultural plants. Hence, the cultivation of garlic and other leek species was well known. In ancient Rome, the consumers of garlic were mainly soldiers, seamen, and slaves. It was a regular item in the diet of Roman legionnaires. Frequently they ate a pottage, named *Moretum*, consisting of the following ingredients (Heinze, 1939): "cracked wheat, water, salt and black pepper, onions, garlic, diced bacon, beef and cheese." Archaeologists who have used this recipe were astonished by the tastiness of the soup.

Many Romans believed that with garlic evil spirits could be banished from their homes. For this purpose, it was not necessary actually to eat the garlic, but sufficient to paint the garlic plant on the wall, as many archaeological excavations from Pompeii have shown. The Roman satirist Juvenal (60–127 AD) designated garlic as a medicine and food of plebeians and soldiers, and did not conceal its use as a remedy for tapeworms. Horatius (65–8 BC) describes it as the food of the Jews, whom he described as *Iudaei foetentes* (stinking Jews) because of their garlic scent. In the third ode the poet pokes fun at the seasoning for his patron Maecenas:

> It appears to me as a poison,
> infested by a witch.
> Give it to the criminals
> instead of the cup of poison!
> It burns my extremities
> like the sun of Apulia,
> like the nettle garment
> the body of Heracles.
> If you ever, Maecenas,
> should have the inclination
> to partake of this plant,
> then may your beloved
> resist your kiss
> and remain far from your hugs
> and flee to the lowest couch.

Pliny the Elder (23–79 AD), the great Roman encyclopedist, observed that the leaves of *Allium sativum* turned black upon prolonged exposure to the air. From this he concluded that garlic attracts all evil and dark powers, and he recommended it for snake bite. The Romans believed that evil demons resided in snakes and that they were transferred to humans through the bite of snakes. Therefore, garlic was considered an effective antidote. Balsam oil in milk mixed with mulberry juice and garlic oil was used as an antidote for monkshood (*Aconitum*) poisoning. Pliny also points to the bad breath that one gives off after garlic consumption, to the aggravation of his fellow beings (Plinius Secundus der Ältere, 1977; Plinius, 1881; Pliny, 1961).

In Virgil's *Second Idyll*, written around the years 42–39 BC, Thestilys crushed wild thyme and garlic to protect the harvesters from snakebite while they relaxed during their noon rest (Virgil, 1880). The physician Pedanios Dioscurides Anazarbeus (1st century AD), who came from Greece and worked in Rome, considered garlic from a medical-scientific point of view. He recommended it as a diuretic and stomachic, for the expelling of intestinal worms, and for the treatment of hemorrhoids. In addition to *Allium sativum*, in his *Materia Medica* he also names *Allium leucoprasum, A. subhirsutum,* and *A. ampeloprasum* as medicinal plants (Dioscorides, 1610; Berendes, 1902). He decisively writes:

> Garlic has a sharp, softening, gas subsiding effect. It causes confusion in the abdomen, dries out the stomach, causes thirst and can cause even abscesses on the skin. When eaten, it expels tapeworms and increases urine's volume. It is also beneficial for dropsy. Like no other remedy, it helps those who have been bitten by the viper and the hemorrhoids, if ground with wine and drunk. With salt and oil it heals skin rashes; mixed with honey as an ointment it also removes white spots, scabies (psoriasis), and liver spots.

This description sounds rather modern. At first, it introduces general characteristics, even including possible side effects; then it lists all therapeutic possibilities.

Scribonius Largus, who lived in Rome under the emperors Tiberius and Claudius and who accompanied Claudius to Britain in 43 AD, presented a collection of prescriptions (*Compositiones*) which can be considered the first "Dispensatory" or "Pharmacopeia" in history. Felix Rinne edited this collection with some comments (Rinne, 1892). Of course, *Allium sativum* is among the listed plant drugs.

Aulus Cornelius Celsus (25 BC–50 AD) is among the most important and most independent medical authors of the Roman Empire. This learned, highly educated man was well versed in medicine by thorough study and practical experience. Of his extensive literary activity, only eight books concerning medicine have survived, collectively entitled *De Medicina Libri Octo*, written around 45 AD (Celsus, 1906).

Here, Celsus recommends garlic to initiate the expulsion of tapeworms. He numbers it among the sharp, bloating, warming, and purgative media and ascribes to it the possibility of keeping bad nutrient fluids from the body. In addition, Celsus emphasizes its effectiveness for breathing difficulties and for coughs. He recommends garlic seeds as an ingredient in suppositories to encourage menstruation.

Marcus Valerius Martialis (40–102 AD), the Roman epigrammatist, regarded garlic as an aphrodisiac (Leclerc, 1918). He wrote:

> CUM SIT ANUS CONJUX ET SIT TIBI MORTUA MEMBRA
> NIL ALIUD BULBIS QUAM SATUR ESSE PODEST.

Claudius Galenus (Galen) (129–199 AD) built his medical-scientific system upon the theories of Hippocrates. Galen was not highly regarded in his lifetime, and in fact was even strongly opposed. However, after his death, his teachings influenced occidental and oriental medicine for almost 900 years. He was born in the Greek city of Pergamon but lived mainly in Rome where he had been appointed by the emperor Marcus Aurelius as his personal physician. Galen summarized the favorable characteristics of garlic and called it the "theriac for the peasant." The knowledge of various leeks, especially garlic, later on became part of numerous Arabic and medieval works through Galen's *Materia Medica* (Galenus, 1948).

1.3 THE IMPORTANCE OF GARLIC FOR THE EUROPEAN NATIONS FROM THE MEDIEVAL ERA TO MODERN TIMES

Allium sativum L. was originally unknown in middle, western, and northern Europe. The Romans brought it to the Celts, who were acquainted with only broad-leaved wild garlic, *Allium ursinum* L., which they called *Kremo*. Garlic was also introduced to Germany by the Romans. It has been said, however, that the ancient Germans and Anglo-Saxons knew about garlic, but apparently other *Allium* species (common designation *leac*) were involved. The word *Lauch*, which is found in several German dialects, is a term that applies to many *Allium* species (Höfler, 1908; Hoops, 1905).

Later, the Benedictine monks were instrumental in the distribution of the plant. Garlic was cultivated in the gardens of the monasteries. It was believed to be a valuable remedy for the plague and many other contagious diseases. Garlic was cultivated in a separate place of the vegetable garden. The *Breviarum rerum fiscalium*, which was begun during the reign of Charlemagne (768–814 AD) and which contains many inventories of the various Carolingian courts, mentions garlic and its use as seasoning and medicine. Reference was always made to classical authors such as Hippocrates, Dioskurides, Pliny, and Galen. Copies of these books usually contained commentaries with religious and mythological interpretations. The *Capitulare de villis*, a regulation for management of the royal estates at the time of Charlemagne, lists the following *Allium* species (Gareis, 1895):

- *Allium cepa, fistulosum* or *margaritaceum* as *"uniones"* or *"cepas"*
- *Allium schoenoprasum* as *"britlas"*
- *Allium porrum* as *"porros"*
- *Allium ascalonicum* as *"ascalonicas"*
- *Allium sativum* as *"alia"*

The famous cloister diagram, which was designed in the 9th century by monks of the monastery Reichenau for the Abbot Gozbert of St. Gallen, Switzerland, and which is now at the Monastery Library of St. Gallen (Fig. 1.2), shows a vegetable garden in addition to an herb garden. However, this was an unused model that did not consider the geographic location of St. Gallen, because laurel, figs, mulberry, and rosemary could certainly not be grown in that climate. Apparently there were also difficulties with garlic, which is mentioned in the vegetable garden. *A. sativum* is found under *"Alias,"* as already mentioned (Fig. 1.3).

Garlic was not received with equal acceptance in middle Europe. The ambassador of King Otto the Great (912–973 AD) at the Byzantine court was disgusted by garlic. He mentioned in a report that the Byzantine emperor smelled strongly of onions and garlic, and that both plants were used extensively for food preparations there (Tergit, 1963).

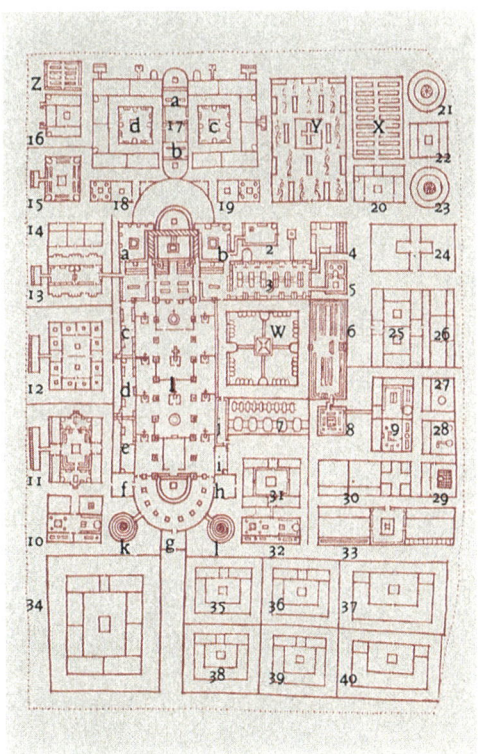

Figure 1.2. Schematic of the monastery of St. Gallen; vegetable garden, item X, near upper right corner (Stiftsbibliothek St. Gallen).

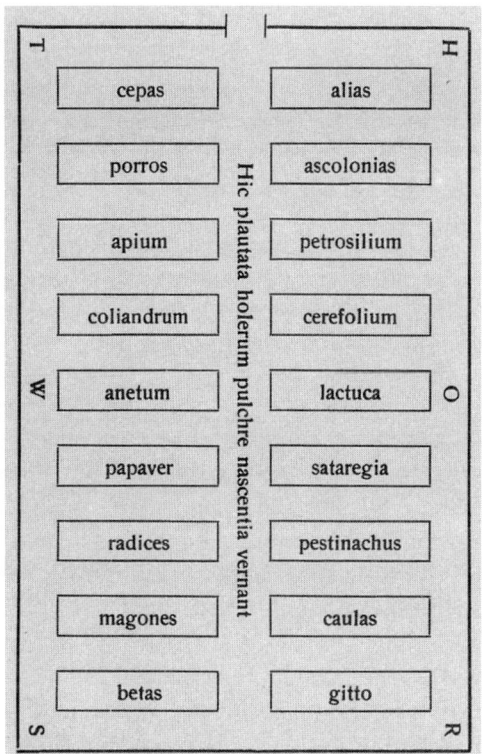

Figure 1.3. *Allium sativum*, shown as "alias" in the vegetable garden of St. Gallen.

Even today, some of this antipathy still remains in northern Europe. Occasionally, opinions were very exaggerated, as expressed by a 19th-century saying from northern Germany:

> He who eats garlic in excess will be full of lice.

As a seasoning garlic was valued chiefly as a sauce that was served with river lamprey or roast beef. In England, garlic was also very little appreciated. Thus, the British did not hesitate to call their archenemies, the Spaniards, the "garlic eaters." Even William Shakespeare said, in *A Midsummer Night's Dream* (act 4, scene 2):

> "And, most dear actors, eat no onions nor garlic, for we are to utter sweet breath."

Nothing has changed in England concerning this matter. Even restaurants that offer French cuisine use garlic only in very small quantities.

It seems somewhat strange to find contradictory opinions in garlic-loving Spain. King Alphonse III of Castilia (866–990 AD) sponsored an order, presumably upon the request of his mistress, to refrain from eating garlic. Such an order existed in a land in which many pilgrims were introduced to a soup known as *olla potrida*. It had been invented by the brotherhood at Santiago de Compostela, the most famous place of pilgrimage. The brotherhood had to take care of the pilgrims, and their soup became a national dish among Spain's middle class. It consisted of cabbage, carrots, onions, pumpkin, pepper, pork, veal, lamb, bacon, and, of course, a considerable amount of garlic. In a cookbook by Marx (Marcus) Rumpott from the year 1604, the soup is called *hollopotrida* and contains not less than 90 ingredients. In a poem by Lord Lytton (Edward George Bulwer-Lytton of Knebworth, 1803–1873 AD), Spain is called "the land of garlic." This appellation is also found in several Spanish proverbs where *los ajos y cebollas* play an important role.

The didactic poem by Macer Floridus entitled *"De viribus herbarum"* proved to be of importance for many later herbal and medicinal books (Behling, 1957). It was written at the end of the 11th century and, in a rather unpolished Latin, describes garlic among 77 medicinal plants. Its true author has not yet been discovered, but the pseudonym he used supposedly refers to the great Roman poet Aemilius Macer.

Duke Robert I of Normandy (ruled 1010–1035 AD) was a great supporter of garlic. In his short, eventful life (he died on a pilgrimage to Jerusalem), he became acquainted with the value of the "aromatic bulb," and he wrote:

> Because garlic has the power to save from death, endure it, though it leaves behind bad breath.

The Medical School at Salerno was one of the most famous colleges of its kind in the Middle Ages. The various fields of the art of healing were treated in didactic poems. Constantinus Africanus of Carthage, a widely traveled man, worked and probably wrote most of his treatises there. He died about 1106 as a monk in Monte Cassino. Occidental medicine owes to him the knowledge of many drugs used by the Arabs, and his works were republished several times. *Allium* is found in the 1536 edition, published at Basle. It is mentioned in the *Alphita*, the drug list of Salerno (13th century) (Thoms, 1926), as well as by Saint Hildegard of Bingen (1098–1179) in her books *Physica* and *Causae et curae* (Hildegard von Bingen, 1955), and by Albertus Magnus in his book *De virtutibus herbarum et animalium* (Albertus Magnus, 1867). Hildegard recommends garlic for jaundice. In both works, theological, mystical, and astrological considerations surround the actual medical statements. The great scholar of the Dominican Order at Cologne gives interesting gardening advice:

> Plant garlic at the feet of sweet-smelling roses. Thus they renew their original odor.

In the 19th century, similar considerations possibly played a role among wine growers who asserted that leeks, especially garlic, grown in the vineyard would yield a stronger aromatic grape. A legend recorded by Sieg points to similar observations on the influences of plants on each other; according to this, garlic was the "star of envy" (Sieg, 1953):

> When St. Mary's flower (Bellis perennis) grew out of the tears of Mary, the devil came and threw a claw (garlic clove) next to each one. Out of these grew the "star of envy." Where it grew, the other plants died. Only St. Mary's flower can prevent further spreading.

The German term *Clobeloch* for *Allium sativum* is found in a Middle High German *Viennese Manuscript* of the 13th century. The Middle Lower German *Gothaer Arzneibuch* (Pharmacopeia of Gotha), presumably written by Eberhard of Wampen (15th century) speaks of *Knoflok* or *Knuflok*. In 1244, Henrik Harpestreng, who worked as physician and canon at the Convent at Roeskilde in Denmark, wrote an herbal book in which he called garlic *Kloflok* (Thoms, 1926). Fischer mentioned in his *Medieval Botany* that a "Weistum" (a legal codex) in 1344 lists the following garden products as subject to taxation: *Reban, Cibölle, Knobloch, Kabaz, Magsam, Lauch*, and *Hanfsam* (Fischer, 1929).

The *Hortus Sanitatis*, in German *Gart der Gesundheit* (Garden of Health), printed in Mainz in 1485 and 1491 by Peter Schöffler, is an important work that was widely circulated. It was edited both in German and Latin (Schöffler, 1485). In this book, woodcuts were used for the illustration of medicinal plants. The draftsman who illustrated garlic used a late gothic form. He transposed the leaves into a twist, such that seven leaves turn toward the left and three toward the right. He shaped the garlic bulb into an ornamental form that simulates a late gothic fish bladder (Fig. 1.4).

The botanists, scientists, and physicians of the 15th–17th centuries frequently used garlic therapeutically, and when they published their observations, they devoted detailed attention to the plant. Paracelsus (1493–1541 AD), whose actual name was Theophrastus Bombastus von Hohenheim, prescribed it as an antidote for the plague, as an expectorant, as a diuretic, for the expulsion of the afterbirth, and externally for treatment of prolapse of the rectum and for abscesses (Paracelsus von Hohenheim, 1928).

Adam Lonitzer (1557), in Latin "Lonicerus" (1528–1586 AD), who was employed as the city physician in Frankfurt-on-Main, authored an herbal book in Latin and German that had over 20 editions (most in Latin). He recommended garlic juice as an external remedy. When applied to the head, it was said to kill lice and nits. When administered internally, it was praised as being effective against worms and poisons. Indeed, this was the belief of many physicians at the time. Again and again, its effectiveness during epidemics of contagious diseases was indicated, although it is not always known whether the term "epidemic" really meant the true plague (*Yersinia pestis*). The most famous saying of the time was:

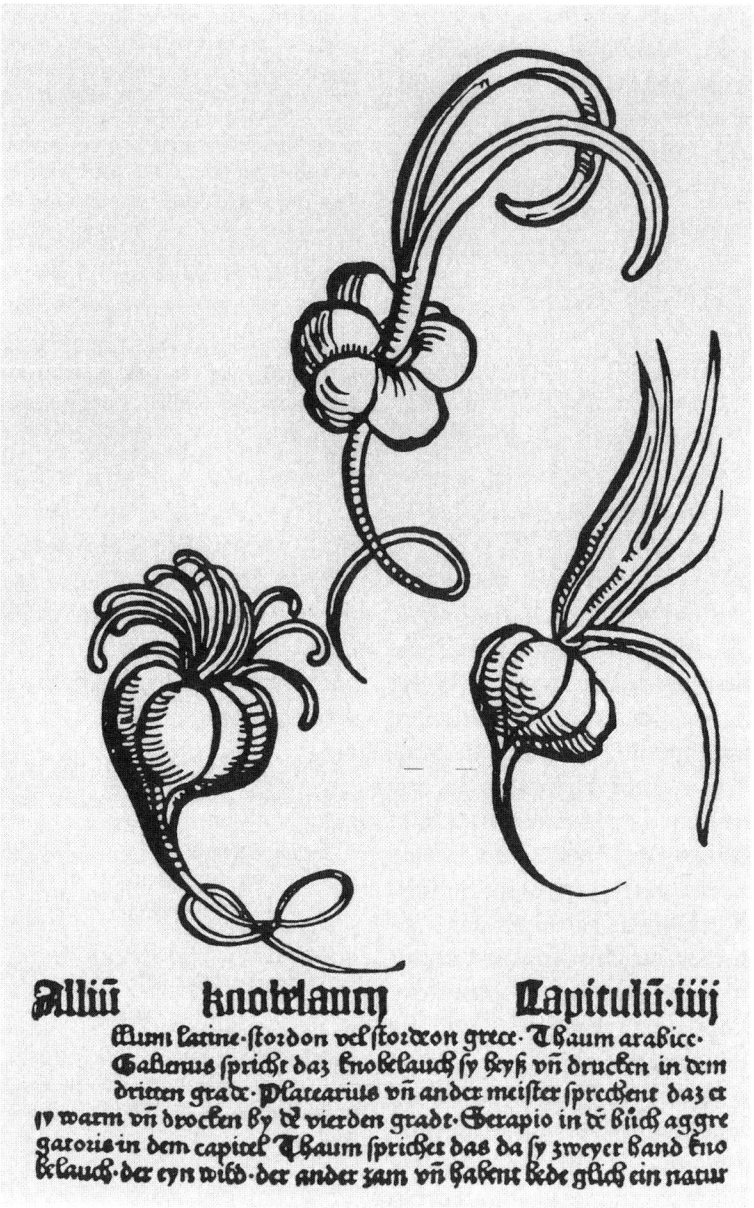

Figure 1.4. Garlic, *Allium sativum,* from the book *Hortus Sanitatis*, Mainz, 1485.

Esst Knoblauch und Bibernell, no sterbet ihr nit so schnell.

(Eat garlic and pimpernel, and you don't die so quickly.)

When the city of Basle was infested with the great plague, it is said that the Jewish population, who was eating garlic regularly, had a much lower fatality rate than the other people. In the year 1721, the plague raged through Marseille. At that time, gangs of thieves passed through the city robbing the sick and dead without infecting themselves. Four of the thieves who were caught by the police offered this explanation: they had regularly consumed a concoction of garlic with wine and vinegar. This remedy soon became rather popular and was called *vinaigre des quatres voleurs* (vinegar of the four thieves).

Sebastian Frank von Wörd (1499–1543 AD), a theologian, originally acted as a priest at Augsburg, then joined the Reformation movement and maintained active relationships with many scholars of his time. He expressed a similar opinion in his pamphlet *Traktat über den Bauerntheriak* (Treatise on Peasant's Theriak) (Frank von Wörd, 1531; Behling, 1964):

> It is warm and dry to the 4th degree, dilutes, impregnates, opens up, disperses, serves against poison, cholic, or cramps caused by bloating, against intestinal worms, poisonous mushrooms if accidentally eaten, or if a lizard ran into the mouth, etc. To prevent the plague, garlic mixed with vinegar can also be used. In the contagious unit at Breslau, the grave diggers used to chew garlic every day in order to feel well, as it is told by Purmann in his Pestbarbier. The Jews take it daily, followed by a mouthful of brandy. The juice of garlic is also a remedy to kill worms, as shown in an example by August Pfeifer concerning a rare heartworm. If the juice is applied externally to the navel, it cures scabies, obstruction of urine, stroke, and stomach ailments. Others mix it with lard applied to the soles of their feet to quiet a cough. If garlic is planted at full moon and also harvested by full moon, it is said to taste sweet. Pharmacies carry an Electuarium de Allio.

Hieronymus Bock, in medieval Latin "Tragus" (1498–1554 AD), a teacher, preacher, and physician from Hornbach in the Wasgenwald, Germany, organized the plants in his *New Kreutterbuch* (the first edition of which was printed in Strassburg in 1539, with several revised editions following) according to plant families (Bock, 1595). About *Allium sativum* he said this:

> Garlic is sweetest in the second
> month after planting,
> before it forms new shoots or seeds....
> The worst is the strong odor which
> may cause headache,
> dull the eyes,
> irritates to anger,
> and promotes sleep....
> Eating garlic prevents all poisoning
> all tired field workers
> who in the heat would drink polluted
> soft water
> and therefore many a germ grow
> which can be overcome by garlic....
> Eating garlic
> gives a clear voice
> and takes away a neglected cough
> is good for dropsy
> promotes urinary action
> serves as laxative
> relaxes abdominal cramps....
> Pressed garlic
> and the juice pressed out
> to be used as ointment
> is good for the skin which is hard
> and dry
> as if leprous.
> The juice removes dandruff from
> the scalp
> and all kinds of spots from the face
> when rubbed in....

Leonhart Fuchs (1501–1566 AD) from Wemding in the Nördlinger Ries, one of the "German fathers of botany," was a professor in Ingolstadt and Tübingen. He made similar statements in his *New Kreüterbuch* (New Book of Herbs), which was published in German in 1543, following *Historia Stirpium*, the Latin edition, which had appeared but one year earlier (Fuchs, 1543; Caesar, 1993). He offers the following saying:

> Garlic warms the belly
> and resolves the churlish and
> stubborn humor.
> Garlic promotes the tendency to sleep
> and lust for sexuality.

The *Kreuterbuch* (Book of Herbs) by "Matthiolus" (1501–1577 AD) contains a rather extensive treatise as well (Mattioli, 1526). Pier Andrea Mattioli (his real name) was a well-known physician and botanist who, from 1554 to 1562, was the personal physician to the emperors Ferdinand and Maximilian II. He writes:

Garlic is not only eaten as spice,
but also as medicine.
It warms and dries the cold
and wet stomach
Relieves constipation
Eliminates flatulence
Offers little nutrition, kills and expels
worms....
For kidney stones: eat garlic
with brandy.
Or cut off three garlic heads into a cup
of white wine, let it come to boiling,
pour through a cloth
and drink warm.
It removes the stone,
increases urination. But use it
with wisdom.
Said potion promotes women to
their time
and drives out the afterbirth....
For the yellow color after jaundice:
eat raw or cooked garlic....
Garlic is not the best thing for hot-
tempered natures.
Those who are plagued by arthritis
or gout
they should not eat garlic
because it repeats and excites the
well-known pain....
Smoke of garlic and its leaves
gives women their coyness
as they lean over it well-covered
and well-exposed to the fumes....
When women faint away
in case of pregnancy
as well as of falling sickness of men
and women.
Also, when children have worms,
a precious medicine. Take a head of
garlic, cut it up,
add *Aloe hepaticum,* or if you
can't get it
a small amount of ox gall;
warm it up
press out the juice
and rub it into the navel
also apply it to pulse and temples
as well as to the nostrils.
A natural animosity exists between
garlic and magnets
this material draws iron unto itself
however, if it is covered with
garlic juice,
it repels the iron....

This last rather curious statement from the *Kreuterbuch* concerning magnetism was followed up in a physical experiment, but it was not confirmed. What the author wanted to express with this sentence remains unknown (Simonis, 1965).

Joachim Camerarius (1555–1574 AD), physician in Frankfurt-on-Main, published Matthiolus' *Book of Herbs* in 1526, and again in 1610 (Camerarius, 1588), using the printing blocks of Conrad Gesner, in Latin "Gesnerus" (1516–1565 AD), a physician from Switzerland. In the 1610 publication, six pages were devoted to a discussion of garlic, together with illustrations (Fig. 1.5) (Gesner, 1555). During that time, herbal books of this kind were distributed in large numbers. But of the Matthiolus book, there were 32,000 copies sold (Mattioli, 1526).

The physician Jacobus Theodorus Tabernaemontanus (1520–1590 AD) from Bergzabern (actually his last name is the Latin translation of his native town) recommended garlic in his *Neuw Kreuterbuch* for the expelling of kidney stones (Tabernaemontanus, 1613):

> Three cloves of garlic cut into a cup
> of white wine
> brought to boiling once
> poured through cloth
> drunk warm will drive the stone
> stimulate urination
> promote the woman's time and expel
> afterbirth.

The botanist, physician, and preacher Otto Brunfels (1488–1534 AD), who is considered one of the "fathers of botany," edited an herbal book with numerous naturalistic pictures of medicinal plants. It was originally published in Latin under the title *Herbarum vivae eicones* and later in German as *Contrafayt Kreuterbuch*. He writes about garlic as *De Allio Teutonice* (German garlic) and comments equally on its use as food and as medicine (Brunfels, 1532).

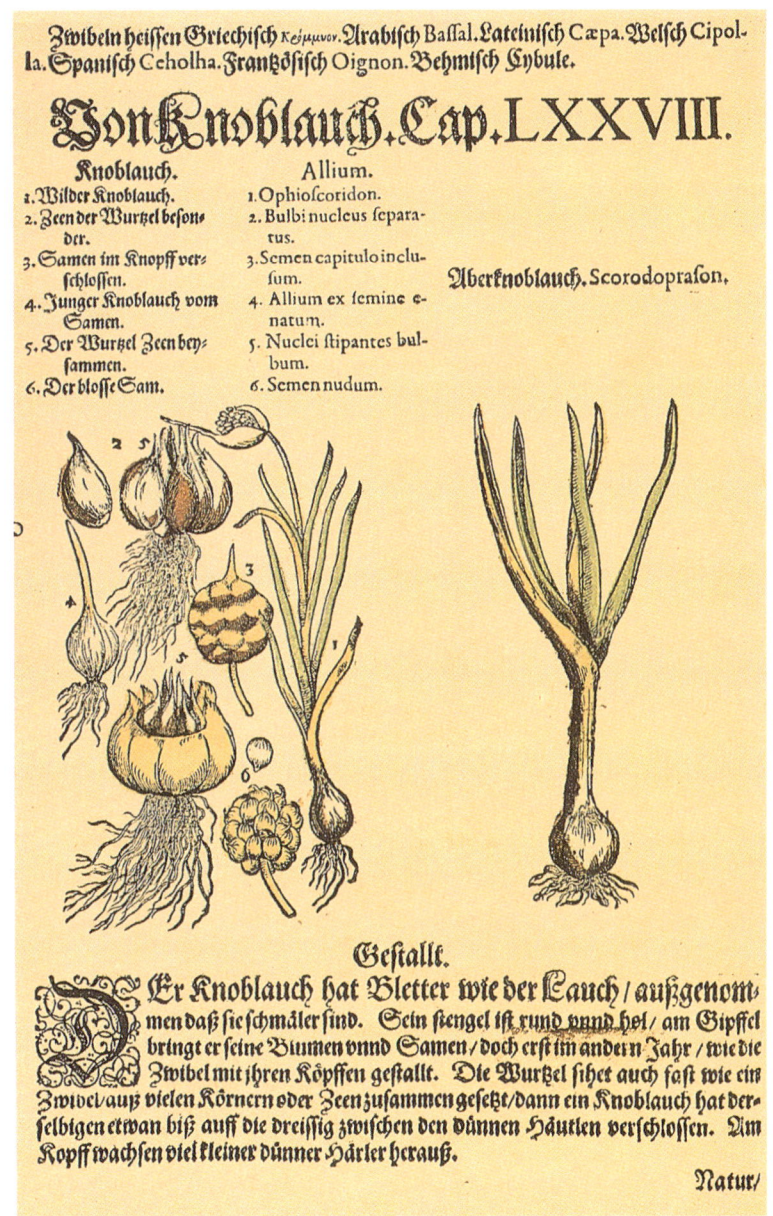

Figure 1.5. Garlic, from the herbal book by Joachim Camerarius, Frankfurt-on-Main, 1610.

Another well-reputed, illustrated book on medicinal plants, also written in Latin, is the *Rariorum plantarum historia* of the Dutch botanist Carolus Clusius from Antwerp, Belgium, whose real name is Charles de l'Escluse (1526–1609 AD). This book describes and comments on the numerous plants of the genus *Allium* in a rather comprehensive manner (Clusius, 1576). The same is true of the book of Clusius' fellow countryman, Junius Rembert Dodoens, latinized as "Dodonaeus" (1517–1585 AD), which appeared in 1557 and is also written in French under the title *Histoire des plantes* (Dodoens, 1557).

In the 17th century, the Englishman Nicholas Culpeper (1616–1654) wrote a very popular book on medicinal plants (Culpeper,

1649). He connected garlic with the astrologic constellation of Aries (Mars) and said that garlic is the plant of the god Mars. This statement is consistent with the accounts of classical antiquity, because during that time soldiers ate garlic for strength and courage. Initially his book received little attention, but after its revision in 1613 by Caspar Bauhin, several further editions were published (Hess, 1860). In spite of the attention given to garlic in herbal books, it was rarely found in pharmacopeias. It remained mainly a remedy of folk medicine, the application of which was very often coupled with superstition (Fig. 1.6).

A nicely illustrated herbal book, the *Parnassus medicinalis illustratus*, was published by Johann Joachim Becher in 1663,

Figure 1.6. Garlic, from *Viridarium reformatum seu regnum vegetabile* (Revised and Updated Herbal Book), Frankfurt-on-Main, 1719.

showing most of the medicinal plants of the genus *Allium* and describing their therapeutical actions, as they were known at this time (Becher, 1663).

Allium sativum is also found in the *Frankfurter Liste,* an inventory of a pharmacy around 1450, and a 1687 pharmaceutical price list from Frankfurt also contains garlic. In a price list from Worms, (1582) *"Radi Allii seu Scorodi"* is listed, and a Würtemberg pharmacopeia of 1741 contained *"Radix Allii sativi vulgaris"* (garlic; diuretic and anthelmintic).

Among the 18th and 19th-century physicians who examined garlic more intensively were Geoffroy, Hecker, Clarus, and Osiander. While Geoffroy holds garlic almost as a panacea (Geoffroy, 1761), Osiander considered it a folk remedy (Osiander, 1829). Clarus prescribed it as a stomachic, an antiscorbutic (i.e., for scurvy), and in the form of an enema, as a remedy for threadworms (*Oxyuris*) (Clarus, 1815). Hecker, who provided the most details, prescribed *Allium sativum* for its ability to aid digestion, stomach sluggishness, abdominal congestion, and chronic diseases developing from these conditions (Hecker, 1815). He also prescribed it, because of its influence on the nervous system, for hysteric or spastic affections; because of its stimulatory effect on the urinary system, for dropsy, sand, and gravel; and because of the expectorant action, for asthmatic problems. He also prescribed it as an anthelmintic and antiscorbutic. For scaly skin he applied garlic externally, like a mustard plaster:

> More frequently it is used for rheumatic deafness. Cotton is soaked with garlic juice and pushed into the ear; this is repeated five to six times the first day. The spot becomes painfully irritated, produces pus, the skin scales off, and often the hearing will return.

1.4 GARLIC IN FOLK MEDICINE AND AS A LEGENDARY AND MAGICAL PLANT

The application of garlic among the common people and in folk medicine is often connected with mystic and religious ideas, a phenomenon already observed in ancient times (Seligmann, 1910; Haltrich & Wolff, 1885). Even then, fishermen and seamen used the cloves as an amulet to protect themselves from evil spirits and diseases. Nothing has changed in this regard in the Orient, Greece, and many coastal areas of the Mediterranean Sea. Little children carry necklaces of garlic to ward off the "evil eye." In Galicia (now in Poland), Jews attached the bulbs to the neck of sick children to protect them from the demons of sickness. The peasants in Bohemia always carried a few garlic cloves in their pockets, hoping that the strongly odorous bulb would extricate any disease from their bodies. The Ruthenians of Bucovina believed that *Allium sativum* would protect them from the jinx. On St. Andrew's Day (November 30) they hung them on doors and windows so that "nobody would take the milk from their cows, and the families would be safe from witches." The Dalmatians and Albanians used it as a vermifuge for children, though not internally, but in the form of necklaces. A pregnant woman protected herself from the witchcraft of evil people by putting parsley and garlic under her linen. In all these relationships with witches and spirits it appears that the memory of witch practices in the 17th and 18th centuries still played a role. During this time, garlic was frequently mentioned, especially in the Comitate (county) of Trentchin (today in Slovakia). At the christening of a child in Estonia, garlic, bread, and money were put into the swaddling cloth of the infant; and the citizens of Königsberg (East Prussia) lauded a healthy child, closing with the sentence: "Garlic, hyacinth bulb, three times white beans" (Pisko, 1896; Seligmann, 1910; Haltrich & Wolff, 1885).

As late as into the 19th century, the superstition still existed among country folk that birds could be kept away from seeded fields by the laying out of garlic bulbs. Some even went so far as to claim that garlic scent dazed the birds so that they could be caught by hand.

The strong-smelling bulbs were used to protect persons of high position, especially princes, from demons and witches. This custom is known from France. Henry IV from Navarre (reigned 1589–1610 AD) was anointed with a mixture of wine and garlic oil in order to keep all evil away. Other crown princes of France were also anointed in this way.

Household animals were also protected with garlic. In Bohemia, dogs, roosters, and ganders were fed garlic on Christmas Eve. The idea of this custom was that all three, especially the dog, would become so vicious and aroused that even the devil would run away. There were similar accounts by the Spanish conquistadors, who fed garlic to their fighting dogs, and who, by the way, introduced garlic to America.

Garlic bulbs were hung up in stables to protect the cows. Opinions varied as to whether the good or bad spirits were influenced. In a legend told in Oostende, Belgium, it was obviously the good ones:

> A farm near the coast flourished under the protection of a helpful spirit. This spirit, whose name was Flexus, did everything that was required of him. All that he desired was a bowl full of good sweet milk every day. An impertinent maid wanted to play a trick on Flexus. Instead of honey, the maid added garlic to his milk. Then the helpful gnome shook with disgust, threw the bowl at the girl's head and said: "Milk and leek, Flexus moves away and good fortune, too." After that everything went wrong on the farm.

In order to prevent the entrance of evil spirits, garlic and old pumpkin stems were cooked in a pot. The content was added to the water, which was then used to clean the room. Following a death, all objects of the deceased were "disinfected" in this fashion. In eastern and southern Europe, it was believed that the dead person would no longer haunt after this procedure. It was believed, too, that the dead might become vampires and could be kept away with garlic. Thus garlic was given the name "bulb of the tree of life" or "stinking rose." In *Dracula*, the old vampire tale by Bram Stoker, the wise professor instructs (in his accented English) the following about garlic (see also Chapter 8 below) (Stoker, 1966):

> "It smell so like the waters of Lethe, and of that fountain of youth that Conquistadores sought for in the Floridas."

In order to protect the beautiful Lucy from being bled by a vampire, the professor covered the entire room with garlic blossoms and decorated her bed with cloves of this spice, so that even the slightest breeze would be saturated with the odor. But the maid removed the badly smelling blossoms, opened windows and doors to air the room, and the next morning it was discovered that Lucy once more had been bled by vampires; she could only be saved by a blood transfusion.

Another story tells of a young mother who had died. She returned frequently to her child until the room was finally sprinkled with garlic broth. It is understandable that among the people of the Balkan nations and in eastern Europe more good effects were attributed to garlic than in areas of middle and northern Europe, where it was detested. That is why there are more folk references found in this area. With the Slovakians, garlic played an important role in magic and folk medicine, even in areas where garlic was not liked (Holuby, 1884). Some uses of garlic can be medically explained: for instance, drinking the pressed juice followed by a glass of gin to cure tapeworm infection, or the mixture of milk and garlic for mawworms. Other uses cannot be medically explained. It is questionable whether a piece of garlic put into the ear will help a toothache, or whether it will cure a boil of the cornea. This use was common in eastern Europe and was carried out by "eye specialists." (This practice was strictly separated by sex; a male therapist was not allowed to treat a female patient.) A peasant named Bosic, living in the Bosnian Sava region at the end of the 19th century, used garlic for epilepsy. His prescription was probably the strangest that ever appeared among garlic applications:

> During 80 days, the patient should fast every Monday and Saturday, and keep

Friday holy. Then, a volunteer hawthorn stem is located, its bark peeled in a strip of a belt; garlic is ground on it and spread to the bast; the entire belt is covered with linen strips and fastened to the body of the patient. More garlic is ground to a pulp which is put into the lining of the patient's cap. Belt and cap should be worn uninterrupted by the patient for 80 days; the garlic pulp in the cap must be moistened freshly every night with ice-cold water.

As an afterthought it is written: "It helps, if God wills it."

Crushed garlic was supposed to inhibit further growth of warts, but the procedure was not to be used for a long period, "since the flesh around the warts would decay." Even the curing of rabies, if it really was rabies, is reported:

> A woman was bitten by a mad dog and was afflicted with rabies herself. Since no one knew what to do with the raving patient, she was locked in a room. There she found a string of garlic, which she thoroughly chewed in her madness. The patient fell into a deep sleep. When the commotion had stopped and everything was quiet, the family went in after a while to see what had happened. They found the woman in a deep sleep, and when she woke up she was healed from the disease.

The same story is also told of a peasant in Dalmatia. Several accounts were given about the use of garlic during the cholera epidemic in 1866. Ground garlic was used to rub the extremities affected by cramps and involuntary contractions. The entire house must have been filled with an overpowering stench. However, nothing is mentioned as to whether cholera itself was cured with this remedy. If it is assumed that garlic has some antibiotic activity, it may be possible. Garlic mixed with fresh butter was supposed to heal difficult boils. This is absolutely demonstrable. However, to heal jaundice by stringing up garlic on a thread and wearing it around the waist is as doubtful as applying the same technique to ward off evil spirits. In Siberia, the inhabitants collected garlic or some other wild growing leek species which they called *Tcheremissa*. The root was credited with so many healing characteristics that, in autumn, pilgrimages of old people and sick people went there to treat themselves with the root right at the source (Lackhovsky, 1932).

In southern areas the field workers ate garlic with their drinking water to protect them from infections by unboiled water. Matthiolus already mentioned this custom in his book, "since the leek withstood all poison." However, the following account, if true, appears rather doubtful scientifically:

> Its poison-repelling power is so great that if a string of garlic is put on a table in a circle, and a spider is placed in the center of the circle, the spider dare not move out of the circle and cross the garlic.

An excellent power of observation is revealed by the following story from the time of King Olaf, the Saint of Norway (995–1030). It is supposed to have happened in 1030 after the battle at Stiklarstadt near Drontheim (Marzell, 1967).

> Several warriors who had been wounded in the battle were treated by a woman skilled in healing (medicine). After she had cleaned the wounds, she gave them leek to eat so that she could possibly detect from the odor of the wounds whether or not they had also injured internal organs.

Such diagnostic application reveals an astounding knowledge of the rapid absorption of the effective constituents of leek. Whether we are dealing with garlic in this story remains unknown (Hoops, 1905). In the older *Edda* which had been written just prior to the above event, the song of *Sigrdrifa* is recited (Döbler, 1975):

> The filling bless and protect you from danger and put leek into your drink.

Here, it is obviously not only a medicine, but a magic potion. *Laece-Boc*, a medical book published by Cocktayne, was based on a manuscript of the 10th century (Thoms,

1926). It contains the following recommended treatment for madness: garlic juice mixed with holy water, to be drunk out of a church bell.

For many years the peasants in the *Erzgebirge* (mountains, now in the Czech Republic) served a spicy garlic soup for supper. A couple of handfuls of well-cleaned and cut garlic cloves were boiled with butter and salt in a pot of water for some time. It tasted similar to beef broth. The soup was served with croutons. This soup, eaten fairly often, kept the family healthy. A farmer's wife related that it made the cheeks red and the limbs fresh and that she rarely had to call for a doctor for her children.

In upper Italy, *Allium sativum* is considered an as excellent remedy for hangover. A tablespoon of ground or pressed garlic, taken by the drunk person, will work like magic. It is also favored in Italy for stomach ailments. Garlic also enjoys great popularity in Russian and Indian folk medicine. The prescriptions have changed but little (Demitsch, 1889; Kloppenburg-Versteegh, 1937; Madaus, 1938).

For short-windedness (asthma), garlic is boiled in water with sugar to a syrup and taken by teaspoon. Garlic should not be disregarded for the treatment of cholera. In its simplest application, the cloves should be eaten, or an alcoholic extract should be used including: 6 tablespoons of Alpinia galanga root, 5 tablespoons of Allium sativum, and 2 teaspoons of salt, allowed to stand in the sun for 14 days. For jaundice, fever, and dropsy, garlic is used in the same way as for asthma. The Indians use garlic as a malaria preventative, for persons who have to work in the marshes. Its extensive external application is remarkable. Insect stings are treated with a poultice of garlic, saltpeter, vinegar, or salt. Bloodshot lesions allegedly disappear quickly by treating them with a compress of garlic and honey. Putrid wounds, boils, skin ulcers, splinters are attended to with garlic roasted in linseed oil. Even the most stubborn skin rash will disappear with a pulp of garlic together with honey and very little water. For toothache a little piece of garlic is stuck into the ears. For bruised nails, fingers, or toes, a dressing with crushed garlic is applied. Also pimples and pustules on the face are cleared by rubbing with garlic juice.

In Russia, crushed garlic with honey is favored for a mucus-producing cough. The mixture is kept on a warm stove overnight and then dispensed to the patient. For a very stubborn cough accompanied by bloody discharge, garlic, finely crushed with *Prunus padus* and honey, is necessary. Garlic and tallow mixed in an ointment and rubbed onto the neck and chest is an external application for angina and whooping cough. Otherwise the indications for garlic in Russian folk medicine are the usual ones like intestinal worms and, formerly, also cholera.

The following summary gives an overview of former usage in Czechoslovakia and other eastern European countries. Garlic healed snake and rabies bites, expelled worms and prevented the stings and bites of poisonous animals, as well as the plague. Garlic stopped coughing, aided in mushroom poisoning, colic, restless sleep, regulated menstruation, was used as a treatment for kidney stones, jaundice, and dropsy; externally it was used for boils, abscesses, and ulcers. Fainting due to pregnancy or epilepsy could be overcome by rubbing garlic onto the pulsing areas of the arteries and by the inhalation of garlic. Urinary inhibition was removed and pain due to podagra was alleviated. In former times, scabies was healed with garlic finely ground with honey. For toothache it was recommended as an infusion of garlic in vinegar with the addition of incense and pine resin; this was held in the mouth. A solution of three garlic cloves cooked in white wine was used to prevent kidney stones and as a diuretic. Six crushed garlic cloves boiled in water and drunk warm was used against passing blood-containing urine. Garlic also alleviated stomach cramps and removed mawworms, where various preparations were used. Here and there garlic infusions were used for tuberculosis. An infusion of garlic and horseradish was taken three times daily for chills. In the area of Podebrade, garlic was given as a condiment with meals to people with rheumatism (Madaus, 1938).

Notable possibilities for its application were also known in French folk medicine. Garlic poultices were applied to festering wounds. Mixed with choke cherries and honey, it relieved coughs.

On the Balkan Peninsula, garlic was and is the symbol of fertility, health, and wealth (Krauss, 1885). It is as unthinkable for it to be removed from the kitchen of these countries as it is from the French kitchen. Every host and farmer stores a considerable amount. Even hermits on the Mount of Athos, who certainly do not care for material values, show their garlic storage with pride. It is even effective in our modern time, especially in Bulgaria, the homeland of a "Methuselah" who celebrated his 110th birthday, because he ate garlic every day. According to Jonathan Swift, "Every man desires to live long; but no man would be old." And garlic is obviously the remedy for it or against it, however you want to look at it.

Garlic odor and lily-of-the-valley odor are certainly an unreasonable combination for the middle European, yet in the Mediterranean region this is perceived differently. Men, especially, use lily-of-the-valley perfume to cover garlic odor. Therefore, it is unexplainable to the proponents of the strong aroma when non–garlic users shrink back. Some emphasize that the aromatic garlic bulb eventually develops a lily blossom, and was for our ancestors the symbol of loveliness and purity. Why, then, not smell of garlic? Alexandre Dumas (1802–1870 AD) said:

> The entire air is characteristically perfumed and penetrated by the fine essences of this mystical and wonderful bulb.

He wrote these words on a journey through the French Provence, which he declared to be the "spiritual home" of garlic, and where, in his opinion, garlic cloves are used as commonly as orange slices are used in other areas. For many, this odor is wonderful and very enticing, almost like a night with a full moon. They think of roasted lamb, snails, and a variety of soups. Raspail, another French writer (Leclerc, 1918), called garlic the "camphor of the poor man."

In conclusion, a book by Georg Friedrich Most, scientifically reworked in the mid–19th century, indicates how many of these folk medicines were still used at that time (Most, 1843). Most was a physician in Rostock, northern Germany. His book, published in 1843, was considered a popular family health book and was widely distributed.

> Garlic (Allium sativum L.). This kitchen herb is very much favored by the Israelites and is similar in its effects to onions, though very much stronger. In various countries, namely southern Germany, France and Italy, garlic is used medicinally for the following cases: 1) for ear and tooth ache: put a fried garlic bulb on the upper arm; the skin will be reddened and thus the pain will be relieved through diversion. 2) Freshly pressed juice used externally: for herpes rashes; soaked in cotton and pushed into the ear canal for nerve deafness; rubbed under the soles of the feet and around the heart area for whooping cough. To eliminate worms in children: garlic cooked in milk and given as an enema is a popular remedy among the country as well as the city population. 3) In former times, garlic was also considered a preventative for infectious diseases. Garlic was boiled in strong vinegar and drunk warm. 4) Daily intake of garlic is much recommended for a prolonged mucous cough and also stomach trouble. Its only drawback is a strong odor upon exhaling which is unpleasant for other people. In order to suppress the odor, chew afterwards 2–3 burnt coffee beans, or green rue, or a fried chard root

(Tabernaemontanus). 5) As a remedy for blushing, one young man took a piece of garlic and inserted it into his anus. The effect was facial paleness and purgative action. 6) According to Tabernaemontanus, if a person is bitten by a mad dog, he should eat much garlic followed by good wine. He also recommended this procedure for snake and viper bite. 7) Galenus prescribed garlic, cooked in milk, to be drunk on an empty stomach by children with worms. Other good vermifuges are all bitter, horrible smelling, mechanically irritating remedies (e.g., creosol). Also, Tabernaemontanus writes that vermin infesting the head are destroyed by the drink. 8) An ointment prepared with garlic juice and honey will remove scales and spots from the face, and will also promote growth of hair. 9) Tabernaemontanus recommends the following for difficulty in urinating: 4 garlic bulbs are roasted in ashes, then ground in a mortar and a small amount of pepper added. The poultice is applied warm to the genitals. 10) For relieving the heavy pain in podagra, Tabernaemontanus recommends garlic infused for 4 days in the urine of a healthy boy and applied externally.

Garlic and folk medicine or folk customs have truly merged!

2
Botanical Characterization and Cultivation of Garlic

GOTTFRIED HAHN

2.1 A BRIEF OVERVIEW OF THE GENUS ALLIUM

The genus *Allium* comprises approximately 600 known species distributed over Europe, North America, North Africa, and Asia. The number of species belonging to the genus has grown steadily over time as botanists have discovered new species throughout the world (Table 2.1). Stearn has provided many valuable references regarding the botanical nomenclature and classification of this genus (Stearn, 1944, 1978, 1980). However, there remains some disagreement concerning the assignment and nomenclature of the individual species. In spite of this, the list contained in Table 2.2 of the most common species will provide the reader with an overview of the diversity of this genus, as well as its economic and pharmaceutical uses.

The general characteristics of the *Allium* species are that the plants are almost exclusively herbaceous, perennials, and usually form bulbs. Some species, however, often form thickened rhizomes. Typical for most species is the leek odor, although this is also known in some plants that do not belong to the genus *Allium*. The leaves of the individual species show a variety of forms, but are generally basal and sessile. The flowers form flat to spherical umbels and are subtended by one or two (sometimes more) spathaceous or leafy bracts. The blossoms are comparatively small with long pedicels. The spathaceous umbels often stand off or bend, forming the shape of a bell. They are either slightly connected at the base or freestanding. The six stamens are fused to the base of the perianth segments. The inner stamens are broadened at the base and sometimes accompanied by a tooth-like formation. The anther is always connected in dorsal position. The style is enveloped at the base by a canal

Table 2.1. Number of Registered *Allium* Species

Author	Year	*Allium* Species	Reference
Samuel Dale	1693	37	Dale, 1693
Albrecht von Haller	1745	24	Haller, 1745
Carl von Linné	1753	31	Linnaeus, 1753; Linne, 1770
John Sibthorp	1823	22	Sibthorp, 1823
George Don	1827	129	Don, 1832
Carl Friedrich Ledebour	1853	73[a]	Ledebour, 1853
Eduard von Regel	1875	262	Regel, 1875, 1887
Edmond Boissier	1882	141[b]	Boissier, 1881
A. Vvedensky	1935	225[c]	Vvedensky, 1946
William T. Stearn	1944	~500	Stearn, 1944
Hamilton P. Traub	1968	>600	Traub, 1968
Per Wendelbo	1971	>600	Wendelbo, 1971

[a] Only in the area of the former Russian Empire
[b] Only in the Near East between Egypt and India
[c] Only in the Caucasus and in Central Asia

Table 2.2 Overview of Some Allium Species

Species	Common Name	Native to and/or Cultivated in	Usage
A. ampeloprasum L. (Porrum ameloprasum Rechb.)	summer leek	Europe, Asia, America	spice
A. ascalonicum L. (Porrum ascalonicum Rechb.)	shallot (Bulbus Alii ascalonici)	Asia minor, cultivated in Europe and Asia minor	tonic and spice, for gastrointestinal disorders
A. atropureum W et K	black-purple leek	from Eastern Europe to Siberia, Hungary, Asia minor, Afghanistan	ornamental
A. bakeri Regel (A. chinense G. Don)	Rakkyo	China, Japan, widely cultivated	spice, vegetable, in folk medicine for cough and diarrhea
A. caeruleum Pall (A. azureum Ledeb., A. coerulescens Don.)	black-purple leek	Asian Russia	ornamental
A. cepa L. (A. esculentum Salisb., Porrum cepa Rechb.)	onion (Bulbus Allii cepae, Bulbus Cepae)	Iran, Belutschistan, cultivated in Europe, Asia, America	spice and vegetable, for cattarh and wound treatment, bactericidal
A. fistulosum L. (A. altaicum Pall., Cepa ventricosa Moench.)	winter onion, two-bladed onion, Welch onion, Japanese bunching onion, ciboule	Korea, eastern Asia, widely cultivated	bulb as spice, seeds as tonic leaves for raising turkeys
A. flavum L.	yellow leek	Alps, Balkans	ornamental
A. giganteum Regel	giant leek	Himalayas, widely cultivated	ornamental
A. karataviense Regel		Turkistan	ornamental
A. moly L.	yellow garlic	southern Europe, widely cultivated	ornamental
A. neapolitanum Cyr. (A. lacteum Sm.)	daffodini garlic Giuseppini (Italy)	Spain, southern France, Italy	ornamental (no garlic odor)
A. nigtum L. (A. magicum L.)	black leek, magic leek	southern Europe, France, Near East	ornamental
A. ophioscorodon Don. (also as var. of A. sativum)	pearl onion	widely cultivated	bulb as spice and vegetable
A. oreophilum C. A. May (A. ostrows-kianum Regel)		Turkistan	ornamental
A. porrum L. (Porrum commune Rechb.)	leek, winter leek (several varieties)	Mediterranean, today cultivated everywhere	spice and vegetable for boils
A. pulchellum Don. (A. montanum Rechb.)	pretty leek	Alps, Balkans	ornamental
A. roseum L.	rose leek	southern Europe	ornamental
A. schoenoprasum L. (Cepa schoenoprasum Moench.)	chives (several articles)	Europe, widely cultivated	spice and vegetable, seeds as vermifuge
A. scorodoprasum L. (considered variety of A. sativum)	snake leek	Europe, Korea, Japan, eastern Asia	spice; eastern Asian varieties rich in leek oils
A. sphaerocephalum	ball leek (Radix Allii spaerocephali)	southern and central Europe	roots in folk medicine
A. triquetum L.		eastern Mediterranean	ornamental; spice
A. ursinum L.. (A. latifolium) Gilib., A nemorale Salisb., Ophioscordon Ursinum Wallr.)	wild garlic, bear's garlic	Europe, northern Asia	herb as tonic, for arteriosclerosis, hypertonia, indigestion, rashes, spice
A. victorialis L. (A. plantagineum Lam., Cepa victoralis Moench.)		Europe, eastern Asia, America	onion as a diuretic and vermifuge; ornamental

which occupies the axis of the ovary. The stigma is obtuse and barely broadened. The fruit, often flattened on top, is a tripartite, trilobate, trilocular capsule with 1–2 (seldom more) seeds in each compartment. The black seeds are spherical to angular, at times also flattened (Fig. 2.1; also Fig. 2.5).

Sometimes bulbils replace the blossoms, especially with *Allium vineale* L. and *Allium oleraceum* L. Bulbils represent ontogenetically compressed lateral shoots. They are often found among Liliaceae. Anatomically they have a short shoot with a sheath containing two fleshy storage leaves and the bud.

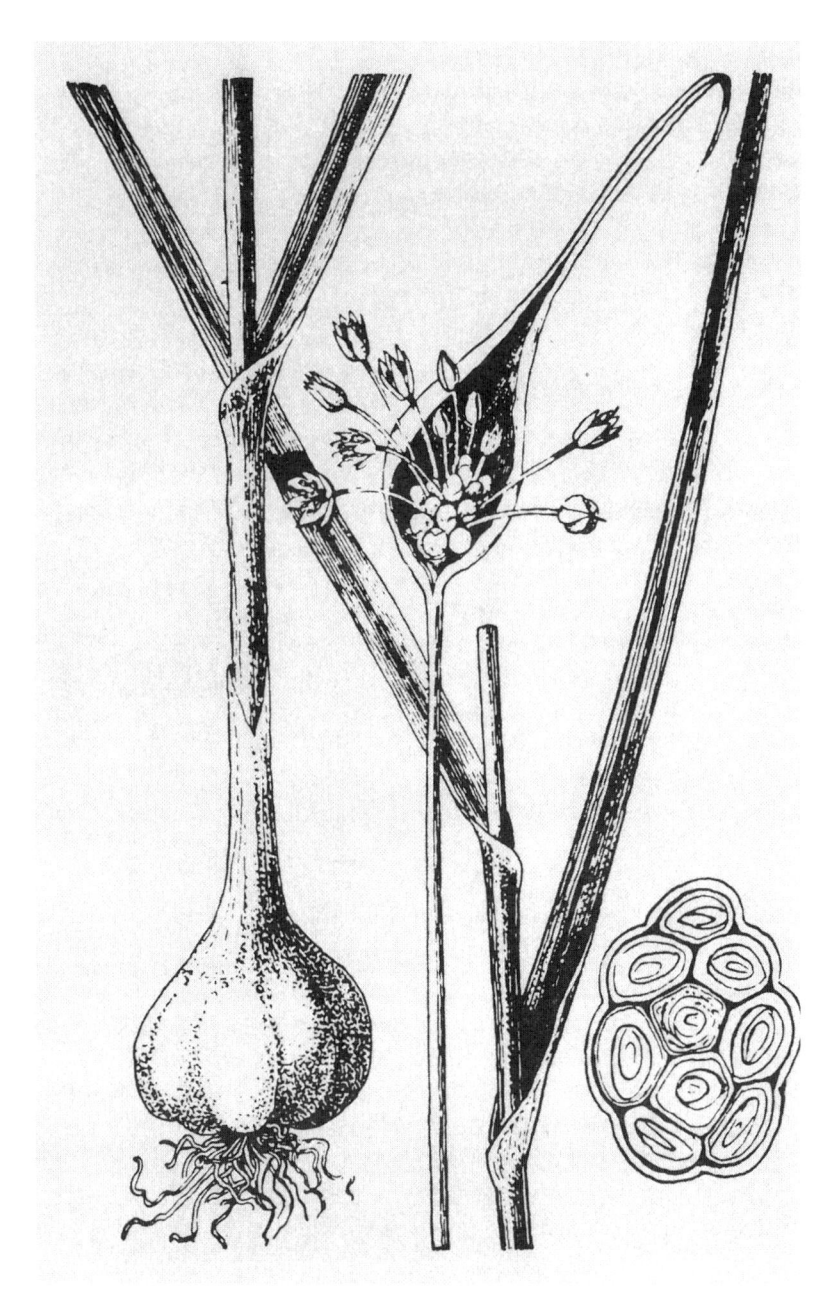

Figure 2.1. *Allium sativum* L. (garlic): flowering plant, bulb, and bulb cross section. (From Weber H. Botanik, eine Einführung für Pharmazeuten und Mediziner. Stuttgart: Wissenschaftl. Verlagsges., 1962.)

Most species of the genus *Allium* are known as wild plants, with little, if any, economic importance. These wild species preferentially inhabit the Northern Hemisphere, where they grow in open meadows and shrub terrain, and in open forests. With the exception of *Allium schoenoprasum* L. or chives, which occurs in Europe, North America, and even in arctic areas, all economically used *Allium* species originate in the Near East or central Asia (Candolle, 1884; Kotlinska et al. 1990; Purseglove, 1988). A difference between the species of the New and Old Worlds appears to be in the haploid chromosome set. As a rule the North American species show n=7, whereas those of the Old World and the cultivated forms have n=8 or multiples thereof (Simmonds, 1976). Some of them are grown in North American gardens for their ornamental blossoms (Harkness & Fox, 1944).

In the older literature, the genus *Allium* was often listed in the Amaryllidaceae family because of its encased inflorescence. Since the blossom has all the characteristics of the lilies, this species was assigned to the Liliaceae family, a classification still maintained in many textbooks and taxonomy manuals (Table 2.3). However, embryologic and phytochemical examinations have led to a new orientation. For a systematic classification, the development of the female gametophyte is without doubt very important. While in the normal type only one megaspore is involved in the development of the gametophyte (monosporic), *Allium* requires two cells (bisporic). Therefore, it is also called "the *Allium* type." For this reason the genus *Allium* now belongs to the Alliaceae family, in the order Asparagales, while earlier authors ascribe the *Allium* species to the Liliaceae family (Purseglove, 1988; Simmonds, 1976; Traub, 1968, 1972; Hermann, 1945; Stearn, 1944; Hegi, 1939; Wendelbo, 1971; Hardman, 1991; Sebastian et al., 1979; De Halacsy, 1904).

2.2 DESCRIPTION OF *ALLIUM SATIVUM*

Garlic, *Allium sativum* L. (2n=16), presumably originated from central Asia and is now known solely in the cultivated form (Hyams, 1971). The wild plants, as described from the Kirghiz grassland, are probably cultivated specimens that have naturalized, or they are distinctly different leek species. Occasionally, *Allium longicuspis* from central Asia is considered as a wild form of garlic (Regel, 1875), but recent isozyme studies strongly indicate that it is the same species as *Allium sativum* L. (Pooler & Simon, 1993; Maass & Klaas, 1995). Helm describes the following varieties (Helm, 1956):

- *Allium sativum* L. var. sativum
- *Allium sativum* L. var. ophioscorodon (Link) Doll
- *Allium sativum* L. var. pekinense (Prokh) Makino

This arrangement is by no means generally accepted. There are, for instance, publications that list the varieties *A. sativum* and *A. ophioscorodon* as separate species or subspecies (Fig. 2.2). *Allium sativum* var. *ophioscorodon* obtained its name from its contorted stalk—"twisted garlic" or "rocambole"—but is also known as "hardneck garlic" or "topset garlic" (in German, *Perlzwiebel*) (Zander, 1993). A very practical discussion of the subspecies and varieties of

Table 2.3 Botanical Classification of the *Allium* Genus

	New Classification	Old Classification
Genus	Allium	Allium
Family	Alliaceae	Liliceae
Order	Asparagales	Liliiflorae
Superorder	Lilianae	–
Subclass	Liliidae	–
Class	Monocotyledoneae	Monocotyledoneae

Artichoke Garlic
A. *sativum*
Subspecies *sativum*
Variety Artichoke

A

Figure 2.2. Examples of the two main subspecies of garlic: **A.** *Allium sativum* L. var. *sativum* (softneck garlic). **B.** *Allium sativum* L. var. *ophioscorodon* (hardneck or topset garlic), shown with its distinguishing curled flower stalk (scape) and umbel capsule. Cross sections of the bulbs show in dashed lines the multiple clove layers or fertile leaves (four in this case, but three to eight is typical) found in softneck varieties, in contrast to the two clove layers and stalk (A) found in hardneck varieties. See Table 3.5 for other examples. (Drawings by Jim Anderson, used with permission from Filaree Productions, Okanogan, Washington.)

Allium sativum, as well as of their anatomy and cultivation, may be found in Engeland's *Growing Great Garlic* (1991). Using isozymes, DNA markers, and morphology, Maass & Klaas (1995) have recently classfied 300 accessions of garlic into four major groups: variety *sativum*, variety *ophioscorodon*, variety *longicuspis* of which *pekinense* is a subgroup, and a subtropical group. In close relationship is *Allium scorodoprasum* L. (in German, *Schlangenlauch*), which occurs in Europe and Asia Minor in many variations. Its rich leaves, containing leek oil, are very popular as a seasoning herb.

Rocambole Garlic
A. sativum
Subspecies *ophioscorodon*
Variety Rocambole
Type; Late

B

Figure 2.2. *Continued*

Garlic does not have a scale-like bulb like *Allium cepa* L. with its leaves enclosed within each other, but has bulbs of many shapes, sizes and colors, most often white to lightly purple. The single parts of the bulb are called cloves; they are formed from auxiliary buds and their collateral daughter buds. These buds are separately enclosed by leaves closely surrounding the compressed stem. Due to the very tight fit, the lateral bulbs or cloves are flattened. The size of the individual cloves varies considerably. About 4–20 cloves are attached to a bulb by a parchment-like white envelope. The base of the bulb, called the "basal plate," has numerous short roots. Each clove possesses a modified leaf or cataphyll that serves as a storage system and surrounds the vegetation point, all protected by an involucral leaf skin. The cloves of *Allium sativum* L. correspond to the daughter bulbs of *Allium cepa* L. Healthy bulbs can become as big as a human fist (Fig. 2.3). Out of these cloves the plant shoots a quill-like, round, hollow, and unbranched stalk or scape, which is encased at the bottom by tubular leaves. The stalk (no

Figure 2.3. *Allium cepa* L. (onion) and *Allium sativum* L. (garlic). **A.** Whole bulbs. **B.** Sliced bulbs showing scale-like structure in onion and cloves in garlic. **C.** Sliced onion bulb and peeled garlic clove. (Photographs by G. Hahn.)

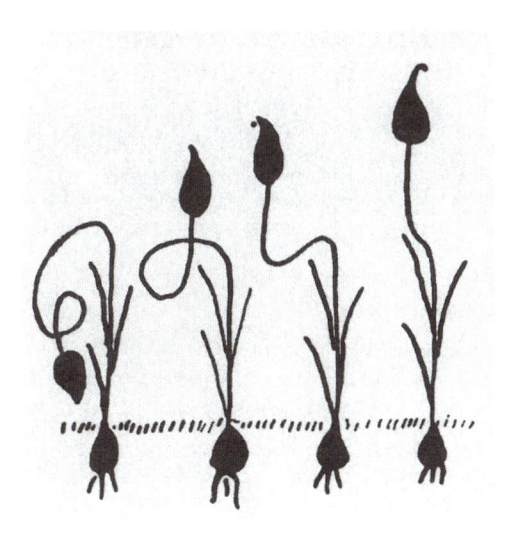

Figure 2.4. Growth changes in a topset garlic plant. (From Simonis WC, Die einkeimblättrige Pflanze. Ulm: Haug, 1965.)

true stem) is 30–90 cm high, depending on location. The approximately 5–9 leaves which encase the stalks are about 12 mm broad, linear, long-tapered, leek-green, entirely margined leaves, keeled at the lower side and grooved on the upper side. Leaves exist only up to the middle of the scape. During growth, the stalk shows curious contortions and spiral formations, which later disappear (Fig. 2.4). The cause is an increased growth at the outer side of the tissue. During leaf development of *Allium sativum* L., Effertz and Weissenböck (1976) have observed a discrepancy in flavonol concentration between the upper and lower epidermis and the mesophyll.

On top of the leafless part of the scape, the loose, spherical inflorescence or flower cluster develops, shrouded by bracts that almost remind one of a dunce cap. In some varieties, the inflorescence develops only after vernalization. The bracts are eventually discarded. The umbellate inflorescence carries only a few greenish-white to whitish-red blossoms (about 5 to 7), which sit on long slender pedicels. The sepals and petals, which are only 3 mm long, surpass the shorter stamens and the even shorter style in the center of the flower. The inner stamens have a broader base and are accompanied on either side by a short, obtuse, tooth-like appendage (Fig. 2.5). The months for flowering are June, July, and August.

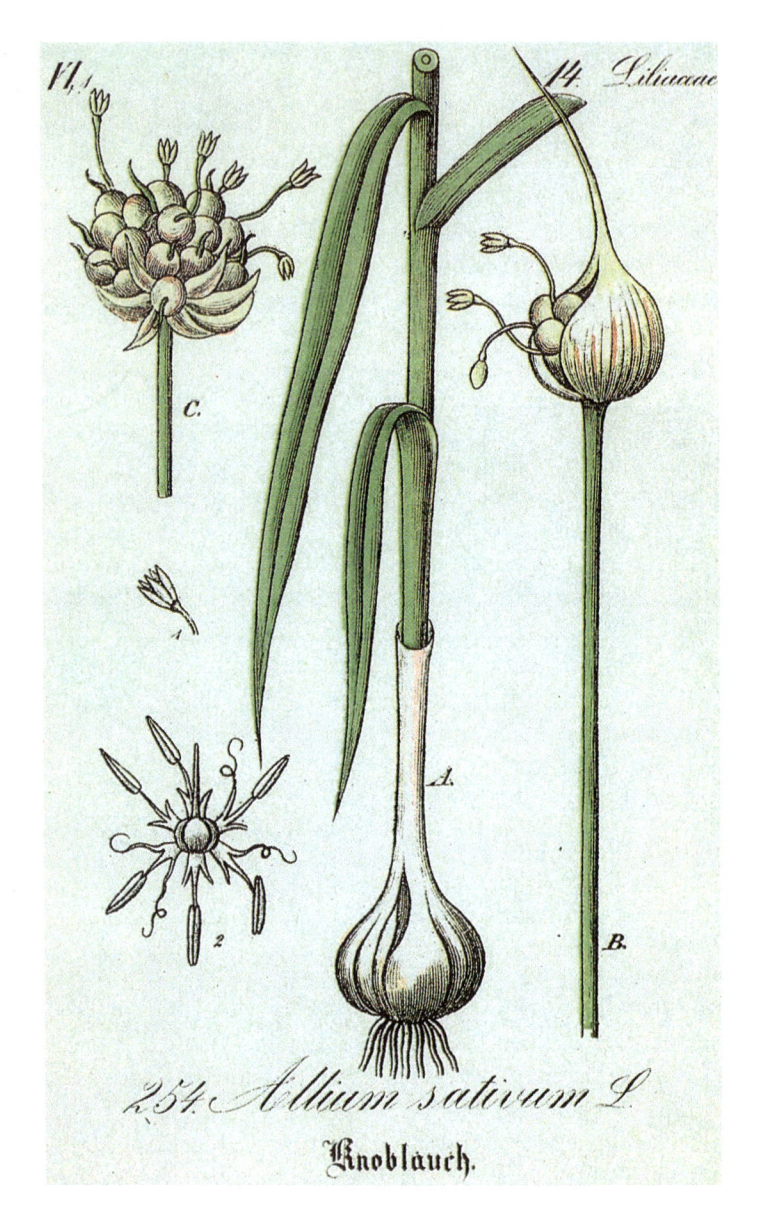

Figure 2.5. Inflorescence (flower cluster) and single flower of *Allium sativum* L. var. *ophioscorodon*. **A.** Roots, bulb, leaves, and lower flower stalk. **B.** Flower stalk (scape) and opened umbel. The umbel consists of the flower cluster, bulbils (vegetative topsets), and an outer cover (spathe or bract), which ends in a pointed beak. **C.** Inflorescence (flower cluster) with flowers and spherical bulbils. *(1)* A single flower with a short pedicel, *(2)* an enlarged flower. (From Schlechtendal DFLV, Langethal LE, Schenk E. Flora von Deutschland. Gera-Untermhaus: Fr. E. Köhler, 1880.)

About 20 to 30 egg-shaped bulbils or "brood bulbs," each 1 cm in diameter, develop from the flowers in the inflorescence. They have an important function in the propagation of the garlic plant, since the blossoms themselves are usually sterile. The ripe bulbils fall off the plant and can be successfully dispersed elsewhere by rain. Every bulbil can become a complete plant. During the first year a "stock bulb" emerges, producing no cloves, however. The usual composite garlic bulb is not developed until the second year.

The bulbils have mistakenly been called "fruits." Since the pollen of today's cultivated garlic is almost always sterile, true fruits are really a rarity (Novak, 1983; Sprecher, 1986). Nevertheless, they are mentioned again and again in botanical descriptions. If real fruits do develop, they are trilocular capsules. Generally, however, propagation occurs by vegetative means.

The obvious variability among garlic specimens suggests an originally genetic propagation of the plant. The sterility of the pollen, however, prevents a targeted breeding project. Earlier attempts to overcome the sterility of common garlic were unsuccessful (Kononkov, 1953; Etoh & Ogura, 1977; Novak & Havranek, 1975; Konvicka et al., 1978; Konvicka, 1973; Etoh, 1985; Koul & Gohil, 1970), except in one case when viable seeds were obtained from a clone of garlic (Etoh, 1983). On the other hand, we know of numerous varieties of the onion plant (*A. cepa* L.) with very high yields of viable seeds (Konvicka, 1984; Etoh et al., 1988). Gori and Ferri (1982) found that the cause for the sterility of garlic plants is a premature atrophy of the microspore. The fine structural changes which are the basis for this process have been described. Novak et al. were able to obtain tetraploid garlic through treatment of the bud tips with colchicine (Novak et al., 1982; Novak, 1978, 1980, 1981, 1983, 1984). Recently, a seed-bearing tetraploid garlic has been found in Italy (Bozzini, 1991). Pooler and Simon (1994) very recently also succeeded in producing a small sample of true seeds from selected garlic clones; this opens the possibility of producing new recombinants through sexual reproduction. On the other hand, Suh and Park (1993) produced new garlic plants through immature bulbil cultures. Possibly, the problem of garlic's sterility will be overcome in the future by the gentechnological approach (Wilkie et al., 1993; Havey, 1992a, 1992b).

Recently, garden centers in Germany have begun to offer *"Knolau"* (linguistic contamination of German *Knoblauch* and *Schnittlauch*)—a hybrid of garlic and chives, which does not produce any bulb. The new product is more aromatic than chives (*Allium schoenoprasum*). Recently, interspecific hybrids have been described in the literature, generated from *A. cepa* L. and *A. sativum* L., and, somewhat earlier, from *A. cepa* L. and *A. fistulosum* L. as well as from *A. fistulosum* and *A. ascalonicum*, respectively, showing different flavor compounds and carbohydrates than the parent plants. (Gupta et al., 1993; Ohsumi et al., 1993a–c; Song & Peffley, 1994; Ulloa-Godinez, 1993; Bark Hur, 1994; Ohsumi & Hayashi, 1994; Ulloa et al., 1994; Hizume, 1994; Cochran, 1950; Nomura et al., 1994; Levan, 1936). The hybrid between onion and garlic shows characteristics intermediate between the two parent plants (Ohsumi & Hayashi, 1994). Recently, transgenic plants expressing a heterologous isopentenyl transferase gene, which improves the tuber yields of the plants, have been patented (Letham et al., 1993).

The anatomic-histologic structure of the garlic bulb displays a certain conformity with *Allium cepa*. Especially characteristic is the occurrence of numerous well-developed oxalate crystals (Fig. 2.6) in the dry white involucres which consist of elongated sclerenchyma fibers. These crystals are absent in the hypodermis of the fleshy leaves. In cross section, the individual cloves show concentric scales which have an epidermis of greatly thickened cells on the inside as well as on the outside. In between is the large lumenal parenchyma tissue into which vascular bundles with oil-containing cells are encased. In addition to the parenchymatous sheath, oil can also be seen in all other tissues, except

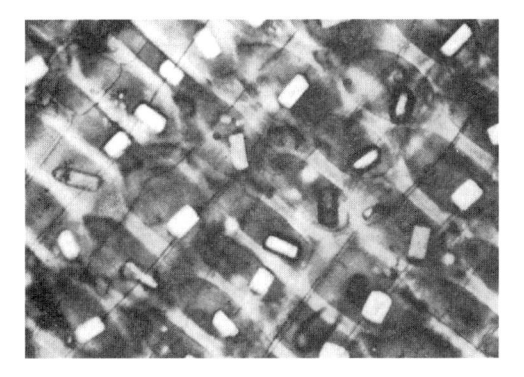

Figure 2.6. Oxalate crystals in the dried white scales of a garlic bulb. (From Schormüller J. Obst, Gemüse, Kartoffeln, Pilze. In: Acker L, ed. Handbuch der Lebensmittelchemie. Berlin: Springer, 1968).

in the vascular cells. The lower part of the bulb, which is called the basal plate, consists of round to elongated, widely differing lignified cells, and hence is rather woody. Since remains of the basal plate are directly connected to the cloves, lignified cells can be found also in some *Allium* preparations. To complete this description, the following are some pharmacognostic data (Anonym, 1985; Hager, 1969; Hörhammer, 1970; Thoms, 1926).

Cut herb: characterized by yellow, horny, strongly hygroscopic pieces and of white to slightly yellow, shiny, paper-like strips of the scales. The strongly hygroscopic pieces have a glassy fissure, the characteristically penetrating odor of garlic, and a burning, pungent taste.

Powder: The off-white to brownish dry garlic powder contains numerous prisms of calcium oxalate from the hypodermis of the outer cataphyll; fragments of tissue with large parenchymal cells and spiral vessels from the mesophyll; thick-walled, elongated epidermis cells of the dried white scales; and fiber-like, elongated, strongly thickened and pitted epidermal cells of the dried cataphylls of the secondary bulbs.

Allium sativum L. is native to the entire Mediterranean region and large areas of Asia, and it is cultivated in many countries. It can be found wild in the Mediterranean area and is collected by the native population, especially during the months of July, September, and October. Presumably, there are other regions in which garlic is cultivated or where it can be found as a wild plant.

2.3 CULTIVATION OF GARLIC IN GARDENS AND AS AN AGRICULTURAL CROP

Garlic used in Europe comes from the Balkan countries, Hungary, Czechoslovakia, France, Spain, and Italy. In Germany, it is cultivated in the area of Schweinsfurt and Nürnberg (the "garlic country"). The little town of Uzès in the backcountry of the Cévennes Mountains has the reputation of being the largest market for garlic in all of France. Large amounts of garlic are produced in Egypt, India, China, and South Korea. During recent years, China has become not only a notable user, but also a major producer of garlic. Argentina is the main producer in South America. In the United States, the area around Gilroy, California, known as the "Garlic Capital of the World," grows about 90% of the U.S.-produced garlic. A great deal of garlic is cultivated in Russia, but it is directly consumed there. Garlic is imported into the Federal Republic of Germany from the following countries (in decreasing order, from more than 3000 tons, but not less than 25 tons each): China, Spain, Italy, France, Mexico, Argentina, Hungary. Some information on consumption, production, and importation of garlic in the U.S. is provided in Table 2.4.

Allium sativum L. prefers a well-fertilized, open ground, and sandy, loamy soil is particularly suitable. However, heavy soil will also yield good crops. Warm, sunny, wind-protected locations are especially good for its cultivation. The plant is very sensitive to excessive moisture; cool, rainy areas should not be employed for its cultivation, as the product would be a low-grade garlic of poor taste quality.

Cultivation of garlic can be done either on an annual or biennial basis, depending on whether cloves (annual) or brood bulbils (biennial) are harvested. Depending on weather and climate, rows of bulbils are set from March to April with 4 dt/ha (deciton/hectare) or 355 lbs/acre in rows 20 cm wide with a planting depth of 5 cm. The planting of cloves is done either in March/April or September/October with 8 dt/ha in a planting arrangement of 20 x 12 cm at a depth of 5 to 8 cm. Fertilization is recommended with 0.30–0.50 lbs/acre nitrogen (nitrate of lime and ammonia), 0.40–0.80 lbs/acre superphosphate, and 1–1.50 lbs/acre potassium salt 40%. Of course, potent fertilization with manure or compost is also feasible. The addition of nitrogen can boost the crop yield of garlic enormously. There are, in addition, chemical growth regulators that can be used to influence the productivity of garlic plants (Souza & Casali, 1991,

Table 2.4. Consumption, Production, and Importation of Garlic in the United States

	1991	1992	1993	1994
Total U.S consumption of garlic [million pounds]	334	328	345	438
% produced in the U.S.	87.0	88.4	87.6	77.6
% from China	1.8	1.1	2.7	14.5
% from Mexico	6.2	6.9	7.3	6.1
% from Argentina	2.4	1.6	1.5	1.3
% from Chile	0.8	0.6	0.7	0.4
% from Taiwan	1.4	0.9	0.3	0.2
% from all other countries	0.4	0.5	0.1	0.1

Source: U.S. International Trade Commission and U.S. Department of Commerce

1992a, 1992b). Usually only a few plants are needed for raising garlic in a garden. Cloves are set 15 cm apart in rows that are 20–25 cm apart. Weeding and tilling must be done, depending on the growth of weeds.

In the present context, the numerous attempts at plant breeding and propagation of good seedlings or bulblets of garlic, as well as of onion and other *Allium* species, by tissue (callus) culture can only be mentioned here (Fujime et al., 1993; Moriconi et al., 1989; Camara et al., 1989; Nagakubo et al., 1993; Munoz & Escaff, 1987; Mohamed-Yasseen & Splittstoesser, 1992; Suh & Park, 1994; Mirghis & Mirghis, 1991; Xue et al., 1991a, 1991b; Vera Lopez et al., 1990; Takagi, 1990; Kohmura et al., 1994; Nagasawa & Finer, 1988a, 1988b; Novak, 1983, 1984; Abo El-Nil, 1977; Kehr & Schaeffer, 1976; Havranek & Novak, 1973; Reichart et al., 1985; Pandey et al., 1992; Inagaki et al., 1992; Viterbo et al., 1992; Novak et al., 1985). Also, pure constituents of the *Allium* species (alliin, allicin, enzymes, etc.) and flavor components (oligosulfides) can be produced in this way (Ohsumi et al., 1989, 1993b; Agrawal et al., 1991; Madhavi et al., 1991; Lee et al., 1994; Zhang et al., 1993, 1994; Chomatova et al., 1990; Mellouki et al., 1994; Collin & Musker, 1988; Selby et al., 1979, 1980; Collin & Britton, 1993; Novak et al., 1985).

Allium sativum L., as all other cultivated plants, can fall victim to diseases, resulting in a greatly diminished yield. Infections with *Peronospora destructor* (Berk.) Casp. occur frequently in colder regions. This fungal disease was first described in 1841 in England; it spread very quickly and occurred in many cultured plants. It infects the bulb, causing soft spots and color changes. Infection with rust (*Sclerotium cepivorum* Berk.) was also seen in England for the first time and was described there for *Allium cepa*, but it may affect *Allium sativum* as well. Likewise "bulb smut" (*Urocystis cepulae* Frost.), observed in the U.S. in the 19th century, is especially harmful. Spores of this fungus remain infectious over long periods of time, and the farmer is forced to refrain from replanting the same ground with garlic for several years.

Many *Allium* species are infected by "gray mold" from *Botrytis* species. The following species have been associated with onion and garlic: *B. allii* Munn., *B. byssoidea* J. C. Walker, and *B. squamosa* J. C. Walker. Since its discovery in 1976 in Germany, this fungal disease has spread widely within 20 years and frequently causes considerable loss of crops. Other infectious agents described are: *Puccinia* and *Uromyces* species ("garlic rust"), *Alternaria*, *Phytophthora* ("herb and bulb rot"), *Fusarium*, and *Aspergillus niger* (Rengwalska & Simon, 1986). Viral diseases, such as *garlic mosaic virus* infections, can also harm the crop yield significantly. Therefore, the possibility of obtaining virtually virus-free garlic plants from meristem cultures is of considerable importance (Havranek, 1974; Yasseen et al., 1995).

Aphids transfer a viral disease, causing a striped, yellow-green discoloration of the leaves, thus causing an impairment of photosynthesis. The most important carrier is the "onion louse" (*Rhopalomyzus ascalonicus* Donc.). The name itself points to the fact that several *Allium* species are among the victims of this parasite.

Bacillus cepivorus Delacr. is one of the bacterial infectants which cause rotting of the bulbs. The metabolic products formed during this process have an obnoxious odor. This infection can also lead to storage losses.

Nematodes, especially *Ditylenchus dipsai* Kühn, cause swellings and deformations in the leaves, growth termination, and damage to the bulbs. The first reports of this disease in onions and garlic came from the U.S. Since the nematodes remain in the soil, the infested ground cannot be planted with *Allium sativum* for several years. *Heterodera marioni* Cornu causes gall-like swellings in the roots, followed by disturbances in water and mineral absorption; the plant withers and dies.

The following biting and sucking insects cause damages and weaken the plant, making it more susceptible to bacterial and viral infections: *Phorbia antiqua* Meig, *Acrolepia assectella* Zell, *Dizygomyza cepae* Her, *Thrips tabaci* Lind., and *Lioceris merdigera* L.

Allium sativum is harvested after the leaves and bulb stems have dried, usually during September or October in the Northern Hemisphere. The bulbs are dug up with wide, broad-pronged digging forks. For larger areas, digging machines are feasible. An average yield is about 60 dt/ha or 5340 lbs/acre (bulbs and side bulbs). Depending on marketing conditions, the bulbs, together with the herbal parts, are bound into braid-like bundles (also called 'braids') and hung up for drying. Garlic is for sale in this form in stores and markets. From large harvests, the bulbs are deleaved and dried on screens. In this case, single bulbs come on the market. The harvested crop has to be well picked over. It is important that the cloves are not dried up under their paper-like scales. Discolored or dried up bulbs or cloves have a bad taste. Good garlic must be firm.

Much of the harvest undergoes processing by specialized companies. There, the garlic bulbs are peeled, then further processed in special cutting machines where they are cut into slices, little strips, or flakes, followed by drying at 60°C. The dried products can then be used as spice or in vegetable mixtures, or they are carefully packaged and put on the market in this form. In the Balkan countries and in Egypt large amounts of *Allium sativum* are ground to dry garlic powder in special garlic mills.

Little needs to be said here about the common use of garlic: it is widely used as a spice. In the Balkan countries, in China, in India, and in the former Soviet Union it is consumed as a condiment in considerable amounts. Its medicinal use is preferentially covered in the scope of this book. Additionally, it may be mentioned that it is not only the bulb that is valued. The young inflorescences or flower clusters are prepared as a tasty vegetable dish in China. The bulbils of *Allium sativum* var. *ophioscorodon* are pickled in vinegar and used as a seasoning.

Finally, some consideration should be given to the question of whether or not *Allium sativum* influences the micro-flora of the soil, which possibly could have an influence on other plants in the vicinity. Agrawal and Rai (1948) contributed an interesting study on this topic. They tested the influence of extracts from bulbs of both *Allium cepa* and *Allium sativum*. They found that extracts which had been obtained by autoclaving stimulated the growth of soil bacteria. Preparations of garlic were more potent than those from onions. Extracts which were prepared in different ways were inhibitory to soil organisms, and again the garlic extracts were more efficient than those from onions. A further observation may be of economic interest: some fungi (*Macrophomina phaseolina, Botryodiplodia theobromae, Colletotrichum corchori*) live as parasites in the seeds of jute (*Corchorus capsularis*). Extracts of the bulbs of *Allium sativum* inhibit the germination of the fungal spores and the growth of the mycelia (Ahmed & Benigno, 1984; Ahmed & Sultan, 1984). To some extent, then, *Allium sativum* does influence the soil flora with its root excretions, a phenomenon about which growers should be informed.

3

The Composition and Chemistry of Garlic Cloves and Processed Garlic

LARRY D. LAWSON

In light of the long historical use of garlic for medicinal purposes, it has become very intriguing and important to determine the composition of garlic so that its active compounds might be revealed. Although garlic is also a commonly eaten vegetable, its health benefits do not lie in its content of classical nutrients, as garlic is normally eaten in relatively small amounts. For example, 5 grams of garlic (a larger clove, and the upper limit that most people are willing to eat) contains only 1.2% of the U.S. Recommended Dietary Allowance (RDA) of vitamin C, its most abundant nutrient, while a 150-gram baked potato contains 50% of the RDA. Discovery and quantitation of the active agents of garlic are also important so that we can assess the level and uniformity of the medicinal quality among the many strains of garlic and to determine any adverse or beneficial effects of agricultural variables, such as soil nutrition, weather, moisture, harvest time, storage conditions, and so on. There are also many changes that can take place when a plant is processed by the various methods that are used industrially for making medicinal food products, and this is especially true for garlic. Furthermore, attempts to remove the odor or odor potential of garlic can cause considerable changes in its composition, including removal of some of the main medicinal principles, since these are also responsible for the odor.

Garlic is sold in several forms: bulbs, pickled cloves, crushed or chopped cloves (bottled with or without vegetable oil), spice powders, and garlic salt. Garlic supplements are sold as powder tablets or capsules with various coatings, as a steam-distilled oil, as a vegetable oil-macerate extract, or as an extract aged in dilute alcohol. As will be discussed later, most of these preparations differ considerably from one another in their composition, due to chemical and enzymatic changes that take place during processing. It is also important to remember that garlic has been effectively used for a variety of ailments. Therefore, a compound which has been discovered to be the active agent for the treatment of one disease may not necessarily be the active compound for another disease. Hence, more than one active compound may need to be sought.

3.1 THE GENERAL COMPOSITION OF GARLIC

All plants contain a large number of compounds. Most of these are necessary for normal metabolism in the plant and are called *primary metabolites*. The content of these compounds is similar among nearly all plants. However, most plants also contain compounds unique to that plant or genus, called *secondary metabolites*, which, among other purposes, serve to provide resistance to a variety of diseases or pests. Therefore, although this chapter will deal to some extent with compounds common to all plants (amino acids, carbohydrates, lipids, etc.), it will deal mainly with compounds that are unique or mostly unique to garlic.

The general composition of garlic is shown in Table 3.1. The water content of garlic (about 65%) is rather low compared to most fruits and vegetables (80–90%). The bulk of the dry weight is composed of fructose-containing carbohydrate, followed by sulfur compounds, protein, and free amino

Table 3.1   **General Composition of Garlic Cloves**

Component	Amount (% fresh weight)	References[a]
Water	62–68	1–4
Carbohydrates (mainly fructans)	26–30	1, 2, 5
Protein	1.5–2.1	Section 3.3.3.
Amino acids: common	1–1.5	Table 3.10
Amino acids: cysteine sulfoxides	0.6–1.9	Table 3.3
γ-Glutamylcysteines	0.5–1.6	Table 3.3
Lipids	0.1–0.2	Table 3.11
Fiber	1.5	6
Total sulfur compounds[b]	1.1–3.5	Table 3.3
Sulfur	0.23–0.37	3, 7–9
Nitrogen	0.6–1.3	3, 7, 10, 11
Minerals	0.7	Table 3.16
Vitamins	0.015	Table 3.16
Saponins	0.04–0.11	Table 3.15
Total oil-soluble compounds	0.15 (whole)–0.7 (cut)	Table 3.14
Total water-soluble compounds	97	12

[a] 1 (Fenwick & Hanley, 1985); 2 (Souci et al., 1986); 3 (Ueda et al., 1991); 4 Lawson et al., 1991c); 5 (Darbyshire & Henry, 1981); 6 (Watt & Merrill, 1963); 7 (Lawson, 1993); 8 (Alfonso & Lopez, 1960); 9 (Pentz et al., 1990); 10 (Parthasarathi & Sastry, 1959); 11 (Sutaria & San Diego, 1982); 12 (Lawson & Wang, 1994).
[b] Excluding protein and inorganic sulfate (0.5%).

acids. Many of these components will be discussed in detail later in the chapter.

3.2 THE SULFUR COMPOUNDS OF GARLIC

The majority of this chapter will deal with garlic's sulfur compounds, as over 90% of investigations on garlic's components have focused on these compounds. One might well ask: why is there so much interest in garlic's sulfur compounds? There are probably three main reasons: the unusually high content of these compounds in garlic compared to other food plants (Fig. 3.1); the long-recognized pharmacological activity of sulfur-containing drugs (e.g., penicillin and sulfonamide antibiotics, probucol for serum cholesterol reduction, thiazide diuretics, hypotensive captopril, and many others); and, most important, research showing that elimination from crushed garlic of a class of volatile sulfur compounds called thiosulfinates, of which allicin is the most abundant, results in the removal of all or most of garlic's antibacterial effects (Cavallito & Bailey, 1944; Cavallito et al., 1944, 1945; Rao et al., 1946; Hughes & Lawson, 1991; Deshpande et al., 1993; Shashikanth et al., 1985, 1986;), its antifungal effects (Barone & Tansey, 1977; Adetumbi, 1985; Yoshida et al., 1987; Ghannoum, 1988; Hughes & Lawson, 1991), its antiarteriosclerotic effects (Silber, 1933), its in vitro antithrombotic effects (Mohammad & Woodward, 1986; Lawson et al., 1992), and its blood lipid–lowering effects (Billau, 1961; Kamanna & Chandrasekhara, 1984; Sitprija et al., 1987; Plengvidhya et al., 1988). Several recent and older reviews on the sulfur chemistry of garlic have been published (Yu & Ho, 1993; Lawson, 1993; Block, 1985, 1986, 1992a, 1992b; Sticher, 1991; Lancaster & Boland, 1990; Sprecher, 1987; Lutomski, 1987; Fenwick & Hanley, 1985b; Whitaker, 1976; Isenberg & Grdinic, 1973; Virtanen, 1965).

3.2.1 Discovery of the Sulfur Compounds of Garlic: Allyl Sulfides, the First Clue

While the odor-causing compounds of garlic have probably been suspected for centuries as being due to sulfur, this was first scientifically demonstrated 150 years ago by the German scientist Wertheim (1844, 1845) who discovered that steam distillation of crushed garlic produced a strongly smelling oil which consisted exclusively of organosulfur compounds. He determined these com-

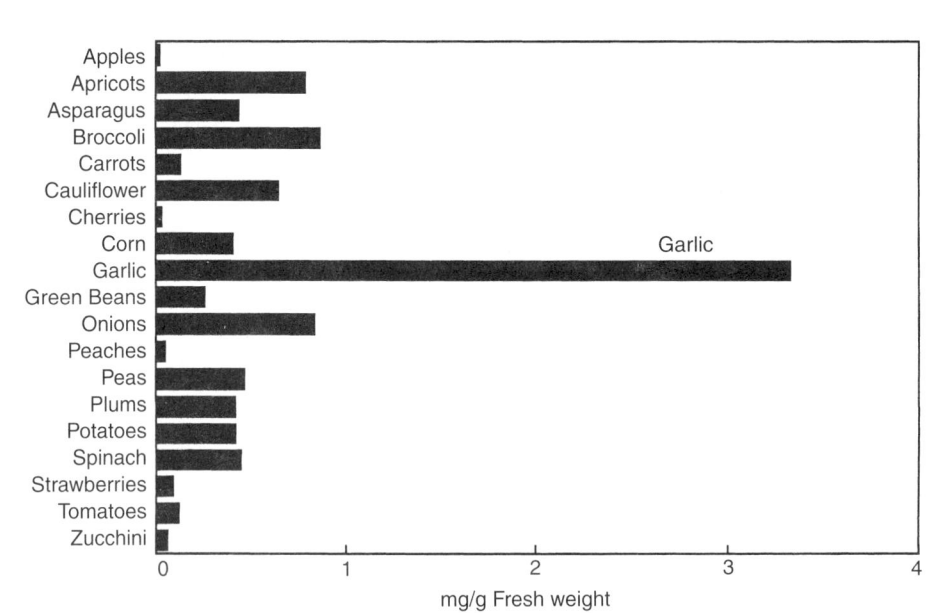

Figure 3.1. Sulfur content of common fruits and vegetables. (Modified from Nielson KK, Mahoney AW, Williams LS, Rogers VC. X-Ray fluorescence measurements of Mg, P, S, Cl, K, Ca, Mn, Fe, Cu, and Zn in fruits, vegetables, and grain products. J Food Comp Analysis 1991;4:39–51.)

pounds to have the basic formula of $C_6H_{10}S$ (his publication says C_6H_5S, but he erroneously calculated hydrogen using an atomic weight of 2) and named the hydrocarbon group *"allyl"* after *Allium sativum*. Semmler named the pure oil of steam-distilled garlic *"Schwefelallyl"* (German for sulfur allyl or thioallyl). However, it required almost 50 years before Semmler (1892) fractionally distilled this type of oil and identified specific compounds. He correctly determined the formula of allyl to be C_3H_5 rather than C_6H_{10} and found the oil to contain 60% *diallyl disulfide*, 20% *diallyl trisulfide*, 10% *diallyl tetrasulfide*, and 6% *allyl propyl disulfide*.

Semmler's results were surprisingly similar to what is found today (Lawson et al., 1991a; Ariga & Kase, 1986), except that the allyl propyl disulfide content was later shown to be an error since *S*-propyl compounds are not produced by garlic (Lawson et al., 1991a, 1991b, 1991c; Block et al., 1992a, 1993; Ziegler & Sticher, 1989). Nevertheless, as will be discussed later in detail, these diallyl sulfides are not actually found in whole or crushed garlic cloves but are produced during steam distillation (a process long used to isolate the *essential oils* of plants) and during the storage of crushed garlic; however, they provided an important clue in identifying the parent sulfur compounds of whole and crushed garlic.

3.2.1.1 DISCOVERY OF ALLICIN

Even though the highly odorous compounds of the oil of distilled garlic had been shown to be sulfur compounds (allyl sulfides), two unsolved mysteries remained: whole garlic cloves did not have an odor and boiled cloves did not have antibiotic activity. Although others had tried to find the antibacterial principle of garlic and precursor of diallyl disulfide (Rundquist, 1910; Peyer, 1927; Vollrath et al., 1936; Dittmar, 1939; Kitagawa & Amano, 1935; Kitagawa & Hirano, 1937; Kitagawa & Noda, 1939; Carl et al., 1939; Tokin, 1943; Yacobson, 1936; Glaser & Drobnik, 1939), it was not until 1944, almost 50 years after Semmler's work, before Cavallito and coworkers (Cavallito & Bailey, 1944; Cavallito et al., 1944, 1945) isolated and identified the antibacterial activity of crushed garlic clove as an oxygenated sulfur compound that possessed the odor of freshly cut garlic, which they named *allicin*. Table 3.2 lists various names for allicin and other thiosulfinates (nomenclature rules are discussed in Block, 1992).

Final proof of the structure of allicin came in 1947 from two laboratories when it was shown that allicin could be synthesized by mild oxidation of diallyl disulfide (Small et al., 1947; Stoll & Seebeck, 1947). However, the actual presence of allicin in uncut garlic was soon doubted, as illustrated by the following statement (Cavallito et al., 1944): "The presence of a chemical substance as unstable as allicin in garlic which has been stored for several months to a year raises the question as to the nature of its state in garlic." A year later, a dried low-temperature (77°C) acetone extract of garlic cloves was used to demonstrate that no diallyl sulfides are present in garlic and that allicin itself is not formed unless water is added to the dried garlic; nor is it formed if enzyme-inhibiting solvents are added (Cavallito et al., 1945).

Table 3.2 Naming of the Thiosulfinates of Garlic

Structure
IUPAC name
Chemical Abstracts name
Historically used names

1. $CH_2=CHCH_2\text{-}\mathbf{SS(=O)}\text{-}CH_2CH=CH_2$
   allyl (or 2-propenyl) 2-propenethiosulfinate
   2-propene1-1sulfinothioic acid S-2-propenyl ester
   allicin or diallyl thiosulfinate
2. $CH_2=CHCH_2\text{-}\mathbf{SS(=O)}CH_3$
   allyl (or 2-propenyl) methanethiosulfinate
   methanesulfinothioic acid S-2-propenyl ester
   allyl methyl thiosulfinate
3. $CH_3\text{-}\mathbf{SS(=O)}\text{-}CH_2CH=CH_2$
   methyl 2-propenethiosulfinate
   2-propene-1-sulfinothioic acid S-methyl ester
   methyl allyl thiosulfinate
4. (E) $CH_2=CHCH_2\text{-}\mathbf{SS(=O)}\text{-}CH=CH_3$
   allyl *trans*-1-propenethiosulfinate
   (E)-1-propenesulfinothioic acid S-2propenyl ester
   allyl *trans*-1-propenyl thiosulfinate
5. (E) $CH_3CH=CH\text{-}\mathbf{SS(=O)}\text{-}CH_2CH=CH_2$
   trans-1-propenyl 2-propenethiosulfinate
   2-propene-1-sulfinothioic acid S-(E)-1-propenyl ester
   trans-1-propenyl allyl thiosulfinate

Note: The IUPAC system will be used throughout the text except for the very common name of allicin. See Figure 3.2 for names and structures of additional garlic thiosulfinates. Since compounds 2 & 3 and 4 & 5 are regioisomers of each other, the names allyl methyl thiosulfinates and allyl trans-2-propenyl thiosulfinates are used when speaking of both isomers together. Sometimes the *trans* is omitted because the *cis* isomer rarely occurs.

From these results they concluded that allicin must be formed upon crushing garlic from an unknown parent compound by the action of an unknown enzyme, and that the diallyl sulfides present in the oil of steam-distilled garlic were formed from allicin.

3.2.1.2 DISCOVERY OF ALLIIN AND ALLIINASE

Although Cavallito's group had provided good evidence that allicin must be formed by the action of an enzyme on a stable precursor when garlic is crushed (Cavallito et al., 1944, 1945;), the isolation and identification of the precursor eluded them. However, a few years after the discovery of allicin, Stoll and Seebeck isolated, identified, and synthesized an oxygenated sulfur amino acid from garlic, which they named *alliin*, and which they found to be the parent compound of allicin (Stoll & Seebeck, 1947, 1948, 1950b, 1951a). Because alliin has two asymmetric atoms, four diastereomeric isomers are possible, but only one isomer was found to occur: (+)-S-allyl-L-cysteine sulfoxide (S-2-propenyl-L-cysteine sulfoxide) (Stoll & Seebeck, 1950b). Interestingly, alliin was the first compound found in nature to possess both an asymmetric sulfur atom and an asymmetric carbon atom. It was later discovered that garlic also contains much smaller amounts of two other compounds similar to alliin: (+)-S-methyl-L-cysteine sulfoxide (*methiin*) (Fujiwara et al., 1958) and (+)-S-*trans*-1-propenyl-L-cysteine sulfoxide (*isoalliin*) (Granroth, 1968).

Stoll and Seebeck (1947) found alliin to have no antibiotic activity unless converted to allicin by a garlic enzyme preparation. They named the enzyme responsible for this transformation *alliinase* (see Section 3.2.5 for complete naming) and two years later thoroughly described its purification, properties, and specific activities (Stoll & Seebeck, 1949a, 1949b, 1950a, 1951c). Of their numerous publications, only one is published in English (Stoll & Seebeck, 1951b), but it is an excellent review of all of their work.

3.2.1.3 DISCOVERY OF THE γ-GLUTAMYLCYSTEINES

Although alliin is the parent compound of allicin and allicin is the parent compound of the diallyl sulfides, a parent compound of alliin also occurs abundantly in garlic cloves, apparently as a reserve pool, that can increase its levels during storage and sprouting (Lawson et al., 1991b). In the early 1960s, both Virtanen's group in Finland, who worked chiefly with onions (Virtanen & Mattila, 1961a, 1961b; Virtanen et al., 1962; Virtanen, 1962b, 1965, 1969; Ettala & Virtanen, 1962; Virtanen & Matikkala, 1960a, 1960b, 1961a; 1961b), and Suzuki's group in Japan, who worked almost exclusively with ^{35}S-labeled garlic (Suzuki et al., 1961a, 1961b, 1962a, 1962b; Sugii et al., 1964b), found that garlic contains about nine different γ-glutamylpeptides, six of which contain the sulfur amino acid, cysteine. Most importantly they found an abundant amount of γ-glutamyl-S-allylcysteine (γ-glutamyl-S-2-propenylcysteine) (Virtanen & Mattila, 1961a; Suzuki et al., 1962a; Sugii et al., 1963b). However, γ-glutamyl-S-trans-1-propenylcysteine, frequently the most abundant γ-glutamylcysteine in garlic, was not discovered until almost 30 years later (Lawson & Hughes, 1990; Lawson et al., 1991b, 1991c; Mütsch-Eckner et al., 1992a).

Lancaster & Shaw (1989) have shown that these two compounds may be the parent compounds of alliin and isoalliin, at least during storage (see Section 3.2.10 for detailed discussion). However, this had earlier appeared to be the case since it had been shown that the content of γ-glutamylcysteines in onions decreased upon sprouting with an accompanying increase in γ-glutamyl transpeptidase activity (Virtanen & Matikkala, 1960a, 1965b). Furthermore, increases in pyruvate levels, which always accompany the enzymatic lysis of cysteine sulfoxides, were also found (Austin & Schwimmer, 1971; Schwimmer & Austin, 1971a, 1971b). These results were further substantiated in recent studies which showed that cool storage of garlic bulbs for a few months resulted in steady decreases of the γ-glutamylcysteines with almost equimolar increases in the cysteine sulfoxides (Lawson et al., 1991b, 1991c).

3.2.2 Total Known Sulfur Compounds in Garlic Cloves

The sulfur content of garlic is very close to 1.0% of its dry weight (0.35% of its fresh weight) (Pentz et al., 1990; Alfonso & Lopez, 1960; Ueda et al., 1991). While alliin, allicin, and the two main γ-glutamylcysteines constitute the majority (about 72%) of the sulfur compounds in whole or crushed garlic, a total of 16 nonprotein organosulfur compounds have been found in whole cloves and 23 in crushed cloves (Table 3.3). Interestingly, of the 16 organosulfur compounds present in whole cloves, only three very minor compounds (methionine, γ-glutamylmethionine, and thiamine) do not contain the amino acid cysteine. The minor organosulfur compounds consist mainly of methyl and 1-propenyl homologs of alliin and allicin, plus γ-glutamyl-S-methylcysteine (which comprise about 13% of the total sulfur), while trace amounts (less than 0.1 mg/g) of a few other related compounds (8 in whole cloves, 11 in crushed cloves) are also found. Protein sulfur (soluble and insoluble) and inorganic sulfate account for 9% and 5% of the total sulfur (Lawson, 1993). The presently known sulfur compounds of garlic constitute about 86% of the total sulfur of garlic cloves and about 98 ± 2% of the total nonprotein organosulfur compounds. Therefore, any as yet undiscovered organosulfur compounds will not be very abundant. It is also important to know what sulfur compounds are not found in garlic, especially those considered as possibilities, and many of these are listed in Table 3.3.

3.2.3 γ-Glutamylcysteines: Important Storage Compounds

The discovery and importance of these sulfur compounds have already been discussed (Section 3.2.1.3). Their abundance is indicated in Table 3.3, and their structures are shown in Figure 3.2. They are present

Table 3.3 Total Known Sulfur Compounds in Whole and Crushed Garlic cloves. (Also included are compounds sought that were not detected.)

Compound	Whole Garlic	Crushed Garlic	Reference[a]
	(mg/g fresh weight)		
S-(+)-Alk(en)yl-L-cysteine sulfoxides			
S-Allylcysteine sulfoxide (alliin)	5–14	nd[b]	1–10
S-Methylcysteine sulfoxide (methiin)	0.5–2	nd	1, 3, 9, 10
S-trans-1-Propenylcysteine sulfoxide (isoalliin)	0.2–1.2	nd	1, 11
S-Propylcysteine sulfoxide	nd	nd	11–13
Cycloalliin	0.5–1.5	0.5–1.5	1, 14
γ-L-Glutamyl-S-alk(en)yl-L-cysteines			
γ-Glutamyl-S-trans-1-propenylcysteine	3–9	3–9	1, 2, 15
γ-Glutamyl-S-cis-1-propenylcysteine	0.06–0.15	0.06–0.15	15
γ-Glutamyl-S-allylcysteine	2–6	2–6	1, 2, 15
γ-Glutamyl-S-methylcysteine	0.1–0.4	0.1–0.4	1, 15
γ-Glutamyl-S-propylcysteine	nd	nd	15
Thiosulfinates			
Allyl 2-propenethiosulfinate (allicin)	nd	2–6	1, 9–11, 13, 16–21
Allyl methyl thiosulfinates[c]	nd	0.3–1.5	1, 9, 11, 13, 16–18
Allyl trans-1-propenyl thiosulfinates[c]	nd	0.05–1.0	1, 11, 13, 16–17
Methyl trans-1-propenyl thiosulfinate[c]	nd	0.02–0.2	1, 11, 16–17
Methyl methanethiosulfinate	nd	0.05–0.1	1, 10, 11, 13, 16–18
Others			
Cysteine	nd	nd	1, 12
Cystine	nd	nd	14
Glutathione, reduced	nd	nd	1
Glutathione, oxidized	nd	nd	1
γ-Glutamyl-S-allylcysteine sulfoxide	nd	nd	14
γ-Glutamyl-S-trans-1-propenylcysteine sulfoxide	nd	nd	15
γ-Glutamyl-S-methylcysteine sulfoxide	nd	nd	22, 24
γ-Glutamyl-methionine	0.02–0.12	0.02–0.12	14, 30
γ-Glutamylcysteine, reduced	nd	nd	1, 14
γ-Glutamylcysteine, oxidized	tr	tr	22
S-2-carboxypropylglutathione	0.09	0.09	23
γ-Glutamyl-S-allylmercaptocysteine	0.01–0.03	0.01–0.03	14, 24
S-Methylcysteine	tr	tr	1, 25
S-1-Propenyl cysteine	nd	nd–0.006	1, 26
S-Allylcysteine	nd–0.026	nd–0.026	1, 12
S-Allylmercaptocysteine[d]	nd	0.002	24
Methionine	0.02	0.02	1, 14
Thiamine	0.002	nd–0.001	14, 27
Allithiamine	nd	nd–0.001	14
Scordinins	0.03 (yield)	<0.03	28
Sulfolipids	nd, 0.01	nd, 0.01	29, 31
Protein, soluble	0.3	0.3	1
Insoluble compounds (prob. mostly protein)	0.6	0.6	1
Inorganic sulfate	0.5	0.5	1

[a] 1 (Lawson, 1993); 2 (Mütsch-Eckner et al., 1992); 3 (Ueda et al., 1991); 4 (Alfonso & Lopez, 1960); 5 (Edwards et al., 1994); 6 (Gaind et al., 1965); 7 (Kappenberg & Glasl, 1990); 8 (Mochizuki et al., 1988); 9 (Sendl et al., 1992); 10 (Sendl & Wagner, 1991); 11 (Lawson et al., 1991b); 12 (Ziegler & Sticher, 1989); 13 (Block et al., 1992); 14 (Lawson & Wang, 1994); 15 (Lawson et al., 1991b); 16 (Lawson & Hughes, 1992); 17 (Block et al., 1993); 18 (Wagner & Sendl, 1990); 19 (Iberl et al., 1990); 20 (Müller, 1990); 21 (Cavallito et al., 1944); 22 (Suzuki et al., 1961); 23 (Tsuboi et al., 1989); 24 (Sugii et al., 1964); 25 (Sugii et al., 1963a); 26 (Sugii et al., 1963b); 27 (Souci et al, 1986); 28 (Kominato, 1969); 29 (Kamanna & Chandrasekhara, 1986); 30 (Virtanen, 1965); 31 (Yang & Shin, 1982).
[b] Not detectable (nd), <0.02 mg/g clove fresh wt. in most cases.
[c] Inclides two isomers: R-SS(=O)-R' and R'SS(=O)-R.
[d] S-AMC (the reaction product of allicin with cysteine) is undetectable in whole garlic and fresh chopped garlic, but as chopped garlic ages, protein hydrolysis yields cysteine, which then reacts with allicin to form S-AMC such that it is present 0.15 mg/g in 12 days and 0.4 mg/g in 90 days (Lawson & Wang, 1994). The small amount (0.002 mg/g) reported is likely the result of the cysteine released by protein hydrolysis within the day of its preparation.

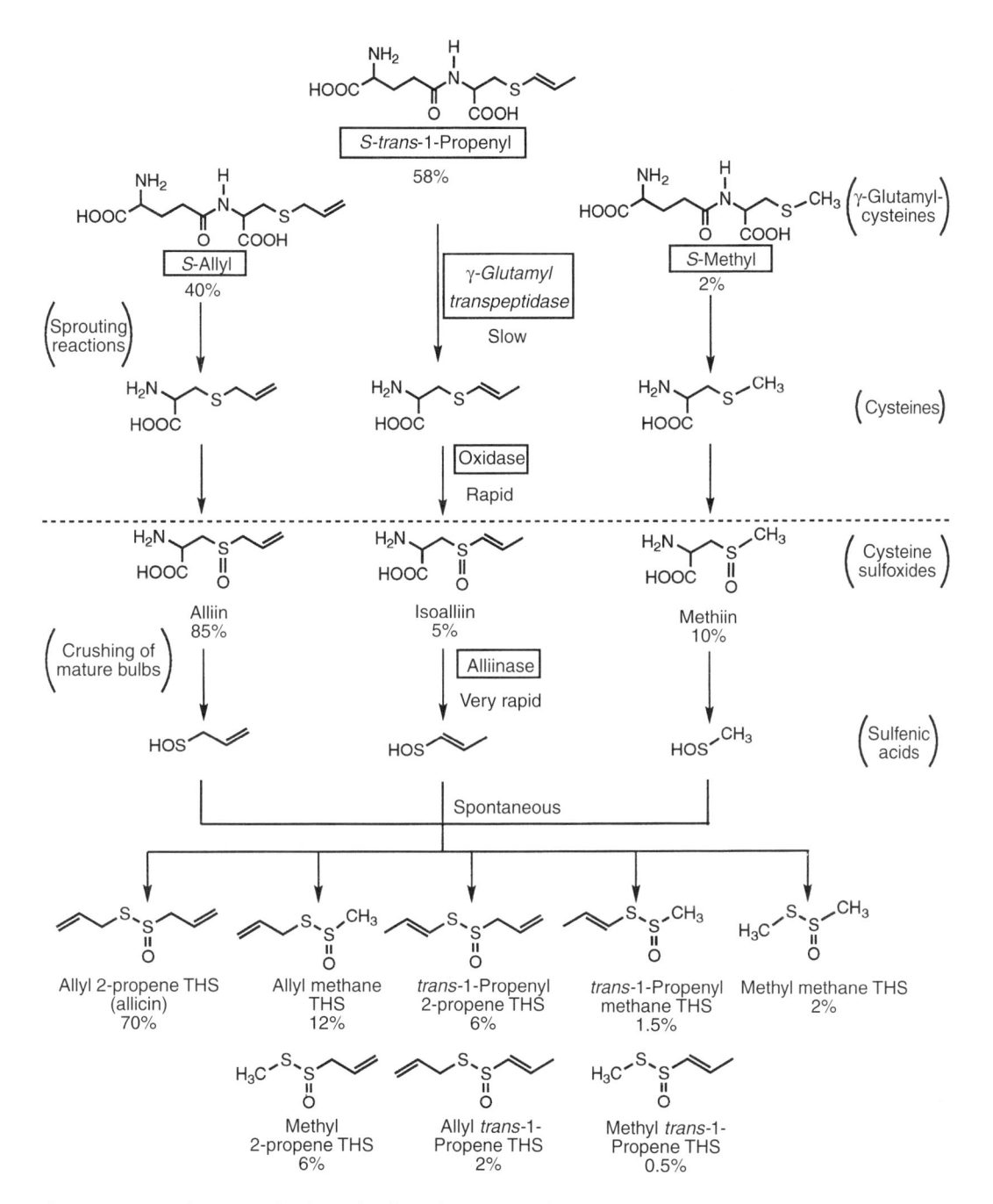

Figure 3.2. Transformation of γ-glutamyl-S-alk(en)ylcysteines and S-alk(en)ylcysteine sulfoxides to thiosulfinates (THS). Includes typical weight percentages for compounds actually found in whole or crushed cloves.

Table 3.4 Whole Plant Distribution of Cysteine Sulfoxides and γ-Glutamylcysteines

	Roots (%)	Bulbs (%)	Above Ground (Leaves and Stems) (%)
Mature plants (harvest time)			
Alliin	2	87	11
Methiin	1	95	4
Isoalliin	6	87	7
γ-Glutamyl-S-allylcysteine	nd[a]	>98	nd
γ-Glutamyl-S-1-propenylcysteine	nd	98	nd
7-week premature plants			
Alliin	6	25	69
Methiin	6	22	72
Isoalliin	20	61	19
γ-Glutamyl-S-allylcysteine	nd	nd	nd
γ-Glutamyl-S-1-propenylcysteine	nd	nd	nd

[a] nd=not detected
From Lawson LD, Wnag ZJ, unpublished data, 1994.

only in the bulbs of the plants and are not present at all until shortly before harvest time (Table 3.4). Although their presence in garlic has been known for 30 years, they have only recently been quantitated in garlic (Lawson et al., 1991b; Mütsch-Eckner et al., 1992b; Sticher & Mütsch-Eckner, 1991). They are nearly as abundant as the cysteine sulfoxides (alliin, etc.) on a weight basis, but because of their larger molecular weight are only about 50% as abundant on a molar basis. The main γ-glutamylcysteines found in garlic cloves are γ-glutamyl-S-trans-1-propenylcysteine and γ-glutamyl-S-allylcysteine. Small amounts of γ-glutamyl-S-methylcysteine are also usually found, but not always (Mütsch-Eckner et al., 1992b). γ-Glutamyl-S-cis-1-propenylcysteine is barely detectable in fresh cloves, but it increases slowly upon clove storage and especially upon boiling of cloves, indicating that it is spontaneously formed from the trans-isomer rather than biosynthesized. Only trace amounts of oxidized γ-glutamylcysteine and of γ-glutamyl-S-allylmercaptocysteine (γ-glu-cys-SS-allyl) are found in garlic, but the latter compound can increase many times over in the preparation of garlic powders (see Section 3.4.1). Neither glutathione (γ-glu-cys-gly) nor γ-glutamylcysteine have been found in garlic (Lawson, 1993), even though they are probably important intermediates in the biogenesis of garlic's sulfur compounds (Lancaster & Shaw, 1989). The γ-glutamyl peptides are closely connected to glutathione metabolism and may have related biological roles in the plant (Rennenberg, 1982).

The sulfoxides of the γ-glutamyl-S-alkenylcysteines have not been found in garlic (Lawson et al., 1991b; Lawson & Wang, 1994). There has been one report of finding the sulfoxide of γ-glutamyl-S-allylcysteine in autoclaved and resin-fractionated garlic, but this is undoubtedly an artifact of the isolation procedure (Ueda et al., 1990). Onion bulbs, however, do contain abundant amounts of γ-glutamyl-S-1-propenylcysteine sulfoxide (Matikkala & Virtanen, 1967; Lawson et al., 1991b). Interestingly, the seeds of onion (but not the bulbs) and several other Allium species do contain large amounts of the γ-glutamylcysteines (mainly S-1-propenyl), but very little of the sulfoxides (Lawson et al., 1991b). Three noncysteine γ-glutamyl-peptides have also been found in garlic: γ-glutamylmethionine (Virtanen, 1965; Mütsch-Eckner et al., 1992b), γ-glutamylglycine (Suzuki et al., 1961a), and γ-glutamylphenylalanine (Suzuki et al., 1961a; Mütsch-Eckner et al., 1992b), which is present at 0.4-1.1 mg/g fresh weight (Lawson & Wang, 1994). Of course, other Allium plants also contain γ-glutamyl peptides (Ali et al., 1991; Lancaster & Shaw, 1991; Matikkala & Virtanen, 1962, 1963, 1964, 1965a; Virtanen & Matikkala, 1961b, 1962; Tsuboi et al., 1989; Kasai & Kiriyama, 1987; Vitalyeva et al., 1968). In onions, γ-glutamyl-arginine and γ-glutamyl-S-(2-carboxy-N-propyl)-cysteine have also been detected (Matikkala & Virtanen, 1967, 1970; Granroth, 1968).

The variation in content of the γ-glutamyl-S-alkenylcysteines among a large number of garlic strains grown on the same three-acre farm during two different years of extreme differences in weather conditions (cool and very wet versus hot and very dry summers) has been examined (Lawson & Wang, 1994; Lawson, 1993). The strains represented both of the well-known subspecies of garlic: *Allium sativum* L. var. *sativum* (softneck garlic) and *Allium sativum* L. var. *ophioscorodon* (hardneck or topset garlic) (see Fig. 2.2). The total γ-glutamylcysteines content was higher in hot, dry 1993 than in cold, wet 1992 for both softneck (35% higher) and hardneck (60% higher) strains; however, the strains with the highest values in 1992 also had the highest values in 1993. The hardnecks consistently produced more γ-glutamyl-*S*-1-propenylcysteine than the *S*-allyl isomer (90% more in wet 1992 and 41% more in dry 1993), but the softnecks produced nearly the same amounts of each in both years (12% and 4% higher, respectively). The content of γ-glutamyl-*S*-allylcysteine tended to be higher in softnecks than in hardnecks (94% higher in 1992, 18% higher in 1993 (Fig. 3.3); however, when the strains of known variety were grouped into individual varieties (Table 3.5), it was found that the porcelain and purple stripe hardneck varieties consistently contained the highest amounts, while rocambole hardneck contained the least. As shown in Figure 3.3, the range in content of the total γ-glutamylcysteines was greater for the hardnecks than for the softnecks.

The γ-glutamylcysteines play an important function as reserve compounds for producing additional alliin and isoalliin during wintering and sprouting, increasing the antibiotic capacity of the young plants (Lancaster & Shaw, 1991). These compounds have been found to be fairly stable when cloves are maintained at room temperature (15% loss in 3 months), but when cloves are stored at 4–6°C, they decrease by 70% each in 3 months with approximately quantitative increases in alliin and isoalliin (calculated from thiosulfinates) (Lawson et al., 1991b, 1991c). The transformation of γ-glutamyl-*S*-alkenylcysteines to *S*-alkenylcysteine sulfoxides is due to increased synthesis or availability of γ-glutamyltranspeptidase (EC 2.3.2.2) and γ-glutamylpeptidase (Ceci et al., 1992a; Matikkala & Virtanen, 1965b; Austin & Schwimmer, 1971; Schwimmer & Austin, 1971a, 1971b) and an oxidase. While specific experiments on the late appearance of this oxidase have not been conducted, it appears to be absent in harvested cloves because the γ-glutamyl-*S*-alkenylcysteine content of chopped cloves decreases about 50% in 1

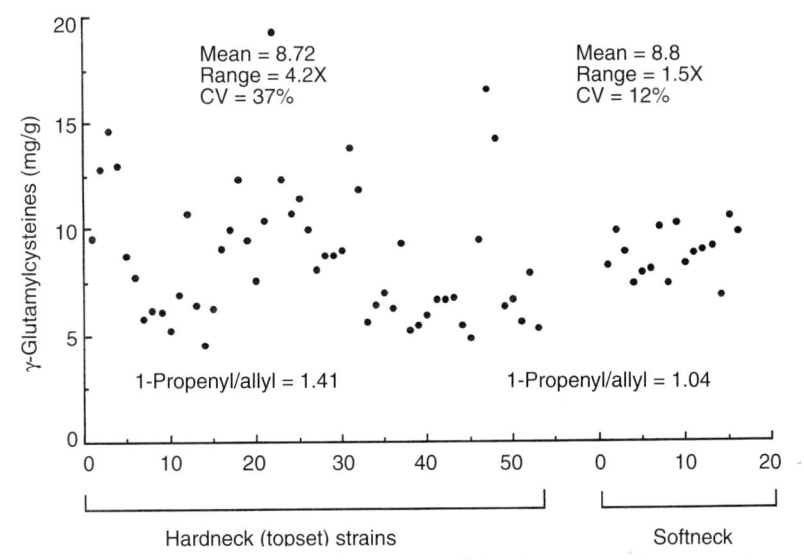

Figure 3.3. Variation in content of γ-glutamylcysteines (*S*-allyl and *S*-1-propenyl) among 69 strains of garlic grown on the same farm in Troy, New York. (From Lawson & Wang, 1994.)

Table 3.5 The Main Sulfur Compounds of 22 Strains of Crushed Garlic[a]

Variety and Strain[b] (State grown, year, number of bulbs	Bulb Weight (g)	Clove Weight (g)	Allicin (mg/g)	Total THS[c] (mg/g)	S-Allyl-γ-GC[c] (mg/g)	S-1-Propenyl-γ-G[c] (mg/g)
Softneck, artichoke						
California Early (CA92-4)	47	2.6	3.2	4.6	2.5	5.1
California Early (UT93-5)	29	2.7	4.6	6.9	—	—
Chinese Sativum (NY93-3)	34	3.3	3.4	5.0	3.6	5.5
Early Red Italian (NY93-3)	29	4.6	3.5	4.7	5.1	4.7
Inchellium (NY93-3)	50	5.4	2.8	3.7	4.0	3.5
Inchellium (NY92-2)	—	—	2.8	3.5	2.0	3.1
Nova Boschaca (NY93-3)	41	5.7	3.5	5.3	2.9	4.0
Oregon Blue (NY93-3)	36	6.9	4.0	5.9	3.2	4.7
Oregon Blue (NY92-2)	—	—	3.2	4.0	3.4	4.3
Softneck, silverskin						
Idaho Silverskin (NY93-3)	40	2.6	4.3	6.1	2.6	6.3
Nichol's Silverskin (NY93-3)	33	3.1	4.0	5.5	4.5	5.5
Nootka Rose (NY93-3)	43	3.2	4.2	6.0	2.7	6.3
Hardneck, rocambole						
Carpathian (NY93-3)	31	5.0	5.4	8.2	2.1	4.2
Carpathian (NY92-3)	—	—	3.5	4.3	1.6	3.3
German Red (NY93-3)	38	6.1	4.9	7.2	2.1	4.0
German Red (NY92-6)	—	—	3.4	4.5	0.9	3.2
German Red (UT93-3)	43	6.7	3.0	4.6	1.2	3.9
Israeli (NY93-3)	41	5.7	5.2	7.4	2.6	4.2
Spanish Roja (NY93-3)	35	11.2	6.7	9.1	3.2	4.8
Hardneck, porcelain						
Brown Rose (NY93-3)	24	3.8	6.1	9.4	6.8	7.8
Georgian Crystal (NY93-3)	35	8.4	4.7	7.6	4.9	5.8
Romanian Red (NY93-3)	28	9.3	7.7	11.8	6.7	7.5
Rosewood (NY93-3)	29	7.0	3.3	5.4	4.7	6.7
Zemo (NY93-3)	31	7.5	5.1	8.4	5.6	6.2
Hardneck, purple stripe						
Brown Tempest (NY93-3)	32	5.2	4.1	7.4	6.0	7.0
Persian Star (NY93-3)	24	2.9	3.5	7.0	4.8	7.5
Red Rezan (NY93-3)	31	3.4	3.1	5.8	4.8	5.9
Yampolski (NY93-3)	29	7.3	4.7	8.4	6.7	7.1

[a] Values given are the means per fresh weight for the number of bulbs shown.
[b] The strains are classified by subspecies (*A. sativum* var. *sativum* = softneck; *A. sativum* var. *ophioscorodon* = hardneck) and varieties as described by Engeland (1991) and in close agreement with the isoenzyme groupings of Pooler and Simon (1993).
[c] Abbreviations: THS, thiosulfinates; γ-GC, γ-glutamylcysteine.

week with a quantitative increase in *S*-allylcysteine and *S*-1-propenylcysteine, but never an increase in *S*-allylcysteine sulfoxide or *S*-1-propenylcysteine sulfoxide, even after 2 years (Lawson & Wang, 1994). On the other hand, the transpeptidase has been added to garlic powder with a resultant increase in flavor and pyruvate release, indicating that the newly formed alkenylcysteines were rapidly oxidized to alliin and isoalliin. The enzymatic turnover of the glutamyl peptides has also been observed in other *Allium* species (Knobloch et al., 1993).

3.2.4 Cysteine Sulfoxides: Precursors of the Thiosulfinates

Garlic contains three *S*-alkylcysteine sulfoxides capable of producing thiosulfinates, *S*-allylcysteine sulfoxide (*alliin* or *S*-2-propenylcysteine sulfoxide), *S-trans*-1-propenylcysteine sulfoxide (*isoalliin*), and *S*-methylcysteine sulfoxide (*methiin*). Alliin constitutes about 85% of these sulfoxides, while isoalliin (5%) and methiin (10%) are considerably less abundant. Their content in garlic is indicated in Table 3.3, and their structures and lysis by alliinase are shown in Figure 3.2. S-Propylcysteine sulfoxide was originally reported to be present in garlic (Fujiwara et al., 1958; Freeman & Whenham, 1975b; Madhavi et al., 1991), but this was probably a misidentification of isoalliin since its absence has been clearly demonstrated (Ziegler & Sticher, 1989; Lawson et al., 1991c; Block et al., 1992a; Edwards et al., 1994a).

Garlic also contains a cyclic cysteine sulfoxide, *cycloalliin* (3-methyl-1,4-thiazane-5-carboxylic acid-1-oxide) (Virtanen & Matikkala, 1959; Ueda et al., 1991; Lawson & Wang, 1994) (Fig. 3.4), which cannot be cleaved by alliinase and hence produces no thiosulfinates (Schwimmer & Mazelis, 1963). Cycloalliin was originally found in onions (Virtanen & Matikkala, 1958, 1959, 1961c; Matikkala & Virtanen, 1958; Virtanen & Spåre, 1961; Virtanen, 1962a) and can be produced chemically from isoalliin by incubation at alkaline pH (Virtanen & Spåre, 1961) or by heating at neutral pH (Ueda et al., 1994), but it appears to be biosynthesized from 1-propenylcysteine (perhaps via γ-glutamyl-S-1-propenylcysteine hydrolysis), since it is not formed from labeled-isoalliin (Müller & Virtanen, 1965), but can be formed from labeled-carboxypropylcysteine (Granroth & Virtanen, 1967a, 1967b). The function of cycloalliin in garlic is unknown.

The cysteine sulfoxides are rare in nature and, except for methylcysteine sulfoxide, which is also found throughout the Cruciferae family, are found almost exclusively in the Alliacae family, especially among the genus *Allium* (Fujiwara et al., 1958). Alliin, however, has only been found abundantly in a few *Allium* plants: garlic, wild garlic *(A. ursinum)*, elephant garlic *(A. ampeloprasum)*, garlic chives *(A. tuberosum)*, *A. moly*, *A. vineale*, *A. victorialis*, and *Allium ampeloprasum*, var. *bulbilliferum* (Freeman & Whenham, 1975b; Fujiwara et al., 1958; Lancaster & Boland, 1990; Lawson et al., 1991c; Boscher & Auger, 1991; Auger et al., 1992). Among non-*Allium* plants, small amounts of alliin have been found in the roots (0.1 mg/g fresh weight) (none in the leaves or flowers) of garlic mustard *(Alliaria officinalis)* (Lawson & Duke, 1994) and in the leaves of the Brazilian garlic bush *(Adenocalymma alliaceum)* at about 1 mg/g, on the basis of the diallyl sulfides found in the steam-distilled oil (Zoghbi et al., 1984).

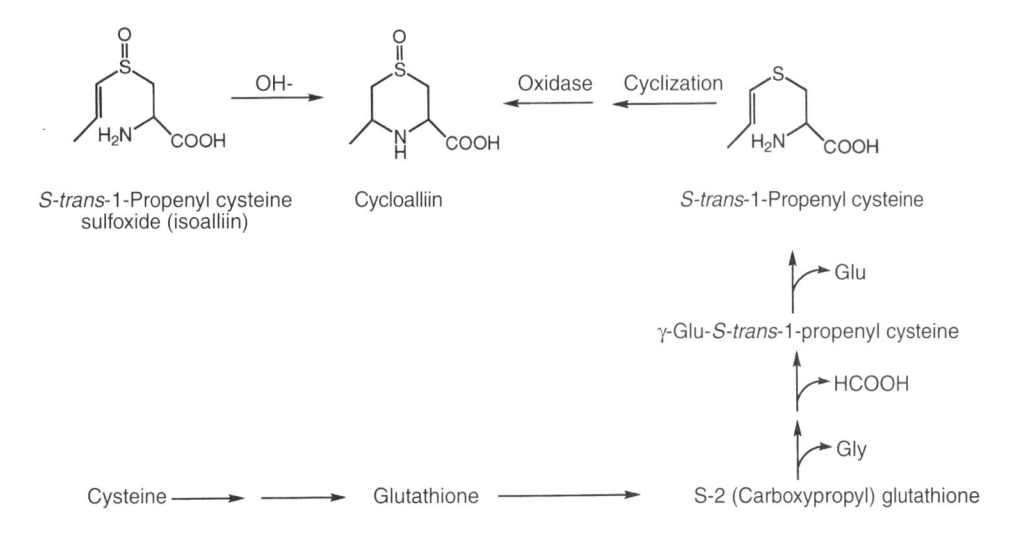

Figure 3.4. Chemical and possible biosynthetic pathways for formation of cycloalliin. (From Granroth B, Virtanen AI. S-(2-Carboxypropyl)-cysteine and its sulfoxide as precursors in the biosynthesis of cycloalliin. Acta Chem. Scand. 1967;21:1654–1656; and Lancaster JE, Shaw ML. Gamma-Glutamyl peptides in the biosynthesis of S-alk(en)yl-L-cysteine sulphoxides (flavour precursors) in Allium. Phytochemistry 1989;34:1229–1235.)

The distribution of garlic's cysteine sulfoxides among the various plant parts is summarized in Table 3.4. About 7 weeks prior to harvest, most of the alliin and methylcysteine sulfoxide are found in the leaves (above ground plant), but by harvest time about 90% of the total plant cysteine sulfoxides are found in the bulb (Lawson & Wang, 1994). Ueda (1991) has also shown a similar trend for alliin. Interestingly, isoalliin was found to move earlier into the bulb than alliin.

The variation of the alliin content (determined directly or from allicin yield) among garlic cloves grown around the world is not large. Results from several laboratories (Lawson et al., 1991c; Mütsch-Eckner et al., 1992b; Lawson, 1993; Ueda et al., 1991; Sendl et al., 1992; Kappenberg & Glasl, 1990) indicate that the greatest variation is 5-fold (3–15 mg/g fresh weight); however, this would include variation due to strain, location, weather, and soil differences as well as variation due to differences in analytical methods. Some of these differences have been minimized by growing a large number of strains on the same one-acre plot in Troy, New York (grown by G. Reynolds) using a single method of analysis. As shown in Figure 3.5, there was a 1.8–2.7-fold variation in alliin content among these strains, which is somewhat less than what has been reported from the many laboratories. Interestingly, the softneck strains gave somewhat lower values and were slightly less variable in their content for each of the cysteine sulfoxides than the hardneck strains. The content of methiin varied more than for alliin or isoalliin.

3.2.5 Alliinase: Lysis of the Cysteine Sulfoxides

The cysteine sulfoxides themselves have no known physiological or pharmacological function other than to be enzymatically transformed into biologically active dialkyl thiosulfinates when garlic cloves are crushed or cut. This transformation is extremely rapid at room temperature, being complete in under 10 seconds for alliin and in 60 seconds for methylcysteine sulfoxide (Fig. 3.6) (Lawson & Wang, 1994). These rates are only moderately increased (30 seconds and 5 minutes, respectively) for oven-dried commercial garlic powders (Lawson & Hughes, 1992). The enzyme responsible for the lysis of alliin is *alliinase* or *alliin lyase* (alliin alkyl-sulfenate-lyase, EC 4.4.1.4), a pyridoxal 5'-phosphate-dependent glycoprotein consisting of two subunits and a carbohydrate content of 5.5–6% (Goryachenkova, 1952; Mazelis & Crews, 1968; Nock & Mazelis, 1986; Jansen et al., 1989; Nock & Mazelis, 1987; Rabinkov et al., 1994). It is about 10 times more abundant in the cloves than in the leaves (Rabinkov et al., 1994). For reviews on its structure and activity, see Lancaster and Boland (1990), Rabinkov et al. (1994), and Whitaker and Mazelis (1991).

For some time there were conflicting reports about the molecular weight of alliinase, with subunit weights varying from 42,000 to 65,000 (Nock & Mazelis, 1986; Brady et al., 1988; Jansen et al., 1989; Kazaryan et al., 1979), but recently garlic alliinase has been cloned and its amino acid sequence determined (Van Damme et al., 1992b; Rabinkov et al., 1994); it has been found to contain 448 amino acids with a protein molecular weight of 51,451, which, with a carbohydrate content of 5.5–6% gives a subunit molecular weight of about 55,000, a value similar to that reported (Jansen et al., 1989) and recently confirmed by immunoblot analysis (Ellmore & Feldberg, 1994). In contrast to an earlier report (Nock & Mazelis, 1989), a great deal of homology is found between garlic alliinase and onion alliinase, since there was 88% agreement in their amino acid sequences (Van Damme et al., 1992b). This homology has been further substantiated by immunological assays (Ho & Mazelis, 1993).

Extensive research has also been conducted on alliinase from onion (*A. cepa*), although this cannot be discussed here in detail (Schwimmer & Mazelis, 1963; Kupiecki & Virtanen, 1960; Schwimmer et al., 1960, 1964; Schwimmer, 1969; Tobkin & Mazelis, 1979; Selby et al., 1979; Freeman,

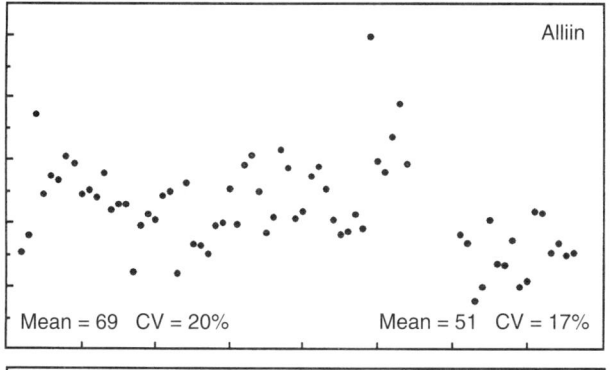

Figure 3.5. Variation in content of cysteine sulfoxides among 53 hardneck (var. *ophioscorodon*) and 16 softneck (var. *sativum*) strains of garlic grown on the same farm in Troy, New York. (From Lawson & Wang, 1994.)

Values are in micromoles/g fresh weight. Scale ranges are 20-130, 0-40, and 1-8.

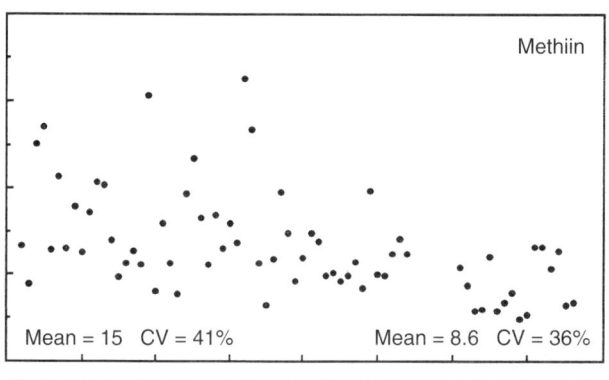

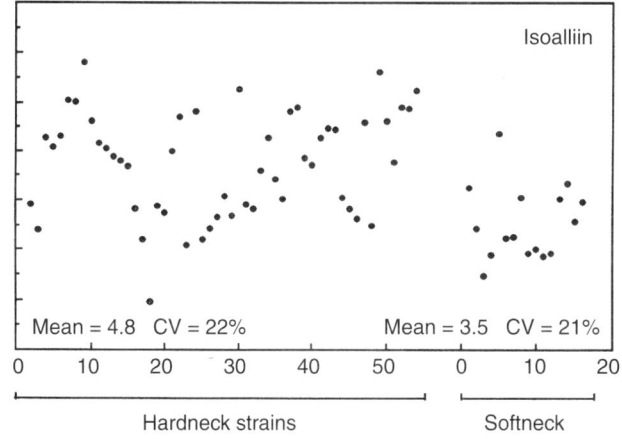

1979; Lancaster & Collin, 1981; Nock & Mazelis, 1989; Thomas & Parkin, 1991; Wäfler et al., 1994). Furthermore, alliinase-like enzymes have also been reported in other *Allium* species (Mazelis, 1963; Tsuno, 1958a; Freeman, 1979; Lohmüller et al., 1994a, 1994b; Fujita et al., 1990; Won & Mazelis, 1989), as well as in non-*Allium* plants (Sweet & Mazelis, 1987; Hamamoto & Mazelis, 1986; Mazelis & Creveling, 1975; Jacobsen et al., 1968; Schwimmer & Kjaer, 1960), and in microorganisms (Murakami, 1960a, 1960b; Saari & Schultze, 1965; Nomura et al., 1963).

A unique feature of alliinase is that it is present in garlic in unusually large amounts for an enzyme, consisting of at least 10% of the total clove protein (Van Damme et al., 1992b; Ellmore & Feldberg, 1994), which means that garlic cloves contain approximately equal amounts (10 mg/g fresh weight) of alliinase and alliin. This may well explain

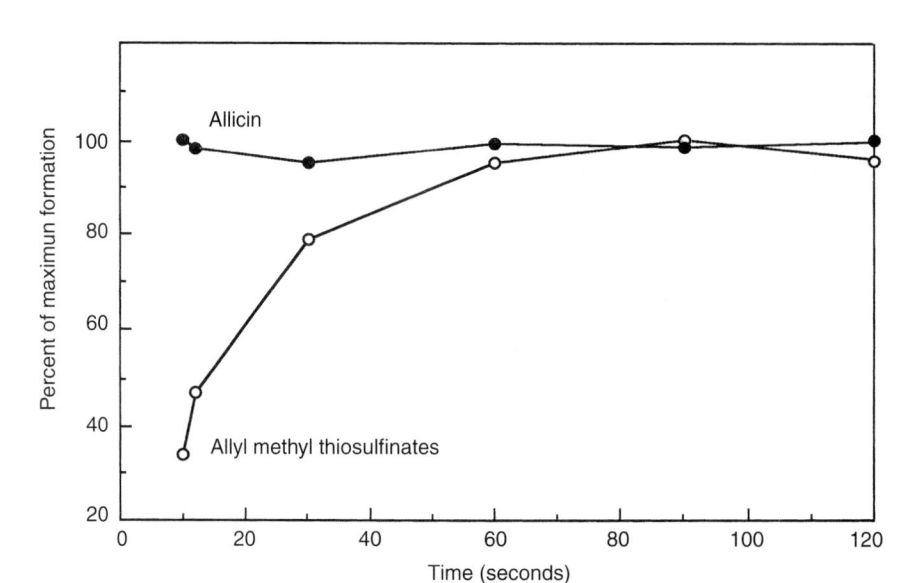

Figure 3.6. Rates of formation of the main thiosulfinates of crushed garlic cloves. (From Lawson & Wang, 1994.)

why alliin is converted so very rapidly to thiosulfinates when garlic is crushed. The stability of alliinase is also quite remarkable, since garlic powders stored for up to 5 years show little loss in ability to produce allicin (see section 3.4.1). Human serum has recently been reported to contain highly specific alliinase antibodies (Tchernychev et al., 1995).

Whole garlic is well-known to be odor free (that is, thiosulfinate free) until the cloves are cut or crushed, indicating that alliinase and the cysteine sulfoxides are stored in separate compartments. In onions, this separation has been shown to be due to location of the cysteine sulfoxides in the cytosol of all the cells, while alliinase is located in the vacuoles of the same cells (Lancaster & Collin, 1981). The same method of compartmentalization was assumed to be true for garlic; however, it has recently been shown that a different form of separation exists in garlic. Using enzyme activity and immunological stains (Fig. 3.7), it was shown that garlic clove alliinase is located only in the relatively few vascular bundle sheath cells located around the veins or phloem (the small yellow spots seen when a clove is cut transversely) (Ellmore & Feldberg, 1994). In agreement with this finding, a major protein of molecular weight 40 kDa (probably alliinase) has been found exclusively in the cytoplasmic globular granules of the same sheath cells (Wen at al., 1995). It was shown in addition that the messenger RNA for alliinase is also located in these cells (Fig. 3.7C), indicating that alliinase is synthesized in the clove rather than being transported to it. More recently, it has been found that alliin is concentrated in the very abundant storage mesophyll cells, with none present near the bundle sheath cells (G.S. Ellmore, personal communication). Hence, separation of alliin and alliinase in garlic cloves involves the location of each in different cell types.

The specificity of garlic alliinase has been studied ever since its discovery by Stoll & Seebeck (Stoll & Seebeck, 1947, 1949b, 1951b). Among the four possible diastereomers of alliin, strong activity is found for the natural L-(+) isomer, with no activity for the D-(+) or D-(−) isomers and greatly reduced activity for the L-(−) isomer (Stoll & Seebeck, 1951b). Less activity is found for the methyl-, ethyl-, propyl-, and butylcysteine sulfoxides than is found for alliin (Stoll & Seebeck, 1951b; Mazelis & Crews, 1968; Perchellet et al., 1990). No activity is found for S-allylcysteine or other S-alkylcysteines, which in fact act as inhibitors of alliinase (Mazelis & Crews, 1968; Jansen et al., 1989;

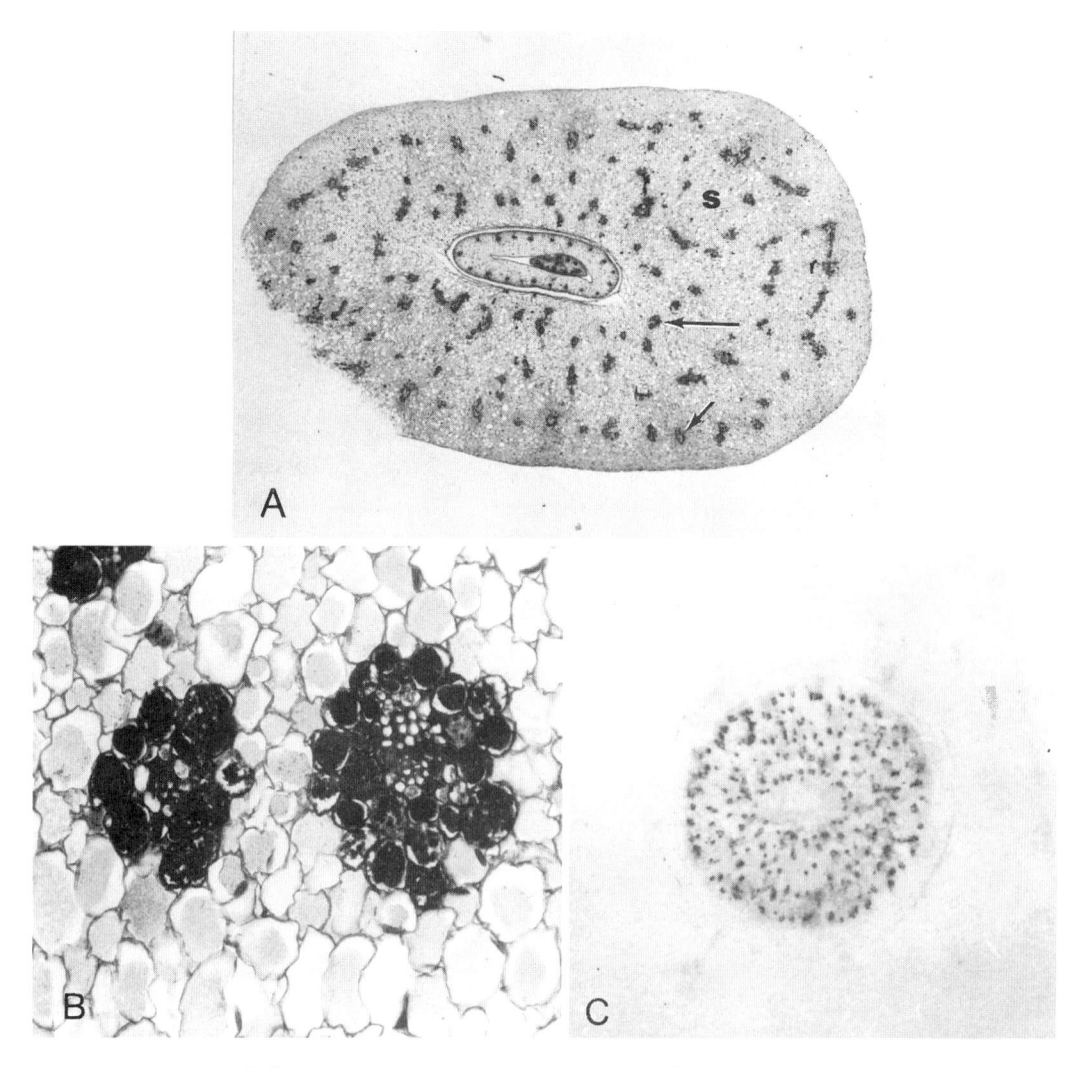

Figure 3.7. Location of alliinase in a garlic clove. **A.** Cross-section of a garlic clove, with arrows pointing to the vascular bundles which occur in rings in the midst of mesophyll storage cells (S). **B.** An enlargement of two vascular bundles reacted with anti-alliinase antibody linked to horseradish peroxidase, showing that alliinase is highly concentrated in the bundle sheath cells (darkest) and absent in the numerous mesophyll storage cells. **C.** Tissue blot of a developing garlic bulb harvested several weeks early and probed for alliinase mRNA using RNA-digoxygenin antisense probe. The distribution of alliinase mRNA mimics the distribution of alliinase in bundle sheaths, suggesting that alliinase expression is at least partly controlled at the level of transcription. (**A** and **B** from Ellmore GS, Feldberg RS. Alliin lyase localization in bundle sheaths of the garlic clove (*Allium sativum*). Am J Bot 1994;81:89–94; **C** prepared by Reddy A, Feldberg RS, and Ellmore GS [unpublished].)

Schwimmer et al., 1964). The activity of onion and garlic alliinases toward alliin and methylcysteine sulfoxide are similar (Mazelis & Crews, 1968; Whitaker & Mazelis, 1991; Schwimmer & Mazelis, 1963) even though no alliin is present in onion. Onion alliinase has much greater activity toward isoalliin than toward methylcysteine sulfoxide (Schwimmer, 1969). The activity of garlic alliinase toward isoalliin has never been reported; however, it has been found that garlic alliinase is at least as active toward isoalliin as it is toward alliin, as the 1-propenyl allyl thiosulfinates are formed faster than allicin in crushed garlic (Lawson & Hughes, 1992).

The optimum pH conditions for maximal activity of alliinase have been found to be pH 5–8 for a crude preparation (Stoll & Seebeck, 1951b) and 6.5 for the highly purified enzyme (Mazelis & Crews, 1968; Jansen et al., 1989), which is considerably lower than the pH optimum of 8.5 for onion or broccoli alliinase (Schwimmer & Mazelis, 1963; Mazelis, 1963). The optimum temperature for alliinase activity is 33–37°C (Jansen et al., 1989; Stoll & Seebeck, 1951b), but there is surprisingly good activity even at 0–2°C (Stoll & Seebeck, 1951b; Lawson & Hughes, 1992).

Garlic appears to have two alliinase enzymes, one which is propenyl-specific for alliin and isoalliin, and one which is specific for methiin or methylcysteine sulfoxide, although a second alliinase has never been isolated. The evidence for this methyl-specific alliinase is threefold: the pH optimum for generating allyl and 1-propenyl thiosulfinates from garlic is 5.0, but the pH optimum for generating the methyl thiosulfinates is 6.5–7; the methyl-specific enzyme is more heat-sensitive, as heat drying of sliced garlic cloves decreases the rate of formation of the methyl thiosulfinates, but not the rate of formation of the allyl or 1-propenyl thiosulfinates; and the methyl-specific alliinase activity is three times more sensitive to inhibition by amino-oxyacetate (Lawson & Hughes, 1992). The presence of a methyl-specific alliinase is not unexpected, since methiin is much more broadly found in the plant kingdom than is alliin. It also appears possible that there may be a 1-propenyl–specific alliinase. This has been indicated by the observation that microwaving 5–6 g whole garlic cloves for 15 seconds followed by homogenization in water results in the formation of only 35% less 1-propenyl thiosulfinates, while allyl and methyl thiosulfinates are decreased by 70% and 97%, respectively (Lawson & Wang, 1994).

Bacteria also appear to have alliinase activity. This was first mentioned by Stoll & Seebeck in 1951, reporting von Euler's finding that *E. coli* incubated with pure alliin develop a garlic odor (allicin). *Bacillus subtilis* has also been shown to possess alliinase activity (Murakami, 1960a, 1960b). These studies, which have not been generally known, are very interesting because they indicate that the intestinal flora may be capable of producing allicin from alliin, assuming that alliin absorption does not occur faster than the likely slower bacterial action. Such a possibility, though untested, is important because much of garlic's medicinal effects have been ascribed to allicin, yet if one eats cooked cloves (cooking inactivates alliinase) or does not chew the fresh clove (stomach acid inactivates alliinase), alliin, not allicin, will enter the intestinal tract.

3.2.5.1 INHIBITION OF ALLIINASE

Acidic conditions (pH 3.5 or below) rapidly destroy garlic's alliinase activity, and hence prevent the generation of allicin and other thiosulfinates, as has been shown under artificial gastric and intestinal conditions (Lawson & Hughes, 1992; Blania & Spangenberg, 1991). Furthermore, neutralization of an inhibiting pH has no effect on restoring alliinase activity when the initial pH is 3 or below (Fig. 3.8). This is of practical importance in the question of how to consume garlic or garlic pills. Since the pH of the stomach is about 1.5 when empty and about 3 after a light meal or as much as 4.5 after a high protein steak meal (Davenport, 1982; Goldschmiedt & Feldman, 1993), unchewed garlic cloves or garlic powder consumed in gelatin capsules that do not have a stomach acid-resistant coating will usually not produce allicin unless consumed with a high protein meal. Although an empty stomach has a very low pH, the volume of acid present is small (about 10 mL) and can be partially neutralized by the modest amount of protein present in garlic powder. Therefore, consumption of garlic powder in uncoated gelatin capsules on an empty stomach does result in significant allicin production, accompanied by considerable gastric pain, as experienced by the author. As is mentioned later on (Chapter 7), gastric pain is easily averted by consumption of garlic with a meal.

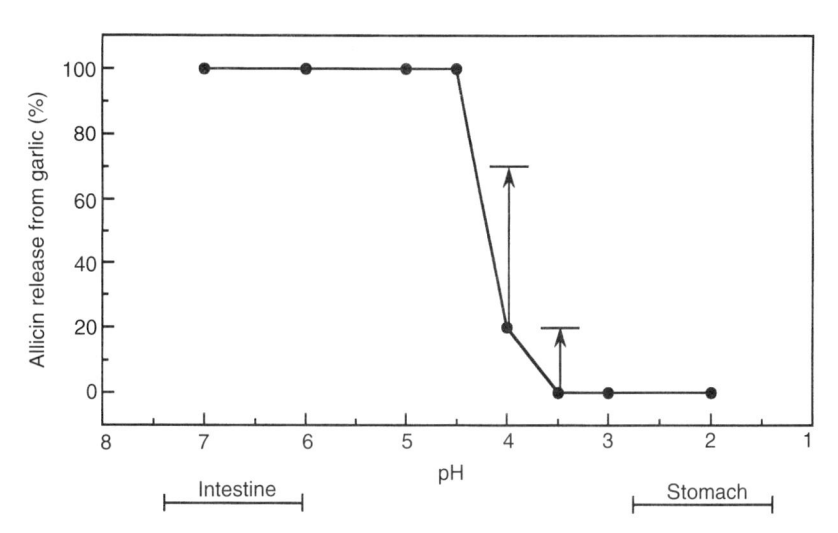

Figure 3.8. Effect of pH on garlic alliinase activity as determined by allicin release. *Arrows* indicate ability of pH neutralization to restore activity. Below pH 3.5, neutralization has no effect. (From Lawson LD, Hughes BG. Characterization of the formation of allicin and other thiosulfinates from garlic. Planta Med 1992;58:345–350.)

Alcohols are general protein denaturants and have been used to inhibit alliinase as well as to prepare various garlic extracts. However, alcohols are not effective inhibitors of alliinase. After 30 minutes of incubation of freeze-dried garlic, no inhibitory effect is found for either ethanol or methanol at a final concentration of 40%. At 70% of either alcohol, there is about 90% inhibition, but even at a final concentration of 99% about 1% of the activity still remains, as determined from allicin formation. Furthermore, static incubation of whole peeled cloves in ethanol (4 mL/g) for 48 hours resulted in conversion of 50–60% of their alliin content to allicin, as found in the ethanol (Lawson & Wang, 1994).

Effective inhibitors of alliinase are important for determining the content of the cysteine sulfoxides in garlic. In addition to general inhibition by protein denaturants, some specific inhibitors have been found, most of which act on the pyridoxal 5'-phosphate moiety of alliinase. While amino-oxyacetate (carboxymethoxylamine, NH_2OCH_2COOH) (Müller, 1990; Lawson & Hughes, 1992) and hydroxylamine (Mazelis & Crews, 1968; Edwards et al., 1994) have been successfully used in garlic preparations, L-cycloserine has been reported to be the strongest inhibitor of purified alliinase, an effect which is irreversible but slow (Kazaryan et al., 1979). The relative activity of each of these inhibitors toward allicin generation from powdered garlic cloves has been compared (Lawson & Wang, 1994). As shown in Table 3.6, L-cycloserine was a very poor inhibitor of this rapid reaction, while amino-oxyacetate was about 10 times more active than hydroxylamine. Inhibition by amino-oxyacetate required 10 mM to be complete and irreversible. At 3 mM, inhibition was 99.7% complete, but activity resumed after 3 hours.

3.2.6 Allicin and Other Thiosulfinates: Odorous But Medicinal

The rapid lysis of the cysteine sulfoxides by alliinase in crushed garlic cloves or in wetted garlic powder proceeds by formation of sulfenic acids which rapidly condense to form dialkyl thiosulfinates. As a primary example of this reaction, Figure 3.9 shows the scheme for the formation of allicin from alliin. This scheme has remained almost unchanged since its first proposal in 1949 by those who discovered alliin and alliinase (Stoll & Seebeck, 1949b, 1951b), but it took 25 years before the existence of the sulfenic acid intermediate could be proven and for

Table 3.6 Inhibitors of Garlic Alliinase

Inhibitor	Concentration to Give 90% Inhibition	Concentration to Give 100% Inhibition
Amino-oxyacetate (carboxymethoxylamine)	0.5mM	10 mM
Hydroxylamine	3 mM	100 mM
L-Cycloserine	200 mM	>1000 mM

Concentrations are based on the amount of allicin formed in 10 minutes after suspension of whole-clove powder in inhibitor solution.

the structure of the sulfenic acid to be shown to be allyl-S-OH rather than allyl-S(=O)H (Block et al., 1984, 1986; Block, 1992b; Isenberg & Grdinic, 1973; Penn et al., 1978).

The low-molecular-weight sulfenic acids have extremely short lifetimes before condensation to form thiosulfinates. Practical evidence for this short life comes from the fact that when water is added to garlic powder, over half of the alliin is converted to allicin in only 6 seconds (Lawson & Hughes, 1992). In addition to the formation of allicin from alliin, pyruvic acid and ammonia are also formed. Hence, pyruvate assays have permitted a rather simple procedure for determining total alliinase-reactive cysteine sulfoxides or total garlic thiosulfinates (Jäger, 1955; Schwimmer & Weston, 1961; Schwimmer, 1971).

Since allyl-, *trans*-1-propenyl-, and methylcysteine sulfoxides are all present in garlic, the action of alliinase produces three different sulfenic acids, all of which condense with each other to give eight of the nine possible thiosulfinates (Fig. 3.2). Because alliin is so much more abundant than the other cysteine sulfoxides, allicin is the main thiosulfinate produced, comprising about 70% of the total, while allyl methanethiosulfinate (12%) is the second most abundant (Fig. 3.2 and Tables 3.3 and 3.5). The missing possible thiosulfinate is di-1-propenyl thiosulfinate, which has been shown by Block and others to be rapidly converted in onion homogenates to *zwiebelanes* (Bayer et al., 1989a, 1989b).

Interestingly, in onions most of the *trans*-1-propenesulfenic acid does not even go to zwiebelanes, but rather forms the well-known tear-producing compound called *lachrymatory factor* ($CH_3CH_2CH=S-O$) (Block, 1992b; Block et al., 1979); however, lachrymatory factor has not been detected in garlic because 2-propenesulfenic acid is produced so rapidly and so abundantly that most of the *trans*-1-propenesulfenic acid is converted to allyl 1-propenyl thiosulfinates and some to methyl 1-propenyl thiosulfinates (Table 3.3 indicates abundance) (Lawson et al., 1991c; Lawson & Hughes, 1990; Block et al., 1992a, 1993).

The methyl thiosulfinates occur mainly (85%) in combination with allyl groups and, due to the slower action of alliinase on methylcysteine sulfoxide than on alliin (Fig. 3.6) (Lawson & Hughes, 1992; Mazelis & Crews, 1968), appear to be formed mostly by a mechanism other than the simple condensation of mixed sulfenic acids depicted in Figure 3.2. Because alliin is depleted before most of the methanesulfenic acid is produced, most of the methanesulfenic acid should self-condense to form dimethyl thiosulfinate, but this is not observed. It has been proposed (Block, 1992b; Lawson & Hughes, 1992) that the more slowly formed methanesulfenic acid reacts directly with the already formed allicin to produce both of the allyl methyl thiosulfinate regioisomers as depicted in Figure 3.10.

Such a mechanism predicts that allyl methanethiosulfinate will be three times more abundant than its isomer, methyl 2-propenethiosulfinate. In practice, a ratio closer to 2:1 is usually found (Lawson et al., 1991c; Block et al., 1992a), probably because some of the allyl methyl thiosulfinates are formed before all of the allicin is generated. Evidence for the formation of the allyl methyl thiosulfinates from already gen-

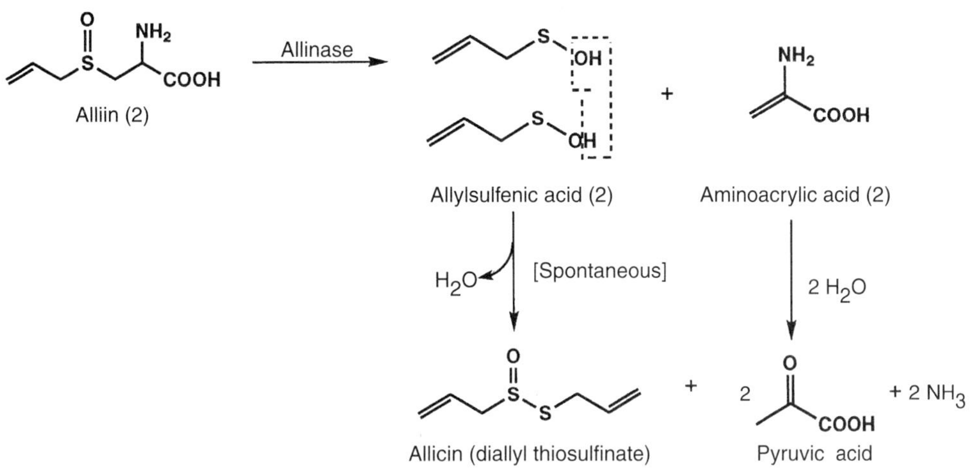

Figure 3.9. Formation of allicin by the action of alliinase (located in the vascular bundle sheath cells) on alliin (located in the mesophyll cells) upon crushing garlic. (Stoll & Seebeck, 1951.)

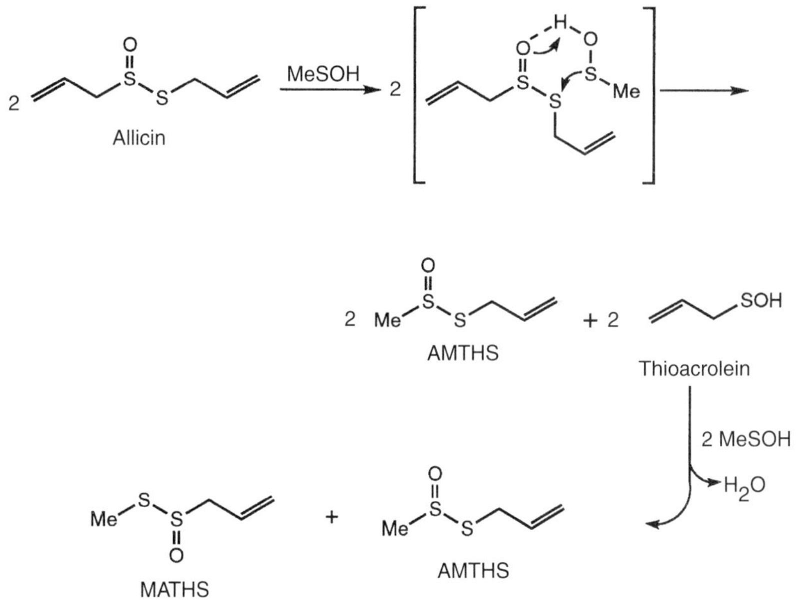

Figure 3.10. Reaction of allicin with methanesulfenic acid (MeSOH) to form 3 moles of allyl methanethiosulfinate (AMTHS) and 1 mole of methyl 2-propenethiosulfinate (MATHS). (Modified from Block, 1992a, 1992b.)

erated allicin comes from the observation that the allicin generated in one minute from garlic powder decreases slightly until all of the methyl thiosulfinates are produced in about 5 minutes (Lawson & Hughes, 1992), and from the observation that homogenization of garlic with a greater volume of water (30 mL/g rather than 10 mL/g) yields about 5% more allicin and correspondingly less allyl methyl thiosulfinates and more methyl methanethiosulfinate (Lawson & Wang, 1994). The greater dilution would make allicin less available to methanesulfenic acid than to another nearby-generated molecule of methanesulfenic acid. The same mechanisms also likely apply to the formation of the minor amounts of 1-propenyl methyl thiosulfinates.

Because the thiosulfinates contain an asymmetric sulfur atom (the S=O), optical isomers would be possible if the condensation of the sulfenic acids proceeded enzymatically rather than spontaneously. So far the only evidence of optical activity in natural allicin comes from the dissertation of Müller (1991), who used a high performance liquid chromatography (HPLC) chiral detector and found a specific rotation of −3.9° mL/dm/g. Bauer (1991) unsuccessfully attempted to separate the enantiomers of allicin and methyl methanethiosulfinate with a chiral HPLC column; however, the enantiomers of methyl methanethiosulfinate and propyl methanethiosulfinate in onion extract have been well separated on a chiral gas chromatography (GC) column, with the finding that the enantiomers were equally abundant (Block & Calvey, 1994). Therefore, the thiosulfinates from onion are racemic (no optical activity), and the same is suspected of garlic. Nevertheless, the issue is probably moot because in the body the two thioallyl groups of allicin are both rapidly converted to either S-allylmercaptocysteine (allyl-SS-cys) or allyl mercaptan (allyl-SH) (Lawson & Wang, 1993; Lawson, 1993).

The allicin yield of fresh garlic bulbs maintained in open air at room temperature has been found (in a study employing a garlic braid) to decrease less than 10% over 6–8 months after harvesting (Lawson et al., 1991c; Lawson & Wang, 1994); however, a more rapid decline has also been reported (Pfaff, 1991).

3.2.6.1 PHYSICAL AND SPECTRAL PROPERTIES OF THE THIOSULFINATES

The thiosulfinates of garlic are colorless, volatile liquids with pungent odors. Allicin is responsible for the usual odor of fresh-cut garlic, while the methyl thiosulfinates have the odor of cooked cabbage. Allicin has a density of 1.112 g/mL, a refractive index of 1.561, and is probably not optically active (see discussion above) (Cavallito & Bailey, 1944). It has a water solubility of about 2%, is moderately soluble in hexane, and is very soluble in organic solvents more polar than hexane (Cavallito & Bailey, 1944; Lawson, 1993; Isenberg & Grdinic, 1973).

Allicin and the other thiosulfinates have been identified by their ultraviolet (UV), infrared (IR), nuclear magnetic resonance (NMR), and mass spectral features. The ultraviolet spectra of the thiosulfinates containing only allyl or methyl groups are characterized by a prominent shoulder at 240 nm in water or 245 nm in methanol, while those containing a 1-propenyl group have a shoulder at 260–265 nm (Jansen et al., 1987; Lawson et al., 1991c; Block et al., 1992a). The extinction coefficients for allicin (2370 M^{-1} or 14.6 mL/mg at 240 nm in water) and most of the other thiosulfinates have been reported (Lawson et al., 1991c). The infrared spectrum for allicin is characterized by a strong S=O stretch at 1080–1090 cm^{-1} (Jansen et al., 1987; Block et al., 1986). The mass spectrum of allicin has been difficult to obtain because of its very rapid degradation in the presence of heat, even for very short periods of time. Although prior attempts have had limited success (Yu & Wu, 1989b; Müller, 1991), a mass spectrum that contains significant molecular ion intensity has only been recently obtained by the use of supercritical fluid chromatography and a low restrictor tip temperature (Calvey et al., 1994a, 1994b). Specific identification of the thiosulfinates has been achieved by the use of proton-NMR couples with regiospecific and stereospecific synthesis (Lawson et al., 1991c; Block et al., 1986, 1992a; Bayer et al., 1989a). The proton-NMR spectra for all of the thiosulfinates of garlic except dimethyl thiosulfinate have been reported (Lawson et al., 1991c); however, some errors exist in this study for the spectra of the methyl 1-propenyl thiosulfinates, and these have been corrected by others (Block et al., 1992a; Bayer et al., 1989a).

3.2.6.1.1 Synthesis and Standardization of Allicin

Because allicin yield is usually used as the measure of garlic quality, preparation of a pure standard is necessary since its instability precludes its commercial availability. Pure allicin can be obtained by isolation from garlic (Lawson et al., 1991c) or by synthesis. Synthesis has been achieved by oxidation of commercially available diallyl disulfide with either hydrogen peroxide/ glacial acetic acid (Stoll & Seebeck, 1947; Mayeux et al., 1988; Iberl et al., 1990a; Lawson et al., 1991c) or with perbenzoic or peracetic acids (Small et al., 1947; Jansen et al., 1987; Block et al., 1986; Freeman et al., 1993). All of these methods employ chromatographic procedures to purify the allicin and do not achieve a purity greater than 95%. A very convenient method to prepare allicin of a purity of 98% or more without chromatographic purification proceeds as follows (Lawson & Wang, 1994). One gram of fractionally distilled (removes diallyl polysulfides) commercial grade (about 80% pure) diallyl disulfide is dissolved in 5 mL cold (4°C) glacial acetic acid, to which 1.5 mL cold 30% hydrogen peroxide is slowly added. After 30 min, the temperature is allowed to increase to 12-15°C and stirring is continued for 4–6 hours until the diallyl disulfide content decreases by only 75–80% (this assures no overoxidation to diallyl thiosulfonate [Freeman et al., 1993]). The reaction is stopped with 15 mL water and extracted with 30 mL dichloromethane. Acetic acid is removed by washing the extract several times with 5% $NaHCO_3$ and then with water to pH 6–7. After evaporation of the solvent, the allicin is redissolved in 500 mL water, and unreacted diallyl disulfide and other low-polar impurities are removed by double extraction with 0.1 volume of hexane. The yield is almost quantitative on the basis of the amount of diallyl disulfide depleted. The purity (weight %) is determined by reversed phase HPLC using 75% methanol at 240 nm, using published extinction coefficients (Lawson et al., 1991a, 1991c) for allicin, ajoene (the main impurity, which elutes about 0.5 minutes after allicin), and other sulfides.

The allicin content of the preparation can be accurately (within 3%) quantitated by sodium borohydride reduction to diallyl disulfide (see Section 4.3.2.2), cysteine depletion (Han et al., 1995; see Section 4.3.2.1), gravimetry after correction for impurities, the extinction coefficient of allicin (accurate within 8%), or other methods (see Section 4.3.2). Gravimetry is conducted by extraction of the aqueous preparation with two volumes of dichloromethane followed by sodium sulfate drying and rotary evaporation at ambient temperature to nearly constant weight and must be conducted on only a portion (enough to give about 30 mg) of the preparation, since allicin rapidly degrades in the absence of solvent. The cysteine depletion method has the advantage of not requiring a standard, but cysteine also reacts with ajoene. The sodium borohydride reduction method has the advantages of being highly specific to allicin and of the commercial availability of diallyl disulfide, although the disulfide must be further purified and stored below 0°C. In our hands, this method has given better reproducibility (± 3%) than the cysteine depletion method (± 8%).

3.2.6.2 STABILITY OF THE THIOSULFINATES

Since its discovery, allicin has been well-known to be a relatively reactive compound, as reflected in both its antibacterial activity and its rapid conversion to other compounds under various conditions. Unfortunately, the instability of allicin has been frequently greatly exaggerated, particularly in the popular literature, with claims of its disappearing in a few minutes to a few hours at room temperature. The earliest known reports of the stability of allicin in water found a half-life (time required for a loss of 50%) of 4 to 6 days (Sreenivasamurthy et al., 1961; Laakso et al., 1989). However, the stability of allicin has long been indicated by the decrease in its antibiotic activity, including in a more recent study which showed the half-life of the anti-candidal activity of garlic to be about 15 days (Hughes & Lawson, 1991).

The stability of allicin in a variety of solvents and under a variety of conditions (Lawson, 1993; Lawson et al., 1991c; Lawson & Wang, 1994), is compared in

Table 3.7. Although allicin is less soluble in water than in organic solvents (see above), its stability increases with increased polarity of the solvent. For example, allicin is most stable in water (especially if acidified), with a half-life of about 30 days at room temperature, and least stable in hexane where its half-life is only 2 hours. Allicin's greater stability in water is due to the hydrogen bonding of water to its oxygen atom, thus reducing the rate of self-reaction which proceeds either by a monomolecular self-elimination reaction (mechanism shown below in Fig. 3.13A) or a bimolecular reaction (Fig. 3.13B). The alcohols also exert a degree of hydrogen bonding to provide some stability.

Table 3.7 Stability of Allicin in Various Solvents

Solvent	Temperature (°C)	Half-life
water (0.1–2 mg/ml)	23°C	30–40 days
water	–20°C	(30% loss in 1 year)
water	–70°C	(no loss in 2 years)
1mM citric acid	23°C	60 days
methanol	23°C	48 hours
chloroform	23°C	48 hours
dichloromethane	23°C	30 hours
ethanol	23°C	24 hours
acetonitrile	23°C	24 hours
diethylether	23°C	3 hours
hexane	23°C	2 hours
none (neat)	23°C	16 hours
none (neat)	–70°C	25 days

The stability of allicin is also very temperature-dependent. Refrigeration increases its stability about 20-fold, but the only sure way to store allicin is in water at –70°C. In the absence of solvent, allicin cannot be stabilized, even at –70°C. Interestingly, allicin is less stable in crushed garlic or aqueous garlic homogenates than it is as a pure compound in water (Fig. 3.11). While the stability of pure allicin in water varies little with its concentration, the stability of allicin in garlic homogenate decreases greatly with increased concentration. However, when sufficiently diluted, the stability of allicin in garlic homogenate is similar to that of pure allicin, indicating that there are water-soluble substances in garlic which decrease allicin's stability (Lawson & Wang, 1994).

Attempts to stabilize allicin by adsorption onto *silica gel* has resulted in its commercial availability by two vendors; however, analysis of these products has revealed no detectable allicin (Jäger et al., 1992; Lawson et al., 1991c). Thus, such "allicin standards" are useless and will lead to greatly erroneous results (Jäger et al., 1992).

Not all thiosulfinates have equal stability. Studies of the stability of each of the thiosulfinates of garlic (Lawson & Hughes, 1992; Lawson, 1993; Block et al., 1992a) have

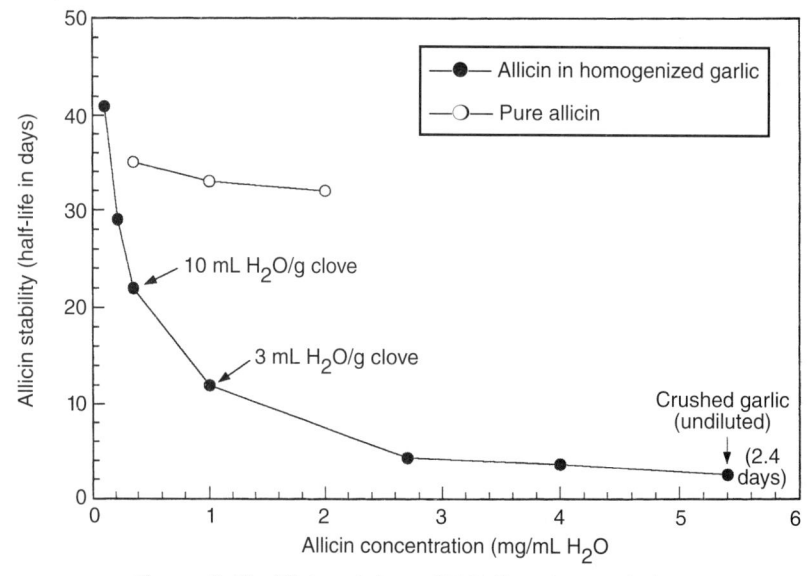

Figure 3.11. Allicin stability at 23°C. (From Lawson & Wang, 1994.)

found that the allyl-S thiosulfinates (allicin, allyl methanethiosulfinate, and allyl 1-propenethiosulfinate) are the least stable of garlic's thiosulfinates because they form thioacrolein ($CH_2=CH-CH=S$) (Block et al., 1984, 1986; Lawson et al., 1991c; Bock et al., 1982a, 1982b) and allyl mercaptan (allyl-SH) (Block et al., 1986) as intermediates for further reactions. However, the allyl-$S=O$ thiosulfinates (methyl 2-propenethiosulfinate and 1-propenyl 2-propenethiosulfinate) cannot form thioacrolein by elimination and are considerably more stable. For example, in ether, allyl methanethiosulfinate has the same half-life as allicin (3 hours), but its regioisomer, methyl 2-propenethiosulfinate, shows little loss in 6 days (Lawson, 1993).

The 1-propenyl-$S=O$ thiosulfinates (allyl 1-propene- and methyl 1-propenethiosulfinates) are the least stable of garlic's thiosulfinates, with allyl 1-propenethiosulfinate being about 50 times less stable than allicin in water (Lawson & Hughes, 1992). Methyl methanethiosulfinate, on the other hand, is about 7 times more stable in water than allicin (Lawson & Wang, 1994). The instability of the allyl thiosulfinates has prevented their analysis by gas chromatography (Brodnitz et al., 1971); however, using very-low-temperature gas chromatography, Block has succeeded in analyzing all other thiosulfinates found in *Allium* plants (Block et al., 1992b).

3.2.6.3 THIOSULFINATE REACTIONS

As already mentioned, the thiosulfinates are very reactive compounds. In addition to reactions with reducing agents, they also undergo a variety of spontaneous reactions in different solvents or media. Because the transformation products of the thiosulfinates are considerably more stable than the thiosulfinates, but still contain the thioallyl (*S*-allyl) or thiomethyl groups, several of these reactions are employed in the production of commercial garlic oil preparations. These reactions are summarized in Figure 3.12 and will now be discussed in detail.

3.2.6.3.1 Thiosulfinate Transformations in Water or Steam: Allyl Sulfides

The thiosulfinates present in crushed garlic are slowly transformed at room temperature principally to *diallyl trisulfide, diallyl disulfide,* and *allyl methyl trisulfide,* listed in order of abundance (Iberl et al., 1990a; Yu & Wu, 1989b). This transformation is greatly accelerated and expanded upon heating, such as during steam distillation of chopped garlic for preparation of commercial garlic oils. As many as 24 different sulfides have been identified in the oil of steam-distilled garlic and include the diallyl (57%), allyl methyl (37%) and dimethyl (6%) mono- to hexasulfides and, occasionally, small amounts of allyl 1-propenyl and methyl 1-propenyl di-, tri- and tetrasulfides (Lawson et al., 1991a; Vernin & Metzger, 1991; Jirovetz et al., 1992; Pino et al., 1991). As previously mentioned (Section 3.2.1), the sulfide transformation products were the first sulfur compounds identified in processed garlic (Semmler, 1892). Allyl sulfides containing four or more sulfur atoms have often not been reported because they are unstable in polar gas chromatography columns (Lawson et al., 1991a; Vernin & Metzger, 1991). The allyl sulfides are rather insoluble in water (0.05 mg/mL for diallyl disulfide and 0.006 mg/mL for diallyl trisulfide), but their solubility can be greatly increased by detergents. For example, in 1% solutions of Triton X-100 or Tween 80, the solubility of diallyl trisulfide is increased to 4.5 and 1.5 mg/mL, respectively (Lawson & Wang, 1994).

The mechanisms proposed by Block for the transformation of allicin to diallyl trisulfide and diallyl disulfide are shown in Figure 3.13 (Block et al., 1986; Block, 1992b). Transformation to diallyl disulfide results in the loss of one-third of the *S*-allyl groups to propene and sulfur dioxide. Transformation to diallyl trisulfide also results in the loss of one-third of the allyl groups to allyl alcohol, but no loss of sulfur. The finding of propene and allyl alcohol in steam-distilled garlic (Yu et al., 1989a, 1989b) confirms these mechanisms. As further evidence, it has been

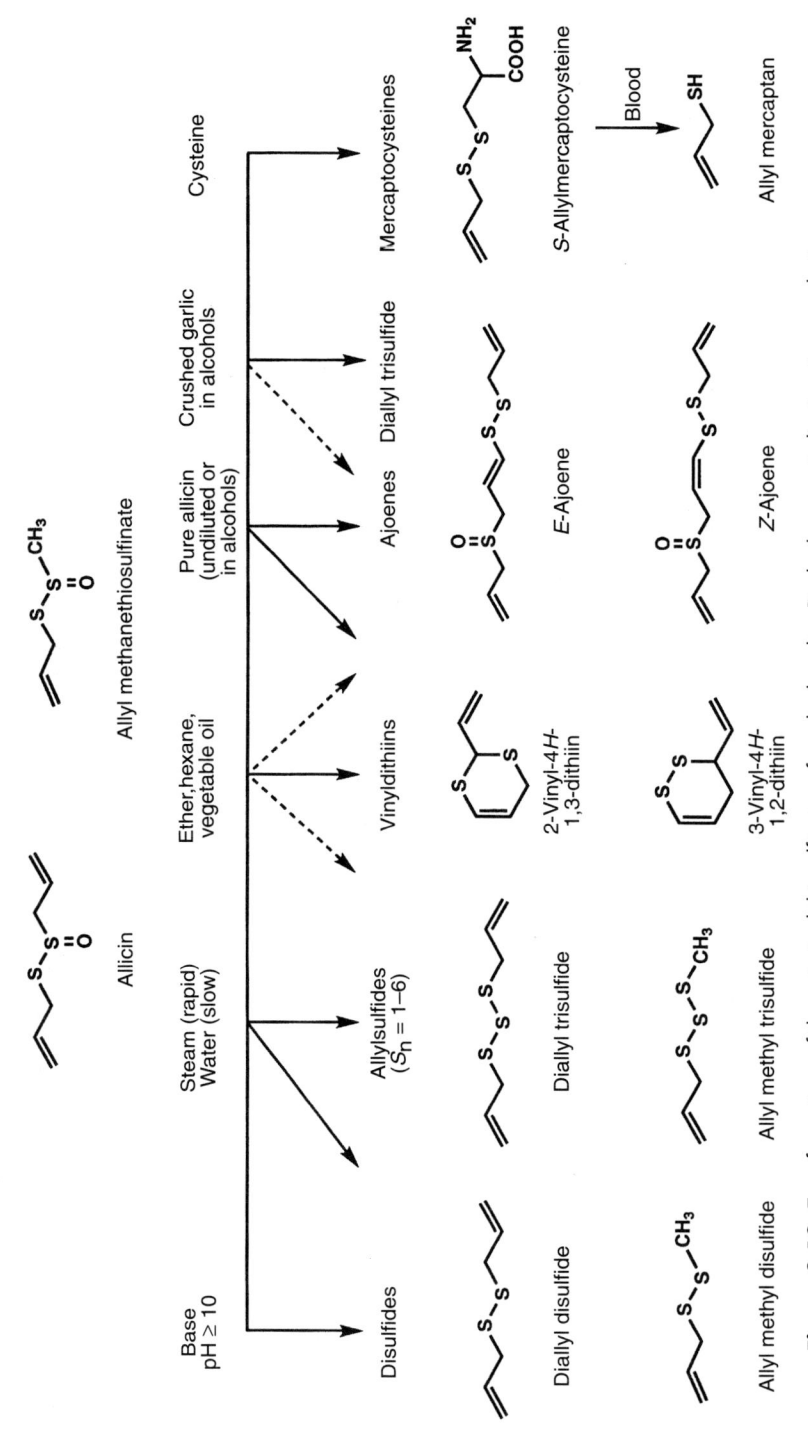

Figure 3.12. Transformations of the principal thiosulfinates of crushed garlic. *Dashed arrows indicate minor products.*

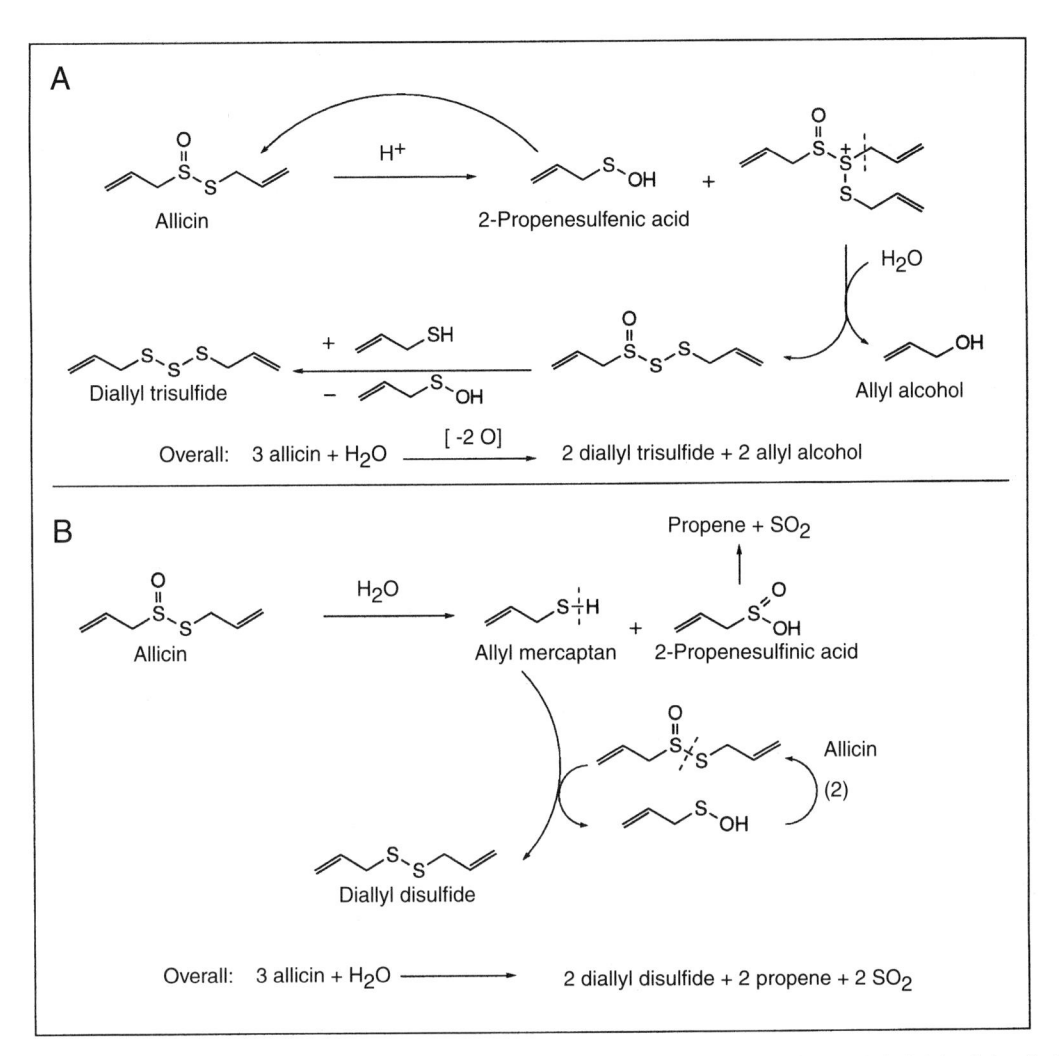

Figure 3.13. Mechanisms for the spontaneous transformation of allicin to diallyl trisulfide and diallyl disulfide (Block et al., 1986; Block, 1992a, 1992b), with addition of overall equations.

shown that heating (120°C) an aqueous allicin solution in a closed container results in a 30% loss of allyl groups to compounds other than allicin or diallyl sulfides (Lawson & Wang, 1994). However, it is possible that other mechanisms also take place, such as those involving the formation of diallyl thiosulfonate (allyl-SSO$_2$-allyl) as an intermediate, particularly if the sulfur dioxide has not escaped (Backer & Kloosterziel, 1954; Block, 1992b).

Both of the mechanisms shown in Figure 3.13 take place simultaneously; however, kinetic studies with heated garlic (Yu & Wu, 1989b; Miething, 1988) and analyses of freshly made steam distillates (Jirovetz et al., 1992; Miething, 1988) show that diallyl trisulfide is clearly the most dominant sulfide formed and is several times more abundant than diallyl disulfide. Nevertheless, commercial steam-distilled products often contain similar amounts of diallyl trisulfide and diallyl disulfide (Lawson et al., 1991a; Vernin et al., 1986; Miething, 1988; Yan et al., 1992), probably as a result of slow disproportionation in the undiluted oil of diallyl trisulfide to diallyl disulfide and diallyl tetrasulfide.

Table 3.8 Fate of Undiluted Allicin at Room Temperature

Time	Allicin	Ajoene	1,3-VDT	1,2-VDT	AAS$_2$	AAS$_3$	Unidentified
Isolated garlic thiosulfinates (87% allicin)							
0 h	100	0	0	0	0	0	0
24 h	32	17	21	7	0	0.5	22
48 h	6	13	12	7	0	1.5	61
Synthesized allicin (98%)							
0 h	100	0	0	0	0	0	0
24 h	19	6	7	4	7	3	53
48 h	3	5	4	3	6	3	77

Values represent weight %, as determined by C18-HPLC. Zero time began when samples were extracted from aqueous preparations and the solvent evaporated. The amount of unidentified compounds was estimated from the difference between amounts of identified compounds and the initial weight, and by assigning the extinction coefficient of diallyl disulfide to each unidentified compound, both methods of which approximately agreed. Abbreviations: VDT, vinyldithiin; AAS$_2$, diallyl disulfide; AAS$_3$, diallyl trisulfide. From Lawson LD, Wang ZJ, unpublished data, 1994.

3.2.6.3.2 Transformation of Undiluted Allicin

Pure allicin is an oily liquid, but this is its most unstable form, and as such, it has been reported to disappear almost completely in less than 20 hours at room temperature (Brodnitz et al., 1971). This original gas chromatography study reported that the allicin is transformed primarily to diallyl disulfide (66%), diallyl sulfide (14%), and diallyl trisulfide (9%). However, using nondestructive HPLC analysis, contrary results have been found (Lawson & Wang, 1994). Table 3.8 shows the results of incubating undiluted allicin, either synthetic or isolated as total thiosulfinates from garlic, in closed vials for 1–2 days. Most of the allicin had decayed in 24 hours; however, the dominant identifiable transformation products were the vinyldithiins and ajoene. Diallyl disulfide was also abundant with synthetic allicin, but was not found with allicin isolated from garlic, indicating that other thiosulfinates present have an influence. The majority of the transformation products, however, especially at 48 hours, were unidentified compounds, of which there were about 40. Incubation in a nitrogen atmosphere rather than in air did not change the results. The reason for the great instability of pure allicin is simply that it easily reacts with itself. As mentioned previously, allicin's stability increases greatly with dilution, especially in solvents capable of hydrogen bonding.

3.2.6.3.3 Thiosulfinate Transformations in Organic Solvents: Vinyldithiins and Ajoenes

Upon incubation in organic solvents, the thiosulfinates undergo different transformations than in water (Fig. 3.14). Furthermore, the distribution of the new compounds formed depends upon the polarity of the solvent and other factors. In low-polarity solvents (e.g., hexane, ether, chloroform, or vegetable oil), allicin and allyl methanethiosulfinate rapidly form mainly *1,3-vinyldithiin* (2-vinyl-4H-1,3-dithiin) (51%), *1,2-vinyldithiin* (3-vinyl-4H-1,2-dithiin) (19%) and lesser amounts of *ajoene* (E,Z-4,5,9-trithiadodeca-1,6,11-triene 9-oxide) (12%) and *allyl sulfides* (18%) (Voigt & Wolf, 1986; Iberl et al., 1990a; Lawson et al., 1991a). The vinyldithiins were first discovered as major degradation products of allicin which were created as artifacts during the gas chromatography of allicin (Brodnitz et al., 1971) and were later shown to be the dominant sulfur compounds found in commercial garlic oils prepared by maceration of crushed garlic in vegetable oil (Voigt & Wolf, 1986). Originally, Brodnitz misidentified the structure of the dominant vinyldithiin, but this has since been corrected (Block et al., 1984, 1986).

Ajoene was originally discovered upon incubation of chopped garlic in methanol as a new compound which had remarkable activity toward inhibition of blood platelet aggregation (Apitz-Castro et al., 1983). Shortly thereafter, the structure of the com-

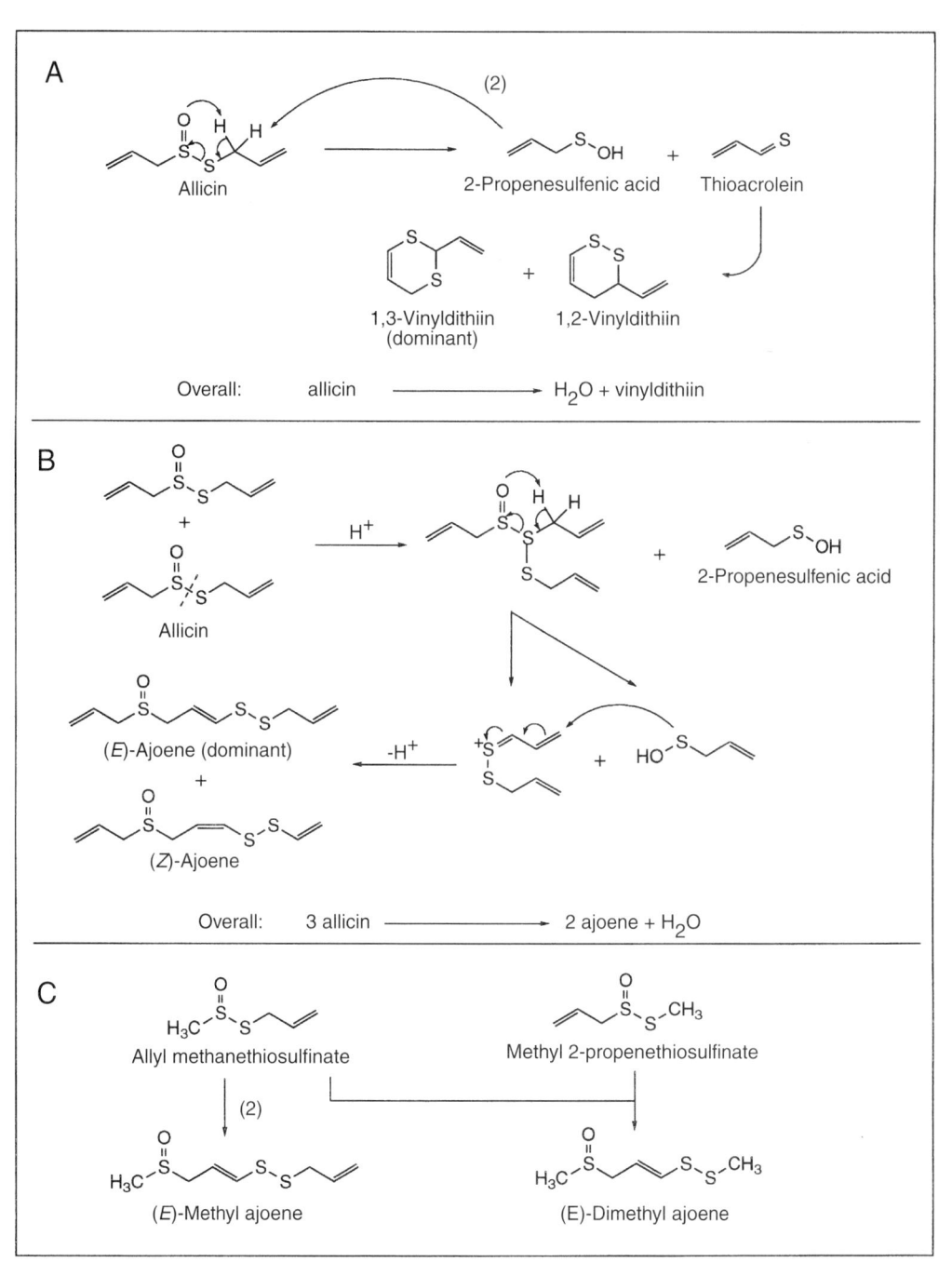

Figure 3.14. Mechanisms for the formation (in organic solvents) of vinyldithiins and ajoene from allicin and the formation of methyl ajoenes from allyl plus methyl thiosulfinates. (Modified from Block E, Ahmad S, Catalfamo JL, Jain MK, Apitz-Castro R. The chemistry of alkyl thiosulfinate esters. IX. Antithrombotic organosulfur compounds from garlic: structural, mechanistic, and synthetic studies. J Am Chem Soc 1986;108:7045–7055.)

pound was determined and the name ajoene ("ajo," pronounced "ah-ho," is Spanish for garlic) was given to it (Block et al., 1984). Ajoene is the only transformation product of allicin which is oxygenated and which has similar water solubility to allicin; however, it is considerably more stable than allicin, showing no decay at all after 6 days at room temperature in the absence of solvent (Lawson & Wang, 1994). For a discussion of its long-term stability in garlic products, see Section 3.4.3. As solvent polarity increases, both the ratio of ajoenes to vinyldithiins (e.g., 0.09 in ether or 1.7 in ethanol) and the ratio of E-ajoene to Z-ajoene (e.g., 0.4 in ether or 7 in ethanol) increase (Iberl et al., 1990a). Increasing the reaction temperature also increases the E/Z ratio, but heat has no effect after the reaction. The ratio of ajoenes to vinyldithiins, however, can be increased to as high as 5:1 when allicin in 85% acetonitrile/water is heated at 80°C for 20 minutes (Lawson & Wang, 1994).

With polar solvents, such as the alcohols, incubation of the thiosulfinates gives variable results. Incubation of allicin, as garlic homogenate, in ethanol produces mainly diallyl trisulfide (73%), diallyl disulfide (8%), ajoene (8%), and no reported vinyldithiins. In contrast, abundant amounts of ajoene (55%) and vinyldithiins (34%), and only minor amounts of allyl sulfides (0–11%), are produced when pure allicin is dissolved in ethanol (Iberl et al., 1990a). Similar results have been found with methanol: incubation of garlic homogenate in methanol yielded diallyl trisulfide (83%), diallyl disulfide (10%), ajoene (1%), vinyldithiins (1%), and about 5% unidentified compounds, while incubation of pure allicin in methanol yielded only 8% diallyl trisulfide and about equal amounts (20–25%) of ajoene and the vinyldithiins, with about 25% unidentified compounds (Lawson, 1993). Apparently, other compounds in garlic, probably the very abundant fructans (see Section 3.3.1), influence the transformation of allicin in alcohols to produce primarily diallyl trisulfide.

Incubation of pure allyl methanethiosulfinate (Block et al., 1986), the second most abundant thiosulfinate released by garlic, or homogenates containing allyl methanethiosulfinate (Sendl et al., 1992; Sendl & Wagner, 1991), in polar solvents such as aqueous acetone has been shown to produce the dimethyl and methyl (less abundant) homolog of ajoene (Fig. 3.14); however, they have less antiaggregatory activity than ajoene (Block et al., 1986). They are particularly abundant when homogenates of *Allium ursinum* L. (wild garlic) are used, because their bulbs contain nearly equal amounts of allyl- and methylcysteine sulfoxides (Sendl & Wagner, 1991; Sendl, 1992). According to the mechanism shown in Figure 3.14, ajoene can form from allicin or from allicin plus allyl methanethiosulfinate; methylajoene can form from allyl methanethiosulfinate or from allicin plus methyl 2-propenethiosulfinate or from allicin plus methyl methanethiosulfinate; and dimethylajoene can form from allyl methanethiosulfinate plus methyl 2-propenethiosulfinate or from allyl methanethiosulfinate plus methyl methanethiosulfinate.

3.2.6.3.4 Thiosulfinate Reactions in Base: Disulfides

Although relatively stable under acidic conditions, thiosulfinates undergo immediate alkaline hydrolysis to form disulfides (Block et al., 1984; Müller, 1989; Yu & Wu, 1989a; Yu et al., 1989a; Kice & Rogers, 1974) and sulfur dioxide (Cavallito et al., 1944). It has been shown that alkaline hydrolysis (sodium hydroxide, pH 11, in 50% methanol) of one mole of allicin gives 0.5 moles of diallyl disulfide, whereas one mole of allyl methanethiosulfinate (allyl-SS(=O)-methyl) gives 0.25 moles each of diallyl disulfide and allyl methyl disulfide, while one mole of methyl 2-propenethiosulfinate (methyl-SS(=O)-allyl) gives 0.25 moles each of dimethyl disulfide and allyl methyl disulfide (Lawson, 1993). No mono- or trisulfides were found. The loss of 50% of the thiosulfinates has not yet been explained, but additional, oxygenated compounds have been observed in the HPLC analyses.

3.2.6.3.5 Thiosulfinate Reactions with Cysteine and Other Thiols

It has been known since its discovery in 1944 that allicin reacts very rapidly with the sulfhydryl group of the amino acid, cysteine, to form two moles of S-allylmercaptocysteine (allyl-SS-cysteine) plus water (Cavallito et al., 1944; Lawson et al., 1992). It has also been shown that allicin inhibits a large number of enzymes in vitro which contain cysteine at their active sites, but very few that do not (Wills, 1956) and that the same reaction takes place with glycylcysteine, N-acetylcysteine (Bailey & Cavallito, 1948), and protein-cysteine (Fujiwara & Murakami, 1960). In fact, many of the explanations given for the biological effects of garlic focus on the ability of allicin to react with sulfhydryl enzymes as well as the sulfhydryl group of acetyl-CoASH, the building block of cholesterol and triglyceride synthesis (Augusti & Mathew, 1974; Papageorgiou et al., 1983; Freeman & Whenham, 1975a). Other thiosulfinates also give the same reaction. For example, methyl and propyl thiosulfinates react with cysteine to produce S-methyl mercaptocysteine and S-propyl mercaptocysteine (Fujiwara et al., 1955). The rate of the reaction is pH-dependent, being maximal at pH 8 and about 50 times slower at pH 5 (Lawson & Wang, 1994). At pH 7, the half-life for the reaction of allicin with cysteine is less than 1 minute (Lawson & Wang, 1993). Recently, a very sensitive method of analysis of the total thiosulfinates of garlic has been developed on the basis this reaction (Han et al., 1995).

Cysteine reacts not only with allicin, but also with all of the allicin transformation products containing a dithioallyl (allyl-SS-) group (Fig. 3.15). It has been known for some time that diallyl disulfide (Papageorgiou et al., 1983) and allyl propyl disulfide (Augusti, 1977) react with cysteine to form S-allylmercaptocysteine and that ajoene probably also does, since it disappears in the presence of cysteine (Winkler et al., 1992a) and since it inhibits the cysteine-dependent enzyme, gastric lipase (Gargouri et al., 1989, 1992). Recently the reaction of cysteine with allicin and its transformation products (0.5 mM) has been studied in some detail under artificial physiological conditions (Lawson & Wang, 1993; Lawson, 1993). It was found that ajoene and diallyl trisulfide react very rapidly (half-life of less than 2 minutes) with cysteine to form S-allylmercaptocysteine, while diallyl disulfide reacts much more slowly (half-life about 30 minutes). One mole of E,Z-ajoene was found to react with two moles of cysteine to form one mole of S-allylmercaptocysteine and a new compound, named *ajocysteine*, which appears to have the structure shown in Figure 3.15. Ajocysteine formation is optimal in the presence of both cysteine and cystine and is most stable at a cystine/cysteine ratio of about 4.5 (3–6). Diallyl trisulfide was found to give 1–2 moles of S-allylmercaptocysteine, depending upon the amount of excess cysteine present. Diallyl disulfide was also found to react with cysteine to give one mole of S-allylmercaptocysteine and one mole of allyl mercaptan. No reaction with cysteine was found to occur for diallyl sulfide, the vinyldithiins, alliin, or the γ-glutamyl-S-alkenylcysteines. These reactions may have physiological significance due to the presence of cysteine, and possibly glutathione, in the intestinal tract during digestion.

3.2.6.4 COMPOSITION OF THE VAPOR (ODOR) OF CRUSHED GARLIC AND GARLIC BREATH

The smell of chopped garlic is a widely known and valued characteristic. Consequently there have been several attempts to determine its composition. These gas chromatography studies, which typically employ heated crushed garlic, called headspace analyses, have usually reported the most abundant volatile compound to be diallyl disulfide, followed by lesser amounts of diallyl trisulfide, allyl methyl di- and trisulfides, allyl 1-propenyl disulfide, and small amounts of the monosulfides (Laakso et al., 1989; Saghir et al., 1964; Pino et al., 1991; Pino, 1992; Oaks et al., 1964; Deruaz et al.,

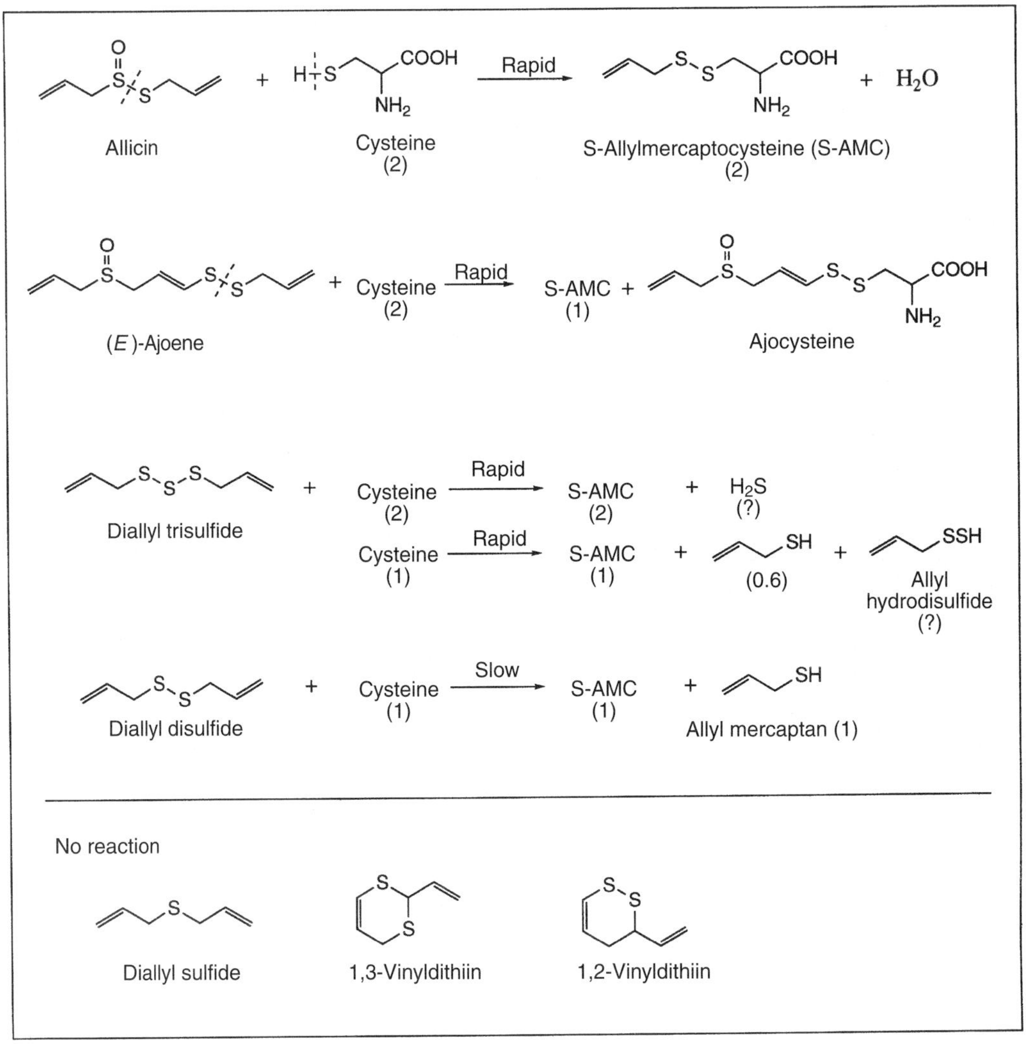

Figure 3.15. Reaction of cysteine with allicin and allicin-derived compounds at neutral pH. Compounds with a (?) beneath them are likely, but have not yet been confirmed. (From Lawson LD, Wang, ZJ. Pre-hepatic fate of the organosulfur compounds derived from garlic (*Allium sativum*). Planta Med 1993;59:A688.)

1994; Cai et al., 1995a). Furthermore, methyl-Se-S-allyl has recently been shown to be the dominant selenium compound in garlic vapor (Cai et al., 1994a; 1994b, 1995a).

The true composition of the odor of crushed garlic, however, is only partially revealed by these studies because allicin and the other thiosulfinates of garlic are also volatile, but yet they cannot be analyzed by normal gas chromatography because of destruction in the heated injection port and column oven (Brodnitz et al., 1971; Block et al., 1992b). Furthermore, heating in the oil bath of the headspace apparatus will convert much of the thiosulfinates to sulfides. When the vapor of crushed garlic is adsorbed onto silica gel followed by HPLC analysis, a method which employs no heat, abundant amounts of the thiosulfinates are found, in addition to diallyl disulfide, and small amounts of allyl 1-propenyl disulfide and allyl methyl disulfide, but no diallyl trisulfide (Table 3.9). The relative abundance of allicin and diallyl disulfide was found to vary depending on the source of the allicin and on the amount of time elapsed. As would be

Table 3.9 Approximate Composition (weight %) of the Vapor (Odor) of Crushed Garlic and Allicin Solution as Determined by HPLC.

Preparation	Allicin (weight %)	Allyl methyl thiosulfinates (weight %)	Diallyl disulfide (weight %)	Allyl 1-propenyl disulfide (weight %)	Diallyl trisulfide (weight %)
Crushed cloves					
30 minutes	39	6	49	6	< 2
15 minutes	47	7	40	5	< 2
7 minutes	55	8	33	4	< 2
1 minute[a]	63	9	25	3	< 2
Rehydrated powder[b]					
30 minutes	69	12	12	5	< 2
15 minutes	74	14	8	3	< 2
1 minute[a]	80	15	4	1	< 2
Pure allicin in water[c]					
30 minutes	90	—	8	—	2

The vapor of samples incubated at room temperature in a small chamber were absorbed onto silica gel for the amount of time shown and extracted with acetonitrile and analyzed by C18-HPLC (Lawson et al., 1991a). Values (weight %) represent the means of 8–10 samples. Not shown is allyl methyl disulfide which was 5–10% as abundant as diallyl disulfide.

[a] The one-minute values were estimated by linear extrapolation.
[b] Garlic cloves (from the same bulbs used to prepare the crushed cloves) were freeze-dried, pulverized, and then rehydrated (2 mL water per gram) to contain the same amount of water as the cloves.
[c] Synthetic allicin in water at 1.8 mg/mL.

From Lawson LD, Wang ZJ, unpublished data, 1994.

expected, the diallyl disulfide content increases with time while allicin decreases; however, for reasons not understood, the appearance of diallyl disulfide is considerably slower in the vapor above wetted garlic powder. The vapor above an allicin solution is almost exclusively allicin, indicating that components in garlic assist in the conversion of allicin to diallyl disulfide. The composition of the vapor of garlic at the time it is cut or crushed was estimated by extrapolation to 1 minute and found to be about 63% allicin and 25% diallyl disulfide (Table 3.9).

The dominance of thiosulfinates in the vapor has been confirmed by Auger et al. (1989, 1990), who showed that the odor of leek is stable and primarily consists of propyl propanethiosulfinate and a much smaller amount of dipropyl disulfide. The odor of cooked chopped garlic, however, is undoubtedly due to the sulfides reported in the headspace studies. These sulfides actually have a much stronger and more repugnant odor than allicin and the other thiosulfinates present in freshly cut garlic.

Another odor of importance to garlic consumers is that of one's breath after consuming fresh garlic. This has recently been examined in good detail (Cai et al., 1995a). Allyl methyl sulfide and diallyl disulfide were found to be the dominant compounds in the breath, and they persisted for several hours. In agreement with earlier studies (Laakso et al., 1989; Minami et al., 1989), allyl mercaptan was initially the most abundant volatile, but was not detectable at 1 hour, indicating that it had been generated in the throat, while the sulfides were expired from the lungs. A selenium compound, dimethyl selenide (methyl-Se-methyl), was also found in the breath at a level around 1% of that of the sulfides, but in sufficient quantities possibly to contribute to the odor of garlic breath. Interestingly, this study also showed that selenium compounds go into the breath about 100 times faster than do sulfur compounds. Another recent and thorough study showed that the breath after consuming crushed garlic contains three groups of compounds: allyl mercaptan, which is initially very high but quickly disappears; the

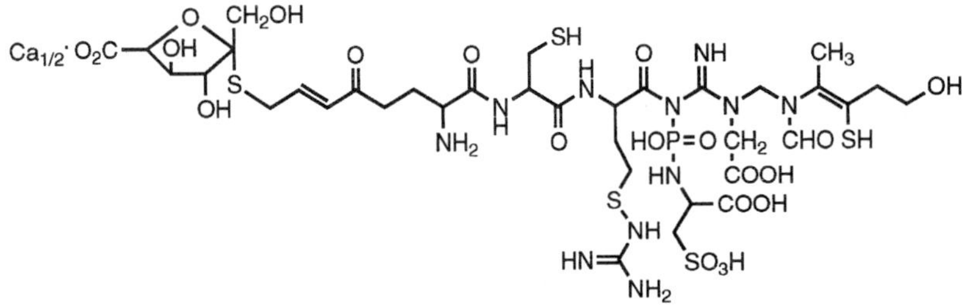

Figure 3.16. The structure of scordinin A (4-thiofructuronic acid-S-allyl-deoxy-γ-glutamyl-cysteinyl-S-guanidinohomocysteinyl-thiamamidine). Thiamamidine structure proceeds from the third peptide bond to the terminal hydroxyl group. Scordinin B has nicotinamide ribose diphosphate esterified to the terminal hydroxyl group. (From Kominato K, Kominato Y. Silkworm attractant, scordinin A, from garlic. Patent No. 73 87 009, Japan, 1973.)

sulfides (allyl methyl sulfide, diallyl sulfide, diallyl disulfide), which are also initially high but gradually decrease over 1–3 hours; and two terpenes (d-limonene and p-cymene), which are initially very minor but increase very rapidly after 1–4 hours to levels similar to those of the sulfides (Ruiz et al., 1994).

In the essential oil of other *Allium* species (e.g., onion, shallot, leek, chives, rakkyo, scallion, and Japanese bunching onion) *alkyl furanones* and *thiophenes* have also been found among the volatile flavor components (Galetto & Hoffman, 1976a, 1976b; Kameoka et al., 1984).

3.2.6.5 COOKING AND ITS EFFECTS ON GARLIC'S MAIN SULFUR COMPOUNDS

Since garlic is more often eaten cooked rather than raw, the effects of cooking on its organosulfur compounds is a common question and an important concern. Because heat inactivates enzymes, it is no surprise that heating garlic cloves inactivates alliinase and hence prevents the formation of allicin and other thiosulfinates. Boiling unpeeled whole cloves for 15 minutes completely inactivates alliinase; however, before the alliinase is inactivated, about 0.5–1% of the alliin is converted to allicin, possibly due to the cloves' bumping into each other, which is then rapidly converted mainly to diallyl trisulfide and to smaller amounts of the di- and tetrasulfides. Thus, even boiled cloves give a garlic breath odor when eaten (Lawson & Wang, 1994). Boiling cloves for 20 minutes also causes hydrolysis of 12% of the γ-glutamylcysteines to S-allylcysteine and S-1-propenyl cysteine, and a similar loss of alliin without production of S-allylcysteine.

When garlic is chopped into tiny pieces or is crushed, most of the cysteine sulfoxides are transformed into thiosulfinates. Heating the crushed garlic at boiling temperature for 20 minutes in a closed container caused complete conversion of the thiosulfinates to diallyl trisulfide and lesser amounts of other sulfides; however, after boiling in an open container for 20 minutes, 7% of the thiosulfinates were retained, but 97% of the sulfides had evaporated. Cooking in whole milk in an open container caused even more rapid loss of allicin (only 0.5% remained after 10 minutes), but 70% of the sulfides formed were retained, even after 40 minutes, presumably by the milk fat. Stir-frying chopped and smashed garlic cloves in hot (near the smoke point) soybean oil for 1 minute in a Chinese wok retained about 16% of the sulfides in the oil, but no allicin remained (Lawson, 1993). Microwaving single 5–6-gram whole cloves at 650 watts revealed that alliinase was completely inactivated in 30 seconds, while methyl-specific alliinase (see Section 3.2.5) was completely inactivated in 15 seconds (Lawson & Wang, 1994).

In attempting to determine the flavor compounds produced when sliced (thiosulfinates formed) or whole garlic cloves are cooked, a series of studies has been published on the volatile compounds formed upon cooking garlic by a variety of methods, such as oil-frying (0.33 hour), baking (1 hour), boiling (1 hour) and microwaving (0.5 hour) (Yu et al., 1993, 1994b, 1994a, 1994c). Most of the studies used a temperature of 180°C (355°F) followed by a 3-hour steam distillation. In all cases using 180°C, 40–60 volatile compounds were identified, with 16 found for boiled garlic. Baking and microwaving garlic slices produced the highest amount of volatiles (0.2%), while boiling and baking whole cloves produced the least (0.0008–0.004%). Diallyl disulfide was found to be the dominant volatile formed in all of the studies. Diallyl trisulfide was abundant only in the baked and microwaved garlic slices. In oil-cooked slices, significant amounts of allyl methyl disulfide, vinyldithiins, and, interestingly, allyl alcohol were also formed.

In studies on heating pure alliin in water (180°C, 1 hour), it was found that 1% volatiles were formed and that they consisted mainly of allyl alcohol (67%) and acetaldehyde (21%) (Yu et al., 1994e). Addition of glucose to the heated alliin generated a meat-like flavor consisting mainly of allyl alcohol and 2-acetylthiazole (Yu et al., 1994d). Heating of aqueous S-allylcysteine produced a larger variety of volatiles, mainly 2-methyl-1,4-dithiepane (a 7-member ring with two sulfur atoms), allylthioacetic acid (allyl-S-CH$_2$COOH), diallyl sulfide, and diallyl disulfide, but almost no allyl alcohol.

Recently, a novel sulfur compound, S-allyl *thiohexanoate*, was identified when chopped garlic and a vegetable oil were heated together, due to the reaction of diallyl disulfide with a dienal (Hsu et al., 1993).

3.2.7 S-Alkylcysteines

Trace amounts of the S-alkylcysteines (S-allylcysteine, S-*trans*-1-propenylcysteine, and S-methylcysteine) have been reported in garlic cloves (Table 3.3). These compounds are probably the immediate, but transitory, biosynthetic precursors of alliin, isoalliin, and methylcysteine sulfoxide (Fig. 3.17). S-

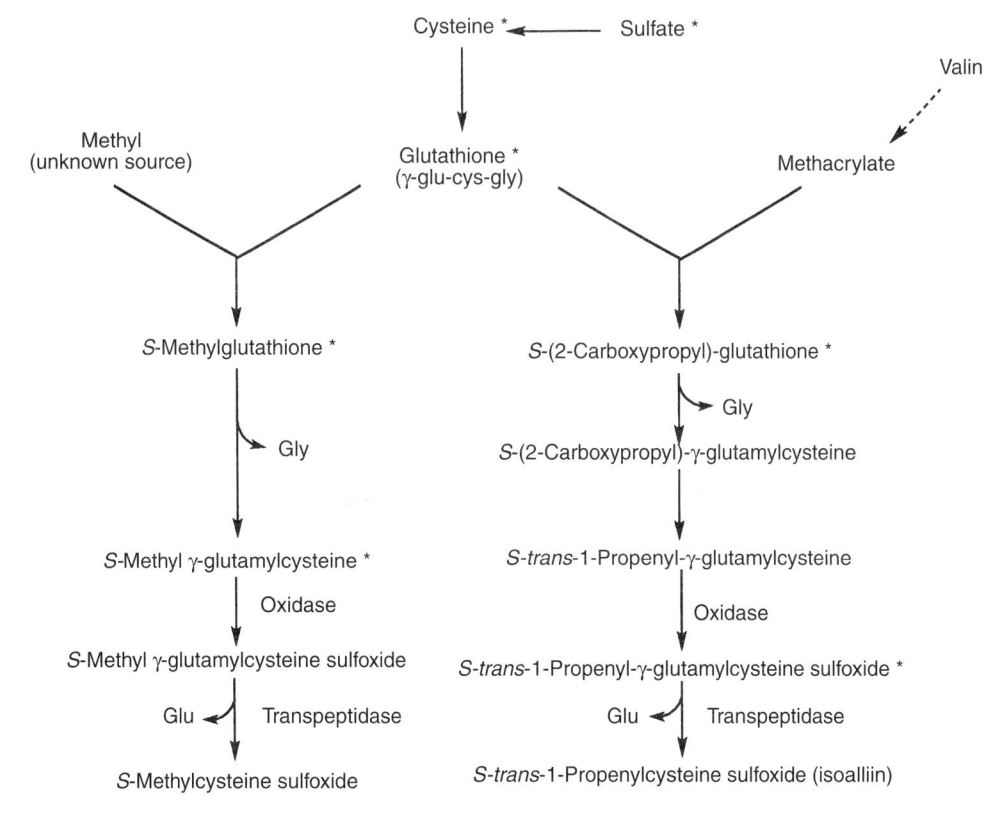

Figure 3.17. Glutathione pathway proposed by Lancaster & Shaw (1989) for biosynthesis of S-methyl and S-*trans*-1-propenylcysteine sulfoxides, based on ^{35}S-sulfate pulse-label studies with onion seedlings. Other studies indicate methionine to be the source of the methyl group (see text). Asterisk (*) indicates proven intermediates.

allylcysteine was first found in garlic at a yield of about 10 ppm fresh weight (Suzuki et al., 1961b). More recently others have found its content in garlic cloves to range from undetectable to 26 ppm (Lawson, 1993; Ziegler & Sticher, 1989). S-trans-1-propenylcysteine has been found in cloves in the range of undetectable to 6 ppm (Lawson, 1993; Sugii et al., 1963b). S-methylcysteine has not been found in garlic (Lawson, 1993; Ziegler & Sticher, 1989) and has only been detected in garlic as the ^{35}S-radiolabeled compound, a method which also detects transitory intermediates (Sugii et al., 1963a).

However, S-allylcysteine and S-trans-1-propenylcysteine increase considerably (to 230–400 ppm for S-allylcysteine and 35–100 ppm for S-trans-1-propenylcysteine) when garlic cloves are dried in the manufacturing of commercial garlic powders (Lawson, 1993; Hansen et al., 1993). These increases are due to hydrolysis of the abundant γ-glutamyl-S-alkenylcysteines while chopped garlic sits prior to drying. Formation of the S-alkenylcysteines proceeds even more efficiently with complete hydrolysis of the γ-glutamylcysteines when chopped garlic is aged in water or dilute ethanol for 3 months or longer (Section 3.4.5). Although such a process typically results in about 7000 ppm (dry weight) S-allyl cysteine and 5000 ppm S-1-propenylcysteine (Lawson & Wang, 1995), no dry commercial aged extract product containing greater than 700 ppm S-allylcysteine has ever been found (Lawson, 1993; Amagase & Milner, 1993; Hansen et al., 1993). It is also likely, though not yet proven, that S-alkylcysteines are rapidly formed in the body by hydrolysis of the γ-glutamylcysteines, particularly in the kidney, which contains an abundant amount of γ-glutamyltranspeptidase activity.

3.2.8 Allithiamine

In the 1950s, Japanese researchers found that the addition of sulfur-containing thiamine (vitamin B_1) to garlic extracts produced a new form of thiamine that was found to be absorbed much better than thiamine itself and was less toxic than thiamine, yet it had all the physiological effects of thiamine (Fujiwara et al., 1954a, 1954b, 1955; Fujiwara, 1976; Fujiwara & Watanabe, 1952; Ikeda, 1969; Matsukawa & Yurugi, 1952a, 1952b, 1952c; Matsukawa et al., 1953; Satoh, 1952g, 1952f; Watanabe, 1953a, 1953b; Yoshimura, 1958a, 1958b; Yurugi, 1954a, 1954b, 1955; Menke, 1994; Bitsch & Bitsch, 1989; Kawasaki, 1963; Ohashi, 1965; Baker et al., 1974). The compound was isolated and found to be thiamine allyl disulfide (allyl-SS-thiamine), which they named *allithiamine* (see also Sections 5.9 and 6.3), and was found to be the reaction product of thiamine with allicin in a reaction similar to the reaction of allicin with cysteine. The optimum conditions for allithiamine formation are pH 8 and 70°C (Fujiwara et al., 1954b). In blood, allithiamine was found to be rapidly converted (50% in 3 minutes, 100% in 30 minutes) to thiamine and allyl mercaptan, which is the reason that allithiamine has the same physiological effects as thiamine (Fujiwara et al., 1954a; Fujiwara & Murakami, 1960; Yurugi et al., 1956).

Allithiamine is now a widely used medical agent and food additive (Okada & Fujiwara, 1987, 1989; Okada & Fuji Sangyo Co. Ltd., 1988; Cheng & Tung, 1981). Its discovery launched a large number of publications at the time and resulted in a commercial product (Alinamin) that was an S-propyl homolog of allithiamine (Thomson et al., 1971; Fujiwara, 1954). The majority of the vitamin B_1 supplements sold in Japan today are still made with this product, although the S-propyl group has been replaced by S-tetrahydrofurfuryl (*Fursultiamine*, INN) or S-benzoyl (*Benfo-tiamine*, INN) (Kawasaki, 1963; Utsumi et al., 1962a, 1962b; Aramaki et al., 1959).

The discovery of allithiamine was important for finding a better-absorbed form of thiamine; however, the compound has little to do with the composition or health benefits of garlic. At physiological pH and temperature, it is probably formed too slowly to be produced before intestinal absorption (Fujiwara et al., 1954b). Even if the reaction did take place, the amount of thiamine in garlic is too small (Table 3.16, below) to be of benefit. Nevertheless, the allithiamine research did expand our knowledge of the composition of garlic because it led to the discovery of some of the homologs of alliin and allicin (Fujiwara et al., 1955, 1958; Tsuno, 1958a, 1958b; Tsuno et al., 1960; Yoshimura, 1958b; Yoshimura et al., 1958; Yoshimura & Arai, 1958).

3.2.9 Scordinin: A Plant Growth Hormone

Scordinin is a complex thioglycophosphopeptide containing five sulfur atoms that was discovered in garlic as a plant growth hormone (coenzyme) by Japanese researchers in the 1950s (Kominato, 1953, 1959, 1960, 1969b; Kominato & Kominato, 1970). It was isolated and studied for pharmacological activity as a possible nonodorous compound from garlic that might be responsible for some of garlic's health benefits. Its isolation is a tedious process involving extraction of boiled cloves with methanol, followed by lead acetate precipitation and binding to activated charcoal (Kominato, 1969a). The work on scordinin is not well described and is found mainly in patents and only a few publications.

Scordinin exists in two main forms in garlic: as the prohormone, *scordinin A*, and as the hormone, *scordinin B*. Scordinin A is the dominant form when garlic is harvested in the summer, while scordinin B, the nicotinamide ribose diphosphate ester of scordinin A, is dominant when the bulbs are stored for several months. The scordinins have also been degraded and fractionated into components A_1, A_2, A_3, B_1, B_2, and B_3, but these are not well defined and may well be isolation artifacts (Kominato, 1970a, 1971; Kominato et al., 1971, 1976a; Kominato & Kominato, 1970, 1972a, 1972b; Michahelles, 1974; Nishii et al., 1970). The mixture of scordinin A and B is also called *oxoamidin* or *oxoreduin* (Kominato et al., 1976b). Scordinin A, shown in Figure 3.16, has the structure of 4-thiofructuronic acid-*S*-allyl-deoxy-γ-glutamyl-cysteinyl-*S*-guanidinohomocysteinyl-thiamamidine. It has a molecular weight of 1147 and is found as a dimer bound to calcium (Kominato & Kominato, 1973; Kominato, 1969a, 1972a). The peptide portion of scordinin A after the allyl group is called scormine (Kominato, 1969b). A compound similar to scordinin was found in *Allium schoenoprasum* var. *viviparum* (Kominato, 1960).

An enzymatic assay has been developed on the basis of the fact that treatment of scordinins with sodium hydroxide produces a consistent amount of lactic acid (Hatanaka & Kaneda, 1980). Scordinin A is found at levels around 0.03% of the fresh weight of garlic, on the basis of extraction yields (Kominato, 1969a, 1972b). A processed garlic has been reported to contain no detectable amount of scordinins (Matsuda, 1990). Using very similar extraction procedures, compounds called *sativin I* and *sativin II* have been isolated from garlic that could well be identical to the scordinins (Deininger & Wagner, 1976). Because the scordinins contain two thiol (SH) groups that would react rapidly with allicin, only garlic in which alliinase has been inhibited (e.g., boiled cloves or cloves pickled in acetic acid) would be expected to contain the unmodified scordinins.

At least one commercial product based on scordinin content is currently marketed, with claims to be an odorless garlic extract containing 100 mg of scordinin per gram powder. Such a product should be considered as a scordinin concentrate rather than as a garlic product because analysis of the product has revealed that it contains very little of other compounds found in garlic cloves: no alliin, 3% of the γ-glutamyl-*S*-alkenylcysteines, and 6% of the arginine (Lawson & Wang, 1994). Various uses for scordinin have been proposed: dietary supplement (Hatanaka & Kaneda, 1980), flavoring agent (Kominato, 1977), preparations with enteric microorganisms (Kominato & Ohira, 1987), metabolic regulator (Kominato, 1970b), nail polish (Ito & Pola Chem. Ind. Inc., 1990), and silkworm attractant (Kominato & Kominato, 1973).

The pharmacological studies on pure or

semipure scordinin A or decomposition products from it have provided some interesting results, but positive effects have only been observed at levels much too great to account for the health benefits seen with garlic cloves. It has been shown to have thiamine sparing activity in animals at 4 mg/kg body weight/day (equivalent to a person consuming 930 g or 300 cloves per day) (Kominato et al., 1976b); to decrease blood pressure and improve muscle tone in rabbits at 5 mg/kg body weight (Suzuki & Motoyoshi, 1966); to reduce serum cholesterol levels in rabbits at 40 mg/kg body weight; to increase swimming endurance in mice at 13–100 mg/kg body weight; and to cause faster growth rates in rats at 0.1% of diet (equivalent to a diet of 13% garlic) (Kominato, 1972a). Furthermore, the aglycon of scordinin A, not present in garlic but produced upon treatment with silver nitrate or mercuric chloride, has been shown to have antibiotic activity (Kominato, 1969a; Kominato & Kominato, 1965; Kominato et al., 1976a; Nishimura & Kominato, 1971). The observed antithrombotic activity of isolated scordinins A and A_1 was found to disappear upon further purification (Michahelles, 1974). (See also Sections 5.1.4, 5.2.6, 5.14, 6.5, and 7.4.)

3.2.10 Biogenesis of Garlic's Sulfur Compounds

The processes by which the garlic plant synthesizes the S-allyl, S-1-propenyl, and S-methyl cysteine sulfoxides and γ-glutamylcysteines, the main sulfur compounds present in a mature garlic bulb, have not yet been well explained. Although proposed pathways for 1-propenyl and methyl compounds have some support, neither the source of the allyl group nor the carrier of the sulfur atom from sulfate to the allyl group have been revealed. Three laboratories have investigated the possible pathways and their findings will now be discussed.

In the early 1960s, Suzuki's laboratory incubated garlic roots for one day in ^{35}S-labeled sulfate and found the label in cysteine (appeared first), methionine (appeared second), and then in methyl cysteine sulfoxide, and alliin (Sugii et al., 1964a; Suzuki et al., 1961a). They also found that feeding labeled methionine to roots resulted in even more rapid formation of labeled S-methylcysteine sulfoxide and S-methylcysteine (Sugii et al., 1963a), indicating that the S-methyl group had been transferred from methionine to serine to form the methylcysteine compounds by the following pathway:

$$sulfate \rightarrow cysteine \rightarrow methionine + serine \rightarrow S\text{-}methylcysteine + homoserine \rightarrow S\text{-}methylcysteine\ sulfoxide$$

Their results also indicated that cysteine may be a precursor of alliin; however, a more thorough study by Granroth (1970), using garlic sprouts incubated three days in ^{35}S-labeled sulfate, found strong labeling of methylcysteine sulfoxide and 1-propenylcysteine sulfoxide (isoalliin), but no labeling of alliin. Feeding the plants ^{35}S-labeled cysteine also did not result in formation of labeled alliin, although isoalliin was rapidly formed; however, feeding ^{14}C-labeled serine did produce labeled alliin as well as methylcysteine sulfoxide, revealing the source of the aminoacyl portion of the alliin molecule, but not the source of the sulfur atom. The conflicting results of Suzuki (Sugii et al., 1961a, 1964a), who reported the presence of labeled alliin after cysteine feeding, is probably due to misidentification of isoalliin as alliin, since isoalliin had not yet been discovered in garlic and since it elutes very close to alliin by chromatography.

The source of the 1-propenyl group of isoalliin was discovered when Granroth fed ^{14}C-labeled valine to sprouted garlic and onion leaves and found rapid labeling of isoalliin as well as some labeled S-(2-carboxypropyl)cysteine ($CH_3CH(COOH)CH_2S$-cys) (2-CPC) and S-(2-carboxypropyl)glutathione (2-CPG), but no labeled alliin. The formation of the carboxypropyl compounds from valine had been reported earlier by Suzuki (Suzuki et al., 1962b), but Granroth (1970) also showed that feeding S-(2-carboxypropyl) cysteine to garlic tissues caused strong labeling of isoalliin, indicating that the 1-propenyl group was formed by the loss of HCOOH from the 2-carboxypropyl group. Since methacrylic acid ($CH_3C(=CH_2)COOH$) had been a well-established product of valine metabolism in other plants, he concluded that in both garlic and onions, valine was the source of the 1-propenyl group and cysteine the source of the sulfur, by the following pathway:

valine → methacrylic acid + cysteine → 2-CPC → S-1-propenyl cysteine → S-1-propenylcysteine sulfoxide (isoalliin)

After finding that decarboxylation of 2-CPC was the source of the 1-propenyl group, he proposed that decarboxylation of 3-CPC would form the 2-propenyl or allyl group, but he found this not to be the case. Even giving labeled acetate to garlic sprouts did not result in labeled alliin. Both Granroth and others in a recent study showed that the addition of allyl-SH to garlic tissue resulted in the formation of alliin, but the physiological significance of this reaction is very doubtful, since garlic has never been reported to produce allyl-SH. Furthermore, it was shown that addition of S-allylcysteine to cultured cells resulted in quantitative formation of alliin, but this only proved the presence of a suitable oxidase (Ohsumi et al., 1993). Thus, the source of the allyl group of garlic still remains a mystery.

Since he had also found a small amount of radioactivity in 2-CP-glutathione, Granroth proposed that glutathione could also combine with methacrylic acid to form 2-CPG, followed by hydrolysis back to 2-CPC. Since he was working with sprouted plants which contain greatly increased levels of γ-glutamyl transpeptidase, it was not possible for him to discover the essential importance of the γ-glutamylcysteine intermediates in the pathway, as was later discovered (Lancaster & Shaw, 1989).

A much-improved understanding of the intermediate steps in the biosynthesis of the 1-propenyl and methyl cysteine sulfoxides of onion has come about as a result of a pulse-label study (Lancaster & Shaw, 1989) in which onion seedlings were incubated briefly (10 minutes) with ^{35}S-sulfate, followed by identification of the labeled sulfur compounds over time, a method which reveals the sequence of appearance and disappearance of intermediates (Parry & Sood, 1989; Freeman & Mossadeghi, 1971). From this study, the scheme shown in Figure 3.17 was proposed (Lancaster & Shaw, 1989; Lancaster & Boland, 1990). This study revealed that glutathione and other γ-glutamylcysteine compounds are essential intermediates in the biosynthesis of the cysteine sulfoxides and account for the known accumulation of γ-glutamyl-S-1-propenylcysteine sulfoxide in onion bulbs (Virtanen & Matikkala, 1961b; Matikkala & Virtanen, 1967; Lawson et al., 1991b; Lancaster et al., 1986; Becker & Schuphan, 1975), as well as the known increase in isoalliin when transpeptidase levels are increased during sprouting (Schwimmer & Austin, 1971a). Lancaster did not observe or propose a source for the methyl group, but the work of Suzuki (1964a) and Granroth (1970) indicates that methionine is probably the main source. Parry and Lii (1991) have confirmed this scheme by showing that addition of γ-glutamylcysteine to methacrylate results in the formation of S-(2-carboxypropyl)-γ-glutamylcysteine. However, Edwards and coworkers (1994b) recently presented evidence that γ-glutamylcysteines may not be the immediate precursors of the cysteine sulfoxides in onions by showing that the ^{35}S in long-term (24 hours) pulse-labeled onion sets was found (at 18 hours) mainly (70–90%) in the cysteine sulfoxides fraction rather than in the γ-glutamylcysteines fraction (20% or less), although the labeling time and intervals between assays were too long to determine precursors.

Lancaster and Shaw (1989) also conducted a less rigorous pulse-label study with garlic and showed similar trends in movement of ^{35}S among the various general polarities of compounds, but unfortunately they looked only at total radioactivity without distinguishing between allyl, 1-propenyl, or methyl compounds. From this study, the authors have proposed that the S-allyl and S-1-propenyl compounds of garlic follow the same pathway of biosynthesis as the S-1-propenyl compounds of onion. Similar results were obtained with callus cultures of garlic (Lancaster et al., 1988).

However, there are four problems with this proposal. First, since the garlic compounds were not separated, the movement of radioactivity could well be due only to 1-propenyl and methyl compounds, not allyl compounds. Second, Granroth (1970) has shown that carboxypropyl compounds cannot be transformed by garlic into allyl compounds and that valine does not efficiently form alliin. Third, there is a basic difference between garlic and onion with respect to the types of 1-propenyl sulfur compounds found in the mature bulbs: in garlic, 90–95% of the S-1-propenyl is found in γ-glutamyl-S-1-propenylcysteine and none is found in the corresponding sulfoxide (GG1PCSO) (nor is any GGAllylCSO found); in onion, however, S-1-propenyl is found about equally in isoalliin and γ-glutamyl-S-1-propenylcysteine sulfoxide, while γ-glutamyl-S-1-propenylcysteine is absent (Lawson et al., 1991b; Matikkala & Virtanen, 1967), indicating that in garlic the transpeptidase probably precedes the oxidase. Fourth, there are some basic differences between the metabolism of allyl and 1-propenyl compounds in garlic: although the content of their γ-glutamylcysteines in cloves are similar, the content of allylcysteine sulfoxide (alliin) is about 17 times greater than that of 1-propenylcysteine sulfoxide (isoalliin) even though the rate of transpeptidase activity toward both γ-glutamyl-S-alkenylcysteines has recently been shown to be very similar (Lawson et al., 1991b, 1991c; Lawson & Wang, 1995), indicating that there is a second source (probably the main source) of alliin that is not γ-glutamyl-S-allylcysteine, such as the serine pathway of Granroth (1970).

Therefore, a new scheme is proposed for the biosynthesis of the allyl and 1-propenyl compounds of garlic (Fig. 3.18), based on the labeling studies of Granroth (serine and lack of valine or 2-CPGSH for alliin), Lancaster (for 1-propenyl pathway down to GG1PCys), and our own analytical work with mature cloves and with premature plants (note in Table 3.4 the complete absence of the γ-glutamylcysteines but abundant presence of the cysteine sulfoxides in plants harvested 7 weeks early). The new scheme depicts two pathways for alliin synthesis: the serine pathway as the main pathway during bulb and clove formation, and a glutathione pathway which operates during storage and sprouting of garlic cloves and which may also operate to a small extent during maturation. Since very little S-allylcysteine has been found, even in premature plants, the oxidase appears to be active throughout development. Another distinction in this proposed pathway from that proposed by Lancaster for onions is that the sequence of the transpeptidase and oxidase reactions are reversed. Placing transpeptidase first allows for the finding that, unlike onions, garlic does not contain γ-glutamyl-S-alkylcysteine sulfoxides (Lawson et al., 1991b; Lawson & Wang, 1994). Furthermore, aging studies have shown that garlic clove homogenate contains considerable, but compartmentalized, transpeptidase activity and no oxidase activity, resulting in complete conversion of the γ-glutamyl-S-alkenylcysteines to S-alkenylcysteines which are not further metabolized (Lawson & Wang, 1995). If oxidase action had been prerequisite, then no S-alkenylcysteines would have formed. The pathway for the formation of methylcysteine sulfoxide (not shown) is probably the same as that proposed by Lancaster, with the possible exception of the reversal of the transpeptidase and oxidase steps. Undoubtedly, time will reveal some flaws in this proposal, but it seems to represent best all the observations currently known.

3.2.11 The Function of Garlic's Sulfur Compounds

In concluding this attempt (Section 3.2) to gather together all current knowledge about the composition, chemistry, and biosynthesis

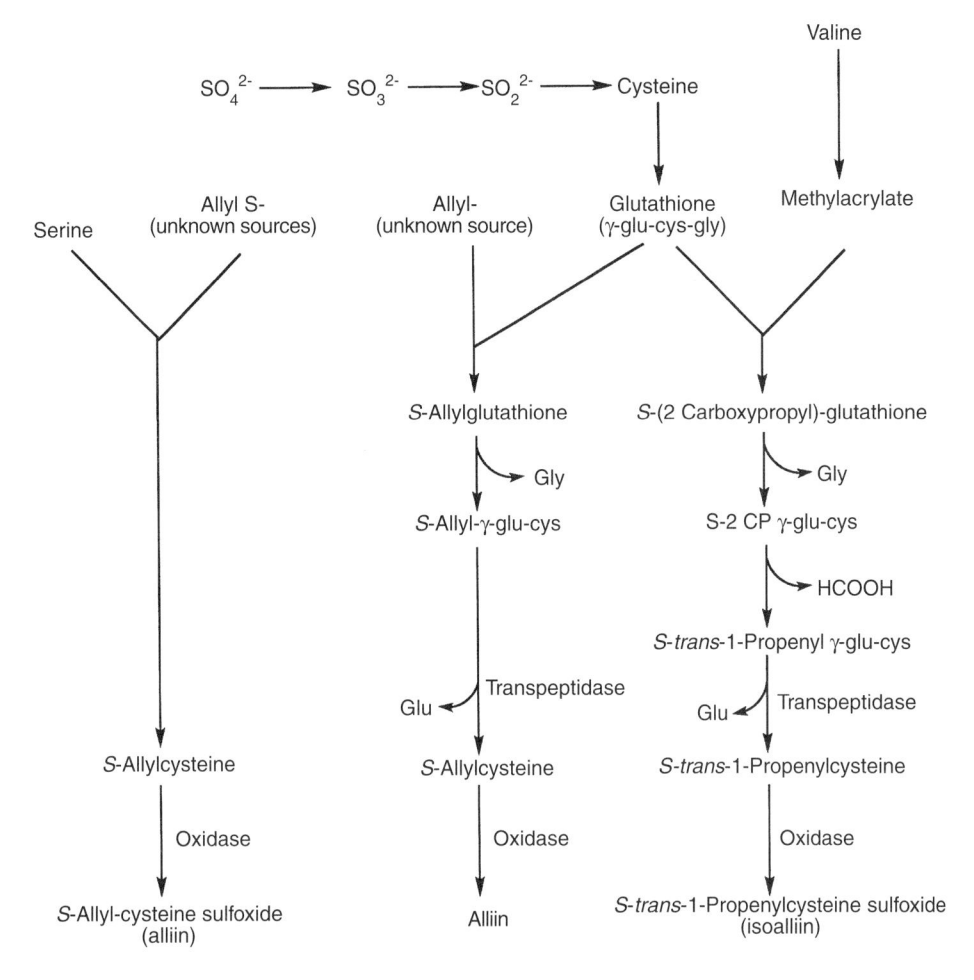

Figure 3.18. Currently proposed pathways in garlic for biosynthesis of alliin and isoalliin. For alliin, the serine pathway is probably dominant during bulb maturation, while the glutathione pathway occurs slowly during storage of mature bulbs and during sprouting. Note that the transpeptidase and oxidase steps have been reversed from those that have been proposed for onions (Fig. 3.17). See text for discussion.

of garlic's unusually abundant sulfur compounds, it may be noted that nature has developed, or has been directed to develop, a rather sophisticated system in garlic for the synthesis and storage of numerous sulfur compounds and abundant amounts of the enzyme alliinase, nearly all of which have the end purpose of producing allicin and other thiosulfinates at an extremely rapid rate upon disturbance of the plant. Such a major effort leads one to wonder what might be the function or benefit of the thiosulfinates for a garlic plant. Since the thiosulfinates are very effective antibacterial and antifungal agents, it has been proposed that they "offer the plant protection against the bulb decay induced by fungi" (Block, 1985). Whether an invading fungus will trigger the release of alliinase, and thus of the production of thiosulfinates, is not yet known. However, allicin is very effective against other invaders of the bulb, such as worms, nematodes, and other parasites (see Section 5.2.5), all of which would initiate its release. Furthermore, since the above-ground plant also has considerable potential to produce allicin, animals and other pests would be quickly repulsed by its irritating taste. Certainly, we do not know all of the ways in which thiosulfinates provide host defense, but defense surely is their main function, a virtue which is also most useful to humanity.

3.3 NONSULFUR COMPOUNDS IN GARLIC

3.3.1 Carbohydrates: Sugars, Fructans, Pectins

Carbohydrates are the most abundant class of compounds present in garlic bulbs and account for about 77% of the dry weight (Fenwick & Hanley, 1985b; Souci et al., 1986; Darbyshire & Henry, 1981; Park & Lee, 1992). Most of the carbohydrate research on garlic has been qualitative, identifying what types of saccharides are present, while only a meager amount of quantitative work has been reported; hence, our understanding of the carbohydrate composition of garlic is far from complete. However, it is safe to say that the majority of the carbohydrate material in garlic cloves, as well as in other *Allium* species, consists of water-soluble fructose polymers called *fructans* or *fructosans* (Kihara, 1929; Khodzhaeva & Kondratenko, 1983, 1984a, 1984b; Belval, 1939, 1943; Belval et al., 1948; Anantakrishnan & Venkataraman, 1941; Srinivasan et al., 1953; Khodzhaeva & Ismailov, 1979; Khodzhaeva et al., 1982; Schmiedeberg, 1879; Archbold, 1940; Darbyshire & Henry, 1979, 1981; Das & Das, 1978; Henry & Darbyshire, 1979, 1980; Rakhimov & Khodzhaeva, 1990; Park et al., 1988; Espagnacq et al., 1988; Bauer et al., 1988; Hirao et al., 1987; Steer, 1982; Lercari, 1982; Darbyshire et al., 1979; Van Wyk, 1967; Bose & Shrivastava, 1961; Sinha & Sanyal, 1959; Bacon, 1959; Chevastelon, 1894). A recent study with six samples revealed that 65 ± 12% of the dry weight (22 ± 4% of the fresh weight) of garlic consists of fructans (measured as total fructose [McRary & Slattery, 1945] minus free fructose [HPLC]) (Lawson & Wang, 1994). Hence, fructans constitute approximately 84% of the carbohydrate content of garlic.

The known mono- and disaccharides of garlic comprise glucose (1.2% dry weight), fructose (1.4%), arabinose (unknown amount), sucrose (7%), and lactose (unknown amount), while maltose has been shown to be absent (Srinivasan et al., 1953; Srinivasan & Bhatia, 1954; Mizuno et al., 1957; Pant et al., 1962; Arime & Deki, 1983; Belval et al., 1948). Trisaccharides found in garlic are raffinose, 1F-fructosylsucrose, and 6G-fructosylsucrose (Pant et al., 1962; Darbyshire & Henry, 1981). A tetrasaccharide called *scorodose*, which contains four fructose units, has been reported to be present in very large amounts (47–53% dry weight) in garlic (Kihara, 1929, 1936, 1939; Arime & Deki, 1983; Deki, 1983). It has the structure of α-D-fructofuranosyl-$(2\rightarrow 6)$-α-D-fructofuranosyl-$(2\rightarrow 2)$-α-D-fructofuranosyl-$(6\rightarrow 2)$-α-D-fructofuranoside, which includes the very unusual $2\rightarrow 2$ linkage (Mizuno et al., 1957). The unusually large content reported for scorodose is in conflict with more recent reports which show that the majority of the fructose polymers in garlic contain 10–60 fructose units, with tetrasaccharides comprising only a small proportion (Darbyshire & Henry, 1981; Koch et al., 1993).

Like other plants of the genus *Allium*, garlic does not store polysaccharides in the form of starch, but rather as the above-mentioned fructans or fructose polymers, which contain an inulin type $(2\rightarrow 1)$ linkage (Khodzhaeva et al., 1982). Very recently it was shown that the soluble polysaccharides of garlic contain 89% fructose and only 11% glucose (Ohsumi & Hayashi, 1994a). Interestingly, all of the garlic fructans contain at least one glucose unit and appear to be synthesized initially from sucrose as a "starter" (Darbyshire & Henry, 1981). Indeed, it has been shown that 4.3% of the composition of garlic fructans is glucose (Das & Das, 1978). Besides their main function as carbohydrate storage compounds, the fructans may play an important part in osmotic regulation and in cold resistance (Darbyshire & Henry, 1981; Park et al., 1988).

Upon more vigorous extraction, the insoluble cell wall polysaccharides have also been obtained and identified in garlic and include a *galactan*, a *galacturonan*, an *arabinan* (Das et al., 1977a), *cellulose* (1.8% dry weight), and a *xyloglucan* (0.3% dry weight) (Ohsumi & Hayashi, 1994b; Das et al., 1977b). Hydrolysis of the insoluble polysaccharides (5.0% dry weight) shows that they contain mainly galactose, galacturonic acid, arabinose, and glucose, with small amounts of xylose and fucose (Ohsumi & Hayashi, 1994a; Krynska & Kawecki, 1970). The content of sugars and polysaccharides in garlic fluctuates throughout the growth of the plant, but reaches its highest levels in the mature bulb (Kawecki & Krynska, 1968; Krynska & Kawecki, 1970).

The dry scales of garlic contain large amounts of carbohydrates called *pectins*. Pectins absorb large amounts of water and are used commercially to make jellies, cosmetics, and pharmaceuticals. In Egypt, large amounts of onion and garlic scales are used to obtain pectins for these purposes. The yield of pectin from garlic scales is 27%, but it is only 0.3% from the cloves. Garlic skin pectins are unbranched polymers (α 1→4 linked) of D-galacturonic acid (polyuronides), and about 50% of the carboxyl groups are esterified with methanol (Alexander & Sulebele, 1973; Abdel-Fattah & Khaireldin, 1970; Abdel-Fattah & Edrees, 1972; Khodzhaeva & Ismailov, 1979; Bonfante-Fasolo et al., 1990; Demkevich, 1985). Of course, pectins and cellulose, hemicellulose, hemicellulose, and lignin are also present in other *Allium* species (Sen & Rao, 1966; Deineko, 1985; Bonfante-Fasolo et al., 1990; Khodzhaeva & Kondratenko, 1985; Zhao et al., 1988; Mankarios et al., 1979; Abdel-Fattah & Edrees, 1971, 1973; Alexander & Sulebele, 1973; Crooke et al., 1960).

Lastly, a novel *amino acid glycoside*, ($\times$)-N-(1'-deoxy-1'-D-fructopyranosyl)-S-allyl-L-cysteine sulfoxide, was isolated from a hydrophilic extract of the leaves (not found in the cloves) of garlic. The glycoside revealed a significant inhibition of in vitro platelet aggregation induced by adenosine diphosphate (ADP) or epinephrine (Mütsch-Eckner et al., 1993).

3.3.2 Enzymes in Garlic

In addition to the unique abundance of the enzyme alliinase, which has already been thoroughly discussed, garlic has been studied for its content of a variety of other enzymes. Of the other enzymes in garlic, *adenosine triphosphatase* (ATPase) should be mentioned first. This enzyme is present in all living organisms. Ultrastructural localization of ATPase in the garlic plant has been subject to intensive investigation and was found in all cells of the phloem, bound to the filaments and protein granula. In addition, ATPase is concentrated in the intercellular channels and in transfer cells (Dong & Zhang, 1986). Other researchers have found ATPase in high concentrations in young, unthickened cells, and that its concentration drops sharply in thickened, dry cells. The presence of *phosphatase* in garlic has been known already for some time (Watanabe et al., 1963).

Polyphenol oxidase, which is also ubiquitous in plants, has a cellular distribution that is exactly reversed from that of ATPase (Li et al., 1986). Polyphenol oxidase obtained from garlic oxidizes triphenols (e.g., pyrogallol, gallic acid) and reacts only weakly with diphenols (e.g., catechin, chlorogenic acid). Maximum activity of the enzyme is achieved at pH 6.5. It is heat stable and is inhibited by cyanide, persulfate, ascorbic acid, and thiourea. Magnesium and copper activate the enzyme (Kim et al., 1981; Kararah et al., 1985).

Phenylalanine-ammonia-lyase (PAL, EC 4.1.1.5) is involved in the synthesis of aromatic compounds in garlic. In differentiating tissue cultures, the enzyme activity increases commensurately with the alliin content (Malpathak & David, 1986). PAL activity has also been studied in onions (Knypl & Janas, 1990).

Peroxidase (EC 1.11.1.7) and *catalase* have been detected in garlic bulbs (Glaser & Drobnik, 1939; Glaser, 1940; Drobnik, 1938), and *superoxide dismutase* (SOD, EC 1.15.1) has been found in bulbs and in cell cultures of garlic (Matkovics et al., 1981; Zhang et al., 1993). Peroxidase in garlic is considered responsible for the discoloring and loss of activity in the dried product. The enzyme has optimal activity at pH 5.05–5.15, with no activity below pH 2.23 or above pH 10.57. It is inhibited by chloride and is fully blocked by cyanide (Cruess, 1944; Sugihara & Cruess, 1945). The isoenzyme pattern of peroxidase in diverse cultivated varieties of garlic was used for their classification in corresponding wild species (Etoh & Ogura, 1981). The activity of SOD within the callus increases with its proliferation and is 1.56 times higher after 15 days than in naturally stored garlic (Zhang et al., 1993). Changes in peroxidase activity in garlic are associated with radiation-induced sprout inhibition (Croci et al., 1991). Peroxidase is also present in onion (Drawert & Görg, 1975).

Alcohol dehydrogenase (ADH), *esterase*, and *phosphoglucose isomerase* have also been determined and their isoenzymes characterized (Siqueira et al., 1985). The existence of two molecular species of *ferredoxin* is very conceivable (Shin & Sakihama, 1986). NAD(P)H-dependent dehydrogenase from onion has been purified and characterized (Serrano et al., 1994). ADH has also been characterized from *A. fistulosum* (Mangum & Peffley, 1994). In *A. porrum* (leek), the peroxidase is bound to the cell wall of the roots (Spanu & Bonfante-Fasolo, 1988). The esterase isoenzyme pattern has been elaborated in *A. cepa* (Hadacova et al., 1981; Drawert & Görg, 1975) and in *A. montanum* from various localities (Hadacova et al., 1985). *Cholinesterase* activity has been found in several species of the *Allium* genus (Hadacova et al., 1983).

A series of specific enzymes is involved in the synthesis and degradation of polymeric carbohydrates in garlic. In the dormant plant, but also in the growing plant, there are carbohydrases such as *polyfructosidase* and *invertase*. A *transfructosidase* is also present (Bhatia et al., 1955). A *pectinesterase* degrades the pectins that are responsible for the viscosity observed in the freshly pressed plant juices. This enzyme is inhibited at pH 3–4 and is inactivated by heating to 72°C for 5 minutes. *Polygalacturonase*-like enzymes are not contained in garlic per se, but activity can occur through contamination with microorganisms (*Penicillium canescens*) (Misekow & Fabian, 1953); it is found, however, in *A. porrum* (Peretto et al., 1992).

The presence of other enzymes involved in carbohydrate metabolism has been reported for other *Allium* plants as well (Kim et al., 1993; Mangum & Peffley, 1994; Serrano et al., 1994; Dumas-Gaudot et al., 1992). *Chitinase* activity has been found in garlic, onion, and leek tissue cultures (Van Damme et al., 1993c; Williams & Leung, 1993; Kim et al., 1992; Dumas-Gaudot et al., 1992). Purification and characterization of *β-galactosidase* from green onions (Kim et al., 1993), *α-glucosidase* from *A. fistulosum* (Suzuki & Uchida, 1984), and *β-glucosidase* from *A. erubescens* (Vardosanidze et al., 1991) have been reported. Partial purification of a ring *β-O-glucosyltransferase* has been achieved from onion bulbs (Latchinian-Sadek & Ibrahim, 1991). An *acyl-CoA elongase* has been partially purified from *A. porrum* (Bessoule et al., 1989).

From germinating garlic cloves, an acidic *invertase* has been isolated which has maximum activity at pH 4.6 and cleaves saccharose and raffinose, but not fructan. Also, an *inulase* has been found with maximum activity at Ph 5.0 and a molecular weight of 76,000. The latter enzyme degrades inulin (a fructan), as well as saccharose and raffinose. Both enzymes increase proportionately with germination, but eventually the activity of invertase greatly surpasses that of inulase (Bhat & Pattabiraman, 1980). A specific *hexokinase*, which phosphorylates D-glucose, D-fructose, D-mannose and D-glucosamine, has been isolated in pure form from garlic and has optimal activity at pH 8.4 (Bhat & Pattabiraman, 1979).

Garlic contains numerous γ-glutamyl peptides (Section 3.2.3) which are considered to be storage products and precursors of the cysteine sulfoxides. The enzymes *γ-L-glutamyl-peptidase* and *γ-L-glutamyl-transpeptidase* are substantially involved in the synthesis and metabolism of these constituents. The first enzyme, a hydrolase, cleaves off the glutamyl residue from the peptide, and sets free the second amino acid or the *S*-alkylcysteine. The second enzyme, an amino acid transferase, transfers the glutamyl residue to an amino acid or a sulfoxide. Hence, it participates in the anabolism of the γ-glutamyl peptides. The transpeptidases are important for flavor development in garlic (Whitaker & Mazelis, 1991; Schwimmer & Friedman, 1973; Ceci et al., 1992b). Closely related enzymes are also present in other *Allium* species (Austin & Schwimmer, 1971; Schwimmer & Austin, 1971a; Matikkala & Virtanen, 1965b; Lancaster & Shaw, 1994). Amino acid synthases have also been characterized in other *Allium* plants (Nakamura & Tamura, 1990; Granroth, 1974; Granroth & Sarnesto, 1974).

Arginase, an enzyme that degrades the basic amino acid arginine, has been found in several *Allium* species, including garlic (Brunel-Capelle & DeSerres, 1967). The content of *lipase* has been examined in connection with the lipid- and cholesterol-lowering effect of garlic (see Section 5.1.1). Ripe (mature) garlic contains two kinds of lipases: an acidic lipase with maximum activity at pH 3.5 and a neutral lipase whose maximum activity occurs at pH 6.5. Garlic lipases are stable against heat up to 65°C and can therefore be present in an active form in a carefully dehydrated garlic powder (Khan & Mahmood, 1984).

A species-specific *deoxyribonuclease*, which causes "chromatinolysis," i.e., degradation of nucleic acids, has been found in garlic. It is easy to presume that this enzyme may have something to do with the mitosis-inhibiting effect repeatedly described in garlic (see Section 5.3) (Hoffmann-Ostenhof & Keck, 1951; Keck & Hoffmann-Ostenhof, 1951; Keck et al., 1951).

A *lysozyme* was discovered in garlic which specifically dissolves cell walls of Gram-positive bacteria (endolysin, muramidase). The leaves are very rich in this enzyme (170 μg/100 g), while the cloves contain relatively little (9.6 μg/100 g). In comparison, the lysozyme content in onions amounts to 30.1 and 3.34 μg/100 g, respectively (Bityukov et al., 1982).

Finally, a protein-containing substance must be mentioned which is not an enzyme, but is a potent inhibitor of an enzyme, namely *guanylate cyclase*. This inhibitor was isolated from fresh garlic. It has a molecular weight of 8000 and, with an isoelectric point of 4.5, has been shown to be heat-stable and acid-stable against proteases. Hence, it is obviously a polypeptide. This inhibitor is also found in several other vegetable plants. Most interesting is the fact that this factor exhibits a tumor-inhibiting activity (Dhillon et al., 1981; Criss et al., 1982, 1983; Fakunle, 1983; Belman, 1983).

3.3.3 Protein, Free Amino Acids, and Dipeptides

The protein content given for garlic in food analysis tables is 5.6–6.3% fresh weight (Fenwick & Hanley, 1985b; Souci et al., 1986; Watt & Merrill, 1963); however, these values are much too high, since they are based on total nitrogen and since garlic contains abundant amounts of nonprotein nitrogen compounds. Analysis of nine strains of garlic using two different assay methods (Smith et al., 1985; Bradford, 1976) has shown that the protein content of garlic cloves is 1.5 ± 0.3% fresh weight, which represents only 30% of the total nitrogen. The remainder of the total nitrogen is present as common free amino acids (25%), cysteine sulfoxides (23%), γ-glutamylcysteines (20%), and γ-glutamylphenylalanine (1.5%) (Lawson & Wang, 1994).

Except for enzymes, few individual proteins from garlic have been isolated. A protein fraction, called F-4, of which nearly all the proteins have a molecular weight near 11 kDa, has been isolated and shown to have immune-stimulating effects in vitro (Hirao et al., 1987; Morioka et al., 1993). Very recently, two proteins of molecular weight 40 kDa (probably alliinase) and 14 kDa (perhaps identical to the 11 kDa protein of the F-4 fraction) have been reported to constitute 96% of the protein content of garlic bulbs, with the 40 kDa protein being exclusively localized in the cytoplasmic globular granules of the vascular bundle sheath cells, as is alliinase, and with the 11 kDa protein being found only in the numerous cortical cells

(Wen at al., 1995). Alliinase and mannose-specific lectin (Section 3.3.12) have been reported as the two major proteins in garlic (Tchernychev et al., 1995).

Lectins are antibody-like proteins found in plant seeds which bind to nonreducing terminal sugar units of cell surface receptors, and can be biologically assayed by their ability to agglutinate red blood cells, yeast cells, tumor cells, and the like, in vitro (Franz, 1988; Mizutani et al., 1981; Rüdiger & Gabius, 1993; Beuth et al., 1994). Lectins from garlic have been shown to have subunit molecular weights of 26,000 and 47,500 and have been found to bind to mannose, while the lectins of onions bind to fucose and lactose (Kaku et al., 1992; Mizutani et al., 1979; Sun & Yu, 1986; Van Damme et al., 1991, 1992a; Sun et al., 1987). Specific lectins are found also in other *Allium* species (Van Damme et al., 1991, 1993a, 1993b; Mizutani et al., 1981, 1985; Mo et al., 1993).

The free amino acid content of garlic is exceptionally high and is about 65% greater than the total protein. The free amino acids of garlic include nearly all of the common amino acids (those present in protein) plus an almost equally abundant amount of S-alkylcysteine sulfoxides, of which alliin is the most abundant, and trace amounts of the S-alkylcysteines. Since these cysteine derivatives have already been thoroughly discussed (Sections 3.2.4 and 3.2.7), this section will deal only with the common amino acids.

The content of individual free amino acids in garlic has been analyzed qualitatively and quantitatively by several laboratories over the past 45 years (Anantakrishnan & Venkataraman, 1940b; Atal & Sethi, 1961; Parthasarathi & Sastry, 1959; Kasai et al., 1984; Jankov, 1961; Demkevich, 1981; Sutaria & San Diego, 1982; Perseca & Parvu, 1986; Ueda et al., 1991; Lawson & Wang, 1994; Tawaraya & Saito, 1994). However, there have been considerable differences between many of the results, making it currently impossible to give a simple list of the abundance of each amino acid. Hence, the results of all four quantitative studies have been included in Table 3.10.

Most of the studies agree that *arginine* is clearly the most abundant free amino acid in garlic. The unusually high content of arginine is interesting, and indicates that it serves as an important nitrogen reserve, as has been proposed for onions (Schuphan & Schwerdtfeger, 1971).

Unfortunately, no two studies agree on the content of the other amino acids. For example, the values for *aspartic* and *glutamic acids* vary from 0.2 to 6.1 and 0.7 to 18.2 mg/g dry weight, while *methionine* and *threonine* vary from 0.03 to 9.9 and 0.6 to 10 mg/g dry weight, respectively. Furthermore, *cystine* varied from not detectable to 1.9 mg/g dry weight. There are at least three reasons for this lack of agreement among investigations: differences in methods of analysis (hopefully, a minor cause), natural variation among garlic strains grown in different parts of the world, and increases due to protein hydrolysis after crushing the cloves. From the rather large standard deviations shown in Table 3.10, it is apparent that considerable differences exist between samples grown in the same country, indicating that even greater differences could exist between different countries. Furthermore, it has also recently been shown that when garlic cloves are cut or crushed the levels of free amino acids immediately begin to increase as a result of protein hydrolysis (Lawson & Wang, 1994, 1995). Although the hydrolysis is slow for most of the amino acids, cystine increases at least tenfold in one day (Section 3.4.5). Cysteine, which contains a free thiol (SH) group, has not been found in garlic. Even if it were present, the allicin formed upon crushing the clove would immediately react with cysteine to form S-allylmercaptocysteine (Section 3.2.6.3.5). However, it has been shown by a specific test for thiol groups (Ellman, 1959) that even alliinase-inhibited garlic contains no detectable level of thiol compounds and, hence, no cysteine (Han et al., 1995).

Garlic also contains an abundant amount of *dipeptides*, all of which have been shown to contain glutamic acid with a γ-peptide linkage to the other amino acid. The γ-glutamyl-

Table 3.10. Free Amino Acids (Common)[a] Content of Garlic Cloves (mg/g dry weight ± standard deviation).

Study[b] Country No. Of Samples	Ueda et al., 1991 Japan 24 (11 strains)	Lawson & Wang, 1994 U.S.A. 6 (6 strains)	Sutaria & San Diego, 1982 Philippines 1	Parthasarathi, & Sastry 1959 India 1
Alanine	0.4 ± 0.3	0.4 ± 0.2	0.6	0.7
Arginine	22.3 ± 8.0	17.4 ± 7.7	21.0	6.0
Asparagine	nr[c]	1.5 ± 1.4	nr	nr
Aspartic acid	3.7 ± 1.6	0.2 ± 0.1	6.1	1.3
Cystine	1.9 ± 0.7[d]	nd[c]	nr	0.3
Cysteine	(see note[d])	nd	nd	nr
Glutamine	(see note[e])	1.4 ± 0.9	nr	nr
Glutamic acid	3.7 ± 1.6	0.7 ± 0.2	18.2	2.7
Glycine	0.04 ± 0.05	0.04 ± 0.01	3.9	1.0
Histidine	0.4 ± 0.2	0.11 ± 0.05	2.0	8.3
Isoleucine	0.15 ± 0.3	0.08 ± 0.04	4.7	0.2
Leucine	0.2 ± 0.15	0.1 ± 0.1	6.7	0.5
Lysine	1.6 ± 0.5	2.0 ± 1.3	3.2	1.1
Methionine	0.03 ± 0.11	0.04 ± 0.02	9.9	1.8
Phenylalanine	0.4 ± 0.2	0.13 ± 0.11	1.7	0.3
Proline	0.6 ± 0.3	nr	5.8	1.1
Serine	1.0 ± 0.4	0.5 ± 0.4	3.5	0.8
Threonine	2.4 ± 0.7[e]	0.6 ± 0.3[f]	1.5	10.0
Tryptophan	nr	0.06 ± 0.08	nr	1.2
Tyrosine	0.4 ± 0.3	0.18 ± 0.17	6.2	0.8
Valine	0.6 ± 0.3	0.2 ± 0.1	nd	0.5
Total	40 ± 11	26 ± 10	94	39

[a]Not included are the S-alkylcysteine sulfoxides (e.g. alliin) and S-alkylcysteines.
[b]Methods of analysis: amino acid analyzer, HPLC of butylthiol-isoinodole derivatives, amino acid analyzer, and paper chromatography, respectively.
[c]nr = not reported, nd = not detected.
[d]Cystine analysis includes cystine and cysteine.
[e]Threonine analysis includes glutamine.
[f]Threonine analysis includes glycine.

S-alkylcysteines mentioned previously (Table 3.3 and Section 3.2.3) account for the majority of these dipeptides; however, γ-glutamylmethionine, γ-glutamylphenylalanine, γ-glutamylglycine, and oxidized γ-glutamylcysteine have also been found in garlic bulbs (Virtanen, 1965; Suzuki et al., 1961a; Lawson & Wang, 1994; Mütsch-Eckner et al., 1992b; Lawson et al., 1991b). Polymers of γ-glutamylcysteine are known *phytochelatins* (bind toxic heavy metals) throughout the plant kingdom, but their content in garlic has not yet been determined (Grill et al., 1987).

Alkaloids, which are biogenetically derived from amino acids, have not been found in garlic except for *trigonellin* (3-carboxy-1-methyl-pyridinium hydroxide) which has been estimated to be present at 0.10 to 0.89 mg/g dry weight, depending on the age of tissue (Khanna et al., 1989). A true *Allium* alkaloid, named *alline*, having a physostigmine skeleton, however, was detected in *A. victorialis*, *A. senescens*, *A. anisoprodium*, *A. odorum*, and others (10 species), which contain up to 0.20% of this substance per dry weight of tissue (Samikov et al., 1986; Antsupova & Polozhii, 1987; Taskhodzhaev et al., 1985). Additionally, some specific proteins have been characterized in several other *Allium* species (Bessoule et al., 1994; Moreno et al., 1994; Basra et al., 1994; Nakamura & Tahara, 1977; Liebe & Quader, 1994).

3.3.4 Lipids: The True Oil in Garlic

Garlic contains only a very small amount of oil-soluble compounds, a fact that may surprise those who have frequently seen "garlic oil" in the stores. This commercial garlic oil is the product of steam distillation of chopped garlic, a process which converts allicin and other thiosulfinates to oil-soluble allyl sulfides. However, undiluted commercial steam-distilled garlic oil has been found to contain only 0.05–0.14% acyl lipids (lipids containing fatty acids), indicating that almost none of the lipids are transported into the steam distillate (Lawson & Wang, 1994).

The total lipid content of garlic (Kamanna & Chandrasekhara, 1980, 1986; Yang & Shin, 1982; Stoianova-Ivanova & Tzutzulova, 1974; Michahelles, 1974; Afzal et al., 1985; Lawson & Wang, 1994; Abdel-Fattah & Edrees, 1972; Huq et al., 1991) and other *Allium* species (Sanchez et al., 1988; Bouthelier & Cattaneo, 1987; Deineko, 1985) have been determined. The results of the quantitative studies on garlic are summarized in Table 3.11. An important consideration in determining the total lipid content is that the cloves must be treated in such a way as to inhibit alliinase and thereby avoid conversion of the cysteine sulfoxides to the more abundant thiosulfinates, which are soluble both in water and in the neutral lipid fraction. The first two studies in Table 3.11 do not state whether this was done; however, the third study inhibited alliinase by treatment of the lyophilized and pulverized cloves (3 strains) with an alliinase inhibitor (amino-oxyacetate) prior to extraction, which could well be the reason for the lower values. Thus, the lipid content of garlic cloves appears to be about 1–2 mg/g fresh weight.

Unlike in many plants, a high proportion of the lipids of garlic are polar lipids (phospholipids and glycolipids). As indicated in Table 3.11, the polar lipids account for 37–60% of the total lipids, and the neutral lipids for 40–63%. The unusually low content of neutral lipids is probably a reflection of the low total lipid content, indicating that garlic's lipids are used primarily for structural purposes rather than for energy storage. The contents of the various lipid subclasses are shown in Table 3.12. There are considerable differences between the results of the two studies, particularly with respect to sterols (Section 3.3.5), phosphatidyl choline, and mono- and diglycerides; however, both agree that triglycerides are the largest single class. The second study (Yang & Shin, 1982) reported the presence of small amounts of sulfolipids, but the amount may be overstated since it was assumed that all of the TLC-origin material was sulfolipids; however, its presence has been confirmed qualitatively (Michahelles, 1974).

The *fatty acid* composition of the total lipids and the lipid classes has been determined with general agreement among the studies (Kamanna & Chandrasekhara, 1980, 1986; Yang & Shin, 1982; Michahelles, 1974; Stoianova-Ivanova & Tzutzulova, 1974; Sanchez et al., 1988; Huq et al., 1991; Bouthelier & Cattaneo, 1987; Deineko, 1985). In the total lipids, the main components are linoleic acid (60–65%), palmitic acid (20–30%), oleic acid (3–10%), and α-linolenic acid (3–6%). Interestingly, small amounts of the long chain saturates 20:0, 22:0, and 24:0 (0.3, 0.9, 0.6%) and the odd-

Table 3.11. Lipid Content of Garlic Cloves

	Kamanna & Chandrasekhara, 1980[a] (n = 1)	Yang & Shin, 1982[a] (n = 6)	Lawson & Wang, 1994[a,b] (n = 3)
Total lipids (mg/g fresh weight)	2.0	3.1–3.4	1.0–1.1
Neutral lipids[c] (%)	62.6	36–44	—
Phospholipids[c] (%)	23.4	36–39	—
Glycolipids[c] (%)	14.0	20–24	—

[a]Method of extraction: chloroform/methanol, according to Bligh & Dyer (1959).
[b]Alliinase was inhibited prior to extraction.
[c]Percent of total lipids.

Table 3.12. Composition of Lipid Classes in Garlic Cloves

	Kamanna & Chandrasekhara, 1986		Yang & Shin, 1982	
	% of Each Lipid Class	% of Total Lipids	% of Each Lipid Class	% of Total Lipids
Neutral lipids				
Triglycerides	41.5	26	8–84	29–36
Monoglycerides	18.5	11.6	nr[a]	
Diglycerides	14.2	8.9	4.4–5.0	1.8–1.9
Free fatty acids	4.4	2.8	1.2–1.6	0.5–0.6
Sterols	16.3	10.2	1–4	0.4–1.6
Sterol esters & hydrocarbons	3.3	2.1	5–7.5	1.8–3.3
Phospholipids				
Phosphatidyl choline	23.5	5.5	60–61	22–24
Phosphatidyl ethanolamine	17.9	4.2	25	9.2–9.8
Phosphatidyl inositol	nr		10–11	3.7–4.5
Lysophospholipids	20.0	4.7	nr	
Phosphatidic acid	3.4	0.8	3.6	1.2–1.4
Glycolipids				
Acyl sterol glycoside	38.6	5.4	46–47	9.6–11
Sterol glycoside	15.6	2.2	22–23	4.5–5.6
Monogalactosyl diglyceride	22.5	3.2	7.6	1.5–1.9
Digalactosyl diglyceride	10.1	1.4	8.0	1.6–1.9
Cerebrosides	nr		13	2.6–3.2
Sulfolipids	nr		2.2–2.5	0.5[b]

[a]Abbreviation: nr = not reported.
[b]The TLC-origin material was assumed to be only sulfolipids.

chain saturates 17:0, 19:0, 21:0, and 23:0 (0.3, 0.04, 0.2, 0.7%) are also found (Lawson & Wang, 1994). Among the lipid subclasses, Kamanna and Chandrasekhara (1980) have found that the triglyceride and the monogalactosyl diglyceride fractions contain large amounts of α-linolenic acid (25–33%), while digalactosyl diglyceride contains little α-linolenic acid, but large amounts of 12:0 and 12:1 (51%). The free fatty acid fraction is made mainly of 12:0, 14:0 and 16:0, while the acylsterol glycosides contain almost exclusively 16:0 (86%). The fatty acid composition of the phospholipids is similar to that of the total lipids (Yang & Shin, 1982; Michahelles, 1974). The presence of arachidonic acid has been reported in one study (Afzal et al., 1985); however, it has been found that this compound co-elutes on a gas chromatograph with undodecanoic acid (21:0), which is the likely identity since garlic lipids contain a series of odd-chain fatty acids and since no other C_{20-22} polyunsaturated fatty acids have been found (Lawson & Wang, 1994).

Prostaglandins, a group of hormones derived from C_{20} polyunsaturated fatty acids, are also present in garlic. Bioassay studies have indicated their presence in garlic (Ali et al., 1990; Rashid et al., 1986), while others have provided analytical evidence for the presence of prostaglandins A_2 and $F_{1\alpha}$ (Al-Nagdy et al., 1988; Pobozsny et al., 1979) and have shown that garlic has the capacity to produce prostaglandins from added arachidonic acid (Ali et al., 1990). Garlic also contains a lipoxygenase system that produces dominant amounts of divinyl ether oxylipins, compounds that have previously been found only in potatoes (Grechkin et al., 1995). Prostaglandins are also abundant in common onions and other *Allium* species (Attrep et al., 1973, 1980; Al-Nagdy et al., 1986; Claeys et al., 1986; Makheja et al., 1981; Ali et al., 1990; Pobozsny et al., 1979; Sun, 1991; Ma et al., 1990; Ubaid, 1989; Gu et al., 1988; Sun et al., 1988; Üstünes et al., 1985; Panosyan, 1981). On the other hand, some garlic-derived compounds (allyl sulfides, ajoenes, vinyldithiins) have been found to inhibit prostaglandin synthetase and lipoxygenase (Wagner et al., 1987; Block et al., 1988).

3.3.5 Sterols and Hydrocarbons: Unsaponifiable Lipids

Sterols (3-β-hydroxylsteroids) and hydrocarbons are nonpolar compounds with solubility characteristics very similar to the lipids, but they do not contain fatty acids. Studies on the content of *phytosterols* in garlic cloves (Table 3.13) have shown that the dominant sterol is β-sitosterol with lesser amounts of campesterol and cholesterol and the absence of stigmasterol. The total sterol content appears to be about 18 µg/g fresh weight (Oka et al., 1973; Huq et al., 1991; Lawson & Wang, 1994). A study by Stoianova-Ivanova et al. (1980) yields similar relative amounts of the sterols, but absolute values 20 times higher than that of the other studies, indicating a calculation error in this study.

The *hydrocarbons* represent the plant waxes of garlic and constitute both saturated paraffins and unsaturated olefins. Both a 1980 study (Stoianova-Ivanova et al., 1980) and a recent study (Lawson & Wang, 1994) have shown that the garlic hydrocarbons constitute about 94% paraffins and 6% olefins and that the dominant chain lengths are odd-numbered (C31>C29>C27>C25>C23=C30=C28=C26=C24>C22). The total paraffin and olefin content of garlic appears to be about 12 ± 5 µg/g fresh weight (n = 6) (Lawson & Wang, 1994), although the study by Stoianova-Ivanova et al. (1980) reported about 430 µg/g fresh weight (n = 1); however, as mentioned with the sterols, the quantitation in this study appears to be in serious error. The *carotenes*, another class of hydrocarbons found in many vegetables, are virtually absent in garlic, with less than 0.01 ppm found (Lawson & Wang, 1994; Watt & Merrill, 1963). *Pristane* (2,6,10,14-tetramethylpentadecane), a hydrocarbon that is presumably derived from the phytol moiety of chlorophyll and may induce a myriad of biological effects, has been found in garlic (0.59 µg/g) and in other *Allium* species (Chung et al., 1989).

3.3.6 Total Lipophilic Compounds

The lipophilic compounds represent the compounds that are extractable with a nonpolar organic solvent or the total amount of oil-soluble compounds present in garlic cloves. As shown in Table 3.14, lipids represent about 98% of the total lipophilic compounds in whole garlic, but only about 21% in crushed garlic due to the formation of thiosulfinates, which constitute about 78% of the total lipophilic compounds.

Table 3.13. Sterols in Garlic (µg/g fresh weight)

Study	β-Sitosterol	Campesterol	Cholesterol
Oka et al., 1973 (n = 1)	13	nd	1.4
Lawson & Wang, 1994 (n = 6)	15 ± 5	1.7 ± 0.6	0.7 ± 0.3
Stoianova-Ivanova et al., 1980 (n = 1)	220	53	28

Table 3.14. Total Oil-Soluble Compounds in Whole and Crushed Garlic Cloves (mg/g fresh weight)

Compounds	Whole Cloves	Crushed Cloves	Reference
Lipids	1.5	1.5	Table 3.11
Sterols	0.018	0.018	Table 3.13
Hydrocarbons	0.012	0.012	Section 3.3.6
Tocopherols	0.002	0.002	Table 3.16
Thiosulfinates	0	5.5	Table 3.3
Total	1.53	7.03	

3.3.7 Steroid and Triterpenoid Glycosides (Saponins)

The saponins are a large and widely distributed group of polycyclic steroid glycosides (C27 based) and triterpenoid glycosides (C30 based), all of which possess a 3β-hydroxyl group, where the sapogenins (or aglycons) are linked to a variety of sugar units. The saponins have strong surfactant (foaming) ability and have a variety of biological activities, although they are only poorly absorbed in humans (Oakenfull, 1981). Several reviews on *Allium* saponins are available (Koch, 1993a; Kravets et al., 1990; Mahato et al., 1982).

A compilation of all the known quantitative data on the saponins of garlic cloves is provided in Table 3.15. The content of total saponins, according to somewhat earlier literature (Müller, 1982; Smoczkiewicz et al., 1977, 1978, 1982; Smoczkiewiczowa & Nitschke, 1975; Nitschke & Smoczkiewiczowa, 1976; Stoianova-Ivanova et al., 1980; Eichenberger & Grob, 1970; Fenwick & Oakenfull, 1983), appears to be around 1 mg/g, which is much smaller than for many other common vegetables (39 mg/g for soybeans, *Glycine max* L. Merill; 14 mg/g for red kidney beans, *Phaseolus vulgaris* L.; 5.8 mg/g for peanuts, *Arachis hypogaea* L.; and 5.5 mg/g for spinach, *Spinacea oleracea* L.) (Fenwick & Oakenfull, 1983). However, because analysis of these compounds is difficult, the actual content of saponins in garlic may be higher (Koch HP, unpublished data, 1995; Matsuura et al., 1989a, 1989b). A comparison of the total identified saponins with the total saponin content indicates that only about half of the saponins in garlic have been identified.

Proto-eruboside B was the first saponin to be fully identified in garlic. It is a furostanol consisting of F-chlorogenin and 5 sugar units: 4 glucose (1 at C_{26}) and 1 galactose (Fig. 3.19). Cleavage of the C_{26} glucose with added β-glucosidase yields eruboside B, a compound which has good anticandidal activity minimum inhibitory concentration (MIC) of 25 µg/mL (Matsuura et al., 1988). *Sativoside B1* (5 glucose and 1 galactose) and *proto-desgalactotigonin* (3 glucose, 1 galactose, 1 xylose) are less abundant furostanols whose sapogenins are F-sativogenin and F-tigogenin, respectively (Matsuura et al., 1989b). Hydrolysis of the saponins yields the sapogenins, which are easier to identify and quantitate than the saponins. Two independent studies have yielded remarkably similar results, indicating that β-sitosterol accounts for the most of the identifiable sapogenin (Michahelles, 1974; Smoczkiewicz et al., 1982; Smoczkiewicz et al., 1977).

Table 3.15. Saponins and Sapogenins in Garlic Cloves

	mg/g fresh weight	Reference
Known saponins		
Proto-eruboside B	0.10	Matsuura et al., 1988
Sativoside B1	0.03	Matsuura et al., 1989
Proto-desgalactotigonin	0.01	Matsuura et al., 1989
β-sitosterol based[a]	0.38	Smoczkiewicz et al., 1982
Total saponins	0.35–0.42	Smoczkiewicz et al., 1977
	0.95	Smoczkiewicz et al., 1982
	1.10	Fenwick & Oakenfull, 1983
Known sapogenins		
β-sitosterol	0.14	Michahelles, 1974
β-sitosterol	0.195	Smoczkiewicz et al., 1982
F-chlorogenin	0.047	Matsuura et al., 1988
F-sativogenin	0.012	Matsuura et al., 1989
F-tigogenin	0.005	Matsuura et al., 1989
Total sapogenins	0.14	Michahelles, 1974
	0.195	Smoczkiewicz et al., 1982

[a]An estimate based on assuming four sugar units per molecule of β-sitosterol.

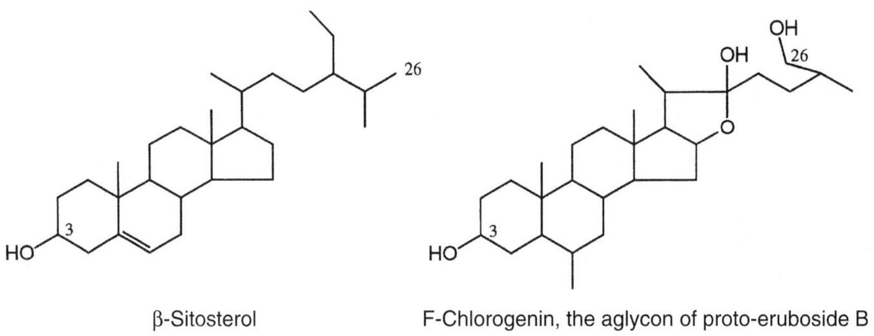

β-Sitosterol

F-Chlorogenin, the aglycon of proto-eruboside B

Figure 3.19. The main known sapogenins (aglycons) of garlic saponins.

These investigators could not detect gitogenin, β-amyrin, or oleanolic acid in garlic and concluded that β-sitosterol was the only detectable sapogenin in garlic; however, more recent studies by Matsuura and colleagues (Table 3.15) have shown the presence of other sapogenins as well. Interestingly, garlic roots have been shown to contain different saponins than the cloves, and include sativosides R1 and R2, F-gitonin, and desgalactotigonin (Matsuura et al., 1989b).

It should be mentioned that sapo(ge)nins are abundant in other *Allium* species such as *A. cepa* (Koczwara, 1949; Eichenberger & Grob, 1970; Sinha, 1959; Nitschke & Smoczkiewiczowa, 1976; Koch, 1993a; Kintya & Degtyareva, 1989; Kintia et al., 1986; Kravets et al., 1986a, 1986b; Smoczkiewiczowa & Nitschke, 1978b; Itoh et al., 1977; Smoczkiewicz et al., 1977; Oka et al., 1973, 1974), *A. porrum* (Harmatha et al., 1987; Smoczkiewiczowa & Wieladek, 1978; Eichenberger & Menke, 1966), *A. fistulosum* (Jung et al., 1993), *A. giganteum* (Mimaki et al., 1994; Kawashima et al., 1991a; Sashida et al., 1991; Kelginbaev et al., 1973, 1975, 1976; Khristulas et al., 1970), *A. tuberosum* (Noda et al., 1988; Oka et al., 1973), *A. bakeri* (Nishino et al., 1990a; Oka et al., 1973), *A. ascalonicum* (Oka et al., 1973, 1974), *A. chinense* (Matsuura et al., 1989c; Oka et al., 1973), *A. macrostemon* (Peng et al., 1992, 1993, 1994; Yao and Peng, 1994; Wu et al., 1992), *A. ampeloprasum* (Morita et al., 1988), *A. vineale* (Chen & Snyder, 1987, 1989), and others (Mimaki et al., 1993; Kawashima et al., 1991a, 1991b, 1993; Vardosanidze et al., 1992; Cherkasov et al., 1985, 1990; Kravets et al., 1990; Vollerner et al., 1983a, 1983b, 1984, 1988, 1989, 1991; Azarkova et al., 1983, 1984, 1985, 1986; Ismailov et al., 1976; Ismailov & Aliev, 1974, 1976; Khristulas et al., 1974; Gorovits et al., 1971, 1973a, 1973b; Kereselidze et al., 1970; Pkheidze et al., 1967). Many of these compounds show remarkable biological effects, including inhibition of platelet aggregation (Yao and Peng, 1994), antitumor activity (Nishino et al., 1990a), antimicrobial action (Matsuura et al., 1989a), and insecticidal (Harmatha et al., 1987) and molluscicidal effects (Chen & Snyder, 1987, 1989).

3.3.8 Vitamins and Minerals

The content of vitamins, minerals, and trace minerals in garlic cloves is summarized in Table 3.16. This table also compares the values found in a large clove of garlic with the U.S. Recommended Dietary Allowance or the typical daily intake for each vitamin or mineral. Although garlic contains levels of nutrients typically found in tuber vegetables (e.g., potatoes), the amounts of garlic normally eaten are so small that even a rather large intake of 5 grams per day provides only 1% or less of the daily need for each of these nutrients. Therefore, nutrients are not important in evaluating the health benefits of garlic consumption. The first nutritional analysis of garlic was conducted over a century ago (König, 1879).

Selenium and *germanium* have often been cited in the popular literature as important elements of garlic's medicinal effects; however, as shown in Table 3.16, the actual amounts of these trace minerals consumable from normal garlic (selenium-enriched garlic is discussed in the next section), especially germanium, are much too small to be of benefit. Nevertheless, selenium is found in garlic at considerably higher levels than in most fruits and vegetables, with the exceptions of cauliflower, spinach, mushrooms, and grains, where it is found in concentrations about equal to those of garlic, and asparagus, where it is three times as abundant (on the basis of fresh weight) (Amer & Brisson, 1973; Ali et al., 1990).

Tellurium, a trace mineral of the same family as selenium and sulfur which is not known to meet any dietary need, has recently been hypothetically proposed to be important to the cholesterol-lowering effect of garlic through inhibition of cholesterol synthesis at the site of squalene epoxidase (Larner, 1995). This theory was based in large part on a report of high tellurium levels in garlic bulbs: 31–73 ppm fresh weight, compared to 0.14 ppm for selenium (Schroeder et al., 1967). This value for tellurium seems much too high, however, since the earth's crust contains about 45 times as much selenium as tellurium. The only other known report of the tellurium content of garlic found a much smaller value, 3 ppm (Liu et al., 1989). Even

Table 3.16. Vitamin and Mineral Content of Garlic Cloves

Item[a]	Average Value[b] mg/100 g Fresh Weight	Recommended Dietary Allowance[c] or Average Daily Intake[d]	% of RDA or Daily Intake in a Large 5 g Clove
Ascorbic acid[1-3,24]	14 ± 4	60 mg[c]	1.2
Thiamine[1,2,4]	0.20 ± 0.05	1.3 mg[c]	.8
Riboflavin[1,2]	0.08	1.5 mg[c]	0.3
Niacin[1,2]	0.55	17 mg[c]	0.2
Pantothenic acid[5]	0.26	4-7 mg[d]	0.2
Vitamin E (10% α, 90% δ)[1]	0.02	9 mg-TE[c]	0.01
Vitamin A	trace	900 RE[c]	—
Calcium[1,6,7,10]	24 ± 11	1000 mg[c]	0.12
Phosphorus[1,6,7,10,25]	177 ± 45	1000 mg[c]	0.9
Potassium[6-8]	440 ± 80	2000 mg[c]	1.1
Sodium[6-8]	11 ± 7	500 mg[c]	0.1
Magnesium[6,8,10]	18 ± 2	315 mg[c]	0.3
Iron[1,6-9]	2.0 ± 1.2	12 mg[c]	0.8
Boron[1,6,10]	0.35 ± 0.2	(see note[e])	—
Copper[1,6,8-10]	0.15 ± 0.1	1.6 mg	0.5
Zinc[1,6,8-10]	1.2 ± 1.1	13.5 mg[c]	0.4
Manganese[1,6,9-10]	0.33 ± 0.17	2.5 mg[d]	0.7
Chromium[8-10]	0.05 ± 0.04	0.10 mg[d]	2.5
Nickel[1,8-10]	0.03 ± 0.02	(see note[e])	—
Molybdenum[1,8,10]	0.016 ± 0.011	0.20 mg[d]	0.4
Cobalt[8-10]	0.022 ± 0.007	(see note[e])	—
Iodine[1]	0.003	0.15 mg[c]	0.1
Tellurium[10,14]	0.003 ± 0.3	(see note[e])	—
Selenium[11-18, 26-31]	0.020 ± 0.023	0.063 mg[c]	1.6
Germanium[19-23]	0.004 ± 0.007	1.5 mg[d]	—

[a]References for each item: 1(Souci et al., 1986); 2(Hanley & Fenwick, 1985); 3(Gusev & Grishina, 1963); 4(Tsuno, 1958); 5(Ishiguro, 1963); 6(Christensen et al., 1968); 7(Watt & Merrill, 1963); 8(Raj et al., 1980); 9(Khan et al., 1985); 10(Liu et al., 1989); 11(Rao & Jones, 1984); 12(Koch et al., 1988); 13(Ip et al., 1992); 14(Lawson & Wang, 1994); 15(Amer & Brisson, 1973); 16(Askar & Bielig, 1983); 17(Goto & Fujino, 1967); 18(Noda et al., 1983); 19(Lawson, 1993); 20(Schleich & Henze, 1990); 21(Tao & Fang, 1993); 22(Lu et al., 1992); 23(Feng et al., 1991); 24(Prins, 1984); 25(Anantakrishnan & Venkataraman, 1940); 26(Kisu, 1985); 27(Zhang et al., 1994); 28(Jirovetz et al., 1993); 29(Suzuki et al., 1982); 30(Matsuzawa et al., 1984); 31(Matysek & Behounkova, 1968).
[b]Mean ± standard deviation for n references. Multiply by 10 to obtain ppm.
[c]U.S. Recommended Dietary Allowance, average value for adult women and men, from National Research Council 1989.
[d]Average daily intake for adults. Values for all except germanium from National Research Council 1989. Value for germaniums, Schauss 1991.
[e]No NRC guideline.

this value, however, may be much too high. Using induced coupling argon plasma (ICAP) and direct acid digestion (to prevent alliinase activation), the tellurium and selenium contents of six freeze-dried garlic bulb samples gathered from several locations (California, Oregon, New York, Mexico, and Xin Jiang, China) were evaluated; tellurium was found to be undetectable (less than 0.1 ppm dry weight or less than 0.03 ppm fresh weight) and selenium was found at levels of 0.05–0.73 (0.24 ± 0.31) ppm dry weight or 0.02–0.24 (0.08 ± 0.10) ppm fresh weight (Lawson & Wang, 1994). Therefore, it is highly unlikely that tellurium could be responsible for any of garlic's pharmacological effects. The report by Schroeder and coworkers (1967) also claims that the tellurium compounds of garlic are volatile; however, it is more likely that any trace amounts of organotellurium present are substituted for sulfur in the cysteine sulfoxides, forming volatile tellurium-substituted thiosulfinates only when the cloves are crushed.

The *boron* content of garlic is important for the plant itself. It has recently been shown that levels of 3 ppm (0.3 mg/100 g) in the bulb or 30 ppm in the above-ground plant are correlated with optimum growth and yield (Chermsiri et al., 1995).

3.3.9 Organoselenium Compounds

The *selenium* content in fresh garlic (0.01–0.6 ppm, average 0.2 ppm, Table 3.16) is about 10,000 times less than that of sulfur; however, because selenium, a nutritionally essential element of the same family as sulfur, can substitute for sulfur in the organosulfur compounds of the sulfur-rich *Allium* species, there has been considerable interest in growing garlic and onion in selenium-enriched soils as less toxic vehicles for increasing the dietary levels of this important component of the body's antioxidant system. In fact, the use of selenium-enriched garlic for pharmaceutical purposes has been patented (Konvicka, 1989).

Remarkably, the level of selenium compounds in garlic has been increased by 2500 times (up to 0.25 mg per 5 g clove, about 4 times the recommended daily intake) by cultivation in selenium-enriched soil (Ip et al., 1992, 1994; Ip & Lisk, 1993, 1994a, 1994b; Wang et al., 1989). The resulting increase in selenium compounds has been shown to provide selenium to selenium-dependent enzymes in animals as well as to improve the anticancer effects of garlic, but without accumulation to excessive levels in animal tissues.

Consequently, an important concern has been to identify the organoselenium compounds present in garlic. Garlic cloves have been shown to possess a selenium-containing polysaccharide and selenoproteins containing *Se-cysteine* and *Se-methionine*, while garlic oil has been found to contain *dimethyl selenide* and *dimethyl diselenide* (Yang et al., 1992; Wang et al., 1989; Bai et al., 1994). Furthermore, it was found 30 years ago that injecting young onion plants with selenite resulted in finding Se-cysteine, Se-methionine, and *Se-methylcysteine sulfoxide* (Spare & Virtanen, 1964). The majority of the selenium, however, appears to be present in *Se-methyl selenocysteine*, as was demonstrated with garlic grown in selenium-enriched soil (Cai et al., 1995b; Block, 1995)

Recently, by use of a selenium-selective detection system, the volatile compounds generated by garlic grown in a selenium-rich soil as well as in normal soil have been identified (Cai et al., 1994a, 1994b, 1995a). Interestingly, of the eight compounds identified, all were methyl-Se or methyl-S-Se compounds, with methyl-Se-S-allyl being the most abundant. Even though allylcysteine sulfoxide (alliin) is about 10 times more abundant than methylcysteine sulfoxide in garlic, no allyl-Se compounds were detected except for the dual case of methyl-Se-allyl. This demonstrates a strong preference of selenium binding to methyl compounds and provides further evidence of a different route for biosynthesis of methylcysteine sulfoxide than for alliin, as was discussed in Section 3.2.10.

Garlic appears to be somewhat unusual among plants for its ability to concentrate selenium as well as to produce its most usable form. Recently, it has been shown that the selenium content of garlic grown on selenium-enriched soil was four times higher (on the basis of dry weight) than that of onions grown on the same soil at the same time, and that when both were given to rats on an equal-selenium basis the garlic had better anticancer activity than onion (Ip & Lisk, 1994b). In most grains and vegetables selenium is found substituted primarily in methionine, an amino acid which has no antioxidant-stimulating effects; however, in garlic, onion, and broccoli most of the selenium is substituted into cysteine (and S-methylcysteine, but not into S-allylcysteine, when grown in Se-enriched soil), which provides selenium to glutathione peroxidase, thus promoting antioxidant activity through increased synthesis of glutathione (Cai et al., 1995b).

3.3.10 Flavonoid Pigments and Phenols

Relatively little work has been reported on the content of *flavonoid* pigments in garlic cloves. Although these highly colorful compounds are very abundant in onions (Omidiji, 1993; Urushibara et al., 1991, 1992; Voldrich et al., 1989; Kiviranta et al., 1986, 1988; Smoczkiewiczowa & Nitschke, 1978a; Herrmann, 1956a, 1977; Weissenböck et al., 1987) and in other *Allium* species (Bilyk & Sapers, 1985; Tan et al., 1994; Voldrich et al., 1989; Yoshida et al., 1987; Daiichi, 1983), they have been shown to be virtually absent in peeled garlic cloves, except for trace amounts of *quercetin* and *kaempferol* (Starke & Herrmann, 1976; Michahelles, 1974; Kaneta et al., 1980; Leighton et al., 1992). However, the reddish purple color of inner garlic scales has been shown to be due to three *anthocyanins:* cyanidin-3-glucoside, which is the main one (Fig. 3.20), and two other cyanidin-3-glucosides which are acylated at the glucose (Du & Francis, 1975; Borukh & Demkevich, 1976). Anthocyanins are also present in the onion (Harborne, 1986; Du et al., 1974; Fuleki, 1971), including malonylated anthocyanins (Bauhin, 1622). For biological actions and technical uses of flavonoids, several comprehensive articles are available (Spilkova & Hubik,

1988, 1992; Asada et al., 1992). Some flavonoids may have *phytoalexin* function (i.e., induction of pathogen resistance) in the plant (Dmitriev et al., 1986; Omidiji & Ehimidu, 1990; Walker & Stahmann, 1955; Grandmaison et al., 1993).

Several *phenolic acids* have been found in garlic, albeit at very small concentrations. Six of these have been determined in peeled garlic cloves whose concentrations total 40 µg/g fresh weight and include *p-hydroxybenzoic acid* (13 µg/g), *caffeic acid* (10 µg/g), *ferulic acid* (7 µg/g), *vanillic acid* (6 µg/g), *sinapinic acid* (2 µg/g), and *p-coumaric acid* (2 µg/g) (Schmidtlein & Herrmann, 1975; Herrmann, 1956c, 1977; Codignola et al., 1989; Okuyama et al., 1986; Bekdairova, 1981). The scales were found to contain a similar total concentration, but *p*-coumaric acid was dominant and *p*-hydroxybenzoic acid was absent. *Salicylic acid* has also been found in trace amounts (1 µg/g) (Swain et al., 1985).

Lignin, a complex polymer of *p*-propenolphenols that serves to provide a rigid cell wall structure, is found at 1.6% in garlic scales (Abdel-Fattah & Edrees, 1972). *Tannins*, which have antioxidant properties, have been detected in the skin of red onions (Akaranta & Odozi, 1986). Other *Allium* species also have phenolic compounds, some of which are even more abundant (Winter et al., 1987; Goda et al., 1987; Rescke & Herrmann, 1982; Runkova & Talieva, 1970; Das & Rao, 1964; Herrmann, 1956b; Hegnauer, 1963). These compounds are also involved in the pathogen resistance of the plants (Walker & Stahmann, 1955; Tada et al., 1988; Patel et al., 1986; Link et al., 1929; Link & Walker, 1933) and in the binding and transport of heavy metals in the roots (Micera & Dessi, 1988, 1989).

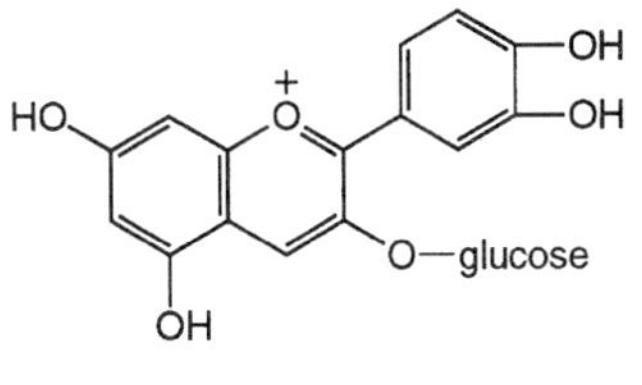

Figure 3.20. Cyanidin-3-glucoside, the main red/purple pigment of garlic scales.

3.3.11 Adenosine, Guanosine, and Nucleic Acids

Adenosine has been reported to be present in relatively high amounts in both garlic and onion, with values as high as 0.2–0.37 mg/g fresh weight for garlic (Michahelles, 1974; Lawson et al., 1991c; Weisenberger et al., 1972). It has also been found in the leaves of *A. tuberosum* and *A. fistulosum* (Choi et al., 1992; Ohga, 1988) and the bulbs of *A. bakeri* (Okuyama et al., 1986, 1989). The effects of garlic compounds on adenosine metabolism are discussed in Section 5.14. The high content of adenosine has caused some to conclude that it may be responsible for garlic's antithrombotic activity (see Section 5.14), even though adenosine is known to be very poorly absorbed and, when absorbed, to be rapidly metabolized by erythrocytes (Reuter, 1980, 1983, 1986).

Recent studies, however, have shown that free adenosine is absent in garlic cloves until it is crushed and that it gradually increases to a maximum value in about 8 hours (Fig. 3.21) (Lawson & Wang, 1994; Lawson et al., 1991c). When the cloves were homogenized in 1N HCl or 50% ethanol, no adenosine could be detected, indicating that it is enzymatically formed upon crushing (probably from adenosine monophosphate), similar to allicin formation from alliin, although 1000 times slower. A very similar time curve was found for *guanosine* release from garlic cloves, although the maximum value achieved was about 40% higher than that of adenosine (Lawson & Wang, 1994).

Therefore, the adenosine and guanosine content of garlic homogenates will vary greatly, depending on how much time has elapsed between crushing and analysis or between crushing and consumption. Although no adenosine is present in garlic cloves, commercial garlic powders, since they are made from oven-dried chopped cloves, have been found to contain adenosine at 0.020–0.170 mg/g dry weight (mean, 0.10 ± 0.06, n = 10), due to adenosine's being formed before the garlic has been sufficiently dried. The lowest value (0.02) was found for a powder prepared by freeze-drying the chopped cloves. Furthermore, no detectable amount (less than 0.005 mg/g) of adenosine has been found in commercial aged extracts of garlic (Lawson & Wang, 1994).

Specific *nucleic acids* have also been characterized, but mainly in other *Allium*

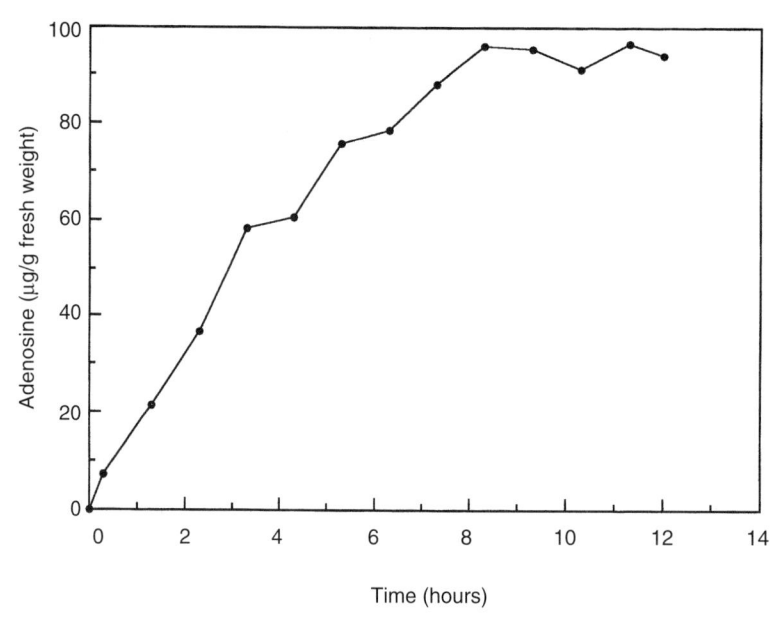

Figure 3.21. Adenosine formation after homogenization of garlic cloves. The zero-time value was obtained by addition of HCl. (From Lawson & Wang, 1994.)

species. Only one study on long-lived mRNA from garlic has been published (Wang et al., 1989). Some genetic work has been done, however: a method for the preparation of high-purity DNA from onions has been patented (Goto et al., 1994); the nucleotide sequence of several tRNAs and the gene sequence have been determined for *A. porrum*; random amplified polymorphic DNA markers have been used for genetic analysis in *A. cepa, A. fistulosum, A. ascalonicum, A. schoenoprasum, A. ampeloprasum,* and a wild relative, *A. roylei* (Wilkie et al., 1993); and restriction enzyme analysis has been conducted on the chloroplast and nuclear DNA sections of *A. cepa* (Havey, 1992a, 1992b).

3.3.12 Miscellaneous Compounds: Phytic Acid, Plant Hormones, Lectins

Phytin, a mixture of calcium and magnesium salts of phytic acid (myo-inositol hexaphosphate) and an important phosphorus storage compound in plants, was first reported to be present in garlic at 0.079% fresh weight, a value which is lower than in many plants such as rice (0.2%) and soybeans (0.28%) (Anantakrishnan & Venkataraman, 1940a). Later, a compound was isolated from garlic that had blood anticoagulant activity in vitro but not in vivo. The sodium salt of the substance was present in garlic at 0.075% fresh weight and had a phosphorus/inositol ratio of 6:1, which strongly suggests that it was phytic acid (Song et al., 1963a, 1963b; Anantakrishnan & Venkataraman, 1940b). Although phytic acid is well known to interfere with absorption of calcium, iron, and magnesium, the amount present in a clove of garlic is much too small to be a therapeutic concern.

A group of ubiquitous *phythormones*, plant growth and senescence agents, have also been found in garlic and related *Allium* species. For instance, *auxin* (indole acetic acid), the primary growth hormone, has been reported (Park & Lee, 1992). *Gibberellins*, another group of growth hormones responsible for secondary growth and bulb formation, have been reported in all parts of the garlic and onion plants, with the amounts varying with stage of development (Rakhimbaev & Olshanskaya, 1981; Rao et al., 1981; Arguello et al., 1983; Nojiri et al., 1993; Park & Lee, 1992; Chang, 1988; Park, 1987, 1988). *Ethylene*, which is a ribosomal cistron regulator in quiescent and senescent plants, has only been found in onion (Karagiannis & Pappelis, 1993, 1994). *Abscisic acid*, which is responsible for senescence of the plants, is found in scapes and seed bulbs of garlic (Wang et al., 1988; Chang, 1988). *Jasmonic acid*, which stimulates shoot and bulb formation, has also been detected in garlic (Ravnikar et al., 1993).

3.3.13 Machado's Garlicin— A Mystery

In 1948, Brazilian investigators reported the discovery of a non-sulfur-containing substance from garlic that was demonstrated to possess strong antibiotic activity, both in vitro and in vivo, including in clinical trials (Machado et al., 1948). They named the substance *"garlicina"* (anglicized to garlicin) and described it as a stable yellow solid (brownish when wet) that is practically insoluble in water but soluble in organic solvents, forms slightly acidic solutions, does not rotate polarized light, and contains an allyl group (according to the Grote reaction) but no sulfur (therefore, not allicin). Unfortunately, the report (in Portuguese and unknown to most garlic researchers) does not give any information on how the material was obtained from garlic, nor does it give its melting point or yield, and it does not say whether it was an extraction mixture or mainly a single compound. Neither the original authors nor anyone else has ever reported again on the finding of such a substance. Because of the unkown and unreproduced nature of this substance, and since there is a well-known U.S. garlic tablet by the same name, Machado's substance will be referred to as "Machado's garlicin" here.

Garlic has been subjected to various treatments which have resulted in the production of antibacterial compounds that are not allicin, but they do not share common properties with Machado's garlicin (Klosa, 1949; Datta et al., 1948). Shortly after the report of Machado (1948), the residue of steam-distilled garlic was reported to contain a compound (an undefined dihydroxycyclohexyl lactone, $C_8H_{12}O_4$) with antibiotic activity, although the compound was not found in unheated garlic (Zwergal, 1952). About 20 years later, another group (Kabelik, 1970) proposed that Zwergal's distillation residue compound might be identical with Machado's garlicin; however, Zwergal's compound was a liquid and did not contain an allyl group, and hence could not be the same as Machado's substance. Furthermore, a decade-long effort to reproduce Machado's garlicin by synthesizing model compounds based on Zwergal's compound failed to produce a compound with similar properties (Koch, 1993b; Jäger et al., 1993; Nour et al., 1992; Rachenzentner & Urban, 1992; Edelsbacher et al., 1988, 1992; Aufreiter et al., 1992; Fleischhacker et al., 1991; Stejskal et al., 1991; Urban et al., 1989; Mostler & Urban, 1989).

The doubtfulness of the existence in normal garlic cloves of a substance with the qualities described by Machado is further supported by the fact that several studies have shown that inhibition of allicin formation (Cavallito et al., 1945; Shashikanth et al., 1985, 1986) or selective removal of allicin (Cavallito & Bailey, 1944; Cavallito et al., 1944; Rao et al., 1946; Barone & Tansey, 1977; Mostler & Urban, 1989; Yoshida et al., 1987; Deshpande et al., 1993; Ghannoum, 1988; Hughes & Lawson, 1991) from crushed garlic also removes all of its antibiotic activity. It is known that some plants produce new compounds (the so-called *phytoalexins*) when exposed to an unusual environmental stress. For example, induction of disease resistance in the closely related onion by biotic elicitors has been demonstrated recently (Perkovskaya et al., 1991; Sato et al., 1979). The compounds responsible for this effect may be the so-called *cibullins* (or *tsibulins*) found by Russian researchers (Tverskoi et al., 1991; Dmitriev et al., 1988a, 1988b, 1989, 1990). Similarly structured compounds can also be found in other plants (Okwute et al., 1986; Lansbury & La Clair, 1993). Therefore, it has been proposed that Machado's garlicin may have been produced in this manner (Koch, 1993b).

Indeed, Japanese researchers have found that damaging garlic cloves with fire, mercuric chloride, hydrogen peroxide, or cellulase causes production of a chloroform-soluble solid compound they named allixin ($C_{12}H_{18}O_4$), which is also different from Machado's garlicin, as it has only very weak antibiotic activity (Kodera et al., 1989; Nishino et al., 1990b; Yamasaki et al., 1991). Since Machado did not report any of his methods, the origin of his substance will probably always remain a mystery.

As a final note, the term "garlicin" has also been used by Chinese authors, although much later, as a mistranslation for aqueous garlic extract, and it is usually used to mean crude allicin (W. Xu, personal communication), although it has been used at least once to mean diallyl disulfide (Xu et al., 1991; Qian et al., 1986; Zhou et al., 1988; Chen et al., 1990; Yang et al., 1987; Lang et al., 1989). For more information on the toxicology, pharmacokinetics, and clinical testing of Machado's garlicin, see Sections 5.2.7, 6.4, and 7.3.

3.4 GARLIC PROCESSING

Now that the analysis of fresh and cooked garlic cloves has been described, it is important to discuss the composition of the many types of commercially processed garlic formulations. These products are usually sold in pill form and have become very popular in recent years, often representing a significant proportion of the garlic consumed. They include powdered dry garlic, oils produced upon treating chopped garlic with steam, vegetable oil, or ether, and aged extracts of chopped garlic. Garlic cloves aged in vinegar ("pickled") are occasionally consumed and will also be discussed. There are several general reviews on the various processing methods (Raghavan et al., 1983; Fenwick & Hanley, 1985a; Koch, 1987, 1988; Sticher, 1991; Gassmann, 1992). There are also several publications on the postharvest treatment of garlic bulbs (de Carvalho et al., 1991, 1993; Srivastava & Mathur, 1956) as well as on the preparation of dried garlic (Stephenson, 1949; Pruthi et al., 1959a, 1959b, 1959c; Lawson & Hughes, 1989; Singh et al., 1959a, 1959b; Pruthi, 1980; Farag, 1991; Müller & Ruhnke, 1993; Madamba et al., 1993).

Table 3.17 provides an overview of the composition of the sulfur compounds in several brands of each of the various types of commercial products available in 1990 and of how they compare to fresh garlic cloves. As would be expected, considerable brand-to-brand variation is found (note the ranges), which reflects both variation in the initial quality of the garlic cloves being used as well as variation in the quality of the processing methods being employed. The composition, stability, and variability of each of these products are the subject of the remainder of the chapter.

3.4.1 Garlic Powder Products

Garlic powders (see also Section 4.2.1) represent the true composition of garlic cloves better than any other type of processed garlic because they are simply dehydrated, pulverized, peeled cloves. However, as will be discussed, some changes do occur during processing, depending on how the cloves are cut and dried. Perhaps the very best way to prepare a garlic powder with virtually no changes in composition is to freeze the cloves in liquid nitrogen followed by cold pulverization and vacuum drying. Although one manufacturer is known to produce such a powder, the procedure is time-consuming and very expensive.

Most commercial garlic powders are prepared by cutting the peeled cloves into smaller pieces, followed by oven drying at 50–60°C and then pulverization. Cutting the cloves before drying is necessary to make drying time practical; however, some conver-

Table 3.17. Comparison of Commercial Garlic Products Available in 1990 for Yield of Total Known Sulfur Compounds (see Lawson, 1993)

	Cysteines[a]		Thiosulfinates		Thiosulfinate Transformation Compounds[d]			Total[e]	S-Allyl[f] (μmole/g)
	γ-Glutamyl[b] cysteines	S-Alkyl[b] cysteines	Allicin	Others[c]	Dialkyl sulfides	Vinyl-dithiins	Ajoenes		
Product Type (n=brands)				(mg/g of clove or product)[g]					
Garlic cloves									
Fresh-picked	14.5 ± 3.1	nd[h]	3.6 ± 0.4	1.1 ± 0.2	nd	nd	nd	20.3	71
Store-purchased	9.4 ± 2.9	nd	3.7 ± 0.9	1.7 ± 0.4	nd	nd	nd	15.9	68
Powder tablets									
Typical (n=4)	12.1 ± 1.8	0.5 ± 0.2	2.9 ± 0.6	0.8 ± 0.5	0.5 ± 0.03	nd	nd	17.5	66
Range (n=9)	2–20		0.2–3.6	0.03–1.3	0.01–0.13	nd	nd		
Steam-distilled oils									
Typical (n=5)	nd	nd	nd	nd	4.4 ± 0.5	nd	nd	4.4	42
Range (n=9)	nd	nd	nd	nd	0.2 – 7.3	nd	nd		
Oil-macerates									
Typical (n=5)	nd	nd	nd	nd	0.16 ± 0.05	0.61 ± 0.1	0.10 ± 0.4	0.9	7
Range (n=9)	nd	nd	nd	nd	0.06–0.22	0.07–0.7	0.02–0.12		
Aged extract[i]	0.5 ± 0.2	0.6 ± 0.1	nd	nd	nd	nd	nd	1.4[i]	5

[a]Not included are the S-alkylcysteine sulfoxides (alliin, etc.), because they (except cycloalliin) are immediately converted to thiosulfinates upon contact with water. Cycloalliin is typically present at 1.1 mg/g in both cloves and powder tablets. See also note i.
[b]Includes the S-allyl, S-1-propenyl, and S-methyl derivatives.
[c]Includes all of the allyl, methyl, and 1-propenyl thiosulfinates shown in Figure 3.2, except allicin.
[d]Includes diallyl, allyl methyl, and dimethyl mono- to heptasulfides; 2-vinyl-4H-1,3-dithiin and 3-vinyl-H-1,2-dithiin; (E)-and (Z)-ajoene.
[e]Total listed compounds, plus the cycloalliin mentioned in notes a and i.
[f]Includes all compounds containing an S-allyl (thioallyl) group. Given as micromoles/g since some compounds have two S-allyl groups (see Section 3.4.7).
[g]Mean ± standard deviation for (n) number of brands. Typical values represent the highest 4–5 values.
[h]Not detected (nd). Limit of detection is 0.02 mg/g for the cysteines, 0.005 mg/g for the thiosulfinates, and 0.001 mg/g for the thiosulfinate transformation compounds.
[i]This product (only one brand found) is sold in liquid form and tablet form. Three lots of each form gave similiar results. Not shown are alliin and cycloalliin, which are present at 0.04 ± 0.03 mg/g and 0.24 ± 0.09 mg/g (n = 6), respectively.

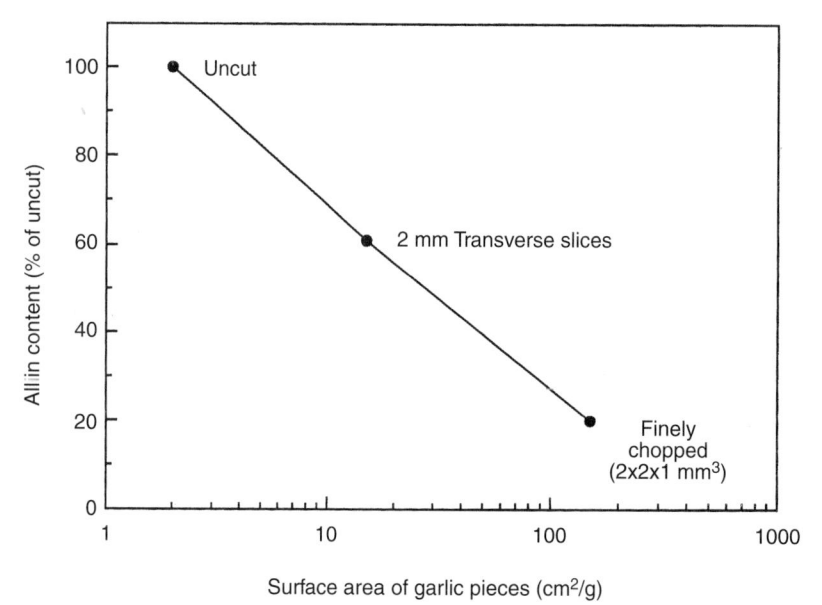

Figure 3.22. Effect of cutting size on the alliin content, or allicin potential, of garlic clove pieces. (From Lawson & Wang, 1994.)

sion of alliin and other cysteine sulfoxides to allicin and other thiosulfinates always occurs when nonfrozen cloves are cut, due to the release of alliinase. As shown in Figure 3.22, the larger the surface area to mass ratio of the chopped cloves (i.e., the smaller the pieces) the greater is the loss of alliin or allicin potential. In fact, a small number of garlic powders are prepared by spray drying of homogenized cloves—a method which causes complete loss of alliin. Hence, the clove pieces should be as large as possible before drying. Furthermore, the more alliin that is lost during cutting, the greater will be the odor of the resulting powder. The odor is due to the spontaneous conversion of the thiosulfinates to diallyl di- and trisulfides, both of which have a much stronger odor than allicin itself. Even though the sulfides are volatile and mostly lost to the atmosphere during the drying process, significant amounts do remain in the powder (Table 3.17).

An important concern in the drying of cut garlic is the effect of the oven temperature, particularly since alliinase activity is so important to the production of allicin. Table 3.18 compares the effects of drying sliced cloves in a freeze-dryer or in an oven at 50, 60, and 70°C on alliinase activity as well as on all the main sulfur compounds found in garlic (Lawson & Wang, 1994). Surprisingly, alliinase activity was found to be unaffected even at the highest temperature, although the activity of methyl-specific alliinase was slightly affected. The cysteine sulfoxides, except for methiin, were affected by oven drying, with a modest effect on alliin and a much larger effect on isoalliin; however, since the abundance of isoalliin is minor, this is not of serious consequence. Furthermore, the γ-glutamylcysteines were unaffected by drying at these temperatures. These results are in agreement with a previous, less thorough, study (Lawson & Hughes, 1992).

The long-term stability of sulfur compounds in garlic powder products has been found to be very high. Well-maintained garlic powder has been reported to lose only 10% of its allicin yield after 5 years of storage at room temperature (Pfaff, 1991; Lawson & Wang, 1994). Probably the most critical factor in retaining the allicin potential of garlic powder is its moisture content, which should not be allowed to go above 7% and should preferably be 4–6% (Müller & Ruhnke, 1993). The allicin potential of gar-

Table 3.18. Drying Temperature and Its Effect on the Composition and Activity of Garlic Clove Slices[a]

Drying temperature (24 hours):	0°C	50°C	60°C	70°C
Alliinase activity[b] (seconds)	< 10	< 10	< 10	< 10
Alliinase (methyl-specific) activity[b] (seconds)	45	45	52	75
Losses (% of 0°C value)				
Alliin (8.9 mg/g fresh weight)[c]		5	15	20 (13[d])
Methiin (1.2)[c]		< 2	< 2	< 2
Isoalliin (0.3)[c]		19	58	76 (38[d])
γ-Glu-S-1-propenylcysteine (4.4)		< 2	< 2	< 2
γ-Glu-S-allylcysteine (4.3)		< 2	< 2	< 2
γ-Glu-S-methylcysteine (0.3)		< 2	< 2	< 2

[a]Cloves from same bulbs were cut transversely into 3-mm slices, slices mixed, and dried on a screen in static air for 24 hours, followed by pulverization. The 0°C slices were lyophilized (freeze-dried).
[b]Time to reach maximum allicin (or methyl thiosulfinates) release. The reaction (20 mL water/g dry) was stopped at various times by 10mM amino-oxyacetate.
[c]Cysteine sulfoxides determined by calculation from content of all thiosulfinates.
[d]Loss at minimum time required to dry (4 hours).

lic powder tablets has also been found to be very stable (36% average loss, range 5–73%, in 5 years for four brands), but more variable than for the powders, probably as a result of different types of excipients employed in the tablets (Lawson & Wang, 1994). The content of γ-glutamylcysteines in tablets and gelatin-encapsulated powders has been found to decline by less than 5% in 4 years; however, in bulk powders there occurs a 20% loss of γ-glutamyl-S-trans-1-propenylcysteine, with no loss of the allyl isomer (Lawson & Wang, 1994), indicating oxidative loss of this compound which has been previously shown to be somewhat unstable (Lawson et al., 1991b).

The powders that are prepared for the spices typically found in grocery stores usually have a smaller allicin yield (3–5 mg/g powder) than those prepared for quality garlic powder tablets (6–14 mg/g), reflecting finer chopping (increases alliin loss) of the cloves to decrease drying time. The garlic salt products have an allicin yield of 0.3–0.7 mg/g. The spice powders have also been found to contain considerably higher amounts of γ-glutamyl-S-allylmercaptocysteine (γ-glu-cys-SS-allyl, reaction product of allicin with γ-glutamylcysteine) (0.4–1.0 mg/g) than those used for higher quality tablets (0.04–0.12 mg/g), which is likely due to the greater amount of allicin produced during processing and possibly to the use of higher drying temperatures. Similarly, the content of S-allylmercaptocysteine (0.2–0.3 mg/g) and S-allylcysteine (Section 3.2.7 and Table 3.17) in garlic powders is much higher than is found in fresh cloves (Table 3.3) (Lawson & Wang, 1994).

The quality of many of the various brands of garlic powder tablets has been compared in several publications (Lawson et al., 1991a; Winkler et al., 1992b; Pentz et al., 1992; Voigt & Wolf, 1986; Lawson, 1993; Hansen et al., 1993; Schardt & Liebman, 1995; Block, 1995). As shown in Tables 3.17 and 3.19, there is considerable variation in quality among brands as judged by the amounts of both allicin and γ-glutamylcysteines released. The quality of garlic powder tablets has been increasing in recent years. While the allicin yield of products available in the U.S. and Germany in 1990 (Table 3.17) did not exceed 3.0 mg/g, several brands now release well over 4 mg/g (Table 3.19). The values for the highest brands compare equally to or better than fresh garlic cloves, keeping in mind that tablets contain 30–60% excipients while cloves contain about 65% water.

An important standard measure of the quality of garlic powder tablets is their ability to form allicin from alliin (called allicin potential) when they become dissolved. Since this reaction is completely dependent upon active alliinase, and since it has been clearly demonstrated that alliinase is imme-

Table 3.19. Garlic Powder Tablets: Effective Allicin Yield and Yield of the Main Sulfur Compounds

Year Purchased -Allicin standardized?[a] -Acid protected?[b] -Tablet weight (g)	Effective Allicin Yield[c] (% of Maximum)	Yield in Water for Crushed Tablets (mg/g tablet)					
		Allicin	Total thiosulfinates	Alliin	Allicin/ alliin ratio	γ-Glutamyl-cysteines[d]	(allyl/1-propenyl ratio)
U.S. Brands							
1 (1994-Y-E-0.45)	4.6 (52)	8.9	12.5	23.8	0.37	17.3	(1.06)
2 (1994-Y-S-0.83)	5.5 (72)	7.7	9.5	19.2	0.40	26.0	(1.10)
3 (1995-U-E-0.80)	3.1 (55)	5.7	8.4	14.2	0.40	7.0	(0.88)
4 (1992-U-E-0.72)	5.2 (99)	5.2	7.3	13.9	0.37	5.9	(0.90)
5 (1995-Y-E-0.55)	4.7 (98)	4.8	7.2	13.4	0.36	15.7	(1.18)
6 (1995-U-E-0.43)	3.4 (91)	3.7	5.4	10.4	0.36	14.0	(1.04)
7 (1993-Y-S-1.00)	0.0 (<1)	3.4	4.6	8.9	0.38	20.2	(1.38)
8 (1995-Y-E-1.25)	2.9 (98)	3.0	4.2	7.8	0.39	14.1	(1.31)
9 (1995-U-S-0.40)	0.0 (12)	0.4	0.5	6.6	0.06	9.5	(1.38)
10 (1995-U-E-0.75)	0.5 (25)	1.8	3.1	4.9	0.37	16.8	(1.02)
11 (1992-U-N-0.90)	0.3 (14)	1.8	2.3	4.5	0.40	18.6	(1.35)
12 (1994-Y-E-0.80)	0.5 (30)	1.8	1.9	9.5	0.19	17.4	(1.18)
13 (1995-U-S-0.25)	1.5 (95)	1.6	2.2	4.2	0.38	3.2	(1.08)
14 (1995-U-N-0.65)	0.0 (<1)	1.6	2.0	3.9	0.41	12.2	(1.98)
15 (1993-Y-N-0.50)	0.0 (<1)	1.3	1.4	15.0	0.09	47.6	(1.37)
16 (1995-N-N-0.80)	0.0 (<1)	0.9	1.1	9.2	0.10	22.0	(1.16)
17 (1993-N-N-0.55)	0.0 (<1)	0.8	0.9	11.8	0.07	30.6	(1.14)
18 (1994-U-S-0.40)	0.0 (<1)	0.5	0.5	5.2	0.10	8.0	(1.11)
19 (1995-U-N-0.80)	0.0 (<1)	0.1	0.1	4.0	0.03	7.4	(0.98)
20 (1992-N-E-0.50)	0.0 (<1)	0.2	0.2	0.4	0.50	1.7	(1.42)
German Brands							
21 (1993-Y-N-0.40)	3.0 (76)	3.9	5.8	10.8	0.36	13.0	(1.55)
22 (1995-Y-S-0.25)	2.4 (81)	2.9	3.5	7.3	0.40	13.3	(2.02)
23 (1995-Y-N-0.53)	2.5 (99)	2.5	3.0	6.2	0.40	11.2	(2.13)
24 (1993-Y-N-0.50)	0.2 (10)	2.4	3.1	6.2	0.39	10.1	(2.88)
25 (1995-U-N-0.40)	3.4 (99)	3.4	4.1	8.4	0.40	12.1	(1.67)
26 (1994-N-N-0.25)	2.0 (99)	2.0	2.6	5.2	0.39	12.3	(2.32)
27 (1993-U-N-0..40)	0.1 (10)	1.0	1.2	2.4	0.42	4.0	(2.08)
28 (1995-Y-N-0.55)	2.4 (97)	2.5	3.0	5.9	0.42	10.1	(2.24)

[a]Y, yes; N, no; U, label claims allicin release but amount not stated.
[b]E, enteric-coating claimed; S, some unspecified type of protection claimed; N, no protection claimed.
[c]Allicin yield (mg/g tablet) under simulated gastrointestinal conditions. This was determined from the amount of unreacted alliin present (partially confirmed by the amount of allicin present, see text) after incubation (in a disintegration apparatus) of six tablets for 1 hour in simulated gastric fluid and 2 hours in simulated intestinal fluid at 37°C following standard disintegration conditions (U.S.P. 1990) that were modified so that the intestinal fluid was added to both the gastric fluid and the tablet remnants rather than adding only the remnants to the intestinal fluid. Hence, 300 mL gastric fluid (0.7% conc. HCl, 0.2% NaCl, 0.32% pepsin, final pH about 1.5), rather than 600 mL, was used, followed by addition of 300 mL concentrated simulated intestinal fluid that contained 2.04% KH_2PO_4, 0.48% NaOH, and 2% porcine pancreatin (to achieve a final pH of 6.7 after addition to gastric fluid).
[d]Sum of S-allyl- and S-1-propenyl-γ-glutamylcysteines.
From Lawson LD, Wang ZJ, unpublished data, 1994.

diately and irreversibly inactivated at the low pH levels usually found in the stomach (although a high protein steak meal can serve to activate alliinase, discussed in Section 3.2.5.1) (Blania & Spangenberg, 1991; Lawson & Hughes, 1992), it is vital that the pills be protected from stomach acid. This has been successfully achieved mainly by coating the tablets with enteric coatings, such as the cellulose esters, which require intestinal enzymes to be removed. Thus, the quality of a garlic tablet will depend not only on its alliin and active alliinase content, but also on the amount of allicin found in the intestines after a tablet has been consumed. While it is not possible to measure such a value, which would surely vary from person to person if it could be measured, an estimate of this *effective allicin yield* can be determined on the basis of standardized simulated gastrointestinal conditions that have been described in the U.S. Pharmacopeia (1990). These conditions and the results for the effective allicin yields of 28 brands of garlic tablets are provided in Table 3.19.

Because allicin is formed enzymatically, it was necessary to modify the U.S.P. method by adding the simulated intestinal fluid to both the remnants and the 1-hour gastric medium, rather than to the remnants alone, because the alliin released into the gastric fluid from many of the tablets could often be converted to allicin in intestinal fluid when tablets fully disintegrated and released additional alliinase. Thus, it was necessary to use an increased buffer concentration by 50% in the intestinal fluid to obtain a final pH near neutral (6.3–6.5). It was also found that the effective allicin yield was best measured from the amount of unreacted alliin (assayed by the method of Ziegler & Sticher, 1989) found in the medium after 1 hour in gastric fluid plus 2 hours in intestinal fluid, compared to the amount of alliin present in the tablets. This was accurate for tablets that completely disintegrated (released all of their alliin) by the end of the 3 hours, which was the case for all brands except two (brands 10 and 11 were only 30% and 50% disintegrated by the end of the 3 hours). Most brands were completely disintegrated within 1 hour after adding the intestinal fluid. The amount of allicin found at the end of the 3 hours was also measured but was found to be a poor representation of the amount of allicin formed, since significant amounts (10–30%) of this fairly volatile compound were lost mainly to evaporation during the hours of agitation in the warm and open environment of the disintegration apparatus.

As shown in Table 3.19, the percentage of effective allicin yields for the tablets, under simulated gastrointestinal conditions, varied from not detectable (less than 1%) to complete release and was not always predictable on the basis of the tablet coatings claimed. Generally speaking, those brands which claimed to be enteric coated (E) or to contain some type of gastric protection (S) and which claimed a specific allicin yield (Y) (brands 1, 2, 5, 8, 22) were the highest quality brands, on the basis of effective allicin yield. Exceptions were brands 3, 4, and 6 among U.S. products and brands 21, 23, 25, 26, and 28 among German products, all of which had moderate to high effective yields even though they met only one of these criteria, and brands 7 and 12, which had very low effective yields even though they met both of the criteria. Brands 14–20 had no effective allicin yield, as they completely disintegrated in the gastric fluid. This was expected for brands 15 and 17, since they were simply garlic powder in gelatin capsules, which easily dissolve in gastric fluid. Brands 8, 14, and 20 were softgel capsules containing garlic powder suspended in an oil; however, brand 8 was effectively enteric coated.

Not shown in Table 3.19 is the amount of alliin found in the gastric fluid after incubation for 1 hour. No alliin was found for most of the enteric-coated tablets (brands 1, 5, 6, 8, 10), but for three of the enteric coated brands (3, 4, and 12) 25%, 24%, and 80%, respectively, of their alliin was found in the gastric medium, indicating that some enteric coatings are of poor quality. Two brands did not claim to be enteric coated (21 and 26) but acted as if they had been, since less than 2% of their alliin was found in the gastric fluid. Interestingly, a lack of enteric coating (the absence of which usually allows entrance of gastric acid into the tablet to inactivate alliinase) does not necessarily mean a poor effective allicin yield if the alliinase is protected in other ways. For example, although brands 11 and 22–25 released about 50% of their alliin into the gastric fluid, brands 22, 23, and 25 had enough active alliinase remaining in the tablet remnants to convert nearly all of this alliin to allicin after addition of the intestinal fluid. The high effective allicin yield of brand 22 is important to note, since this brand has been extensively used in clinical trials on serum cholesterol reduction (see Section 5.1.1.2).

Of course, the actual amount of allicin released in vivo will depend upon how much time the tablets actually spend in the stomach, upon whether the tablets are consumed with or without food, and upon the protein (a buffer) content of the meal; the pH of an empty stomach, about 1.5, increases to about 3.0 (alliinase inactive) after consumption of a light meal and to as high as 4.5 (alliinase active) with a steak meal, within 30 minutes of consuming the meals (Davenport, 1982; Goldschmidt & Feldman, 1993). Consequently, a few of the brands in Table 3.19 were also tested for effective allicin yield with a gastric pH that had been adjusted to 3.0. It was found that if the allicin release after a gastric pH of 1.5 was moderate to high (50% or more) (e.g., brands 1, 21, and 22), then a gastric pH of 3 made little difference. However, if the effective allicin release was very low (less than 15%) (e.g. brands 11 and 24), then a gastric pH of 3 resulted in a major improvement (to about 50%); however, much of this improvement may be due to the fact that the dissolved tablets caused the unbuffered gastric pH to increase to 4.4–5.0, a range where alliinase is very active (Fig. 3.8). In the stomach, however, this much of an increase in pH due to the dissolved tablets is not likely to occur, since the stomach continually secretes hydrochloric acid during digestion. Therefore, it is recommended that many of the brands be consumed with or just after a meal, and it is recommended that manufacturers prepare tablets on the basis of effective allicin release, using the standard gastric pH of about 1.5 (USP XXII, 1990).

Another cause for poor allicin formation, aside from the action of gastric acid, is the initial lack of active alliinase in the tablets. Alliinase activity is indicated by the ratio of allicin to alliin, determined when crushed tablets are incubated in water (30 minutes) or in a solution of alliinase inhibitor (Section 3.2.5.1). If all of the alliin were converted to allicin, this ratio would be 0.46, but since allyl methyl and allyl 1-propenyl thiosulfinates are also formed from alliin, the actual ratio for garlic cloves is found to vary from 0.32 to 0.42 (the range found for 90 strains of garlic). Ratios below this range indicate damaged alliinase. For the 28 brands analyzed and shown in Table 3.19, six brands (9, 15–19) had severely inactivated alliinase (allicin/alliin ratio of 0.03–0.10), while a seventh (12) was moderately inactivated (ratio of 0.19). The cause of alliinase inactivation in these particular brands is not known, but may be due to excessive drying temperatures, perhaps well above 70°C, or excessive drying times. The inactivation of alliinase appears not to be due to any interaction with excipients present in the tablets, since the addition of active alliinase to the tablet solutions resulted in the rapid and complete formation of allicin.

With respect to the accuracy of the allicin yield claims, the values found (based on crushed tablets) for the 12 brands (Y) making claims for specific amounts of allicin yield were found to meet or exceed claims, except for brands 12 and 21 which were deficient by about 40%. When the effective allicin yield is considered, only brands 5, 8, 22, 23, and 28 were able to meet their claims. For the 12 brands (U) that made unspecific claims of high allicin yield ("maximum allicin potential," "allicin rich," "allicin guaranteed," etc.), only four (brands 3, 4, 6, and 25) had at least moderate (2.0 mg/g or more) allicin yields for crushed tablets, all four of which also had at least moderately effective allicin yields.

The content of γ-glutamylcysteines (*S*-allyl plus *S*-1-propenyl) in garlic powder tablets was also found to be highly variable and to be approximately correlated with the alliin content (Table 3.19). The ratio of γ-glutamylcysteines to alliin ranged from 0.4 to 3.4 (1.8 ± 1.0) in the U.S. products, but was more consistent in the German products, where the ratio ranged from 1.2 to 2.4 (1.7 ± 0.4). Because of such variability, there is little likelihood of being able consistently to standardize a garlic product on the basis of the γ-glutamylcysteines content while also standardizing the allicin yield.

3.4.2 Oil of Steam-Distilled Garlic ("Essential Oil")

Steam-distilled garlic oils (see also Sections 3.2.1, 3.2.6.3.1, 3.3.4, 4.2.4) have been prepared for at least 150 years. For historical reasons, the oil produced by steam distillation of a plant is usually designated as its essential oil, assuming that the distillation did not cause any chemical changes. For garlic, however, this is an erroneous term because the compounds present in the oil of steam-distilled garlic, allyl sulfides, are not present in whole or crushed garlic cloves but are produced from the thiosulfinates upon steam treatment of crushed garlic. As mentioned previously, garlic contains only minute amounts of oil (Section 3.3.4). While the term "essential oil of garlic" usually refers to the product of steam distillation, it has also been occasionally used to refer to ether extracts of crushed cloves (Bordia et al., 1974, 1975a, 1977; Bordia, 1981).

True steam-distilled oil of garlic contains exclusively (about 98%) allyl, methyl, and 1-propenyl mono- and polysulfides. However, commercial products sold for human consumption are diluted about 200-fold with various vegetable oils such that the final composition of sulfides represents about the same amount of allicin and other thiosulfinates present in the same weight of crushed garlic. This dilution also stabilizes the polysulfides and greatly decreases the extremely strong odor of the undiluted oil. The composition of typical commercial steam-distilled garlic oil products is provided in Table 3.20 (see also Fig. 4.3). The dominant compounds are the diallyl di-, tri-, and tetrasulfides and the allyl methyl di-, tri-, and tetrasulfides, which comprise 86% of the total sulfides. As shown by the small standard deviations, surprisingly little variation was found in the percentage composition among the various brands tested; however, as shown in Table 3.21, a 50-fold variation was found in the total content of the sulfides among 15 brands, reflecting mainly various degrees of dilution.

The stability of steam-distilled garlic oil products is an important concern, especially since it is known that undiluted allyl sulfides gradually disproportionate to form higher sulfides, which is the reason that commercial diallyl disulfide typically contains about 10–15% diallyl tri- and tetrasulfides. A 5-year stability study on four brands of steam-distilled oil products in gelatin capsules (Table 3.22) showed virtually no change in the amount of the main allyl sulfides. The largest decrease was an 11% decrease for allyl methyl disulfide, which may be due to diffusion through the capsule rather than to instability, since it is the smallest of these sulfides.

3.4.3 Oil of Oil-Macerated Garlic

Maceration or grinding of garlic cloves in a common vegetable oil, followed by isolation of the clear oil, is used to prepare a product that is much more common in Europe than in the United States (see also Sections 3.2.6.3.3 and 4.2.5). However, a similar product, bottled chopped or crushed garlic suspended in an oil, is not uncommon in U.S. grocery stores, and as such has probably been used for centuries, particularly where plant or fish oils were available. As mentioned elsewhere, this oil contains some unique transformation compounds of allicin and other allyl thiosulfinates that are not found in any other type of garlic product: the ring-structured vinyldithiins and the oxygenated ajoenes.

The composition of typical commercial oil-macerates of garlic is provided in Table

Table 3.20. Composition of Typical Commercial Steam-Distilled and Oil-Macerated Garlic Products

Compounds	Steam-Distilled Products μg/g[a]	%[b]	Oil-Macerated Products μg/g[a]	%[b]
Diallyl monosulfide	87 ± 64	2.0	n[c]	
Diallyl disulfide	1134 ± 94	25.8	34 ± 17	3.9
Diallyl trisulfide	810 ± 96	18.4	65 ± 16	7.5
Diallyl tetrasulfide	356 ± 43	8.1	n	
Diallyl pentasulfide	94 ± 2	2.1	n	
Diallyl hexasulfide	18 ± 2	0.4	n	
Allyl methyl monosulfide	39 ± 32	0.9	n	
Allyl methyl disulfide	549 ± 97	12.5	n	
Allyl methyl trisulfide	666 ± 146	15.2	58 ± 19	6.7
Allyl methyl tetrasulfide	263 ± 54	6.0	n	
Allyl methyl pentasulfide	74 ± 17	1.7	n	
Allyl methyl hexasulfide	14 ± 3	0.3	n	
Dimethyl monosulfide	n			
Dimethyl disulfide	56 ± 10	1.3	n	
Dimethyl trisulfide	147 ± 56	3.3	n	
Dimethyl tetrasulfide	59 ± 19	1.3	n	
Dimethyl pentasulfide	16 ± 3	0.4	n	
Dimethyl hexasulfide	5 ± 3	0.1	n	
Allyl 1-propenyl disulfide	24 ± 4	0.5	n	
Allyl 1-propenyl trisulfide	13 ± 3	0.3	n	
2-Vinyl-4H-1,3-dithiin	n		435 ± 69	50.4
3-Vinyl-4H-1,2-dithiin	n		167 ± 28	19.4
E-ajoene	n		68 ± 28	7.9
Z-ajoene	n		36 ± 16	4.2
Total sulfides	4390 ± 502		845 ± 98	

[a] Mean ± standard deviation per gram of product for five-six brands.
[b] Weight percent of total sulfur compounds.
[c] n = not detected. Limit of detection is 5 mg/g product.
From Lawson LD, Wang ZJ, Hughes BG. Identification and HPLC quantitation of the sulfides and dialk(en)yl thiosulfinates in commercial garlic products. Planta Med 1991; 57: 363–370.

3.20 (see also Fig. 4.4). Little variation has been found in the percentage composition among the various brands; however, considerable variation has been found in the concentration of these allicin transformation products among the many brands tested (Table 3.17, with some more recent values in Table 3.21). On the other hand, Winkler et al. (1992b) analyzed six German brands and found much less variation in the amounts of total sulfur compounds (0.72–1.7 mg/g product). Of course, the concentrations of these compounds will depend not only on the quality of the garlic cloves used, but also very much on the ratio of vegetable oil to garlic used in the processing, in addition to any further dilution. If a garlic bulb is homogenized with an equal weight of oil, which probably represents the highest concentration achievable, one would expect to find a concentration of total thiosulfinate transformation products of about 3–5 mg/g; however, such a high concentration has been found only in two recent brands of which the author is aware. An excellent study on the changes that take place when garlic is macerated in oil or other media under a variety of conditions has been conducted by Iberl et al. (1990a). This study showed that the yield of transformation products could be increased about 35% by crushing the cloves first, to produce thiosulfinates more efficiently, before the addition of the oil (Table 3.23). It is interesting that alliinase still has reasonably good activity even when the oil is added prior to maceration, which probably reflects the great speed with which alliinase operates once a clove is chopped, rather than the ability of alliinase to function in an oil environment.

Table 3.21. Content Comparisons Among Brands of Commercial Garlic Oil Products (update of Lawson et al., 1991a)

	Composition (mg/g product)			
	Vinyl dithiins	Ajoene	Allyl sulfides[a]	Total
Steam-distilled garlic oils				
A (U.S.)	n[b]	n	10.9	10.9
B (U.S.)	n	n	7.4	7.4
C (U.S.)	n	n	6.5	6.5
D (U.S.)	n	n	4.9	4.9
E (Germany)	n	n	4.7	4.7
F (Germany)	n	n	4.6	4.6
G (Germany)	n	n	4.0	4.0
H (U.S.)	n	n	3.7	3.7
I (U.S.)	n	n	3.5	3.5
J (India)	n	n	1.9	1.9
K (England)	n	n	1.8	1.8
L (England)	n	n	1.4	1.4
M (U.S.)	n	n	0.6	0.6
N (U.S.)	n	n	0.3	0.3
O (U.S.)	n	n	0.2	0.2
Oil-macerated garlic				
A (U.S.)	4.7	1.06	0.26	6.0
B (U.S.)	3.3	0.54	0.30	4.1
C (Sweden)	0.69	0.12	0.16	0.96
D (Germany)	0.67	0.09	0.17	0.93
E (Switzerland)	0.63	0.11	0.15	0.89
F (Germany)	0.50	0.12	0.15	0.76
G (Austria)	0.49	0.08	0.17	0.74
H (Germany)	0.43	0.09	0.10	0.62
I (Germany)	0.28	0.05	0.23	0.55
J (U.S.)	0.28	0.02	0.06	0.37
K (Austria)	0.07	0.02	0.15	0.24

[a] Also includes the very small amount of dimethyl sulfides.
[b] n = none detected (< 0.005 mg/g).

Table 3.22. Five-Year Stability of the Main Allyl Sulfides in Commercial Steam-Distilled Garlic Oil Products

	AAS2	AAS3	AAS4	AMS2	AMS3	AMS4
Initial Values (µg/g)	400–2100	350–1470	145–540	140–610	300–1200	130–400
Decrease (%)	< 2	< 2	4	11	< 2	5

All samples (four brands in gelatin capsules) were stored in the dark at 23°C and analyzed at time of purchase and again 5 years later. Abbreviations: AAS2, AAS3, AAS4 (diallyl di-, tri-, and tetrasulfides); AMS2, AMS3, AMS4 (allyl methyl di-, tri-, and tetrasulfides). Samples were analyzed according to Lawson et al., 1991a. From Lawson LD, Wang ZJ, unpublished data, 1994.

The stability of the compounds present in oil-macerates has been tested. The general stability of one product was reported to be at least 3 years (Voigt & Wolf, 1986). The results of a recent study of the stability of the individual vinyldithiins as well as of ajoene for several brands for up to 5 years is presented in Table 3.24. Overall, the vinyldithiins were very stable, especially the 1,2-isomer, while ajoene was considerably less stable. Protection from air in a gelatin capsule greatly increased the stability of both ajoene and 1,3-vinyldithiin, compared to storage in a tightly sealed half-full bottle. Nevertheless, even in gelatin capsules, ajoene consistently decreased at a rate of about 1.5%/month at room temperature until none remained by 5 years. Therefore, it is recommended that oil-macerate products which are intended to retain their ajoene concentration for more than 12–18 months be kept refrigerated. The loss of ajoene over time is surely due to degradation, since it is much less volatile than the vinyldithiins; however, no increase in any identified or unidentified compounds in the C18 HPLC chromatograms has been observed.

3.4.4 Ether-Extracted Oil

Extraction of the thiosulfinates of crushed garlic with ether, followed by evaporation of the ether, has been used to produce an oil from garlic that has a composition very similar to that of the oil-macerates, consisting chiefly of vinyldithiins, ajoene, and diallyl sulfides. Although this type of product is probably not commercially available, it has been used in several animal studies and several clinical studies to lower serum cholesterol levels. It has an advantage over the oil-macerates in that the solvent is removable, thus allowing for far greater concentrations of the relevant compounds. In practice, such a product is usually diluted with glycerin or a vegetable oil.

Following the same procedure described by Bordia et al. (Bordia et al., 1975a, 1975b) for his many animal and clinical studies with this type of oil, a study was conducted on the time course of the transformation of allicin in ether (Lawson & Wang, 1994). As shown in Figure 3.23, approximately half of the allicin is converted in 10 hours to vinyldithiins and lesser amounts of ajoene, while complete conversion required about 4 days. In contrast to incubation in oil (Iberl et al., 1990a), no diallyl sulfides were formed. The actual composition of this type of oil, however, depends on how long it is maintained in the ether. Once the ether is removed and the undiluted oil obtained, production of ajoene from any previously unreacted allicin will proceed at a greater rate (equal to that of 1,3-vinyldithiin), as discussed in Section 3.2.6.3.2, along with production of a small amount of diallyl trisulfide. Analysis of actual capsules used in Bordia's studies revealed that the content of vinyldithiins was 5.7 mg/g, ajoenes 1.0 mg/g (after correcting for loss in storage), and allyl di- and trisulfides 1.4 mg/g, for a total of 8.1 mg/g, which is considerably higher than for most oil-macerates.

Table 3.23. Maceration of Garlic in Linseed Oil: Percent Transformation of Allicin

Time (hours)	0	6	24
Allicin (crushed cloves)	0.45	0.23	0
Allicin (whole cloves)	0.33	0.13	0.02
1,3-vinyldithiin (crushed)	0	0.10	0.17
1,3-vinyldithiin (whole)	0	0.08	0.13
Ajoenes (crushed)	0	0.07	0.10
Ajoenes (whole)	0	0.03	0.05
Diallyl disulfide (crushed)	0	M	0.04
Diallyl disulfide (whole)	0	0.04	0.05

Cloves were either crushed 30 minutes prior to addition of the oil or were macerated as whole cloves in the oil. Values represent percent of fresh weight. From Iberl B, Winkler G, Knobloch K. Products of allicin transformation: ajoenes and dithiins, characterization and their determination by HPLC. Planta Med 1990; 56: 202–211.

Table 3.24. Stability of the Vinyldithiins and Ajoene in Commercial Garlic Oil-Macerates

	1,3-Vinyldithiin	1,2-Vinyldithiin	Ajoene
Study 1 (single high sulfur brand)			
Initial values (µg/g)	2400	720	550
Gelatin capsules			
Decrease in 18 months (%)	3	0	22
Decrease in 37 months (%)	8	1	63
Bottle (half-full)			
Decrease in 18 months (%)	31	1	86
Decrease in 37 months (%)	53	4	92
Study 2 (4 typical brands)			
Initial values (µg/g)	150–490	95–190	45–115
Decrease in 60 months (%)	24 (range, 5–64)	0–2	> 98

All samples were stored in the dark at 23°C. Study 1 used the same oil in both capsules and a half-full bottle and was compared to samples stored at −70°C. Study 2 employed four European brands in gelatin capsules. Samples were analyzed by C18HPLC according to Lawson et al., 1991a. From Lawson LD, Wang ZJ, unpublished data, 1994.

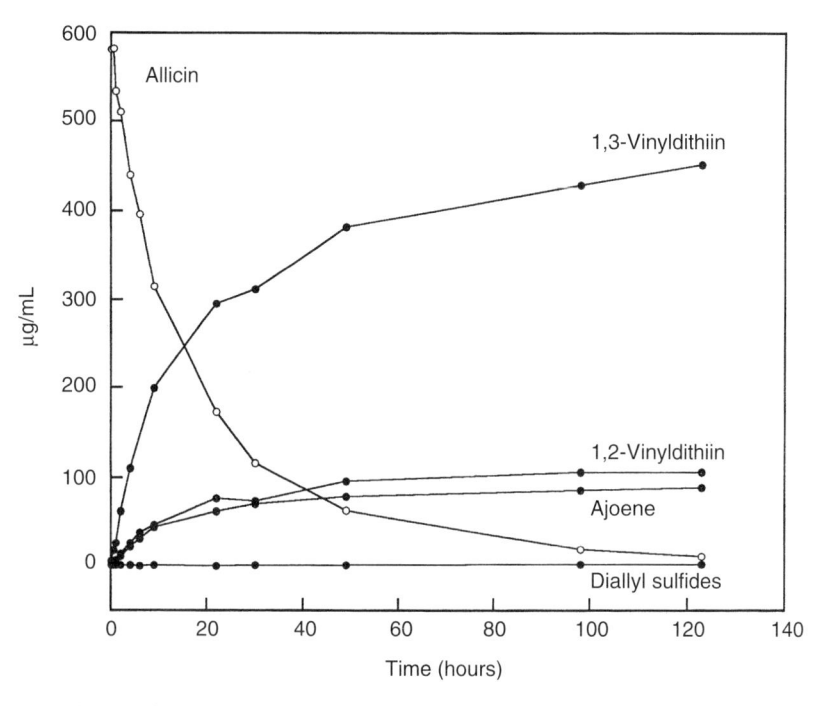

Figure 3.23. Incubation of garlic thiosulfinates (mainly allicin) in ether at 23°C.

3.4.5 Extract of Garlic Aged in Dilute Alcohol

A garlic product that is commonly sold in Japan, the U.S., and parts of Europe is chopped garlic that has been incubated or aged in 15–20% ethanol for 18–20 months at ambient temperature (Nakagawa et al., 1989; Hirao et al., 1987; Amagase & Milner, 1993). The incubation medium or extract is then filtered and concentrated to dryness and is sold in both dry (tablets and powder capsules) and liquid forms. The liquid form contains 10% (w/v) ethanol, as confirmed in several recent (1994–1995) lots.

Only recently, however, have the chemical changes been studied that occur when garlic extracts are aged for such long periods of time (Lawson & Wang, 1995). This study determined the content of sulfur compounds and free amino acids present in both 20% ethanol and water extracts of finely chopped (2x2x1 mm pieces) or transversely sliced (2 mm thick slices) garlic cloves when incubated up to 24 months. The results for chopped garlic in 20% ethanol (12 mL/g) are provided in Table 3.25. Nearly identical results were found at 3 mL/g, as well as for incubation in water, indicating that the alcohol has no effect, except perhaps to inhibit bacterial growth. As shown in the table, allicin and other thiosulfinates were nearly depleted by 90 days, having been transformed to diallyl and allyl methyl tri-, di- and tetrasulfides; however, only trace amounts of the sulfides were actually found in the extract, since they had diffused into the atmosphere through the polyethylene lids that were used, but they were found (45% of expected yield) in the containers of an auxiliary study that used less permeable Teflon-lined Bakelite lids.

Unexpectedly, considerable amounts of alliin were also found in the extracts. This was the result of rapid diffusion of alliin (complete in 1 day) through the pieces into the medium, with little diffusion of the much larger alliinase. At about 30 days all alliinase activity ceased and the alliin content remained stable. In the study using transverse slices, where the surface area to mass ratio was much smaller (resulting in much less alliinase release), four times as much

Table 3.25. Compositional Changes During Aging of a 20% Ethanol Extract of Chopped Garlic[a]

Compound	1	5	30	90	360	720	Commercial Aged Extract Products[b]
			(mg/g dry extract)				
Allicin	8.3	7.9	4.1	0.43	0	0	0
Allyl methyl thiosulfinates	2.1	1.9	1.3	0.35	0	0	0
Alliin	5.0	4.7	3.2	2.8	2.9	2.7	0.32 ± 0.13
Cycloalliin	3.5	3.6	3.6	3.5	4.0	3.6	0.34 ± 0.08
γ-Glutamyl-S-allylcysteine	12.7	11.7	5.8	1.10	0	0	0.25 ± 0.12
γ-Glutamyl-S-1-propenylcysteine	15.9	13.2	3.4	0.5	0	0	0.09 ± 0.07
S-Allylcysteine	0.2	0.9	5.9	7.2	7.1	7.2	0.62 ± 0.07
S-1-Propenylcysteine	0.5	1.4	6.7	8.1	6.5	4.4	0.37 ± 0.04
S-Allylmercaptocysteine	0.01	0.16	0.64	1.2	1.7	1.9	0.14 ± 0.03
Cystine	0.07	0.18	0.47	0.83	0.94	1.2	0.01 ± 0.01
Total sulfur compounds[c]	48.3	45.6	34.1	26.0	23.1	21.0	2.14
Glutamic acid	1.1	3.2	9.7	14.2	15.8	16.2	1.2 ± 0.2
Arginine	25	26	27	28	30	33	2.2 ± 1.1
Total common free amino acids	39	45	54	58	66	70	6.4 ± 2.3

Age of Garlic Extract (days): 1, 5, 30, 90, 360, 720

[a]Whole cloves were chopped into small pieces (2x2x1 mm) and placed into a 20% ethanol solution (12 mL/g) in a closed container and stored at room temperature, with samples being removed at the indicated times and analyzed by published methods (Ueda et al., 1990, Ziegler & Sticher, 1989, Lawson et al., 1991a, 1991c).
[b]Mean ± standard deviation (mg/g product) for six lots (purchased 1994 at local stores) of powder capsules or tablets. The product labels state that they were aged for 20 months (600 days) and contain only the extract, whey, and magnesium stearate.
[c]Not included are protein and trace amounts of γ-glutamyl-S-allylmercaptocysteine and trace amounts of allyl sulfides.
From Lawson LD, Wang ZJ, unpublished data, 1994.

alliin was found in the extract than is shown in Table 3.25. Cycloalliin, which is not affected by alliinase and is not contained in protein, remained constant throughout the study.

Perhaps the most dramatic change that occurred was the complete and quantitative hydrolysis of γ-glutamyl-S-allylcysteine and γ-glutamyl-S-1-propenylcysteine to S-allylcysteine and S-1-propenylcysteine in about 90 days (Fig. 3.24). The hydrolysis appears to be caused by a γ-glutamylpeptidase, since the process is inhibited by 5% acetic acid and since there is a much larger increase in free glutamic acid than in the other amino acids (Table 3.25). The S-allylcysteine formed was stable throughout the remainder of the 2 years, but S-1-propenylcysteine gradually declined to about 55% of its maximum level. Two other sulfur compounds that were initially absent in the fresh cloves were also found to increase greatly: cystine and S-allylmercaptocysteine (allyl-SS-cysteine). They are likely the result of cysteine's being released from protein hydrolysis, which occurred throughout the study, followed by oxidation to cystine or reaction with allicin to form S-allylmercaptocysteine, respectively.

Table 3.25 indicates the approximate values of sulfur compounds and amino acids that can be expected in the dried aged extracts of a typical strain of garlic, depending upon the amount of time the extracts have been aged. While the values for alliin will depend greatly upon the size of the garlic pieces, this is not a factor for the other compounds. Since S-allylcysteine and S-1-propenylcysteine are the most abundant sulfur compounds in the extracts, and since the S-allylcysteine content is used to standardize commercial aged extracts, the quality of these products will depend upon the content of γ-glutamyl-S-allylcysteine initially present in the garlic cloves. The content of γ-glutamyl-S-allylcyteine in 21 strains of garlic (Table 3.5) was found to be 3.7 ± 1.6 (range, 1.6–6.8) mg/g fresh weight. These will produce an S-allylcysteine content of 6.1 ± 2.7 (range 2.7–11.3) mg/g dry weight (92% of the dry weight of garlic is soluble in 20% ethanol or water), which approximately represents the amount that should be found in commercial products that are not diluted with excipients.

However, as shown in Table 3.17 (note i) for 1990 purchases (0.30 mg/g) and in Table

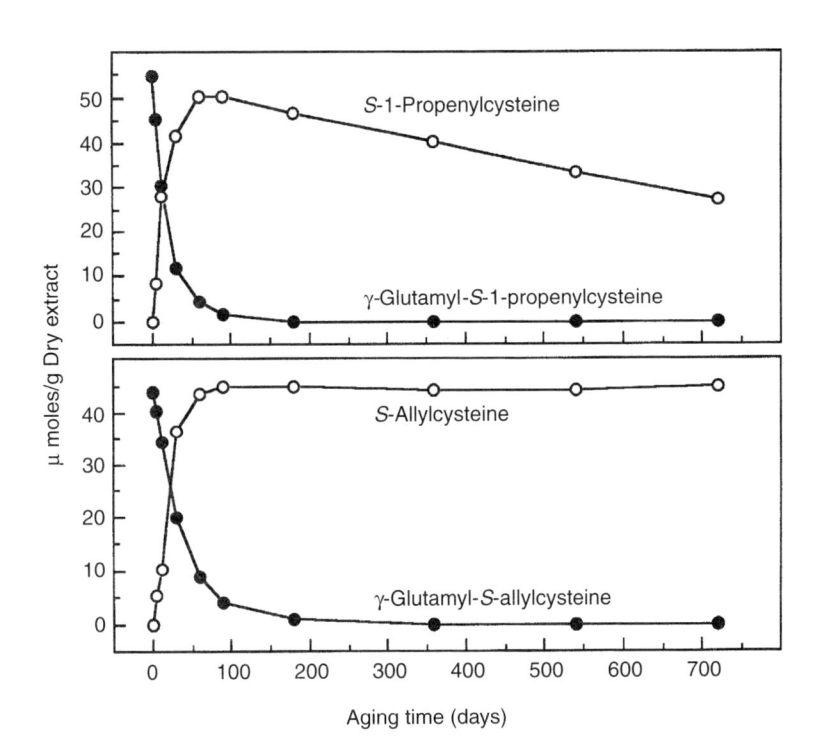

Figure 3.24. Transformation of γ-glutamyl-S-alkenylcysteines to S-alkenylcysteines during aging of an extract (20% ethanol) of chopped garlic. (After Lawson LD, Wang ZJ. Changes in the organosulfur compounds released from garlic during aging in water, dilute ethanol, or dilute acetic acid. J Toxicology 1995;14:214.)

3.25 for 1994 purchases (0.62 mg/g), the amounts of S-allylcysteine found in commercial dry aged garlic extracts are much smaller than expected. Others have reported similar and lower values: 0.46 mg/g powder (Amagase & Milner, 1993); 0.13 mg/capsule (0.26 mg/g powder) and 0.37 mg/mL liquid (0.33 mg/g) (Hansen et al., 1993); and 0.25 mg/capsule (0.53 mg/g) (Schardt & Liebman, 1995; Block, 1995). The alliin content of commercial aged extracts has been found to vary from less than 0.02 mg/g (Hansen et al., 1993) to 0.04 mg/g (Table 3.17) for 1990 purchases, to 0.32 mg/g (Table 3.25) for 1994 purchases. Small amounts of γ-glutamyl-S-allylmercaptocysteine (0.04 ± 0.02 mg/g powder) have also been found (Lawson & Wang, 1994). Only trace levels of the allyl sulfides have been found in commercial extracts (Table 3.17)—about 0.9 ppm diallyl sulfide, 0.3 ppm diallyl disulfide, and 0.5 ppm diallyl trisulfide, which are in agreement with published values for a special aged extract concentrate (Weinberg et al., 1992, 1993).

The total sulfur content of the liquid form of the commercial aged extract was found to be 0.091 ± 0.003% for three recent lots, in contrast to the 0.35% (1.0% dry weight) typically found in garlic cloves or garlic powders (Alfonso & Lopez, 1960; Pentz et al., 1990; Ueda et al., 1991). However, the dry aged extract product was found to contain much more sulfur (0.36 ± 0.03% for seven 1994 lots) than the liquid form, but most of this sulfur probably originates from added excipients, since both the liquid and dry forms contain very similar amounts of all the compounds listed in Table 3.25. The total nitrogen content of the liquid aged product was found to be only 0.23 ± 0.01% (three 1994 lots), compared to 1.1% (3.3% dry weight) for garlic cloves or garlic powders (Ueda et al., 1991; Lawson, 1993). However, again the dry extract product was found to contain much more nitrogen (5.5 ± 0.6% for the seven lots) than the liquid form, indicating that most of this nitrogen came from the added excipients. The fructan content

(Section 3.3.1) of the aged products was found to be 8.3 ± 1.2% and 9.2 ± 1.0% (four lots each) for the dry and liquid forms, compared to 22 ± 4% (65 ± 12% dry weight) for garlic cloves and powders. The high fructan content of garlic appears to decrease by hydrolysis during aging, since the commercial aged extracts (both forms) had a fructose/fructan ratio of about 0.33, compared to the ratio of 0.011 found in cloves and powders (Lawson & Wang, 1994).

3.4.6 Garlic Aged ("Pickled") in Dilute Acetic Acid

One of the oldest methods of preparing and preserving garlic is that of storage of cloves in vinegar (about 5% acetic acid), a method that is called pickling. This is still a very common food form of garlic in various parts of the world, especially in China and Russia, and is not uncommonly found in U.S. grocery stores. Pickling garlic decreases its pH to about 3.5, resulting in inactivation of most of its enzymes. Hence, pickled garlic yields no allicin upon chewing and has a content of S-allyl compounds similar to what one finds in fresh cloves. In a recent study on pickling whole cloves in 2% acetic acid, allinase activity, measured as allicin release upon homogenization, decreased by 70% in 10 days and was absent after 60 days of storage. The pungency of the cloves also declined in a similar pattern (Kim et al., 1994).

Figure 3.25 shows the changes that take place over time in the sulfur compounds when garlic powder is incubated in 5% acetic acid (Lawson & Wang, 1995). Because of enzyme inactivation, the γ-glutamylcysteines, especially the S-allyl homolog, are much more stable in 5% acetic acid than in dilute alcohol or water (compare Fig. 3.24). On the basis of half-lives, the γ-glutamyl-S-allylcysteine is about 20 times more stable, while γ-glutamyl-S-1-propenylcysteine is only about five times more stable, again showing that S-1-propenyl compounds are more subject to chemical reactions. Further

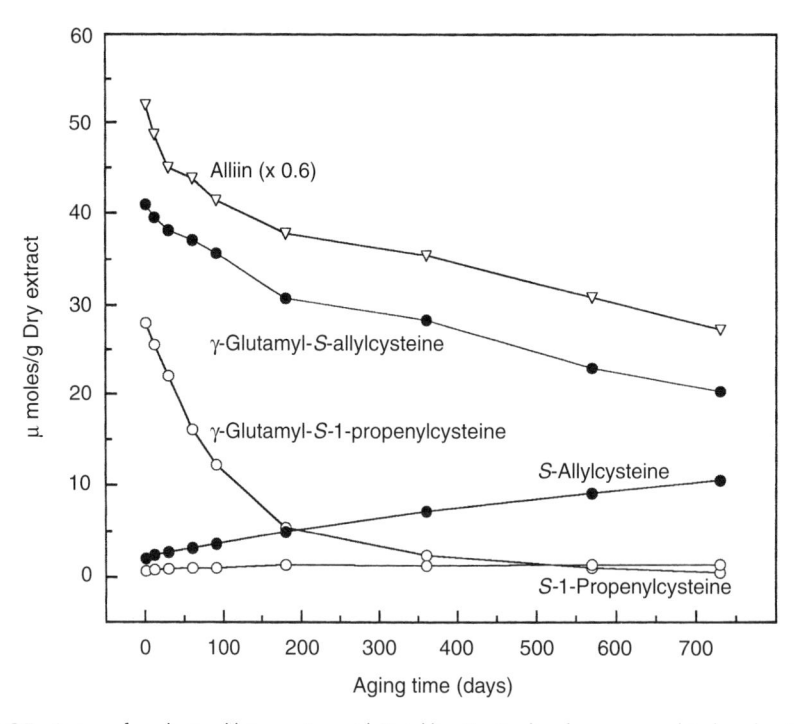

Figure 3.25. Aging of garlic in dilute acetic acid ("pickling"). Garlic cloves were dried, pulverized, and incubated in 5% acetic acid (12 mL/g clove) and the medium analyzed as described in Figure 3.24. Similar amounts are found in pickled whole cloves. (From Lawson & Wang, 1994.)

evidence that the acetic acid is causing some unique changes is the fact that no S-1-propenylcysteine was found and only around half of the expected amount of S-allylcysteine was found. This is in contrast to the results upon aging in dilute alcohol or water, where the content of S-alkenylcysteines was found to correspond quantitatively to the losses in γ-glutamylcysteines.

Interestingly, alliin was found to decline at a rate very similar to that of γ-glutamyl-S-allylcysteine to unknown degradation products, although allyl alcohol and cysteine would be the expected products. However, cycloalliin remained constant throughout the 2 years, while glutamic acid increased only threefold and total common free amino acids increased only about 8% (not shown), indicating very little protein hydrolysis activity. While the results shown in Figure 3.25 are for the incubation medium, the same results apply to the content of pickled garlic cloves because these water-soluble compounds equilibrate fairly quickly with their environment. For example, when commercial pickled whole cloves were analyzed 60 days after they were prepared (155 g cloves with 85 mL vinegar), the same concentrations of sulfur compounds were found in the cloves as in the vinegar. Thus, the true content of these compounds in pickled garlic will depend not only upon the age of the preparation, but also upon the relative volume of vinegar in which the cloves sit.

3.4.7 Total Thioallyl Compounds

It has been proposed that the quality of all types of garlic products should be compared by their content of total thioallyl (S-allyl) compounds (Lin, 1991; Lee et al., 1994). Although some support for this proposal is found in the observations that the thioallyl moiety has been shown to have important specific activity (e.g., S-allyl compounds are more anticarcinogenic than S-propyl or O-allyl compounds [Sundaram & Milner, 1995]) and that several of the S-allyl compounds from garlic products are metabolized in blood to allyl mercaptan (Lawson & Wang, 1993), the proposal nevertheless has some serious faults. Most importantly, the biological activity of each of the thioallyl compounds of garlic is not known, and for those where it is known, there are vast differences in activity. For example, the most abundant S-allyl compounds of crushed garlic are allicin and γ-glutamyl-S-allylcysteine, yet allicin is a very strong antibiotic while γ-glutamyl-S-allylcysteine has no antibiotic activity. On the other hand, γ-glutamyl-S-allylcysteine has been shown to have good activity in vitro toward inhibiting angiotensin-converting enzyme, while allicin has no such activity (Elbl, 1991); and diallyl trisulfide is five times as active as diallyl disulfide toward the inhibition of platelet aggregation in vitro, while S-allylcysteine has no activity here (Lawson et al., 1992).

However, for interest's sake the content of total thioallyl compounds among the various types of garlic products is compared in Table 3.17. Generally, the thioallyl content parallels the content of total sulfur compounds for the various types of products, although the proportion for steam-distilled oils compared to garlic cloves is considerably higher because a large portion (about 30%) of the sulfur compounds in garlic cloves is present as γ-glutamyl-S-1-propenylcysteine, yet only about 1% of the compounds in the steam distillates contain the 1-propenyl group.

4
Garlic Preparations: Methods for Qualitative and Quantitative Assessment of Their Ingredients

REINHARD PENTZ AND CLAUS-PETER SIEGERS

4.1 INTRODUCTION

Garlic has been used for spicing food, as a vegetable, and for therapeutic purposes in folk medicine for thousands of years. The benefits it conveys for health and health care have led to numerous commercial garlic preparations over the last decades. Fresh garlic, of course, can offer the same advantage as commercial preparations when eaten on a regular and sufficient basis, but disadvantages include sprouting or rotting with prolonged storage, variable quality depending on different cultivation areas and climate conditions, stomach complaints after consumption, and, last but not least, an unpleasant odor.

Garlic preparations are manufactured by the pharmaceutical and natural products industries in various forms and are widely available in pharmacies, drug stores, and health food stores. The amount of garlic processed annually is considerable (Table 4.1). The most important garlic preparations made from raw garlic, as well as their applications, are summarized in Table 4.2.

In order to deliver the potential advantages of garlic preparations over fresh garlic, they should contain sufficient and standardized amounts of biologically active compounds. The main problems in this regard arise from the fact that correlations between specific ingredients and their beneficial effects have not been clearly identified. In fresh garlic, the main "prodrug" principles are believed to be the alliin/alliinase and the cysteine-containing γ-glutamylpeptide systems, which are substantially preserved only in dry powder preparations, whereas in other garlic products tertiary alliin-derived sulfur-containing compounds would be the essential active ingredients (Section 4.2). Depending on the type of processed garlic, different substances may serve as the main identifying compounds for evaluating the quality, or at least the consistency, of the preparations.

Table 4.1. Statistics of Garlic Production for 1974/76 and 1983

	Amount Produced (1000 tons)		Area Cultivated (1000 hectares)	
	1976/76	1983	1974/76	1983
Industrial countries	492	574	96	102
Developing countries	1358	1828	216	296
Asia	1035	1480	175	252
Europe	420	455	83	85
Africa	162	160	7	7
South America	129	141	28	31
North America	87	134	10	12
Former Soviet Union	17	32	9	11
Total World	1850	2402	312	397

From Hanley AB, Fenwick GR. Cultivated Alliums. J Plant Foods 1985; 6:211–238. (Source: *Yearbook of the Food and Agriculture Organization of the United Nations.*)

Table 4.2. Commercial Garlic Products and Their Uses

Product	Appearance	Main Usage
Pressed juice (stabilized)	Viscous liquid	Medicine and manufacturing
Garlic oil	Liquid with strong odor	Medicine and spice mixtures
Dry powder	Powder, granules, pellets	Medicine and foods
Liquid extracts	Aqueous or alcoholic	Spice (e.g., with vinegar/5% NaCl)
Dried extracts	Powder	Medicine and foods
Oil macerates	Oily solution	Medicine
Aged extracts	Liquid or powder	Medicine
Pickled garlic	Garlic cloves in vinegar/salt	Foods
Crushed or chopped garlic	Sauce	Foods

4.2 THE MAIN INGREDIENTS OF DIFFERENT GARLIC PREPARATIONS

Several procedures are used to obtain certain garlic-derived products for pharmaceutical purposes. In the processing of the fresh garlic bulb, attempts are made to preserve the genuine substances (e.g., alliin), mainly by avoiding the activation of alliinase, or to produce secondary compounds by maceration, thus initiating rapid conversion of alliin to allicin, combined with further treatment with steam, oil, or other solvents.

For comparative purposes, several properties and substance compositions of intact garlic cloves shall be described. Fresh garlic contains about two-thirds water, along with a variety of medicinally interesting compounds, most of which are organosulfur substances, with some probably less important nonsulfur compounds (e.g., saponins, polysaccharides, etc.). An overview of the components in fresh and dried garlic is presented in Tables 3.1, 3.3, and 3.16.

Garlic cloves typically contain 0.35% (1.0% dry weight) total sulfur and 1.1% (3.1% dry weight) total nitrogen (Alfonso & Lopez, 1960; Pentz et al., 1990). On the basis of alliin analysis, alliin-derived sulfur has been determined to be 0.55% of the dry weight (calculated from Table 3.3). On the basis of hydrolysis-released SH, alliin-derived sulfur has been determined to be 0.2% of the dry weight, and total organosulfur 0.5–1.0% (Pentz et al., 1990). All of the important organosulfur compounds contain the amino acid cysteine and can be subclassified into the compounds listed in Table 3.3 (Lawson, 1993).

γ-Glutamylcysteines are the precursor compounds of cysteine sulfoxide moieties and are gradually hydrolyzed during wintering and sprouting, as well as when stored at room temperature, or even faster when stored cool. The differences between fresh-picked and store-purchased garlic are thus lower γ-glutamylcysteines content and higher cysteine sulfoxide content in the latter.

4.2.1 Dry Garlic Powders

Dehydration of garlic bulbs without damaging the cell structure, followed by grinding to garlic powder, should completely preserve the composition and alliinase activity of fresh garlic. This is indeed true when the cloves are frozen in liquid nitrogen, pulverized, and then vacuum-dried. For efficient drying, however, the cloves must be chopped to a certain extent, thus causing a partial but immediate release of alliinase-generated thiosulfinates which are lost upon drying. When garlic cloves are dried at raised temperatures (e.g., 60°C), neither alliinase nor alliin are affected, but a loss of 20–50% γ-glutamyl-S-$trans$-1-propenylcysteine may occur (see Section 3.4.1) (Lawson & Hughes, 1992; Lawson, 1993). Above 60°C, marked losses of active ingredients are inevitable (Fischer, 1936; Pruthi et al., 1960; Raghavan et al., 1986; Sumi et al., 1987). Surprisingly, room temperature drying, as well as lyophilization, due to the stability of alliin and alliinase, does not yield a better quality of dry powder than hot air drying at normal pressure (Kononko et al., 1984).

An appropriately processed garlic powder, especially when produced from good quality crops, will contain an approximately 2.5-fold higher concentration of ingredients than fresh garlic. Small amounts of S-alkylcysteines (S-allyl, S-1-propenyl, and S-methyl) and even lower levels of dialkyl sulfides, neither of which are found in fresh garlic, are generated during processing and can be detected in garlic powder.

During storage of garlic powder, humidity must be kept low to prevent molding processes and the activation of alliinase, which will result in reduction of quality, discoloration ("browning"), and possibly even spoilage. Additives to prevent lumping (e.g., silicates, phosphates, or starch) are sometimes applied (Stephenson, 1949). Pesticides are also sometimes added in order to prevent spoilage (Lorenzato, 1984). Of course, these additives should be declared.

Garlic bulbs, as well as dried garlic powder, can also be preserved by irradiation with gamma rays (^{60}Co, 0.1–10 kGy). Garlic can then be stored up to 10 months without loss of quality, and in this manner early sprouting of garlic bulbs can also be prevented (Cho et al., 1984; Curzio & Ceci, 1984; Kwon et al., 1984; Nunez et al., 1985; Croci et al., 1987). Previous exposure to gamma rays can later be detected by chemiluminescent measurements (Boegl & Heide, 1985; Heide & Boegl, 1985).

In addition to containing appropriate levels of active ingredients as determined by analysis (Section 4.3), a properly processed dry powder from garlic should meet various other criteria (Table 4.3). The powder must flow freely, and the odor should be pure garlic-like, not musty or rancid. Furthermore, the grain size should not be lower than 100 µm, since finer powder easily lumps and offers a larger surface area for oxidation reactions. The color of the powder should be white to cream-like and must not contain brown discolored spots. Extensive investigations on the various factors involved in production, packing, and storage of garlic powder were performed by Pruthi and colleagues (Pruthi et al., 1959a, 1959b, 1959c, 1959d, 1959e, 1959f; Singh et al., 1959a, 1959b, 1959c).

In garlic powder from carefully dehydrated garlic bulbs or cloves, stored under adequate conditions, the ingredients such as alliin/alliinase and γ-glutamylpeptides are stable over long periods of time (Pfaff, 1991; Sreenivasamurthy et al., 1961). In order to circumvent the inherent danger of premature enzymatic reaction of alliin with alliinase, attempts have been made to manufacture and apply alliin and alliinase as separate formulations for drug application (Hess et al., 1987).

4.2.2 Dry Extracts with Conservation of Alliin

Other methods employed to avoid alliinase-influenced decomposition of alliin include the treatment of whole cloves with certain extraction media such as ethanol or hot ethanol steam. Of course, this extract or the evaporation-prepared powder derived from it will contain no alliinase activity, which ultimately is necessary to produce the beneficial compounds derived from alliin. Therefore, when garlic preparations are manufactured with this type of garlic powder, addition of a certain amount of "active" (i.e., alliinase-containing) garlic powder, sufficient to bring about complete conversion of alliin, is necessary.

Commercial alliin-containing garlic powder preparations are sold in capsules or coated tablets and should have an intact alliinase system which is activated upon rehydration, for alliin itself will not produce beneficial health effects. Pharmaceutical formulation design should also take into account the potential inactivation of alliinase by the acidic gastric juice. Although Aye (1989)

Table 4.3. Limiting Values for Dry Powders of Garlic

Parameter (maximum)	Weight%
Moisture	5.0
Total ash	6.5
Acid insoluble ash	0.5
Foreign matter	3.0

From Raghavan B, Abraham KO, Shankaranarayana ML. Chemistry of garlic and garlic products. J Sci Ind Res 1983; 42:401–409. (Source: *"Indian Standard Specification for Dehydrated Garlic."*)

described this system as being only reversibly suppressed at low pH in the stomach, others have observed a complete irreversible inactivation of alliinase (Blania & Spangenberg, 1991; Lawson & Hughes, 1992). Coating the preparation with a substance resistant to gastric juice will not only avoid this negative effect but will also provide a means by which gastric complaints can be avoided.

4.2.3 Garlic Juice

Obtained as a viscous fluid by cold pressure extrusion from garlic cloves, garlic juice quickly turns brown. The viscosity can be effectively lowered by the addition of pectinase, and discoloration can be prevented by adding 10% acetic acid and 5% sodium chloride or by brief heat treatment (5 minutes around 90°C). This kind of pressed juice is only used in the food industry for seasoning. For pharmaceutical purposes, allicin as the main active component may be somewhat stabilized by the addition of lemon juice, sodium benzoate (Bockmann et al., 1969), or other compounds such as sodium bisulfite, citric acid, or ascorbic acid (Hirano & Hirano, 1984). Upon storage without additives, allicin rapidly decays (up to 50% loss in 2 days, Figure 3.11), and after prolonged storage the mixture consists mainly of diallyl trisulfide and smaller amounts of diallyl disulfide, methyl allyl trisulfide, and ajoenes (Iberl et al., 1990b).

The pressed juice also contains the water-soluble γ-glutamylcysteines and adenosine (see Fig. 3.21), as well as other constituents, which are not present in oil-macerates or steam-distilled garlic oils. As a result of insufficient information provided by manufacturers concerning preparation and storage procedures, it is often impossible to determine whether a preparation is obtained from pure garlic juice. This can only be ascertained by analytical measurements of allicin or diallyl trisulfide levels (the main component of allicin transformation).

4.2.4 Steam-Distilled Garlic Oil

Steam-distilled garlic oil is frequently referred to as the "essential oil of garlic," which may erroneously be confused with pure garlic juice or other oil preparations. Processed from crushed garlic in water, followed by hot water steam distillation, an oily liquid separates from the aqueous part of the condensate. The amount of volatile water-insoluble compounds was originally quoted as 0.1–0.2% (Wertheim, 1844, 1845; Semmler, 1892a, 1892b). Distillation of the entire plant yields 0.05–0.09% volatile oil (Guenther, 1952). A modern method of combining distillation and extraction should yield 0.410–0.575% of the raw material as water-free garlic oil (Raghavan et al., 1983). The physicochemical properties of garlic oil have been described by Zwergal (1952): density at 20°C, 1.092; refractive index at 20°C, 1.5795; viscosity at 20°C, 1.73 cSt; iodine number, circa 900.

Steam-distilled garlic oil is utilized for the formulation of commercial garlic products upon dilution with a vegetable oil and is by far the most common garlic oil found in the European and U.S. markets (Miething & Thober, 1985; Koch & Hahn, 1988; Reuter, 1988; Koch & Jäger, 1990). None of the ingredients in this oil is present in garlic cloves, since the oil contains almost exclusively di- and polysulfides. The composition of a typical commercial preparation is 26% diallyl disulfide, 19% diallyl trisulfide, 15% allyl methyl trisulfide, 13% allyl methyl disulfide, 8% diallyl tetrasulfide, 6% allyl methyl tetrasulfide, 3% dimethyl trisulfide, 3% monosulfides, 4% pentasulfides, and 1% hexasulfides (Lawson et al., 1991a). By gas chromatography/mass spectrometry (GC/MS) analysis, Vernin and colleagues, who only determined mono-, di-, and trisulfides, found an even higher proportion of diallyl disulfide, in the range of 30–50% (Vernin et al., 1986; Vernin & Metzger, 1991). Freshly distilled garlic oil, however, contains predominantly diallyl trisulfide (Miething, 1988; Jirovetz et al., 1992a; Yan et al., 1992; Kim, 1995). Mazza and colleagues (1992) detected 22 volatile compounds on GC analysis of aqueous and methanolic extracts of garlic bulbs, but these are artifacts produced in the GC from the thiosulfinates (Brodnitz et al., 1971; Saghir et al., 1964; Toohey, 1986).

Propyl sulfides were originally thought to be present in steam-distilled garlic oil (Semmler, 1892a; Saghir et al., 1964; Fenwick & Hanley, 1985), although this was never verified by mass spectroscopy. More recent studies have shown the absence of propyl sulfides (Vernin et al., 1986; Vernin & Metzger, 1991; Lawson et al., 1991a), propyl thiosulfinates (Lawson et al., 1991c; Block et al., 1992a), and propyl cysteine sulfoxide (Ziegler & Sticher, 1989) in garlic or its preparations, but have shown the presence of 1-propenyl compounds which were previously not thought to be present in garlic.

During storage without dilution or without additives, the di- and polysulfide pattern changes toward compounds with longer polysulfide chains. Because commercial steam-distilled garlic oils are diluted (more than 99%) with vegetable oils, the sulfides composition is much more stable. The finished product is either prepared with soft gelatin capsules or microencapsulated with dextrin before preparation of tablets or dragées. The stability of this product is discussed in Section 3.4.2.

4.2.5 Oil-Macerated Garlic Products

This type of processed garlic is rare in the United States but is very common in Europe. The production of oil-macerates is accomplished by grinding garlic cloves and mixing the product with vegetable oils (such as soybean oil, wheat germ oil, peanut oil, and others) before the oil-insoluble components are separated. Chopping and grinding of garlic induces the enzymatic conversion of alliin to allicin (which is much more soluble in oil than in water) more or less according to this reaction's stoichiometry, but allicin and other thiosulfinates are unstable in lipophilic solvents and are transformed to vinyldithiins (70%), sulfides (18%), and ajoenes (12%) (Brodnitz et al., 1971; Voigt & Wolf, 1986; Iberl et al., 1990a; Lawson et al., 1991a). These percentages represent a typical mixture, but different amounts may result, depending on the preparation procedure (Iberl et al., 1990b).

Summaries of the constituents of finished products available on the German market have been published (Winkler et al., 1992; Lawson et al., 1991a; Sarter et al., 1991; Koch & Jäger, 1990). The report of Sarter and colleagues (1991) includes only the vinyldithiins. The report of Koch and Jäger (1990) has been criticized by Sarter's group as being erroneous due to the destruction of most of the main vinyldithiin by the use of high GC oven temperature. The *vinyldithiins* consist of two isomers, which are described as (*a*) 2-vinyl-(4H)-1,3-dithiin and (*b*) 3-vinyl-(4H)-1,2-dithiin. The relative amounts of the two in commercial products have been found to range between 2.0 and 3.9 (Voigt & Wolf, 1986; Winkler et al., 1992; Lawson et al., 1991a; Sarter et al., 1991; Iberl et al., 1990a). The main *sulfides* in commercially available garlic oil-macerates are diallyl trisulfide, diallyl disulfide, and methyl allyl trisulfide (Iberl et al., 1990a; Lawson et al., 1991a; Winkler et al., 1992). The stability of the commercial oil-macerates is discussed in Section 3.4.3 (Voigt & Wolf, 1986).

Ajoenes present in oil-macerates can be subdivided into the *trans* isomer *(E)*- and the *cis* isomer *(Z)*-ajoene, first identified by Block and colleagues (1984). The *E/Z* ratio and yield depend on the polarity of the solvent system and the reaction temperature during processing (Iberl et al., 1990a). In finished oil-macerated garlic products the *E*-ajoene has always been found to be dominant over the *Z*- isomer, usually about twofold. The stability of ajoenes in garlic brands was shown to be rather low on storage at room temperature, with the content decreasing by about 50% over 6 months (Lawson et al., 1991a); however, more careful studies have found a 22% loss of ajoenes in gelatin capsules stored 18 months, but an 86% loss if the oil was stored in a half-full glass bottle, indicating air sensitivity (see Section 3.4.3).

4.2.6 Aged Alcoholic Garlic Extract

This type of processed garlic is prepared by storing sliced garlic in 15–20% aqueous ethanol for 20 months, followed by filtration and concentration (Hirao et al., 1987; Nakagawa et al., 1989). The final liquid extract has been reported to contain 10% (w/v) ethanol and small amounts of several water-soluble sulfur compounds (0.30 mg/g S-allylcysteine, 0.15 mg/g S-1-propenylcysteine, 0.11 mg/g S-methylcysteine, 0.28 mg/g γ-glutamyl-S-allylcysteine, 0.17 mg/g γ-glutamyl-S-1-propenylcysteine, 0.04 mg/g S-allylmercaptocysteine, 0.01 mg/g cystine, and less than 0.02 mg/g alliin (Lawson, 1993; Hansen et al., 1993). See Section 3.4.5 for more recent data.

4.2.7 Comparison and Value of Different Garlic Preparations

When the quality of different commercial garlic preparations is to be compared, problems arise from our incomplete knowledge concerning the ways in which specific garlic components and/or their transformation products convey health benefits. Even the calculated values of "allicin-equivalents" and the "amount of processed fresh garlic represented by a certain preparation" may be misleading. For instance, several compounds derived from allicin using specific preparative processes will never appear in vivo when fresh garlic is consumed, although garlic extracts could be claimed to represent a comparatively high amount of fresh garlic, albeit consisting of ingredients not occurring naturally. Moreover, there are many garlic brands which are prepared as mixtures from different types of processing (e.g., oil-macerates combined with steam-distilled products).

4.2.8 Odorless Garlic Preparations

Another problem has to do with the question of whether the unpleasant odor of garlic is directly related to its medicinal value (Koch, 1989). Is it possible to avoid olfactory annoyance by adding odor inhibitors (see Table 8.1) or by special manufacturing procedures (e.g., "aged garlic extract" or "odorless garlic extract"—see Section 5.13) while still retaining the beneficial health effects?

The threshold concentrations for olfactory perception of various odor carriers that can be found in processed garlic are summarized in Table 4.4. Shankaranarayana (1981) reported the following odor thresholds in water: methanethiol, 0.02 ppb (0.0002 ppb in air); dimethyl sulfide, 0.3 ppb; dimethyl disulfide, 3 ppb; and allylmethyl disulfide, 6.3 ppb. According to Laakso and colleagues (1989) and Minami and colleagues (1989), allyl mercaptan was shown to be the dominant sulfur-containing compound in human breath after consumption of garlic cloves (see Section 3.2.6.4).

Interestingly, the odor perception caused by onions and garlic can be enhanced by 10–40 times by the addition of cysteine or homocysteine. The reasons for this are unknown, and no influence on the alliinase system could be shown (Schwimmer & Guadagni, 1967; Granroth, 1968). There is, however, a correlation between the amount of enzymatically produced pyruvate and olfactory perception of sulfur compounds, since pyruvate is produced at the same time as the thiosulfinates (Schwimmer & Guadagni, 1962).

Many manufacturers of garlic preparations claim that their product is "odorless" or "odor minimized" because they do not cause odor problems when consumed. This perhaps may be true of preparations made with "aged garlic," because of its low content of water-soluble sulfur compounds, or of "stomach

Table 4.4. Threshold Concentrations of Odor Perception of Some Components of Processed Garlic

	Steam-distilled Garlic Oil ppb	Diallyl disulfide ppb	Diallyl sulfide ppb
Absolute odor threshold	0.7	30	100
Recognition of garlic odor	3	80	300
Range of pleasant garlic odor	20 upwards	100–7500	1000–20,000
Sharp, nauseating odor	750	10,000	40,000

From Teleky-Vamossy G, Petros-Furza M. Evaluation of odor intensity versus concentrations of natural garlic oil and some of its individual aroma compounds. Nahrung 1986;30:775–782.

acid-resistant garlic powder tablets" as long as not too many are consumed. The sources of odor from garlic powder tablets are twofold. First, there is the odor from a small amount of allyl sulfides present in the powder itself as a result of transformation of the thiosulfinates released when the cloves are chopped prior to drying; the allyl sulfides can be reduced by cutting the cloves into larger pieces or by freeze-drying the cloves. The appearance of this type of odor in the breath can be greatly reduced by using acid-resistant tablets that do not disintegrate in the stomach. The second type of odor is the result of the formation of allicin in the intestines, which is converted to the very odorous allyl mercaptan in the blood, a portion of which makes its way to the lungs. This type of odor is usually not noticeable with smaller doses and cannot be avoided in preparations that provide the health benefits of allicin.

Several attempts have been made to suppress the odor carriers, for example, by addition of charcoal, titanium dioxide, milk sugar, potato starch, gallic acid, or other substances (Sandoz, 1926; Sächsisches Serumwerk AG Dresden, 1936), but none of these have been successful (Hotzel, 1936; Oppikofer, 1947). Odor-minimized garlic powder can supposedly be obtained either by inhibition of alliinase—and hence of allicin formation—through addition of compounds such as fumaric acid, phytic acid, ascorbic acid, or by adsorption, solution, or dilution with silicic acid, germanium oxide, saponins, fatty oils, and various plant extracts, or by microwave treatment (Bockmann et al., 1969; Lonyai et al., 1973; Yokozawa & Onaga, 1981; Kikkoman Shoyu Co. Ltd. 1982; Sakai, 1981, 1985; Araki, 1983; Morinaga & Tomoda, 1984; Osaka Yakuhin Kenkyusho Kk., 1984). Many of the methods for the production of "odorless garlic" have been patented (see Table 8.1). Perhaps the easiest method of preparing odorless garlic pills is simply to put a high quality (low sulfides) garlic powder into gelatin capsules. When the capsules dissolve in the stomach, alliinase is immediately inhibited by the acidic conditions. However, in some individuals the gastric pH may be higher than normal, and a large meal can increase the gastric pH to the point of activating alliinase. Therefore, the most sure way of preparing an odorless powder is to inactivate alliinase by cooking (boiling or microwaving) the whole cloves prior to powder manufacturing (see Section 3.2.6.5). Of course, if alliin is prevented from being transformed to allicin there will be minimal odor development (Moser, 1937, 1948), but alliin itself has not been shown to have beneficial effects on human health.

Another attempt to produce odorless garlic preparations has been accomplished by the addition of chlorophyll, from parsley, for example. This method was recommended as early as in the 16th century by Jacobus Theodorus Tabernaemontanus (1520–1590) in his *Neuw Kreuterbuch* (first edition, 1588). The "odor blocking" potency of chlorophyll has not been studied. However, through the addition of other strong-smelling volatile oils such as thyme, peppermint, or others, "blockage" of scent may be accomplished by masking the strong garlic odor (Koch, 1989).

The odorous compounds can be reversibly bound by complex formation with cyclodextrins (cyclic glucopyranose with 6, 7, or 8 units). In the central cavity of the cyclodextrin molecule, lipophilic "guest molecules" with appropriate dimensions can be enclosed, such as allicin or diallyl sulfide (Lindner et al., 1981; Lindner, 1982; Riken Chem. Industry Co. Ltd. 1983, 1984; Kawashima, 1986). The enclosed compounds are released easily and remain unchanged upon degradation of the oligosaccharides by digestive enzymes.

New manufacturing procedures of "odorless garlic preparations" are in special demand in countries of the Far East, as indicated by the numerous patent applications in Japan (see Table 8.1). There are, for instance, registered garlic extracts with sweet *sake* (rice wine) and an "odorless garlic wine" (Hakamada, 1982; Fujiwara, 1986). In China, garlic preserved with sugar (candied garlic) is highly valued and satisfies the consumer in odor, taste, and color (Mie, 1983).

A modern, galenical form of medicine is the so-called microcapsules. These are tiny spheres or pearls made of an inert material, into which the medicinal ingredients can be incorporated. Upon contact with liquid, the encapsulated substances are slowly released. Microcapsules have already been manufactured with allicin in order to stabilize this active compound and mask its odor (Shen et al., 1983).

Even when odorless brands of garlic preparations are manufactured, taken in appropriate amounts, most of them will produce a typical garlic odor by the liberation of sulfur compounds upon swallowing and absorption. Thus olfactory effects will appear from volatile compounds either in expired air or through perspiration from the skin.

4.3 ANALYTICAL DETERMINATION OF CONSTITUENTS IN GARLIC AND GARLIC PREPARATIONS

The quality of garlic or garlic preparations is considered to be conveyed by certain substances which are responsible for its health effects and its flavoring characteristics. Although whole garlic bulbs contain several primary constituents regarded as precursors of beneficial substances (e.g., alliin or γ-glutamylpeptides), constituents in processed garlic may be even more complex. This is due to enzymatic and chemical cleavage, as well as to rearrangements and degradations that arise when different procedures are used for the preparation of commercial products. Furthermore, products and ingredients may change during prolonged storage, with volatile substances in particular partially or even completely lost.

Depending on the category of commercial garlic preparation, there can be several compounds (mostly considered to be biologically active principles, but not all proved to be) which may serve as parameters to allow qualitative and quantitative characterization of the product. All of the substances of interest in this respect are sulfur-containing compounds. As such, several methods have been developed and suggested for the analysis and standardization of garlic and processed garlic products. An overview of the published methods for evaluating the most important compounds is given in the following sections.

4.3.1 Alliin and Related Cysteine Sulfoxides

These compounds are present only in the intact garlic bulbs and in appropriately produced garlic powders (and garlic powder tablets). Analytical procedures for determining alliin content have to avoid the influence of alliinase, for this would result in the rapid conversion of alliin to secondary products. Inactivation of alliinase can be achieved by applying extraction media such as methanol/water (4/1, v/v), methanol followed by aqueous 0.1% formic acid (1/2, v/v), or by using aqueous solutions of alliin inhibitors (see Section 3.2.5.1).

4.3.1.1 CYSTEINE SULFOXIDES: EARLIER APPROACHES

An indirect method based on the release of pyruvic acid (see Fig. 3.9) by the enzymatic reaction with alliinase was published 40 years ago (Jäger, 1955). Pyruvate reacts with 2,4-dinitrophenylhydrazine (DNPH) to form a product which can be spectrophotometrically determined. The DNPH method has been extensively examined and its reliability improved (Lutomski et al., 1968, 1969b). Further modifications of the DNPH procedure have also been published by other authors (Alfonso & Lopez, 1960; Schwimmer & Weston, 1961; Freeman & Whenham, 1975b; Zelikoff & Belman, 1985; Bekdairova & Klyshev, 1982; Rao & Rao, 1982; Örsi, 1976; Pharmacopoea Helvetica VI, 1971). The DNPH method has been applied to measuring similar substances in the onion (*A. cepa*) and in other species of the genus *Allium* (Freeman & Whenham, 1975a, 1975b; Tewari & Bandyopadhyay, 1975, 1977).

Thin-layer chromatography (TLC) of garlic ingredients, including alliin and other cysteine sulfoxides, has been carried out by Hörhammer and colleagues (1968). A procedure for qualitative TLC assays on silica gel plates (Table 4.5) has also been published in which detection was achieved with ninhydrin reagent (Wagner et al., 1983; Müller, 1989). The homologous S-alkylmercaptocysteines were chromatographed by Michahelles (1974) (Table 4.6). Molnar and colleagues (1991) determined alliin by densitometry after separation by TLC, while Ostrowska (1987) employed the ninhydrin colorimetric method for its quantification. Lancaster and Kelly (1983) developed an elaborate method for measuring isomeric and homologous S-alkyl- and S-alkenyl-L-cysteine sulfoxides in *Allium cepa*, which would presumably be applicable for measurement in garlic as well. After purification of the raw extract by electrophoresis, a two-dimensional separation on silica gel plates was performed.

TLC or a modified high performance system (HPTLC) can be used for quality control of fresh garlic and garlic powder, and this method has certain advantages over high performance liquid chromatography (HPLC) applications (e.g., low price, simple equipment, flexible, fast), although an accurate quantification of alliin is not easily obtained. Quantitative TLC has been used for the standardization of garlic and garlic preparations (Kappenberg & Glasl, 1990; Hoppen et al., 1989).

Paper chromatography (PC) has been used for the detection of several garlic and onion constituents, including alliin (Gaind et al., 1965; Virtanen & Matikkala, 1959; Virtanen, 1965; Kajita, 1957; Yoshimura, 1958a; Yoshimura et al., 1958). A paper electrophoretic method has also been employed (Yoshimura et al., 1958).

In earlier days, biological assays for the determination of effective constituents in garlic were usually performed to evaluate the antibiotic activity in microbiological test systems (Cavallito & Bailey, 1944; Small et al., 1947; Kupinic, 1979). Klein and Souverein (1954) reported a biological assay for the enzyme-substrate system alliin/alliinase, and Kedzia and colleagues (1969), as well as Lutomski and colleagues (1969a; 1969b) quantitated the antibacterial activity of lyophilized garlic powder. Other methods of this kind are only of historical interest (Hintzelmann, 1935a; 1935b; Orzechowski, 1933; Orzechowski & Schreiber, 1934; Platenius, 1935; Martius, 1951; Scheibe, 1958; Whitaker, 1976; Kuhn, 1932).

The methods mentioned above represent separation and determination techniques which have various shortcomings for quantitative evaluations. Modern analytical methods, which combine separation and detection devices, are described in the following section.

4.3.1.2 CYSTEINE SULFOXIDES: GC AND HPLC METHODS

Column chromatography can be considered to be the forerunner of HPLC methods, and has been applied by Michahelles (1974) for fractionating and determination of garlic constituents. In the last decade, this method has been almost completely replaced by high performance liquid chromatography (HPLC). Several papers have been published describing the evaluation of alliin in garlic (Ziegler

Table 4.5. Thin-layer Chromatography of Alkylcysteine Sulfoxides and Alkylcysteines

Compounds	Rf-values[a]	
	S-alkylcysteine sulfoxides	S-alkylcysteines
methyl-	0.18	0.56
ethyl-	0.27	0.68
allyl-	0.50	0.70
iso-propyl-	0.30	0.74
n-propyl	0.38	0.76
n-butyl	0.47	0.79

[a]Solvent system: n-butanol/n-propanol/acetic acid/water (30/10/10/10).
From Wagner H, Bladt S, Zgainski EM. Drogenanalyse. Dünnschichtchromatographische Analyse von Arzneidrogen. Berlin: Springer, 1983; and Wayne H, Bladt S, Zgainski EM. Plant drug analysis. Berlin: Springer-Verlag, 1984.

Table 4.6. Thin-Layer Chromatography of S-Alkylmercaptocysteines

Compounds	Rf-value[a]
methyl-	0.50
allyl-	0.70
propyl-	0.79
butyl-	0.91

[a]Solvent system: n-butanol/acetic acid/water (40/10/50).
From Michahelles E. Über neue Wirkstoffe aus Knoblauch (Allium sativum L.) und Küchenzwiebel (Allium cepa L.). Dissertation, University of Munich, 1974.

& Sticher, 1988, 1989; Mochizuki et al., 1988; Ueda et al., 1991; Kitada et al., 1988; Ziegler et al., 1987; Iberl et al., 1990b; Velisek et al., 1993; Thomas & Parkin, 1994).

In principle, these methods are based on analytical amino acid procedures. Most of the common amino acids and nonprotein amino acids (e.g., alliin) do not absorb sufficiently for direct detection after chromatographic separation and thus have to be coupled to chromophoric groups. One of the preferred methods is the reaction with o-phthaldialdehyde (OPA)/thiol, which results in isoindole derivatives with both strong ultraviolet-absorbing and fluorophore properties (Ziegler & Sticher, 1989; Auger et al., 1993; Mütsch-Eckner et al., 1992; Velisek et al., 1993; Ziegler et al., 1991; Meier et al., 1989). In order to improve the stability of the substituted isoindole, tertiary-butylthiol is usually employed for the building reaction. After this precolumn derivatization, HPLC separation is performed and alliin detected by UV (340 nm), fluorescence (excitation 230 or 335 nm, emission 410 or 420 nm), or electrochemical devices (750 mV versus Ag/AgCl). HPLC columns for this analysis are of the C18 reversed phase type, and the composition of the mobile phase consists of aqueous phosphate buffer with acetonitrile and small amounts of 1,4-dioxane and tetrahydrofuran.

An HPLC method for the quantification of alk(en)yl-L-cysteine sulfoxides (ACSO) and related amino acids in *Allium* plants presumed to be involved in thiosulfinate biosynthesis was developed by Thomas and Parkin (1994) quite recently. The method is based on fluorescent detection of 9-fluoromethyl chloroformate–derivatized ACSO adducts. The detection limit is about 2.5 mg/100 g fresh weight. By this procedure total ACSO levels of 300–500 mg/100 g fresh weight were found in garlic cloves. Similar results were obtained with other *Allium* species on a dry weight basis.

Gas chromatographic (GC) analysis of derivatized alliin and other sulfur amino acids in the callus tissue of *A. sativum* has recently been reported (Hayashi et al., 1993), as well as GC determination of alliin in garlic products (Saito et al., 1988).

Alliin can also be calculated from analysis of total allyl thiosulfinates. Total allyl thiosulfinates can be determined by conversion of the allyl groups to allyl mercaptan (Koch et al., 1989), allyl ethyl disulfide (Tahara & Mizutani, 1979), or diallyl disulfide (Müller, 1990), standards of which are readily available but always need further purification by fractional distillation. They can also be determined by HPLC analysis of all the thiosulfinates, using synthetic or isolated allicin as the standard and the relative extinction coefficients for the determination of the allyl methyl and allyl 1-propenyl thiosulfinates (Lawson et al., 1991c). This method also allows calculation of the content of the 1-propenyl- and methylcysteine sulfoxides.

Alliin, in the absence of active alliinase, is a stable compound, and solutions of it can therefore be used for external standard preparations. Alliin may either be isolated by preparative chromatography from garlic extracts or prepared by synthesis from L-cysteine, allyl bromide, and hydrogen peroxide (Müller, 1989; Iberl et al., 1990b).

4.3.2 Allicin and Other Thiosulfinates

Allicin is rapidly generated (in less than 1 minute) when the cell structure of garlic cloves is disturbed and alliinase, otherwise contained in separate cells, activates the transformation reactions of alliin. On a molar basis, two molecules of alliin can be transformed to one molecule of allicin, or 1.000 g alliin can produce 0.458 g allicin; however, the theoretical amount of allicin is never produced because allyl methyl and allyl 1-propenyl thiosulfinates are also formed from alliin. Using a method in which all allyl thiosulfinates are converted to diallyl disulfide, Müller (1990) has shown that all of the alliin is converted to allyl thiosulfinates.

4.3.2.1 THIOSULFINATES: EARLIER APPROACHES

Various methods for indirect estimation of allicin from alliin measurements have been described and were mentioned in Section 4.3.1.1. Carson and Wang (1959a) introduced N-ethylmaleimide (NEM) for the determination of total thiosulfinates. Schwimmer and Mazelis (1963) used this reaction for the analysis of alliinase activity. Other authors improved this procedure (Watanabe & Komada, 1966; Nakata et al., 1970; Freeman & McBreen, 1973). Sekine and colleagues (1972) applied N-[p-(2-benzimidazolyl)phenyl]-maleimide (BIPM) instead of NEM to achieve a fluorescent reaction product (Fig. 4.1). Khaletski and Reznik (1957) detected the volatile constituents from garlic by complexing them with $HgCl_2$.

Various other methods involving sulfoxide, thioether, or thiosulfinate reactions that yield products for spectrophotometric measurement have been reported by Safronov (1963) (reaction with indole derivatives and glyoxylic acid) and Shannon and colleagues (1967b; 1967a) (reaction with glycine and formaldehyde). Kajita (1957) describes a coupling reaction of allylsulfenic acid with phenylpropylamine to produce allyl sulfenamide, which results in reaction products that can be detected in the urine. However, all of these procedures have shortcomings in selectivity, basically because they fail to differentiate between the several thiosulfinates.

Japanese authors were the first to use paper chromatography (PC) for the separation or identification of allicin and its homologs (Fujiwara et al., 1955, 1958; Yoshimura, 1958b), a method that was applied by others as well (Barone & Tansey, 1977). According to Michahelles (1974), the reaction products with cysteine are very suitable for the identification of dialkylthiosulfinates by chromatography (Table 4.6). As Cavallito's group (1944) observed long ago, each allicin molecule reacts with two molecules of cysteine, yielding two S-thioallylcysteine moieties (see Fig. 3.15). Lukes (1971), and later Augusti (1977), used these derivatives for analyzing thiosulfinates from onions.

Thin-layer chromatography (TLC) was applied to garlic ingredients for the first time by Schultz and Mohrmann (1965b), who separated the components of the "volatile garlic oil"—especially the oligosulfides but not the thiosulfinates. A systematic study of the ether-extractable components by means of TLC on silica gel plates was presented by Güven and colleagues (1972), with detection being performed through the use of several spray reagents which produced colored spots. Augusti (1976) described the chromatographic identification of sulfoxides from *Allium* species, and Granroth (1970) employed autoradiography using radioactively labeled compounds for detection. Michahelles (1974) used TLC to analyze thiosulfinates in ether extracts of garlic samples. The thiosulfinate bands were eluted with isopropanol, followed by NEM derivatization, and the absorption measured spectrophotometrically. By this procedure, the content of allicin and other thiosulfinates from several garlic samples were determined; the values obtained were not greatly different from those found today using more sophisticated methods (Tables 4.7 and 4.8).

In a recently published paper, Müller and Ruhnke (1993) described the TLC analysis of allicin on silica gel plates using ninhydrin reagent detection for identity assurance of garlic powders. Blania and Spangenberg (1991) reported on a high performance thin-layer chromatographic (HPTLC) method for allicin (and ajoene) determination. After chromatographic separation, thiosulfinate-related compounds (e.g., the lachrymatory factor from onion) can also be detected directly on TLC plates by the glycine-formaldehyde reaction (Tewari & Bandyopadhyay, 1975).

A new method for the determination of allicin, as well as for total garlic thiosulfi-

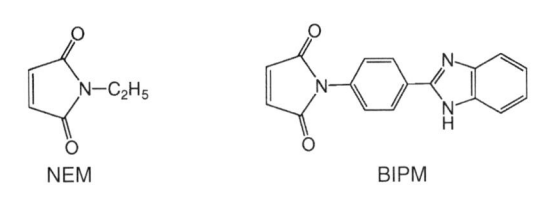

Figure 4.1. NEM and BIPM reagents for the determination of allicin.

nates, was also published recently (Han et al., 1995). This method is based on one of the earliest reactions ever reported for allicin (Cavallito et al., 1944) and involves the rapid reaction of one molecule of allicin with two molecules of cysteine, resulting in the formation of S-allylmercaptocysteine and the subsequent decrease of the standardized amount of cysteine. The decrease in cysteine concentration is measured spectrophotometrically using 5,5'-dithio-bis(2-nitrobenzoic acid) (DTNB) or 4,4'-dithio-pyridine (DTDP), the latter of which increased the sensitivity of the assay by about 40%. Other garlic thiosulfinates also react with cysteine in the same manner as they do with allicin. Thus, this method can be employed to measure the total garlic thiosulfinate concentration or the allicin concentration in a purified preparation. Unlike GC and HPLC methods, this procedure does not require a standard to quantify allicin or total thiosulfinates, since it requires only the extinction coefficient of DTNB or DTDP.

4.3.2.2 THIOSULFINATES: GC AND HPLC METHODS

Gas chromatography methods (GC) have not been successful because of the thermal instability of allicin when it is chromatographed on columns, even at very low temperatures. Oaks and colleagues (1964) were the first to attempt the separation and identification of allicin in an extract (hexane) of fresh garlic by means of GC (carbowax column), but no allicin was found. According to Brodnitz's group (1971), only degradation products of allicin could be detected, which was confirmed by coupling the GC procedure with mass spectrometry (MS). By this method, for instance, nine sulfides could be identified in onion (Brodnitz & Pollock, 1970). Aye and colleagues described the difficulties of direct determination of allicin by GC and proposed a *direct* analytical GC procedure for allicin, assuming a consistent loss (Tschirch & Lippmann, 1933; Müller & Aye, 1991; Müller, 1989). Block's group (1992b, 1993) successfully used cryogenic GC/MS analysis (i.e., at 0°C) of thiosulfinates and other compounds for several *Allium* species except for those yielding allyl thiosulfinates. Similar attempts were made also by other authors (Tsuji et al., 1991; Jirovetz et al., 1992b; Artacho et al., 1992; Sinha et al., 1992; George et al., 1991; Kallio et al., 1990; Mubarak & Kulatilleke, 1990).

Indirect methods for the measurement of allicin by GC have also been published. Koch and Jäger determined two vinyldithiins as the presumed single thermal degradation products of allicin (Koch & Jäger, 1989; Koch, 1991). Müller doubted this assumption and showed that vinyldithiins gave inconsistent results (Müller, 1989). Another attempt was made by J. Koch and colleagues (1989), who reduced allicin to allyl mercap-

Table 4.7. Allicin Yield from Several Garlic Samples

Sample	Allicin (%)	Harvest (year)	Storage (months)
A	0.16	1971	10
B	0.18	1971	10
C	0.19	1971	10
D	0.23	1972	3
E	0.24	1972	3

From Michahelles E. Über neue Wirkstoffe aus Knoblauch (Allium sativum L.) und Küchenzwiebel (Allium cepa L.). Dissertation, University of Munich, 1974.

Table 4.8. Thiosulfinates from a Garlic Sample

Thiosulfinate	% of Total Thiosulfinates	% of Fresh Garlic
dimethyl-	1.7	0.004
methylallyl-	5.1	0.012
diallyl and methylpropyl-[a]	78.0	0.180
dipropyl[a] and propylallyl-[a]	15.2	0.035

[a]Note: What was originally thought to be propyl thiosulfinates were later shown to be 1-propenyl thiosulfinates (Lawson et al., 1991c; Block et al., 1992a).
From Michahelles E. Über neue Wirkstoffe aus Knoblauch (Allium sativum L.) und Küchenzwiebel (Allium cepa L.). Dissertation, University of Munich, 1974.

tan and determined the latter product by head space analysis. Müller (1989) utilized a different reaction product of allicin, diallyl disulfide (DADS), which is rapidly and quantitatively produced (1 mole per 2 moles allicin—see Section 3.2.6.3.4) in alkaline medium (pH 11), followed by GC analysis of DADS. Nevertheless, all of these methods measure total allyl thiosulfinates rather than allicin itself.

However, Tahara's group (1979a, 1979b) determined the true allicin yield from garlic by reducing the thiosulfinates with sodium borohydride to disulfides and measuring allicin as DADS (not other thiosulfinate will reduce to DADS), using allicin generated from allyl thiomethylhydantoin-S-oxide as the standard. They found an allicin content of 3.4 mg/g fresh weight, which agrees well with other studies (see Tables 3.3 and 3.17). Lawson and Wang (1994) have recently verified the accuracy, reproducibility, linearity, and stability of this method with excellent results using aqueous extracts of crushed garlic containing 0.02–1.0 mM allicin and freshly prepared sodium borohydride at 4 mM in a final concentration of 33% acetonitrile, measuring DADS by HPLC (75% methanol). However, some caution must be used since 100–300 mM sodium borohydride is commonly used to reduce disulfide bonds to thiols (Jocelyn, 1987). Extraction into an organic solvent would make GC analysis possible. Another advantage of this method is that it offers the ability to analyze the other thiosulfinates of garlic as combined regioisomers, yielding allyl methyl disulfide, dimethyl disulfide, allyl *trans*-1-propenyl disulfide, and methyl *trans*-1-propenyl disulfide (see Fig. 3.2).

In summary, direct determination of garlic samples by GC methods cannot be applied for specific analysis of allicin or the other allyl thiosulfinates, whereas indirect analysis should prove more suitable, though it is rather elaborate and depends on various parameters which must be controlled precisely. Furthermore, several assumptions concerning reproducibility and quantitative aspects of allicin reactions have to be confirmed.

High performance liquid chromatography methods (HPLC), which to a certain degree represent a modern development of column chromatography, are currently the most useful means for the determination of allicin and/or other thiosulfinates, both in raw garlic and in dry garlic powders. The advantage of HPLC over GC is that it is operated at room temperature, allowing allicin itself to be measured, while artifacts or reaction products only occur if they were formed prior to HPLC analysis (Pfaff, 1991; Matikkala & Virtanen, 1967; Michahelles, 1974; Barone & Tansey, 1977).

An indirect method for allicin determination by HPLC was proposed by Voigt and Wolf (1986). Here, allicin is determined as the sum of vinyldithiins, sulfides, and ajoene by a procedure analogous to the preparation of garlic oil-macerate (Table 4.9).

Miething (1985) was the first to determine allicin by applying a normal phase HPLC method (Si-60 column, n-hexane as eluant), but his results were highly variable due to the instability of the allicin standard, stored in ether.

Jansen and colleagues (1987) published a comprehensive investigation of allicin determination using reversed phase HPLC (ODS-2 column, eluant was 60% methanol and

Table 4.9. Compounds Isolated by HPLC and Their Content in Various Garlic Samples

Sample	I	II	III	IV	V	VI
Garlic, fresh	0.1	0.4	0.10	03	0.01	0.03
Oil macerate 1:1	0.1	0.3	0.1	0.05	0.05	0.03
Garlic powder 1	0.07	0.06	0.02	0.03	0.05	0.1
Garlic powder 2	0.07	0.8	0.2	0.1	0.1	0.05

I-ajoene; II-2-vinyl-1,3-dithiin; III-3-vinyl-1,2-dithiin; IV-diallyl disulfide; V-allyl-methyl trisulfide; VI-diallyl trisulfide.
From Voigt M, Wolf E. Knoblauch: HPLC-Bestimmung von Knoblauchwirkstoffen in Extrakten, Pulver, und Fertigarzneimitteln. Dtsch Apoth Ztg 1986;126:591–593.

40% water containing 1% formic acid). Results from measurements of different garlic samples by this group, using modified analytical conditions, were published later (Iberl et al., 1989, 1990b; Lawson & Hughes, 1989, 1990; Müller, 1989; Winkler et al., 1992; Müller & Ruhnke, 1993).

Lawson's group used C18 HPLC and Si HPLC to identify and quantitate all eight of the thiosulfinates present in a variety of garlic homogenates and garlic products, and reported the extinction coefficients for most of them so that they could be quantitated universally without the need to resynthesize them (Lawson & Hughes, 1989; Landshuter et al., 1992; Lawson et al., 1991a, 1991c). Wagner's group used HPLC to analyze allicin and the methyl thiosulfinates in garlic and wild garlic (*Allium ursinum* L.) (Wagner & Sendl, 1990; Sendl & Wagner, 1990). Block and colleagues (1992a, 1993) extended the HPLC method also to various other *Allium* species and noted the trends regarding the various proportions of thiosulfinates in each species. Chromatograms showing the HPLC separation of the thiosulfinates of crushed garlic are presented in Figure 4.2.

A special problem concerning quantitative allicin determination arises from the instability of pure allicin (see Section 3.2.6.2), which makes it necessary to prepare standard solutions in water that are stored at −70°C. Allicin can be isolated by extraction of homogenized garlic cloves (Cavallito & Bailey, 1944; Miething, 1985; Lawson et al., 1991c). Purification is performed by TLC, column chromatography, or preparative HPLC and identification by mass spectrometry, infrared (IR), or proton nuclear magnetic resonance (NMR).

Allicin is also easily synthesized (see Section 3.2.6.1.1) by oxidation of purified diallyl disulfide with H_2O_2 (Iberl et al., 1990b). Purification can then be performed by separating the crude product on a Sephadex LH-20 column. Stabilization of allicin by adsorption to silica gel at a concentration below 0.1% has been reported (Jansen et al., 1987). It was claimed that, when stored at −24°C, no significant decrease in content and purity of this "allicin standard" was observed over a period of 3 months; however, Koch and colleagues (1992) clearly showed that this is not true, with adsorbates of allicin on silicate carriers being practically useless in this respect. Tahara and Mizutani (1979, 1979) proposed to use another suitable compound with adequate stability, namely the hydantoin derivative of allicin, or of its homologs.

A method of standardization of allicin-releasing potential is provided by the addition of defined amounts of alliin to the garlic samples. Differences in measurements of allicin values in samples with and without added alliin can be used in order to quantify the allicin content (Pentz et al., 1990, 1992; Müller & Ruhnke, 1993; Grünwald, 1989). Alliin can be prepared according to the procedures cited in Section 4.3.1.2., and the synthetically produced L-(±) isomers may be used in place of the naturally occurring L-(+)-alliin, although synthetic alliin reacts about 20-fold slower (Stoll & Seebeck, 1951). Recently, however, L-(+)-alliin has become commercially available (Extrasynthese, Lyon-Nord, France).

An important recent development in the analysis of allicin has been the use of *supercritical fluid chromatography* (SFC). Although the method has not yet been shown to be quantitative, due to some thermal destruction in the MS detector, it will probably be successful in time and will uniquely permit analysis of all of the thiosulfinates of all *Allium* species, including regioisomers and geometric isomers (Calvey et al., 1994b). Furthermore, supercritical CO_2 extraction of thiosulfinates from homogenized garlic has been shown to be virtually complete, as determined by LC-MS (Calvey et al., 1994a).

4.3.3 Polysulfides

These sulfur compounds, including disulfides and higher homologs, are the exclusive components of steam-distilled garlic oil, but they also appear in oil-macerates as well as in garlic preparations made by mixtures of

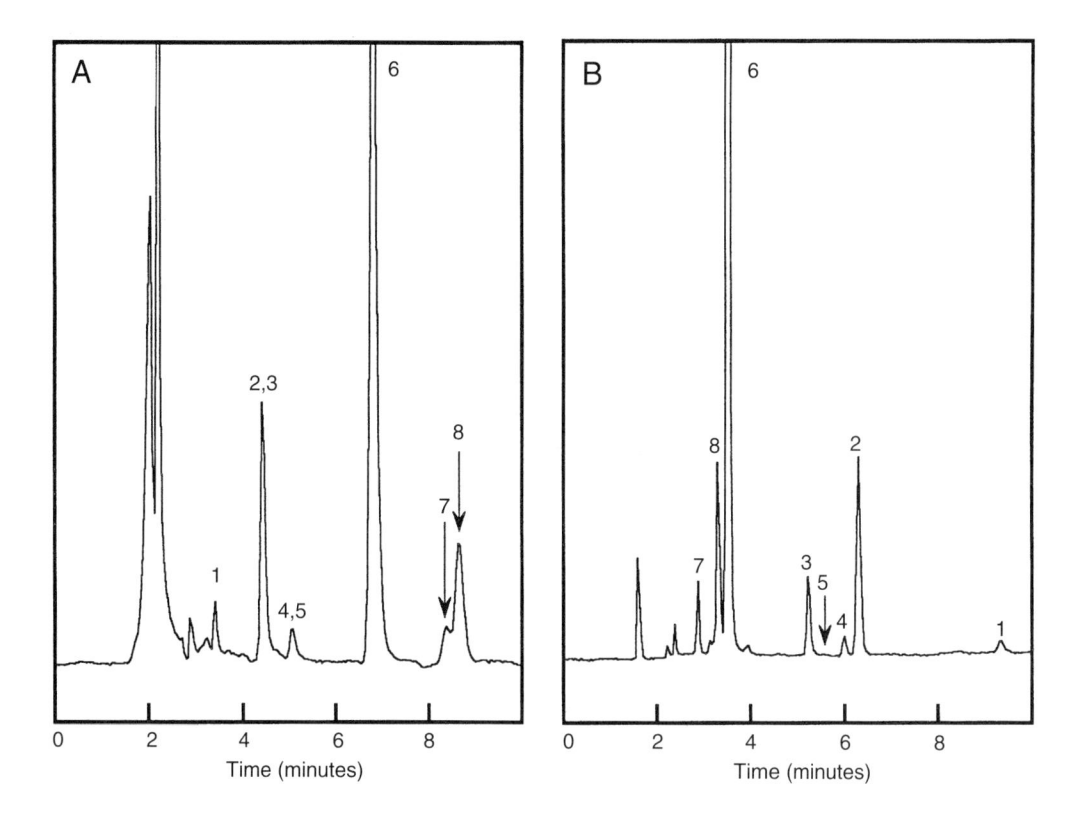

Figure 4.2. Separation of garlic clove thiosulfinates, **A.** By C18 HPLC (methanol/water, 1:1) of an aqueous extract, or **B.** By Si HPLC (hexane/isopropanol, 95:5) of a chloroform extract of garlic homogenate. Identified as follows:

1. Methyl methanethiosulfinate
2. Allyl methanethiosulfinate
3. Methyl 2-propenethiosulfinate
4. trans-1-Propenyl methanethiosulfinate
5. Methyl trans-1-propenethiosulfinate
6. Allyl 2-propenethiosulfinate (allicin)
7. Allyl trans1-propenethiosulfinate
8. trans-1-Propenyl 2-propenethiosulfinate

(From Lawson LD, Wood SG, Hughes BG. HPLC analysis of allicin and other thiosulfinates in garlic clove homogenates. Planta Med 1991;57: 263–270.)

both. Monosulfides are only of minor significance. These components were discussed above, in Sections 3.2.6.3.1 and 4.2.4.

4.3.3.1 POLYSULFIDES: EARLIER APPROACHES

The first analysis of the sulfides of steam-distilled garlic was achieved by Semmler (1892a) using fractional distillation. Peyer, as early as 1927, tried to determine the sulfur-containing ingredients of several garlic preparations (pressed juice, oily extract, Tinctura Allii, and others) according to the method used at that time for analyzing the isothiocyanates ("mustard oil") in the seeds of mustard (*Semen Sinapis*). However, he emphasized that not all sulfur-containing constituents can be determined in this way (Peyer, 1927). Somewhat later, Breinlich (1950) applied this method as well, with the results summarized in Table 4.10.

Lehmann (1930) exploited a chromogenic reaction between sulfides and sodium nitroprusside in alkaline solutions which results in the generation of a red color. This reaction, which is very sensitive, occurs only with dialkyl disulfides and polysulfides, but not with monosulfides. Using this method, Kuhn (1932) examined the content of total volatile compounds in a number of garlic preparations (Table 4.11). The color reaction was modified by Markow (1964) to obtain a

Table 4.10. Determination of "Volatile Garlic Oil" (%) by Four Different Methods

Preparation	Argentometric	Gravimetric (KMnO^4)	Bromine Water	"Mustard Oil" (DAB 6)
Garlic clove	0.036	0.035	—	—
Garlic juice	0.032	0.034	0.029	—
Garlic powder 1	0.051	0.052	0.043	0.13
Garlic powder 2	0.040	0.040	0.030	0.07

From Breinlich J. Die Bestimmung des ätherischen Ölschwefels in schwefelölhaltigen Pflanzen unter besonderer Berücksichtigung des Knoblauchöls, Oxydation des Allyl-Senföls mit alkalischer Jodlösung. Pharm Zentralhalle 1950;89:217–223.

Table 4.11. Content (%) of "Volatile Garlic Oil" in Some Older Preparations

Garlic juice, raw	0.144
Garlic juice, filtered	0.083
Garlic decoction	0.056
Garlic decoction, pasteurized	0.062
Tinctura Allii sativi	0.063
Oleum Allii sat. via frig. parat.	0.061
Garlic juice, preparation unknown	0.034
Garlic juice, preparation unknown	0.070

From Kuhn A. Wertbestimmung Knoblauchhaltiger Produkte. In: Jahrbuch Madaus, n.p., 1932:41–43.

"micro method" for the determination of sulfur compounds in garlic.

Lehmann (1930) also measured the reducing compounds in garlic by potassium permanganate titration. The results were in fairly good agreement with those achieved by the nitroprusside method. Farber (1957) used this method for analyzing volatile compounds of garlic and onion. Indian scientists developed a method of analysis in which the volatile components of garlic are oxidized by chloramine T (Shankaranarayana et al., 1981).

A quantitative method for determination of isolated diallyl trisulfide (erroneously translated as allicin in the Chemical Abstracts) was developed for the Chinese Pharmacopeia of 1979. The method involves oxidation in a stream of oxygen, followed by titration of the obtained sulfuric acid (Xu, 1981).

The specificity of these various procedures are obviously dependent on the separation techniques employed prior to analytical determination. Although the procedures correctly describe total sulfur measurements (Section 4.3.7), they do not exclusively represent levels of the specific di- or polysulfides.

Schultz and Mohrmann (1965b) were the first to report the TLC separation of garlic-derived sulfides. Several liquid phase systems and methods of detection on silica gel plates were applied. A study by Hörhammer and colleagues (1968) deals with TLC determinations of garlic ingredients as well. A review article on TLC of sulfides and glucosinolates presumed to be present in garlic was presented by Danielak (1973). Differentiation and quantification of the polysulfides is also possible by means of ^{1}H-NMR analysis (Martin & Pearce, 1966). Block's group (1980b, 1980a) used the analogous ^{13}C-NMR and ^{17}O-NMR methods for this purpose.

4.3.3.2 POLYSULFIDES: GC AND HPLC METHODS

A very convenient method for determining volatile sulfur compounds from garlic preparations is *gas chromatography* (GC). Methodological approaches and applications for the separation and identification of garlic constituents by GC and later by GC/MS have been published by many authors, the work of whom cannot be discussed in every case. Volatile flavor compounds of garlic and other spices mainly consist of polysulfides (Carson & Wong, 1959b, 1961a, 1961b; Jacobsen et al., 1964; Saghir et al., 1964; Wahlroos & Virtanen, 1965; Schwimmer & Guadagni, 1967; Brodnitz et al., 1969; Freeman & Mossadeghi, 1971, 1973; Boelens et al., 1971; Freeman & Whenham, 1974a, 1974b, 1975a, 1975b; Legendre et al., 1980; Ariga & Hideka, 1985; Tokitomo & Kobayashi, 1992; Pino, 1992; Pino et al., 1991; Jirovetz et al., 1992b; Sheen et al., 1992; Skapska & Karwowska, 1990; Guo & Cui, 1990; Saito et al., 1989).

Table 4.12. Sulfides in the Oil of Steam-distilled Garlic and Their Relative Retention Times by Gas Chromatography

Methylpropyl sulfide	0.07
Dimethyl sulfide	0.10
Dipropyl sulfide	0.13
Dimethyl disulfide	0.16
Diallyl sulfide	0.22
Methylpropyl disulfide	0.40
Dipropyl disulfide	0.64
Dimethyl trisulfide	0.67
Diallyl disulfide	1.00
Methylallyl trisulfide	1.68

This table is shown for historical purposes and contains several errors. Propyl compounds are now known to be absent in garlic and its oils. Missing are diallyl trisulfide and diallyl tetrasulfide, which are abundantly present.
From Balrati G, Cagna D, Giannone L. Contributo alla conoscenza dell' olio essenziale d'aglio. Ind Conserve 1970; 45:125–130.

Oaks and colleagues (1964) separated 18 sulfides in the vapor of fresh garlic. Schultz and Mohrmann (1965a) conducted GC analysis on garlic oil, and Bernhard (1968) identified six sulfides in onion oil by this method. Baldrati's group (1970) determined 10 sulfur compounds in fresh garlic and in two samples of garlic oil (Table 4.12). Brodnitz and colleagues (1971) successfully used the GC method, as well as Yu's group (1989) and Yan's group (1993). Vernin and colleagues (1986) were able to identify eight different oligosulfides by GC/MS, and Tokarska and Karwowska identified 19 individual sulfides using the same method (Tokarska & Karwowska, 1983). Block's group (1984) applied GC methods as well for fresh garlic and garlic powders. Recently, Deruaz and colleagues (1994) applied an analytical strategy of coupling headspace GC, atomic emission spectrometric detection, and MS to the volatile sulfur compounds from garlic.

However, since higher polysulfides are thermally unstable (i.e., those with four and more atoms in the sulfur chain), GC methods are restricted to mono-, di-, and trisulfides present in garlic samples when using polar GC columns (with nonpolar columns tetrasulfides can also be analyzed) (Lawson et al., 1991a; Vernin & Metzger, 1991). The mono-, di-, and trisulfides account for about 80% of total sulfides in steam-distilled garlic oil, but during storage a conversion to higher sulfide polymers may occur (Lawson, 1993). Thus, GC analysis of garlic sulfides is the preferred method for the majority of the volatile compounds (e.g., for flavor validation). However, quantitative GC evaluation of garlic products containing pentasulfides or larger (only 5% of the total sulfides) is somewhat limited.

HPLC methods for qualitative and quantitative determination of all sulfides in garlic products seem to be the most promising procedures, since they are nondestructive when operated at room temperature. In 1986, Voigt and Wolf reported the analysis of di- and trisulfides by HPLC, as did others (Winkler et al., 1992; Wu & Wu, 1981; Miething, 1988; Iberl et al., 1990b). Recently, Lawson and colleagues (1991a) published a comprehensive investigation of the HPLC separation, spectrophotometric properties, retention times, and extraction behavior of 20 sulfides from steam-distilled garlic. Extinction coefficients were found to increase according to the number of sulfur atoms, and a linear relationship between the logarithm of absolute C18 HPLC retention times and the number of sulfur atoms was also demonstrated. These results permit identification and quantitation (via relative extinction coefficients) of nearly all the sulfides, with only a few standards that can be purchased or easily prepared.

Complete extraction of the sulfides of steam-distilled garlic oil from oil-macerates can be achieved with acetonitrile. Through the use of an eluant consisting of acetonitrile/water/tetrahydrofuran (70/27/3) and a detection wavelength of 240 nm, separation of diallyl, allyl methyl, and dimethyl sulfides with one to seven sulfur atoms was achieved by C18 reversed phase HPLC (Fig. 4.3). However, allyl methyl sulfide and dimethyl disulfide could only be separated from one another when methanol/water/tetrahydrofuran (50/45/5) was employed as the eluant (Lawson et al., 1991a). Similar investigations on garlic preparations using HPLC methods were performed by Winkler's group (1992), and on shallot by Wu and Wu (1981).

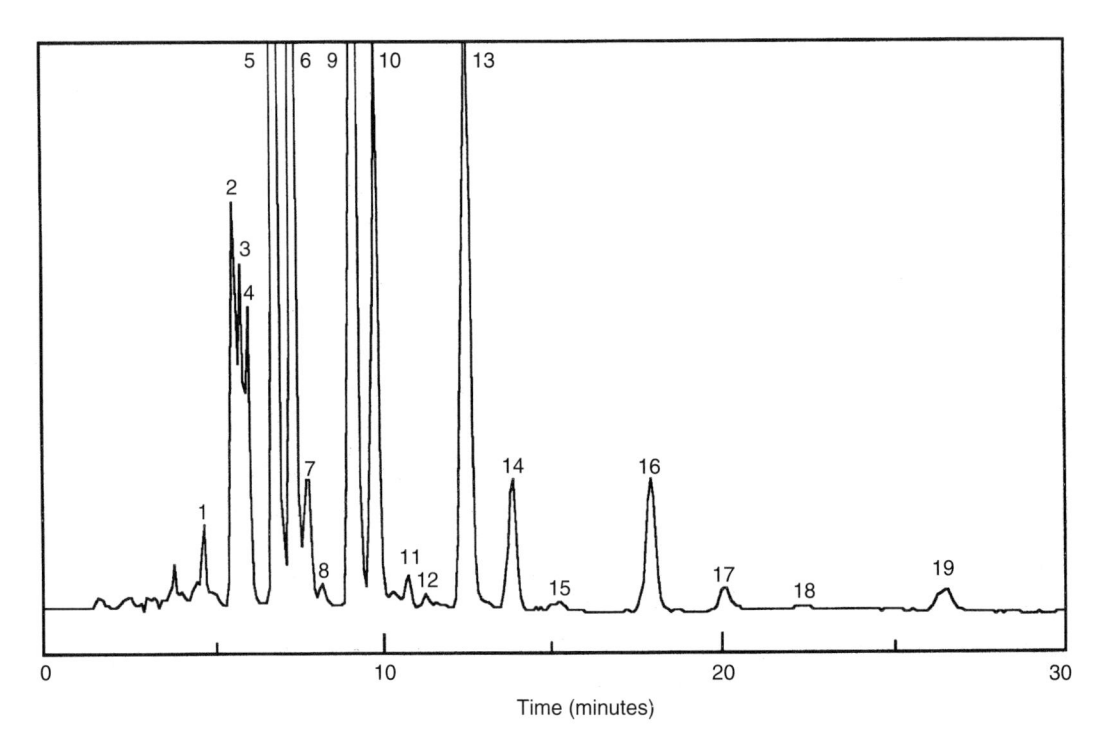

Figure 4.3. Analysis of a typical commercial oil of steam-distilled garlic by C18 HPLC (acetonitrile/water, 70:30). Identified as follows:

1. Allyl methyl sulfide and dimethyl disulfide
2. Allyl methyl disulfide
3. Diallyl sulfide
4. Dimethyl trisulfide
5. Diallyl disulfide
6. Allyl methyl trisulfide
7. Dimethyl tetrasulfide
8. Allyl trans-1-propenyl disulfide
9. Diallyl trisulfide
10. Allyl methyl tetrasulfide
11. Dimethyl pentasulfide
12. Allyl trans-1-propenyl trisulfide
13. Diallyl tetrasulfide
14. Allyl methyl pentasulfide
15. Dimethyl hexasulfide
16. Diallyl pentasulfide
17. Allyl methyl hexasulfide
18. Dimethyl heptasulfide
19. Diallyl hexasulfide

(From Lawson LD, Wang ZJ, Hughes BG. Identification and HPLC quantitation of the sulfides and dialk(en)yl thiosulfinates in commercial garlic products. Planta Med 1991;57:363–370.)

Standard samples of mono- and polysulfides can be prepared from commercially available compounds, either by fractional distillation of crude diallyl disulfide, preparative HPLC, synthesis or semisynthesis of several compounds, or by combinations of these procedures (Lawson et al., 1991a; Carson & Wong, 1959b).

4.3.4 Vinyldithiins

Vinyldithiins are presumed to be the main active ingredients of garlic oil-macerates, since they are the most abundant sulfur compounds in these preparations (see Section 3.2.6.3.3). They can be determined by using either GC or HPLC methods.

Brodnitz and colleagues (1971) reported on the thermal decomposition of allicin when injected into a GC system. About 78% of the total chromatogram peak areas represented two compounds, which were identified, on the basis of NMR and MS data, as 3-vinyl-[4H]-1,2-dithiin (23.5%) and 2-vinyl-[4H]-1,3-dithiin (55.4%). Block's group (1984) described GC measurements of allicin transformation products, including the vinyldithiins. Koch and Jäger (1989) determined these two dithiins by a GC method, using dipropyl disulfide as the internal standard, in order to trile/water/methanol; 50/41/9) on C18 columns which separates vinyldithiins, allicin, ajoenes, and sulfides. They used this method to study the kinetics of formation of these compounds as well as the influence of solvent polarity on the amounts of each produced. They also determined the amounts of these compounds in several types of European oil-macerate products with vinyldithiins ranging between 0.5 and 1.2 mg/g (Winkler et al., 1992). Lawson and colleagues (1991a) applied a similar method of analysis to vinyldithiins in commercial oil-macerated garlic products (Fig. 4.4), with vinyldithiins ranging between 0.07 and 0.7 mg/g of product.

Pure preparations of both vinyldithiins have been achieved by heating crude allicin in acetone/methanol (6/4), followed by purification by column chromatography or preparative HPLC (Block, 1986; Block et al., 1986; Iberl et al., 1990a, 1990b). These compounds have also been prepared by incubation of purified allicin in hexane for 1.5 hours at 45°C, followed by TLC purification, and they were identified by GC/MS and found to have a molecular weight of 144. The 1,3-isomer is distinguished from the 1,2-isomer by its consistent predominance (Lawson et al., 1991a).

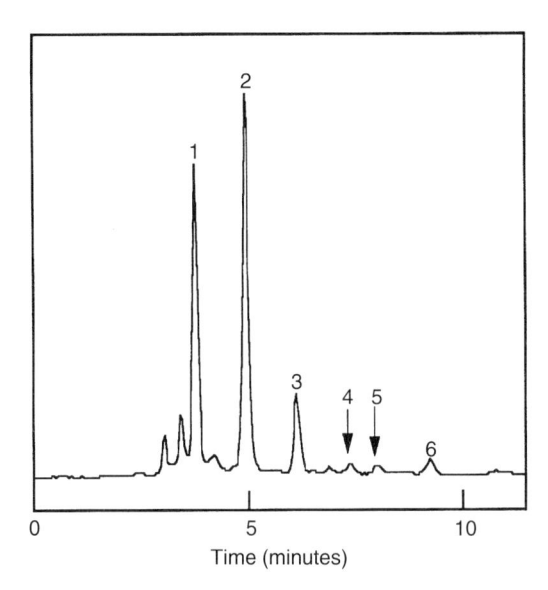

Figure 4.4. Analysis of a typical commercial oil-macerate of garlic by C18 HPLC (acetonitrile/water, 70:30). Identified as follows:

1. *E*- and *Z*-ajoene
2. 1,3-vinyldithiin
3. 1,2-vinyldithiin
4. diallyl disulfide
5. allyl methyl trisulfide
6. diallyl trisulfide

(From Lawson LD, Wang ZJ, Hughes BG. Identification and HPLC quantitation of the sulfides and dialk(en)yl thiosulfinates in commercial garlic products. Planta Med 1991;57:363–370.)

measure allicin indirectly. Sarter and colleagues (1991) reinvestigated this application and changed the analytical conditions (mainly by applying lower temperatures), thus avoiding thermal decomposition of 2-vinyl-[4H]-1,3-dithiin in particular (Bock et al., 1982). The improved procedure was used to analyze vinyldithiins in garlic oil-macerates. Iberl and colleagues (1990a) used GC analysis to check the purity of synthesized vinyldithiins.

HPLC determination of vinyldithiins circumvents the possibility of thermal cleavage of these compounds, since this method can be applied at ambient temperature. Voigt and Wolf (1986) were the first to analyze vinyldithiins in oil-macerated garlic by this method.

Iberl's group (1990a, 1990b) described an isocratic HPLC elution system (acetoni-

4.3.5 Ajoenes

Block's group (1984; 1986) was the first to report on a new group of oxygen-containing transformation products from allicin (see Section 3.2.6.3.3). These compounds, called *ajoenes*, are predominantly formed when allicin or allicin-containing garlic extracts are stored or heated in moderately polar solvents like acetone/water, acetone, ethanol, and the like (Iberl et al., 1990a), and are commercially found only in oil-macerated garlic products, in which they account for about 12% of the sulfur components.

The ajoenes consist of two isomers, *(Z)*- and *(E)*-4,5,9-trithiadodeca-1,6,11-triene-9-oxide (see Figs. 3.12 and 3.14). The *Z*-isomer is dominant when prepared in nonpolar solvents (vegetable oil, hexane, ether), whereas the *E*-isomer is dominant when prepared in polar solvents (alcohol, acetone/water) and can be increased by heating the reaction medium (Iberl et al., 1990a). Commercial products typically have an *E/Z* ratio of approximately 2, with a range of 0.6 to 2.5 (Iberl et al., 1990a; Lawson et al., 1991a). The ratio of ajoenes to vinyldithiins also increases with solvent polarity, with the lowest for ether (0.1) and the greatest for isopropanol (2.0) (Iberl et al., 1990a). In addition, dimethyl-substituted ajoenes and about half that level of monomethyl-substituted ajoenes may be present in garlic preparations, albeit to a lesser extent. These are transformation products of allyl methane thiosulfinate (Block et al., 1988), the second most abundant thiosulfinate produced in garlic clove homogenates (Lawson et al., 1991a).

The only analytical method which has been reported for the determination of ajoenes is the HPLC procedure (Fig. 4.4), due to the instability of the ajoenes in a heated GC column (Block et al., 1986; Lawson et al., 1991a). Voigt and Wolf (1986) published a reversed phase HPLC method for ajoene and other sulfur compounds, but without separation of the isomeric or homologous ajoene components. Iberl's group (1990b) and Lawson's group (1991a) achieved the separation of *E*- and *Z*-ajoene with normal phase (Si) HPLC using hexane/isopropanol as the eluant. Sendl's group (1992) separated the methyl ajoenes by reversed phase HPLC.

For standard sample preparations, ajoenes can be synthesized and purified (Block, 1986; Iberl et al., 1990a; Lawson et al., 1991a). According to Iberl and colleagues (1990a), pure or crude allicin, when stored in isopropanol for a period of 12 days at room temperature, yields a maximum amount of *E*- and *Z*-ajoenes which can be separated by preparative Sephadex LH-20 column chromatography.

4.3.6 γ-Glutamyl Peptides

So far, three γ-glutamylcysteines have been identified as genuine components of garlic cloves (see Section 3.2.3). They consist of *S*-allyl, *S*-1-propenyl and much smaller amounts of *S*-methyl derivatives of γ-L-glutamyl-L-cysteine. Most of the studies on these peptides (some of which are present in onions as well) were conducted prior to 1970 and involved separation by electrophoresis and chromatography (Virtanen, 1965, 1969; Matikkala & Virtanen, 1967; Granroth, 1968, 1970). With modern HPLC methods, comprehensive and more quantitative investigations have been performed (Lawson & Hughes, 1990; Lawson et al., 1991b, 1991c; Lawson, 1993; Mütsch-Eckner et al., 1992).

In principle, GC analysis of these low molecular weight peptides could be applied after derivatization, but determinations using this method have not been described so far. It is more convenient and less destructive to separate and detect the γ-glutamylcysteines by HPLC (Fig. 4.5). These compounds can be assayed directly in aqueous extraction solutions of fresh garlic, garlic powder, garlic powder preparations, and seeds of other *Allium* species. Appropriate procedures were published by Lawson and colleagues (1991b) using C18 HPLC, the composition of the eluant being 0.05 mol/L KH_2PO_4 (pH 4.5)/MeOH (97.5/2.5). Variations of the pH value (2.5 and 3.0) and application of gradient elution improves the resolution of these and several other constituents of garlic. The γ-glutamylcysteines have also been analyzed by HPLC after derivatization with *o*-phthaldialdehyde/*tert*-butylthiol (Mütsch-Eckner et al., 1992). Pure γ-glutamyl peptides for standards can be obtained either by preparative HPLC methods or by peptide synthesis procedures.

The γ-glutamylcysteines can also be assayed indirectly by HPLC analysis of the *S*-alkenylcysteines (or *S*-methylcysteine) after treating the garlic with γ-glutamyl-transpeptidase (Lawson et al., 1991b). For a standard in determining the content of γ-glutamyl-*S*-allylcysteine, *S*-allylcysteine is easily synthesized by reacting allyl bromide with cysteine. Because *S-trans*-1-propenylcysteine is difficult to prepare, the content of γ-glutamyl-*S-trans*-1-propenylcysteine can be determined on the basis of its extinction

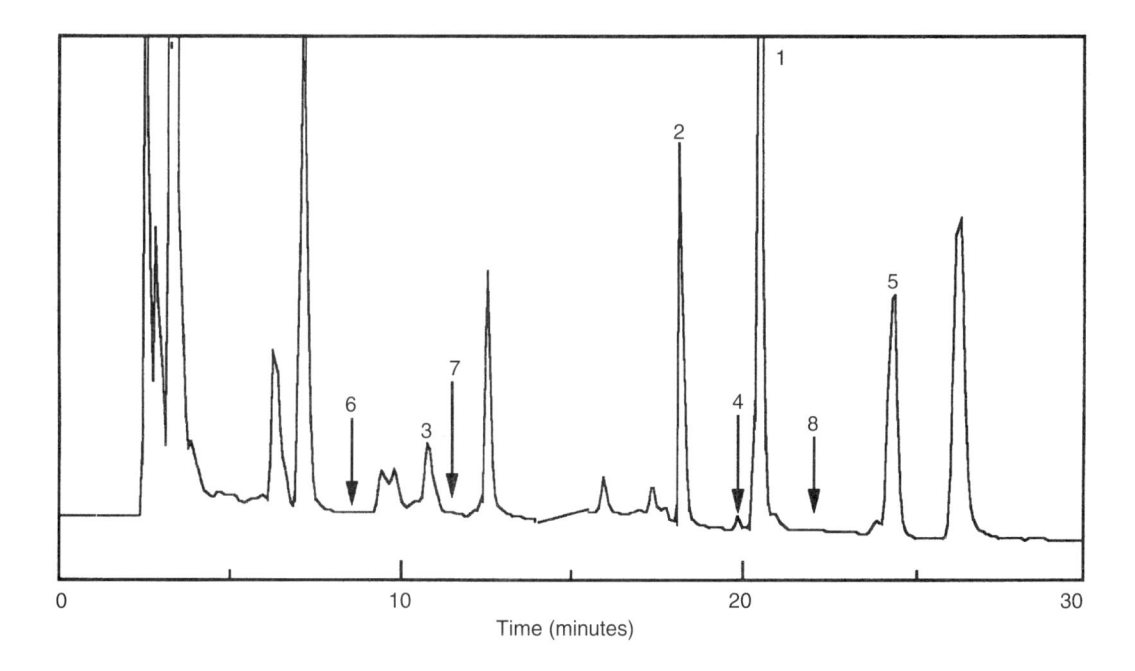

Figure 4.5. Analysis of the γ-glutamylcysteines and related compounds in a garlic clove homogenate. Identified as follows:

1. γ-Glutamyl-S-trans-1-propenylcysteine
2. γ-Glutamyl-S-allylcysteine
3. γ-Glutamyl-S-methylcysteine
4. γ-Glutamyl-S-cis-1-propenylcysteine
5. γ-Glutamylphenylalanine
6. Sulfoxide of 1 (absent)
7. S-Allylcysteine
8. S-Allylmercaptocysteine (absent)

(From Lawson LD, Wang ZYJ, Hughes BG. Gamma-glutamyl-S-alkylcysteines in garlic and other Allium spp.: precursors of age-dependent trans-1-propenyl thiosulfinates. J Nat Prod 1991;54:436–444.)

coefficient relative to that of γ-glutamyl-S-allylcysteine.

4.3.7 Elemental Sulfur

The sulfur compounds of garlic can be subdivided into organosulfur and inorganic sulfur components. As potentially biological active constituents, only the organosulfur compounds are of special interest. Since the organosulfur compounds have already been discussed in detail, this section will deal only with methods for determination of total *elemental sulfur*.

4.3.7.1 ELEMENTAL SULFUR: EARLIER APPROACHES

Determination of total sulfur content in garlic cloves and in analytical fractions were historically among the first attempts to estimate the amount of active ingredients. Peyer in 1927 tried to determine the value of different garlic preparations according to the method applied for "mustard oil" (Peyer & Remund, 1928). Lehmann (1930) used the nitroprusside reaction to generate a red color with di- and polysulfides, and Kuhn (1932) examined the content of "volatile garlic oil" (Table 4.11). Markow (1964) used a micromethod application of this procedure for determining the sulfur compounds of garlic. Zhang (1985) in China oxidized the sulfur compounds either to SO_4^{2-}, which was then precipitated as $BaSO_4$ and measured gravimetrically, or reduced them to S^{2-} by the $HgNO_3$ precipitation method with subsequent titration of the superfluous Hg^{2+} with Fe^{3+} and SCN^-. However, gravimetric and argentometric methods have long been used for this purpose (Gstirner, 1955; Ebert, 1982).

Three methods, not restricted to sulfur components, have been applied by several

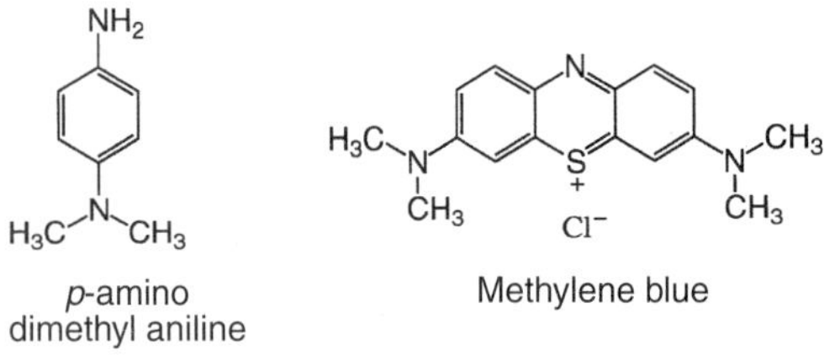

Figure 4.6. Reagent and reaction product for the methylene blue determination of total water-soluble sulfur.

authors (Lehmann, 1930; Farber, 1957; Abramjan & Sarkisjan, 1963). Potassium permanganate titration has been used; measured the iodine number of garlic samples has been measured (Zwergal, 1952); and both bromine water oxidation and alkaline permanganate solution have been used together with $BaSO_4$ precipitation of the produced sulfate (Breinlich, 1950).

Breinlich (1950) as well as Guillaume and Wadie (1950) used another method for the estimation of total sulfur in garlic oil, applying silver nitrate that had been heated with aqueous garlic samples in order to form Ag_2S. The excess of $AgNO_3$ was determined by titration with rhodanide. The results of the different procedures are summarized in Table 4.10.

Currier (1945) described a colorimetric determination of water-soluble sulfur compounds. This involves acid hydrolysis, reduction with zinc dust, and reaction with p-amino-N,N-dimethyl aniline. The dye methylene blue is formed (Fig. 4.6), which is spectrophotometrically determined. Kosyan (1985) mineralized the organic sulfur compounds to inorganic S^{2-} at 200°C and determined this by means of the color reaction with p-amino-N,N-diethyl aniline (absorbance at 669 nm). Researchers in Egypt employed the acidimetric silver method and the cupric butyl phthalate method for this purpose (El-Hadidy et al., 1981).

Jäger (1955) obtained about 0.8 g steam-distilled oil from 100 g garlic cloves with 176 mg total sulfur. The distillation residue still contained 138 mg sulfur. Thus only a fraction of the element could be obtained in the volatile portion.

4.3.7.2 ELEMENTAL SULFUR: MODERN METHODS

Nowadays, determination of total sulfur in foodstuffs and plant material is mainly performed by measuring sulfate concentrations in solutions obtained after complete oxidation of the samples. The sulfate is then determined by using instrumental methods like atomic absorption and induced-coupled plasma spectroscopy. In practice, wet incineration with $HNO_3/HClO_4$ or combustion in oxygen and completion with H_2O_2 (preferably in autoclaved systems) is applied to promote the generation of sulfate. Decomposition with oxidizing acids at elevated temperatures can conveniently be achieved by application of microwave treatment in commercial devices with Teflon digesters. As such, several methods for determination of sulfate alone are commonly employed. Precipitation as $BaSO_4$ in a solution with surface active additives (e.g., Tween 80) and measurement of turbidity can be applied (Hunt, 1980).

Decoloration of a Ba-methylthymol blue solution was described by Madsen and Murphy (1981), where interference by calcium and other metal ions was avoided with an in-line cation exchange tube in the analytical part of the instrumentation. This method can be automated, and a flow-injection system can measure up to 180 samples per hour. Ion chromatography (IC) can be used as well for determining sulfate, with detection achieved either with a conductive cell or by indirect UV absorption.

The total sulfur content of garlic is about 1.0% of its dry weight or 0.35% of its fresh weight (Alfonso & Lopez, 1960; Pentz et al., 1990; Ueda et al., 1991). The cysteine sulfoxides and γ-glutamylcysteines account for about 82% of the total sulfur, with most of the remainder present in soluble protein (3%), sulfate (5%), and insoluble compounds (6%) (Lawson, 1993).

4.3.8 Hydrolysis to Thiols

Beneficial effects of garlic ingredients in biological systems are quite unequivocally dependent upon sulfur-containing compounds. As mentioned previously, various garlic preparations with different organosulfur substances are claimed to show positive health effects. In order to achieve a common scaling standard for these preparations, the potency of SH group liberation (thiol formation) upon alkaline hydrolysis was proposed as one method for standardization (Pentz et al., 1990, 1992). Although garlic contains no SH compounds (Han et al., 1995), their formation may be the fate of some of the organosulfur compounds in humans after ingestion of garlic and garlic preparations (Koch, 1992; Lawson & Wang, 1993).

Hydrolysis is typically performed by incubation of garlic samples with 4 M NaOH at 95°C for 2 hours in closed tubes. Oxidation of the thiols generated is avoided by using an argon atmosphere. The reaction mixture is then transferred into a buffer solution and treated with 5,5'-dithio-bis(2-nitrobenzoic acid), whereafter the reaction product is measured photometrically at 412 nm for the determination of SH groups (Ellman, 1959).

Several pure compounds (cysteine derivatives, di- and polysulfides) appearing in garlic samples have been analyzed, and formation of thiols upon hydrolysis has been calculated to be about 30% (cysteines) and 90% (sulfides) of the theoretically possible liberation. With garlic preparations, good correlations have been established between amounts of SH groups and total sulfur, alliin, allicin, and γ-glutamylpeptides. In general, SH group determination provides a convenient method to estimate total biologically active sulfur compounds in garlic samples without expensive instrumentation (Pentz et al., 1992); however, some precautions are mentioned in Section 3.4.7.

4.4 OTHER GARLIC COMPONENTS

There are several sources of information available on the total content of common constituents of garlic such as water, protein, fat, carbohydrates, vitamins, minerals, and so on (see Table 3.1) (Souci et al., 1986; Liu et al., 1989). Here, we restrict ourselves to several additional compounds that are among the presumed active substances of garlic cloves.

Isoalliin (S-trans-1-propenylcysteine sulfoxide) was first characterized in 1961 (Virtanen & Spåre, 1961), and it is the major S-substituted cysteine sulfoxide found in onion. Its presence in garlic was first indicated by the discovery of trans-*1-propenyl thiosulfinates* in garlic homogenates (Lawson & Hughes, 1990; Lawson et al., 1991b, 1991c). Isolation and structure elucidation revealed its identity with the component present in onion (Mütsch-Eckner, 1991). Separation of alliin and isoalliin has been achieved by HPLC of the derivatized compounds (Thomas & Parkin, 1994; Mütsch-Eckner et al., 1992). The summation of thiosulfinates containing 1-propenyl groups has been used to determine isoalliin. The same applies to *methiin (methylcysteine sulfoxide)*, which can also be measured indirectly by determining total *methyl thiosulfinates* using HPLC procedures (Lawson et al., 1991c; Lawson, 1993).

S-Allyl cysteine (SAC) is claimed to be the main biologically effective compound, or at least the quality-determining component, in "aged garlic extract" (see also Sections 3.2.7, 3.4.5, and 5.13.1). SAC can be determined with the same HPLC system that is used for γ-glutamylcysteines measurements (Fig. 4.5 and Section 4.3.6.). When fresh garlic homogenates were analyzed, SAC was only found in very small amounts in some of the strains and not at all in others (Lawson, 1993; Pentz R, unpublished data, 1994). However, in garlic powder and powder preparations this component can be detected (0.2–1.1 mg/g dry weight). In samples of "aged garlic extract" formulations, about the same small amount of SAC could be measured, as reported in an analysis of 55 garlic brands (Hansen et al., 1993).

Several other sulfur-containing ingredients of fresh garlic and processed garlic have been described, but they are of low concentration and are not likely to contribute toward the beneficial health effects to any significant extent.

Adenosine has been proposed as an important antithrombotic compound from garlic (see also Section 3.3.11), but the adenosine content of fresh garlic is small (0.02 mg/g) (Lawson et al., 1991c). The quantitative determination of this nucleoside can be performed using various methods. Enzymatic measurement with adenosine deaminase (adenosine aminohydrolase), which produces inosine, has been described by Möllering & Bergmeyer (1974). More recently, HPLC methods using reversed phase columns with a detection wavelength of 254 nm have been widely applied for nucleoside analysis in biological material (Ontyd & Schrader, 1984).

Scordinins are complex thioglycosides of uncertain structure which have been suggested to be important natural ingredients of garlic (see also Section 3.2.9). According to Hatanaka and Kaneda (1980), an enzymatic determination of scordinins can be performed. A sample is hydrolyzed in 10% NaOH for 2 hours at 100°C, as a result of which the alanine fraction of scordinins is converted to lactic acid. The lactate concentration is determined with lactate dehydrogenase (LDH) and is related to the genuine scordinin amount.

Saponins of garlic consist of several sapogenins and glycoside moieties, with oligosaccharide chains containing five different types of monosaccharides (see also Section 3.3.7). The sapogenins themselves are derivatives of either pentacyclic triterpenes or steroids of spironostanol and furostanol type. The complex mixture of saponins can be separated by TLC on silica plates, and detection can be achieved by spray reagents (Oakenfull, 1981). Semiquantitative measurements can be performed by application of densitometry (Smoczkiewicz et al., 1977; Fenwick & Oakenfull, 1983).

Other chromatographic methods (e.g., HPLC) may also be helpful in the separation and quantification of saponins, although none has yet been published. A more convenient procedure is achieved by determining the saponins enzymatically, using the adenosine deaminase (ADA) suppression assay (Koch et al., 1992a, 1992b, 1993, 1994). A comprehensive description of results on saponin evaluation in garlic has been provided by Koch (1993).

Separation and identification of the complex *carbohydrates* of garlic bulbs (scorodose, etc.) has been accomplished by HPLC (Deki, 1983). The chemical composition of the *fatty acids, phospholipids*, and *sterols* in garlic cloves has been investigated by TLC and HPLC (Huq et al., 1991).

4.5 PROPOSALS FOR STANDARDIZATION OF GARLIC PREPARATIONS

Basically, two different kinds of garlic preparations must be considered in the problem of standardization: (*a*) garlic preparations that more or less represent fresh garlic in a stabilized condition—i.e., appropriately processed garlic powder preparations); and (*b*) formulations with ingredients not naturally present in fresh garlic, representing artificially generated transformation products of genuine garlic components—i.e., garlic oil from oil-macerated or steam-distilled preparations and aged garlic extract.

4.5.1 Garlic Powder Preparations

The main biologically active agents of garlic powder are believed to be the organosulfur compounds alliin (as the precursor of allicin) and, possibly, the γ-glutamyl-S-alkylcysteines. However, alliin as such has been assumed to provide no or low beneficial health effects, whereas γ-glutamyl peptides may have some, although few studies have been published on them so far. Alliin in garlic powder has to be converted into allicin by means of alliinase activity when it comes into contact with water. Garlic powder preparations, therefore, should contain an active alliinase system, sufficient to convert rapidly all of the alliin present.

Quality control of dry garlic powder has recently been discussed by Müller and Ruhnke (1993), involving TLC, GC, and HPLC methods for the identification and determination of alliin and allicin. As for the finished garlic powder preparations, we would propose a standardization based on the yield of allicin upon contact with water (allicin-releasing potential). Using pure and stable alliin as the standard addition compound, quantification can easily be achieved. Allicin is a less stable standard, but is much easier to isolate or synthesize than alliin, and it is stable in aqueous solution when stored at −70°C. The appropriate instrumentation for the analytical procedure would be an HPLC device with a reversed phase C18 column and an ultraviolet light detector as described in Section 4.3.2.2.

A more complete standardization should take into account, in addition to allicin or total thiosulfinates, the quantitative determination of two dominant γ-glutamyl peptides, γ-glutamyl-S-trans-1-propenylcysteine and γ-glutamyl-S-allylcysteine. HPLC again is the method of choice (Section 4.3.6.), but one problem which remains even now is the availability of pure compounds for preparing standards (Jäger et al., 1992). Procedures for their isolation from garlic or from seeds of other species have been described (Lawson et al., 1991b). However, they can be assayed indirectly as discussed in Section 4.3.6.

The results of standardization measurements should not be expressed as percentages (e.g., allicin yield 0.6%) if no clear relation is obvious (i.e., with no statement as to whether this applies to the powder itself or to the tablet). An unambiguous description could be provided by expressing values in mg alliin or allicin/tablet. Also, when describing allicin, words like "contains" or "content" should be replaced by "yields" or "releases" or "potential," since, like fresh garlic, no garlic product contains allicin until active alliinase is released.

Typically, a daily dosage of 3.6 to 5.4 mg allicin is recommended in order to achieve sufficient beneficial health effects, which corresponds to 0.6–1.8 g fresh garlic (containing 0.3% allicin releasing potential) or 0.3–0.9 g garlic powder (yielding 0.6% allicin). German authorities have determined an amount of 4 g fresh garlic/day as necessary to confer health effects (Spinka & Stampfer, 1956), although this figure is considered by some to be too high (Sticher & Mütsch-Eckner, 1991).

4.5.2 Garlic Oil Preparations

Preparations containing garlic oil are more difficult to standardize. As far as pure oil-macerates are concerned, the main ingredients (approximately 70% organosulfur compounds) are represented by two vinyldithiins. Assuming, then, that the other biologically active sulfur compounds (sulfides and ajoenes) are always present in a more or less constant fraction related to vinyldithiins, measurement and declaration of the vinyldithiins will be sufficient for standardization purposes (Koch & Jäger, 1989; Koch, 1991). Methodologically, GC procedures can be applied for the measurement of vinyldithiins or, more conveniently, HPLC determination according to the procedures described in Section 4.3.4. Of course, pure vinyldithiins are required in order to prepare standard samples, but if the analytical detection response for vinyldithiins and a commercially available compound like diethyl sulfide or diallyl disulfide are established first, the latter may serve as a reference standard for further determinations.

As for garlic oils produced by steam distillation, about 20 allyl and methyl sulfides represent the organosulfur ingredients. To circumvent the separation and quantification of these specific compounds, a determination of total sulfur would be appropriate and sufficient. However, devices that perform elemental analysis are not common. As a more convenient alternative for routine analyses can be established by employing a linear correlation between total sulfur (Section 4.3.7.2) and hydrolysis-releasable SH compounds (Section 4.3.8).

Of course, a more sophisticated measurement of the dialk(en)yl sulfides using GC (taking care to avoid loss of polysulfides by thermal decomposition) or HPLC methods can be performed. Again, the availability of pure reference compounds is restricted, but commercially available diallyl disulfide may be used after purification as a standard.

Several garlic oil formulations consist of mixtures from oil-macerated and steam-distilled products. For these preparations, standardization is rather complicated, and instead one might propose that either some leading substances (vinyldithiins and diallyl disulfide) or total sulfur be measured (Koch & Jäger, 1990). Claims made for commercial oil preparations can be related either to total sulfur content or to amounts of specific organosulfur compounds; however, it is not appropriate to back-calculate to the amounts of garlic used to produce a specific quantity of oil from measurements made on ingredients present in these processed oils (Koch & Jäger, 1990).

4.6 SUMMARY

Garlic, as a medicinal plant and a natural remedy, is indexed in many contemporary pharmacopeias (Table 4.13), and a variety of preparations are still derived from it. In recent years, several approaches have been taken to evaluate the quality of raw garlic and its preparations. Depending on the type of processing, a number of different compounds are claimed to deliver the health-beneficial properties of garlic products. With the increasing capabilities of modern analytical methods, most of the compounds have now been identified and quantified. Isolation of various substances from fresh and processed garlic in reasonable amounts has made it possible to administer these compounds to laboratory animals and investigate their biological effects. From these studies it has been evident that some compounds revealed rather specific effects, whereas others acted in a manner more typical of garlic as a whole. However, in humans, investigations with garlic samples of defined analytical composition have mainly been performed with garlic powder preparations. Hence, the measurements of various other garlic preparations with regard to their beneficial health effects remain to a certain extent arbitrary. In practice, the quality of garlic formulations is related to the content of marker compounds or suspected active compound groups. In powder preparations, the role of alliin as the principal active constituent remains unchallenged, and its level may be taken as a measure of the quality of the product. Obviously, allicin as the primary transformation product of alliin is a better marker, because its generation indicates an active alliinase system, which is necessary for the formation of allicin and other thiosulfinates. These secondary products, though not identified under physiological conditions, are believed to play an important role in biological systems.

Other garlic preparations, such as oils of steam-distilled garlic homogenates and garlic macerated in vegetable oils, reveal a variety of allicin (and probably other thiosulfinates) transformation products. Measurement of vinyldithiins, ajoenes, or other organosulfur compounds may be performed for quality characterization and standardization.

Nowadays, it is not the lack of instrumentation or analytical procedures that restricts an elaborate standardization; it is the unavailability of pure standard compounds that prevents a convenient quantification to describe the quality of the ranges and types of garlic products available on the market.

Table 4.13. Garlic in Various Pharmacopeias

Bulbus Allii sativi	Deutsche Arzneibuch (DAB) 6, Erg. B.
Allii Bulbus	Pharmacopoea Hispanica (Hisp.) IX
Allium	Indian Pharmaceutical Codex (Ind. P.C.) 1953
Garlic	British Pharmaceutical Codex (B.P.C.) 1949
Ail	Codex Français (C.F.) 65
Allium sativum	Homöopathische Arzneibuch (HAB) 1934, 1953, 1978
Garlic	Homeopathic Pharmacopoeia of the United States (HPUS) 1964, 1979
Garlic	British Herbal Compendium 1992
Garlic	British Herbal Pharmacopoeia 1990

5
Therapeutic Effects and Applications of Garlic and Its Preparations

HANS D. REUTER, HEINRICH P. KOCH, AND LARRY D. LAWSON

Garlic has captured a secure position in modern medical science, especially in the years following World War II. It had formerly been popular—and still is, to some extent—as a carminative for dyspeptic problems and diarrhea, as an antimicrobial for bacterial, fungal, and viral infections, as well as a vermifuge for intestinal parasites. In recent years, garlic has become highly valued because of its excellent effectiveness in arteriosclerosis, its ability to lower elevated serum cholesterol and triglyceride levels, its hypotensive effects, its anticarcinogenic effects, and its antidiabetic effect on moderately elevated blood sugar levels. Garlic also inhibits thrombocyte aggregation and activates fibrinolysis. Numerous reviews on the medicinal effects of garlic have been written, and new studies are continually being published (Reuter & Sendl, 1994; Koch & Hahn, 1988; Kritchevsky, 1991; Fulder & Blackwood, 1991; Kabelik, 1970; Pros, 1979; Lutomski, 1983, 1980; Smoczkiewiczowa et al., 1981; Koch, 1985a 1985b; Becker, 1985; Orth-Wagner, 1986; Sprecher, 1986a, 1986b, 1986c, 1989; Weiss, 1986; Kendler, 1987).

In this chapter the most important indications for garlic and its preparations are treated in greater detail. Primary focus is placed on the question of which of the effects of garlic can be pharmacologically or clinically documented and which constituents of the plant are responsible for these effects. While fresh garlic is not tolerated in larger amounts by many people, reasonable amounts and many pharmaceutical preparations of garlic are well tolerated. Contraindications for garlic therapy are practically unknown. A reduced dosage is recommended when irritation or inflammation of the gastrointestinal tract are manifest (Koch, 1992c; Scheibe, 1958). See Chapter 7 for a discussion of tolerable and adverse levels of garlic consumption.

5.1 EFFECTS ON THE HEART AND CIRCULATORY SYSTEM

Of all the effects of garlic that have been reported over the years, perhaps the most interesting are those on the heart and circulatory system. By appropriate application, garlic may protect the blood vessels from the deleterious effects of free radicals, exert a positive influence on blood lipids, increase capillary flow, and lower elevated blood pressure levels. This means that development of arteriosclerosis can be prevented or an already existing condition favorably influenced.

The cardiovascular effects of garlic were essentially rediscovered in the late 1960s (Srinivasan, 1969; Papayannopoulos, 1969; Kendler, 1987), although they were also documented in several older publications (May, 1926; Schlesinger, 1926; Taubmann, 1934; Weiss, 1957; Mansell & Reckless, 1991). However, garlic is an old constituent of folk medicine for treating cardiovascular problems in Asian countries (Nadkarni, 1954; Sreenivasamurthy & Krishnamurthy, 1959; Lau et al., 1983; Kritchevsky, 1991), and has also been used by many people in European countries for some time. According to recent studies, some of the antiatherosclerotic effects are based on the reduction of thrombocyte adhesiveness and aggregation. The tendency of the platelets to aggregate and to form thrombi is significantly decreased by the effective constituents of garlic. In addition, fibrinolysis is enhanced, resulting in more rapid dissolution of coagulated blood, plaques, and clots. The decrease of lipoproteins circulating in the blood in the form of LDL (low density lipoprotein), and of cholesterol, occurs through increased formation of antiatherogenic HDL (high density lipoprotein) at the expense of LDL (Reuter, 1980, 1983b, 1991; Ernst, 1983, 1984; Lau et al., 1983). Furthermore, the effect of garlic on free radicals (Section 5.4) must be considered in connection with the decreased levels of blood lipids and decreased deposition of cholesterol into the vessel walls (Kourounakis & Rekka, 1991; Popov et al., 1994; Phelps & Harris, 1993).

The fluidity of the blood can be maintained by eating fresh garlic or by corresponding doses of pharmaceutically manufactured garlic products. This is especially important in older people. It has been repeatedly shown that garlic strengthens the heart and stimulates well-being (Keys, 1980; Buck et al., 1982; Bordia, 1986; Kiesewetter et al., 1991a).

5.1.1 Cholesterol and Lipid-Lowering Effects

Although there have been controversial discussions concerning the significance of high cholesterol levels for the incidence of arteriosclerosis, several recent studies clearly show that a correlation exists between the concentration of blood lipids and the narrowing of coronary vessels. Some studies, including the large Framingham study, have revealed a significant correlation between serum cholesterol and the risk for heart disease, in both men and women (Castelli, 1988). Furthermore, a major 25-year follow-up study in the United States, Europe, and Japan has recently shown that increased serum total cholesterol levels are directly associated with increased coronary heart disease in all cultures (Verschuren et al., 1995). In numerous studies and reports dealing predominantly with garlic and onion preparations, it has been documented that this process can be stopped or even reversed by lowering the cholesterol level (Brown et al., 1990; Ornish et al., 1990; Tatami et al., 1992; Buchwald et al., 1990; Cashin-Hemphill et al., 1990; Watts et al., 1992; Kendler, 1987; Ernst, 1987, 1989; Barrie et al., 1987; Grünwald et al., 1991; Reuter, 1990, 1993b; Bongartz & Grünwald, 1992; Kiesewetter & Jung, 1992; Mirhadi & Singh, 1988; Walper, 1992a, 1992b; Chen & Tang, 1991; Betz & Weidler, 1989; Hobbs, 1993; Anonym, 1976). Two meta-analyses of the clinical studies on the effect of garlic on total serum cholesterol have been recently published (Warshafsky et al., 1993; Silagy & Neil, 1994), and a critical review of the less recent evidence on humans with emphasis on commercial preparations is also available (Kleijnen et al., 1989).

As mentioned earlier, the most important risk factors for developing arteriosclerosis with its secondary effects, such as myocardial infarction, stroke, and occlusive arterial disease, are hyperlipidemia and hypercholesterolemia, in addition to obesity, high blood pressure, diabetes, and nicotine and alcohol abuse. Other risk factors are a lifestyle with little exercise and frequent stress situations. Patients who have been confirmed as having peripheral, coronary, or cerebral arteriosclerosis certainly also suffer a disorder in fat metabolism (Sogani & Katoch, 1981; Soh, 1960). Something can and must be done in this situation. Timely prevention is, as usual, the best therapy, and garlic is an almost ideal remedy.

5.1.1.1 CHOLESTEROL AND LIPID LOWERING: ANIMAL AND IN VITRO STUDIES

The lipid-lowering effects of garlic have been clearly shown in animal experiments. The first experiments on experimental ergosterol-induced arteriosclerosis in cats clearly demonstrated an antiarteriosclerotic effect of fresh pressed garlic juice and the ether extract. No effects were seen upon feeding of old pressed garlic juice, with greatly decreased thiosulfinates, or upon feeding the residue of steam-distilled garlic (Silber,

1933; Orzechowski, 1933). Also, the alimentarily developed atheromatous changes in the vessels (wall thickening) disappeared partially or completely after feeding with fresh garlic, garlic powder, pressed juice, or garlic oil (Tempel, 1962; Kritchevsky, 1975; Jain, 1975a, 1975b, 1978; Jain & Konar, 1976a, 1976b, 1978; Jain & Vyas, 1975a; Bordia et al., 1977; Bordia & Verma, 1980; Sarkar & De, 1981; Kaur et al., 1982; Mirhadi et al., 1983; Mand et al., 1985; Auer et al., 1989). Even in chickens, a drop in cholesterol content in blood was observed after feeding with garlic (Abdo et al., 1983; Horton et al., 1991; Qureshi et al., 1987). Only in pigs was there no noticeable effect on the development of atheromatous conditions (Lee & Lee, 1980).

The cholesterol and lipid–lowering effects of the ether-extractable active substances from garlic (1 g garlic, or about 4 mg ether extract/kg) are of the same magnitude as those of clofibrate (33 mg/kg) (Bordia et al., 1975a). The thiosulfinates of garlic are easily extracted with ether and are subsequently rapidly transformed to vinyldithiins (70%), sulfides (18%), and ajoenes (12%) (Lawson, 1993). Other authors have demonstrated the hypolipidemic effect of garlic oils (both steam-distilled oils and ether-produced oils) in rats and rabbits that were fed ethanol and an especially fat-rich diet for the development of hyperlipidemia (Bobboi et al., 1984; Shoetan et al., 1984; Sodimu et al., 1984; Bordia & Verma, 1978; Ikpeazu et al., 1987; Nagai & Osawa, 1974). Likewise, distilled garlic and onion oils exerted a small, but significant, decrease in serum lipids when given to rats made hyperlipidemic with a high glucose diet (Adamu et al., 1982; Wilcox et al., 1984). A significant decrease in total lipids and cholesterol in plasma of rats was also observed after administration of a mixture of 98% diallyl disulfide and 2% diallyl trisulfides (allitin, 100 mg/kg intraperitoneally) (Pushpendran et al., 1980a). A Chinese research group studied the prevention and treatment of atherosclerosis with diallyl trisulfide (allitridi) in rabbits with equally good results (Zhao et al., 1982). Onion has also shown similar effects (Vatsala et al., 1980; Sainani et al., 1979d; Augusti, 1974a).

A garlic oil as well as diallyl disulfide (80–160 mg/kg) have been shown to reduce the activity of liver and serum lipase (but not of kidney) and of glutathione reductase in all tissues in animals with enzyme activities that had been elevated with a hyperlipidemic diet of glucose and alcohol. It was proposed that diallyl disulfide reacts with the SH-groups of certain enzymes (Adoga, 1987; Huh et al., 1986b).

Other authors fed lyophilized garlic and various garlic fractions to rats and reported similar results. The animals had been previously fed a diet rich in lard and cholesterol, whereupon the lipid parameters rose as expected. After garlic administration, cholesterol, triglyceride, and very low density lipoprotein (VLDL) levels fell significantly, while HDL levels increased. Biliary excretion of the neutral steroids increased greatly, while that of bile acids increased only slightly (Chi et al., 1982; Chi, 1982). Aside from a slight neogenesis of the liver and increased fecal excretion, decreased activity of cholesterol-7-hydroxylase in the liver has been observed (Kritchevsky et al., 1980). Furthermore, dietary protein has been found to modify garlic-induced changes in phospholipid and cholesterol contents in rat brain (Vimala & Sharma, 1988).

Oral administration of petroleum ether extracts of *Allium sativum* (about 3 mg extract oil/kg body weight) and *Allium cepa* to albino rats significantly inhibited the rise of serum cholesterol and serum triglyceride levels (by 90%, $P < 0.001$) caused by an atherogenic diet. Both agents were found to confer significant protection against atherosclerosis induced by the atherogenic diet (Lata et al., 1991).

The effectiveness of garlic has been compared to that of onion (*A cepa*). Rats fed the corresponding powders (20% in feed) for 12 weeks showed significantly lower serum and liver cholesterol values than at the onset of the experiment, whereas the fecal sterols and prothrombin time had increased. Garlic was clearly more effective than onion (Bakhsh & Chughtai, 1985).

Indian authors tested a low dose of commercial steam-distilled garlic oil in normal dogs and, after an initial increase in serum cholesterol, found a significant lowering of cholesterol values, which even remained for 1 week after conclusion of therapy and then returned to their initial levels (Das et al., 1982). Subsequent analysis of the garlic oil capsules indicates that the dogs received 0.1–0.2 mg/kg body weight allyl sulfides (Lawson & Wang, 1994).

The lipid-lowering effect of garlic is observed not only in mammals, but also in chickens. Connected with this effect was the significant inhibition on the enzymes involved in the biosynthesis of cholesterol in the liver of the animals (Qureshi et al., 1983a, 1983b, 1983c; Ahmad, 1986).

Billau (1961) conducted an intensive study of how various garlic preparations influence hyperlipidemia that had been produced in rabbits by high fat and cholesterol diets. She observed that the lipid-associated turbidity of serum decreased distinctly in almost all cases where the active ingredients of garlic had been given orally or intramuscularly. The total extract (i.e., containing all constituents of garlic) was the most effective. Extracts prepared with urea and/or alcohol reacted similarly, though more weakly. On the other hand, aqueous extracts which had been boiled, resulting in loss of the thiosulfinates, did not produce well-defined results.

Garlic may reduce serum lipids by decreasing fat absorption. Human gastric lipase (HGL) is a sulfhydryl enzyme involved in the digestion and absorption of dietary fats and is inhibited by sulfhydryl-binding agents. Gargouri and colleagues (1989, 1992) found that ajoene, a component of garlic oil-macerates, has the in vitro ability to inactivate HGL, presumably because of its sulfhydryl-binding activity. Since ajoene, allicin, and diallyl trisulfide have all been found to bind rapidly to sulfhydryl compounds (Lawson, 1993; Lawson & Wang, 1993), these results probably also apply to allicin-yielding garlic products as well as to steam-distilled garlic oils.

Mirhadi and coworkers observed that feeding a cholesterol-rich diet to male rabbits increases collagen biosynthesis and accumulation in the aorta, liver, kidneys, heart, and lungs. This effect was partially suppressed by garlic supplementation (homogenized cloves), particularly in the aorta. The cholesterol content of plasma, aorta, and liver, which was greatly increased by the atherogenic diet, was reduced as well. It is suggested that garlic reduces the accumulation of collagen through increased mobilization of lipids and/or decreased biosynthesis and maturation of collagen. Histopathological studies also support this view. Also, the activity of phospholipase in cell-free supernatants of aorta and liver was increased, as was the activity of dehydrogenase (Mirhadi et al., 1986, 1991; Mirhadi & Singh, 1988).

Some controversy has existed concerning the effective compounds of aqueous extracts of garlic (AQGE) on plasma lipids. The reduction of total lipids, phospholipids, and cholesterol in serum and in the liver of rats through administration of AQGE (Augusti & Mathew, 1973) was proposed to be due to the action of adenosine (Koch & Hahn, 1988). The authors also refer to empirical data showing that adenosine lowered the cholesterol level, probably through stimulation of excretion and/or inhibition of intestinal absorption of cholesterol (Yamasaki, 1973). From the observation that the most effective principle is stable to autoclaving at 120°C (converts allicin to diallyl trisulfide and diallyl disulfide, which are also active) it was proposed that the responsible substance could only be adenosine, since allicin was assumed to be insoluble in water (Chi et al., 1982; Chi, 1982; Qureshi et al., 1983a, 1983b, 1983c; Agarwal, 1992; Yamasaki, 1973). However, allicin has been clearly shown to be both lipophilic and hydrophilic. It is soluble in water at 10–20 mg/mL (see Section 3.2.6.1), while the most concentrated AQGE used contains 2 mg allicin/mL. Thus allicin can be very abundant in AQGE. Furthermore, allicin is quite stable in aqueous solution, with a half-life of 2–40 days, depending on dilution (see Fig. 3.11). It can therefore be assumed that allicin is present in AQGE and is probably responsible for the lipid-lowering effects. Adenosine might also contribute to the lipid-lowering effects; however, the adenosine content of crushed garlic that has sat long enough to reach maximum levels (absent in freshly cut cloves—see Fig. 3.21) is considerably smaller (about 0.1 mg/g) than that of allicin (about 4 mg/g). Adenosine is also poorly absorbed and is rapidly metabolized by erythrocytes (Reuter, 1980, 1983a, 1986). However, the presence of allicin in AQGE may improve the absorption of adenosine. Only in the presence of substances with lipophilic and hydrophilic areas in the same molecule, e.g., allicin, can adenosine be absorbed into the blood (Reuter, 1980, 1983b). There are some effects which could be in some way related to the action of adenosine, such as the rise in amino acids after the administration of an AQGE to rats (Augusti & Mathew, 1973).

In order to test the cholesterol-lowering principle in garlic, several fractions were prepared and evaluated on rats: the distilled oil (mainly diallyl trisulfide and diallyl disulfide produced from allicin), the distillation residue (all other water-soluble compounds), and a defatted garlic powder (alliin not removed). The distilled oil and garlic powder lowered the cholesterol content in serum and liver as expected; the distillation residue, however, was completely inactive (Kamanna & Chandrasekhara, 1984). The same authors had earlier observed a decrease of cholesterol and LDL levels, and an increase in HDL levels in the serum of rats fed with lyophilized garlic powder (0.5–3% in feed) (Kamanna & Chandrasekhara, 1982). Therefore, the active agents appear to be allicin and allicin-derived allyl sulfides.

Fujiwara and colleagues (1972), in a targeted experiment with S-methyl-L-cysteine sulfoxide (a homolog of alliin) on artificially induced hypercholesterolemic rats, demonstrated a significant drop in blood and liver cholesterol values when administered at the very high daily dose of 200–400 mg/kg body weight. Augusti & Mathew (1974) gave freshly prepared allicin from garlic (100 mg/kg body weight) to rats for 2 months, and found thereafter a significant reduction of the serum and liver triglyceride and cholesterol levels in comparison to untreated controls. Unfortunately, a lower dose was not tested. At a dose of 12 mg/kg, garlic and onion oils (steam-distilled), dipropyl disulfide, and diallyl sulfide were all found to have significant hypolipidemic effects in both normal and hyperlipidemic rats, although the garlic oil consistently gave the best results (Abo-Doma et al., 1994). An ether-extract oil of garlic (100 mg/kg) was also found to lower total lipid levels in mice (Gupta, 1988). Finally, the development of induced arteriosclerosis was prevented and other existing conditions reversed with a garlic oil (Münch, 1966). An increased synthesis and excretion of bile acids was observed after feeding garlic or onions to rabbits (Jain, 1975a, 1975b, 1976).

The mechanism of action of the hypolipidemic efficacy of garlic has been proposed to be due to the reaction of allicin with coenzyme A, the SH-group of which is needed for the biosynthesis of fatty acids, triglycerides, phospholipids, and cholesterol. Through the blocking of the SH-group, the acetyl transfer is inhibited (Augusti & Benaim, 1975; Reuter, 1980). However, this is probably not the case, since allicin has been shown to be immediately converted to allyl mercaptan (allyl-SH) when in contact with blood (Lawson & Wang, 1993; Lawson, 1993). It is possible that allicin lowers serum cholesterol by inhibiting cholesterol absorption, since it has recently been demonstrated in rats fed for 6 weeks a commercial steam-distilled garlic oil (0.002% of diet or 1.6 mg/kg body weight) resulted in a 23% reduction of serum cholesterol, a 14% decrease in cholesterol absorption, and a 26% increase in bile acid secretion (Srinivasan & Srinivasan, 1995).

Next to allicin, alliin and S-methyl-L-cysteine sulfoxide could have the strongest lipid-lowering activity, as has been indicated when fed to rats at the high level of 1% of the diet (Itokawa et al., 1971, 1973) or at a high dose of 200 mg/kg (Sheela & Augusti, 1992, 1995); however, these compounds would only be present when alliinase has been inactivated, such as in cooked garlic. The mechanism of action for these compounds has been proposed to be the inhibition of cholesterol biosynthesis (Qureshi et al., 1983b, 1983c). The effect is not coupled to hormones of the thyroid gland (Chaudhuri et al., 1984). On the other hand, garlic oil (5 mg/kg) clearly stimulates the activity of dopamine hydroxylase and elevates the catechol level in hypercholesterolemic rabbits, which influences the lowering of cholesterol (Srivastava et al., 1984).

Recently, Gebhardt and coworkers could show that exposure of primary rat hepatocytes and human HepG2 cells to water-soluble garlic extracts resulted in a concentration-dependent inhibition of cholesterol biosynthesis (Gebhardt, 1991, 1993; Gebhardt et al., 1994). At low concentrations, inhibition was exerted at the level of β-hydroxy-β-methylglutaryl-CoA reductase (HMG-CoA reductase). At high concentrations (above 0.5 mg/mL), inhibition was also seen at later steps, resulting in the accumulation of the precursors lanosterol and 7-dehydrocholesterol. Alliin had no effect upon HMG-CoA reductase. Investigations of the effect of allicin and ajoene on HepG2 cells showed a significant inhibition of sterol biosynthesis (IC_{50} = 7 and 9 μM respectively) at the level of HMG-CoA reductase (Jäger et al., 1992). At somewhat higher concentrations allicin and ajoene also inhibited the late steps of cholesterol biosynthesis. In the case of allicin, small amounts of dihydrolanosterol and 7-dehydrocholesterol were formed at concentrations of 5–10 μM. From these results, the authors concluded that a major point of inhibition occurs at the level of lanosterol-14-demethylase. In contrast, nicotinic acid and adenosine caused moderate inhibition of HMG-CoA reductase activity and cholesterol biosynthesis, suggesting a participation of these compounds, at least in part, in the early inhibition of sterol biosynthesis by garlic extracts (see also Platt et al., 1992; Grünwald, 1992; Brosche & Platt, 1991; Qureshi et al., 1987). Nevertheless, as was previously mentioned, both allicin and ajoene are immediately converted to allyl mercaptan in blood and never reach the liver to affect cholesterol biosynthesis. However, it has been recently shown with isolated rat hepatocytes that allyl mercaptan (50 μM), and especially diallyl disulfide (5 μM), are capable of enhancing palmitate-induced inhibition of cholesterol synthesis (Gebhardt, 1995).

In another in vitro study with cultured hepatocytes, it was shown that high concentrations of water, methanol, and petroleum ether extracts of garlic cloves and of an aged garlic extract (containing 0.05 mM S-allylcysteine) decreased the synthesis of cholesterol, while the clove extracts also inhibited fatty acid and triglyceride synthesis. The aged extract increased fatty acid synthesis at lower concentrations, but inhibited it at the highest concentration. Furthermore, S-allylcysteine alone did not inhibit cholesterol synthesis, except at a very high concentration (2 mM), which is 40 times higher than the amount contained in the similarly inhibiting level of aged extract, indicating that S-allylcysteine has little to do with the cholesterol-lowering activity of the aged extract (Yeh & Yeh, 1994). (Section 5.13.1 covers further cardiovascular studies on this aged extract.)

Lipids, being fundamentally water-insoluble substances, do not exist in free form in blood, but are coupled to certain protein components of blood plasma. In the pathogenesis of arteriosclerosis, pathological modifications in the composition of plasma proteins play an important part. Jain & Konar (1976a, 1976b) therefore studied the influence of crushed garlic (juice) on serum proteins of arteriosclerotic rabbits. They detected a shifting of the ratio between albumins and globulins after electrophoretic fractionation. The garlic ingredients distinctly counteracted the rise of globulins, which normally can be found in arteriosclerotic animals.

Some critical remarks should be made concerning experimental hyperlipidemia in animals. The effects of a cholesterol-rich diet in animals such as rabbits, guinea pigs, and dogs clearly differ from the effect that is exerted by cholesterol on human arteries. Rabbits are up to 3000 times more sensitive to cholesterol load in developing arteriosclerosis than are human subjects. Their serum cholesterol amounts to 46 ± 8.8 mg/dL. Guinea pigs and dogs do not show arteriosclerotic deposits in the arteries, despite an increase in serum cholesterol (Holtmeier, 1993). Hypocholesterolemic and antiatherosclerotic effects of garlic have been observed also in goats (Kaul & Prasad, 1990). Onion has a comparable hypolipidemic effect in rabbits (Sebastian et al., 1979; Sharma et al., 1975).

5.1.1.2 CHOLESTEROL AND LIPID LOWERING: HUMAN STUDIES

In addition to the animal studies on the effectiveness of garlic on impaired fat metabolism, a whole series of human studies has been conducted. These studies were performed in healthy persons, as well as in patients with hyperlipidemia and hypercholesterolemia (Tables 5.1 and 5.2). It is known that people who regularly eat larger amounts of garlic and onions have lower lipid and cholesterol levels in their blood than people who refrain from eating these vegetables (Sainani et al., 1976, 1979a). For quite some time, Russian physicians have treated arteriosclerotic patients with garlic preparations (e.g., brand Allifid) (Bojko, 1962).

As a measure for the clinical efficacy of garlic and its preparations, the changes in concentrations of total cholesterol, total triglycerides, and, more recently, the proportions of the different lipoproteins, have been determined in numerous controlled studies, most of which have used garlic tablets standardized to a specific alliin content and allicin yield. Table 5.1 summarizes the results of 40 studies on the effect of different garlic preparations on the total cholesterol levels of 43 groups of patients and volunteers. For all of the studies taken together, the mean decrease in serum cholesterol levels was 10.6%, with the treatment lasting from 3 weeks to several months. The seven studies with 3–10 g fresh garlic on 301 people gave an average decrease of 16%. The 13 placebo-controlled studies with alliin/allicin–standardized garlic powder tablets (0.3–1.2 g powder, mostly 0.6–0.9 g) on 427 treatment patients showed an average decrease of 10.3%, while the 10 open studies (not placebo-controlled) on 4179 people with the powder tablets gave a similar average decrease of 11.3% per study, or a 12.9% decrease per person. The average decrease in the four studies with oils containing predominantly allicin-derived vinyldithiins (oil of ether-extracted garlic (OE) and oil of oil-macerated garlic (OM); OM is similar to OE but is highly diluted) was 14.3%. The average decrease for the four studies with steam-distilled oils was only 4%, although the amounts used varied considerably (4–120 mg) and the effect increased as the dose increased. A single study on an aged garlic extract showed a significant increase in serum cholesterol in the first 2 months followed by a 14% decrease in 6 months. (Section 5.13.1 covers other cardiovascular studies with this product.)

Serum triglyceride levels were also measured in 32 of these studies (34 groups of patients and volunteers) and were found to decrease an average of 13.4% (Table 5.2). Only two studies with fresh cloves reported triglyceride values, which decreased an average of 30%. In the 12 placebo-controlled studies with garlic powder tablets, there was an 8.5% decrease in the 406 treatment patients, while in the 10 open studies on 3262 people there was a decrease of 15% per study or 20.2% per person. The three ether extract studies (OE) gave an average decrease of 30%, but the single oil-macerate study showed a nonsignificant increase of 16%. The only two studies with steam-distilled oils gave results that reflect the amounts used.

Seven long-term studies on the consumption of 3–15 g of uncooked garlic cloves or the juice therefrom have been conducted, all of which found substantial decreases in serum cholesterol levels. Kerekes and coworkers (1974, 1975) demonstrated a lowering of serum cholesterol by 29 mg/dL in healthy persons (lipid levels not stated) who received 3 grams fresh garlic daily for 4 weeks. A very recent study from Kuwait found that subjects who ate 3 g finely

Table 5.1. Clinical Studies: Effects of Garlic Preparations on Serum Total Cholesterol (mg/dL)

Author, Year	Cases[a]	Dose/Day	Prep.[b]	Days	Before	After	Diff.%[c]
Ali, 1995	8/0 N	3 g	F	112	236	186	−21*
Augusti, 1977b	5/0 HL	10 g	F	56	305	218	−29**
Bhushan, 1979	25/0 N	10 g	F	56	223	190	−15***
Gadkari, 1991	50/0 N	10 g	F	56	213	180	−15***
Jain, R. C., 1977	6/0 N	5 g	F	21	227	199	−12
Kerekes, 1975	7/0 N	3 g	F	30	?	?	−29 mg
Sucur, 1980	200/134 HL	5 g	F	25	272	244	−10*
Auer, 1989, 1990	24/23 BP	0.6 g	P	84	268	239	−14*
Beck, 1993	1997/0 HL	0.9 g	P	112	290	242	−16***
Bimmermann, 1988	49/49 CHD	1.2–1.8 g	P	126	294	?	+8 ns
Brewitt, 1991	1024/0 HL	0.6–1.2 g	P	168	297	248	−16***
Brosche, 1989, 1990	29/0 HC	0.6 g	P	84	260	240	−8***
	11/0 N	0.6 g	P	84	163	161	−1 ns
De A Santos, 1993	25/27 HL	0.9 g	P	168	267	243	−5*
Ernst, 1986	10/10 HL	0.6 g	P	24	?	?	−22**
Grünwald, 1992	45/0 HC	0.6 g	P	126	330	304	−8***
Harenberg, 1988	20/0 HL	0.6 g	P	28	278	258	−7*
Holzgartner, 1992	47/47 HL*	0.9 g	P	84	282	210	−25***
Jain, A. K., 1993	20/22 HL	0.9 g	P	84	262	247	−6**
Kandziora, 1988b	20/20 BP*	0.6 g	P	84	314	294	−6
Kandziora, 1988a	20/20 BP	0.6 g	P	84	280	252	−10*
Kiesewetter, 1993a	32/32 PVD	0.8 g	P	84	267	234	−12***
König, 1986	53/0 PVD	0.6 g	P	28	321	278	−14**
Luley, 1986	85/85 HL	0.6–1.4 g	P*	42	284	270	−5 ns
Mader, 1990a	111/110 HL	0.9 g	P	112	266	235	−12***
Rinneberg, 1989	44/0 HL	1.2 g	P	126	298	264	−11*
Sas, 1988	20/52 HL	0.6 g	P	42	330	298	−10*
Semm, 1987	30/0 HL	0.3 g	P*	28	298	251	−16**
Simons, 1995	28/28 HL	0.9 g	P	84	251	253	+1 ns
Vorberg, 1990	20/20 HL	0.9 g	P	112	295	233	−21***
Walper, 1994	917/0 HL	0.6 g	P	42	266	253	−3***
Zimmermann, 1990b	23/0 HL	0.9 g	P	21	287	260	−9*
Plengvidhya, 1988	30/30 HL	0.7 g	SP	56	280	290	+3 ns
Sitprija, 1987	17/16 DB	0.7 g	SP	30	219	229	+5 ns
Bordia, 1981	20/0 N	15 mg[d]	OE	180	233	201	−14*
	33/29 CHD	15 mg[d]	OE	300	298	228	−23*
Bordia, 1982	20/0 NB	15 mg[d]	OE	21	174	165	−5 ns
Bordia, 1986	30/30 HC	1.5 mg[d]	OM	90	272	230	−15**
Arora, 1981	25/0 CHD	4 mg[d]	OS	84	269	279	+4 ns
	17/0 N	4 mg[d]	OS	84	247	248	0 ns
Barrie, 1987	20/20 N	18 mg	OS	28	195	180	−7**
Zhejiang, 1986	141/0 BP	120 mg	OS	30	261	225	−14***
Lau, 1987	15/12 HL	4 mL	AE	60	306	324	+6*
	15/12 HL	4 mL	AE	180	306	262	−14*

[a]Cases: (treatment/placebo) N = healthy persons; NB = healthy persons fed additionally 75 g butter daily; HL = hyperlipidemic patients; HL* = bezafibrate, a cholesterol-lowering drug, was the control rather than a placebo; HC = hypercholesterolemic patients; CHD = patients with coronary heart disease; PVD = patients with arterial occlusive disease; BP = patients with hypertension; BP* = resperine, a diuretic drug, was the control; T = patients with increased platelet function parameters.
[b]Preparations: F = fresh cloves (uncooked); P = garlic powder tablets, standardized on alliin/allicin; P* = powder tablets, not standardized; SP = spray-dried garlic (alliin and alliin-derived compounds removed); OE = oil of ether-extracted garlic; OM = oil of oil-macerated garlic; OS = oil of steam-distilled garlic; AE = aged garlic extract.
[c]Significance of the difference, when given: ns = not significant; *, $P < 0.05$; **, $P < 0.01$; ***, $P < 0.001$.
[d]Value represents analyzed (Lawson, 1993) amount of garlic-derived compounds in the oil products.

chopped garlic cloves each morning for 16 weeks showed a 21% decrease ($P < 0.02$) in serum cholesterol, although no significant decrease was found after 4 weeks (Ali & Thomson, 1995). In another experiment, with hypercholesterolemic patients (237–350 mg/dL) who drank the juice from 10 g of garlic daily for 2 months, serum cholesterol was lowered 28.5% (Augusti, 1977). Bhushan and colleagues (1979) employed healthy subjects (initial cholesterol, 160–250 mg/dL) who had never previously ingested garlic and

Table 5.2. Clinical Studies: Effects of Garlic Preparations on Serum Triglyceride (mg/dL)

Author, Year	Cases[a]	Dose/Day	Prep.[b]	Days	Before	After	Diff.%[c]
Jain, R. C. 1977	6/0 N	5 g	F	21	110	83	−25**
Sucur, 1980	200/134 HL	5 g	F	25	264	174	−34**
Auer, 1989, 1990	24/23 BP	0.6 g	P	84	171	140	−18***
Beck, 1993	1997/0 HL	0.9 g	P	112	235	186	−21***
Bimmermann, 1988	49/49 CHD	1.2–1.8 g	P	126	203	?	+5 ns
Brewitt, 1991	1024/0 HL	0.6–1.2 g	P	168	235	189	−20***
Brosche, 1989, 1990	29/0 HC	0.6 g	P	84	207	174	−16*
	11/0 N	0.6 g	P	84	145	131	−10 ns
De A Santos, 1993	25/27 HL	0.9 g	P	168	130	134	+3 ns
Ernst, 1986	10/10 HL	0.6 g	P	24	?	?	−35**
Grünwald, 1992	45/0 HC	0.6 g	P	126	168	150	−11 ns
Harenberg, 1988	20/0 HL	0.6 g	P	28	231	240	+4
Holzgartner, 1992	47/47 HL*	0.9 g	P	84	218	144	−34***
Jain, A. K., 1993	20/22 HL	0.9 g	P	84	151	165	+9 ns
Kandziora, 1988b	20/20 BP*	0.6 g	P	84	213	197	−8***
Kandziora, 1988a	20/20 BP	0.6 g	P	84	209	193	−8*
Kiesewetter, 1993a	32/32 PVD	0.8 g	P	84	184	177	−4
König, 1986	53/0 PVD	0.6 g	P	28	238	199	−16***
Luley, 1986	85/85 HL	0.6–1.4 g	P*	42	170	161	−5 ns
Mader, 1990a	111/110 HL	0.9 g	P	112	226	188	−17***
Rinneberg, 1989	44/0 HL	1.2 g	P	126	212	184	−14**
Sas, 1988	20/52 HL	0.6 g	P	42	382	357	−7
Semm, 1987	30/0 HL	0.3 g	P*	28	200	158	−21**
Simons, 1995	28/28	0.9 g	P	84	129	128	−1 ns
Vorberg, 1990	20/20 HL	0.9 g	P	112	206	156	−24**
Zimmermann, 1990b	23/0 HL	0.9 g	P	21	253	206	−19
Plengvidhya, 1988	30/30 HL	0.7 g	SP	56	355	335	−6 ns
Sitprija, 1987	17/16 DB	0.7 g	SP	30	201	195	−3 ns
Bordia, 1981	20/0 N	15 mg[d]	OE	180	110	70	−36*
	33/29 CHD	15 mg[d]	OE	300	170	130	−29*
Bordia, 1982	20/0 NB	15 mg[d]	OE	21	100	74	−26*
Bordia, 1986	30/30 HC	1.5 mg[d]	OM	90	95	110	+16 ns
Arora, 1981	25/0 CHD	4 mg[d]	OS	84	107	113	+3 ns
Zhejiang, 1986	223/0 BP	120 mg	OS	30	273	193	−29***

For notes and abbreviations, see Table 5.1.

found a significant drop in blood cholesterol (15%, $P < 0.001$) after consuming 10 g garlic daily for 2 months. Daily consumption of 5 g crushed fresh garlic for 3 weeks by healthy individuals (average initial cholesterol, 228 mg/dL) and patients with arteriosclerosis caused distinct decreases in serum cholesterol (10%) and triglyceride (24%) levels, an increase in fibrinolytic activity (24%), and no change in blood coagulation time (Jain, 1977). A study with 30 students fed 10 g garlic after breakfast daily for 8 weeks reported a 15% decrease ($P < 0.001$) in serum cholesterol levels, while no decrease was found in 20 control subjects (Gadkari & Joshi, 1991). In a crossover trial with 200 patients (average initial cholesterol, 272 mg/dL), consumption of 15 g fresh garlic for 25 days resulted in a 12% decrease in serum cholesterol levels and a 35% decrease in triglyceride levels (Sucur, 1980). In a much shorter, but higher dosed study, the increase of serum lipids caused by a fat-rich diet given for 7 days was reversed and even lowered beneath the prestudy values by the addition of 40 g fresh garlic to the diet during the 2nd week (Bakhsh & Chughtai, 1984).

The large majority of the clinical trials with garlic powder tablets have been conducted with the brand Kwai, which has been standardized on alliin/allicin–yield (1.3%/0.6% of the garlic powder content) for many years. The earliest of these studies was conducted by Ernst and colleagues (1985, 1986) and involved patients with cholesterol values of more than 260 mg/dL. Group A was fed a cholesterol-free diet, while group B received 600 mg/day garlic powder in tablet form in addition to the diet. The various lipid parameters were assayed after 12 and 24 days. While group A showed only significant reduction in total cholesterol on both assay days, group B showed even greater reduction in total cholesterol, triglyceride, and LDL values ($P < 0.01$) (Fig. 5.1). The HDL values remained constant in both groups (Ernst et al., 1985, 1986).

In 40 geriatric volunteers aged 70 and over, the effect of 600 mg/day for 3 months of a dried garlic powder preparation, standardized to 1.3% alliin, on the lipid and lipoprotein composition of plasma and the lipid content of erythrocyte membranes was investigated (Brosche et al., 1990; Brosche, 1989). Participants with normal plasma cholesterol levels (less than 200 mg/dL; n = 11) showed little or no changes, but those with elevated plasma cholesterol (higher than 200 mg/dL; n = 29) showed a 16% ($P < 0.05$) decrease in plasma triglycerides, a 7.7% ($P < 0.001$) decrease in total cholesterol, and a 12% ($P < 0.001$) decrease in cholesterol ester, with no change in free cholesterol. A significant reduction (5%) in LDL cholesterol levels was found, but not in the LDL/HDL ratio. At the same time, there was an 8% ($P < 0.001$) increase in the cholesterol concentration of the erythrocyte membrane, indicating transfer of cholesterol from the plasma.

In a Danish multicenter study on Kwai tablets, the LDL and HDL cholesterol values were also determined and showed a 5% decrease in LDL cholesterol and a 5% increase in HDL cholesterol, while the HDL/LDL cholesterol ratio improved by 12% (Grünwald et al., 1992; Morck & Schulz, 1992). In a noncontrolled drug monitoring study, the efficacy and tolerance of 900 mg of these garlic powder tablets taken daily for 16 weeks was examined in 1997 patients with elevated serum cholesterol levels (higher than 200 mg/dL) (Beck & Grünwald, 1993). Serum cholesterol, LDL cholesterol, HDL cholesterol, and triglyceride values were all significantly ($P < 0.0001$) and positively influenced (−16.4%, −15.1%, +9.4%, and −21%, respectively). The amount of decrease of LDL cholesterol levels was closely correlated to the total cholesterol levels before treatment. In patients with initial cholesterol levels of 201–250 mg/dL, 251–300 mg/dL, and more than 300 mg/dL, the mean decreases in LDL cholesterol levels were 8.4%, 13.7%, and 20.4%, respectively.

A 12-week study comparing the effect of standardized garlic powder tablets (900 mg daily) with that of bezafibrate (600 mg daily), the most commonly prescribed blood lipid–lowering drug in Germany, has also been conducted. The multicenter, double-blind study was performed with 94 patients having cholesterol and/or triglyceride values exceeding 250 mg/dL. After 4 weeks of treat-

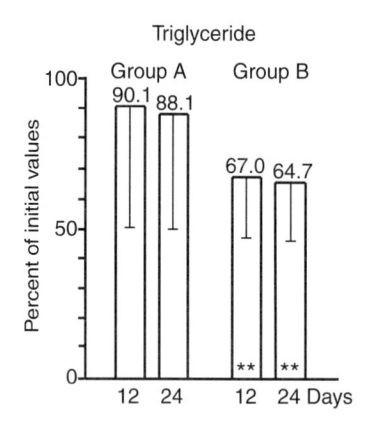

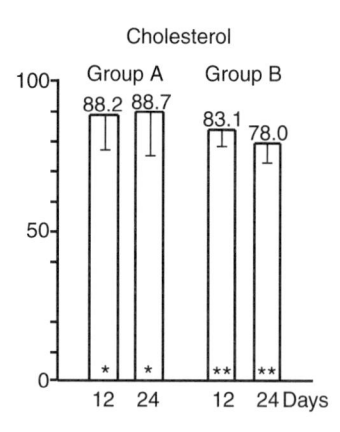

Figure 5.1. Lipid and cholesterol–lowering effects in hyperlipidemic patients. (From Ernst E, Weihmayr T, Matrai A. Knoblauch plus Diät senkt Serumlipide. Ärztl Prax 1986;38:1748–1749.)
*$P < 0.05$; **$P < 0.01$.

ment, the decreases in cholesterol, LDL cholesterol, and triglyceride levels were all statistically highly significant, and there were no differences between the effects of garlic and bezafibrate. HDL cholesterol values in the course of 4 weeks also increased significantly, again without any differences between the two regimens (Holzgartner et al., 1992). A controlled clinical study conducted in the U.S. with standardized dry garlic powder tablets (300 mg three times daily for 12 weeks) found a significant fall in total cholesterol and LDL-cholesterol, but no change in triglycerides and HDL cholesterol (Jain et al., 1993; Wehr, 1993). In summary, the influence of garlic on atherogenic LDL has been investigated in a total of nine clinical studies, seven of which reported a significant decrease, with a mean decrease of 16%.

Rotzsch and colleagues (1992), in a six-week, placebo-controlled, double-blind study, investigated the efficacy of a standardized garlic powder preparation (brands Kwai and Sapec, 900 mg daily) on reducing induced hypertriglyceridemia in 24 volunteers with low plasma HDL_2 cholesterol concentrations (less than 10 mg/dL for men, less than 15 mg/dL for women) after intake of a standardized high-fat test meal containing 100 g butter on days 1, 22, and 43. The increase in plasma triglycerides (measured 3–5 hours after consuming the butter) was clearly reduced by up to 35% after garlic consumption in comparison with the placebo group, with the lowering effect increasing from day 1 to day 22 to day 43. The fasting plasma triglyceride levels also decreased consistently with time in comparison with the placebo group. Similar studies on the same preparation have also been performed by others (Mader, 1990a; Vorberg & Schneider, 1990; Zimmermann & Zimmermann, 1990a; Brosche, 1989). Under garlic supplementation, there was also a tendency to higher HDL_2 cholesterol concentrations (Rotzsch et al., 1992; Mader, 1990a; Morck & Schulz, 1992; Kiesewetter et al., 1991a; Jain et al., 1992).

Five studies showing no effect on serum lipids from garlic powder tablets, performed under double-blind test conditions, have been reported which are in contrast to the numerous clinical studies in which the effectiveness of garlic against hyperlipidemia and hypercholesterolemia has been substantiated. A commercial garlic dry powder preparation (Zirkulin) did not show any significant effect (Luley et al., 1986; Luley & Schwartzkopff, 1985). However, as was later established, the study had been carried out with a production lot that was manufactured from a low-quality (low allicin yield) garlic powder (Anonym, 1992, 1988; Koch, 1987), although a recent lot of the same brand was found to have a good allicin yield. A study by Bimmermann and coworkers (1988) with the brand Carisano also found no effect on serum cholesterol, although there was a very significant 30% increase in HDL_2 after 12 weeks; however, two subsequent studies with the same brand did report significant decreases in serum cholesterol (Sas et al., 1988; Brewitt & Lehmann, 1991; Lehmann & Brewitt, 1991; Morck, 1988). Very recently a cross-over study using an allicin-standardized tablet for 12 weeks with 28 subjects placed on an isocaloric, fat-restricted diet (less than 30% of energy as fat, less than 300 mg/day dietary cholesterol) found no change in levels of serum cholesterol, triglycerides, or lipoproteins (Simons et al., 1995). There have also been two clinical studies conducted with the powder of spray-dried garlic in which hyperlipidemic patients consumed garlic capsules containing 700 mg of the powder daily for 4 to 8 weeks (Plengvidhya et al., 1988; Sitprija et al., 1987). Both studies failed to find any significant effect on serum cholesterol or serum triglyceride levels. Although the studies failed, they provide important implications that allicin may be responsible for the lipid-lowering effects of garlic since the spray drying procedure causes the unique loss of alliin upon homogenization as well as loss of allicin upon drying.

In a study with a garlic-ginkgo combination product (brand Allium Plus) (200 mg garlic extract containing 5.7 mg alliin; 120 mg ginkgo extract) for 8 weeks including the Christmas/New Year's season when the consumption of fat is unusually high, no decrease in serum cholesterol was found in these hyperlipidemic patients; however, there was a decrease in serum cholesterol in 35% of the treatment patients compared to a decrease in only 20% of the placebo patients, a difference which was statistically significant ($P < 0.05$) (Kenzelmann & Kade, 1993).

One trial which tested an ether-extract garlic oil (15 mg undiluted oil per person daily) on patients suffering from coronary heart disease (average serum cholesterol, 298 mg/dL) is of particular interest (Bordia, 1981). The placebo-controlled study resulted in an initial nonsignificant increase in serum total cholesterol and LDL cholesterol levels prior to declining below the pretreatment values by 8 months of treatment. After 10 months of treatment, total cholesterol levels had decreased by an average of 23%, with no significant changes in the control patients. In the same study, patients with lower serum cholesterol levels (average of 233 mg/dL) showed no increase in serum cholesterol before achieving a significant decrease in 6 months. The authors proposed that the initial increase of plasma cholesterol in the first group was due to a garlic-induced mobilization of lipids from deposits in the vessel walls. This interpretation is also supported by data from animal experiments (Bimmermann et al., 1990); however, an initial increase in serum cholesterol has never been found in clinical studies using garlic powder tablets.

In a clinical study with both healthy subjects and patients with ischemic heart disease, Arora and colleagues (1985) tested a commercial preparation of steam-distilled garlic oil, using a daily dose equivalent to 50 g garlic (as claimed by the manufacturer). After 20 weeks there was no significant lowering of lipid and cholesterol values and only a small increase in fibrinolytic activity, leading the authors to conclude that garlic oil has no effect. However, subsequent analysis of the brand of capsules used in the study revealed that the 250 mg capsules contained only 0.5 mg undiluted garlic oil (allyl sulfides) (Lawson, 1993). The subjects in the study would have had to consume 500 capsules per day to achieve the equivalence of 50 g garlic; however, since 3–9 capsules per day were more likely consumed, this would have been equivalent to 0.3–0.9 g garlic per day.

From China comes a report of an extensive field test with garlic on 308 patients who suffered from hyperlipidemia, hypercholesterolemia, high blood pressure, coronary heart disease, diabetes, and liver ailments. The patients were given 120 mg garlic oil (presumably steam-distilled oil and claimed to be equivalent to 50 g garlic) daily for 20–30 days. The results showed significant reduction of lipid parameters, blood pressure, plasma fibrinogen values, platelet aggregation time, and blood sugar values, although the latter were elevated at the start of the experiment. The subjective symptoms of the patients were also improved. In no case were harmful side effects registered (Zhejiang, 1986).

Several single-dose studies (not shown in Table 5.1) with unusually large amounts of garlic have been conducted to observe any changes in serum lipid levels after only a few hours. No drop in the serum cholesterol level was observed in 4 hours when healthy males with normal serum cholesterol levels (199 mg/dL) consumed an aqueous extract from 50 g garlic or 50 g onions (Sharma & Sharma, 1979). However, when similar subjects were also fed a high-fat meal containing 100 g butter to induce elevated serum cholesterol levels, both 50 g crushed fresh garlic (4% decrease) or 50 g boiled garlic (1% increase) prevented the rise that was found at 4 hours in controls (19% increase) (Sharma et al., 1976). From the results of this study, it appears that garlic cloves also contain non–alliin-derived compounds with hypocholesterolemic activity when consumed at a very high dose. Such compounds, aside from other possibilities, could be saponins (Koch, 1993b). On the other hand, both the freshly pressed juice from 50 g garlic and an ether extract (contains only the allicin and other thiosulfinates and their transformation products) from the same amount of juice caused significant ($P < 0.001$) decreases in cholesterol levels (7%

decrease for the juice and 9% decrease for the ether extract, compared to a 7% increase in the controls) at 3 hours in subjects whose serum cholesterol levels were dietetically elevated by simultaneous consumption of 100 g butter (Bordia et al., 1975a). Other single-dose studies, involving healthy individuals as well as patients with coronary heart disease and increased serum cholesterol values, have revealed equally good results (Bordia & Bansal, 1973; Bordia et al., 1974; Jung & Kiesewetter, 1991; Jung et al., 1989b; Vorberg & Schneider, 1990; Jain & Vyas, 1977).

On the basis of the results of two epidemiological studies, one with 200 subjects with coronary heart disease and one with 168 normal subjects and 51 subjects with coronary artery disease, Indian authors concluded that they have to dispute the principal effectiveness of garlic for this disease (Katoch & Sogani, 1979; Sogani & Katoch, 1981; Hardman, 1991); however, the studies did not say whether the garlic was eaten primarily cooked (no allicin potential) or fresh, although it is likely that most of the garlic eaten was cooked.

As was shown more recently in cell models, garlic exerts a direct antiatherosclerotic and atherogenic effect on the arterial cell wall (Orekhov, 1992, 1995). For these experiments, smooth muscle cells cultured from the intima of human aorta were used. Addition of 40% atherogenic serum to the cultured normal cells for 24 hours caused about a 1.5-fold increase in free cholesterol, about a 3-fold increase in cholesteryl ester, and about a 5-fold elevation in the rate of DNA synthesis. Garlic powder extract at 0.1–1000 µg/mL reduced cholesteryl ester accumulation and inhibited (^{3}H)-thymidine incorporation stimulated by atherogenic serum. At a concentration of 1000 µg/mL, the extract significantly decreased free cholesterol accumulation. Furthermore, the addition of serum from four patients with angiographically assessed coronary sclerosis to cultured cells from uninvolved intima of human aorta caused a 46 ± 5% ($P < 0.05$) increase in total cell cholesterol. Blood serum taken from the same patients 2 hours after oral administration of 300 mg garlic powder tablets resulted in substantially less (19 ± 6%) accumulation of cholesterol in cultured cells.

5.1.1.3 CHOLESTEROL AND LIPID LOWERING EFFECTS: ACTIVE COMPOUNDS

Although the compounds responsible for the blood cholesterol and triglyceride lowering effects of garlic have not yet been absolutely proved, there is considerable in vivo evidence that allicin (at a dose of 0.05–0.10 mg/kg body weight on the basis of clinical studies with allicin-standardized garlic powder tablets) is responsible for most of the lipid lowering that occurs at a daily dose representing 2–3 g fresh garlic for 1–2 months. At much higher doses (50 g) or for much longer periods of time (more than 5 months), other, as yet unidentified, compounds also appear to have activity. While the evidence from these studies has been previously mentioned, it will be summarized here.

1. Of the 33 human studies conducted with garlic cloves and garlic powder tablets, the only studies (with one exception—Simons et al., 1995) in which no effect on lipid lowering was found employed tablets which contained either no allicin potential (spray-dried) (Sitprija et al., 1987; Plengvidhya et al., 1988) or low allicin potential (Luley et al., 1986). One additional negative study with an allicin-standardized tablet has been reported (Bimmermann et al., 1988), although subsequent studies with the same brand demonstrated positive effects (Sas et al., 1988; Brewitt & Lehmann, 1991).

2. Clinical trials with garlic oils, which contain only allicin-derived garlic compounds plus added vegetable oils, have lowered serum lipids. Three of these used an ether-extract at a dose of 0.25 mg sulfur compounds per kg body weight (Bordia, 1981, 1989; Bordia et al., 1982); one used an oil-macerate at 0.03 mg sulfur compounds per kg (Bordia, 1986); and one used a steam-distilled garlic oil at 2 mg sulfur compounds

per kg (Zhejiang, 1986). The effects with allicin-derived oils are good evidence that allicin is also active since allicin and several of its derived compounds are all metabolized to allyl mercaptan in whole blood or liver (Lawson & Wang, 1993; Egen-Schwind et al., 1992b).

3. In animal studies, fresh garlic, but not boiled garlic (in which alliinase is inactivated), clears the lipid turbidity in blood of rabbits fed a high-fat diet (Billau, 1961). Fresh garlic juice, but not old garlic juice (in which allicin has evaporated or converted to less active sulfides) decreased atherosclerotic lesions in cats (Silver, 1933). Garlic oil (allicin-derived) and garlic powder, but not the residue of steam-distilled garlic (which contains everything in garlic except alliin, allicin, and related thiosulfinates), was effective in rats (Kamanna & Chandrasekhara, 1984).

4. In animal studies, the allicin-derived garlic oils are effective in reducing serum cholesterol at low doses: 0.2 mg steam-distilled oil per kg body weight in dogs (Das et al., 1982); 12 mg, or 1.6 mg steam-distilled oil per kg, in rats (Abo-Doma et al., 1994; Srinivasan & Srinivasan, 1995); and 0.5 mg ether-extract oil per kg in rabbits and rats (Bordia & Verma, 1980; Lata et al., 1991).

5. Evidence for active compounds not related to allicin at larger or longer doses come from two human studies. Consumption of a single dose of 50 g of either fresh or boiled (inactivates alliinase) garlic was found equally to lower a diet-elevated serum cholesterol level (Sharma et al., 1976). Commercial aged garlic extract (contains no allicin potential and only trace amounts of allicin-derived compounds—see Section 3.4.5) at 4 mL per day did significantly lower serum lipids, but the effect was not seen until after 5–6 months (Lau et al., 1987).

More sure proof that allicin is responsible for most of the lipid-lowering effects of garlic could be provided in clinical studies in which the effects of allicin-standardized garlic powder is compared to those of powders which have been prepared from heated garlic (to inhibit alliinase) or from spray-dried garlic (causes complete loss of both alliin and allicin).

5.1.2 Effects on Blood Pressure, Vascular Resistance, and Heart Function

The effects of garlic preparations on blood pressure have been investigated in numerous animal experiments, as well as in several clinical studies. The hypotensive effect has been attributed to the vasodilative constituents of garlic (Rashid & Khan, 1985; Rashid et al., 1986).

5.1.2.1 BLOOD PRESSURE EFFECTS: ANIMAL STUDIES

As early as 1921, Loeper reported the effect of wild garlic on the circulatory system in dogs (Loeper & Debray, 1921; Loeper et al., 1921). He injected preparations of wild garlic intravenously and found, along with a distinct decrease of the blood pressure, an increase of the amplitude and a retardation of pulse frequency.

In 1929 Lio and Agnoli observed that garlic extracts first elevate, and then diminish the tone of the smooth muscles. They had assumed already that this effect correlated with blood pressure–lowering activity. In 1931 Swetschnikow and Bechterewa noticed different effects of garlic on circulation and the cardiovascular system (Swetschnikow & Bechterewa, 1931). An alcoholic extract at various dilutions produced a concentration-dependent dilation of the peripheral vessels (rabbit ear veins) and an improvement in circulation. Likewise, a widening of the kidney vessels occurred in the isolated organ. In the perfused heart, an enlargement of the coronary vessels could be observed, while the amplitude and frequency of the heart action was rather diminished. The results of this early study might be explained by the action of adenosine, if sufficient quantities were present.

An alcoholic garlic extract (2.5–25 mg/kg body weight) caused a blood pressure drop of 10 to 50 mm Hg in dogs, which was not influenced by the prior application of atropine or antihistamines or by vagotomy. The effect of garlic was partially reversed by propranolol, an antiarrhythmic drug. Upon administration of higher dosages (25–50 mg/kg) the hypotensive action lessened, and at the end a slight increase in blood pressure was noticed. With a further increase in dosage (100 mg/kg and above), a blood pressure decrease of 20 to 30 mm Hg was observed, which lasted 2 to 4 hours (Chandorkar & Jain, 1972). Petkov (1966) also reported that the systolic blood pressure in experimentally hypertensive dogs can be reduced to normal values by garlic consumption.

Upon intravenous administration of garlic extract in dogs (25 mg/kg) the diastolic, systolic, and mean blood pressure levels were reduced by an average of 23.7, 28.9 and 24.3%, respectively, but the effect was not dose-dependent. This action was affected neither by atropine nor by bilateral vagotomy, while brompheniramine maleate (Dimetane) counteracted the effect (Sial & Ahmad, 1982).

Experiments with cats also resulted in a marked hypotensive effect (Malova, 1951). The mean effective dose of lyophilized pressed garlic juice, given intravenously to dogs (calculated as 48.5 ± 4.3 mg/kg), prompted a decrease of blood pressure by 25% (Malik & Siddiqui, 1981).

An aqueous extract of fresh garlic (15 mg/kg of an extract of 56.6 g garlic in 25 mL saline, intraperitoneally) caused a marked and lasting decrease in blood pressure in normotensive rats (still significant at 24 hours) and an increase in heart rate, while in hypertensive rats the decrease in blood pressure was not accompanied by any change in heart rate (hypertension induced with desoxycorticosterone acetate) (Banerjee, 1976). In another experiment, spontaneously hypertensive rats which had been given garlic extract (0.1–0.5 mL/kg, orally) reacted within 30 minutes with a significant blood pressure drop (by 51–56 mm Hg) which lasted up to 24 hours (Foushee et al., 1982). Ruffin and Hunter (1983) repeated these experiments and confirmed the hypotensive effect. However, they also observed serious side effects in those animals that received extremely high doses, some of them several times a day.

A standardized dry garlic powder (1 g/kg) fed to spontaneously hypertensive rats for 9 months distinctly inhibited the rise of blood pressure (150 versus 205 mm Hg in hypertensive controls) (Jacob et al., 1991). A decreased blood pressure was also seen after 200 mg garlic powder and was accompanied by bradycardia. Garlic also inhibited the tendency to form an eccentric cardiac hypertrophy seen very early in spontaneously hypertensive rats. Furthermore, the hypertension in the left ventricle was significantly reduced. Garlic powder also decreased the hydroxyproline concentration in the myocardium, suggesting a reduction of myocardial fibrosis.

In another experiment, a dry garlic powder preparation (1–3% in the diet for 6 weeks) prevented the increase of blood pressure in stroke-prone spontaneously hypertensive rats (Ogawa et al., 1993). The content of apo E in the LDL and HDL fractions tended to increase, although changes in the levels of serum lipids and apolipoproteins were not observed.

The effect of the water-soluble portion of an ethanol extract of garlic on the blood pressure–lowering properties on the isolated rat fundus and colon (concentration-dependent contraction of the fundus and relaxation of the colon) are similar to those exerted by prostaglandin E_2 (PGE_2), suggesting a prostaglandin-like activity of garlic extract (Rashid & Khan, 1985; Agarwal, 1992). However, it is also known that allicin attenuates the formation of PGE_2 by inhibiting the PGH_2 to PGE_2 isomerase (Shalinsky et al., 1989). Contrarily, ethanolic extracts (allinase-inhibited) of garlic and onion fed to spontaneously hypertensive rats at 1–2 and 4–8 g/day caused no decrease in blood pressure (Kiviranta et al., 1989).

Garlic juice inhibited the contractions of rabbit and guinea pig aortic rings induced by norepinephrine in calcium-free and calcium-containing Krebs-Henseleit solutions, as well as the contractions of rabbit and guinea pig tracheal smooth muscles induced by acetylcholine and histamine. The spontaneous movements of rabbit jejunum and guinea pig ileum and the contraction of isolated rabbit hearts were also reversibly inhibited in a concentration-dependent manner. These data suggest that the hypotensive action of garlic juice may be due, at least in part, to a direct relaxant effect on smooth muscle (Aqel et al., 1991; Ozturk et al., 1994).

Freshly pressed garlic juice (corresponding to 2 mL/kg body weight/day) or ether-prepared garlic oil (2.0 and 4.0 mg/kg) were both found to protect rats from myocardial necrosis induced by isoproterenol (Saxena et al., 1979, 1980). Furthermore, the influence of fresh garlic juice on the electrocardiogram (ECG) has been studied in rats (Tongia, 1984). The juice (0.1–0.5 mL) was injected into the tail vein and caused a definite change in the ECG, a positive inotropism (increased cardiac contraction power) and a positive chronotropism (enhanced heart rate). The effect started 10 seconds after the injection, and lasted 10–15 minutes, depending on concentration.

The external dialysate of homogenized garlic has been shown to cause a drop in diastolic blood pressure (from 113 ± 4 to 70 ± 3 mm Hg) and a decrease in heart rate (from 198 ± 10 to 164 ± 17 beats/minute) in a dose-dependent manner (Martín et al., 1992). The ECG showed a regular sinus bradycardiac rhythm. Addition of garlic dialysate to isolated left rat atrium evoked a decrease in tension development. The frequency, as measured by the spontaneous beating of the right atrium, was also reduced. Both effects were dose-dependent. In addition to these effects, the positive inotropism and chronotropism, induced by addition of isoproterenol 10^{-9} M, were partially antagonized by preincubation of the rat atrium with garlic dialysate. The above findings can be explained by a depressant effect on automaticity and tension development in the heart, suggesting a β-adrenergic receptor blocking action produced by the garlic dialysate.

Further investigations of the cardiovascular system were carried out with intact frog heart and with the isolated, perfused heart. An aqueous garlic extract reversibly lowered the heart frequency and increased the contraction power. The positively inotropic effect of low garlic concentrations was reversed at high concentrations. The active factor is heat-stable, as the effect is obtained with both boiled and unboiled extracts. In addition to the heart action, a muscle contraction at the rectum of the frog was observed (Rao et al., 1981).

Aside from the blood pressure–lowering effect, an increased diuresis was observed in dogs after oral and intravenous garlic administration (Loskutov, 1950b; Sharafatullah et al., 1986). The hypotensive factor is contained in the aqueous fraction of the garlic extract. It is believed that the effect is based upon release of histamine. Stimulation of muscarinic adrenal receptors or a depression of the vasomotor center could be excluded (Sial & Ahmad, 1982; Ribeiro et al., 1986).

Intragastric administration of garlic powder to anesthetized dogs induced the maximum dose-dependent (2.5–15 mg/kg) natriuretic and diuretic response after 30–40 minutes. Basal levels were reached after 100–150 minutes. A simultaneous decrease in arterial blood pressure was observed which lasted for more than 4 hours. High garlic doses (15 and 20 mg/kg) provoked bradycardia and T wave inversion during the first 10–15 minutes of the experiment, with recordings returning to normal and staying normal throughout the remainder of the experiment (Pantoja et al., 1991).

Isensee and coworkers (1993), in experiments on hearts isolated from rats that had been fed a 1% garlic powder diet for 10 weeks, showed that garlic reduces the incidence of ventricular tachycardia and fibrillation (following ligation of the descending branch of the left coronary artery for 20 minutes) significantly as compared to untreated controls. The size of the ischemic zone was significantly smaller. Reperfusion experiments (for 5 minutes after 10 minutes of ischemia) revealed similar results. The period of time until the extrasystole and ventricular tachycardia or ventricular fibrillation occurred was prolonged in most cases, and the duration of arrhythmia was shortened. No significant alterations of cardiac membrane fatty acid composition was found. Inhibition of cyclooxygenase by acetylsalicylic acid caused a moderate increase in arrhythmias and in size of the ischemic zone in the garlic group as well as in the untreated controls. Thus, the prostaglandin system does not seem to play any significant role in the cardioprotective action of garlic.

An antiarrhythmic profile of a garlic external dialysate was assayed recently in dogs and isolated atrial preparations (Martin et al., 1994). The preparation decreased the positive inotropic and chronotropic effects of isoprenaline in a dose-dependent manner. This suggests that garlic has also significant antiarrhythmic effects in both ventricular and supraventricular arrhythmias. Other authors confirmed this observation in the isolated rat heart (Jacob et al., 1993a). The antiarrhythmic effect was related to the size of the ischemic zone, pointing to a vascular component of the protective action. The effect of garlic treatment has been visualized also in the electrocardiogram of rabbits (Gupta et al., 1987).

5.1.2.2 BLOOD PRESSURE EFFECTS: CLINICAL STUDIES

Successful clinical use of garlic for treating elevated blood pressure and arteriosclerosis has been known since the early part of this century. It has been reported that regular garlic intake causes both a prolonged lowering of hypertension and an improved sense of well-being in patients. As early as 1928, Schwahn achieved definite blood pressure decreases as well as increases in productive heart power with garlic therapy, not only in older patients, but also in younger hypertonic patients. It has also been known for a long time that onion, which has active constituents similar to garlic, possesses cardioactive and diuretic properties (Kreitmair, 1936).

Petkov and colleagues (1966) reported that the clinical administration of garlic to patients with essential hypertension and arteriosclerosis leads to a lowering of systolic blood pressure by 8 to 33 mm Hg and, in most cases, of diastolic blood pressure by 4 to 16 mm Hg. Of 21 hypertensive patients, 5 regained normal blood pressure and 10 had a distinct reduction of hypertension. It was only in six patients, those in the final stage of arteriosclerosis, that blood pressure could not be lowered. However, an increased subjective feeling of well-being was nevertheless achieved. Petkov has summarized his extensive studies over many years on the various effects of garlic in several review articles (Petkov, 1962, 1966, 1979, 1986). Tchilov and coworkers (1951, 1956) mentioned similar results for treatment of hypertension with garlic. Russian authors have also described a lowering of blood pressure with garlic preparations by 5 to 20 mm Hg (Korotkov, 1966).

Garlic has also been shown to decrease the sensitivity of the peripheral vessels to adrenaline (Bogatskaya, 1953). Fresh garlic, or garlic juice stabilized with alcohol, causes a strong muscle contraction in certain organisms, for instance, in the tapeworm. Furthermore, garlic inhibits intestinal peristalsis, increases the tone of the bronchial musculature, and contracts the portal vein and coronary arteries (Malova, 1950, 1951).

Erken (1969) treated patients suffering from hypertension and arteriosclerosis with a commercial preparation containing oil-macerates of garlic, *Viscum album* L. (European mistletoe) and *Crataegus oxyacanthum* (hawthorn berry). Apart from the normalization of the serum cholesterol and blood pressure levels observed in the less severe cases, there was a favorable influence on symptoms such as headache, buzzing in the ears, and sensation of pressure and dizziness (at systolic pressures higher than 200 mm Hg).

Lutomski (1984), when testing a combination preparation (60% European mistletoe and hawthorn berry) corresponding to 300 mg fresh garlic (brand Ilja Rogoff), in a double-blind, placebo-controlled clinical study, observed significant improvement of the secondary effects of arteriosclerosis (headache, dizziness, insomnia, indigestion) in the patients of the experimental group, as well as an improvement of their general sense of well-being. The initial hypertension was lowered by an average of 10 mm Hg systolic and

20 mm Hg diastolic. Semm (1987) found similar results with the same preparation. In an open study with 50 moderately hypertensive patients, Lutomski and Mrozikiewicz (1989) found an 8–15% decrease in blood pressure and achieved significantly ($P < 0.05$) decreased total cholesterol (9%), decreased LDL cholesterol (18%), and increased HDL cholesterol (46%) using an undescribed commercial garlic juice preparation.

Along with decreased blood pressure, an improvement in blood fluidity is also generally observed. In the scope of an "open doctor's office study," König and Schneider (1986) treated patients suffering from peripheral arterial circulatory disturbances with daily doses of 1.8 g dry garlic powder (brand Kwai) over 4 weeks. Excessive differences between the blood pressure in the large toe and that in the upper arm were lowered toward the norm, and both systolic and diastolic blood pressures were decreased (Table 5.3). All changes were statistically very significant ($P < 0.001$) (König & Schneider, 1986; Jung et al., 1989b). A similar pilot study with comparable results was performed with a combined garlic preparation (Fodor, 1993).

Jung and coworkers (1989a, 1991) measured a distinct decrease of blood pressure, which did not occur in the control subjects, in healthy subjects 6 hours after intake of a dry garlic powder preparation. The blood pressure–lowering effects were accompanied by a decrease of hematocrit level and viscosity and an increase in capillary flow rate (erythrocyte velocity), suggesting that these hemorheological parameters depend on dilatation of peripheral vessels.

The effect of garlic on blood pressure that was seen after intake of different garlic preparations obviously exhibits large interindividual differences. The data from 15 studies on the blood pressure–lowering effect of garlic are presented in Table 5.3. As can be seen, garlic has a slight hypotonic effect in hypertonic patients. In the course of the 1-month and 6-month treatments, the systolic and diastolic blood pressures decreased by an average of 6.7 and 7.9%, respectively. The largest effect was seen in patients with high pretreatment values. In subjects with normal blood pressure, garlic failed to exert any effect. A recent critical meta-analysis of garlic's effect upon blood pressure and on its cardioprotective properties, including eight of the studies presented in Table 5.3, as well as three additional unpublished studies, suggests that garlic powder preparations may be of considerable clinical value in subjects with mild hypertension (Silagy & Neil, 1994; Neil & Silagy, 1994; Kandziora, 1988b).

Table 5.3. Clinical Studies: Effects of Garlic Preparations on Blood Pressure

Author, Year	Cases[a]	Dose/Day	Prep.[b]	Days	Before	After	Diff.%[c]
					(Systolic/Diastolic, mm Hg)		
Auer, 1989, 1990	24/23 BP	0.6 g	P	84	171/102	152/91	−11*/−11**
Brewitt, 1991	119/0 HL	0.6–1.2 g	P	168	166/102	156/93	−6***/−15***
De A Santos, 1993	25/27 HL	0.9 g	P	168	145/90	120/80	−17***/−11**
Grünwald, 1992	23/0 HC	0.6 g	P	126	158/96	147/92	−7*/−4
Harenberg, 1988	20/0 HL	0.6 g	P	26	137/86	126/81	−8**/−6**
Holzgartner, 1992	47/47 HL*	0.9 g	P	84	143/83	135/79	−6**/−5*
Jain, A. K., 1993	20/22 HL	0.9 g	P	84	129/82	130/81	0/−1
Kandziora, 1988b	20/20 BP*	0.6 g	P	84	178/100	167/85	−6/−15
Kandziora, 1988a	20/20 BP	0.6 g	P	84	174/99	158/83	−9***/−16***
Kiesewetter, 1990	32/32 PVD	0.8 g	P	84	152/85	150/82	−2/−3*
Kiesewetter, 1991b	30/30 T	0.8 g	P	28	116/74	108/67	−7/−9*
König, 1986	53/0 PVD	0.6 g	P	28	167/107	156/97	−6**/−9***
Mader, 1990b	53/51 HL	0.9 g	P	112	151/92	143/85	−5*/−8
Simons, 1995	28/28 HL	0.9 g	P	84	122/76	119/76	−2/0
Vorberg, 1990	20/20 HL	0.9 g	P	112	144/90	138/87	−4***/−3*
Barrie, 1987	20/20 N	18 mg	OS	28	ave. 94	ave. 89	−5**

For notes and abbreviations, see Table 5.1. From Reuter HD, Sendl A. Allium sativum and Allium ursinum: chemistry, pharmacology, and medical applications. In: Economic and medicinal plant research. New York: Academic Press, 1994:59–113.

In a drug-monitoring study with 1997 patients suffering from hyperlipidemia (given 900 mg garlic powder as tablets daily for 16 weeks), a decrease of the average systolic blood pressure from 146 to 140.5 mm Hg and of the average diastolic blood pressure from 85.5 to 82.5 was seen (Beck & Grünwald, 1993). However, for the patients with mild hypertension only and who did not take other blood pressure–lowering drugs, a much more distinct effect was seen (systolic from 161.9 to 151.4 mm Hg and diastolic from 97.6 to 89.7 mm Hg; $P < 0.001$).

In an open-design study with nine patients having severe hypertension (diastolic greater 115 mm Hg), the acute effect of a garlic preparation, standardized at 1.3% alliin (0.6% allicin release), was investigated for the effect of a large single dose (2400 mg garlic powder). In sitting patients, the blood pressure fell by 7 ± 3 mm Hg (systolic) and by 16 ± 2 mm Hg (diastolic). The decrease of the diastolic blood pressure was still statistically significant ($P < 0.05$) for 5–14 hours after receiving the dose (McMahon & Vargas, 1993).

5.1.2.3 BLOOD PRESSURE EFFECTS: POSSIBLE ACTIVE COMPOUNDS

There is, as yet, no final clarity about the actual *effective substances* that influence blood pressure. However, good beginnings for elucidation of this problem exist. In 1946, de Torrescasana (1946) treated cats with an alcohol extract and two fractions of the extract and found that there are at least two effective principles in garlic: (*a*) a compound, soluble in chloroform, which exerts a tonic action on the heart muscle as well as a slight blood pressure–lowering action, and (*b*) at least one chloroform-insoluble compound, which has a strong blood pressure–lowering action, as well as a tonic effect on the intestines. Allicin appears not to be involved, since other animal studies showed that removal of allicin by cooking the aqueous extract (Rao et al., 1981) or the inactivation of alliinase by stomach acid exposure of gelatin capsules of garlic powder to dogs (Pantoja et al., 1991) does not prevent the activity. Furthermore, a water-soluble, ethanol-insoluble extract of uncrushed garlic cloves (allicin removed with ethyl acetate) effectively reduced blood pressure in dogs (Sial & Ahmed, 1982; Rashid & Khan, 1985).

One of the active constituents may be adenosine, if it is sufficiently present (see Section 3.3.11). While in general oral adenosine is only poorly absorbed from the gastrointestinal tract, in the presence of polar lipophilic substances, such as allicin, absorption might be significantly increased. As suggested by Reuter, it seems reasonable to assume that the polar sulfur-containing transformation products of alliin and allicin, respectively, serve as a carrier for adenosine and facilitate the otherwise poor absorption of this compound from the gastrointestinal tract. For dimethyl sulfoxide, a compound with a structure corresponding to that of allicin, this effect is well documented (Reuter, 1987).

Perhaps more important than the content of adenosine in garlic is the possibility that garlic may increase blood adenosine levels by inhibition of adenosine deaminase, a property that has been demonstrated in vitro with the pure enzyme (Koch et al., 1992b) as well as with cultured aortic endothelial cells (Melzig et al., 1995). These studies have also demonstrated that allicin is not responsible for the effect and that the activity decreases slowly upon storage of cloves, a characteristic of the γ-glutamylcysteines content (Section 5.14). Adenosine enlarges the peripheral blood vessels, allowing the blood pressure to decrease. However, adenosine is also involved in the regulation of blood flow in the coronary arteries. The adenosine concentration increases when the oxygen content in the blood decreases, as well as when the need for oxygen increases. Caffeine and theophylline, the purine alkaloids in coffee and tea, block the adenosine receptors, and consequently trigger a rise of blood pressure (Daly, 1982).

Recent investigations of changes in the physical state functions of the membrane potential of vascular smooth muscle cells suggest another possible mechanism of the blood pressure–lowering effect of garlic. Aqueous garlic extract, as well as ajoene, was found to cause membrane hyperpolarization in cells of isolated blood vessel strips (Siegel et al., 1991a, 1991b, 1992). Hyperpolarization causing vasodilation is probably the result of the closing of the potassium and calcium channels. In vitro, the uptake of Ca^{2+} and HPO_4^{2-} of the sheep aorta matrix was inhibited by garlic extract (Mirhadi & Singh, 1987).

Sendl and coworkers (1992) demonstrated that extracts from garlic and wild garlic (*Allium ursinum*) inhibit angiotensin-converting enzyme, an enzyme which is involved in the mechanism of increasing hypertension. The highest inhibitory activity was found for an aqueous extract from the leaves of wild garlic containing γ-glutamylcysteines, followed by that from garlic leaf, and then the bulbs of both plants (see also Reuter, 1993b). Elbl (1991) found that all three of the γ-glutamylcysteines of garlic have equal activity toward inhibition of angiotensin-converting enzyme in vitro, but that allicin has no activity.

Very recently, a new possible mechanism of action for the hypotensive effect of garlic has been reported. It was found that both water and alcohol (inhibits most of the allinase activity) extracts of garlic or garlic powder tablets significantly increased the production of nitric oxide (which is associated with decreased hypertension) in vitro as determined by increased calcium-dependent nitric oxide synthetase activity (Das et al., 1995a). This possibility has also been confirmed in vivo, when consumption of 4 g fresh garlic was found to double the nitric oxide synthetase activity of blood platelets in 3 hours (Das et al., 1995b).

5.1.3 Effects on Blood Coagulation, Fibrinolysis, and Blood Flow

Normally the blood coagulation system is in equilibrium with a second, opposite system that breaks down the final product of the coagulation cascade, fibrin, into fibrin degradation products. An imbalance between the two systems toward coagulation causes the thromboembolic complications of arteriosclerosis. In general, arteriosclerosis is accompanied by decreased fibrinolytic activity. Thus, one of the goals of effective antiarteriosclerotic therapy is to increase fibrinolysis to a normal level by appropriate drug therapy.

Aqueous garlic extracts and garlic oil influence plasma fibrinogen level, coagulation time, and fibrinolytic activity, in humans as well as in animals (Ogston, 1983, 1985; Roser, 1990; Anonym, 1976). In experiments with laboratory animals that were artificially rendered arteriosclerotic, the fibrinolytic activity in the serum was increased with dry garlic or garlic extract, as well as with garlic and onion oils (Bordia et al., 1977; Bordia & Verma, 1980; Harenberg et al., 1988; Sharma & Nirmala, 1985; Kumar et al., 1981; Srivastava & Mustafa, 1993). Also, in patients suffering from arteriosclerosis, the same effect was obtained (Gupta et al., 1966; Menon et al., 1968; Jain & Andleigh, 1969; Bordia et al., 1977, 1978; Barrie et al., 1987; Sharma et al., 1978; Bordia & Joshi, 1978). Garlic also inhibits blood coagulation induced by Staphylococci through lowering of the activity of bacterial coagulase (Fletcher et al., 1974). Furthermore, a case has been reported in which the blood clotting time (8–12 minutes) of a woman who consumed heavy amounts of garlic decreased to 6 minutes when she stopped eating garlic (Burnham, 1985).

Sainani and colleagues carried out an epidemiological study on three groups of patients: group I with weekly consumption of 50 g garlic plus more than 600 g onions); group II with weekly consumption of 10 g of garlic and 200 g of onion; and group III, the control group, which, for religious reasons (members of the Jain sect in India), followed a strictly vegetarian diet, but without any garlic and onions. Groups I and II showed a significant decrease in fibrinogen levels within 4 weeks as compared to group III ($P < 0.001$) (Sainani et al., 1976, 1979a, 1979b, 1979c).

Another study by Bordia and coworkers involved the intake of the juice or the ether-extracted oil from 50 g garlic together with the consumption of 100 g butter. The effect observed was similar for both garlic preparations, namely decreased concentration of plasma fibrinogen ($P < 0.01$), increased fibrinolytic activity ($P < 0.001$) and increased blood coagulation time ($P < 0.01$). Again, a definite influence of garlic on all three parameters was observed (Bordia et al., 1975a, 1975b; Bordia, 1989). Similar studies in healthy subjects with much smaller amounts (0.9–1.0 g) of garlic powder preparations have also been reported (Legnani et al., 1993; Jain et al., 1977).

Table 5.4 summarizes the results of 11 clinical studies with 14 groups of subjects, including normal subjects, geriatric patients, and patients with hypercholesterolemia, hyperlipidemia, or coronary heart disease. The studies objectively show that activation of endogenous fibrinolysis by small doses of garlic is real and occurs rapidly. The activating effect is detectable within several hours (Jung et al., 1989a) and increases even further when garlic is taken regularly for several months (Bordia et al., 1977; Chutani & Bordia, 1981; Olschok, 1981). The average increase of fibrinolytic activity in all the groups presented in Table 5.4 was 56% as compared with the pretreatment values.

Investigation of the acute hemorheological (blood flow) effect of garlic was done with single doses of 600–1200 mg dry garlic powder preparation. The most pronounced changes were found for plasma viscosity, tissue plasminogen activator (TPA) activity, and hematocrit level. The maximal decrease in hematocrit level was in the range of 5–10%; this is interpreted as a "rheoregulatory effect" in the sense of an "internal hemodilution," resulting from garlic's peripheral vasodilative effect (Harenberg et al., 1988).

Wolf and coworkers (1990) investigated the hemorheology in conjunctival vessels in a randomized, placebo-controlled, double-blind, crossover trial. In the first phase, measurements were carried out in a characteristic blood vessel area before and 5 hours after administration of 900 mg pure dry garlic powder or placebo. The second phase was conducted after a washout phase of 14 days. Garlic powder administration resulted in a significant ($P < 0.002$) increase of the mean diameter of the arterioles by 4.2%. Similar observations were made in the venules (increased 5.9%, $P < 0.0001$). During the placebo phase, there were no significant changes. Capillaries were not significantly influenced, either during the placebo or the verum phase.

Kiesewetter and Jung also investigated the hemorheological changes in patients with stage II peripheral arterial occlusive disease after a daily dose of 800 mg standardized dry garlic powder given for 4 weeks in a double-blind, placebo-controlled study. Significant ($P < 0.05$) changes were seen in increased

Table 5.4. Clinical Studies: Effects of Garlic Preparations on Fibrinolytic Activity

Author, Year	Cases[a]	Dose/Day	Prep.[b]	Days	Before	After	Diff.%[c]
Arora, 1981	25/0 CHD	4 mg[d]	OS	84	55	53	−4 ns
	17/0 N	4 mg[d]	OS	84	56	52	−7 ns
Bordia, 1975b	10/0 N	20 mg[d]	OE	1	43	92	+114***
Bordia, 1977	10/0 N	20 mg[d]	OE	90	90	205	+128***
	10/0 CHD	20 mg[d]	OE	90	70	129	+84***
	20/0 CHD	20 mg[d]	OE	20	55	107	+95***
Bordia, 1982	20/0 N	15 mg[d]	OE	21	73	127	+74***
Bordia, 1986	30/30 HC	1.5 mg[d]	OM	90	53	66	+25
Harenberg, 1988	20/0 HL	0.6 g	P	28	2.5	2.9	+15**
Jain, R. C., 1977	6/0 N	5 g	F	21	77	95	+23**
Jung, 1989a	10/0 N	1.2 g	P	1	0.78	1.27	+63*
Jung, 1991	10/10 N	0.9 g	P	1	1.2	2.3	+86*
Kiesewetter, 1990	10/0 N	0.6 g	P	1	1.4	2.1	+51***
Olschok, 1981	24/6 G	1.5 g	P	28	35	45	+38**

For notes and abbreviations, see Table 5.1. From Reuter HD, Sendl A. Allium sativum and Allium ursinum: chemistry, pharmacology, and medicinal applications. In: Economic and medicinal plant research. New York: Academic Press, 1994:54–113.

capillary erythrocyte flow rate and decreased plasma viscosity. Plasma fibrinogen levels also decreased ($P < 0.1$) while the hematocrit values and erythrocyte aggregation did not change significantly (Kiesewetter & Jung, 1991; Kiesewetter et al., 1990, 1991a, 1991b, 1993a, 1993b, 1993c; Jung et al., 1990a, 1990b; Jung & Kiesewetter, 1991).

As a rule, it can be stated that an existing hyperlipidemia leads to a reduction of coagulation time, which is expressed as an increased tendency for development of arteriosclerosis. The lipid-induced shortening of coagulation time can clearly be avoided by garlic treatment (Arora & Arora, 1981).

There are only speculations as to which ingredients might be responsible for the increase of fibrinolytic activity. Some authors believe that the sulfur compounds are responsible, as indicated by the studies employing the ether extracts of garlic (Bordia et al., 1975a, 1975b); however, studies with single compounds from garlic have only been conducted with cycloalliin, a compound which was found to elevate fibrinolytic activity (Augusti et al., 1975; Agarwal et al., 1977).

Finally, Song and colleagues isolated a substance from garlic (GEF, garlic extract factor) that increases the prothrombin time and precipitates calcium ions. It was shown that this compound is identical with phytic acid (see Section 3.3.12) (Song et al., 1963a, 1963b, 1963c).

Inhibitors of platelet aggregation and specific effects on blood coagulation and fibrinolysis are found not only in garlic but also in other *Allium* species, especially in onion (Kawakishi, 1991; Ariga & Kase, 1986; Morimitsu & Kawakishi, 1991; Phillips & Poyser, 1972; Nishimura et al., 1988; Yg et al., 1988; Nagda et al., 1983; Chauhan et al., 1982; Mittal et al., 1974; Jain, 1971a, 1971b, 1971c; Menon, 1969, 1970; Jain & Andleigh, 1969; Carson, 1968; Menon et al., 1968; Gupta et al., 1966).

5.1.4 Antithrombotic (Platelet Antiaggregatory) Effects

Apart from the action of lipoproteins, blood platelets (thrombocytes) also play a significant role in atherogenesis. Activated platelets adhere to pathologically changed vessel walls. One of the factors that render the vessel wall susceptible to the adhesion of platelets is the deposition of cholesterol into the vessel endothelium (Section 5.1.1). The adhesion of platelets is followed by an aggregation of the cells, together with a release of certain platelet constituents. Mitogenic factors in particular are released into the endothelial cells, which proliferate further under their influence, thus constricting the lumen and slowing down the flow of blood. These changes can lead to a total occlusion of the vessel, causing myocardial infarction, stroke, or occlusive arterial disease.

Numerous pharmacological studies have been conducted on the inhibition of the adhesion or aggregation of platelets by garlic preparations, garlic compounds, and transformation products of these compounds. As a consequence of these findings, platelet function inhibitors are now part of standard therapy for the prophylaxis of obliterative arteriopathies.

It is well known now that certain garlic constituents affect platelet aggregation by inhibiting the cyclooxygenase-catalyzed formation of cyclic endoperoxides from arachidonic acid, by inhibiting the release of platelet constituents, and by increasing the intracellular cAMP levels. The inhibitory action of factors in onions on the arachidonic acid cascade has recently been reviewed (Kawakishi and Morimitsu, 1994).

Furthermore, garlic also influences the processes preceding aggregation, such as activation (i.e., the transition of platelets from the discoid into the influences the activated form with pseudopods) (Bordia, 1986; Apitz-Castro et al., 1988), and the adhesion of the cells (Bordia & Verma, 1980; Bordia et al., 1978; Sharma & Nirmala, 1985; Sharma & Sunny, 1988). The aggregation-inhibiting effect of garlic at the level of the formation of PGG_2/PGH_2 occurs independently of the respective aggregation inducer (thromboxane A_2 [TXA_2], collagen, thrombin, platelet activating factor, adenosine diphosphate (ADP), serotonin, epinephrine, and vasopressin) (Bordia et al., 1978; Makheja et al.,

1979, 1981; Apitz-Castro et al., 1983a; Morimitsu et al., 1992). Fresh garlic (100–150 mg/kg) leads to complete inhibition of thrombocyte aggregation for 1 to 2 hours after consumption (Boullin, 1981). On the other hand, garlic oil (steam-distilled) preparations at a dose of 0.07 mg/kg can be completely without effect in this regard (Samson, 1982). Onion exerts a similar effect (Baghurst et al., 1977; Makheja et al., 1979; Doutremepuich et al., 1985).

Rabbits that received a lethal dose of collagen and arachidonic acid intravenously were found to develop severe thrombocytopenia, indicative of in vivo platelet aggregation and hypotension. These changes were associated with an increase in the plasma level of thromboxane B_2 (TXB_2) and 6-keto-$PGF_{1\alpha}$, as measured by radioimmunoassay. Pretreatment of the animals with an aqueous extract of garlic (500 mg/kg) provided protection from thrombocytopenia and hypotension. TXB_2 synthesis was significantly reduced in animals pretreated with garlic and then injected with a lethal dose of either collagen or arachidonic acid. The amount of TXB_2 synthesized in these animals apparently was not sufficient to induce thrombocytopenia or hypotension (Ali et al., 1990).

Sendl and coworkers investigated the effect of solvent extracts of wild garlic (*Allium ursinum*) and ordinary garlic with defined chemical compositions for their in vitro inhibitory potential on 5-lipoxygenase (5-LO), cyclooxygenase (CO), and thrombocyte aggregation (TA). The inhibition rates, expressed as IC_{50} values, of both extracts for 5-LO, CO, and TA correlated well with the percent content of the major sulfur-containing compounds (thiosulfinates and ajoenes) of the respective extracts. In the 5-LO and CO tests, the garlic extracts were slightly superior to the extracts from wild garlic, whereas in the TA test no difference was found (Sendl et al., 1992; Wagner & Sendl, 1990; Sendl & Wagner, 1990; Wagner et al., 1991).

An aqueous extract of garlic, thin-layer chromatography (TLC) fractions of the extract, and an ether-chloroform extract of the aqueous extract were examined for their effect on platelet aggregation and platelet synthesis of thromboxane and lipoxygenase products from endogenous arachidonic acid (AA). The aqueous extract and the ether-chloroform extract inhibited the aggregation induced by all agents. Two TLC fractions inhibited epinephrine- and arachidonic acid–induced aggregation. The solvent extract inhibited the incorporation of labeled AA into platelets of platelet-rich plasma; and, at higher doses, it inhibited the degradation of platelet phospholipids and reduced the formation of TXB_2 and lipoxygenase-derived products from labeled platelets. At low dosages, the solvent extract inhibited the cyclooxygenase and lipoxygenase enzymes, but it did not affect the degradation of platelet phospholipids (Srivastava & Justesen, 1989).

Weissenberger and colleagues (1972) showed that platelet aggregation in platelet-rich plasma can be inhibited in vitro by an onion extract. The authors considered *adenosine* to be the effective substance in this plant. Garlic can produce amounts of adenosine comparable to onion and several other medicinal plants (see Section 3.3.11 for content). Michahelles, as well as Reuter, later demonstrated conclusively that adenosine is contained in the aqueous extracts of garlic and that it may contribute to aggregation inhibition by preventing ADP binding to the platelet membrane in a competitive manner (Michahelles, 1974; Reuter, 1983a, 1983b). However, as will shortly be discussed, adenosine exerts no effect in whole blood (Lawson et al., 1992).

Makheja and colleagues (1990) investigated the in vitro effects of adenosine, allicin, and allyl sulfides from both garlic and onion on platelet-rich plasma. Both adenosine and allicin inhibited platelet aggregation without affecting cyclooxygenase and lipoxygenase metabolites of arachidonic acid. The trisulfides inhibited platelet aggregation as well as thromboxane synthesis along with induction of new lipoxygenase metabolites. The authors proposed that adenosine is probably more active in vivo than allicin or the allyl sulfides. However, it should be noted that adenosine acts by competitive blocking of the binding side for ADP at the platelet membrane. Adenosine has multiple biological functions, among them vasodilative, antihypertensive, and spasmolytic actions which are especially significant. These effects are explained by the vascular smooth muscle–relaxing influence of adenosine. In addition, adenosine inhibits lipolysis in fat cells, promotes glucose oxidation, and increases the release of the hormone glucagon from the pancreas. The antidiuretic effect can be explained by vasoconstriction in the kidney. Adenosine also stimulates the synthesis of steroid hormones in the adrenal glands. Finally, as already mentioned, it inhibits or prevents the aggregation of platelets (Born & Cross, 1963; Born et al., 1965; Clayton et al., 1963; Lutomski & Mrozikiewicz, 1989). All these reactions are mediated by reciprocal interaction of the biological transmitter with specific adenosine receptors in the specific organ (Daly, 1982).

The corresponding effect of an alcoholic garlic extract on the smooth muscles has also been demonstrated by the influence on the movement of the uterus, and it may be that the adenosine contained in garlic is responsible for these effects (Tinao & Terren, 1955). In addition, garlic influences the metabolism of other purines in a significant manner. For instance, the enzyme xanthine oxidase from liver is inhibited by garlic. Since this effect is greater with boiled garlic than with fresh garlic juice, it is obvious that a compound other than allicin is involved (Huh et al., 1986a).

The results of several recent investigations demonstrate, however, that *allicin* (natural and synthetic) also acts as a potent aggregation-inhibiting substance in vitro. Allicin at 10 µM completely inhibits platelet aggregation (Mohammad & Woodward, 1986; Mohammad et al., 1980). Aqueous garlic extracts (AQGE) containing allicin also inhibit induced platelet aggregation. After organic solvent removal of the thiosulfinates, the inhibiting effect of AQGE disappears (Mohammad & Woodward, 1986). However, during the extraction process, another aggregation-inhibiting agent may have also been removed, although most of the solvent-soluble (lipophilic) compounds of crushed garlic are thiosulfinates (see Table 3.14).

The pharmacological effects of synthetic allicin have been examined (Mayeux et al., 1988). It was found to inhibit human platelet aggregation (IC_{50} = 0.3 mM) without affecting cyclooxygenase or thromboxane synthase activity or cAMP levels, and without influencing the activity of the vascular prostacyclin synthase. However, it did inhibit lysosomal enzyme release and enhanced the mesenteric circulation in vivo, independently of prostaglandin release or β-adrenergic mechanism.

Further evidence that platelet aggregation in vitro is due to the thiosulfinates results from experiments on whole blood platelet aggregation induced by collagen. This was inhibited by AQGE, the chloroform-extracted thiosulfinates of AQGE, pure allicin at the same concentration as in the AQGE, and other allicin-derived compounds, but not by the chloroform-extracted AQGE, alliin, S-allylcysteine, or adenosine (Lawson et al., 1992).

It should be mentioned that the experiments cited above are in vitro studies and that allicin immediately forms allyl mercaptan when it contacts whole blood. The platelet aggregation–inhibiting ability of allyl mercaptan has not yet been published.

To elucidate further the relative role of garlic constituents on platelet aggregation in vitro, the inhibitory effects of adenosine and 16 quantitatively determined organosulfur compounds derived from garlic cloves or commercial garlic preparations have been studied against collagen-stimulated platelet aggregation in whole blood and in platelet-rich plasma (PRP) (Lawson et al., 1992). An estimation of the antiaggregatory activity of several brands of the major types of commercial garlic preparations was determined from the activities of the individual compounds present in each sample. In PRP most of the antiaggregatory activity of garlic clove homogenates was due to adenosine. However, in whole blood neither adenosine nor the polar fraction had any effect, and all of the antiaggregatory activity was due to allicin and other thiosulfinates. Allicin was equally active in whole blood and PRP. Among the various brands tested, there was a several-fold variation in content of the organosulfur compounds and activity for all types of garlic products tested. The best dry garlic powder tablets were equally active as clove homogenates, whereas steam-distilled oils were only 35% as active and oil-macerates (due to low content) only 12% as active. A garlic product aged many months in dilute alcohol had no activity. For steam-distilled oils, most of the activity was due to diallyl trisulfide. For the oil-macerates, most of the activity was due to the vinyldithiins. Ajoene, a compound unique to the oil-macerates, had the highest specific activity (slightly more active than 1,2-vinyldithiin, diallyl trisulfide, or allicin, and considerably more active than diallyl disulfide, allyl methyl trisulfide, or 1,3-vinyldithiin) of all the compounds tested, but because of its low concentration it accounted for only 19% of the total activity of the oil-macerates. The in vitro activity of steam-distilled garlic oil has also been confirmed by Apitz-Castro and colleagues (1983b).

Besides diallyl trisulfide, other polysulfides, such as dimethyl trisulfide and allyl methyl trisulfide, have also been found to be active. Simple mono- and disulfides, however, are only slightly effective or without any activity (Gaffen et al., 1984). An aqueous garlic extract at pH values above 8.5 does not inhibit induced aggregation, because under these conditions allicin is transformed into the poorly inhibiting diallyl disulfide (Mohammad & Woodward, 1986). Allyl methyl trisulfide, which is present in garlic oil at concentrations of 4–10%, already induces this effect at a concentration of about 10 µM, and the effect lasts up to 2 hours. In the meantime, this compound has been synthesized and is being clinically tested in Japan on a large scale (Ariga et al., 1981; Ariga & Oshiba, 1981a, 1981b; Oshiba et al., 1981; Makheja et al., 1981; Ariga & Kase, 1986).

Inhibition of normal arachidonate metabolism is one of the mechanisms by which the various garlic constituents and their metabolites affect platelet aggregation, since the inhibition of thrombocyte aggregation by garlic or onion oil goes hand in hand with a decrease in synthesis of TXB_2 and 12-hydroxyheptadecatrienic acid (HHT) from arachidonic acid. The lipoxygenase pathway of the arachidonic acid cascade is also inhibited. Additionally, garlic and onion inhibit fatty acid oxygenase. In thrombocytes this enzyme is inhibited 10 times more than in other tissues. The consequence of this is a stagnation of prostaglandin and/or thromboxane synthesis (PGE_2, PGD_2, PGI_2, TXB_2) (Makheja et al., 1980, 1981; Ariga et al., 1981; Wagner et al., 1987; Shalinsky et al., 1989; Agarwal, 1992; Ali et al., 1990; Srivastava & Mustafa, 1989).

Inhibition of cyclooxygenase results in a diminished synthesis of prostaglandins (PGF_2, PGE_2, PGD_2) and TXA_2 in the thrombocytes (Doutremepuich et al., 1985; Wagner et al., 1987; Ali & Mohammed, 1986; Shalinsky et al., 1989). Ali and Mohammed (1986) found that, besides dose-dependent inhibition of thromboxane production during blood clotting, synthesis of prostacyclin in rabbits was not affected by an aqueous garlic extract. The synthesis of thromboxane by the aorta was also completely suppressed. A similar pattern was found upon the enzymatic synthesis of thromboxane and prostacyclin in ex vivo tissues after intraperitoneal administration of aqueous garlic extract (1 mL/kg) for 1 week. Aortic synthesis of prostacyclin was significantly increased in the garlic-treated rabbits as compared to the controls. Also, an inhibition of calcium action by thrombocytes under the influence of garlic was considered as a possible mechanism. Similar observations were made with the active ingredients of onion (Srivastava, 1984a, 1984b, 1986a, 1986b).

Special attention has been given to *ajoene*, one of the transformation products of allicin (see Section 3.2.6.3.3), because of its strong antiaggregatory activity. Apitz-Castro and coworkers have shown that ajoene can inhibit platelet activation induced ex vivo by all known agonists, and that it significantly reduces thrombocytopenia associated with the circulation of blood through extracorporeal devices ex vivo (circulation of fresh heparinized human blood through a dialyzer or oxygenator), as well as in vivo (subjection of dogs to extracorporeal circulation). Under all experimental conditions, ajoene proved to be very efficacious in preventing platelet loss (60–65% loss in controls versus 15–20% loss in the presence of ajoene, $P < 0.01$). Moreover, recuperation of platelet function was achieved after 3 to 4 hours in the in vivo experiments. Recent in vivo animal experiments have also shown that ajoene prevents platelet binding to damaged blood vessels and prevents thrombus formation (Apitz-Castro et al., 1986a, 1988, 1991, 1992, 1994; Jain & Apitz-Castro, 1987; Srivastava & Tyagi, 1993).

By spectral measurements, ajoene was found to interact cooperatively with a purified hemoprotein by modifying the binding interaction of the protein with the ligands, deemed to be physiologically relevant as effectors. The characteristics of the modifications were found to parallel those of the ajoene-induced modifications of agonist-induced aggregation kinetics in gel-filtered calf platelets. Hemoprotein is also implicated in platelet activation (Jamaluddin et al., 1988).

Ajoene, at least, affects the fibrinogen-supported aggregation of washed human platelets (ID_{50} = 13 μM) and inhibits binding of ^{125}I-fibrinogen to ADP-stimulated platelets (ID_{50} = 0.8 μM). In both cases, the inhibition is of the mixed noncompetitive type. Furthermore, the fibrinogen-induced aggregation of chymotrypsin-treated platelets is also inhibited by ajoene in a dose-dependent manner (ID_{50} = 2.3 μM). Other membrane receptors, such as the ADP or epinephrine receptors, are not affected by ajoene. Ajoene strongly quenches the intrinsic fluorescence emission of purified glycoproteins IIb–IIIa (ID_{50} = 10 μM). These results indicate that the antiaggregatory effect of ajoene is caused by direct interaction with the putative fibrinogen receptor (Apitz-Castro et al., 1983a). Furthermore, it was found that the fibrinogen-induced aggregation of ADP-stimulated platelets was inhibited by ajoene in a dose-dependent manner with an ID_{50} of 3.3 μM. The inhibition is of the mixed noncompetitive type (Apitz-Castro et al., 1986b).

Rendu and colleagues (1989) investigated the effect of ajoene on the earliest steps of aggregation and found that ajoene inhibits platelet release by decreasing the viscosity of the inner membrane of the platelets, which then prevents fusion to granules or to plasma membrane.

Block synthesized a number of ajoene homologs and tested them for their capability of inhibiting human thrombocytes (Block et al., 1986; Block, 1986). The results show that the central double bond of ajoene is of decisive importance, since reduction of the bond abolishes its activity. Stepwise oxidation of the sulfur atom to the sulfone and to thiosulfonate also reduces the activity, or even reverses it. Based on previous results, the authors concluded that ajoene enters a disulfide exchange reaction with sulfhydryl groups of membrane proteins, which then modifies the thrombocyte membrane in a manner that renders it incapable of adhesion. Diallyl mono-, di-, and trisulfides were found to have less than half the activity of ajoene.

Other authors, however, believe that alliin is the active principle that prevents platelet aggregation (Liakopoulou-Kyriakides et al., 1985; Liakopoulou-Kyriakides, 1985). This assumption, however, could not be confirmed by other authors (Mütsch-Eckner, 1991; Lawson et al., 1992), who found that alliin, isolated from garlic, did not influence aggregation induced by collagen, ADP, or epinephrine. A significant inhibition of ADP- and epinephrine-induced aggregation by (more than 30%), however, was achieved by 1 mM (–)-N-(1'-deoxy-1'-β-D-fructopyranosyl)-S-allyl-L-cysteine sulfoxide, a component of garlic leaves but not of garlic cloves (Mütsch-Eckner, 1991).

The more complex scordinins, first described by Japanese researchers, display antimicrobial activity and supposedly also have a beneficial effect on fatty livers (Kominato, 1969a, 1969b). In addition, these compounds do have a regulating influence on smooth muscle, heart rhythm, and blood pressure. A scordinin fraction of garlic was found to have antiaggregatory activity in test on rats, but further purification revealed that adenosine was the responsible agent (1974).

In another *Allium* species from China (*A. bakeri*), both adenosine and phenolic compounds, structurally similar to flavonoids, were identified as effective components (Okuyama et al., 1986, 1989). This suggests that the flavonoids present in garlic may participate in the aggregation-inhibition effect as well. *A. bakeri* is native to China and has been manufactured into Chinese drug preparations ("xiebai" and "dasuan") that inhibit thrombocyte aggregation (Okuyama et al., 1989).

Clinical studies demonstrating the platelet-inhibiting effects of garlic or garlic preparations have been comparatively few. This fact, in view of the positive results of numerous pharmacological investigations, is surprising and may be explained by methodological difficulties. Recent investigations (Reuter HD, unpublished data, 1993) suggest that in aggregation tests using collagen, epinephrine, ADP, and thromboxane as inducers, the inducers may have been used at concentrations that were too high. High concentrations of aggregating agents can abolish platelet inhibition caused by low-dosed garlic preparations. In contrast to the induced aggregation, spontaneous aggregation seems to be more sensitive.

Bordia (1978) was the first who systematically investigated the effect of garlic oil (an ether-extract oil) on in vitro human platelet aggregation induced by ADP, epinephrine, and collagen (Bordia, 1978). Positive effects were found both in vitro and after oral consumption (0.4 mg/kg body weight). The results of these early experiments are presented in Table 5.5. The results of aggregation tests clearly show that the inhibitory effect of garlic can be overcome by increasing the concentration of the aggregating agent. In the scope of a clinical study with a commercial oil-macerated garlic preparation (brand Aktiv-Kapseln, four to six 260-mg capsules daily for 1 to 3 months, 0.5 mg garlic-derived oil/kg body weight), platelet adhesiveness and platelet aggregation were tested against placebo (Bordia, 1986). The results of this experiment are summarized in Table 5.6. The difference between verum and placebo therapy was significant for both adhesion and aggregation ($P < 0.01$ and 0.05, respectively).

Table 5.5. Effect of Garlic Oil (Ether Extract) on ADP-, Epinephrine- and Collagen-induced Aggregation In Vitro (%) and Ex Vivo

Garlic Oil	ADP		Epinephrine		Collagen
	1 µg	0.5 µg	1 µg	0.5 µg	1 mg/mL
0	74.2 ± 3.5	40.2 ± 2.1	77.4 ± 6.2	48.2 ± 5.5	88.2 ± 4.9
2.5 µg	50.0 ± 3.1	24.1 ± 2.3	41.1 ± 2.3	18.1 ± 1.2	17.1 ± 2.3
5 µg	28.0 ± 2.2	12.0 ± 1.0	24.2 ± 1.6	8.0 ± 1.0	12.0 ± 2.3
10 µg	5.2 ± 0.5	2.1 ± 0.2	5.0 ± 0.6	1.1 ± 0.1	3.1 ± 0.2
Orally[a]	54.2 ± 6.1	31.1 ± 4.2	57.8 ± 7.1	24.2 ± 2.6	25.2 ± 4.1

[a]After consuming 25 mg (0.4 mg/kg body weight) daily for 5 days.
From Bordia A. Effect of garlic on human platelet aggregation in vitro. Atherosclerosis 1978;30:355–360.

Table 5.6. Effect of Garlic Preparation (Oil-Macerate) on Adhesion and Aggregation of Human Platelets Ex Vivo

Test parameter	Initial value	After 1 month	After 2 months	After 3 months	After placebo 4 months
Adhesion (%)	33.1 ± 3.8	27.1 ± 3.9	25.5 ± 4.1	23.1 ± 3.1	27.0 ± 3.3
Aggregation[a]	0.77 ± 0.05	0.90 ± 0.06	1.00 ± 0.07	1.00 ± 0.09	0.87 ± 0.04

[a]The higher the value, the lower the aggregation rate. From Bordia A. Klinisch Untersuchung zur Wirksamkeit von Knoblauch. Apoth Magazin 1986;4:128–131.

In a clinical study in China on 34 hyperlipidemic patients treated with steam-distilled garlic oil (2 mg/kg body weight) for 20 days, there was a very significant ($P < 0.001$) 30% decrease in ADP-induced platelet aggregation (Zhejiang, 1986). Boullin (1981) reported on an acute effect after administration of 100 mg fresh garlic per kg body weight (0.4 mg allicin/kg) to three normal subjects, with a 58% decrease in aggregation induced by 0.1 µM ADP. Very recently, it was shown that consumption of 3 g freshly chopped garlic daily by normal volunteers for 4 to 16 weeks resulted in a 20 to 90% reduction, respectively, in serum thromboxane production, indicating that platelet aggregation had been inhibited in vivo (Ali & Thomson, 1995). On the other hand, a controlled study with 14 men given an ether-extracted garlic oil adsorbed onto calcium carbonate at 1.2 mg/kg for 5 days found no effect on induced platelet aggregation (Morris et al., 1995). However, the procedure in preparing this oil has been repeated and the oil was found to consist mainly of unidentified compounds, similar to what is shown in Table 3.8, which makes this a very different oil from the oil macerates or the oil used by Bordia's group (both of which contain mainly vinyldithiins) and from the steam-distilled oils which consist mainly of allyl sulfides.

Kiesewetter and coworkers investigated the effect of garlic on platelet aggregation in patients with an increased risk of juvenile ischemic attack (Kiesewetter et al., 1991b, 1993a; Jungmayr, 1994; Kiesewetter & Jung, 1992). In a double-blind, placebo-controlled study on 60 volunteers who had cerebrovascular risk factors and constantly increased platelet aggregation, it was demonstrated that the ingestion of 800 mg daily (0.1 mg allicin/kg) standardized garlic powder tablets for 4 weeks leads to a significant reduction of circulating platelet aggregates, as well as of spontaneous platelet aggregation. The ratio of circulating platelet aggregates decreased by 10.3% ($P < 0.01$) and spontaneous platelet aggregation decreased by 56.3% ($P < 0.01$) during the period of garlic intake, with no significant change in the placebo group.

Several *animal studies* have also indicated the in vivo activity of garlic and garlic products. In rabbits, a cholesterol-rich diet fed for 3 months significantly increased the adhesiveness of platelets, by as much as 180%. After a period of 9 months without the cholesterol-rich diet, the adhesiveness was still elevated by 54%. Treatment of the animals with an ether-extract garlic oil (0.5 mg/kg) after the regression period resulted in adhesion values of 88% below the initial values and 44% below the values of the controls (Bordia & Verma, 1980). In an experimental thrombosis model, it was shown that a garlic product with a composition similar to that of diluted steam-distilled garlic oil (0.4 mg garlic-derived oil per kg) prevents acute platelet thrombus formation in stenosed canine coronary arteries (DeBoer & Folts, 1989). When diallyl trisulfide (allitridi) was administered intravenously to rats at 15 mg/kg body weight, ADP-induced platelet aggregation was decreased by 80% (Zhaosheng et al., 1984). Mice given garlic powder (50 mg/kg body weight intraperitoneally, 0.3 mg allicin/kg, equal to a 10-g clove for a person) showed a significant decrease (50%) in the number of hyperthermia-induced thrombi (Zaghloul et al., 1995).

In conclusion, the numerous in vitro studies on inhibition of platelet aggregation have been substantiated by both human and animal studies which show similar in vivo anti-aggregatory effects for garlic cloves, garlic powders, and allicin-derived garlic oils at similar levels of allicin yield or content of allicin-derived compounds. This indicates that allicin is probably responsible for most of the in vivo antithrombotic activity of garlic at moderate doses. Further evidence for allicin's role can be established by comparisons between allicin-yielding and non–allicin-yielding garlic powders.

5.2 ANTIBIOTIC EFFECTS

The various effects of garlic on bacteria, fungi, protozoa, and viruses have been shown in vitro as well as in vivo. The antibiotic activity is mainly due to allicin. Alliin has been shown to have no activity (Stoll & Seebeck, 1947, 1951; Hughes & Lawson, 1991). Apparently, the S(=O)S (thiosulfinate) structure plays an important role, because upon reduction of allicin to diallyl disulfide, the antibacterial effect is greatly abolished. Inhibition of certain SH-containing enzymes in the microorganisms by the rapid reaction of thiosulfinates with SH-groups is assumed to be the mechanism involved in the antibiotic effect (Cavallito & Bailey, 1944; Cavallito et al., 1944; Cavallito, 1946; Bailey & Cavallito, 1948; Small et al., 1949; Saratikov & Plakhova, 1950; Wills, 1956; Barone & Tansey, 1977; Szymona, 1952; Tynecka & Szymona, 1972; Ionescu et al., 1954).

Hughes & Lawson (1991) showed that the antibacterial and antifungal effects of aqueous garlic extracts are completely abolished as soon as the thiosulfinates (allicin, etc.) are removed with chloroform. The rate of loss of antifungal activity of stored extracts paralleled the decreasing allicin content. Since the antibiotic activity parallels allicin stability, it can be increased through chemical stabilization, such as when garlic is stored in 2% vinegar, after which the activity was found not to diminish until after 4 months (Matysek & Behounkova, 1968).

The antibiotic activity of allicin is quite remarkable. Even at dilutions of 1:85,000 to 1:125,000, it completely inhibits a variety of Gram-positive as well as of Gram-negative bacteria. The antibiotic activity of one milligram of allicin equals that of 15 IU of penicillin (10 μg penicillin G), which is approximately 1% of penicillin's activity (Cavallito & Bailey, 1944; Zwergal, 1952).

Prolonged heating of cloves at high temperatures causes loss of the antimicrobial activity. Therefore, dry garlic powders may lose much of their effectiveness if heated too severely (see Table 3.18). Under certain circumstances, such a product may even be stimulating to the growth of bacteria and fungi. For instance, substances (probably peptides) have been isolated from boiled garlic, which greatly stimulate the growth of *Streptococcus faecalis* (Shashikanth et al., 1981).

There may be considerable differences in the antimicrobial activities among the many varieties of garlic and particularly among garlic preparations, in which the allicin is converted to either diallyl sulfides (steam-distilled oils) or vinyldithiins and ajoenes (ether-incubated or oil-macerated oil preparations), or in which S-allylcysteine is the main sulfur compound remaining (aged extracts). This is illustrated in Table 5.7, which compares the antibacterial activity of allicin with the activity of these compounds and with garlic-derived oils. Although allicin is clearly the most active compound, the activity of the diallyl sulfides increases with each additional sulfur atom. Ajoene, which is the only other compound listed that contains an S=O group, was almost as active as allicin, particularly against *S. aureus*; however, there is only a small amount of this compound present in most commercial products (see Table 3.21). The activity of steam-

Table 5.7. Antibacterial Activity (In Vitro) of Garlic-Derived Sulfur Compounds and Oils

	Staphylococcus aureus	Escherichia coli
Allicin	27	27
Ajoene (E/Z)	55	150
Diallyl tetrasulfide	130	1000
Diallyl trisulfide	250	1900
Diallyl disulfide	900	1900
Diallyl sulfide	> 2500	> 2500
Allyl mercaptan	> 4000	> 4000
S-Allylcysteine	> 4000	> 4000
S-Allylmercaptocysteine	2000	> 4000
Garlic oil (steam-distilled)[a]	80	2000
Garlic oil (ether-incubated)[b]	300	300

Values represent the average minimum inhibitory concentrations (MIC, μg/mL) (the smaller the value, the greater the activity) for at least three experiments, conducted by the method of Hughes and Lawson (1991), except that 0.5% Triton X-100 detergent was added to the incubation medium to solubilize the sulfides and garlic oil. At up to 4%, the detergent had no effect on bacterial growth. The concentrations of the sulfides in the incubation medium were verified by HPLC analysis.
[a]Composition of this pure (undiluted with other oils) commercial steam-distilled oil: diallyl pentasulfide (3%), diallyl tetrasulfide (9%), diallyl trisulfide (27%), diallyl disulfide (21%), diallyl sulfide (9%), allyl methyl trisulfide (18%), allyl methyl disulfide (10%).
[b]Composition of this ether extract of crushed cloves: 1,3-vinyldithiin (58%), 1,2-vinyldithiin (15%), allyl methyl trisulfide (7%), ajoene (1%), unidentified compounds (20%).
From Lawson LD, Wang ZJ, unpublished data, 1994.

distilled garlic oil is significant against *S. aureus* and is about four times as great as expected from the activities of its components, indicating synergistic activity. It is interesting that these compounds and oils, with the exception of allicin and the ether-extract oil, are much more active against *S. aureus* than against *E. coli*, indicating that the inhibitory activities of these compounds can be selective.

5.2.1 Antibacterial Effects

The famous French researcher and physician, Louis Pasteur (1822–1895), was the first to describe the antibacterial effect of onion (Pasteur, 1858) and garlic juices (no known reference, but historically believed). The German physician Albert Schweitzer (1875–1965) first treated amoebic dysentery in Africa only with garlic (cited in Simon, 1932). During the Great Plague of London in 1665, garlic is said to have protected many people from the infection. Likewise, it was used in France during the epidemic of 1721 (*"vinaigre des quatre voleurs"*) (Binding, 1970). In India, garlic is an ancient folk remedy for various diseases and is regularly used there, both internally and externally, as a medicine (Chopra, 1933).

During World War I, while stationed in the Balkans, Austrian and German military physicians became acquainted with the high antiinfective effect of garlic (Kretschmer, 1935b; Aust, 1931). Later, they successfully used a combination of dried, powdered garlic with phenyl salicylate (allphen) for various infectious intestinal diseases, including cases of cholera and dysentery (Marcovici, 1915; Marcovici & Schmitt, 1915; Marcovici & Pribram, 1915; Aust, 1931; Kretschmer, 1935a, 1935b, 1935c). Since then, publications have repeatedly appeared reporting the bactericidal and antiseptic activities of garlic and some of its substituents (Vlaykovitch, 1924; Vollrath et al., 1936; Tynecka & Gos, 1973; Abdou et al., 1972; Dombray & Vlaicovitch, 1924; Kubota & Ken, 1927; Chopra, 1933; Walton et al., 1936; Kitagawa & Hirano, 1937; Egger, 1937; Böcker, 1938;

Kitagawa & Noda, 1939; Osborn, 1943; Small et al., 1947; Rao & Verma, 1952; Szilagyi, 1952; Kourennoff, 1970; Grzybowski & Lewicka, 1987; Warner, 1994; Ahmad et al., 1993). Even acid-fast bacteria, such as *Mycobacterium tuberculosis* and *Mycobacterium leprae*, are not resistant to the antibiotic action of garlic (McKnight & Lindegren, 1936; Jain, 1993; Abbruzzese et al., 1987; Chaudhury et al., 1962; Fitzpatrick, 1954; Sreenivasamurthy et al., 1962; Kathe, 1905). The effectiveness of garlic also includes phytopathogenic bacteria (Kharkina, 1951).

In modern times, Glaser and Drobnik (1939) were the first to conduct a scientific investigation of various garlic preparations, and they established that garlic oil manifested an especially high bactericidal effect. Since then, many publications have appeared confirming high antimicrobial activity against Gram-positive and Gram-negative bacteria and fungi by freshly pressed juice, aqueous and alcoholic extracts, lyophilized powders, and other preparations of garlic and onion. Raw juices were also found to be highly effective against *Escherichia coli, Pseudomonas, Salmonella, Candida, Klebsiella, Micrococcus, Bacillus subtilis,* and *Staphylococcus aureus* (Abdou et al., 1972; Kitagawa & Amano, 1935; Tynecka & Szymona, 1972; Gonzalez Fandos et al., 1994; Xiguang, 1986; Kupinic et al., 1980; Kupinic, 1979; Mano, 1962a, 1962b, 1962c). It is remarkable especially that enterotoxic coli strains and other pathogenic intestinal bacteria, which are responsible for diarrhea in humans and animals, are more easily inhibited by garlic than those which constitute the normal intestinal flora (Sharma et al., 1977; Kumar & Sharma, 1982; Rees et al., 1993; Caldwell & Danzer, 1988; Weiss, 1941).

Garlic exhibits a broad antibiotic spectrum against Gram-positive and Gram-negative bacteria, including *Klebsiella, Pasteurella, Corynebacterium,* and *Mycobacterium* strains (Kabelik & Hejtmankova-Uhrova, 1968). The IC_{50} value of a garlic extract preparation against 30 mycobacteria strains was determined to be in the range of 1.34–3.35 mg/mL. The most sensitive germ was *Mycobacterium bovis* with an IC$_{50}$ value of 1.34 mg/mL, while for six strains of *M. tuberculosis* the value was 1.67 mg/mL (Delaha & Garagusi, 1985).

Jezowa and colleagues (1966) tested the activity of a vacuum-dried garlic preparation against various microorganisms and compared it with a number of commonly used antibiotics. Garlic was effective in all cases, and even worked against those strains that had become resistant to antibiotics. Petricic and Lulic (1977b) tested the effectiveness of garlic oil and garlic extract on the growth of 22 strains of microorganisms and found a notably high antibiotic activity in vitro.

Srivastava and colleagues (1982) observed similar effects on 21 strains of pathogenic bacteria. Abbruzzese (1987) showed that aqueous garlic extract inhibits the growth of *Mycobacterium tuberculosis, M. avium-intracellulare complex,* and *M. kansasii* at concentrations ranging from 0.98 to 2.94 mg/mL. Synergism could not be demonstrated when garlic extract was added to various concentrations of four commonly used antituberculous drugs. A recent reinvestigation of the antitubercular activity of an ether-extract garlic oil (17% allicin) showed equally good activity both in vitro and in vivo in guinea pigs (0.5 mg/kg body weight, intraperitoneally) (Jain, 1993).

Garlic, onion, and shallot have been tested for antimicrobial activity against pathogenic aerobic and anaerobic bacteria. Garlic showed the greatest activity, while the combination of garlic with antibiotics was synergistic against *Acinetobacter calcoaceticus*. No activity was found against anaerobic bacteria (Didry et al., 1987). Studies on the active principles of garlic revealed that the oligosulfides, transformation products of allicin, do have some antibacterial activity as well (Lang & Chang, 1981). Even the hydantoin derivatives of some thiosulfinates show antimicrobial properties (Tahara et al., 1979).

The combination of garlic extracts with antibiotics leads to partial or total synergism, mainly against aerobic bacteria (Didry et al., 1992). This is true also for other *Allium* species, such as for wild garlic (*A. ursinum*) (Tynecka et al., 1993), elephant garlic (*A. ampeloprasum*) (Hughes & Lawson, 1991), shallots (*A. ascalonicum*) (Didry et al., 1987), Chinese chives (*A. tuberosum*), and Japanese bunching onions (*A. fistulosum*) (Akema et al., 1987), and others (Rudat, 1969; Matysek & Behounkova, 1968).

Aqueous extracts of garlic contain the water-soluble thiosulfinates (allicin, etc.) and therefore have strong antimicrobial activity, as has been shown by the results of numerous studies (Al-Delaimy & Ali, 1970; Saleem & Al-Delaimy, 1982; Elnima et al., 1983; Dababneh & Al-Delaimy, 1984; Rode et al., 1989; Maruzella & Sicurella, 1960; Karaioannoglou et al., 1977; Mantis et al., 1978; Paszewski & Jarosz, 1978; Rao & Rao, 1978; Anussorn-Nitisara et al., 1979; Inouye et al., 1983; Khan et al., 1985; Patel et al., 1986; Lutomski & Mrozikiewicz, 1992). The extraction of the antimicrobial compounds from garlic with water and/or lower alcohols has been patented (Tamura, 1994).

Both an aqueous extract of garlic and allicin isolated from garlic were shown to have good in vivo activity at nontoxic doses in treating rabbits for experimental intestinal shigellosis (Chowdhury et al., 1991). The treated rabbits were fully cured in 3 days, while 80% of the untreated animals died within 2 days. Furthermore, the blood of the treated animals was found to have good short-term antibacterial activity. The only flaw in the study is the fact that the allicin was stored in the absence of solvent, a condition which causes rapid conversion of allicin to vinyldithiins, ajoene, and a variety of unknown compounds (see Table 3.8). Shashikanth and coworkers (1985, 1986) also showed that feeding fresh garlic decreased the intestinal bacteria of rats, but that feeding boiled garlic (allicin formation prevented) had no effect, indicating that allicin is also responsible in vivo for the intestinal antibiotic effects of garlic.

In addition to the observation of strong antibiotic properties, the complete lack of resistance of germs to garlic has been found repeatedly (Dankert et al., 1979; Singh & Shukla, 1984). Following intragastric administration of garlic extract (corresponding to 8 µM allicin) to rats, 0.4 µM was recovered in the intestines and 2.4 µM in the appendix after 4 to 6 hours. At the same time, the intestinal microflora was reduced by 50–60%, while the microflora of the appendix was reduced by 80%. Generally, aerobic bacteria are more susceptible to allicin than are anaerobic bacteria (Shashikanth et al., 1984, 1985).

Comparison of the antibacterial effect of garlic extract and 12 antibiotics against bacterial strains including *Escherichia*, *Proteus*, *Salmonella*, *Providencia*, *Citrobacter*, *Klebsiella*, *Hafnia alvei*, *Aeromonas*, *Staphylococcus*, and *Bacillus* species showed that all strains except *Pseudomonas aeruginosa* were sensitive to garlic. Sensitivity to garlic was also exhibited by some strains that were already resistant to antibiotics (Kabelik & Hejtmankova-Uhrova, 1968).

As a consequence of the bactericidal activity of garlic, toxin production by the microorganisms is also prevented. Thus, it was demonstrated that the addition of 1500 µg garlic oil per gram of sausage meat practically prevented the formation of toxins from *Clostridium botulinum* type A. This is a plausible explanation for the conserving effect of garlic spice in the food industry (Sanick, 1974; Sanick, 1975; DeWitt et al., 1979). The gas production by *Clostridium perfringens* under laboratory conditions is also prevented by garlic (Savitri et al., 1986).

In human medicine, garlic oil (steam-distilled) is used for external and internal infections, especially of the mucous membrane of the mouth cavity, the upper respiratory system, and the lungs. Various diseases of the larynx, such as tonsillitis, bronchitis, and pussy aphthae are clear indications. Garlic even shows good activity against the "problem germs" of the typhus-paratyphus-enteritis group and *Pseudomonas aeruginosa* (Fortunatov, 1952; Rudat, 1957). Allicin also enhances, in a synergistic manner, the effectiveness of antibiotics such as streptomycin or chloramphenicol, against, for instance, *Mycobacterium tuberculosis* (Gupta & Viswanathan, 1955).

Before the era of sulfonamides and modern antibiotics, garlic preparations were used in epidemics of typhus, paratyphus, cholera, dysentery, amebic dysentery, diphtheria, tuberculosis, influenza, and poliomyelitis (Kubota & Ken, 1927; Fleury, 1932; Babushkin, 1952; Latsinik & Kuperman, 1955). Madaus (1976) reported some older clinical studies on the antiinfective activity of garlic. Zwergal (1952) dealt intensively with the garlic constituents with antibiotic activity in vitro, as did Kabelik and Petkov (Kabelik, 1970; Petkov & Kushev, 1966). In addition, detailed summaries of this topic have been presented (Lutomski, 1980; Hiller, 1964; Virtanen, 1958a, 1958b; Petkov, 1986; Adetumbi & Lau, 1983; Vonderbank, 1950). A dry garlic powder preparation (brand Wang 1000) has been claimed to have exceptional antimicrobial activity (Lutomski et al., 1988).

Plant antibiotics, the so-called "phytoncides," are frequently used, even today, in the former Soviet Union for disinfecting wounds and for the promotion of wound healing, as well as for other purposes (Toropzev & Kamnev, 1946; Belohvostov, 1949; Erlichman, 1949; Dubrova, 1950; Yakovlev & Zvyagin, 1950; Berchenko & Trifonova, 1951; Margolina et al., 1951; Fortunatov, 1952; Kamenetskaya, 1952; Kardashev et al., 1952; Korotkov, 1953; Logacheva, 1953; Bedrintseva, 1954; Ejzer, 1954; Levitin, 1954; Novikov, 1954; Latsinik & Kuperman, 1955; Linde & Rodina, 1956; Abdullaeva, 1959; Zimina et al., 1963; Tagiev, 1967; Nikolaeva, 1979; Peterssone, 1958; Sykow, 1954; Epshtein & Frolkis, 1953; Petroff, 1963). Summaries of the numerous studies originating from the former Soviet Union on the various pharmacological effects of garlic, as well as on its clinical applications, have been amply reviewed (Müller-Dietz & Rintelen, 1960; Müller-Dietz et al., 1969).

Case reports and clinical studies on the antibiotic effect of garlic are also available. An observation from World War I indicates that an antiinfective effect occurs in vivo, since it was found that soldiers whose diet included garlic suffered less frequently from dysentery than their comrades who did not eat garlic (Madaus, 1976; Ernst, 1981; Reuter, 1983a).

In a children's hospital in Poland, multiple cases of gastroenterocolitis, dyspepsia, pneumonia, sepsis, and nephrosis in infants of ages 6 months to 3 years were successfully treated with garlic preparations (Jezowa et al., 1966). In addition, the effect of garlic extract on the mouth flora was examined. A mouthwash that contained 10% garlic extract in Ringer's solution caused a drastic reduction of total bacterial counts in the mouth cavity (Elnima et al., 1983). An alcoholic garlic extract was used in the former Soviet Union for pulpitis and periodontitis (Vitalyeva et al., 1968).

Fortunatov (1955) studied the influence of various secretions and body fluids, such as saliva, blood, gastric fluid, gall, and urine, on the antimicrobial activity of garlic constituents. No inactivation occurred in saliva or blood, while the other fluids seemed actually to enhance the activity of garlic constituents. Excretion of the active constituents in urine does not occur before 6 hours after oral administration.

Johnson and Vaughn (1969) determined the death rate kinetics of bacterial populations (*Salmonella typhimurium* and *Escherichia coli*) in the presence of rehydrated dry garlic and onion powder. Maximum effect was observed in 1–5% dispersions. The average reduction times were 1.1–1.2 hours in resting (static) cultures and 1.8–2.1 hours for those in the active growth phase. The bactericidal effect corresponded to a 0.1% solution of dipropyl disulfide or allylpropyl disulfide in resting cultures. Bacteria that were actively growing were only inhibited for a short time, since they overcome the effect of bacteriostasis in 2 to 6 hours.

Lactic acid bacteria isolated from fermented Chinese cabbage (Kimchi), such as *Lactobacillus plantarum*, *L. brevis*, *Leuconostoc mesenteroides*, and *Pediococcus cerevisiae*, were inhibited by garlic during the fermentation process (Lee & Kim, 1988).

The growth of *Lactobacillus casei*, however, was stimulated by garlic extract (Park et al., 1980).

For the sake of completeness, it should be mentioned here that the antibacterial effect of garlic applies not only to bacteria that exist in the human or animal body in a symbiotic or parasitic manner, but also to soil bacteria—bacterial populations living in the soil and rhizosphere or outside of it (Agrawal & Rai, 1948).

The activity of SH enzymes is strongly influenced by allicin and other thiosulfinates (Section 5.2), but is not influenced by disulfides, thioethers, or sulfoxide compounds. Reduction of allicin to diallyl disulfide reduces the antibacterial effect considerably (Hörhammer et al., 1968; Kabelik & Hejtmankova-Uhrova, 1968; Schultz & Mohrmann, 1965a, 1965b). Besides the reaction of allicin with the SH groups of various bacterial enzymes, which causes reduced growth, SH groups can also be oxidized by the labile oxygen of allicin. The S(=O)S structure seems to be essential for the bactericidal effect of allicin.

The assumption that SH enzymes are the targets for the attack of allicin is further confirmed by the structural differences of the various bacterial strains (Tynecka & Gos, 1975). The various bacterial species differ in the lipid content of their cell membranes. For example, the cell membrane of *E. coli* contains 20% lipid, while that of *Staphylococcus aureus* contains only 2% lipid (Salton, 1964). These structural differences probably cause varying membrane permeability to allicin and other garlic constituents.

Another investigation into the mechanism by which allicin inhibits bacterial growth used *Salmonella typhimurium* as its model (Feldberg et al., 1988). It was found that allicin caused only a delayed and partial inhibition of DNA and protein synthesis, but that RNA synthesis was completely and immediately inhibited, suggesting that allicin acts primarily by inhibiting RNA synthesis.

Finally, Bielzer (1992) showed that the in vitro inhibitory activity of freshly prepared garlic juice decreases rather quickly during storage at room temperature. When the juice samples were stored in sealed test tubes at 4°C, the decrease of the antibacterial activity was slowed down. Deactivated garlic juice could not be reactivated by means of cysteine and/or hydrogen peroxide. These effects are expected on the basis of the instability and chemistry of allicin. The antibacterial property of garlic juice was also diminished by serum or blood. This effect may be explained by a promoting effect on the bacteria by the body fluids and by the fact that allicin is rapidly converted to allyl mercaptan in blood (Lawson & Wang, 1993) (Table 5.7).

5.2.2 Antifungal Effects

As already mentioned, garlic also has appreciable activity against pathogenic fungi and yeasts. It is said that it has antifungal or antimycotic (i.e., against the fungal mycelium) properties. The antifungal properties of garlic have long been used in folk medicine for the treatment of *Candida* infections, especially those of the skin.

Schmidt and Marquardt (1936) first established the extraordinary fungistatic and fungicidal action of freshly pressed garlic juice and dried garlic with epidermophyte cultures. Later, American and Russian authors reported similar findings almost simultaneously (Small et al., 1947; Lesnikov, 1947). Since then, numerous studies have appeared in which the inhibition of fungal growth by garlic and/or its constituents, mainly allicin, is described (Rao et al., 1946; Sasaki, 1957; Yamada & Azuma, 1975a, 1975b, 1975c, 1977; Petricic & Lulic, 1977a, 1977b; Hitokoto et al., 1978; Paszewski & Jarosz, 1978; Elnima et al., 1983; Louria et al., 1989; Prasad & Sharma, 1980; Tynecka & Gos, 1975; Maleszka et al., 1991).

Kabelik (1970) was among the first to call attention to the unusually strong effect of garlic against pathogenic yeasts. However, the therapeutic use of garlic against yeasts is limited mainly to candidiasis (Kabelik & Hejtmankova-Uhrova, 1968). However, Walker and Stahmann (1955) had already pointed out the antifungal action of onions.

Adetumbi (1989) found that the antimycotic activity of an aqueous garlic extract is drastically lowered by the addition of sulfhydryl compounds, such as cysteine or glutathione. This indicates that allicin is the active compound. In addition, garlic reduced the oxygen uptake by the microorganisms by as much as 70%. Biosynthesis of protein, nucleic acids, and lipids were all found to be seriously inhibited by the extract (Adetumbi & Lau, 1986; Adetumbi et al., 1986).

Further evidence that allicin is responsible for the in vitro anticandidal activity of garlic has been demonstrated in a study where pure allicin was found to be highly active—with a minimum inhibitory concentration (MIC) of 7 µg/mL—yet solvent removal of the thiosulfinates also removed all the activity. Allicin was found to be about twice as active as the allyl methyl thiosulfinates and about 10 times as active as ajoene. The study also showed that several varieties of onion had much less anticandidal and antibacterial activity than garlic (Hughes & Lawson, 1991).

Growth and respiration are also inhibited by garlic juice in *Candida albicans*, *Trichophyton cerebriforme*, and *T. granulosum*. Inhibition of succinate dehydrogenase is assumed to be the mechanism of action (Szymona, 1952). At a dilution of 1:1000, garlic juice had no harmful effect on tissue cultures, such as chicken embryo fibroblasts or kidney cells; however, it completely inhibited the growth of yeast. Microorganisms are much more sensitive to the active constituents of garlic than are higher organisms (Tynecka & Skwarek, 1974).

DiPaolo and Carruthers (1960) carried out some model experiments with various yeast strains (*Saccharomyces cerevisiae*, *S. ellipsoideus*, *S. carlsbergensis*) and enzymatically-produced allicin. Total growth inhibition occurred with the original strains, but not with some mutants of the yeasts. Grzybowski and coworkers (1988) tested the effect of garlic against the pathogenic fungi *S. cerevisiae*, *Kloeckera apiculata*, *Candida utilis*, *Oospora lactis*, and *Penicillium notatum* with good results.

Some of the dermatophytes against which garlic is effective are *Microsporum gypseum, Trichophyton terrestre, Malbranchea pulchella,* and *Chrysosporium tropicum* (Shrivastava & Singh, 1982). Effectiveness against *Aspergillus parasiticus, Aspergillus ochraceus, Penicillum patulum, P. roqueforti,* and *P. citrinum* has also been reported (Azzouz & Bullerman, 1982). *Saccharomyces cerevisiae, Candida albicans, Microsporum canis, Trichophyton mentagrophytes,* and *T. rubrum* are further species responding to garlic (Guevara et al., 1983). Yamada and Azuma (1977) have determined the MIC (minimum inhibitory concentration) of allicin against a series of dermatophytes to range from 1.57 to 6.25 µg/mL culture medium.

Good activity was also observed against *Candida albicans, Aspergillus parasiticus, A. flavus,* and *A. ochreus* (Barone & Tansey, 1977; Hitokoto et al., 1978). At dilutions of 1:200 to 1:1600, garlic juice exhibits fungistatic activity to *Aspergillus niger, A. oryzae, Rhizopus nigricans,* and *Mucor racemosus.* At higher concentrations it is effective against *Didium lactis, Penicillium glaucum,* and *P. notatum* (Dubrova, 1950). Garlic oil (100 µg/mL) also inhibits the production of ethanol and CO_2 by brewer's yeast (*Saccharomyces cerevisiae*) and uncouples respiration in the cells (Conner et al., 1984).

American researchers tested 139 species of zoopathogenic fungi and yeasts for their in vitro sensitivity to aqueous garlic extract. The results were quite variable, since many microorganisms did not grow at all while others grew well. Some of them had established even more mycelia after 21 days than the controls. From these results, it can be seen that the antifungal effect is specific and not general (Appleton & Tansey, 1975; Tansey & Appleton, 1975). On several yeasts which grow on foods or are used industrially, it was observed that even very small amounts (25 ppm) of garlic oil have a strong inhibitory effect (Conner & Beuchat, 1984a, 1984b, 1984c; Conner et al., 1984).

Aqueous garlic extract was found to inhibit the growth of certain pathogenic fungi (*Coccidioides immitis, Auxarthron zufiianum, Uncinocarpus resii*) at a minimum concentration of 1:32 to 1:40. Annually, about 120 Americans fall victim to coccidiosis ("valley fever"), an intestinal disease caused by these organisms (Awasthi & Leathers, 1984; Adetumbi & Lau, 1986).

Moore and Atkins (1977) examined the effectiveness of an aqueous garlic extract against a series of human pathogenic fungi and yeasts, which originated from patients with infectious vaginitis containing *Candida, Cryptococcus, Rhodotorula, Torilopsis,* and *Trichosporon.* All these organisms were inhibited or killed by this extract at a dilution of 1:1024. Activity was highest at 37°C. The fungistatic activity of the extract remained stable for 4 weeks, but the fungicidal activity declined by 1 week unless stored frozen. Fromtling and Bulmer (1978) studied the activity of an aqueous garlic extract on 18 strains of *Cryptococcus neoformans* and found that an amount of 20 µL extract per agar plate was sufficient to be inhibitory or lethal for most strains. Sandhu and colleagues (1980) also observed high sensitivity of yeasts isolated from cases of vaginitis to aqueous extracts of garlic.

Amer and coworkers (1980) tested the activity of an aqueous garlic extract on various medically important dermatophytes (various *Microsporium, Trichophyton,* and *Epidermophyton* species). Inhibitory concentrations were in the range of 130–200 µg/mL. The tests, which were carried out on guinea pigs and rabbits, resulted in complete healing of the experimentally induced infections within 14 days. Prasad and colleagues (1982) tested a garlic extract (1:10 in distilled water) as a topical application on rabbits with experimentally induced dermatitis (*Microsporium canis*) and found complete healing of the infection without any side effects. Additionally, garlic was successfully used to treat a case of infection with *M. canis* in humans contracted from dogs (Rich, 1982). *M. gypseum* is also highly sensitive to the volatile components of garlic (Singh & Deshmukh, 1984). Finally, there is also a report about the successful treatment of a case of sporotrichosis with fresh garlic juice (Tutakne et al., 1983).

Upadhyay and colleagues (1980) tested in vitro the susceptibility to fresh garlic juice of a large number of fungi and yeasts isolated from the tears of patients with eye diseases. A remarkable antimycotic activity was observed, and the authors recommended the application of garlic in ophthalmology. Mahajan (1983), however, could not detect growth inhibition on 13 fungi, which are known to cause keratosis in humans, with garlic (form not stated) concentrations up to 10 mg/mL.

Caporaso and colleagues (1983) tested the antifungal activity of active components of garlic that occur in blood and urine after consumption of fresh garlic extract. Some activity could be demonstrated against *Candida* and *Cryptococcus* in the blood, but not in the urine. The authors believe that garlic has only limited value for treatment of infections of the urinary tract.

Both aqueous garlic extract and preparations containing diallyl trisulfide (allitridi) have been shown in China to have good success in curing patients of life-threatening cryptococcal meningitis (Hunan Medical College, 1980; Cai, 1991). Furthermore, the anticryptococcal titers of the blood and cerebrospinal fluid of diseased patients increased twofold when given the diallyl trisulfide preparation intravenously (Davis et al., 1990). In a further in vitro study of this effect, it was shown that a garlic preparation containing mainly allicin and vinyldithiins inhibited *Cryptococcus neoformans* with an MIC of 6–12 µg/mL and was synergistic with amphotericin B (Davis et al., 1994).

In poultry farming, the addition of 2–5% garlic (chips or extract) to the feed is used for the prevention of mycosis in the animals (Prasad & Sharma, 1981). Garlic was also effective against *Candida albicans* and *Aspergillus fumigatus* as well as against a strain producing aflatoxin, *Aspergillus parasiticus*. In veterinary practice, garlic extract is used for treatment of infected wounds, in calves, for example, and for promotion of wound healing (Singh et al., 1984; Singh, 1984; Kasakowa, 1953a; Titta, 1904). In India, a garlic preparation is used for scabies in pigs (Dwivedi & Sharma, 1985). Onion extracts also have sporicidal activity against aflatoxin-producing fungi (Bilgrami et al., 1992).

Soni and coworkers (1992) showed that garlic at concentrations of 5–10 mg/mL caused over 90% inhibition of aflatoxin production by *Aspergillus parasiticus*. Onion is of comparable effectiveness against this organism (Sharma et al., 1981). Furthermore, garlic will not support the growth of *Aspergillus flavus* strains that produce aflatoxins (Mabrouk & ElShayeb, 1980).

Fliermans (1973) found that aqueous extracts of garlic strongly inhibit growth of the mycelium of *Histoplasma capsulatum*, with an IC_{50} of 1:4,000,000 and an LC_{50} of 1:127,000. He suggested planting garlic in cultivated fields of other plants to keep the soil free from *Histoplasma* when a health risk exists there. Similar results and suggestions have been published by others as well (Timonin & Thexton, 1950; Coley-Smith & King, 1969; Tansey & Appleton, 1975; Coley-Smith, 1986).

Comparative studies on the effects of garlic juice, nystatin, griseofulvin, and amphotericin B against *Candida albicans*, *Cryptococcus neoformans*, *Geotrichum candidum*, *Aspergillus fumigatus*, and *Epidermophyton mentagrophytes* have shown that the antimycotic activity of garlic exceeds that of all the drugs investigated (Tynecka & Gos, 1975).

Investigations of aflatoxin-forming *Aspergillus flavus* strains revealed that minced garlic as well as aqueous garlic extracts, in concentrations of 1–5 mg/mL, inhibit aflatoxin formation on a rice-corn broth for 6 days. A fungicidal effect was seen only at higher concentrations (50–100 mg/mL). Fresh garlic (10–50%) in a sesame and sunflower seed medium completely inhibited aflatoxin formation for up to 30 days, while autoclaved garlic exerted only a 40% inhibition. With 10% garlic after 3, 6, 12, 18, and 24 days, 35%, 49%, 42%, 35%, and 40% inhibition, respectively, were found (Mabrouk & ElShayeb, 1980; Hitokoto et al., 1980).

Ghannoum (1988) investigated the mechanism for growth inhibition of *Candida albi-*

cans by aqueous garlic extract by several means. He found that the garlic extract caused damage to the outer surface of the cells and caused several alterations in the lipid content. Furthermore, the effect of garlic could be eliminated by the addition of thiol compounds. He concluded that garlic extract acts by oxidizing thiol groups of essential enzymes, an effect which is well-known for allicin. Ghannoum (1990) also showed that garlic may reduce the adhesion of *Candida* cells to the mouth, since the adhesion of the cells to isolated buccal epithelial cells was inhibited by brief exposure of the buccal cells to the garlic extract, an effect that was probably due to allicin, since the addition thereafter of thiol compounds eliminated the reduced adhesion.

Further elucidation of the antimycotic principle was achieved by Yoshida and coworkers (1987), who prepared six different fractions from garlic and examined them for antifungal activity on two test organisms (*Aspergillus niger* and *Candida albicans*). The strongest activity was seen with ajoene, which inhibited both organisms at a concentration of less than 20 µg/mL; however, ajoene demonstrated little antibacterial activity. San-Blas and colleagues (1993, 1989) showed that ajoene inhibits the growth of *Paracoccidioides brasiliensis*, a fungal pathogen for humans, by affecting the integrity of the fungal cytoplasmic membrane. Ajoene was also found to act against phytopathogenic and epiphytic microorganisms (Reimers et al., 1993).

A chloroform-extract concentrate from garlic, which appeared to be nontoxic when used as inhalant, proved to be very potent for the elimination of *Candida albicans* from the respiratory tract of pediatric patients (Alkiewicz & Lutomski, 1992). An allicin-glycerol semisolid mixture (Insoles) is claimed in a patent to be effective for the control of trichophytosis in feet (Ko, 1991).

Tests for possible application of garlic oil and synthetic diallyl disulfide to control various phytopathogenic fungi in agriculture have revealed a considerable activity against these pests. In a field study with peanut and mung bean plants, garlic oil (50–10,000 ppm) was found to be protective (Murthy & Amonkar, 1974). *Gibberella zeae*, a parasite of maize, is inhibited by garlic oil (8,000–10,000 ppm) more effectively than by captafol (Mabrouk & ElShayeb, 1980). The larvae of *Trogoderma granarium* Everts, another parasite on maize kernels, can be controlled by 1–2% garlic oil (Jood et al., 1993). Other *Allium* species, especially onion and leek, inhibit a variety of rot fungi (Favaron et al., 1993). Aqueous extracts of garlic and Chinese chive (*A. tuberosum*) have good fungicidal properties but do not inhibit plant growth (Fujii et al., 1991). There is evidence that certain plant phenols are involved in pathogen resistance of endomycorrhizal roots (Grandmaison et al., 1993).

Garlic juice and several synthetic oligosulfides and mercaptans of the *Allium* plants cause a stimulating effect on root growth of the fungus *Sclerotium cepivorum* Berk. This effect occurs with most *Allium* plants except those which contain predominantly S-methylcysteine sulfoxide rather than the allyl or 1-propenyl homologs. However, it is possible that this growth effect is due to an inhibition of the soil bacteria, which normally inhibit fungal growth (Coley-Smith & King, 1969; Esler & Coley-Smith, 1983; Coley-Smith, 1986). Other authors have reported antifungal properties for the oligosulfides (Jirousek & Jirsak, 1956).

Aged garlic extract, which contains no allicin or allicin-derived compounds, was found in vitro to have no anticandidal or other antifungal activity (Hughes & Lawson, 1991; Pai & Platt, 1995); however, when injected into mice (0.1 mL, equal to 230–350 mL for a person) which had been preinjected with *Candida albicans*, the blood and kidneys were found to contain 80% fewer organisms (Tadi et al., 1990). Hence, at large doses garlic appears to have compounds other than allicin capable of antibiotic activity.

Last, but not least, it should be mentioned that garlic oil (steam-distilled) is also an effective remedy against wood-destroying fungi (*Lenzites trabea, Plyporus vesicolor,*

Lentinus lepideus) (Maruzella et al., 1960). Therefore, it has been suggested that garlic be planted between railroad ties and along fences, in order to prevent premature decay of the wood (Tansey & Appleton, 1975). Of course, the antifungal components of garlic also protect garlic itself from fungal infestation (Durbin & Uchytil, 1971).

5.2.3 Antiprotozoal Effects

In the late 1920s, Russian researchers observed that the volatile constituents of fresh garlic and onion strongly inhibit and even kill various microorganisms. Ensuing experiments with ingredients of these plants, and later of other plants, were mainly performed in protozoa, especially *Paramaecium caudatum*, but also with eggs and larvae of mollusks and frogs. The favorable results prompted Tokin to coin the descriptive word "phytoncides" for the antibiotically active components of higher plants. The protistocidal activity of garlic has since then been well documented (Tokin, 1943). The same effect was also described by Bulgarian researchers (Petkov, 1986).

Soviet physicians applied garlic successfully for the therapy of lambliasis, an intestinal disease caused by protozoa (*Lamblia intestinalis*) (Kramarenko, 1951). In the former Soviet Union experiments were carried out on animal-pathologic protozoa which live as parasites in frogs (*Opalina ranarum, O. dimidiata, Balantidium entozoon*). Here, garlic juice proved to be an agent of exceptional protistocidal activity (Kurnakov, 1952).

More recently, Israeli researchers rediscovered the germicidal action of allicin on *Entamoeba histolytica*, the causative agent of amoebic dysentery. Even at extremely low concentrations (30 µg/mL), allicin destroys the amoeba cultures grown in the laboratory. Other, nonpathogenic strains of the *Entamoeba* genus are also killed by garlic (Mirelman & Varon, 1987; Mirelman, 1987).

However, certain oligosulfides also have antiparasitic activity. For instance, diallyl trisulfide has been found in vitro to kill human and animal pathogenic protozoa, such as *Trypanosoma* species, *Entamoeba histolytica*, and *Giardia lamblia* at concentrations that are nontoxic to mammalian cell lines (Lun et al., 1994). Another secondary constituent of allicin, ajoene, inhibits the proliferation of *Trypanosoma cruzi*, possibly by inhibition of phosphatidylcholine biosynthesis (Urbina et al., 1993).

5.2.4 Antiviral Effects

Relatively few publications exist that deal with the activity of garlic or its components against viruses. Hanley & Fenwick (1985) report that, during an influenza epidemic, the former Soviet Union once imported 500 tons of garlic cloves for the acute treatment of the disease ("Russian penicillin"). In eastern Europe and in the South and East Asian countries (e.g., India, China), garlic is used as a substitute for the more effective, but often much too expensive, western medications.

Before the development of vaccines against poliomyelitis, garlic was used successfully as a prophylactic against polio (Mayerhofer, 1934; Huss, 1938). Yakovlev and Zviagin (1950) were the first to observe an efficacy of garlic against viral influenza A, but its activity against the influenza B virus was not determined until much later (Frolov & Mishenkova, 1970).

In China, garlic preparations containing mainly diallyl trisulfide (allitridi) have been shown to have in vitro and in vivo activity against herpes simplex type 1 and influenza B viruses, but not against the coxsackie B virus (Tsai et al., 1985; Cai, 1991). This same product at a dose of 60–120 mg/day has been shown to prevent viral pneumonia in bone marrow transplant patients (Lu et al., 1988; Lu, Guo, et al., 1990; Lu, 1994) and to have in vitro activity against human cytomegalovirus (Guo et al., 1993; Meng et al., 1993).

An allicin-urotropin preparation for parenteral use has been claimed in a patent to be effective against viral infections, among them AIDS (Holzhey et al., 1992). A recent paper from Russia reports on the blockade by ajoene of integrin-dependent processes in the HIV-infected cell system (Tatarintsev et al., 1992).

In veterinary medicine, garlic extract is said to be active against foot-and-mouth disease. This application has even been protected by a patent (Schiefer, 1953).

When mice were fed a dilute alcohol extract of garlic for 15 days prior to experimental infection with an influenza virus strain, statistically significant antiviral activity was found. However, no therapeutic effect was achieved when garlic was given at the time of infection or when the mice were infected with a Japanese encephalitis virus (Nagai, 1973a, 1973b).

Significant activity against influenza viruses was also shown with intranasal and intramuscular application of an aqueous garlic extract in mice. The treatment led to a distinct decrease of the hemagglutination titer. Prophylactic application prolonged the survival time of the animals. Sodium fluoride enhanced the antiviral effect of garlic extract (Esanu & Prahoveanu, 1983).

A single report is available on the effectiveness of garlic against rickettsia. Kumar and coworkers (1981) conducted an experiment with *Coxiella burnetii*, the causative agent for Q fever, in chickens. Among the results were delayed and weaker antibody reaction for the prophylactically treated chickens (daily fed 2 g chopped cloves per animal) and faster recovery than the animals of the control group. From these results, the authors inferred that garlic is also effective against the whole group of viruses of this type.

Several compounds from garlic as well as several types of commercial garlic products have been investigated for in vitro virucidal activity toward several viruses (Weber et al., 1992; Hughes et al., 1989). Ajoene was somewhat more active than allicin or other thiosulfinates, while no activity was found for alliin, *S*-allylcysteine, or the diallyl sulfides, although solubility may be a problem in the latter case. Only commercial products capable of producing allicin showed virucidal activity.

A recent in vitro study has shown that allyl alcohol and diallyl disulfide can selectively kill HIV-1–infected cells, although propanol, diallyl sulfide, dipropyl disulfide, and allyl mercaptan did not (Shoji et al., 1993). Another in vitro study has shown that an aqueous garlic extract can kill a rotavirus without affecting host mammalian cells (Rees et al., 1993).

Finally, Yacobson (1936) isolated a bacteriophage from ultrafiltered pressed garlic juice, which was successful in neutralizing *Escherichia coli*, *Salmonella typhi*, and other bacteria. The titer of the virus was $1:10^5$ to $1:10^7$.

5.2.5 Antiparasitic Effects

The defensive effect of garlic against intestinal parasites and other endo- and ectoparasites has been known since antiquity, although garlic can hardly be classified as an effective anthelmintic. Garlic alone is not sufficient as a vermifuge; however, it should be of value as an adjuvant and prophylactic to prevent the establishment of intestinal parasites. In this sense, the formerly common application against oxyuriasis, orally as tincture and rectally as an enema, must be understood (Marcovici, 1915; Brüning, 1931a, 1931b; Erdmann, 1931; Mayerhofer, 1934).

Garlic is effective for roundworm (*Ascaris strongyloides*) (Rico, 1926; Hofmann & Held, 1953; Kempski, 1967), *Oxyuris* (Standen, 1953; Erdmann, 1931; Brüning, 1931a), as well as hookworm (*Ancylostoma caninum* and *Necator americanus*), and the eggs of the parasites (Hahn & Koh, 1949; Soh, 1960). According to some experiments, dosages of garlic oil and enemas of garlic extracts are as effective as treatment with various drugs, such as piperazine, yatren, or hexylresorcine (Gessner, 1953; Standen, 1953; Kempski, 1967). At normal dosages, garlic treatment is completely harmless, so it can also be used for pregnant patients (Flamm, 1935). Overdosing, however, can result in intestinal inflammation and diarrhea (Flamm, 1935; Hofmann & Held, 1953).

Allicin is probably the active compound for the anthelmintic effect of garlic. Diallyl disulfide is ineffective in this regard (Araki et al., 1952). The early literature proposed that diallyl sulfide is the active compound (Rico, 1926); however, this was before allicin had been discovered and before the composition of steam-distilled garlic oil was more generally known.

In veterinary medicine, garlic can be successfully applied for treating worm infestation. Thus, for instance, a new preparation of garlic powder and garlic oil is used for filarial infection in dogs in the form of an additive to the feed (0.1–0.2%). No microfilariae were detected in the blood after 3 to 4 months of treatment (Riken Chem. Ind. Co. Ltd. 1982). In dogs infested with *Necator americanus* and *Ancylostoma caninum,* raw garlic mixed with the food for 5 days (11–18 g/day) produced a significant reduction of larvae and fertile eggs 2 days after termination of intake (Bastidas, 1969). Even in carps infested with *Capillaria* species, a hexane extract from garlic was used successfully as an anthelmintic (Pena et al., 1988).

Finally, garlic extract and garlic oil, as well as allicin, are also very effective in the destruction of nematodes in the soil, which infect and destroy certain cultivated plants. Other *Allium* species also have comparable nematicidal activity (Tada et al., 1988). This finding is of special interest for developing countries, since garlic can be used there as a cheap nematicide, instead of expensive synthetics, for pest control (Nath et al., 1982; Gupta & Sharma, 1993).

5.2.6 Insecticidal and Repellent Effects

Garlic oil also has the properties of an insecticide and insectifuge. It was repeatedly reported in the older literature that garlic and onions can kill or repel insects. Several years ago, investigations were conducted for the purpose of discovering plant constituents that could be used for mosquito control in lieu of DDT (Hills, 1972). Both steam-distilled garlic oil and synthetic diallyl disulfide were applied. It has been shown that diallyl disulfide and diallyl trisulfide (allitin) have insecticidal properties, while diallyl sulfide is inactive (Banerji et al., 1978; Lutomski, 1980; Venugopal & Narayanan, 1981).

Moreover, garlic kills mosquito larvae (Amonkar & Banerj, 1971; George & Eapen, 1973; George et al., 1973). The methanol extract as well as garlic oil have been found to be extremely effective against insects carrying various infectious diseases (*Culex* and *Aedes*). The mean lethal dosage for larvae of *Culex tarsalis* has been determined to be 25 ppm for extracts and 2 ppm for garlic oil (Amonkar & Reeves, 1970). The active principles of the garlic oil were later identified as diallyl disulfide and diallyl trisulfide (Amonkar & Banerj, 1971). These oligosulfides exert their toxic effects by blocking the synthesis of important proteins in the insect larvae or by inhibiting the incorporation of amino acids into larval proteins (George & Eapen, 1973; George et al., 1973).

The possibility of using garlic extracts for the control of plant-damaging aphids (*Myzus persicae*) was studied under laboratory conditions. A crude extract was effective on larvae and pupae of some natural enemies to aphids, such as *Syrphus corollae, Chrysopa carnea,* and *Coccinella septempunctata.* It was, however, observed that the useful insects were clearly less affected than the aphids, which were negatively affected in their growth, life span, and reproductive capability (Venugopal & Narayanan, 1981; Nasseh, 1982). It appears that garlic oil is as well suited for the control of plant pests as synthetic insecticides, especially in the developing countries (Krishnaiah & Mohan, 1983). Garlic juice and pulp relieve and disinfect insect stings.

Other insects are repelled by the alliaceous odor ("repellent effect"). Occasionally their larvae are killed (Pandey et al., 1976). Grain aphids can be killed by garlic extract as well (Nasseh, 1983). Oddly enough, however, there are some insect species that are attracted by the odor of the *Allium* species (Ishikawa et al., 1978; Soni & Finch, 1979). For instance, silkworms are strongly attracted by scordinin, a compound in garlic (see Section 3.2.9) (Kominato & Kominato, 1973). The most attractive substance for the leek moth (*Acrolepiopsis assectella*) is propyl propanethiosulfinate, which is produced by

Allium porrum (Auger et al., 1989). Dimethyl trisulfide, a component of onion oil, is the strongest attractant for larvae of the onion fly (*Delia antiqua*), and it may be used to attract newly hatched larvae away from onion seedlings (Soni & Finch, 1979).

Catar (1954) reported that the volatile ingredients of garlic have an extremely strong repellent effect on *Ixodes ricinus*, a tick that carries the encephalitis virus. Kabelik (1970) observed, furthermore, that garlic juice even kills the ticks, within a few minutes. Other authors report similar findings (Reznik & Imbs, 1965). This property of garlic has not yet been practically applied, but considering the wide spread of meningoencephalitis in spring and summer, it may be an important avenue of research.

An insect repellent comprised of garlic oil and vitamin B_1 has been patented for protection of humans and animals against blood-feeding pests (Weisler, 1989). Insecticidal bait preparations for cockroaches, which contain garlic powder, boric acid, and glycerol, have been described in another patent (Takizawa, 1989). The fruit fly (*Drosophila melanogaster*) is strongly repelled by garlic and Chinese chive (*A. tuberosum*) (Iwanami et al., 1985). Garlic extract has been shown to repel blackfly and mosquito (Bhuyan et al., 1974). An allicin-containing pesticide has even been developed for golf course greens (Sakai, 1992).

Renapurkar and Deshmukh (1984) looked into the possibility of combating fleas, which function as carriers of many human and animal pests, with plant extracts ("pulicide effect"). Extracts of crushed garlic cloves proved to be exceptionally effective. The mean lethal concentration (LC_{50}) with a 1-hour exposure time to the parasites was found to be 0.25% for the hexane extract and 0.50% for the acetone extract, compared to 0.31% for DDT and 0.44% for malathion.

An original experiment on larvae of *Lohita grandis* should be mentioned, because it provides some insight into the effect of garlic on the cholesterol supply in these animals at this stage of development (Mandal & Choudhuri, 1982). The insects need cholesterol for their own steroid hormones (e.g., ecdysone), but are dependent on dietary cholesterol, obtained as food or through the action of symbiotic intestinal bacteria, since they cannot synthesize this steroid themselves. The mortality rate sharply increased upon addition of fresh garlic juice to the glucose solution which served as feed, or through the injection of garlic extract into the larvae. This effect is explained by the antibiotic effect of garlic (allicin). The drastic reduction of symbiotic microbial population in the larvae caused death due to the acute lack of steroids.

The intense smell of diallyl disulfide serves as a repellent not only for lower animals such as insects, but has also been recently utilized to drive away birds, which can cause a great deal of damage to some cultivated crops (Mizutani et al., 1979).

Another aspect related to insect development and behavior in connection with *Allium* plants is the influence of the sulfur compounds in phytophage-parasite relations. Some insect species, like the onion fly or the leek moth, are not only attracted by *Allium* flavor compounds, but also need them for their normal development. These compounds mediate host searching, oviposition, phagostimulation, and other functions. The sulfur compounds are present in phytophagous larvae, in feces, and in soil. Symbiotic microorganisms seem to interfere with their production and transformation, particularly when amino acid precursors are converted into volatiles (Auger et al., 1993; Dindonis & Miller, 1980; Ishikawa et al., 1978).

5.2.7 Machado's Garlicin

The antibiotic garlicino (Machado's garlicin), a sulfur-free substance, has not yet been sufficiently chemically defined, nor has its isolation ever been repeated (Koch, 1993a) (see Section 3.3.13). Nevertheless, it was clinically tested in a multicenter study (31 hospitals in Brazil) in patients with shigellosis, salmonellosis, and amoebiasis as well as with worm infestations and protozoan infections of the intestines. Machado's garlicin given for 9 days had a satisfactory therapeutic effect on all diseases (Machado et al., 1948).

5.3 ANTICANCER EFFECTS

In ancient times, garlic was used for the treatment of cancer of the uterus (Hartwell, 1960; Essman, 1984; Doetsch, 1989; Konvicka, 1983). Numerous reports, including several important epidemiological studies, have entered the scientific literature ever since, asserting that garlic has a favorable effect on various forms of cancer. The following provides an overview of the current research and points of view concerning this very interesting special area of medicine.

Six decades ago, several statistical studies indicated that cancer occurs the least in those countries where garlic and onions are eaten regularly—for instance, in the French Provence, Italy, the Netherlands, the Balkans, Egypt, India, and China (Lackhovsky, 1932; Lorand, 1934; Caspari, 1936; Güntzel-Lingner, 1941; Yang et al., 1994b; Wang et al., 1985; Dausch & Nixon, 1990; Seelert, 1989; Harris, 1979; Dorant et al., 1994b). In a review article, Auler (1936) referred to the connection between nutrition and cancer, especially to the cancer growth–inhibiting effect of leek plants (*Allium* plants). The practicing physicians of the time were very good observers, but almost nothing was known about the real background of this phenomenon. It was thought that the inhibitory action of garlic on putrefaction in the intestines, together with the secretion-stimulating effect, brought about detoxification and an increase in resistance.

Stimulation of gastric juice secretion and restoration of the intestinal flora, combined with the resulting prevention of gastrointestinal autointoxication, may help to remove at least one of the possible causes of cancer. Garlic may therefore be useful as a cancer preventive agent, and its application as an anticancer drug is based on this assumption (Stöger, 1967, 1968, 1970, 1976; Caldwell & Danzer, 1988; Anonym, 1980). More recently, this idea has again been pursued, not only in Europe, but also in the Third World countries, where the favorable effects of garlic for cancer are well known (Krishnamurthy & Sreenivasamurthy, 1956; Duris et al., 1981; Schimmer et al., 1994; Abdullah et al., 1988; Sagmeister, 1987). For instance, the consumption of black or green tea, as well as of garlic, is known to be a culinary practice which inhibits tumorigenesis in the lung, forestomach, and esophagus (Yang et al., 1993, 1994b).

The only known study in which garlic has been used to treat patients with advanced stages of cancer was conducted by Spivak (1962). A garlic juice preparation was administered in doses of 0.2–2 mL intravenously or 1–5 mL intramuscularly daily for 3–7 days. Of 35 patients with cancer at various sites (lung, cervix, stomach, lower lip, mammary gland, larynx, and leukemia), 26 showed positive treatment results of differing degrees, though complete healing was not achieved in any case. There is a single-case report, however, of a man whose pituitary tumor shrank by 50% during the 5 months in which he ate 5–7 grams fresh garlic daily. This was the first case ever reported of reduction of this type of tumor without chemotherapy or surgery (Rainov & Burkert, 1993).

5.3.1 Anticancer Effects: Epidemiological Studies

Epidemiological studies comparing cancer incidence with consumption levels of individual or grouped foods are currently the most important evidence that garlic may significantly reduce the risk of cancer, especially cancers of the gastrointestinal tract. Table 5.8 lists the results for all known epidemiological studies related to garlic clove consumption. In nearly all the studies, the decreased cancer incidence associated with garlic consumption was statistically significant. Most studies did not distinguish between raw and cooked garlic.

Stomach cancer is the cancer most clearly associated with an effect by garlic consumption. This is true for both raw garlic and cooked garlic (Mei et al., 1982; Buiatti et al.,

Table 5.8. Epidemiological Studies on Cancer and Garlic Clove Consumption

Country (Reference)	Cancer Type	Food	Cancer Association (odds radio)[a]
Argentina (Iscovich et al., 1992)	Colon	Garlic	0.2–0.4
China (Mei et al., 1982)	Stomach	Raw garlic (20 g/day)	0.08[b]
China (You et al., 1988, 1989)	Stomach	Garlic (> 4 g/day)	0.7
		Garlic (0.3–4 g/day)	0.8
		All *Allium* (44–64 g/day)	0.5
China (Zheng et al., 1992)	Larynx	Garlic	0.5 – 0.6
Italy (Buiatti et al., 1989)	Stomach	Cooked garlic	0.4 – 0.6
Italy (Cipriani et al., 1991)	Stomach	Garlic and onions	0.8
Iran (Cook-Mozaffari et al., 1979)	Esophagus, men	Raw garlic (≥ 1 serving/month)	1.1[c]
		Pickled garlic (ever)	0.6
Switzerland (Levi et al., 1993a,b)	Breast	Garlic	0.6 –0.7
	Endometrium	Garlic	0.6
U.S.A.	Colon	Garlic (≥ 1 serving/week)	0.7
(Steinmetz & Potter, 1994)	Colon (distal)	Garlic (≥ 1 serving/week)	0.5

[a]Values below 1 indicate decreased risk of cancer.
[b]Cancer incidence was only 8% of those consuming less than 1 g/day.
[c]Not statistically significant, but all other studies are.

1989). A particularly noteworthy study in China was a comparison between two counties in the same province which have very different garlic-eating habits (Mei et al., 1982; Han, 1993). The incidence of stomach cancer in Cangshan County, where the average person consumes 20 g fresh garlic daily, was only 8% of that in Qixia County, where less than 1 g/day is eaten. Furthermore, the concentration of gastric nitrite was found to be only 23% as high in the Cangshan residents as in the Qixia residents, an effect associated with the killing of nitrate-reducing bacteria by thiosulfinates (allicin, etc.) (Mei et al., 1982, 1985). The inhibitory effect of garlic on the formation of carcinogenic nitroso compounds was further substantiated when the same investigators fed 5 g chopped garlic plus proline and nitrate to volunteers and then measured urinary levels of nitrosoproline. The urinary levels did not increase at all, whereas major increases were found when the same individuals were fed the same diet without garlic (Mei et al., 1989).

Another study in China, involving 564 patients with stomach cancer and 1131 controls in an area of China with a high incidence of gastric cancer, revealed a significant reduction in gastric cancer risk with increased consumption of *Allium* vegetables (You et al., 1988, 1989). Subjects in the group of the highest quartile of intake (more than 64 g/day) experienced only 40% of the cancer incidence of those in the lowest quartile (less than 32 g/day). Protective effects were also found for individual *Allium* foods: garlic, scallions, and Chinese chive, but not for onions.

A very important epidemiological (prospective cohort) study for Americans has recently been published in which the intake of 127 foods (including 44 vegetables and fruits) was determined in 41,387 women (ages 55–69), followed by a five-year monitoring of colon cancer incidence (Steinmetz et al., 1994). The most striking result of this "Iowa Women's Health Study" was the finding that garlic was the only food which showed a statistically significant association with decreased colon cancer risk. For cancers anywhere in the colon, the modest consumption of one or more servings of garlic (fresh or powdered) per week resulted in a 35% lower risk, while a 50% lower risk was found for cancer of the distal colon. Both a critique of this study and a good reply by the authors have been published (Ballard-Barbash et al., 1995; Steinmetz & Potter, 1995).

Although this study of 127 foods did not include onions, several other epidemiological studies have shown that onions and other *Allium* species are usually associated with decreased gastrointestinal cancer risk, although the results have been less consistent than with garlic (Steinmetz & Potter, 1991, 1993; Haenszel et al., 1972; You et al., 1989; Cook-Mozaffari et al., 1979; Tajima & Tominaga, 1985; Shu et al., 1993; Tuyns et al., 1992; Graham et al., 1994; Levi et al., 1993a, 1993b). Furthermore, no effect of onions or leeks has been associated with lung cancer risk (Dorant et al., 1994b). An excellent review of nearly all of the epidemiological cancer studies for the various *Allium* vegetables has been published recently (Dorant, 1994). Some of the garlic studies have also been reviewed (Dorant et al., 1993).

In a recent series of epidemiological studies from the Netherlands, no association was found between garlic supplement (pills) consumption and risk for stomach, colon, rectum, lung, or breast cancers (Dorant, 1994; Dorant et al., 1994a, 1994b, 1995). However, these studies should be viewed with caution, since the composition of garlic supplements varies tremendously with both type and brand (see Tables 3.17, 3.19, and 3.21).

5.3.2 Anticancer Effects: Animal and In Vitro Studies

Very early reports are available on the experimental verification of the tumor-inhibiting action of garlic and its constituents. Königsfeld and Prausnitz (1913) tested various allyl compounds, among them diallyl sulfide, and observed some inhibition in the development of tumors in mice. They considered the allyl structure to be responsible for this activity. Dittmar (1939) also observed a slight inhibition of inoculated tumors by diallyl disulfide but not by diallyl sulfide. Kubo and Kishida (1956) observed an inhibition of Ehrlich carcinoma cells in vitro by diallyl disulfide and other allyl derivatives. Keck and Hoffmann-Ostenhoff (1951; Kech et al., 1951) found mitosis-inhibiting effects when using aqueous extracts from other *Allium* species.

Caspari (1936) fed mice prophylactically with garlic before inoculation with tumor cells, and continued to treat them with garlic. In this fashion he achieved an increase in immunity in the test animals. In addition to fresh garlic, he also used stabilized (no allicin yield) commercial garlic juice and allyl mustard oil (which at that time was considered to be an ingredient of garlic). The latter two products were ineffective. Makowsky (1936), who repeated these experiments, was also able to inhibit the growth of inoculated tumors with garlic.

Von Euler and Lindeman (1949) carried out an experiment with alliin (which had just been isolated from garlic bulbs [Stoll & Seebeck, 1948]) in rats with sarcoma (Euler & Lindeman, 1949). Reduction and dissipation of the tumor was initiated not only by direct injection of alliin (5–10 mg) into the 2–4 cm sarcomas, but also by intramuscular injection of 1–3 mg alliin, in 11 of 22 animals. Novicova and coworkers (1957), on the other hand, could not detect any activity for alliin against various transplanted and induced tumors.

Ehrlich ascites tumor cells, incubated with fresh garlic extract (containing 2.8 mM allicin) for 1 hour and then injected into mice, were no longer lethal to the animals. On the other hand, cells treated with an extract of boiled garlic (containing 5.5 mM alliin) had normal carcinogenic activity and were lethal to the animals. Also, cells of Yoshida sarcoma, a spontaneous mammary tumor, and of methyl cholanthrene–induced sarcoma lost their carcinogenic properties upon treatment with allicin (more than 2.8 mM) (Nakata, 1973). A sugar solution of the ethanol extract from garlic showed adjuvant action in animals immunized with Ehrlich ascites tumor cells attenuated with allicin (Nakata & Fujiwara, 1975).

Weisberger and Pensky (1957, 1958) experimented with S-ethylcysteine sulfoxide, a synthetic analog of alliin, which they incubated with alliinase (to obtain the respective thiosulfinate) prior to application. Growth of the tumor was suppressed in mice with sarcoma-180 ascites tumors. Without enzymatic treatment, the cysteine derivative had no effect. This shows that the effective principle must be allicin and the other thiosulfinates from garlic.

A series of synthetic dialkyl thiosulfinates was tested for cytostatic activity in Ehrlich ascites tumor cells. Derivatives with isobutyl, *t*-butyl, and amyl groups proved to be the most effective (Kametani et al., 1959). In another study, it was found that the diethyl derivative caused 100% inhibition of inoculated tumors (sarcoma-180) in mice (Rocchietta, 1958). Other synthetic derivatives of natural thiosulfinates also show antitumor activity (Hirsch et al., 1965; Zhou et al., 1988). Even allithiamine (see Sections 3.2.8 and 5.9) proved to be effective as an antitumor agent in the same model (Cheng & Tung, 1981).

Di Paolo and Carruthers (1960) performed an experiment in mice with solid and ascites tumors (sarcoma-180). They administered allicin to tumor-bearing animals and compared its action with that of an aqueous garlic extract. Both test samples were administered in increasing amounts (2–280 µg/0.2 mL) intravenously, intraperitoneally, and orally as single and multiple doses for up to 7 and 14 days, respectively. None of these treatments exerted significant inhibition of tumor growth as compared to the untreated controls. In contrast, pretreatment of tumor cells with allicin before implantation led to total inhibition of tumor development; the aqueous garlic extract, however, was ineffective. Other *Allium* plants have similar effects on ascites sarcoma of rats (Kimura & Yamamoto, 1964). Kröning (1964) fed fresh garlic and an alliinase-inhibited garlic preparation to two mouse strains with hereditary tumors. He used AKR mice, which spontaneously develop virus-induced leukemia, and C3H mice, which usually die of breast cancer. Although neither of the two diets produced any effect in the AKR mice, the C3H mice were completely protected from tumor development by fresh garlic. The garlic preparation without alliinase, however, was ineffective. This result again confirms that not all kinds of tumor respond equally, and that the thiosulfinate in garlic must be the active principle against certain forms of cancer. Methylcysteine sulfoxide (methiin) has been shown to be only 10% as effective as the thiosulfinate derived from it, methyl methanethiosulfinate, in reducing the genotoxicity in mice treated with benzo-[*a*]pyrene (BP) (Marks et al., 1993).

Fujiwara and Natata (1967) contributed additional interesting information on this topic. If Ehrlich ascites tumor cells are treated with fresh garlic juice (contains allicin) before injection, they lose their capacity to induce tumors in mice. Not only do animals injected with garlic-treated cells develop no tumors, but they even become immune to untreated tumor cells. Thus, the virulence of cancer cells can be neutralized with fresh garlic, while their ability to induce an antibody response remains intact. This opens up completely new perspectives for a potential antitumor therapy with garlic. Other authors have reported similar results (Aboul-Enein, 1986; Lau et al., 1986a, 1986b).

Aqueous garlic extract has been shown to exert anticancer effects on cultured normal human skin cells as well as to increase their longevity (Svendsen et al., 1994). In mice, the number of tumor cells injected into the peritoneal cavity was significantly decreased after feeding garlic, while the initiation of tumors was delayed and the survival time of the animals prolonged by 50% (Choy et al., 1983).

Criss and colleagues (1983) treated rats carrying a transplantable hepatoma with a diet containing 5% fresh garlic and varying amounts of casein. The animals fed with garlic showed 10–25% less tumor growth than the controls. In a second series of experiments, a garlic extract was injected subcutaneously into rats, showing a 30–50% reduction in tumor growth afterwards. These animals, however, also had toxic symptoms. It was discovered that garlic contains an inhibitor for guanylate cyclase (a protein-like substance of unknown structure—see Section 3.3.2). Thus, it is assumed that tumor inhibition is affected by a decrease of cyclic-GMP in the tumor tissue. Generally, cyclic-GMP is considered a key substance in cancer growth regulation (Criss et al., 1982, 1983; Fakunle, 1983). Similar results were reported by other authors (Dhillon et al., 1981; Belman, 1983b).

El-Mofty and colleagues (1994) have recently found good antitumor effects by low doses of allyl sulfides and minced garlic in aflatoxin-treated toads. When fed minced garlic, yielding about 1.8 mg allicin/kg toad, there was an 80% decrease in liver tumor incidence. When fed a commercial steam-distilled garlic oil (analyzed by Lawson & Wang, 1994), at 2.3 mg allyl sulfides (44% diallyl trisulfide, 27% diallyl disulfide, 22% allyl methyl trisulfide) per kg body weight, there was a 50% decrease in tumor incidence. The results indicate that alliin-derived volatile sulfur compounds account for much of the anticancer effects seen with garlic at this dose level.

When a commercial steam-distilled garlic oil (brand Ranbaxy; 10 mg, about 20 μg of allyl sulfides) was applied topically to albino mice during the initiation phase of BP-induced skin carcinogenesis, there was a decline in the number of tumor-bearing mice, as well as in the mean number of tumors per mouse affected (Sadhana et al., 1988). Meng and Shyu (1990) demonstrated a strong inhibitory effect on both tumor number and tumor volume of topically applied aqueous garlic extract (0.5%, about 20 μg allicin/mL) on 7,12-dimethylbenz[a]anthracene (DMBA)–induced oral carcinogenesis in Syrian hamsters (Shyu & Meng, 1987). Zhang and coworkers (1989a, 1989b, 1990) found that both aqueous garlic extract and diallyl trisulfide suppress the mutagenic activity induced by 4-nitroquinoline-1-oxide and N-methyl-N'-nitro-N-nitrosoguanidine (MNNG) in $E.$ $coli$.

An inhibitory effect on papilloma (skin tumors) production in Sencar mice was induced by a single topically applied 2-mg dose (about 65 mg/kg) of garlic oil (undiluted steam-distillate), administered 30 minutes before a carcinogenic dose of DMBA. Furthermore, a single 5-mg dose of garlic oil maximally inhibited the DMBA-induced epidermal DNA synthesis by 86% when applied 2 hours before the carcinogen administration. Both onion and garlic oils (5 mg) inhibited the 12-O-tetradecanoylphorbol-13-acetate (TPA)–stimulated DNA synthesis, while 1 mg of each oil inhibited the first and second stages of tumor promotion. The authors concluded that these oils inhibit all stages of tumorigenesis (Perchellet et al., 1990). Topically applied diallyl sulfide and diallyl disulfide have also been shown to reduce greatly DMBA-induced skin tumors in mice (Dwivedi et al., 1992). Others have found that garlic oil, onion oil, ajoene, and various allyl and 1-propenyl sulfides, but not dipropyl disulfide, inhibit tumor promotion and lipoxygenase activity (Belman, 1983b; Belman et al., 1987, 1989, 1990).

Weekly topical application of DMBA for 25 weeks at two dose levels resulted in 74% and 100% tumor incidences, respectively. When aqueous garlic extract was topically applied for 3 days every week prior to DMBA administration, the incidences of tumors were reduced to 32% and 43%, respectively ($P < 0.01$) (Rao et al., 1990).

Oral administration of a garlic extract was found to increase the life span of mice containing transplanted Ehrlich ascites tumors by 41% and to increase longevity by 70% after administration of a lethal dose of cyclophosphamide. The garlic extract also inhibited the two-stage chemical skin carcinogenesis induced by DMBA and croton oil (Unnikrishnan & Kuttan, 1990; Unnikrishnan et al., 1990).

Furthermore, garlic (aqueous extract) prevented the carcinogenesis induced by 3-methyl-cholanthrene in the uterine cervix of virgin albino mice, when administered daily at 0.4 g/kg body weight for 2 weeks before and 4 weeks following carcinogen administration. The cervical carcinoma incidence, as compared with that of the positive control (73%), was only 23%, a highly significant ($P < 0.01$) effect (Hussain et al., 1990).

A 20% ethanol extract of garlic (allicin containing) was also found to inhibit the

TPA-induced enhancement of phospholipid metabolism. The change in phospholipid metabolism is one of the earliest phenomena caused by this tumor promoter in vitro. Even the first stage of tumor promotion in the two-stage mouse skin carcinogenesis in vivo was suppressed by treatment with the garlic extract (Nishino et al., 1989).

Favorable modification of clastogenicity of three known clastogens, namely mitomycin C, cyclophosphamide, and sodium arsenite, by aqueous garlic extract (25–100 mg/kg) was demonstrated in bone marrow cells of mice in vivo (Das et al., 1993b).

Numerous publications have attempted to explain the mechanism of action of the anticarcinogenic activity of garlic. Romanyuk (1954) expressed the opinion that the antibiotic substances (thiosulfinates) of garlic inhibit the activity of proteolytic enzymes (e.g., cathepsin) in human malignant tumors. Vainberg (1952) found that garlic extract inhibits the enzyme activity of pyruvate dehydrogenase, wherein thiol groups are targets of certain garlic components, and hence proposed the anticancer effect to be due to an interruption of cellular respiration. Similarly, it has been postulated that an interaction between thiosulfinates (e.g., allicin) and the sulfhydryl groups of the cellular enzymes is the mechanism of action (Weisberger & Pensky, 1957; You et al., 1988). This appears to be confirmed by several of the experiments described above. Enzyme inhibition may thus be generally considered as the mechanism whereby the active compounds in garlic exert their tumor-inhibiting effect.

Another plausible explanation for the inhibitory action of garlic and onion oil on the development of skin cancer in the presence of tumor promoters has recently been presented (Perchellet et al., 1986). It is the stimulation and elevation of the natural, antioxidative, protective functions within the organism that are directly coupled to the redox system of glutathione (GSH) and oxidized glutathione (GSSG). As support for this proposal, garlic oil, onion oil, and diallyl sulfide were all found to enhance the activity of glutathione peroxidase in epidermis cells that had been pretreated with the tumor promoter TPA. Inhibition of the peroxidase by TPA was prevented by the oils. Additionally, garlic oil was found to block ornithine decarboxylase (ODC) activity in epidermis cells. ODC-induction, caused by phorbol ester and other tumor promoters, is prevented by the enhanced activity of GSH-peroxidase. The same research team has published additional results in this special topic of cancer research (Belman, 1983a, 1983b; Zelikoff et al., 1986; Zelikoff & Belman, 1985).

Chinese authors proposed a different mechanism of reaction for stomach cancer, on the basis of the antimicrobial effects of garlic's thiosulfinates. An aqueous garlic extract was found to inhibit the growth of *Fusarium moniliforme* in pure culture. In addition, it lowered the concentration of nitrate and prevented formation of dimethylnitrosamine in the medium. Since nitrosamines are carcinogenic, prevention of their formation, as has been shown with garlic, can be considered as further indication for its antitumor activity (Liu & Lin, 1985; Liu et al., 1985, 1986a, 1986b, 1991; Lee & Jang, 1991; Kim et al., 1987; Lin et al., 1986; Teel, 1993).

The existence of antimutagenic factors in garlic was demonstrated with the use of the familiar Ames test for mutagenicity with *Salmonella typhimurium* (Kimm & Park, 1982; Schimmer et al., 1994; Park et al., 1991; Kim et al., 1991; Knasmüller et al., 1989; Stacchini & Mangegazzini, 1987). In the Ames test, the addition of aqueous garlic extract also caused a definite decrease in the frequency of mutations induced by various peroxides or high-energy radiation, as long as the garlic preparation was applied as a preventative (de Martin et al., 1986). The antimutagenic effect in the Ames test of garlic extract was also observed with nitrosomethylguanidine (Park et al., 1981; Knasmüller et al., 1989).

Ten organosulfur compounds from garlic and onion were studied for their modifying effects on diethylnitrosamine-induced neoplasia in the liver of rats (Takada et al.,

1994; Matsuda et al., 1994). It was found that glutathione S-transferase is significantly enhanced by methyl, propyl, and allyl mono-, di-, and trisulfides and that the effect is interrelated with polyamine synthesis. Deng and coworkers (1993, 1994) found that diallyl trisulfide (a major component of steam-distilled garlic oil) modulates drug-induced unscheduled DNA synthesis in rat hepatocytes. Thus, it has been suggested that these sulfides be used in cancer therapy, either as active agents or to alleviate the adverse side effects of chemotherapeutic agents (Dausch & Nixon, 1990).

People who eat garlic regularly have a low pH and low concentrations of nitrite in their gastric fluid. In addition, fewer nitrate-reducing bacteria are found in their stomach than in people who rarely or never consume any garlic. This is taken as proof for the well-known fact that garlic eaters rarely develop cancer of the stomach. In vitro, it was shown that ethyl ethanethiosulfinate (a homolog of allicin) inhibits the growth of and the nitrite formation by nitrate-reducing bacteria (Mei et al., 1982, 1985). Fresh garlic extract, as well as diallyl trisulfide, also showed strong cytotoxic effects in two lines of human stomach cancer cells, and the effect is comparable to that of 5-fluorouracil and mitomycin C (Pan, 1985; Pan et al., 1988).

Konvicka (1983, 1984) investigated the histologic appearance of mitosis inhibition by garlic extract in a plant model system (root tips of onion and *Secale cereale*). The extent of mitotic inhibition varied depending on concentration (even 0.1% garlic extract was effective). The microscopic picture revealed chromosome agglutination (pyknosis effect), chromosome rupture, and chromatin loss, up to complete karyolysis and karyorrhexis. Contrary to the inhibition of mitosis by colchicine, where only the spindle formation is altered, garlic affected all phases of mitosis. These findings show that the inhibitory effect on cancer by garlic, which has been observed in many cases, may be due to an interference with cell division.

Similar results were reported on experiments with Chinese hamster ovary (CHO) cell cultures. Radiation with high-energy gamma rays induced mutation in CHO cells that show resistance, for example, to thioguanine. An aqueous garlic extract diminished the mutation by as much as 80%. When bacteria were used as the indicator organism (e.g., strains of *Salmonella typhimurium*), garlic revealed an antimutagenic action in 20–30% of cases. As an explanation for this effect, it was proposed that the sulfur compounds in garlic act as radical scavengers. Garlic also successfully protected the DNA of the cell nucleus against active radicals formed during high-energy radiation (Knasmüller et al., 1986; Baer & Wargovich, 1989).

Selenium and germanium have also been proposed as components that may be responsible for the cancer inhibitory effects of garlic (Bolton et al., 1982). However, their levels in normally grown garlic are much too small to have any effect, especially when one considers the amount of garlic typically eaten (see Table 3.16). Nevertheless, enrichment of the soil with selenium has been shown to increase greatly the amount of selenium compounds in garlic (see Sections 3.3.8 and 3.3.9) and to improve the anticancer effects of garlic (Ip et al., 1992, 1994; Ip & Lisk, 1994a, 1994b).

Diallyl sulfide (diallyl monosulfide) has been investigated as a garlic-derived anticarcinogen in a large number of animal studies. Although it is only a minor component of steam-distilled garlic oil (see Table 3.20), it has been very popular to study because it is commercially available in pure form, while other diallyl sulfides are not (commercial diallyl disulfide is only 70–85% pure). Consistently positive effects have been found with large doses of diallyl sulfide; however, some caution is recommended in extrapolating the results with diallyl sulfide to the larger diallyl sulfides, since it has been shown that the much more abundant diallyl disulfide and diallyl trisulfide are metabolized in the blood much differently than diallyl sulfide (Lawson & Wang, 1993; Lawson, 1993). Studies up to 1989 dealing with the anticarcinogenic effects of various allyl sulfides have been reviewed (Lau et al., 1990).

Gudi and Singh (1991) investigated the mechanism of the antineoplastic effect of diallyl sulfide against BP-induced carcinogenesis in mice. Diallyl sulfide treatment caused a significant dose-dependent increase of glutathione-S-transferase (GST) activity in the mouse stomach (1.6-fold increase by 8.5 mg, about 300 mg/kg body weight) as a result of increased synthesis of the enzyme. Pulmonary and renal GST activity was also increased, but not in a dose-dependent fashion, while the hepatic enzyme was not influenced. Treatment of animals with diallyl sulfide increased stomach GSH peroxidase activity (2.5-fold increase by 8.5 mg) as compared to the control. On the other hand, GSH peroxidase activity in liver and kidney was not changed by the treatment. These results suggest that diallyl sulfide may exert an antineoplastic effect by modulating GSH-dependent detoxification enzymes.

Cytochrome P-450 and P2E1 enzymes have been proposed as targets for chemoprevention by diallyl sulfide against chemical carcinogenesis and toxicity, as demonstrated by both in vitro and in vivo evidence in rats (Yang et al., 1993, 1994a; Wargovich, 1992; Brady et al., 1988, 1991; Pan et al., 1993; Reicks et al., 1993, 1994; Chen et al., 1994; Steele et al., 1990). Diallyl sulfide (50–200 mg/kg) has also been shown to decrease the amount of carcinogen-induced DNA methylation (Ludeke et al., 1992). The antimutagenicity of both diallyl sulfide and ajoene has been demonstrated by their ability to inhibit aflatoxin-induced mutations in *Salmonella typhimurium* and to inhibit binding of the aflatoxin to DNA (Tadi et al., 1991). Corresponding effects have also been reported by other authors (Maurya & Singh, 1991; Scharfenberg et al., 1990).

Wargovich's group has investigated the protective effects and possible mechanisms of action of diallyl sulfide (200 mg/kg body weight) on carcinogen-induced gastrointestinal cancer in animals, especially esophageal cancer (Sumiyoshi & Wargovich, 1989; 1990; Wargovich & Goldberg, 1985; Wargovich, 1987, 1988, 1992; Wargovich et al., 1988a, 1988b; Wargovich & Eng, 1989; Baer & Wargovich, 1989; Wargovich & Imada, 1993; Hu & Wargovich, 1989). Similar positive results have been found with DMBA-induced toad liver tumors, where diallyl sulfide (125 mg/kg) was effective if given prior to, but not after, administration of the carcinogen, indicating an effect on tumor initiation (Sadek & Abdul-Salam, 1994).

Diallyl sulfide has also been investigated for its ability to reduce injury to the colon after gamma ray exposure to mice (Baer & Wargovich, 1989). The sulfide (200 mg/kg) was administered to mice 3 hours prior to a single whole body dose of radiation from a ^{60}Co source and was found significantly to inhibit nuclear aberrations due to radiation in the dose range from 0.5 to 10 Gy. The degree of protection was correlated with the dose of diallyl sulfide, but it was not effective if given only after irradiation. The radiation exposure was also found to increase DNA synthesis and ornithine decarboxylase (a DNA synthesis regulator) activity; however, this effect was also found to be significantly reduced by prior administration of diallyl sulfide. The protection by diallyl sulfide was found to involve a polyamine-dependent pathway.

Diallyl sulfide was found to inhibit the metabolic activation of the tobacco-specific 4-(methylnitrosamino)-1-(3-pyridyl)-1-butanone (NNK) in mouse lung (Hong et al., 1992). Mice were pretreated with diallyl sulfide (200 mg/kg) for 3 days, followed by a single 2-mg dose of NNK. In comparison to the control group, diallyl sulfide pretreatment significantly decreased the incidence of NNK-induced lung tumors by 62%, and the tumor multiplicity by 92%. The formation of NNK metabolites in both the lungs and liver was also greatly decreased by the diallyl sulfide pretreatment, indicating that the sulfide probably functions by inhibiting the metabolic activation of NNK.

Hayes and coworkers (1987) showed that diallyl sulfide inhibits 1,2-dimethylhydrazine (DMH)–induced hepatocarcinogenicity in rats. At doses above 50 mg/kg, administered 1 hour before DMH (50 mg/kg), diallyl sulfide partially reduced the numbers of foci positive for γ-glutamyl transpepsidase and glutathione-S-transferase-P. By comparison, all doses of diallyl sulfide (50–100 mg/kg) completely prevented liver necrosis by DMH (200 mg/kg). The results indicated that diallyl sulfide inhibits the promotion stage rather than the initiation stage of carcinogenesis. Diallyl sulfide (300 mg/kg) has also been shown to reduce greatly the tumor incidence and the amount of damaged DNA of aristolochic acid–induced forestomach tumors in rats, indicating that the sulfide prevented conversion of papilloma into malignant tumors (Hadjiolov et al., 1993). When fed to carcinogen-treated mice at 0.5% of the diet, diallyl sulfide was shown to reduce the incidence of benign tumors of the lung and thyroid (Jang et al., 1991).

Diallyl disulfide and, to a lesser extent, other allyl polysulfides have also been studied for anticancer effects in animals. For example, a preparation containing mainly diallyl disulfide has been shown to decrease BP-induced bone marrow genotoxicity (Marks et al., 1992). At 0.01% of the diet (about 4 mg/kg), diallyl disulfide decreased azoxymethane-induced invasive tumors of the colon and increased glutathione transferase activity (Reddy et al., 1993). An in vitro study has shown that preincubation of mouse macrophages with diallyl trisulfide (10 µg/mL) significantly increases the ability of the macrophages to kill three tumor cell lines (Feng et al., 1994).

Wattenberg and Sparnins have published a number of studies on the effects of several sulfides found in steam-distilled garlic oil in protecting against carcinogen-induced neoplasia (early tumors), especially in the forestomach of mice (Wattenberg et al., 1986, 1989; Sparnins et al., 1986, 1988; Wattenberg, 1985, 1992). They found that the allyl group of the sulfides was particularly important, since protection against BP-induced neoplasia by allyl methyl trisulfide, allyl methyl disulfide, diallyl trisulfide, and diallyl sulfide were nearly abolished if propyl groups were substituted for the allyl groups. Furthermore, diallyl trisulfide, having two allyl groups, was more active than allyl methyl trisulfide. In a comparison of diallyl sulfide to diallyl trisulfide, the trisulfide was found to be more active in preventing forestomach neoplasia, but it was less active in preventing lung adenomas (Sparnins et al., 1988). In another study in which forestomach carcinogenesis was induced by N-nitrosodiethylamine (NDEA), diallyl trisulfide was not tested, but diallyl disulfide and allyl mercaptan were found to be considerably more effective than diallyl sulfide. At about 150 mg/kg, diallyl disulfide reduced forestomach tumor formation by 90% and lung adenomas by 30% (Wattenberg et al., 1989). When compared at equal weights (this also compares equal moles of allyl groups), allyl mercaptan was found to be even more active than diallyl disulfide, especially when administered to the mice shortly before administration of the NDEA. The results with allyl mercaptan are particularly important, since allicin and diallyl trisulfide are now known to be rapidly converted (diallyl disulfide also, but much more slowly) to allyl mercaptan in fresh whole blood (Lawson & Wang, 1993).

Sundaram and Milner (1992, 1993, 1994, 1995) have shown that diallyl disulfide inhibits tumor growth both in vitro and in vivo. Diallyl disulfide was found to inhibit the growth of both cultured mammary tumor cells and cultured human colon tumor cells in a dose-dependent manner, with a concentration of 0.05 mM providing 30% inhibition of growth. Diallyl trisulfide was more inhibitory and diallyl sulfide was slightly less inhibitory to cell growth than the disulfide. In an in vivo experiment, the injection of 1 mg of the disulfide into mice (around 30–50 mg/kg) inhibited growth of transplanted human colon tumor cells by 60%. The importance of both the allyl group and the sulfur atoms in diallyl disulfide was revealed when these investigators found no inhibitory activity for dipropyl disulfide, allyl chloride, or diallyl glycidyl ether (allyl-O-CH_2CH_2-O-allyl). A possible mechanism for the effect of diallyl disulfide was found when it was shown that the disulfide caused a dose-dependent rise in intracellular free calcium.

Another possible mechanism for the effect of diallyl disulfide has recently been demonstrated in vitro where it was shown that the disulfide inhibited growth of a human colon cancer cell line and also decreased membrane association of the p21 ras gene. It was proposed that diallyl disulfide inhibits isoprenylation of the p21 ras, which then prevents its ability to localize to the appropriate membranes (Bowman et al., 1995).

S-allylcysteine, a trace component of garlic that increases during garlic product manufacturing (0.03 mg or less per gram garlic clove, or up to 0.5 mg/g in garlic powders or aged extracts—see Tables 3.3 and 3.17, and Section 3.2.7), has also been examined in several investigations for possible anticancer effects. Compared to diallyl disulfide (present at 260 mg/g in undiluted steam-distilled garlic oil or about 1.1 mg/g commercial garlic oil capsules—see Table 3.20), S-allylcysteine is about 20 times less active in inhibiting the in vitro growth of human colon tumor cells (Sundaram & Milner, 1994), has no effect on inhibiting the growth of mammary tumor cells up to 1 mM (Sundaram & Milner, 1992), and does not increase intracellular calcium levels (Sundaram & Milner, 1995). However, it has been shown in vivo to decrease the binding of the carcinogen DMBA to the DNA of rat mammary tumors (Amagase & Milner, 1993a) and to inhibit significantly ($P < 0.05$) the incidence of dimethylhydrazine-induced colon tumors in mice at a large dose (200 mg/kg), although neither S-propylcysteine nor S-allylmercaptocysteine at the same doses had an effect (Sumiyoshi & Wargovich, 1990).

The growth of both cultured human neuroblastoma cells and human melanoma cell lines was found to be inhibited by S-allylcysteine in a dose-dependent manner, with 50% inhibition occurring at 4–5 mM; differentiation was also affected in the melanoma cells, but not in the neuroblastoma cells (Welch et al., 1992; Takeyama et al., 1993). Another in vitro study showed that S-allylcysteine did not affect the growth of human carcinoma cell lines, but that it enhanced the growth inhibition caused by the drug cisplatin by suppressing glutathione levels (Yellin et al., 1994). The growth of cultured normal smooth muscle cells was not affected by up to 60 mM S-allylcysteine, but a disulfide homolog, S-allylmercaptocysteine, inhibited growth by 30% at 0.3 mM (Lee et al., 1994).

Aged garlic extract, a commercial product which contains S-allylcysteine as its most abundant sulfur compound (see Tables 3.17 and 3.25), has been the subject of a number of anticancer studies. Most of the animal studies with the aged extract have shown positive effects with large doses and have important implications for whole garlic, which contains approximately 10 times more total thioallyl compounds.

In tests on mice inoculated with tumor cells, aged garlic extract (25 mg, equivalent to 60–80 g for a person) was directly injected into the bladder. Normally, the tumors develop within 14 days, but after 21 days the animals treated with the garlic preparation clearly showed fewer and smaller bladder tumors than the untreated controls (Marsh et al., 1987; Lau et al., 1985, 1986a, 1986b). At 12 mg/mL, the aged extract was found in vitro to decrease significantly both BP-induced and aflatoxin-induced mutagenesis and DNA binding in *Salmonella typhimurium* (Tadi et al., 1990, 1991; Lau et al., 1991a). It was also shown in vitro to prevent carcinogen-induced phospholipid synthesis in cultured HeLa cells (Nishino, 1993).

When fed to rats at 2–4% of the diet (equivalent to 10–20 g for a person), the aged extract was found significantly to decrease carcinogen-induced (DMBA) DNA adducts and DMBA-DNA binding in the mammary tissue. There was also a 50% decrease in incidence of, and a 38% decrease in number of, mammary tumors. The effect required the presence of 20% corn oil in the diet and was enhanced by the presence of 1 ppm selenium (Liu & Milner, 1990; Liu et al., 1992a; Zoumas et al., 1992; Amagase & Milner, 1993b). Substantial increases in mammary and liver glutathione (twofold increase) and glutathione-S-transferase activity (42–87% increase) were also observed (Liu et al.,

1992b). In a comparative study, the feeding of an aqueous garlic extract, an aged extract, and a commercial garlic powder to DMBA-treated rats at 1.5–2% of the diet resulted in 46%, 50%, and 56% decreases in mammary tissue DMBA-DNA binding, respectively. Pure S-allylcysteine was also found to have good inhibitory activity, but the activity of the three garlic products was not proportional to their content of this compound (Amagase & Milner, 1993a). Furthermore, the aged extract caused a 50% decrease in DNA adducts caused by feeding rats sodium nitrite or nitroso compounds (Lin et al., 1992, 1994).

A cancer prevention study with aged garlic extract was recently conducted on humans. Sixteen men were given 10 mL extract daily for 12 weeks, along with monthly doses of acetaminophen, a common drug that undergoes a similar metabolic route as many of the carcinogens. The aged extract was found to have no effect on the levels of the plasma or urinary metabolites of acetaminophen, which led the authors to conclude that the extract has limited potential as a chemopreventive agent (Gwilt et al., 1994).

While most authors have concentrated on the sulfur-containing constituents of the *Allium* plants, Nishino and colleagues (1990a) found another remarkable substance with anticarcinogenic properties in *Allium bakeri*: laxogenin, a steroid containing four oxygen atoms. Allixin, a sulfur-free phytoalexin isolated from stressed garlic (amounts not reported, absent in normal garlic), has been shown in vitro and in vivo to have antitumor activity and to be able to decrease aflatoxin-induced mutagenesis and DNA binding. It also inhibited the formation of aflatoxin-glutathione conjugates (Yamasaki et al., 1991; Nishino, 1990b, 1993; Kodera et al., 1989).

In addition, it seems of interest that the volatile ingredients of garlic can completely inhibit the germination of pollen from flowers (e.g., *Camelia sinensis*). This effect is similar to that of irradiation with gamma rays (200 kR) (Iwanami, 1981). Finally, it should be mentioned that the garlic plant itself, namely the root tips, has been used as a test object for the determination of clastogenic and mitoclastic effects of other test substances (Valadaud-Barrieu, 1983).

5.3.3 Anticancer Effects: Active Compounds

From the many publications that have just been reviewed, it is apparent that the anticancer effects of garlic are likely due, perhaps equally, to both allicin or allicin-derived compounds as well as to unidentified compounds not related to allicin. The following is a summary of the evidence for possible active compounds.

1. Epidemiological studies from six different countries have consistently shown that garlic consumption is associated with decreased risk of gastrointestinal cancer. Since garlic is mainly eaten cooked (alliinase inactivated) in most of these countries, allicin is not necessary to achieve significant cancer reduction.

2. A major decrease in incidence of gastric cancer in China, particularly where large amounts of allicin-yielding fresh garlic are eaten, is associated with the antibiotic effects of garlic and its thiosulfinates (allicin) toward decreasing the amount of nitrate-reducing bacteria in the stomach and hence the amounts of carcinogenic nitrosamines formed (Mei et al., 1981, 1982, 1985). Therefore, allicin does appear to have an important role in prevention of gastric cancer.

3. Animal studies have indicated the importance of allicin, since dietary fresh garlic, but not alliinase-inhibited garlic, greatly decreased breast cancer incidence in C3H mice (Kröning, 1964). Furthermore, at similar doses of volatile sulfur compounds, both chopped garlic (1.8 mg allicin/kg body weight) and steam-distilled garlic oil (2.3 mg allicin-derived allyl sulfides/kg) similarly decreased liver tumor incidence in toads (El-Mofty et al., 1994).

4. A large number of animal studies with allicin-derived diallyl disulfide and diallyl sulfide, most using very large doses (100–200 mg/kg) (with the exception of

Reddy and coworkers (1993), who showed positive effects with 4 mg diallyl disulfide/kg), have shown positive effects toward decreasing carcinogen-induced cancer. Although allicin itself has not been tested, these studies indicate that allicin-derived compounds have the ability to affect cancer incidence. Studies with allicin-derived compounds are important to understanding the cancer effects of fresh garlic, since diallyl disulfide and other allicin-derived compounds are known to produce the same metabolite, allyl mercaptan, in blood or liver as does allicin (Lawson & Wang, 1993; Egen-Schwind et al., 1992b). Interestingly, alliin, which would be abundant in cooked garlic, may also be metabolized in the liver to diallyl disulfide and then to allyl mercaptan (Egen-Schwind et al., 1992b; Pentz et al., 1991).

5. Evidence from animal studies that compounds in garlic other than allicin also have an important anticancer role comes from the many studies that have been conducted with commercial aged garlic extract (contains little alliin, no active alliinase, and only trace levels of allicin-derived compounds) fed to rats at 2–4% of the diet, which have shown a consistent decrease in carcinogen-induced tumors and DNA adducts (Liu & Milner, 1990; Amagase & Milner, 1993; Lin et al., 1992, 1994). The effective compounds in this extract are not known. S-allylcysteine has been proposed as the possible active agent, but its activity as a pure compound and its content in the extract are far too small to account for a significant portion of the activity of the aged extract.

Future studies on determining the anticancer activity of garlic and its active compounds should ideally include human intervention trials with both fresh and cooked garlic and possibly spray-dried (no alliin) garlic. Animal studies on the allyl sulfides or other garlic-derived compounds should be conducted with considerably lower doses than have been typically used. Studies on the allyl sulfides would benefit by the use of steam-distilled garlic oil, which contains a variety of allyl sulfides, including the abundant diallyl tri- and tetrasulfides, which are more reactive than the mono- or disulfides. Furthermore, studies with garlic fractions should use fractions of known composition, particularly with respect to total sulfur, γ-glutamyl-S-alkenylcysteines, S-alkenylcysteines, alliin, fructans, etc.

5.4 ANTIOXIDANT EFFECTS

The natural antioxidant effect of garlic is of utmost practical interest, especially in connection with its antiarteriosclerotic (Section 5.1.1), antihepatotoxic (Section 5.10.2), and antitumor properties (Section 5.3). Concerning the development of arteriosclerosis, it is now generally accepted that free radicals play a major role in the deposition of cholesterol into the vessel wall. Sources of free radicals, aside from high energy radiation, include pollutants such as nitrogen oxide and metabolites formed in the absence of sufficiently high reductive potential. Free radicals affect the cell at the level of DNA, protein synthesis, and membrane function. In the arterial vessels, the main substrates of the free radicals are the unsaturated fatty acids of the low density lipoproteins (LDL). Oxidative damage to the LDL is accompanied by the loss of their normal receptor function. Oxidatively modified LDL is consumed by resident macrophages, which swell and perish, resulting in a mass of dead macrophages, the so-called foam cells. The foam cells push the endothelial cell layer outward and initiate the first step of atherosclerosis. Garlic constituents inhibit the formation of free radicals, support the endogenous radical scavenger mechanisms, and protect LDL against oxidation by free radicals. This explains the overall cardioprotective effect of garlic (Jacob et al., 1993a, 1993b), as well as of other *Allium* species (Rietz et al., 1993; Nishimura et al., 1988).

5.4.1 Antioxidant Effects: In Vitro Studies

Dhar (1951) was the first to demonstrate an antioxidant effect in garlic, by showing that the Indian *ghee*, or clarified butter (a liquid butter fat used in cooking), could be protected from peroxidation by the addition of an alcoholic extract of crushed garlic. The active compounds were found in the petroleum ether fraction of the extract. A reproduction of the methods used in preparing the petroleum ether fraction of the alcoholic extract revealed that it consisted mainly of ajoene and some diallyl trisulfide (Lawson & Wang, 1994). The biochemical redox potential of an aqueous garlic extract has been determined polarographically (Chagovets & Zalesskii, 1951).

Garlic and onion extracts have been used for the preservation of oil or lard, with a duration of the antioxidative effect for 3 to 6 months (Zalewski, 1960; Pizzocaro et al., 1985; Sethi & Aggarwal, 1957). Zalewski (1960) found that crushed garlic prevents rancidity of lard and attributed the antioxidant activity to the ether fraction, which contains vinyldithiins, ajoene, and diallyl sulfides. On the other hand, others have found a neutral or slightly pro-oxidative effect when using methanol-extracted garlic powder or garlic mixed in oil, neither of which method will activate alliinase (Huang et al., 1981; Gerhardt & Blat, 1984). Naito and coworkers (1981a) reported that alliin and S-alkylcysteines have good antioxidant activity, but that the activity of alliin is decreased after treatment with alliinase, indicating that allicin may be less active.

Studies with an allicin-containing aqueous garlic extract on human serum (Domjan et al., 1988) and with an allicin-yielding garlic powder on human granulocytes (Siegers, 1992) showed a decrease in lipid peroxidation and oxygen radicals, respectively. Specific lipoxygenase inhibitors have been identified in garlic and onion oil (Block et al., 1988; Vanderhoek et al., 1980).

Experiments with liver microsomes have demonstrated the in vitro activity of specific compounds. Hikino and colleagues (1986) showed that steam-distilled garlic oil (diallyl sulfides) inhibited CCl_4-induced free radical formation at 1 mg/mL, and CCl_4-induced lipid peroxidation at 0.01 mg/mL, but that alliin did not inhibit either, even at 1 mg/mL.

Horie and coworkers (1989) showed that a 20% ethanol extract of garlic at 10–20 mg/mL significantly protected the microsomes from induced lipid peroxidation. Although the authors did not analyze their extract, a reproduction of their extract method revealed that it contained high amounts of allicin (5 mg/g) as the only significant ether-soluble compound, as well as the water-soluble γ-glutamyl-S-alkylcysteines (17 mg/g) (Lawson & Wang, 1994). In a further study, it was shown that purified diallyl sulfides, especially those containing four to seven sulfur atoms, reduced lipid peroxidation by 80–90% at 25 µg/mL, thus revealing the active compounds in the steam-distilled garlic oil (Horie et al., 1992). This finding was also patented (Awazu et al., 1992). Kourounakis and Rekka (1991) recently reported the effects of purified alliin and a garlic powder of established allicin potential on induced lipid peroxidation and free radical scavenging. Alliin (25 mM) did not inhibit the lipid peroxidation but did cause a 50% decrease in hydroxyl formation. However, at allicin levels of 0.2 mM and 3 mM, the garlic powder caused 50% inhibition of both lipid peroxidation and hydroxyl radical formation.

The in vitro antioxidant effect of an aqueous extract from a garlic dry powder preparation (Kwai) was investigated by Popov and coworkers, using the method of photochromoluminescence (Popov et al., 1994; Lewin & Popov, 1994). Garlic extract obtained from 1 mg garlic powder was found to be antioxidatively equivalent to 30 nmol ascorbic acid or 3.6 nmol α-tocopherol. A dose-related oxidation-inhibiting effect of the garlic extract was also established in a test system that employs the Cu^{2+}–initiated oxidation of low density lipoproteins.

Several recent in vitro studies have demonstrated garlic's antioxidant activity as a free radical scavenger. Protection against lipid oxidation provided by fresh garlic extract and commercial garlic products, as well as by synthetic antioxidants (BHA, BHT, TBHQ, propyl gallate), was evaluated using a new chemiluminescence technique for the determination of lipid hydroperoxides. The effects were found to be concentration-dependent and similar to that of the synthetic antioxidants. Furthermore, the garlic extract was found to scavenge hydroxyl radicals (Yang et al., 1993). An allicin-standardized garlic powder product was recently shown to effectively scavenge hydrogen peroxide–derived hydroxyl free radicals at an allicin-equivalence as low as 0.0018 mg/mL (Prasad et al., 1995). Another research group investigated the radical scavenging effectiveness of garlic extract (prepared from Kwai dry garlic powder) in vitro in several radical scavenging systems by electronic spin resonance (ESR) and low level chemiluminescence measurements. Significant radical scavenging capacity was detected. An additional interesting finding was that radicals present in cigarette smoke were reduced by garlic (Török et al., 1994).

Chinese workers studied the antioxidant action of garlic oil (steam-distilled) and diallyl trisulfide (allitridi) using lipid peroxidation and chemiluminescence in mouse liver mitochondria induced by a Vc/FeSO$_4$ reaction system and by a hematoporphyrin-induced photohemolysis, respectively. Both garlic oil and diallyl trisulfide inhibited lipid peroxidation and photohemolysis and enhanced O$_2$ production. Moreover, allitridi effectively prevented inactivation of red cell membrane acetylcholine esterase (Fu et al., 1993).

The molecular mechanism of the antioxidant activity of certain garlic-derived compounds, such as diallyl mono-, di-, and trisulfides, as well as dry garlic powder (standardized to 0.6 mg allicin per 100 mg) and caffeic acid as positive control, was investigated by their effect on superoxide (O$_2^-$) quenching and their ability to scavenge peroxyl or alkoxyl radicals. The observed antioxidant activity of the investigated compounds may be due to multiple mechanisms of action, but appears to be due mainly to an inhibition of the chain reaction induced by hydroxyl radicals (Rekka & Kourounakis, 1994). There is also evidence that selenium-containing proteins and polysaccharides may participate in the scavenging of active oxygen radicals (Lu et al., 1992).

The antioxidant activity of aged garlic extract and some of its compounds has been the subject of some recent in vitro investigations. Yamasaki and coworkers (1994) showed that the extract as well as S-allylcysteine (both at 2 mg/mL) could similarly and significantly inhibit lipid peroxidation and increase cell viability in bovine artery endothelial cells that were subsequently treated with 50 μM hydrogen peroxide, a strong oxidant. Since the concentration of S-allylcysteine in the aged extract is about 0.6 mg/g (see Table 3.25; see also Yeh & Yeh, 1994; Amagase & Milner, 1993a), only about 0.1% of the activity of the aged extract could be accounted for by S-allylcysteine. In a study that compared the chemiluminescence reducing activity (a measure of lipid peroxidation) between an aged garlic extract (not the same as the commercial aged extract, since it was aged for only 10 months), fresh garlic, and boiled garlic (0.15% each) to rat liver microsomes exposed to the oxidant t-butyl hydroperoxide, Imai and colleagues (1994) found that only the aged extract was effective. Unfortunately, an alcohol control was not studied, an important concern because liquid aged extracts contain about 10% (w/v) alcohol (see Section 3.4.5), an amount that would be 100 times greater than the S-allylcysteine content. Furthermore, in vitro comparisons of various garlic products in liver cells is subject to considerable error, since the metabolic influences of the blood and other organs are being omitted—and this is a particularly important issue for allicin, which is very rapidly metabolized in the blood to allyl mercaptan. S-allylcysteine, S-1-propenylcysteine, alliin, and several glutamylcysteines were also examined and found to have similar activity to each other, but the concentration needed to achieve an effect similar to the aged extract was very high, 5 mM. On the basis of the concentration of S-allylcysteine reported in the study for the extract, S-allylcysteine could have accounted for only 0.4% of the activity of the aged extract. The activity of pure S-allylmercaptocysteine (concentration in the tested extract not reported) was twice that of the other compounds tested, but on the basis of the low amount of this compound found in the commercial extracts (0.14 mg/g—see Table 3.25), S-allylmercaptocysteine could only account for about 0.004% of the activity of the extract. A study in mice with a non-commercial aged garlic extract (100 mg/kg) showed that the extract could offer protection against CCL_4–induced lipid peroxidation (Kagawa et al., 1986).

5.4.2 Antioxidant Effects: In Vivo Studies

The antioxidant effect of allicin in vivo has been demonstrated by its influence on several enzymes involved in antioxidative processes such as catalase and glutathione peroxidase, as well as the concentration of lipid peroxides in blood. Allicin (0.46 mg/kg, orally, equivalent to 8 g fresh garlic for a person) significantly increased the activity of these enzymes and decreased the lipid peroxide content, obviously by a free radical scavenging mechanism (Han et al., 1992). Siddiqui and coworkers (1988) found that rats fed for 1 week a small dose of allicin-derived commercial steam-distilled garlic oil (20 mg diluted oil or 0.1 mg allyl sulfides per kg body weight, representing 1.5 g garlic for a person) resulted in a 14% decrease in liver lipid peroxidation. Furthermore, orally administered steam-distilled garlic oil has been found to inhibit ethanol-induced mitochondrial lipid peroxidation, which suggests the possible usefulness of garlic for treatment of alcoholism (Umalakshimi & Devaki, 1992; Devaki et al., 1992; Venmadhi & Devaki, 1992).

Only two studies have looked at the possible antioxidant effects of garlic in humans. In a double-blind, placebo-controlled, crossover trial, 10 people consumed garlic powder tablets (Kwai) containing 600 mg/day (corresponding to 3.6 mg allicin) (Phelps & Harris, 1993). After 2 weeks, a serum lipoprotein fraction (mainly LDL) was found to be 34% less susceptible to copper-induced oxidation than that of the placebo group. However, another study with 900 mg/day of the same brand found no effect on LDL oxidation in 28 hypercholesterolemic subjects placed on a fat-restricted diet (less than 30% of energy as fat, less than 300 mg dietary cholesterol per day) for 12 weeks (Simons et al., 1995).

Concerning enzymes involved in preventing oxidation, it has been found that when rats are given a cholesterol-rich diet together with garlic powder for 3 weeks, the activities of glutathione-disulfide reductase and glutathione-peroxidase increase by 87% and 100%, respectively, in comparison with the garlic-free controls (Betz et al., 1992; Heinle & Betz, 1994; Siegers, 1993). Sklan and colleagues (1992) studied the effect in chickens (2% of diet as garlic or onion powder) and found somewhat conflicting results: although there was a 40% increase in liver glutathione and an increase in Mn-dependent superoxide dismutase (SOD) activity, there were decreases in Cu/Zn-SOD and glutathione peroxidase. In a study comparing alliin and an ethanol fraction of garlic to the saponins of ginseng, it was found that the garlic components inhibited lipid peroxidation in vitro, increased liver and brain SOD activity in vivo, and sometimes gave better effects than the ginseng saponins (Choi & Byun, 1986). Thus garlic can activate the endogenous protective mechanisms against free radical formation. Further in vivo evidence of this comes from a recent study in which cholesterol-fed rats were also fed a very large dose of alliin (200 mg/kg body weight) for 2 months, resulting in significantly decreased tissue peroxides and increased tissue glutathione, SOD activity, and catalase activity (Sheela & Augusti, 1995).

5.4.3 Antioxidant Effects: Active Compounds

The compounds responsible for the antioxidant activity of garlic or garlic products have not been thoroughly proved; however, on the basis of studies that have been conducted, allicin and allicin-derived allyl sulfides appear to be responsible for most of the in vivo effects observed at low doses (0.1–0.5 mg aged garlic extract per kg body weight) (Kagawa et al., 1986). In vitro studies with liver microsomes or endothelial cells also indicate that allicin-derived allyl sulfides are active at low concentrations (0.01–0.025 mg/mL) (Hikino et al., 1986; Horie et al., 1992), while S-allylcysteine and alliin are active only at higher concentrations (1–4 mg/mL) (Yamasaki et al., 1994; Imai et al., 1994; Kourounakis & Rekka, 1991). The activities found for the allyl sulfides are important in evaluating the antioxidant potential of allicin, since both allicin and allyl sulfides are metabolized in blood and liver to the same metabolite, allyl mercaptan, which is a strong antioxidant (Egen-Schwind et al., 1992b; Lawson & Wang, 1993). Allyl mercaptan generation by the blood cells appears to account for the in vivo antioxidant activity observed for allicin, a compound that is normally considered a prooxidant. However, at much higher concentrations than could be achieved from eating fresh garlic, the ability of blood to reduce allicin to ally mercaptan becomes saturated and allicin displays its prooxidant effects. For example, it has been shown that when very high concentrations of allicin are added directly to blood (53 mg/mL, equivalent to a person consuming 70 kg of garlic in a single meal), the iron in hemoglobin becomes oxidized, resulting in dark-colored methemoglobin, which does not bind oxygen (Freeman & Kodera, 1995).

5.5 IMMUNOMODULATORY EFFECTS

There is a growing body of evidence that garlic may have significant enhancing effects on the immune system. While most of the work has been conducted on animals or in vitro, the human studies that have been conducted are encouraging.

Preliminary studies in humans, using an alliin/allicin standardized garlic powder preparation, have demonstrated positive effects on immunoreactions and phagocytosis (Brosche & Platt, 1993, 1994). In geriatric subjects, the administration of 600 mg garlic powder per day for 3 months induced significant ($P < 0.01$) increases in the percentage of phagocytosing peripheral granulocytes and monocytes when tested ex vivo for their ability to engulf *Escherichia coli* bacteria. The cell counts of lymphocyte cell subpopulations were also increased. Another human study was conducted with an undefined garlic extract (5–10 g/day) which was given to AIDS patients. For the seven patients who completed the 12-week study, there was a major increase in the percent natural killer cell activity from a seriously low mean value of $5 \pm 4\%$ to a more normal mean value of $36 \pm 15\%$ (Abdullah et al., 1989). The natural killer cell activity of blood was approximately doubled in volunteers who consumed either raw garlic or aged garlic extract (allicin absent), indicating that allicin is not needed for the effect (Kandil et al., 1987, 1988).

Animal studies with garlic or garlic extracts have shown increased activity of B and T lymphocytes (Brosche & Platt, 1993), decreased antibody titers (Gupta & Godhwani, 1984), increased lymphocyte phagocytotic activity (Lau, 1989), and increased hemolytic plaque-forming cells (Yokoyama et al., 1986).

Arzamastsev (1993) investigated the effect of standardized garlic powder at doses of 0.1–1000 mg/kg body weight on the primary immune response and the number of antibody-forming cells in mice. Doses of 10 to 1000 mg/kg garlic induced a significant twofold increase in the titers of hemagglutinins and hemolysins in the blood as well as of the antibody-forming cells in the spleen ($P < 0.05$). A maximum intensification of the immune response and an increase of the number of antibody-forming cells was seen at 10 and 100 mg/kg. The value of 10 mg/kg corresponds to the usual therapeutic dose in humans.

An antiarthritic effect of commercial steam-distilled garlic oil (fed at 2.5 mg/kg body weight) was found for formaldehyde-induced arthritis in rats, with a 26% decrease in arthritis. Interestingly, the effect of garlic was synergistically increased by the addition of boron to the diet (Shah & Vohora, 1990).

A comparative in vitro study of various garlic fractions on the immune function of human peripheral blood mononuclear cells has recently shown that the thiosulfinate fraction (allicin, etc.) increases natural killer cell activity, while the whole homogenate and the polar fraction (only the thiosulfinates removed) increased interleukin-1 production. All fractions increased interleukin-2 production (Burger et al., 1993). The effects on interleukins are important, since they stimulate several immune responses. Another recent in vitro study has shown that diallyl trisulfide at a modest concentration (3 µg/mL) can significantly activate mouse T cells and that the activation is due to a decreased production of nitric oxide, which is inhibitory to T cells, by macrophages. At the same time, the ability of the macrophages to kill three tumor cell lines was significantly improved (Feng et al., 1994).

A protein fraction (F4) isolated from an aged garlic product has been shown to have interesting immune effects. It has been found in vitro to increase the activity of mouse peritoneal macrophages, to increase T lymphocyte blastogenesis, and to increase the cytotoxicity and proliferation of human peripheral blood lymphocytes (Hirao et al., 1987; Lau et al., 1991b; Morioka et al., 1993). The amount of this protein fraction in fresh garlic or in aged garlic has not yet been reported. Whether or not the activity from this protein could survive the digestive processes also remains to be determined.

The identity of the active compounds for the effects thus far observed on the immune system with garlic and garlic products is far from conclusive. Since both allicin-derived garlic oils as well as garlic extracts not containing allicin are effective in vivo at moderate doses, it appears that both allicin and other unidentified compounds are responsible for the effects. Both types of compounds may be important to the overall effects of garlic, since the immune system involves several types of cell, each of which may be affected differently, as has been indicated in the in vitro studies.

5.6 ANTIINFLAMMATORY EFFECTS

Prasad and colleagues (1966) reported the antiinflammatory effects of a garlic extract (allisatin). The preparation was given in dosages of 200 g/kg/day to rats which had been made arthritic by formaldehyde injections. Betamethasone (0.5 mg/kg/day) was used as the positive control. The garlic preparation, as well as the glucocorticoid, reduced swellings and necroses in the joints, but the volume of the exudate remained unchanged. Allisatin also had no effect on the adrenal glands. While the steroid increased the frequency of stomach ulcers, the garlic preparation did not show any ulcerogenic effect.

The effects of garlic extracts and pure garlic constituents on the metabolism of arachidonic acid have been studied with respect to platelet aggregation. The leukotrienes (LT), which are formed from arachidonic acid through the action of the enzyme 5-lipoxygenase, play an essential role as mediators of the inflammatory reaction. LTB_4 (leukotaxin) is a mediator of leukotaxis, while the slow reacting substances (LTC_4, LTD_4, and LTE_4) affect vascular contraction and permeability. It was shown that chloroform and acetone/chloroform extracts of garlic, containing allicin as the major active ingredient, strongly inhibited 5-lipoxygenase in vitro (Table 5.9) (Sendl et al., 1992; Block et al., 1988).

In another experiment, garlic extract, as well as allicin and diallyl disulfide, exhibited a necrosis-inhibiting effect. Injuries at the mucous membrane of the stomach, induced by absolute alcohol in rats, were alleviated by these substances, depending upon the dose administered. The minimal effective dose was 10% for the extract and 10 mM for the pure substances. Indomethacin reversed the favorable effect (Chung, 1985).

Some authors have also detected a prostaglandin-like effect of garlic. A corresponding effect was, e.g., experimentally demonstrated on the isolated stomach fundus and colon from rats. The inhibition of thrombocyte aggregation also indicates the

Table 5.9. Inhibition of 5-Lipoxygenase by Solvent Extracts of Garlic

Concentration µg/mL	Acetone/Chloroform-Extract % Inhibition	Chloroform-Extract % Inhibition
9.6	—	100.0 ± 0
4.8	—	88.1 ± 8.3
2.4	—	40.8 ± 13.2
1.9	100.0 ± 0	—
0.96	96.7 ± 4.1	—
0.57	70.5 ± 4.1	—
0.45	20.2 ± 5.8	—

From Sendl A, Elbl G, Steinke B, Redl K, Breu W, Wagner H. Comparative pharmacological investigations of Allium ursinum and Allium sativum. Planta Med 1992;58:1–7.

involvement of prostaglandins (Makheja et al., 1980, 1981; Ariga et al., 1981; Ariga & Oshiba, 1981b; Rashid et al., 1986). Recently, a concentrated form of an aged garlic extract (10 mL/day) was given to eight women for 90 days in order to determine its antiinflammatory activity as indicated by inhibition of prostaglandin synthesis; no activity was found, however, as the amount of prostaglandin metabolites in the urine remained unchanged throughout the 3 months (Chiou et al., 1995).

5.7 HYPOGLYCEMIC EFFECTS

Elevated blood glucose concentrations can be lowered by the consumption of onions and garlic (Collip, 1923a, 1923b, 1923c; Brahmachari & Augusti, 1962a, 1962b, 1962c, 1962d; Mathew & Augusti, 1973b; Jain & Sachdev, 1971; Sharma et al., 1977), as has long been known in folk medicine, for instance, in Bohemia and Russia (Kit & Orlik, 1960). This prompted Mahler and Pasterny (1924) to conduct a clinical experiment with two diabetics, and the result was surprisingly positive. The patients received 10–15 g fresh garlic, sliced and distributed into three portions over the day. Urine sugar excretion was reduced from 3.2% to 1% in one case, and 3.8% to zero in the other case. Blood sugar decreased correspondingly.

Janot and Laurin (1930) confirmed this blood sugar–lowering effect in rabbits. They prepared an alcoholic extract of 30 or 40 g garlic and injected it subcutaneously. The blood sugar level clearly fell within 1 to 2 days after a slight temporary increase. Using a sophisticated preparative method, Laland and Havrevold (1933) succeeded in isolating a blood sugar–lowering substance from garlic, which they erroneously assumed to be an alkaloid. Nevertheless, the hypoglycemic effectiveness of the compound was unequivocally established in rabbits and in pancreatectomized dogs.

The hypoglycemic principle is a lipophilic compound, being soluble in ether. An ether extract of garlic bulbs showed significant blood sugar–lowering activity in normal and in alloxan-diabetic rats (Jain & Vyas, 1975b). Garlic oil was also found to increase insulin and urea levels and to change the activities of key liver enzymes in ethanol-fed rats (Devaki et al., 1992; Venmadhi & Devaki, 1992). On the other hand, other authors have found diallyl disulfide, a major component of steam-distilled garlic oil, to be hyperglycemic when given in large doses (100 mg/kg) (Pushpendran et al., 1982).

In experiments comparing 0.1 g/kg tolbutamide with 0.1 mg/kg allicin, tolbutamide reduced blood glucose levels by 24.1% and allicin by 15.4%. Simultaneously, insulin activity was increased 18% by allicin and 26% by tolbutamide. The effectiveness was substantial (Brahmachari & Augusti, 1962c; Mathew & Augusti, 1973c). Allicin, isolated from garlic, also lowered blood glucose levels and increased insulin activity in alloxan-diabetic rats (Augusti, 1975). Doses of 0.05, 0.10, and 0.25 g/kg, after 2 hours, resulted in a decrease of blood glucose levels of 16.7, 20.0, and 24.3%, respectively.

Chang and Johnson (1980) studied the incorporation of ^{14}C-acetate and ^{14}C-glucose into lipids and glycogen, as well as the glucose and insulin concentration in the serum of rats, after feeding the animals with a standard diet with or without the addition of garlic (5 g/100 g feed). The results showed that the glucose level in blood was lowered by garlic, while the concentrations of insulin in the blood and the glycogen level in the liver increased. It seems reasonable to suppose that the hypoglycemic effect is due to an enhanced release of insulin, the blood sugar–lowering hormone.

Jain and coworkers investigated the hypoglycemic effect of freshly pressed garlic juice (25 mL/kg body weight) in rabbits kept on a diet supplemented with 1 g glucose. The increase of blood glucose levels was markedly less than in the controls that were fed a garlic-free diet (increase from 110 mg/dL to 164 mg/dL and from 102 mg/dL to 180 mg/dL, respectively) (Jain et al., 1973; Jain & Vyas, 1975b; Jain & Konar, 1977). In alloxan-diabetic rats a similar effect was observed with onions (Jain & Vyas, 1975a; Augusti, 1974b; Augusti et al., 1974).

Moreover, the authors found that the juice of 25 g garlic has about the same effect as the standard dosage of 0.25 mg/kg tolbutamide. A comparable effect is brought about by onions (Jain & Vyas, 1974). The authors (Jain & Konar, 1977) then prepared extracts of carefully dehydrated garlic with various solvents—95% alcohol (A), petroleum ether (P), diethyl ether (E). They tested the extracts for their blood sugar–lowering activity in alloxan-diabetic rabbits. Again, the group receiving a standard dosage of tolbutamide (T) and an untreated control group (C) were used for reference. The results are summarized graphically in Figure 5.2. It can be seen that garlic has a highly significant ($P < 0.001$) blood sugar–lowering capacity that is 60–80% as effective as tolbutamide and that the effective principle is obviously lipophilic in nature.

A comparative study in rabbits using an aqueous extract of fresh garlic (equal to 3 g garlic/kg/day) also resulted in a strong blood sugar and blood lipid–lowering effect (Zacharias et al., 1980). Similar results were also reported by other authors, who used rats that had been made diabetic with adrenalin or streptozocin (Osman, 1984; Farva et al., 1986; Adoga & Ohaeri, 1991), while others could not confirm this effect (Mossa, 1985).

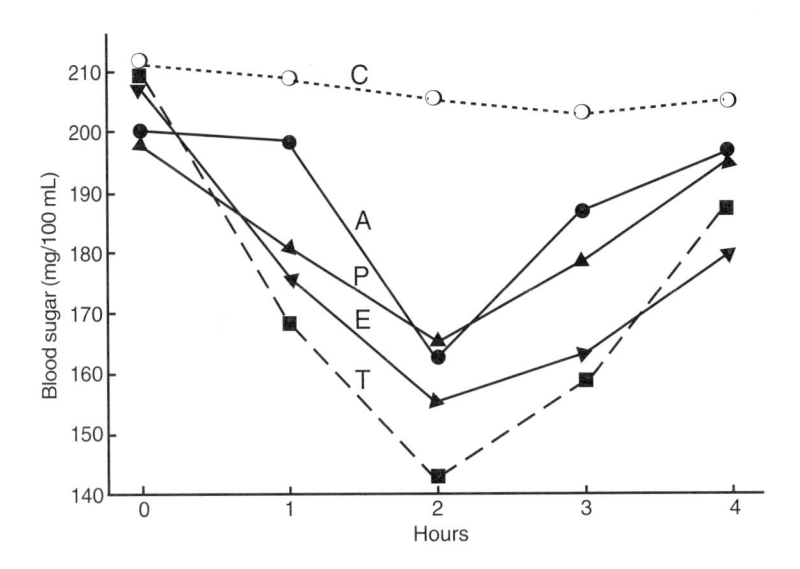

Figure 5.2. Blood sugar–lowering effects of garlic preparations extracted with (A) ethanol, (P) petroleum ether, (E) diethylether, in addition to (T) a standard dose of tolbutamide and (C) a control group. (From Jain RC, Vyas CR. Effect of onion (Allium cepa) and garlic (Allium sativum) on experimentally induced hypercholesterolemia in rabbits. Artery 1975;1:363.)

Augusti and Mathew (1975) examined the influence of allicin (100 mg/kg/day) on the activity of various rat liver enzymes. After 15 days of treatment, the activities of α-glucan-phosphorylase and lipase increased, while the activity of glucose-6-phosphatase decreased, and the activity of hexokinase remained unchanged. The blood sugar level was not altered in this experiment. This effect is obviously dependent on the insulin status. Simultaneously, the insulin level rises, but the expected increase in fatty acids fails to appear (Augusti & Benaim, 1975). The concentration of 2,3-diphosphoglycerate also drops after the administration of allicin (Banerjee et al., 1982). Allicin also produced a dose-dependent hypoglycemia in alloxan-diabetic rabbits (Augusti, 1975).

Steam-distilled onion oil (erroneously described as containing allyl propyl disulfide) has also been found to lower blood sugar levels (Wilcox et al., 1984; Augusti, 1974b). Diphenylamine, which was demonstrated to be a blood sugar–lowering principle in the onion (see Section 7.1), has not yet been found in garlic (Opdyke, 1978; Karawya et al., 1984; Smoczkiewiczowa et al., 1990).

Alliin (S-allylcysteine sulfoxide) has also been found to show significant antidiabetic effects in alloxan-diabetic rats. Administration of 200 mg/kg body weight decreased significantly the concentration of serum lipids, blood glucose, and the activities of serum enzymes, such as alkaline phosphatase, acid phosphatase, lactate dehydrogenase, and liver glucose-6-phosphatase, whereas it increased significantly liver and intestinal HMG-CoA reductase activity and liver hexokinase activity (Sheela & Augusti, 1992).

Morphological and biochemical studies of the pancreas of guinea pigs with alloxan– and streptozocin–induced diabetes revealed a significant regeneration of the islets of Langerhans after application of 128 mg/kg garlic oil, which exceeded the effect obtained by application of 250 mg/kg tolbutamide (Begum & Bari, 1985; Brahmachari & Augusti, 1962b).

To explain the mechanism of the blood glucose–lowering capacity of garlic extracts, it was proposed that allicin, which rapidly reacts with sulfhydryl groups, might protect insulin against its inactivation by endogenous cysteine, glutathione, and albumin fractions that are rich in sulfhydryl groups (Mathew & Augusti, 1973a).

The results of Mathew and Augusti (1973a) and of Chang and Johnson (1980) indicate that garlic extract probably has a direct effect on insulin production. Compared to the relatively large number of animal experiments on the blood sugar–lowering effect of garlic or garlic constituents, only a few human studies with normal subjects have been reported (Augusti & Benaim, 1975; Kiesewetter & Jung, 1992; Kiesewetter et al., 1991b; Sitprija et al., 1987). Augusti and Benaim (1975) tested the effect of a single dose of onion oil, 0.125 g/kg (n = 6), against placebo (n = 6) on blood glucose levels and found a significant decrease in the verum group from 75 to 59 mg/dL ($P < 0.05$), compared to a change from 72 to 66.5 mg/dL in the placebo group (Augusti & Benaim, 1975). Sitprija and colleagues (1987) in Thailand gave diabetic patients a spray-dried garlic powder (contains no alliin) at 700 mg daily for 1 month and found no significant changes in blood glucose or serum insulin levels. The failure of this study may have been due to the absence of alliin or to low dosage.

In an important placebo-controlled clinical study, test subjects received tablets containing 800 mg/day of alliin/allicin standardized garlic powder (n = 10) or placebo tablets (n = 10) for 4 weeks. The blood sugar levels during the treatment period decreased significantly from 90.6 to 77.8 mg/dL ($P < 0.05$) in the garlic group, but no significant change (from 88.6 to 85.8 mg/dL) occurred in the control group (Kiesewetter & Jung, 1992; Kiesewetter et al., 1991b).

Although garlic cannot be considered to be a real antidiabetic, with consistent intake, it can help patients with slight diabetes by lowering the blood sugar concentration and thus curb insulin requirements.

Finally, in a related topic but an unrelated effect, garlic extract has been patented as a substitute for peroxidase in an assay for measuring urine glucose levels (Kitahara, 1974).

5.8 HORMONE-LIKE EFFECTS

Hormone-like effects of plants on animals and humans are not as unusual as it might at first appear. There are many interrelations between plant and animal organisms. Although some have been known for a long time (e.g., plant estrogens), even since antiquity, most are yet to be discovered (Koch, 1992a; Keller, 1988; Vorwahl, 1923). In the literature, one frequently finds implications that garlic also contains hormone-like active components (Glaser & Drobnik, 1939; Glaser, 1937; Berger, 1960; Hager, 1969; Ikram, 1972; Weiss, 1974; Koch, 1992a). Here, it is stated that garlic stimulates the male as well as the female sex hormones. Furthermore, aphrodisiac properties have been attributed to garlic since ancient times.

According to an investigation by Glaser and Drobnik (1937, 1939), aqueous garlic extracts were injected into castrated male fish (*Rhodeus amarus*), whereupon the fish temporarily developed their "mating attire." In addition, using the Allen-Doisy test, the authors tested the extract in mice for female hormone effect. With an extract from 2 g garlic, they obtained the effect of 10–30 "fish units" of the male hormone. The extract from 3 g garlic yielded 13 "mouse units" of female sex hormones. The authors mention that this effect occurs at "physiologically equipotential" concentrations. As no studies were performed to elucidate the chemical nature of the constituents that caused these hormone-like effects, it would make little sense to cite any of the speculations made in this paper. Nevertheless, the need for further investigation of this interesting phenomenon remains undisputed.

Velazquez and Rodriguez (1955a, 1955b) have readdressed the issue of hormone-like effects of garlic. They administered an alcoholic garlic extract to ovariectomized rats, whereupon the animals assumed an almost normal sex cycle. Thereafter, the animals were always 1 day in estrous cycle and 5 days in the anestrous period (normally the period is 2 and 3 days, respectively). With rabbits, after administration of a garlic extract (20 mg/day for 4 days) obtained with ethanol and ethyl acetate, the authors observed a fourfold increase of 17-ketosteroids over the normal values in urine. From these findings, it can be concluded that an effect occurs which is similar to the action of adrenocorticotropic hormone (ACTH). An increased amount of steroid hormones is released, and the degradation products subsequently appear in the urine. They also found that a dosage of 2 mg garlic extract has the same effect on the uterus of guinea pigs as 0.001 IU Pituitrin (posterior pituitary hormone) (Velazquez et al., 1958).

Tinao and Terren (1955) found that garlic extract enhances the rhythmic movements of the uterus, both in the frequency and the amplitude of contractions. In addition, garlic promotes the stimulating effect of estradiol and antagonizes the depressing effect of high dosages of steroid hormones, so that an increased movement of the organ occurs. The paralyzing effect of sodium glycerophosphate is annulled by garlic (see also Sokolov, 1954).

In this context, it must also be mentioned that garlic is assumed to have a pronounced antifertility effect. Qian and colleagues (1986a) reported an in vitro spermicidal effect of a substance which they denominate as "Allitridum" (mostly diallyl trisulfide), the putative active principle of steam-distilled garlic oil. Dixit and Joshi (1982) found that chronic administration of garlic powder to rats results in inhibition of spermatogenesis, while Al-Bekairi and Shah (1990) observed androgenic, estrogenic, and antiestrogenic properties in a water extract from garlic bulbs in male and female mice. Antigonadenosine tropic effects, however, had already been reported over 40 years ago for the closely related onion plant (Klosa, 1951), as have been garlic's estrogenic properties (Glaser & Drobnik, 1939).

An extensive neurological study by Japanese authors (Hinoki et al., 1981) indicates an influence upon central nervous system functions by the volatile, odor-intensive components of garlic through a stimulation of the olfactory receptors. Certain subjects, who upon such olfactory stimulations reacted with dizziness, visual disturbance, uncoordinated motor skills, and the like, displayed the symptoms very markedly when "garlic fumes" (1 mL allinamine solution, equal to 5 mg diallyl sulfide, on a cotton ball, held to the nose for 60 seconds) were administered to them. It was shown that the stimulation is transmitted from the olfactory receptors over centripetal pathways to the limbic system and hypothalamus. The disturbances occur through viscerosomatic reflexes. "Normal"-reacting subjects do not show this reaction to garlic. It is conceivable that corresponding influences on other, neurally and/or hormonally directed systems, possibly including the sexual organs, are brought about by garlic in a similar reflective way. Thus, an aphrodisiac effect of garlic and other *Allium* species (Al-Bekairi et al., 1991) cannot be disregarded and should be clarified by modern neurophysiological testing methods.

Petkov (1986) reported experiments that were performed in rabbits in the 1940s, the results of which indicate an influence on the thyroid gland by certain garlic constituents (see also Virtanen, 1969). A series of alkyl and alkenyl disulfides have indicated an antithyroid effect in rats (Cowan et al., 1967; Salji et al., 1971; Saghir et al., 1966, 1967, 1968; Abdo & Al-Kafawi, 1969). A similar effect on the thyroid in humans has not yet been observed. In other sources, the opinion was expressed that garlic stimulates the pituitary gland, which influences other glands and thus controls fat and carbohydrate metabolism (Anonym, 1983; Anonym, 1984; Greenstock, 1982).

Conditioned aversion in animals to the alliaceous odor was effected by injecting an oily solution of diallyl disulfide into rats; the animals then avoided any food and water to which garlic had been added (Maruniak et al., 1983). The odorous constituents of garlic can cross the placental barrier and thus appear in the fetal environment (Nolte et al., 1992). Furthermore, juvenoid activity (juvenile hormone, the sloughing hormone of insects) has been reported for certain garlic components (Srivastava et al., 1985).

More recently, Winterhoff and Egen-Schwind (1987, 1991) investigated the effect of fresh garlic (up to 1.5 g/kg), dry garlic powder (up to 1 g/kg), vinyldithiins (2.2 mg/kg), and allicin (6 mg/kg) on thyroxine synthesis and thyroid glands in rats. The decrease of thyroidal secretion and of thyroid stimulating hormone (TSH) levels, which was seen after a single dose of fresh garlic or dry garlic powder, also occurred after oral application of vinyldithiins. Pretreatment with fresh garlic and dry garlic powder for 4 days resulted in similar, though less pronounced changes. In contrast to these effects, after treatment for 11 to 14 days, an increased activity of the thyroid gland was observed. Simultaneously, the thyroxine levels increased significantly. It is suggested that the garlic constituents directly affect the pituitary gland. Other authors also found significant inhibition of ^{131}I uptake in the rat thyroid gland with garlic and onion extracts and a number of sulfides (Artacho et al., 1992).

Finally, for the sake of completeness in a discussion of hormone-like effects, it must be mentioned that garlic and other *Allium* plants also have phytohormone-like activities. Among the various compounds having such effects are ethylene (Karagiannis & Pappelis, 1993, 1994), gibberellins (Rao et al., 1981; Arguello et al., 1983, 1986; Park, 1987, 1988), abscisic acid (Chang, 1988; Wang et al., 1988), and others, all of which markedly influence the plant's development. However, since they do not have significant influence on human or animal organisms, there is no need to discuss them here.

5.9 ENHANCEMENT OF THIAMINE ABSORPTION

The effectiveness of garlic for scurvy, which is related to its content of vitamin C, was confirmed in 1924 by the Medical Academy in Paris. In the 19th century, Japanese researchers had already established the effectiveness of garlic for beriberi, which occurs because of vitamin B_1 deficiency. They observed an interesting characteristic of garlic in regard to the absorption of vitamin B_1: allicin forms an adduct with thiamine (allithiamine), which improves the solubility and absorption of thiamine (see Section 3.2.8). Allithiamine has the same physiological effects as vitamin B_1, since the adduct is rapidly reduced by the blood.

Fujiwara and coworkers in Japan showed that, after administration of allithiamine, the absorption of the vitamin from the alimentary tract (Fig. 5.3), as well as its elimination in the urine (Fig. 5.4), is considerably higher, even when equivalent amounts (on a molar basis) of thiamine were given. More vitamin B_1 was also detected in the central nervous system after allithiamine intake than after thiamine intake. The physiological absorption of the natural vitamin is limited, with a threshold of 5–10 mg, beyond which no more is absorbed. However, appreciably more can be supplied to the body by consuming allithiamine than the vitamin itself (Fujiwara & Watanabe, 1952; Ikeda, 1969; Fujiwara, 1976).

Satoh, in a series of publications (Satoh, 1952a, 1952b, 1952c, 1952d, 1952e, 1952f, 1952g, 1952h), demonstrated that garlic, fed to animals as well as to humans, stimulates thiamine synthesis by intestinal bacteria. A combination of vitamin B_1 and alkyl disulfides together with garlic extract is also used as an addition to feed in fish farming (Okada & Fujiwara, 1987). Animal experiments showed that garlic-fed mice experienced weight gain, greater agility, and better defense against infections than the controls without garlic in the diet (Makowsky, 1936).

Regular consumption of garlic plays an important role in the absorption of vitamin B_1, which is necessary for the function of the nervous system and the coronary vessels, a process that is often disturbed in elderly persons. Could the striking longevity of garlic eaters in some areas of the world be related to this fact?

5.10 EFFECTS ON ORGANIC AND METABOLIC DISTURBANCES

Garlic has been a popular folk remedy for many years. People in about their mid-40s, with incipient aging problems, especially like to use it (Weiss, 1982; Ressin, 1985; Kourennoff, 1970). Therefore, it is mainly consumed for self-medication. When mental and physical capacities start to decline, gar-

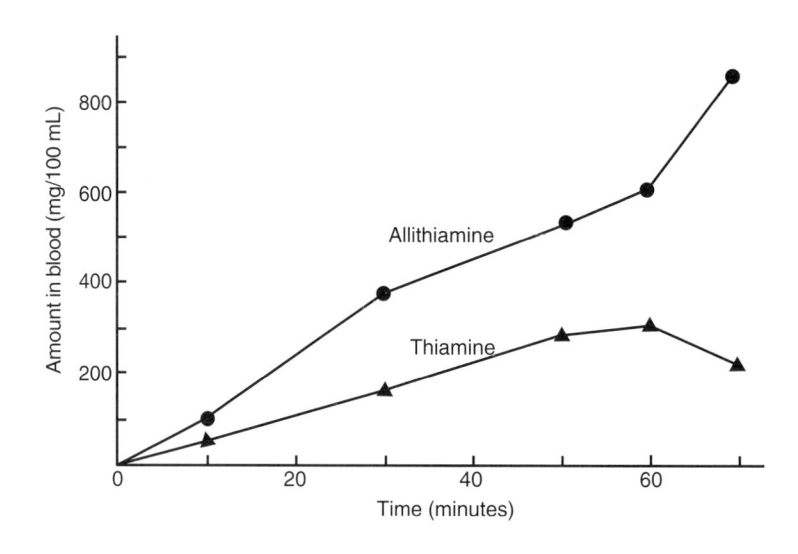

Figure 5.3. Thiamine levels in the blood of rabbits after intrajejunal application of 1 mg thiamine and allithiamine (1 mg thiamine + 2 mL garlic extract). (From Fujiwara, M. Allithiamine and its properties. J Nutr Sci Vitaminol 1976;22:57–62.)

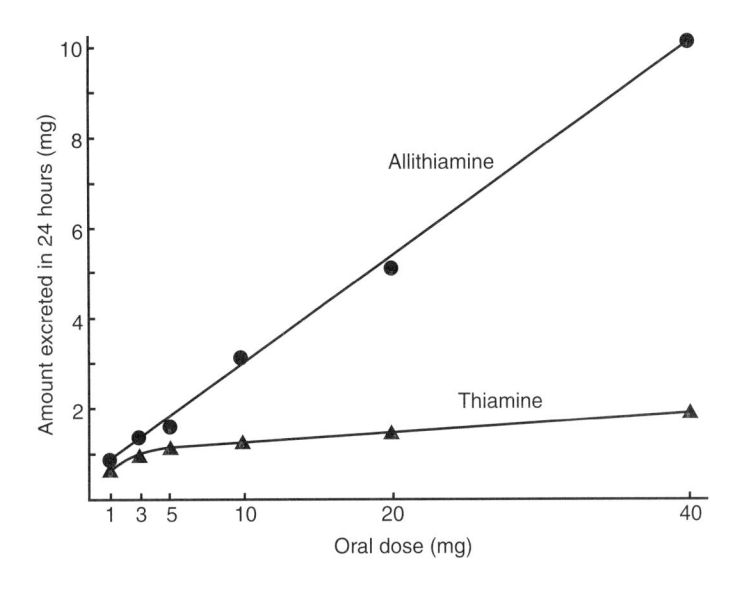

Figure 5.4. Excretion of thiamine into the urine after administration of increasing doses of thiamine and allithiamine. (From Fujiwara M, Watanabe, H. Allithiamine, a newly found compound of vitamin B1. Proc Imp Acad (Tokyo) 1952;28:156–158.)

lic is well suited as a tonic for weakness and for strengthening the defense mechanisms. Its exceptional suitability for the prevention of gastrointestinal infections and for the stimulation of the secretion of gastric juice, alimentary enzymes, and bile is very helpful. Garlic is frequently given for symptoms indicating incipient arteriosclerosis, such as peripheral circulation disorders, high blood pressure, hyperlipidemia, hypercholesterolemia, disorders of blood coagulation time and thrombocyte aggregation, light forms of diabetes, and general decline of memory capacity. Prophylaxis of arteriosclerosis and infarction, including normalization of values of the above-mentioned parameters, have been treated in detail in the previous sections. Effects on endocrine functions and vitamin balance have also been mentioned. The present section will discuss the further influence of garlic on organic and metabolic disorders, so far as this information is needed for the understanding of the effectiveness of this natural medicine. Again, experimental data with animals are of importance, since they reveal the manner and mechanism of action of garlic's effects.

5.10.1 Effects on Enzyme Activities

An aqueous extract of garlic cloves in vitro strongly inhibits the clinically important serum and liver enzymes, glutamic-oxaloacetic transaminase (SGOT), glutamic-pyruvic transaminase (SGPT), lactic dehydrogenase (LDH), and cholinesterase (CE). Inhibition by up to 90% was observed, and the inhibiting factor proved to be heat-stable and not dialyzable. On the other hand, the extract increased adenosine triphosphatase (ATPase) activity in intact liver mitochondria, but not after destruction of the organelles. From these results it is inferred that garlic contains several unknown factors which block enzymes and modify the functions of biological membranes (Bogin & Abrams, 1976). Other enzymes that are influenced by garlic oil and garlic extract are serum aspartate aminotransferase and serum alkaline phosphatase; while the activity of the first enzyme was increased, that of the latter decreased (Joseph & Sundaresh, 1989).

Wills (1956) tested the effect of pure, synthetic allicin on a total of 28 different enzymes (Table 5.10). He demonstrated that strong inhibition occurs especially in

Table 5.10. Effect of Allicin (0.5 mM) on Various Enzymes In Vitro

Inhibited by Allicin	Not Inhibited by Allicin
Succinate dehydrogenase	Cytochrome oxidase
Urease	Lipase
Papain	Rennin
Xanthine oxidase (milk)	Pepsin
Xanthine oxidase (liver)	Trypsin
Choline oxidase	Invertase
Hexokinase	α-Amylase
Cholinesterase	Esterase (serum)
Glyoxalase	D-Amino acid oxidase
Triosephosphate dehydrogenase	Ascorbic acid oxidase
Alcohol dehydrogenase	Catalase
Lactate dehydrogenase	Carboxylic acid anyhdrase
Tyrosinase	Carboxylase
Alkaline phosphatase	Adenosine triphosphatase
	β-Amylase

From Wills Ed. Enzyme inhibition by allicin, the active principle of garlic. Biochem J 1956;63:514–520.

enzymes which contain an SH-group in their active centers (all enzymes in the left column of Table 5.10). Particularly strong was the inhibition of succinate dehydrogenase, triosephosphate dehydrogenase, and xanthine oxidase. Unlike allicin, the compounds diallyl sulfide, diallyl disulfide, and diallyl sulfoxide have no inhibitory effect on the enzymes, and alliin, at a concentration of up to 2 mM, was practically inactive.

Garlic oil influences oxidative phosphorylation in liver mitochondria of mice. The blocking occurs at coupling centers I, II, and III. The extent of inhibition of respiration depends on the respective substrate. Inhibition of oxidation of nicotinamide adenine dinucleotide (NAD)–dependent substrates, such as glutamate, is stronger than that of succinate or ascorbate. Diallyl disulfide has an effect similar to garlic oil, but the effect of dipropyl disulfide is much smaller (George & Eapen, 1974).

Similar findings were found by Bogin (1973), who showed that in erythrocytes and mitochondria, which served as models, the biological membranes are somehow modified in their integrity by the active principles of garlic. Garlic extract acted as an uncoupler of oxidative phosphorylation, inhibited the oxidation of succinate by 50%, and increased the activity of mitochondrial adenosine triphosphatase (ATPase) by 100%. Consequently, the membranes are clearly more permeable to certain substrates (e.g., ATP) and more sensitive to hypotonic solutions.

The activity of microsomal NADPH-cytochrome-P-450–reductase and of NADH-cytochrome-b$_5$-reductase is inhibited as soon as an aqueous extract of garlic (contains no alliin) is added to buffer-suspended microsomes. Incubation of garlic extract with isolated pig liver microsomes lowers the activity of cytochrome P-450–dependent ethoxycoumarin deethylation. These effects, however, are not related to lipid peroxidation. From the fact that equal results are obtained with pure alliin, it is suggested that alliin might also be an active principle for the inhibitory action observed in vitro (Oelkers et al., 1992).

A change in the activity of certain liver enzymes was also observed in vivo after administration of allicin (100 mg/kg/day) to rats for 15 days. After this treatment, the activity of phosphorylase and lipase increased significantly, while the activity of glucose-6-phosphatase decreased. The activity of hexokinase and the blood sugar level remained unchanged (Augusti & Mathew, 1975). According to another source, glucose-6-phosphatase activity, as well as microsomal oxygenase and adenosine triphosphatase activities, are increased by diallyl disulfide (Devasagayam et al., 1982). The sulfane structure in the oligosulfides from garlic are believed to be responsible for the regulatory role of the various biochemical systems, especially enzymes (Toohey, 1989).

An extremely high activity of alkaline phosphatase and alcohol dehydrogenase was found in serum, liver, and kidneys of rats which had been on a prolonged diet of high sugar and alcohol. Simultaneous doses of garlic oil significantly diminished the activity of both enzymes, which indicates a reduction of the synthesis of fatty acids and cholesterol and/or NADPH. In rats fed with lyophilized garlic, the cholesterol, triglyceride, and LDL levels decreased, while biliary elimination of neutral steroids increased. The de novo synthesis of steroids in the liver was also decreased, and this effect correlated well with an inhibition of cholesterol-7α-hydroxylase (Adoga & Osuji, 1986).

Concentration-dependent inhibition of β-hydroxy-β-methylglutaryl-CoA (HMG-CoA) reductase was found on preincubation of rat liver microsomal preparations with diallyl disulfide. Formation of protein internal disulfides, inaccessible for reduction by thiol agents, but not of protein dimer, is likely to be the cause of this inactivation (Kumar et al., 1991).

Russian authors reported an influence of garlic substances on invertase and alkaline phosphatase (Morozov, 1950; Danilenko & Epshtein, 1953). No influence, however, was observed on amylase and pepsin, but the activity of peroxidase in blood is reduced (Vinokurov et al., 1947), and even the metal concentration of the enzymes is changed (Saratikov & Khomjakov, 1952).

Vainberg (1952) detected an inhibition of pyruvate dehydrogenase in nerve tissue of guinea pigs. Huh and colleagues (1985a) examined the effect of garlic on xanthine oxidase in liver. Pure allicin and the heat-treated allicin-containing fraction from garlic cloves reduced enzyme activity in vitro; however, alcohol extracts elevated it. Extracts from leaves and roots were without effect. Agarwala and coworkers (1952) observed a strong inhibition of urease by allicin, an effect that was prevented by H_2S or glutathione.

The general reactivity of allicin and oligosulfides in garlic upon sulfhydryl groups in proteins was intensively studied by Japanese researchers. Many garlic effects can be explained by an interaction of thiosulfinates with specific proteins or enzymes, usually resulting in their final denaturation (Utsumi et al., 1962). However, garlic has also been found to be a potent activator for some enzymes, such as protease of yeast, lipase of *Bacillus subtilis*, amylase from malt, and several enzymes of *E. coli*. This discovery has even been patented (Mitsui, 1980). A liver enzyme whose activity is greatly increased in vitro by allicin is fructose 1,6-bisphosphatase, a key regulatory enzyme of gluconeogenesis, since the activity of this enzyme is stimulated by oxidation of its four sulfhydryl groups (Han et al., 1993).

Since most of the studies cited above were in vitro experiments, they should be viewed with caution with respect to the effects of allicin and aqueous garlic extracts because allicin is now known to be metabolized very rapidly in blood to allyl mercaptan. Similar blood metabolism is also found for some of the allyl sulfides abundantly present in garlic oils (Lawson & Wang, 1993; Lawson, 1993).

5.10.2 Antihepatotoxic Effects

Adequate liver protection by several defined constituents of garlic was shown with freshly prepared rat hepatocytes that had been treated with carbon tetrachloride (CCl_4), a strong liver toxicant. Six substances were tested: *S*-methylcysteine, *S*-ethylcysteine, *S*-propylcysteine, *S*-allylcysteine, *S*-allylmercaptocysteine, and alliin. The antihepatotoxic effect was measured by the SGPT values in the incubation medium, and visualized through the morphology of the cells. At the concentration of 0.5 μg/mL, *S*-allylcysteine, *S*-propylcysteine, and *S*-allylmercaptocysteine revealed complete neutralization of cytotoxicity. Alliin was also effective, although it prevented only the excretion of SGPT from the cells. *S*-methylcysteine and *S*-ethylcysteine had no effect at all (Nakagawa et al., 1985).

Two enzymes that control the detoxification of xenobiotic agents in rat liver, glutathione-*S*-transferase (EC 2.5.1.18) and glutathione peroxidase (EC 1.11.1.9), are activated by pressed garlic juice (Huh et al., 1985b).

Belman and coworkers (1986, 1987, 1989) determined the IC_{50} values, as well as the kinetic inhibition constants, for most of the active constituents of garlic and onion. Di-1-propenyl sulfide was the only irreversible inhibitor observed, with K_i = 59 μM and k_3 = 0.53/minute. The inhibition in the presence of substrate was noncompetitive at 88 and 132 μM linoleic acid, with K_i = 129 μM. At 173 μM linoleic acid, however, inhibition was competitive, with K_i = 66 μM. Diallyl trisulfide, allyl methyl trisulfide, and diallyl disulfide were competitive inhibitors, while 1-propenyl-propyl sulfide and ajoene were mixed inhibitors. Nordihydroguaiaretic acid (NDGA), the most potent lipoxygenase inhibitor, is a competitive inhibitor with K_i = 0.29 μM.

These results were also confirmed by data obtained from in vivo experiments on mice poisoned with CCl_4, wherein fatty liver and lipid peroxidation were measured. Garlic extract was given 6 hours after liver poisoning at doses of 10, 100, and 500 mg/kg, and vitamin E (25 mg/kg) was given as a positive control. The high amount of lipid peroxides and accumulation of triglycerides in the liver observed in the afflicted animals was significantly lowered in animals treated with garlic extract. Vitamin E inhibited lipid peroxidation, but not fat deposition (Kagawa et al., 1986).

According to Japanese workers, *S*-allylcysteine and *S*-allylmercaptocysteine, which are proposed to be the main active constituents of an aged garlic extract, are also effective in controlling hepatopathy and liver damage induced by hepatotoxins during acute hepatitis (Nakagawa et al., 1989; Kodera, 1991; Naito et al., 1981b; Blakely et al., 1993). Possibly, ajoene is also involved in the antihepatotoxic effect of garlic products that contain it (Kasuga et al., 1988).

Tolerance to both heat and cold could be increased in CCl_4-poisoned rats with doses of garlic oil (0.006 mL in peanut oil, prophylactically over 3 days), similar to the effect of α-tocopherol (450 mg/kg) and glucose (300 mg) (Ahujarai & Bhatia, 1984; Bhatia & Ahujarai, 1984). Additionally, *S*-allylmercaptocysteine and *S*-methylmercaptocysteine show remarkable protection in vitro to hepatocytes against the toxic effects of CCl_4 and galactosamine (Hikino et al., 1986).

Another model used for the testing of the antitoxic effects of garlic is the pathologic change in organs of rats that had been poisoned by cadmium. The animals received 100 ppm Cd with or without garlic addition (6.67%) in the feed over 12 weeks. The group fed with garlic had significantly fewer lesions in the testes, liver, and kidneys. It is assumed that there are components in garlic that chelate the metal (D. G. Lee et al., 1985; H. S. Lee et al., 1985). A similar detoxifying effect of garlic was observed in rats exposed to methyl mercuric acetate (Park & Cha, 1984).

Fields and colleagues (1992) performed a study to determine whether a reduction in hepatic lipogenesis would be beneficial in the amelioration of copper deficiency. They found that garlic oil extract did ameliorate the severity of copper deficiency, although hepatic lipogenesis was not affected.

Similarly, the activity of aniline hydroxylase, glutathione peroxidase and the hepatic xanthine oxidase is increased by constituents of garlic oil (diallyl disulfide) (Huh et al., 1983, 1986a, 1986b).

5.10.3 Dyspepsia and Indigestion

The use of garlic for stomach and intestinal troubles has been known since ancient times. The consumption of garlic by the people of the Balkans is considered to be the reason for their exceptional resistance to all kinds of intestinal diseases (Glaser, 1940; Petkov, 1986).

The stimulating effect of garlic oil on the mucous membrane of the stomach causes a stimulation of the glands, which then secrete an increased amount of hydrochloric acid and digestive enzymes into the gastric juice. Garlic is among the strongest acid stimulants in cases of anacidity or hypoacidity (Varga, 1938). Crushed garlic stimulates movement of intestinal villi even at a dilution of 1:1000. At the same time, glucose absorption increases by 15–20% (Kokas & Ludany, 1933). Bile secretion is also enhanced, and the constituents of garlic oil are excreted from the liver and become enriched in the bile. As a result, there is a choleretic and cholekinetic effect (Schindel, 1934). Possibly, there is also a spasmolytic effect on the smooth muscles of the intestine (Kretschmer, 1935a, 1935c; Gessner, 1953). When added to meat dishes as a spice, garlic eases the salt-free diet of hypertonics in a pleasant way (Christoffel, 1930a, 1930b).

During World War I, Austrian physicians successfully used a product of dried garlic and phenyl salicylate (mixing ratio 0.5:0.15, brand Allphen) for various infectious intestinal diseases. Such diseases occurred regularly at that time, among them numerous cases of dysentery and cholera. Animal experiments were set up to confirm the effectiveness of the preparation; carefully dried garlic proved affirmative, but not distilled garlic oil. Under this therapy, the test animals (rabbits) survived the otherwise lethal challenge of dysentery toxin and injections with highly virulent *Shigella dysenteriae*. Beyond this, good curative effects of the product were observed (Marcovici, 1915; Marcovici & Schmitt, 1915; Marcovici & Pribram, 1915). Similar results were reported on another garlic preparation (allisatin), used as treatment for various intestinal diseases, including dysentery, intestinal tuberculosis, celiac in infants, and similar complaints (Lebinski, 1927; Wolf, 1927; Bachem, 1928; Bonem, 1928; Erbach, 1928; Schultze-Heubach, 1928; Ortner, 1929; Tilger, 1929; Becher & Fussgänger, 1930; Strecker, 1930; Aust, 1931; Geithner, 1931; Schubert, 1931; Mayerhofer, 1934; Fischer, 1936; Barowsky & Boyd, 1944; Burks, 1954; Noether, 1925; Roos, 1925).

Damrau and Ferguson investigated the effect of dried garlic on gastrointestinal motility by means of roentgenography, and they demonstrated the elimination of flatulence and meteorism, gaseous colic, and nausea in clinical experiments (Damrau & Ferguson, 1949, 1950). Good results were achieved in fermentation dyspepsia and nervous dyspepsia. The spasmolytic effect on stomach and intestine, and on hyperperistalsis, was made visible by x-ray control. The authors designated the (unknown) effective principle of garlic as "gastroenteric allichalon." It was said that bromination of the active components in garlic (e.g., diallyl disulfide) not only removed the unpleasant odor, but also enhanced its carminative effect (Ferguson, 1949).

Contrary to the sedation of intestine motility in general, garlic has a special stimulating effect upon the intestinal villi. Even at a dilution of 1:1000, garlic juice stimulates the motility of villi twofold to threefold over the resting motility state. At the same time, absorption (e.g., of glucose) is accelerated (Kokas & Ludany, 1933).

The effect of garlic oil on gastrointestinal propulsion was studied in mice with experimental diarrhea. It was found to depress this parameter significantly. Thus, it was concluded that garlic oil may act as relaxant of smooth muscle in the gastrointestinal tract and that it should be evaluated in patients with hypermotile intestinal disorders (Joshi et al., 1987). Diallyl sulfide was found to have a protective effect on the glandular stomach mucosa (Hu et al., 1990; Hu & Wargovich, 1989; Titta, 1904).

The relaxing effect on the smooth musculature of the mucous membrane in the stomach was confirmed in the isolated fundus of rat stomach in vitro. Here, garlic extract, prepared with chloroform, strongly antagonized the contractions caused by prostaglandin E_2 or acetylcholine. The (unknown) active principle cannot be an oligosulfide, because the pure compounds like diallyl sulfide, diallyl disulfide, and dimethyl disulfide showed only rather weak effects when tested similarly (Gaffen et al., 1984). Russian authors also reported on the successful treatment of dysentery with garlic, especially in children (Strunk, 1951; Babushkin, 1952; Latsinik & Kuperman, 1955).

The antibacterial property of garlic is highly effective in the intestines, especially to pathologic intestinal flora. The foreign organisms are suppressed and the normal coli vegetation is favored (Ozek, 1945; Shashikanth et al., 1986; Noda et al., 1985). This explains the repeatedly observed defense effects of garlic against pathogenic organisms causing typhus, paratyphus, cholera, dysentery, and the like. The simultaneous antiseptic and fermentation inhibitory effects of steam-distilled garlic oil reduce or prevent gastrointestinal autointoxication, especially by phenols and indoles. Finally, the antibacterial effect expresses itself in a general increase of resistance and weight gain in patients (Gessner, 1953). Clinical efficacy for gastrointestinal problems has also been confirmed by several other investigations (Barowsky & Boyd, 1944; Zimmermann, 1985; Singh, 1984; Damrau & Ferguson, 1949, 1950).

Garlic is also widely used in veterinary medicine, especially in the countries of the former Soviet Union. Fresh garlic and steam-distilled garlic oil are highly effective in the therapy of gastrointestinal catarrh in horses, cattle, and goats, and they are equally effective in the treatment of the atony and hypertony of the forestomach of ruminant animals. The average oral single dose of fresh garlic or garlic oil is 20–30 g or 40–80 mL for large animals and 3–8 g or 5–15 mL for small animals, respectively (Loskutov, 1950a). In raising poultry, the addition of garlic to the feed (3% for 8 weeks) leads to higher survival rates and faster growth of the broilers, an effect that is supposedly related to the suppression of pathologic intestinal organisms (Haenel et al., 1962).

5.10.4 Respiratory Diseases

In former times, garlic preparations were also used for diseases of the respiratory organs, such as bronchitis and asthma. This indication has long been abandoned; however, recent reports have appeared concerning the favorable influence of garlic and onions for respiratory dysfunction, as well as for complaints due to high altitude (Schinzel & Graf, 1979).

The treatment of allergies and inflammatory conditions with extracts of garlic and onions has been claimed in a patent (Lichtenstein et al., 1985). Experimentally, garlic extract protected mice from ultraviolet B radiation–induced suppression of contact hypersensitivity (Reeve et al., 1993). The surprising activity of alcoholic extracts of garlic and onion in treating chronic asthma was explained by the inhibition of histamine release from basophils and mast cells, and/or by the inhibition of lipoxygenase in neutrophils (Dorsch, 1986). Russian physicians used garlic for some time in the treatment of various inflammatory diseases of the respiratory apparatus (Bulatov et al., 1965). For whooping cough, children received garlic ingredients by way of inhalation (Atroshenko, 1954).

In the Western world, few people are aware that garlic is regularly used for various diseases of the respiratory tract in the Third World countries. Garlic preparations are used successfully in these countries as expectorants in tuberculosis of the lungs, bronchiectasis, gangrene of the lungs, and whooping cough, as well as for tuberculosis of the trachea, lupus, and other diseases (Krishnamurthy & Sreenivasamurthy, 1956). In the former Soviet Union, preparations of garlic and onions are used to soothe sore throat and inflammation of the mouth cavity, as well as for the treatment of influenza. Garlic ("Russian penicillin") is purported to be as effective as the common antibiotics (Fortunatov, 1952). Also, in Bulgarian folk medicine, garlic is valued as a remedy for chronic bronchitis and other diseases (Petkov, 1986). Kempski (1967) described using garlic therapy for the treatment of helminthic bronchitis, a disease of the respiratory tract caused by larvae of intestinal parasites which infest the lungs.

Last, but not least, a case of severe hepatopulmonary syndrome (hypoxia related to diffuse intrapulmonary shunting) that was treated with garlic must be mentioned (Caldwell et al., 1992). The etiology of this syndrome is unknown, but it may result from a disorder of peptide metabolism in the gut. The symptoms may be ameliorated by somatostatin and reversed by successful liver transplantation. A female patient who failed somatostatin therapy and declined liver transplantation took large daily doses of powdered dry garlic on her own. In the follow-up, she experienced partial relief of her symptoms, as well as objective signs of improvement, over 18 months of continuous self-medication.

5.10.5 Antidote for Heavy Metal Poisoning and Other Toxins

According to Petkov (1966), the sulfur compounds of garlic are useful as antidote for "saturnism," chronic lead poisoning. The feeding of homogenized garlic cloves to lead-fed chickens has been recently shown to decrease significantly the lead content of the muscle and liver tissues, particularly if the garlic was fed after the lead intoxication (Hanafy et al., 1994). In a clinical study, garlic was given to workers in a lead mine who suffered from lead intoxication. It was found that the symptoms of poisoning were considerably diminished by this treatment (Petkov et al., 1965). Garlic and onion juice have similar influence on lead toxicity (Sheo et al., 1993; Geier & Woitowitz, 1988). The toxic effects of arsenic in mice have also been shown to be significantly reduced by feeding the animals with aqueous garlic extract (Roychoudhury et al., 1993; Das et al., 1993a).

Petkov and Kushev (1966) investigated the influence of garlic juice on the excretion and accumulation of gold (^{198}Au) in rats. It was shown that garlic accelerated the excretion of the radioactive isotope and decreased the accumulation in various organs.

Garlic exercised a surprisingly good protective effect against experimental intoxication with other heavy metals (e.g., cadmium and mercury), as well as with organometallic compounds (e.g., methylmercury and phenylmercury). The effect is comparable to that of 2,3-dimercaptosuccinic acid (DMSA) or 2,3-dimercapto-1-propanol (BAL, British anti-Lewisite) and other antidotes (Kitahara & Fujinaga Pharm. Co. Ltd., 1977; Park & Cha, 1984; D. G. Lee et al., 1985; H. S. Lee et al., 1985; Lee et al., 1986; Rhee et al., 1985; Ohm et al., 1986; B. D. Kim et al., 1987; D. S. Kim et al., 1987; Kim et al., 1984; Song et al., 1987; Cha, 1987; Hwang et al., 1986).

The detoxifying effect is connected with the activity of certain enzymes involved in biotransformation of xenobiotic agents, especially glutathione-S-transferase in the liver. Huh and colleagues (1985b, 1986b) demonstrated in rats that the hepatic GSH-S-transferase is greatly enhanced by subcutaneous injections of diallyl disulfide (80–160 mg/kg/day, over 3 days). The microsomal enzymes and serum transaminase are not affected.

In rats, oral administration of garlic, simultaneously with cadmium, methylmercury and phenylmercury, resulted in a decreased accumulation of heavy metals in liver, kidneys, bone, and testes. Histopathological damages and the inhibition of serum alkaline phosphatase activity by these heavy metals was reduced. This effect was not found in the 1.7% garlic-treated group, but was most remarkable in the 6.7% garlic-treated group. The protective effect of garlic was superior to those of BAL and D-penicillamine (PEN), and nearly similar to those of 2,3-dimercaptosuccinic acid and N-acetyl-DL-penicillamine. However, garlic was not effective as a curative agent in heavy metal poisoning. The excretion of cadmium was enhanced by garlic, and the effect was more pronounced in the feces than in the urine (Cha, 1987).

Allithiamine, which has already been discussed (Sections 3.2.8 and 5.9), also proved effective as an antidote in cyanide poisoning. Natural vitamin B_1 (thiamine) is without any effect in this regard. The cyanide detoxifying effect relies on the stimulation of rhodanese enzyme activity by sulfur compounds (Tazoe, 1960a, 1960b, 1960c, 1960d; Murakami & Tazoe, 1960). An aged garlic extract given to mice at 5 mL/kg (equivalent to 350 mL for a person) was shown to decrease the toxic effects of a lethal dose of doxorubicin (Kojima et al., 1994).

In mice, the intraperitoneal administration of aqueous garlic extract (50 mg for 14 days) reduced considerably the toxicity of a simultaneously applied chronic lethal dose of cyclophosphamide (50 mg/kg body weight for 14 days) with an increase in life span of more than 70%. Administration of the garlic extract did not improve the lymphopenia produced by cyclophosphamide or liver alkaline phosphatase, but there was a significant reduction in liver glutamic-pyruvic transaminase. Moreover, there was a reduction in the level of lipid peroxidation induced in the liver by cyclophosphamide administration. Administration of the garlic extract did not interfere with the tumor-reducing activity of cyclophosphamide (Unnikrishnan et al., 1990).

Diallyl sulfide has been shown to block the cyclophosphamide-induced nuclear aberrations in the urinary bladder and hair follicles in mice (Goldberg & Josephy, 1987). The effect was not mediated through inhibition of mitosis, as determined by ^{3}H-thymidine autoradiography in the hair follicles. Diallyl sulfide pretreatment decreased the amount of radioactivity excreted in the urine in the first 24 hours following cyclophosphamide treatment, and it blocked the appearance of acrolein, a cytotoxic metabolite of cyclophosphamide, in the urine over this time period. These results suggest that diallyl sulfide acts by conjugating the toxic metabolites of cyclophosphamide, thereby limiting their systemic circulation and diverting their route of excretion from the urine.

Pretreatment with garlic alone, or in combination with *Crataegus*, has resulted in a dose-dependent protective effect on isoprenaline-induced damage of heart, liver, and pancreas in rats (Ciplea & Richter, 1988). Furthermore, garlic is considered an excellent remedy for chronic nicotine poisoning. Toxic damage causing distress to the heart and arteries, smoker's catarrh, and disturbed intestinal functions are reportedly eliminated, or at least relieved, by treatment with garlic (Madaus, 1976).

5.10.6 Other Effects and Applications

Garlic has also been recommended for accelerated healing of infected wounds, burns, and ulcerated skin surfaces. In clinical studies, it was found that local burns heal faster, because the active components of garlic increase the regeneration of the skin and prevent secondary infections (Chopra, 1933; Lutomski, 1980; Kasakowa, 1953a). A mixture of garlic extract and vitamin B_1 has been used in bath preparations for the treatment of atopic dermatitis (Aki, 1993).

Even in burns of the cornea of the eye, a garlic emulsion is said to boost the formation of epithelial cells of the injured cornea without leaving any cloudiness in the eye (Safarli, 1955). Garlic has also been reported to influence favorably the arteriosclerotic changes of the retina (Weiss, 1984a, 1984b). There is evidence that allicin, but not its precursor, alliin, may be involved in some manner in the lowering of intraocular pressure (IOP), as was demonstrated unilaterally in rabbit eyes (Chu et al., 1993). Doses of 10 µg allicin administered topically reduced the IOP by 6 ± 1 mm Hg (n = 4). The molecular mechanism of this effect, however, remains to be determined.

The parchment-like skins of garlic bulbs have found a curious application in otology: ruptured ear drums have been surgically repaired by overlaying the injured part with a layer of garlic cells. The healing occurred without problems in 14 such injuries (Perrin et al., 1983).

Ulcus cruris (ulcers of the lower leg) has been successfully treated with an ointment that contained lyophilized garlic. The preparation had a strong bactericidal effect and stimulated granulation and growth of epithelial tissue in wounds (Zabel et al., 1979; Kasakowa, 1953a, 1953b). Some authors have reported favorable effects of garlic for treating paradentosis and inflammation of the tooth root (Jummel, 1940; Kruczkowska, 1958; Melkozerowa, 1978). Positive results were also seen in the complications after tooth extractions (Madaus, 1976; Lutomski, 1980). Garlic is even used in a preparation for brittle fingernails and in hair tonics to revitalize the hair and nails (Knudsen & Mavala S.A., 1968; Andreotta, 1967).

Reports are also occasionally found on the neurotropic effects of garlic (Anonym, 1984). Razumovich (1979) examined the influence of volatile garlic substances on the physiological activity of frogs, guinea pigs, rabbits, and dogs, and observed remarkable effects on conditioned reflexes and other functions. Usually, the first short phase of excitation was followed by a second, longer phase of inhibition. The stimulating effects are similar to those of sympathicotropic compounds.

In a patent (Matsuda, 1982), alliin and allicin are claimed as suitable for the treatment of neuralgia. Experimentally, in dorsal horn cells of the cat, it was shown that garlic ingredients can indeed induce neural sensitization reactions (Hong et al., 1992). Indian physicians suggested garlic for the treatment of acute lepromatous neuritis (Sreenivasamurthy et al., 1962).

In a review article, garlic is listed among the medicinal plants with antirheumatic and antineuralgic effects (due to the allicin content), as they are used by the people of northern Italy (Cappelletti et al., 1982). Also, garlic is used in Poland as an antirheumatic in folk medicine (Borkowski, 1970). Garlic has relatively strong antiinflammatory activity in comparison with other plant drugs (Mascolo et al., 1987). A commercial garlic combination (with ginkgo) generally improves the well-being of patients, as was shown in a placebo-controlled, double-blind study (Kade & Miller, 1993).

5.11 GARLIC IN HOMEOPATHY

Although the theory and methods of Samuel Hahnemann's homeopathy (Hahnemann, 1833) are highly questionable from a scientific standpoint (Koch, 1993c) and are therefore rejected by serious scientists (Marburger Erklärung, 1992; Hopff & Prokop, 1992), as well as by most doctors, we nevertheless refer to it here, simply as a matter of objectivity; however, we address the topic only as far as it is concerned with the use of garlic. No comments will be made on homeopathic theory or practice.

Allium sativum was mentioned in Hahnemann's dispensary records, which were compiled during his lifetime (1755–1843) or shortly after his death. Garlic is also mentioned in Hager's *Medicamenta homoeopathica et isopathica omnia* (Hager, 1861). An "essence" from freshly ground garlic cloves was prepared by allowing the ground garlic to stand for 10 days in a cool place, after which alcohol was added in equal volumes.

In the *Pharmacopoea homoeopathica polyglottica* by Schwabe in 1872 (PHP 72, 1872), the origin of today's official edition of the German *Homoeopathic Pharmacopeia* (HAB 78, 1978), *Allium sativum* is mentioned in the appendix, and the preparation from fresh garlic cloves described according to a general prescription (Section 3 of HAB 78). Indications in the literature allow the assumption that garlic had already been introduced into homeopathy in the first years of its practice.

Allium sativum is also cited in Schwabe's 1934 *Homoeopathic Pharmacopeia* as a regular article (HAB 34, 1934). The HAB 34 was accepted in Germany as the official collection of prescriptions for homeopathy until 1980. The article on *Allium sativum* was transferred into the third edition of the pharmacopeia without major changes (HAB 53, 1953). The current official HAB 78 also contains an article on *Allium sativum* with the usual description for preparation (according to Section 3) and the details for the testing of identity and purity. It is practically a verbatim transfer of Section 3 from the HAB 34.

Another pharmacopeia that contains an article on garlic is the eighth edition of the *Homeopathic Pharmacopoeia of the United States* (HPUS 79, 1979). Here, the "urtincture" (standard tincture) is systematically adjusted to one-tenth of the juice content of the drug. Here, also, a definite statement can be found that garlic was introduced into homeopathy by a certain Dr. Petroz in 1852. Furthermore, garlic is recorded in the *Homeopathic Pharmacopoeia of India* (HPI

71, 1971). The preparation is made in the same way as in HPUS 79. The *Mexican Homeopathic Pharmacopeia* (FHM 61, 1961) gives the preparation method under regulation 3, which preparation is equivalent to a garlic concentration of one-sixth. This method is consistent with that of *Hager's Handbook* (Hager, 1969).

A collection of prescriptions with the designation *Homéopathie: Pharmacotechnie et monographies des médicaments courants*, is used in France. It is published by the Syndicat des Pharmacies et Laboratoires Homéopathiques, an association of companies producing homeopathic preparations that does not have official status. In the first volume of this compendium is an article on *Allium sativum*. The official collection of prescriptions is based on a general article, "*Préparation Homéopathique*," in the official French pharmacopeia. Garlic is processed in a manner similar to that of the American pharmacopeia, so that the concentration of the standard extract is one-tenth of the initial drug.

As mentioned above, garlic was introduced into therapy in 1852 by Petroz in France. His specifications were completely transferred into Allen's 1874 *Encyclopedia of Pure Materia Medica*, where garlic is listed as a remedy for diseases of various organs. Testing of the drug on healthy people primarily manifested a relaxation of the gastric and intestinal tonus with elimination of dyspepsia and flatulence, a favorable influence on chronic inflammation of the mucous membranes of the respiratory tract, and finally, relief of pain in the hip joints and lower back musculature. From this, the main areas of garlic application in homeopathy have been derived: irritation of the alimentary tract, chronic bronchitis, and coxofemoral problems. The dosage is relatively low: D3 to D6 (diluted by 1,000 to 1,000,000). Voisin lists *Natrium sulfuricum, Rhododendron, Drosera*, and *Causticum* in a comparison (Voisin, 1984). Garlic is also considered as a blood sugar–lowering remedy in homeopathy, though rarely used in practice (Münch, 1966; Hager, 1969; Voisin, 1984; Leeser, 1984; Mezger, 1984).

In conclusion, it should be noted that garlic in homeopathy, though not listed among the so-called major remedies, is a regular part of the *Materia Medica Homeopathica*. The traditional use of garlic in homeopathy as a stabilized fresh plant extract corresponds to that of modern "phytopreparations." The substantial effect of garlic components is not to be excluded, though administered at dosages of such low potencies, there is little hope of cure available to the patient.

5.12 DOSING OF GARLIC AND ITS PREPARATIONS

The 1988 monograph of the "Kommission E" at the Bundesgesundheitsamt in Berlin (BGA, German Health Department, equivalent to the U.S. Food and Drug Administration), as the basis for the admission of a pharmaceutical garlic product to the market, literally recommends "4 g fresh garlic cloves or 8 mg volatile oil" as the daily dose for reducing elevated blood lipid levels and preventing age-dependent vascular changes. This is approximately one clove of garlic, although clove sizes vary considerably (see Table 3.3). According to the monograph, the application form is "the ground drug and its preparations for internal use; garlic oil in the form of oil macerate or obtained by water steam distillation." At the time of publication, there was no standardization of garlic. Unless other compounds come to be identified as having antiarteriosclerotic effects, it seems reasonable to standardize garlic powder on the basis of its concentration of alliin or allicin yield and to standardize other preparations on their content of various allicin transformation products, including ajoenes, vinyldithiins, and polysulfides, depending on the preparation conditions employed. Allicin at 0.45 mg is equivalent to 1.0 mg alliin. Concerning the results of the pharmacological and clinical studies performed since 1988, a significant lipid-lowering effect of garlic is to be expected by a daily dose of about 600–900 mg garlic powder, standardized to 1.3% alliin or 0.6% allicin, corresponding to 3.6–5.4 mg allicin.

In some older pharmacopeias and codices, various directions are given for the dosing of garlic. The *German Pharmacopeia* (DAB 6, 1926) states 5 g for a single dose. The *Extra Pharmacopoeia* of 1958 declares 2–8 g. We have already reported the dosages of garlic and garlic oil in veterinary medicine. For *"Tinctura Allii sativi"* and *"Sirupus Alli sativi"* (DAB 6), 5.0 g or 30.0 g, respectively, are given as the medium single dose (DAB 6, 1926; Braun, 1949; Seel, 1952). Details on the dosing of various preparations, though obsolete today, have been reported by Piotrowski (1948)—alcoholates, dialysate, maceration, garlic vinegar, garlic syrup, and others.

5.13 DEODORIZED GARLIC EXTRACTS

Currently, two commercial extracts are available that do not contain active alliinase or significant amounts of allicin or its transformation products and are hence considered odorless. These products and their known or possible biological effects will be presented in this section. For prior discussions on garlic odor and its control or reduction in garlic products, see Sections 3.2.6.4, 3.4.1, and 4.2.8.

5.13.1 Aged Garlic Extract

A garlic extract from Japan that is widely sold under various brand names (Kyolic, Kyoleopin, Leopin-Five) is garlic aged in dilute alcohol ("Aged Garlic Extract" or AGE). The product is manufactured by storing sliced garlic in 15–20% ethanol for 20 months, followed by filtration and concentration (Hirao et al., 1987; Horie, 1978; Nakagawa et al., 1989). During the long incubation period, the thiosulfinates are completely lost, as well as most of the alliin and γ-glutamylcysteines (see Section 3.4.5). Only trace amounts of the odorous oligosulfides (diallyl and allyl methyl mono-, di-, and trisulfides) derived from allicin and the other thiosulfinates remain (Weinberg et al., 1993). The key substance in AGE, however, has been proposed to be *S*-allylcysteine (see Table 3.25 for content), a stable organosulfur compound that is generated by hydrolysis of γ-glutamyl-*S*-allylcysteine (see Fig. 3.24).

To date, a total of approximately 90 research papers and meeting abstracts have been published on all aspects of AGE. Most of the research on AGE has been conducted in the areas of cancer effects, immune effects, cardiovascular effects, and stress relief. Positive effects in animal or isolated cell experiments have been seen mostly only with high intakes (2–4% of diet, corresponding to 10–20 g AGE/day for a person) or high concentrations of AGE. The area in which there are the most publications (17) is that of the anticancer or cancer-preventive activities of AGE, including both pharmacological and case-control studies. These have been described in some detail at the end of Section 5.3.2. The cardiovascular studies are discussed below. Investigations on possible immune effects (Section 5.5) (Kandil et al., 1987, 1988; Abdullah et al., 1989; Hirao et al., 1987; Lau et al., 1991b; Morioka et al., 1993), antimicrobial effects (Section 5.2.2) (Tadi et al., 1990; Hughes & Lawson, 1991; Pai & Platt, 1995), antiviral effects (Section 5.2.4) (Weber et al., 1992; Nagai, 1973a, 1973b), antioxidant effects (Section 5.4) (Imai et al., 1994; Yamasaki et al., 1994; Kagawa et al., 1986; Ohnishi & Kojima, 1994), antiinflammatory effects (Section 5.6) (Chiou et al., 1995), antihepatotoxic effects (Section 5.10.2) (Nakagawa et al., 1985, 1989; Blakely et al., 1993), and hypoglycemic effects (Nagai et al., 1975) were described earlier.

Antistress effects at high doses have been found in mice (Yokoyama et al., 1986; Takasugi et al., 1984, 1986), while reasonable doses in humans have been reported to result in relief of fatigue, depression, various complaints, and recovery from cancer therapies (Hasegawa et al., 1983; Kajiyama, 1982; Tanaka, 1982; Kawashima et al., 1986; Miyoshi et al., 1984; Okada & Miyagaki, 1983). A recent study showed that a 2% diet of AGE improved memory in mice and increased longevity in a strain of mice with a very short life span (Moriguchi et al., 1994). Not everything that has been reported in this literature, however, is absolutely convincing, but the readers are invited to look into the original publications to make their own evaluation. Furthermore, the in vivo effects of AGE have rarely been compared directly with those of cooked garlic (which also does not yield allicin) or fresh garlic, both of which contain about 10 times as much total sulfur compounds and total nitrogen compounds as AGE (see Tables 3.17 and 3.25) and are therefore likely to result in significantly better therapeutic effects.

Cardiovascular studies with animals have indicated variable effects. Feeding AGE to chickens or rats has been shown to reduce blood cholesterol and triglyceride levels (Abuirmeileh et al., 1991; Yeh & Yeh, 1994). Furthermore, intravenous injection of a liquid AGE containing added thiamine (10 mg/mL) into rats at 0.25 mL/kg was shown to increase significantly the peripheral circulation for 5–10 minutes (Yokoyama et al., 1988). On the other hand, other studies with rats fed up to 4.8% AGE for 8 weeks showed no effects on levels of serum cholesterol, serum triglyceride, hemoglobin, or hematocrit (Chanderbhan et al., 1993; Wiesenfeld et al., 1993).

Clinical studies concerning the cardiovascular effects of AGE in humans have been few and less conclusive than those with fresh garlic and garlic powder products (Section 5.1.1.2). A study on the short-term peripheral capillary blood flow in volunteers after consumption of four capsules of AGE revealed no change in plasma viscosity, erythrocyte flow rate, hematocrit, or fibrinogen content, while consumption of 900 mg of alliin/allicin–standardized garlic powder tablets produced significant improvements in all four parameters (Koscielny et al., 1991a, 1991b; Jung et al., 1990b). In a study on serum lipids, 4 mL liquid AGE/day was given to hyperlipidemic (cholesterol 226–440 mg/dL) patients for 6 months (Lau et al., 1987). Significant decreases in serum cholesterol (Table 5.1) and triglyceride levels were found, but not until after 5 to 6 months, although HDL did increase by 4 months; however, a significant rise in serum cholesterol and triglycerides was found in the first 2 months of consumption, along with a significant decrease in HDL after 1 month. These results have been partially confirmed in a preliminary report of a second study in which 45 hypercholesterolemic (serum total cholesterol 230–290 mg/dL) men on a fat-restricted diet consumed nine capsules (6300 mg) of AGE for 6 months. After elimination of those who showed the least changes (one-third of the subjects), an 8% decrease in serum total cholesterol levels was found, with no effect on HDL cholesterol (Steiner & Lin, 1994). In the same study, significant reductions in platelet adhesion to fibrinogen and platelet aggregation were also found, although a prior in vitro study showed that even high concentrations of AGE had no effect on inhibition of platelet aggregation in whole blood (Lawson et al., 1992).

Although the authors of the first study (Lau et al., 1987) proposed that the initial increases in serum lipids observed during the first 2 months were due to the removal of cholesterol from the tissues into the circulation, no such initial increases have ever been found in human studies where fresh cloves or garlic powder tablets caused decreases in serum lipid levels. Furthermore, normolipidemic subjects (cholesterol 150–200 mg/dL) in the same study also showed significant increases in serum cholesterol and triglyceride levels in the first 2 months, along with a significant decrease in HDL after 1 month. In contrast to the 5–6 months needed for the AGE to lower blood lipid levels, numerous studies with garlic cloves and standardized garlic powder tablets have shown significant decreases in 1 to 2 months (Table 5.1). The differences between the two types of garlic products may be due to the low amounts of sulfur compounds in AGE as well as the lack of allicin potential (see Tables 3.17 and 3.25 and Section 5.1.1.3).

5.13.2 Odorless Garlic Extract

Quite recently, a novel garlic preparation (brand Tölsat) has been introduced (available only in Germany) which claims to be odorless and therefore does not possess the odor disadvantage of most of the commercial preparations. While the genuine plant material is practically odorless, the odorous material (allicin and other thiosulfinates) are not formed until the cloves are chewed, cut into pieces, or mashed into a purée. The chemical processes that take place at this stage have been extensively discussed in Chapter 3. They involve the formation of odorous allicin, from its odorless parent compound, alliin, through the catalytic action of the enzyme alliinase, followed by the gradual formation of malodorous allicin decomposition products. Thus, odor formation is blocked as soon as the activity of alliinase is inhibited by a suitable means, such as by thermal denaturation of the enzyme. This is, simply speaking, the clue to the whole problem. The technical problems in inhibiting alliinase have only been recently resolved, as discussed at length in a review article (Koch, 1989).

This novel product, a dry powdered extract, is made from raw garlic bulbs that are microwave-irradiated and deep-frozen prior to crushing and extraction with aqueous alcohol, followed by removal of the solvent under reduced pressure and then lyophilization of the residue. This treatment effectively eliminates the active alliinase without damaging the plant tissue (Wäfler et al., 1994), while the freeze-thaw injury to the cell membranes eases the efflux of ions and organic solutes (Arora & Palta, 1991) from the cell compartments during the subsequent extraction step. No transformation of the genuine alliin or its homologs occurs during the whole processing. The final product is a dry, scentless powder that can be directly compressed into tablets (Koch, 1993d, 1994a).

5.14 OTHER ACTIVE COMPOUNDS IN GARLIC

As has been discussed in previous sections of this chapter (Sections 5.1.1.3, 5.1.2.3, 5.1.4, 5.3.3, 5.4.3, and 5.5), compounds other than those from the alliin/alliinase/allicin system also contribute to some of garlic's therapeutic effects, particularly with respect to blood pressure reduction, anticancer effects, and immune effects. Among these compounds, the more hydrophilic or polar compounds probably play an important role (allicin is also reasonably water soluble, but it is much more soluble in ether and therefore not considered to be polar). This has long been suspected, but little attention has been paid to other compounds because they are more difficult to isolate and because of the great amount of interest in organosulfur compounds derived from the cysteine sulfoxides of the genus *Allium*, as was pointed out by Koch (1993e).

The realization that additional active principles are contained in the water-soluble, ether-insoluble polar fraction was the second basis for the development of the "odorless garlic extract" mentioned in Section 5.13.2. Among the candidates for therapeutic effects are these well-known and much less known constituents of garlic and onion bulbs:

- alliin and its homologs
- γ-glutamylpeptides
- scordinins
- steroids and triterpenoids
- flavonoids and other phenols
- fructans

Some of these have been found to possess properties of active drug compounds. Thus, it is certain that the natural drug "garlic" is not a monosubstance product, but a rather complex mixture of active agents which probably act synergistically to produce a clinical effect.

One of the effects of the polar fraction of garlic has been proposed and also experimentally verified by Koch and coworkers during recent years (Koch et al., 1992b, 1993; 1994; Koch, 1992b). Accordingly, at least some of garlic's clinical effects may be due to a purinergic mechanism of action. It was shown that aqueous/alcoholic garlic and onion extracts in vitro strongly inhibit the enzyme adenosine deaminase (ADA, EC 3.5.4.4). A clear dose-dependent inhibition of enzyme activity, leading to an accumulation of the substrate adenosine in the system, was observed. For instance, 10 μL fresh cell sap (supernatant of homogenized and centrifuged garlic bulbs) exerts up to more than 90% inhibition of ADA in vitro (Koch et al., 1992b). Adenosine, on the other hand, is known to regulate a number of important physiological processes in vivo, among them heart rate, contraction force, rate of blood flow, vasodilation, rennin release, rate of lipolysis, platelet aggregation, respiration, and neurotransmitter release (Koch et al., 1992b). Surprisingly, many of these effects are among the empirically found activities of garlic.

The degradation of adenosine by ADA deamination is irreversible, leading to the elimination of adenosine. Therefore, inhibition of ADA in vivo will increase the actual adenosine concentration in the human body, thereby triggering the effects which would normally follow an exogenous dose of adenosine through the interaction of the nucleoside with purinergic receptors. Thus, garlic (and other *Allium* species) can be considered natural ADA enzyme inhibitors. Furthermore, pure samples of alliin, allicin, diallyl disulfide, and ajoene do not display any inhibitory action on ADA in vitro or are only slightly inhibitory. This clearly suggests that they are probably not the active principles in garlic preparations having the assumed therapeutic effects as listed above (Koch et al., 1992b). Additional evidence that allicin does not inhibit ADA comes from a recent study in which three different fresh or dried garlic preparations of similar allicin yield showed greatly varying activities toward inhibition of ADA in cultured aortic endothelial cells (Melzig et al., 1995). Furthermore, the ADA inhibition activity decreased slowly during 3 months of storage of the cloves, an indication that γ-glutamyl-cysteines may be involved, since they are known to decrease upon storage while alliin does not (see Section 3.2.3).

However, what additional compounds then could also be responsible for garlic's therapeutic effects? The short list of candidates for this function (as enumerated in the list above, which can easily be expanded to include many more natural constituents of the garlic bulb) provides some suggestions for future research. Triterpenoids (Koch et al., 1994), flavonoids and related plant phenolics (Koch et al., 1992a), and, surprisingly, certain fructans which were isolated from garlic bulbs (Koch et al., 1993), were identified as potent inhibitors of ADA enzyme activity. Obviously, the steroid-like triterpenoids (saponins of the oleanolic acid type) that have been identified in garlic and onion (see Section 3.3.7), are favored as the most likely active agents in the polar fraction (Koch et al., 1992b, 1994; Koch, 1993b).

This does not necessarily exclude the variety of alliin-derived organosulfur compounds from participation in the therapeutic effects of garlic extracts. Certainly, alliin and other cysteine derivatives, γ-glutamyl peptides, sulfur glycosides, and the like, are included in the extracted polar fraction; although enzymatically active alliinase is no longer present in this extract (Section 5.13.2), there are other enzymes available, either in the human body (Egen-Schwind et al., 1992; Lachmann et al., 1994) or in symbiotic microorganisms (Murakami, 1960a, 1960b; Nomura et al., 1963; Saari & Schultze, 1965), that may be capable of converting the inactive precursors into active agents within the human organism. By no means is it required that the malodorous components be present, which opens the possibility of getting rid of the unwanted alliaceous odor, while retaining at least some of the favorable therapeutic properties of the *Allium* plants.

6
Biopharmaceutics of Garlic's Effective Constituents

Heinrich P. Koch

Today it is generally assumed that the intensity and duration of a drug effect are in causal relationship to the amount of effective compound (administered dosage) in the patient's body. The pharmacological effect caused by the drug is furthermore dependent on how much of the given effective compound (the dosage) is available at a given time at the actual site of action (the biophase). The *bioavailability* of a drug is determined by a series of sequentially and/or simultaneously occurring processes; it is the responsibility of researchers in *biopharmaceutics* and *pharmacokinetics* to investigate these processes (Koch & Ritschel, 1986).

These processes can be only briefly outlined here. First, the effective ingredient, contained in a drug dosage form (tablet, film tablet, sugar-coated tablet, capsule, etc.) is dissolved in the juices of the stomach and/or small intestine (*liberation*), then absorbed by the mucous membrane and transferred into the blood (*absorption*). The compound is distributed throughout the body by the circulating fluid (*distribution*) where a greater or lesser part of the molecules reach their specific receptor site, at which they trigger a specific reaction. During this time, it can happen that the drug substance is attacked by the body's own enzymes and biochemically transformed, creating new active or inactive compounds, the metabolites (*metabolism* and *biotransformation*). Lastly, the residues of the drug and its metabolites are again removed through the function of the excretory organs (*elimination*).

It makes sense that the rate at which these events proceed determines the intensity and duration of the drug effect. For this reason, today it is required that a drug comply with the above described principles. A compound that is not absorbed, for instance, usually cannot exert an effect in the organism; one that is degraded (assuming that the degradation products are inactive) or eliminated too quickly may only have a short effect, if any. Bioavailability plays a decisive role in the realization of a therapeutic effect and should be determined for every compound thought to be effective.

6.1 PHARMACOKINETICS OF THE SULFUR-CONTAINING ACTIVE COMPONENTS OF GARLIC

Systematic tests concerning the action and the catabolism of the effective components of garlic in the human body or in animals are scarce. However, there are a few isolated indications in the literature which give some information on the absorption and elimination of single compounds or related groups of compounds. A comprehensive article summarizing the state of the art in the field of metabolism and pharmacokinetics of the constituents of garlic has appeared recently (Koch, 1992).

Although allicin is generally accepted as an important active compound released from garlic, its mechanism of action in vivo, except as an intestinal or dermal antibiotic, has not yet been determined. An important clue in the identification of its mechanism of action is the recent finding that the addition of allicin (0.5 mM) directly to fresh whole blood causes immediate conversion of allicin to two moles of allyl mercaptan (Lawson & Wang, 1993; Lawson et al., 1991). Diallyl trisulfide and ajoene, two possibly active compounds produced from allicin during commercial processing of garlic, were also shown immediately to form allyl mercaptan in blood. Therefore, in vitro or ex vivo studies on the mechanism of action of these compounds should not use the parent compounds, but rather should use allyl mercaptan, or possibly a metabolite of allyl mercaptan. The fate of allyl mercaptan in the body is not yet known. The fact that allicin is rapidly converted to other metabolites in the blood and other organs does not mean that it does not have important indirect in vivo systemic effects. There are many examples from both nutrients and drugs of compounds that need to be metabolized to other compounds in order to have systemic pharmacological effects: linoleic acid must be converted to prostaglandins and other eicosanoids; retinol (vitamin A) must be reduced to retinal and combined with opsin; pyridoxine (vitamin B_6) must be converted to pyridoxal phosphate; N-acetylcysteine and Procysteine must be hydrolyzed to cysteine.

It is understandable that the conspicuous garlic odor in the breath has given reason for speculation about the elimination of the odor carrier. It was suspected that mercaptans occurred, as well as hydrogen sulfide, but this could not be verified (Leeser, 1984). The fact alone that the odor carrier is eliminated through the breath, regardless of whether it was administered orally or rectally, proves that absorption of the basic components has occurred (Chopra, 1933). Lehmann concluded that the absorption rate of the sulfur compound in the gastrointestinal tract must be rather high, since the expiration occurs only a few minutes after oral application of garlic. He also conducted quantitative evaluations and estimated that up to 10% of the administered garlic oil is eliminated by expiration (presumably as allyl mercaptan and diallyl sulfide) (Leeser, 1984; Laakso et al., 1989).

In a more recent study, diallyl sulfide, a trace component of steam-distilled garlic oil, in an oily solution (25% corn oil, 0.1 mL intraperitoneally) was injected into rats. Three minutes after the injection, diallyl sulfide could be detected in the breath of the animals by means of gas chromatography (GC), and it was still detectable after 5 hours (Maruniak et al., 1983).

Noether measured the elimination duration of the odor carrier in human urine and observed that the odor is still very distinct 76 hours after ingestion of 2 g raw garlic, and that even after 96 hours traces could be perceived. The exhaled air no longer had an odor after 24 hours. Thus, the persistence of oligosulfides in the human body is considerable. In a cat, however, no odor could be detected in the urine or feces after repeated administration of pressed juice (Noether, 1925). Fortunatov states that the excretion of the antimicrobial components ("phytoncides") of garlic does not occur until 6 hours after internal application (Fortunatov, 1955).

The odor components have been found also in the breast milk after ingestion of garlic by the mother, and the smell could be traced even in the newborn baby's expired air (Mennella & Beauchamp, 1991, 1993).

Kretschmer (1935) presented further evidence on the catabolism of the active constituents of garlic in the human organism. It can be presumed that garlic's constituents are transferred from the stomach through the liver into the bile, and this stimulates bile secretion. Enrichment occurs in the gallbladder, and elimination follows with bile secretion into the small intestine. With this action, the bacterial flora of the intestines is also favorably influenced through an increase of the *E. coli* population.

A pharmacokinetic study with a defined garlic-derived compound was performed with diallyl disulfide in mice (Pushpendran et al., 1980). The animals were given a sublethal dosage (100 mg/kg intraperitoneally) of ^{35}S-labeled diallyl disulfide, which was emulsified in physiological saline with Tween. At various times, the animals were sacrificed, the liver removed, and the accumulated radioactivity measured (Fig. 6.1). In addition, various subcellular fractions were prepared 2 hours after application and the contained radioactivity determined. Most of the radioactivity was found in the cytosol

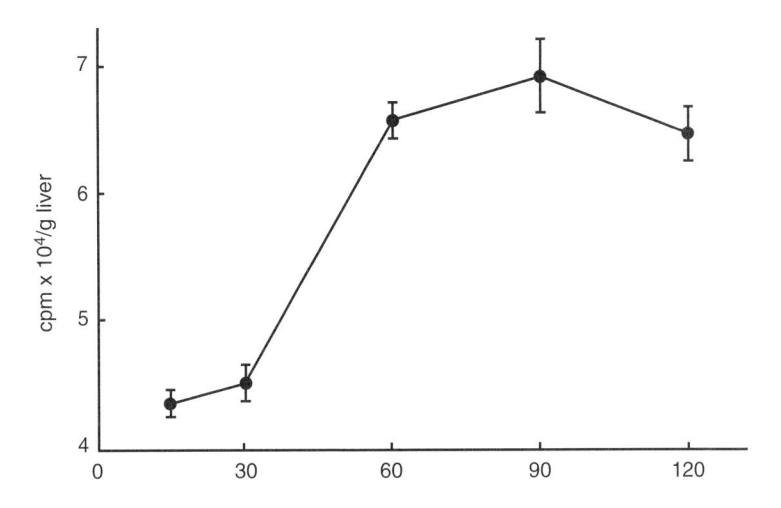

Figure 6.1. Radioactive sulfur in mouse liver at various times after administration of ^{35}S-diallyl disulfide (100 mg/kg intraperitoneally). (From Pushpendran CK, Devasagayam, TPA, Chintalwar, GJ, Banerji A, Eapen J. The metabolic fate of [35S]-diallyl disulphide in mice. Experientia 1980;36:1000–1001.)

fraction (of which 90% was in the nonlipid, nonprotein supernatant) with much less in the mitochondria. It was also observed that the largest part of the labeled sulfur found in the liver was totally degraded after a short time, being converted by more than 80% into sulfate with only a small amount (9%) still organically bound.

A study on the bioavailability of diallyl trisulfide (in Chinese, *dasuansu*) was performed in rabbits after administration of suppositories containing 15 mg dasuansu in palmityl acetate as a base (total weight 2.2 g). The half-life of the compound in plasma was found to be 1.46 hours by suppository and 2.32 hours by injection. The bioavailability was calculated to be 98.6% by the rectal route (Wang et al., 1989).

It is sometimes easier to study the metabolism of drug compounds in isolated organs than in vivo. Thus, the metabolic and kinetic behavior of some garlic constituents, such as alliin, allicin, and diallyl disulfide, were investigated in the isolated perfused rat liver. Allicin showed a remarkable first pass clearance effect and passed the liver unmetabolized only at concentrations high enough that it caused considerable cell injury. Diallyl disulfide and allyl mercaptan were identified as metabolites of allicin, with diallyl disulfide being rapidly metabolized to allyl mercaptan. Alliin and diallyl disulfide were found in the perfusate after liver passage of alliin, but no allicin was found.

Qualitative detection of sulfur compounds in human urine after garlic consumption is possible by a color reaction (Kajita, 1957). This reaction was not used for pharmacokinetic studies, however. By gas chromatography/mass spectroscopy (GC/MS), on the other hand, the isolation and identification of garlic-derived sulfur compounds from human urine after the ingestion of garlic oil was achieved. Diallyl disulfide, diallyl sulfide, and dimethyl disulfide were detected by this technique from as little as 10 mL urine (Bartzatt et al., 1992).

Possible marker compounds have been proposed to detect compliance in patients during clinical trials. An analytical study demonstrated that nine well-defined organosulfur compounds from garlic extracts can be used for this purpose (Weinberg et al., 1993).

Quite recently, a pharmacokinetic study of the garlic constituents alliin and allicin, as well as of the vinyldithiins, was performed (Lachmann et al., 1994). The ^{35}S-labeled compounds were each administered as oily solutions at oral doses of 8 mg/kg body weight to rats, and the activity levels were monitored for 72 hours in blood, urine, feces, and exhaled air. The urinary metabolite pattern was also determined by thin-layer chromatography (TLC), and whole-body autoradiography was performed. The blood activity profile of alliin differed considerably from those of allicin and the vinyldithiins, both for absorption and elimination, which were distinctly faster with alliin than with the other compounds. Absorption of alliin was complete after 10 minutes and elimination after 6 hours, whereas for allicin the absorption values were 30–60 minutes and for the vinyldithiins 120 minutes. The mean total urinary and fecal excretion after 72 hours was 85.5% of the dose for allicin and 92.3% for the vinyldithiins. The expired air showed only traces of activity, while autoradiography demonstrated enrichment of activity in the mucosa of the airways and pharynx and deposits in the vertebral column and ribs. No difference was found in organ distribution between allicin and the vinyldithiins. From the urinary metabolite pattern, it could be seen that there was no unchanged alliin, allicin, or vinyldithiins present, suggesting rapid and extensive metabolism of the three test compounds in the organism. Conjugates with sulfuric acid or glucuronic acid were not detectable either.

S-Allylcysteine (SAC) has recently been examined in a pharmacokinetic study for distribution and bioavailability in the body organs and urine of three animal species (Nagae et al., 1994). Although SAC is present only in trace amounts in garlic, it is assumed that some systemic enzymes may hydrolyze garlic's abundant γ-glutamyl-SAC to SAC. Kidney, for example, contains a high amount of γ-glutamyl transpeptidase activity, while liver contains activity only in the tissue that produces bile. Intestinal enzymes, however, probably do not hydrolyze γ-glutamyl-SAC itself, since glutathione, a similar compound, is absorbed without hydrolysis (Hagen et al., 1990). Absorption was found to be remarkably rapid and was maximal in 15 minutes in several tissues, and was more abundant at 15 minutes in kidney and liver than in plasma. Bioavailability of SAC in rats was found to be dose-dependent; at doses of 50, 25, and 12 mg/kg, 98, 77, and 64% of the dose was absorbed, respectively. Since consumption of a 5-g clove of garlic (1.7 g dry weight) by a person would result in only 0.2 mg SAC/kg (after transpeptidase hydrolysis) or 0.05 mg SAC/kg after consumption of 5 g commercial aged garlic extract, the bioavailability of SAC at practical levels of consumption may be very small. The half-life of SAC varied greatly between rat, mouse, and dog (2.1, 0.8, 10.3 hours), as did the excretion form, N-1-acetyl-SAC (30–50, 7, and less than 1%); such large variations among these three species show that animal results for this compound cannot be extrapolated to humans.

Alliin, which is seen as the key substance for the pharmacological effects of garlic, was investigated for its metabolic fate in rats. It was found that there is only little activity of alliinase-like enzymes in the organism of rats. Nevertheless, alliin was very quickly transformed by isolated liver, kidney, and mucous membrane of small intestine, creating the secondary product diallyl disulfide (DADS), which was also metabolized quickly into as yet unknown tertiary products (Pentz et al., 1991; Egen-Schwind et al., 1992b). The resorption-excretion balance of sulfur-containing components of garlic was made possible by determination of hydrolyzable SH-groups (Pentz et al., 1990, 1991).

In connection with the pharmacokinetics of the sulfur-containing constituents of garlic, a study on the stability of these compounds in gastric and intestinal fluids is of interest. Rao and coworkers studied the inactivation of allicin in gastric juice (pH 1.7) and pancreatic juice (pH 9.3) and found that, in the former, virtually no activity is lost within 24 hours, while in the latter, the activity gradually decreases and reaches zero after 24 hours. Also, the presence of blood does not influence the antimicrobial activity of a high concentration of allicin (6.6 mg/mL) (Rao et al., 1946).

Diallyl disulfide (DADS), the main constituent in commercial garlic oil, has a strong stimulating effect on various enzymes involved in the metabolism of xenobiotic agents. It enhances the detoxification of poisonous substances in the organism. Thus, the activity of the NADH- and NADPH-dependent microsomal oxygenase in rats increased after intraperitoneal injection of DADS at a high dose (100 mg/kg). After 4 hours, glucose-6-phosphatase and adenosine triphosphatase were also increased (Devasagayam et al., 1982). In contrast, other researchers found an inhibition of hepatic drug-metabolizing activities by DADS given at 0.04 mg/kg orally, and they suggest that this effect is due to a change in the cytochrome P-450 moiety (Siddiqui et al., 1988).

In another study, it was found that DADS (single dose, 500 mg/kg intraperitoneally) exerts a significant depression of hepatic cytochrome P-450, aminopyrine N-demethylase, and aniline hydroxylase, while microsomal protein content, NADPH-cytochrome c reductase, benzphetamine N-demethylase, and cytosolic glutathione S-transferase remain unaffected 24 hours after treatment. Daily administration of 50 mg/kg for 5 days of DADS produced a significant increase of hepatic cytochrome P-450, aminopyrine N-demethylase, and benzphetamine N-demethylase activities, but not in the aforementioned parameters of biotransformation reactions. These data indicate that the effect of garlic oil on the hepatic drug-metabolizing enzyme system is dose-dependent (Dalvi, 1992).

In vitro inhibition of cytochrome P-450 reductases from pig liver microsomes by aqueous garlic extracts containing allicin or by purified alliin was demonstrated through the decrease of the activity of cytochrome P-450–dependent ethoxycoumarin deethylation. As measured by malondialdehyde release, the effect on the enzyme system was not due to lipid peroxidation. No loss of cytochrome P-450 was observed either. Alliin or its transformation products are believed to be the active principles for the inhibitory effect in vitro (Oelkers et al., 1992).

Specific inhibition of certain cytochrome P-450 isoenzymes was observed with diallyl sulfide (DAS), a minor component of steam-distilled garlic oil, and with its putative metabolites diallyl sulfoxide (DASO) and diallyl sulfone ($DASO_2$). It was found that DAS inactivates P-450 2E1 and induces P-450 2B1. Each compound displayed competitive inhibition of p-nitrophenol hydroxylase activity in liver microsomes of rats. Preincubation of the microsomes with $DASO_2$ inhibited the enzyme in a process that is time- and NADPH-dependent and saturable with pseudo–first-order kinetics. The K_i value for $DASO_2$ is 188 µM and the maximal rate of inactivation 0.32 min^{-1}. $DASO_2$ is ineffective in the inactivation of ethoxyresorufin dealkylase, pentoxyresorufin dealkylase, or benzphetamine demethylase activities. Purified P-450 E1 in a reconstituted system is inactivated in a time- and NADPH-dependent manner by $DASO_2$. The metabolic conversion of DAS to DASO and $DASO_2$ was observed in vitro and in vivo. The results suggest that DAS inhibits the metabolism of P-450 2E1 substrates both by competitive inhibition mechanisms and by inactivating P-450 2E1 via a suicide-inhibitory action of $DASO_2$ (Brady et al., 1991a, 1991b).

In a follow-up study of the same research group, the mechanism of P-450 2B1 induction by DAS was investigated. Following a single dose of DAS (200 mg/kg, intragastric), liver pentoxyresorufin dealkylase, which is representative for P-450 2B1 activity, was induced 3-, 16-, 26-, and 43-fold at 6, 12, 18, and 24 hours after the treatment. A corresponding increase in the level of P-450 2B1 protein was observed by immunoblot analysis. In contrast, the level of P-450 2E1 mRNA in the liver of DAS-treated rats was not changed. These and other findings clearly demonstrate that the induction of P-450 2B1 by DAS is due to transcriptional activation. Besides the increase of enzyme activity in the liver, P-450 2B1 mRNA was also markedly increased in the stomach (23-fold) and in the duodenum (66-fold), but not in the lung and nasal mucosa (Pan et al., 1993).

The most comprehensive study on the modification of hepatic drug-metabolizing enzymes in rats fed large amounts of diallyl sulfide (DAS) or diallyl disulfide (DADS) (2000 ppm of diet, i.e., 200 mg/kg, for 15 days) was performed by Haber and colleagues (1994, 1995). This treatment resulted in a modification of both enzymatic activity and/or microsomal level of cytochrome P-450 forms (Table 6.1).

A special study examined the possible transmammary modulation of hepatic xenobiotic metabolizing enzymes in mouse pups whose lactating mothers were fed crushed garlic (200 and 400 mg/kg orally for 14 and 21 days). No alteration in hepatic cytochrome P-450 and cytochrome b_5 content and GSH S-transferase activity occurred in either dams or pups, while GSH reductase and GSH peroxidase activities were significantly reduced in both dams and pups that were exposed to the highest dose for 21 days (Chhabra & Rao, 1994).

In humans, a model compound, acetaminophen, was used to study the influence of the constituents of a commercial aged garlic extract (brand Kyolic) on the drug-metabolizing enzymes. The commonly used drug was chosen because it forms a reactive electrophilic metabolite after oxidative metabolism. The subjects undergoing the test ingested daily doses of the aged garlic extract for 3 months; at the end of each month and 1 month after termination, a 1 g oral dose of acetaminophen was given, and plasma and urine were measured for the test compound as well as its glucuronide, sulfate, cysteinyl, mercapturate, and methylthio metabolites. It was found that aged garlic extract had no discernible effect upon oxidative metabolism but was associated with an increase of sulfate conjugation of the drug (Gwilt et al., 1994).

6.2 PHARMACOKINETICS OF THE VINYLDITHIINS

The pharmacokinetic behavior of 2-vinyl-4H-1,3-dithiin (I) and 3-vinyl-4H-1,2-dithiin (II), two main components of the oil-macerate preparations of garlic, was investigated after oral administration of 27 and 9 mg of the pure compounds, respectively, to rats (Egen-Schwind et al., 1992a). The pharmacokinetic parameters were calculated using either the one-compartment (1COMEX) or the two-compartment open model (2COMEX) (Tables 6.2 and 6.3) (Koch & Ritschel, 1986). Compound I was less lipophilic and thus was rapidly eliminated from serum, kidney, and fat tissue, whereas II was more lipophilic and showed a tendency to accumulate in fat tis-

Table 6.1. Cytochromes P-450 and Microsomal Enzymes of Rats as Influenced by Feeding (200 mg/kg body weight) with Diallyl Sulfide or Diallyl Disulfide

Cytochrome/Enzyme	Symbol	Influenced by	
		DAS	DADS
Cytochrome P-450 1A2		+	+
Cytochrome P-450 2B1/2		+	+
Cytochrome P-450 3A1/2		+	+
Cytochrome P-450 2E1		−	−
NADH cytochrome c reductase		+	+
P-Hydroxybiphenyl UDP-glucuronyltransferase	UDPGT2	+	+
P-Nitrophenol hydroxylase	PNPH	−	−
N-Nitrosodimethylamine demethylase	NDMAD	0	0
Laurate ω–hydroxylase	LAH	−	−
Glutathione S-transferase	GST	+	+
Epoxide hydrolase	EH	+	+
Ethoxyresorufin O-deethylase	EROD	+	+
Aryl hydrocarbon hydroxylase	AHH	+	+
Methoxyresorufin O-demethylase	MROD	+	+
Ethoxycoumarin O-deethylase	ECOD	+	+
Pentoxyresorufin O-depentylase	PROD	+	+
Benzoxyresorufin O-debenzylase	BROD	+	+
Erythromycin N-demethylase	ERDM	+	0

+ increase/enhancement, − decrease/depression, 0 not influenced.
From Haber D, Siess MH, de Waziers I, Beaune P, Suschetet M. Modification of hepatic drug-metabolizing enzymes in rat fed naturally-occurring allyl sulfides. Xenobiotica 1994;24:169–182.

Table 6.2. Pharmacokinetic Parameters of 2-Vinyl-4H-1,3-dithiin

Parameters Model	Liver 1COMEX	Serum 2COMEX	Kidney 2COMEX	Fat tissue 1COMEX
$t1/2\ \alpha$ [h]	3.7	0.2	0.2	3.2
$t1/2\ \beta$ [h]		5.1	8.1	
α	4.56	3.46		
K_{el} or β [h^{-1}]	0.19	0.14	0.09	0.22
Cl_{tot}	0.87 kg/h	61 L/h	20.6 kg/h	0.33 kg/h
AUC [µg/mL.h]	0.4			
$V_{d\beta}$ [L]	447			

Abbreviations: 1COMEX or 2COMEX, one or two compartment model; $t1/2\alpha$ or $t1/2\beta$, elimination half-life during α-phases or β-phase; α, initial distribution rate constant; K_{el} or β, terminal disposition rate constant; Cl_{tot}, total clearance; AUC, area under serum concentration time curve; $V_{d\beta}$ or V_d, apparent volume of distribution. From Egen-Schwind C, Eckard R, Jekat FW, Winterhoff H. Pharmacokinetics of vinyldithiins, transformation products of allicin. Planta Med 1992; 58: 8–13.

Table 6.3. Pharmacokinetic Parameters of 3-Vinyl-4H-1,2-dithiin

Parameters Model	Serum 1COMEX	Kidney 1COMEX	Fat tissue 1COMEX
$t1/2$ [h]	6.7	13.1	5.6
K_{el} [h^{-1}]	0.10	0.05	0.12
Cl_{tot}	85 L/h	6.0 kg/h	0.13 kg/h
AUC [µg/mL.h]	0.1		
V_d [L]	814		

Abbreviations: 1COMEX or 2COMEX, one or two compartment model; $t1/2\alpha$ or $t1/2\beta$, elimination half-life during α-phases or β-phase; α, initial distribution rate constant; K_{el} or β, terminal disposition rate constant; Cl_{tot}, total clearance; AUC, area under serum concentration time curve; $V_{d\beta}$ or V_d, apparent volume of distribution. From Egen-Schwind C, Eckard R, Jekat FW, Winterhoff H. Pharmacokinetics of vinyldithiins, transformation products of allicin. Planta Med 1992;58:8–13.

sue. Experiments with liver homogenate confirmed the in vivo findings. Allicin, the precursor of I and II, was metabolized more rapidly in liver homogenate than were the vinyldithiins (Egen-Schwind et al., 1992b).

6.3 PHARMACOKINETICS OF ALLITHIAMINE

In the past, allithiamine (Sections 3.2.8 and 5.9) has raised great interest because of its vitamin B activity. In this connection, its behavior was intensively investigated in vitro and in vivo (Itokawa, 1963a–d).

The pharmacokinetics and metabolism of a radioactively labeled model substrate, ^{35}S-thiamine propyl disulfide, were studied in rats and rabbits. For comparison, identically labeled thiamine was used in parallel experiments. After intravenous injection of the test substance, the excretion of ^{35}S in urine was lower than after comparable amounts of thiamine; upon oral administration, however, it was considerably higher. In all cases, the urine contained only free thiamine. The preparation did not accumulate in the erythrocytes, as compared to thiamine. It was bound to plasma albumin. Distribution studies showed that the test substance is absorbed from the gastric intestinal tract within 30 minutes, and that its absorption occurs faster than that of thiamine (Itokawa, 1963a–d). Other authors reported on the absorption of lipid-soluble allithiamine and thiamine propyl- and tetrahydrofurfuryl-disulfide, and found significant increased thiamine activity in whole blood, red blood cells, and cerebrospinal fluid (Baker et al., 1974; Thomson et al., 1971).

The conversion of allithiamine to thiamine and to further degradation products has been thoroughly studied as well. The excretion of thiamine, as well as of pyruvic and lactic acids, was determined in blood, urine, and stool after oral and parenteral administration (Yano, 1958a–e).

6.4 PHARMACOKINETICS OF MACHADO'S GARLICIN

It is highly surprising that a remarkably thorough study on the absorption and elimination as well as on the diffusion and retention in the tissue was conducted with the antibiotic garlicin (Section 3.3.13), although its chemical composition has never been determined to satisfaction (Machado et al., 1948), and this at a time when "pharmacokinetics" had not yet been "invented" (Koch & Ritschel, 1986).

Garlicin was found to be quickly absorbed from the gastrointestinal tract. After oral administration of 50 units of the antibiotic, it could be extracted with chloroform from the urine after only 2 hours. It was eliminated by the renal route and with the bile. The excretion was maximal in 6 hours; the urine concentration gradually decreased to zero by 24 hours. Elimination proceeded in a similar manner after three repeated dosages on consecutive days (60, 50, and 50 units). It is therefore possible to maintain a constant concentration in the body by multiple dosing (Machado et al., 1948).

Garlicin passes the blood-brain barrier, as activity was detected in the spinal fluid of patients who had received 600 units orally. The crossing into the central nervous system was observed in rats after the administration of 20 units/day. Active garlicin was also found in the liver and bile of the animals. It was present in plasma, but did not enter the blood cells (Machado et al., 1948) (compare also Koch, 1993).

6.5 DOES GARLIC ACT AS AN "ABSORPTION ENHANCER"?

In the case of vitamin B_1 (thiamine) it was shown that sulfur-containing compounds of the allicin type accelerate the uptake of the vitamin from the gastrointestinal tract and, in general, positively influence its availability. Of course, in this case the improvement of the absorption is due to the formation of a new chemical compound, allithiamine, with changed physicochemical properties (Section 5.9).

In the absorption of adenosine, it has been speculated that certain sulfur-containing constituents of garlic serve as true absorption enhancers. Adenosine is normally not absorbed from the gastrointestinal tract. Since garlic extracts contain, in addition to adenosine, compounds (thiosulfinates) similar to dimethyl sulfoxide (DMSO), which is known to serve as carrier for a series of nonpolar substances, transporting them across the membrane border, the absorption of garlic's adenosine may be enhanced by them (Reuter, 1980, 1983). This would be another new function of the sulfur-containing constituents of garlic.

In 1933, Kokas and Ludany observed in experiments with dogs that garlic juice greatly increases the absorption from the intestine. Absorption of glucose is increased twofold to threefold at a dilution as low as 1:1000. The authors speculated that this effect may be due to stronger movements of the intestinal villi under the influence of garlic.

Bogin found that garlic extract increases the permeability of certain compounds (e.g., ATP) through erythrocyte and mitochondrial membranes, which serve as models for biological membranes. Uncoupling of oxidative phosphorylation and subsequent modifications of the membranes are assumed to be mechanisms of this phenomenon (Bogin, 1973).

Recently, the gastrointestinal absorption enhancement in the presence of a special high scordinin–containing garlic extract was observed and protected by a patent. The study of Kominato's group suggests that such an extract can improve the uptake of various essential amino acids which are used as food additives and for liver therapy (Kominato et al., 1986).

Lastly, not only does garlic act as an absorption enhancer in animal organisms, but it also does so in plants. Garlic oil was mixed with other oils and added to herbicides in order to facilitate the absorption of the latter by the plants (Heyn, 1978).

7
Toxicity, Side Effects, and Unwanted Effects of Garlic

HEINRICH P. KOCH

Where there is light, there is also shadow! This familiar saying is also true for phytopharmacons: "A drug that is considered to have no side effects must also be suspected of having no main effects." The frequently occurring adverse reactions, euphemistically called *side effects* (a better expression would be *adverse effects*), which almost regularly occur with highly effective pharmaceuticals, are as much to be expected with effective components of plants as with synthetic drugs. Garlic is no exception. It should not be surprising, then, that the various active ingredients of garlic occasionally cause side effects, as can be seen from various case reports in the literature. A comprehensive review of garlic toxicity has appeared recently (Koch, 1992).

This chapter presents a discussion of the possible side effects that may result from consuming unusually large amounts of garlic. However, before doing so, it is also prudent to discuss safe levels of garlic consumption. Garlic is ingested daily both fresh and cooked by many people throughout the world, and this has been the case for thousands of years. Obviously, it is far from being a toxic food, when consumed at reasonable levels. Although little has been published on how much garlic various people consume, there have been a few reports from studies on the medicinal use of garlic. In Cangshan county of Shandong Province in northern China, the average daily consumption of fresh garlic has been reported to be 20 g (about 4–5 cloves), an amount which has been associated with greatly decreased stomach cancer (Mei et al., 1982). This is probably near the upper limit of a safe level of garlic consumption, although it is expected that cooked garlic, which contains no or little allicin, could be safely eaten in considerably higher amounts. Several clinical studies on the reduction of serum lipid values have had the subjects consume 3–15 g fresh garlic daily for several weeks without any reported side effects. A summary of these studies is presented in Table 7.1. It can be recommended, then, that the ingestion of 10 g fresh garlic per day represents a safe level when eaten with a meal (an important requirement), although somewhat higher levels may also be reasonably tolerated if gradually achieved. Certainly, the adage of "a clove a day," about 3–5 g garlic, is well within the safe limit (Koch, 1988).

Table 7.1. Levels of Fresh (Uncooked) Garlic Consumed in Clinical Studies

Daily Amount Consumed (grams)	Duration of the Study (weeks)	Number of Persons	Side Effects	Reference
3	4	4	none	(Kerekes & Feszt, 1975)
3	16	8	none	(Ali & Thomson, 1995)
5	3	6	none	(Jain, 1977)
10	8	5	none	(Augusti, 1977)
10	8	15	none	(Bhushan et al., 1979)
10	8	50	none	(Gadkari & Joshi, 1991)
15	3.5	200	none	(Sucur, 1980)

7.1 ACUTE, SUBACUTE, AND CHRONIC TOXICITY OF GARLIC

Absorptive poisoning by the plant itself is not suspected. In certain plant extracts or in individual active compounds, as well as in garlic oil, however, if they are administered in a highly concentrated form or as pure compound, toxic effects have to be expected (Siegers, 1989; Dalvi & Saluhnke, 1993).

Testing of the acute toxicity of allicin in mice yielded LD_{50} values of 60 mg/kg intravenous injection and 120 mg/kg subcutaneous injection (Cavallito & Bailey, 1944). Recent tests of acute toxicity of an aged garlic extract on Wistar rats and ddY mice in which the lethal effect of oral, intraperitoneal, and subcutaneous application was determined, gave LD_{50} values greater than 30 mL/kg (Nakagawa et al., 1984).

Allicin is strongly *hepatotoxic* when administered over a long time in very large doses (approximately 100 mg/kg, the equivalent of a person eating 1750 g or 500 cloves of raw garlic every day). The activity of certain liver enzymes decreases, that of others increases, the content of glycogen and RNA is lower, and the fat content is higher than in untreated animals (Augusti & Mathew, 1975; Lee, 1967).

For two synthetic homologs of allicin, di-*tert*-butyl-thiosulfinate and di-*n*-amyl-thiosulfinate, the LD_{50} values in mice were determined to be 392 and 346 mg/kg, respectively (Kametani et al., 1959).

Oral consumption of garlic oil (10, 100, and 200 mg/day) by humans does not affect the erythrocytes, but it does if it is administered parenterally to cats (100 mg/kg). In this case, serious anemia occurs (Lebinski, 1927; Satoh, 1952). Leukocytes are also changed by allicin; phagocytic activity of polymorphonuclear leukocytes of rabbits in vitro was diminished in the presence of allicin (6.25 to 50 µg/mL). The germicidal activity of the leukocytes, however, was increased after the treatment (Yamada et al., 1976).

In a more recent study on the hemolytic activity of the alkenyl disulfides of the Alliaceae, also given in very high doses, it was found that di-1-propenyl disulfide and diallyl disulfide were more powerful hemolytic agents than the saturated dipropyl disulfide. On the other hand, no degenerative changes were found with all three agents in liver, spleen, or kidney (Munday & Manns, 1994; Williams et al., 1941). Other compounds toxic to erythrocytes were identified in the onion, namely sodium *cis*- and *trans*-1-propenyl-thiosulfate and *n*-propyl-thiosulfate (Yamato, 1994).

Studies concerning the effect of garlic extracts on erythrocytes revealed changes in the hemoglobin spectrum. It is assumed that the responsible factor, which in its native form is heat-stable and nondialyzable, acts on the membranes of the blood cells. It could be responsible for the occurrence of anemia in animals after ingestion of various *Allium* plants (Farkas & Farkas, 1974; Spice, 1976; Bogin et al., 1984).

Simple sulfur compounds are considered to be among the active principles in garlic. Uemori (1946) studied intensively the toxicity of diallyl sulfide (in alcoholic solution). The intravenous toxic dosage for rabbits was calculated to be 0.755 mL. The isolated frog heart stops after high dosages. The reaction is only partially counteracted by caffeine and atropine; adrenaline has no effect at all. It was believed that the point of action was at the vagus nerve ends and the smooth musculature. The motility of isolated rabbit intestine was decreased by diallyl sulfide; in live rabbits, myosis, blood pressure drop, and increase in heart rate were observed after intravenous applications of the extract (4 to 10 mL) (Uemori, 1929).

Toxic effects on the liver or the function of liver enzymes have been repeatedly reported, especially when the *Allium* plants have been ingested in very large amounts. Garlic and onion oils can be hepatotoxic (Adamu et al., 1982), as well as extracts of garlic (Bogin & Abrams, 1976; Burks, 1954; Ciplea & Richter, 1988; Huh et al., 1983, 1985a–c, 1986a–b). Aqueous garlic extracts, given to rats intraperitoneally, increased lactate dehydrogenase activity in liver (Ali et al., 1991).

Diphenylamine, contained in onions and considered to be responsible for the blood sugar–lowering effect, has not yet been found in garlic. Diphenylamine is toxicologically questionable because of its nephrotoxic properties. It has been calculated that at a concentration of 1%, the intake of 100 g onions (or 35 mg diphenylamine) per day can become dangerous for humans (Opdyke, 1978; Matsuzawa et al., 1984; Siegers, 1987; Koch, 1994).

Toxicity studies were also undertaken with an aged garlic preparation containing vitamins (brand Kyoleopin, from Japan). LD_{50} for rats after oral and subcutaneous applications were 30 mL/kg. No toxic symptoms were observed within 7 days. Sub-cutaneous and chronic toxicity testing of the same preparation generally revealed no toxic effects. Only in some groups of animals that had received extremely high dosages were the number of erythrocytes and hemoglobin values slightly reduced and the spleen enlarged. Otherwise, there were no histopathological changes found in various organs and tissues (Kanezawa et al., 1984). The same research team also presented data of a study on chronic toxicity of aged garlic extract. After the administration of 2000 mg/kg five times per week over 6 months, no change in weight was observed, although the food intake of the treated animals was less than that of control animals. No other differences were observed in the urological, hematological, and serological parameters. Histological examination did not reveal any indication of pathological changes in organs and tissues (Sumiyoshi et al., 1984).

The adverse effects of an aqueous extract of 300 and 600 mg/kg/day of garlic bulbs (the dry weight concentrated sixfold) were studied in mice that received this treatment for 21 days. The extract adversely affected weight growth, biologic parameters, and histologic structures of the animals (Fehri et al., 1991). In another study, rats that were treated with either raw or boiled aqueous garlic extracts were assayed for lactate dehydrogenase and transketolase activities in liver and red blood cells, respectively. Both enzyme activities were elevated, the increase being more prominent with the raw garlic extract treatment (Ali et al., 1991). A significant rise was also observed in urea and D-aspartate aminotransferase, but alkaline phosphatase was inhibited in the serum of rats fed aqueous garlic extract. Histological examination of the liver also revealed changes. Garlic oil (100 mg/kg), on the other hand, was lethal (death was due to acute pulmonary edema) when fed to fasted rats, but no effect was seen when given to fed rats (Joseph et al., 1989).

In a targeted study for the establishment of general side effects of antihypertensive therapy with garlic extracts on rats, serious, lethally toxic effects were at times observed. Of course, these animals had received especially high daily dosages (0.25 to 1.5 mL/kg) of a highly concentrated extract. The observed effects were diarrhea, strong drop in blood pressure, and increase in heart rate, as well as a general state of weakness which caused the death of some animals (Ruffin & Hunter, 1983).

On a lower organism (*Oxytricha*, an infusorian) that was used as a test object for toxic effects of allicin, a concentration of 45 mg allicin per 100 mL incubation medium caused total inhibition of reproduction, and 0.18 mg/100 mL showed 20–30% growth inhibition (Petroff, 1963). Solutions of 0.25 to 2.5% pressed garlic juice kill fish and leeches (Malova, 1950). Using *Cynops pyrrhogaster*, an amphibian, as test organism, signs of abnormal behavior and disturbances were noted in the pigmentation of the epithelium when influenced by the volatile ingredients of garlic. Upon prolonged exposure, the injuries led to the death of the animals (Iwanami et al., 1982). Aqueous extracts of garlic are effective antifungal agents, but do not impair the growth of higher plants (Fujii et al., 1991).

In testing the *genotoxicity* of orally fed garlic (7.5 g/kg) on bone marrow cells of mice ("micronucleus test"), no significant changes in comparison to untreated controls were found (Abraham & Kesavan, 1984). In a study for *mutagenicity* and *cytotoxicity* using the Ames test, the micronucleus test, and the Chinese hamster embryo (CHE) test, freshly pressed garlic juice and alcoholic garlic extract were tested. No mutagenicity was observed in the Ames test (Schimmer et al., 1994; Stacchini & Mangegazzini, 1987; Rockwell & Raw, 1979), but the other tests showed concentration-dependent increases in modified bone marrow cells of mice and hamsters. Freshly pressed juice was considerably more cytotoxic than an aged extract (Yoshida et al., 1984). A certain general toxic effect of fresh garlic (20% of diet) on rats and mice had already been observed earlier (Carl et al., 1939).

Hepatocytes are generally employed for toxicity testing. In this context, a method for separating hepatocytes from rat liver by using collagenase together with garlic extract, which was recently disclosed in a patent, may be of some interest (Hirao et al., 1994).

7.2 TREATMENT OF ACUTE TOXICITY

The treatment of a patient after an accidental overdose of garlic must be performed primarily etiotropically, i.e., the poison must be removed as quickly as possible. In practice, this is done through washing and flushing. In addition, antiirritant poultices can be applied externally. Internally, if the rectum is injured, mucilage can be applied rectally (Gessner, 1953).

7.3 TOXICOLOGY OF MACHADO'S GARLICIN

Garlicin, an antibiotic that may be present in some treatments of garlic (Koch, 1993) (see Section 3.3.13), which was not further defined chemically at the time of the early experiments, has nevertheless been intensively tested for chronic toxicity (Machado et al., 1948).

No lethal, nor even toxic effects could be induced in rats with increasing amounts of garlicin in aqueous-alcoholic solution up to 1000 units/kg, i.e., 700 times the therapeutic dosage, injected intraperitoneally. Prolonged application (10 units orally twice daily) over a prolonged time did not affect weight increase, blood coagulation capacity, or histopathologic parameters compared to untreated controls. The antibiotic also had no influence on gestation or postpartum (Machado et al., 1948).

7.4 TOXICOLOGY OF SCORDININ

General pharmacologic and toxicologic studies have also been conducted with scordinin, a structurally somewhat complicated compound that is present in garlic at approximately 0.03% of fresh weight (see Section 3.2.9). The LD_{50} in mice was determined to be 12.5 mg/kg after subcutaneous and intraperitoneal application, and 15.1 mg/kg after oral application (Suzuki & Motoyoshi, 1966).

The effect on the smooth musculature, heart rhythm, blood pressure, and respiration were studied in vitro and in vivo. The isolated ileum of rabbit showed an increase in tonus and peristalsis with 600 µg/mL scordinin (scordinin C_b). Isolated guinea pig intestine contracted at 64 µg/mL. The increased tonus was again normalized by diphenhydramine hydrochloride (Benadryl), but not by atropine. The heart rate of the isolated rabbit auricle was decreased at 1 mg/mL, and this effect was increased by eserine salicylate. Blood pressure, measured on dog and rabbit, decreased temporarily after intravenous injection of 5 mg/kg. This blood pressure effect was neutralized by Benadryl and atropine. Pretreatment with scordinin had no effect on the hypotensive effect of acetylcholine or on the lowering of the blood pressure after electrical stimulation of the vagus nerve. Scordinin caused vomiting and a definite drop in blood pressure at a concentration of 50 mg/kg. The effect was reversed by metoclopramide (100 µg/kg). Respiration was clearly enhanced with 10 mg/kg scordinin intravenously in dogs as well as rabbits (Suzuki & Motoyoshi, 1966).

7.5 TOXICOLOGY OF A COAGULATION INHIBITOR FROM GARLIC

Song and coworkers isolated a compound from garlic that inhibited the coagulation of blood. It could be *phytic acid,* according to the chemical analysis (Section 3.3.12) (Song et al., 1963a, 1963b, 1963c). This active compound was examined pharmacologically and toxicologically. Mouse LD_{50} was found to be 222 mg/kg. Intravenous injection of 70 mg into rabbits caused muscular hyperactivity, which finally led to the death of the animals. The substance showed a strong hypocalcemic action in vivo as well as in vitro, which could explain this effect (Song et al., 1963a, 1963b, 1963c).

A similar observation, reported by another author, could be attributed to the same active compound (Sanfilippo, 1946). Daily administration of garlic juice to guinea pigs over 3 weeks resulted in a twofold to threefold increase of the normal calcium concentration in blood. After two months of continuous treatment, the calcium level had again normalized. Calcium was also greatly increased in bones and teeth, but the phosphate concentration was reduced.

7.6 LOCAL IRRITATION

Fresh garlic ingested on an empty stomach may cause irritation of the mucous membrane with the sensation of heartburn and eventually long-lasting pains. For individuals with a sensitive stomach or gastric ulcer, this can be a serious complication (Varga, 1938). Allicin and the homologous thiosulfinates are believed to be the causative factor of this irritation (Koch, 1992; Koch & Hahn, 1988).

The mechanism of action of this side effect (for the ingestion of onions) has been explained recently (Block, 1992a, 1992b; Koch, 1994). The organosulfur compounds may inhibit the enzyme activity of lipoxygenase and cyclooxygenase in the gastric mucosal membrane; consequently, smaller quantities of prostanoid compounds are formed and relaxation of the esophageal sphincter occurs, leading to an increased upward reflux of the acidic gastric fluid. Heartburn is the initial symptom, followed by mucosal lesions with extended duration of the esophagitis (Block & Purcell, 1992).

Garlic oil can cause severe inflammation after local application of large amounts and prolonged exposure, which can lead to tissue necrosis. Such accidents have formerly been observed on occasion, when indiscriminate applications, such as enema into the rectum and poultices on the skin were done, especially on infants (Mayerhofer, 1934). Large internal dosages of garlic oil can cause violent vomiting and diarrhea. It can even lead to kidney irritation with severely decreased diuresis (Flamm, 1935).

Schwahn (1928) observed in an experiment on himself, after ingestion of a larger amount of garlic infusion, nausea and a collapse-like condition (Schwahn, 1928). In an experiment to establish the tolerance threshold for garlic oil, this was administered at 625 mg three times per day. The symptoms observed thereafter were anorexia, nausea, diarrhea, weight loss, metrorrhagia, and menorrhagia (Arora et al., 1981).

A 78-year-old patient who suffered from an itching eczema on his legs treated this by self-medication with a cataplasm of smashed fresh garlic; this treatment, however, caused a severe dermatitis which could only be brought under control with corticosteroids (Bojs & Svensson, 1988). In another reported case of self-medication with detrimental results, a 65-year-old patient swallowed a whole garlic clove. The result was an intestinal occlusion; he was saved only by an emergency surgery (Szybejko et al., 1982).

An incident of sustained skin burns from a garlic–petroleum jelly plaster that had been applied to the feet of a 17-month-old child for 8 hours at the direction of a naturopathic physician was reported (Parish et al., 1987). A second case of skin burn occurred when the wrists of a 6-month-old infant were wrapped in crushed garlic for 6 hours to treat (successfully) a high fever (Garty, 1993). A third case of skin burn on the foot of a 6-year-old girl occurred when crushed garlic was placed under a bandage for 2 days (Canduela et al., 1995). A fourth case report describes an epidural hematoma with associated platelet dysfunction which occurred after excessive garlic ingestion. This patient had an extraordinary fondness for garlic and consumed an average of four cloves daily as a precautionary measure against heart disease (Rose et al., 1990).

Local irritation of respiratory organs can occur upon inhalation of garlic powder or dust, which can develop into serious asthma attacks (Falleroni et al., 1981). The occasionally observed diarrhea after garlic consumption could be caused by adenosine, contained in the plant (see Section 3.3.9). It has been reported elsewhere that adenosine irritates the intestine and causes secretory diarrhea (Turnheim, 1985).

Consumption of sandwich spreads made of a mixture of chopped garlic, water, and soybean or olive oils has caused two serious outbreaks of botulism, the first involving 36 cases (St. Louis et al., 1988), the second involving three cases (Morse et al., 1990). In both incidents, the containers had been kept unrefrigerated for an extended period of time, and had high enough acidity (pH 4.5 and 5.7, respectively) to allow growth of *Clostridium botulinum*. Other foods prepared in oils with water present, such as potato salad and sautéed onions, have also caused outbreaks of botulism (St. Louis et al., 1988). Onions, though not directly toxic, can also be an unusual vehicle for serious poisoning. A case of poisoning has been reported in which botulism was transferred into a restaurant through sautéed onions (MacDonald et al., 1985).

The local irritation caused by garlic was also confirmed in an animal experiment. Injection of an aqueous extract (0.1 to 0.2 mL) into the rat paw led to severe local inflammation. Administration of antihistamines prior to this treatment could not prevent the response, suggesting that there is not a histaminergic mechanism involved in this particular event (Banerjee, 1976).

7.7 ALLERGIES TO GARLIC

Some of the ingredients of the *Allium* species, especially garlic (*Allium sativum* L.) and onion (*Allium cepa* L.) (rarely other leek plants), can cause allergic contact eczema upon frequent and intensive contact. It is mainly a concern for housewives and professional groups, such as chefs, workers in the food industry, produce dealers, and gardeners. The location of the eczema on the tips of the fingers of the left hand is typical, because onions and garlic cloves are predominantly held with the left hand, while the right hand holds the knife for cutting (Burks, 1954).

Frequently, there are cross-allergies between garlic and onion. In addition to allicin, other constituents may be causative factors. The typical compounds are contained in both species or are formed from precursors in the plant. Of course, there are patients who react only to garlic and not to onion and vice versa, so that several inducing factors must be considered (Papageorgiou et al., 1983).

Diagnosis of such an allergy is done by cutaneous reaction tests. Fresh garlic or onion pieces are applied to the skin, usually the inside of the forearm, and held in place with impermeable tape. Some authors suggest freshly pressed juice or ether, acetone, or ethanol extracts for the test (Sinha et al., 1977). In the case of allergy, i.e., the presence of specific antibodies, a local inflammation or eczema occurs within 48 hours.

Numerous cases of contact allergies and asthma caused by *Allium* species are described in medical literature (Table 7.2). Three books (Hausen, 1988; Fisher, 1986; Benezra et al., 1985) and several review articles about "plant dermatitis" are available (Rook, 1960, 1962; Schulz & Hausen, 1975).

The first report on asthma attacks after inhaling garlic dust was published in 1940 (Razumovich, 1979). Two additional cases were reported more recently on hypersensitivity against the dust occurring when garlic skins were powdered. Here also, it manifested itself through asthma and rhinal conjunctivitis (Couturier et al., 1980). A similar case occurred during bagging and packaging of dry garlic powder (Falleroni et al., 1981).

In another case, a patient who worked in a spice factory suffered asthma attacks every time garlic powder was processed. The causal relationship was secured through various allergy tests. In this case, a blind test with 1.6 g garlic powder in a gelatin capsule prompted an attack (Lybarger et al., 1982). At the present time, this is the only known

Table 7.2. Reports on Allergies from *Allium* Species

Year	Number of Patients	Allium sativum	Allium cepa	Source
1938	1	+	−	(Sepulveda, 1938a,b)
1940	1	+	−	(DiPaolo & Carruthers, 1960)
1950	1	+	−	(Edelstein, 1950)
1952	1	+	+	(Burgess, 1952)
1954	14	+	+	(Burks, 1954)
1955	14	+	−	(Cueva & Duran, 1955)
1961	6	+	+	(Borda & Bozzola, 1961)
1961	2	+	−	(Daitsch et al., 1961)
1964	4	+	+	(Neves, 1964)
1972	1	+	+	(Bleumink et al., 1972)
1973	8	+	−	(Bleumink & Nater, 1973)
1976	17	+	+	(Hjorth & Roed-Petersen, 1976)
1977	45	+	+	(Sinha et al., 1977)
1978	10	+	−	(Strobel et al., 1978)
1978	3	+	+	(Van Ketel & De Haan, 1978)
1979	13	+	−	(Pasricha & Guru, 1979)
1979	5	+	+	(Yoshikawa et al., 1979)
1980	1	+	−	(Mitchell, 1980)
1980/82	2	+	−	(Couturier et al., 1980, 1982)
1981	1	+	+	(Falleroni et al., 1981)
1981	1	+	−	(Martinescu, 1981)
1982	1	+	−	(Campolmi et al., 1982)
1982	1	+	−	(Lybarger et al., 1982)
1982	4	+	−	(Saito et al., 1982)
1983	1	+	−	(Molina et al., 1983)
1983	7	+	+	(Papageorgiou et al., 1983)
1983	4	−	+	(Veien et al., 1983)
1983/84	11	+	−	(Fernandez de Corres et al., 1983)
1984	1	+	−	(Haenen, 1984)
1984	1	+	−	(Kramer, 1984)
1985	1	+	−	(Kirsten & Meister, 1985)
1985	1	+	+	(Lautier & Wendt, 1985)
1987	2	+	+	(Cronin, 1987)
1988	1	+	−	(Bojs & Svensson, 1988)
1991	8	+	−	(Lembo et al., 1991)
1992	34	+	−	(McFadden et al., 1992)
1993	3	+	−	(Seuri et al., 1993)
1994	1	+	−	(Burden et al., 1994)
1994	3	+	−	(Räsänen et al., 1994)

+ positive test reaction; − negative test reaction

case where garlic precipitated such an incident through oral intake. Of course, we are dealing here with a highly sensitive patient. Cross-sensitization occurs also between various garlic varieties (Haenen, 1984).

7.8 INTERACTION WITH OTHER DRUGS

So far only one example of a specific interaction between garlic products and a common drug substance, warfarin, has been reported in the literature. Patients who were stabilized by anticoagulant therapy with warfarin started intake of garlic pearls or tablets; in both cases blood clotting times were roughly doubled. This is considered a potentially serious interaction (Sunter, 1991). Very recently, a warning was promulgated on heavy garlic consumption as a possible risk for postoperative bleeding (Burnham, 1995; Petry, 1995).

7.9 ECOLOGICAL HAZARDS

A quite different aspect of common toxicity is based on the uptake of potentially toxic contaminants, present in soil, water, or air, by medicinal plants. During the processing of the plant material, the toxic agents may be transferred into the products manufactured from such plants and become hazardous to humans and animals.

Toxic heavy metals have been reported to be absorbed by plant roots and may become enriched in other parts of the plant, causing cytotoxic effects. This includes lead (Lerda, 1992; Wierzbicka, 1984, 1986, 1987a, 1987b, 1988, 1994), arsenic (Srinivas & Rao, 1993), aluminum (Berggren & Fiskesjö, 1987; Liu et al., 1993; Wheeler & Follett, 1991), manganese (Joardar, 1988; Singh & Singh, 1994), nickel (Liu et al., 1994), chromium (Liu et al., 1992; Micera & Dessi, 1988; Srivastava et al., 1994), vanadium (Micera & Dessi, 1989), and many others (Baycheva et al., 1988; Fiskesjö, 1985, 1988, 1993; Ilyushenko & Shchegolkov, 1990; Jirovetz et al., 1993). Also, radioactive fallout can, of course, be absorbed by the plants. Toxic effects of cesium chloride on chromosomes of *Allium sativum* have been reported (Ghosh et al., 1992), but nothing is known about the influence of this mineral in garlic or onion on humans.

The growth of the plants themselves is also affected by various inorganic and organic contaminants. It was reported that boron (Francois, 1991), nicotine (Sopova et al., 1985), salicylate (Briand & Kapoor, 1989), tubulosine (Farjaudon et al., 1988), and chemical mutagens (Böhmova et al., 1988) have cytological effects on *Allium* plants. Even urea, a natural metabolite that is usually considered nontoxic, can induce mitotic and amitotic anomalies in the roots of garlic plants (Chaurasia, 1979). Antibiotics, such as penicillin, cloxacillin, and streptomycin, cause cytological effects in meristematic cells of *Allium* plants (Dobrzynska & Podbielkowska, 1991). The growth yield of *Allium* plants is highly sensitive to increased salinity of the soil (Singh & Abrol, 1985).

Pest control in *Allium* crops is not solely an agricultural problem; it creates also potential ecological hazards to the consumers, since residues of the preservatives can contaminate the end products and thus induce toxicological effects in humans and animals, as well as in the crop itself. Residue analysis is therefore an important measure against the danger of poisoning (Waliszewski & Waliszewski, 1986; Blum & Gabardo, 1993; Bugaret & Marin, 1992; Cessna, 1991a, 1991b, 1991c; DeResende et al., 1987; Duhova et al., 1986; Malet, 1993; Ragab et al., 1984; Ramos et al., 1987; Sediyama et al., 1992; Stan & Christall, 1991; Gulati et al., 1994). Some of the pesticides used in garlic plantations can cause genetic disturbances in the plant, as well as in the parasites (Choi et al., 1992, 1993; Sumi et al., 1993). The same is true for herbicides used for weed control (Igue et al., 1982; Nortje & Henrico, 1985).

Unclear as yet is the influence on human health by measures that are usually taken to preserve crop plants, either by chemical treatment, e.g., with ethylene oxide (Farkas & Andrassy, 1988) or fungicides (Lammerink, 1990), or by γ-irradiation (Ceci et al., 1992; Croci et al., 1987, 1990, 1994; Curzio et al., 1986; El-Oksh et al., 1971; Kwon et al., 1984, 1988, 1989; Malpathak & David, 1986; Nunez et al., 1985; Narvaiz, 1995; Kobayashi et al., 1994), microwave treatment (Riva et al., 1993; Yu et al., 1993), freezing (Farkas, 1987; Park et al., 1988; Wäfler et al., 1994), and others (see Federal Register, 1985). It is well-known that the parasites, mainly fungi and viruses which normally dwell on *Allium* plants, can be efficiently brought under control by such treatment; however, it is also true that the plant material and/or its chemical composition can be significantly altered by these conservation methods (Arguello et al., 1983, 1986; de Carvalho et al., 1987; Kalra, 1987; Zhao & Wang, 1983). The toxicologic consequences of this processing have not yet been satisfactorily investigated.

8
Nonmedicinal Applications of Garlic and Curiosities

HEINRICH P. KOCH

Under "nonmedicinal" applications, the reader will naturally think of the usage of garlic in the kitchen as a spice or condiment for meals. These aspects should in no way be underestimated. The position of garlic as a spice is equal to that as a medicine. The feeling of well-being that everyone knows after a delicious meal is, after all, an effect that greatly contributes to our health. This book, though, is devoted to the therapeutic importance of the plant *Allium sativum*, and so the usage of garlic in the art of cooking should be excluded here. Nevertheless, several books have been published that properly extol garlic as a vegetable and spice. These books also contain numerous recipes from the German-Austrian region (Zeller, 1985), the cuisine Provençale (Lichtner, 1984), and the annual Gilroy Garlic Festival in California (Gilroy, 1987). With this brief comment, then, we leave the subject. The only fact that must be mentioned in this context is this: garlic is also a remarkably effective preservative for meat products, even if they are made from such extraordinary starting material as camel meat (Heikal et al., 1972).

Garlic is a truly remarkable means of preserving well-being, even under highly stressful conditions. This is confirmed by the report of a 60-year-old man who toured through all of Europe on his bicycle, taking nothing other than garlic and red beets as his medicine (Nassauer, 1990).

Let us dwell for a moment on a medicinal, or rather pseudomedicinal aspect of garlic. Perhaps the reader has wondered why Count Dracula, vampires, and similar figures exist in fictional works and folklore. Even a serious scientific work, if it claims to discuss all interesting information about garlic, cannot ignore the theme of Dracula in aesthetic literature. Ever since the classic *Dracula* novel by Bram Stoker, the theme has occurred again and again in many variations, in films on television and movies as well (Stoker, 1966; Ludlam, 1962; Masters, 1972; Sturm, 1968). It is surprising that the vampire Dracula originated with a historically documented person from "Transylvania," which, according to the novel, is a land of the eastern Carpathian mountains (probably Wallachia). (More details can be found in Sturm, 1968, and Weslowski, 1910). The curiosity is that the creatures of horror in the Hollywood films may have really existed and still do live among us, though not in the described versions. These are in no way supernatural beings, but simply humans who suffer from a rare disease called *porphyria*.

This disease, for which there is still no known cure, is caused by a genetic enzyme defect, or an acquired metabolic disturbance, in which the porphyrin synthesis in the blood-generating organs and/or the liver is affected. It leads to abnormal formation and deposition of the pigment and its precursors in urine (porphyrinuria). The increased porphyrin, moreover, causes extreme light sensitivity in the afflicted, who can acquire severe skin lesions through exposure to sunlight. That should explain why Dracula avoided the sun! (Jain & Apitz-Castro, 1987; Hall, 1986)

In addition, these patients suffer from other highly undesirable symptoms, such as colic, pareses, circulation irregularities, respiratory paralysis, and depression. Another symptom of this disease is the deficiency of iron-containing hemoglobin, the red pigment of the blood, which is of vital importance for oxygen transport. The sulfur-containing ingredients of garlic exacerbate the disease. This would be a very plausible explanation for the legendary preference of Count Dracula and his "Undeads" for fresh blood and his abhorrence of garlic odor. The blood-sucking could possibly be interpreted as a primitive attempt to treat this disease. All in all, it is an astonishing story. An Italian saying goes: *"Si non è vero, è ben trovato!"* (if it is not true, it is well composed).

The molecular basis of the putative "vampire-repellent" action of garlic and other matters related to heart and blood have been amply discussed by Jain and coworkers (Jain & Apitz-Castro, 1987, 1993; Apitz-Castro & Jain, 1987; Jain et al., 1988; Allais, 1987; Masters, 1972).

To complete this topic, a real case of auto-vampirism was recently reported: a 21-year-old prisoner showed repeatedly typical symptoms of gastrointestinal bleeding with tar-like feces and low hemoglobin values. After a thorough medical examination and interrogation, he confessed to have cut his forearm with a razor blade and sucked the blood during the night. With the resulting symptoms, he was able regularly to escape the prison and land in the hospital (Halevy et al., 1990). However, vampirism of this kind has long been known as a clinical phenomenon (McCully, 1964; Prins, 1984; Van den Bergh & Kelly, 1964).

Another garlic phenomenon with medical-scientific background which aroused much attention in recent years is "the case of the black-spotted dolls" ("The garlic touch") (Anonym, 1986a, 1986b). Harris reported a case of a patient who made imitations of antique porcelain dolls. It was observed that whenever the patient had touched the porcelain head of the doll with her hands as it was painted, black spots appeared upon firing. The doll heads became worthless. Closer investigation revealed that the patient consumed large amounts of garlic. The volatile sulfur compounds were eliminated through perspiration on the hands. Upon firing, they had reacted with the iron contained in the porcelain clay, forming black iron sulfide; hence, the black spots (Harris et al., 1986).

The phenomenon is probably based on a rare metabolic anomaly of the patient. It could be shown that she had only a limited capacity to degrade sulfur compounds via sulfoxidation and to eliminate them through the urine. The (technical) problem was solved with a garlic-free diet. Inquiries at other companies that manufacture such dolls revealed that approximately 10% of the trained doll makers produced rejects with black spots (Anonym, 1986a, 1986b).

In this context, the discolorations of garlic (and onion) preparations, like chopped slices, powders, purées, condiments, and the like deserve attention, at least from the manufacturers of such products. Various measures are undertaken to avoid the problems resulting from such discolorations. While garlic preparations undergo a greening reaction followed by browning (Bae & Lee, 1990; Kim et al., 1992; Lukes, 1986; Pruthi et al., 1960; Wu, 1992), onion products show a reddish discoloration, finally leading to a browning as well (Joslyn & Peterson, 1958, 1960; Joslyn & Sano, 1956; Li et al., 1967; Mizoi et al., 1992; Shannon et al., 1967a, 1967b; Butz et al., 1994).

Still another peripheral medical problem, which is only indirectly related to garlic, should be mentioned. Every anesthesiologist knows the phenomenon that patients who are treated intravenously with the basic narcotic *thiopental* perceive the taste (or is it an odor perception?) of garlic or onion in the mouth following injection. There is no explanation yet for this phenomenon. The English anesthesiologist Body thinks the answer lies in the similarity of structure between the molecules of thiopental and allicin (Body, 1986).

Another remarkable observation is that the characteristic odor components of garlic permeate into the breast milk after ingestion of garlic by the mother (Mennella & Beauchamp, 1991, 1993). It was reported that babies of women who were treated with 1.5 g garlic extract in a capsule consumed up to 140% more milk than those of the control group, whose mothers had no intake of garlic. Thus, the offspring of mothers fond of garlic become "garlic lovers" themselves, already in their earliest days of life (Koch, 1992b; Anonym, 1993).

Not only are humans fond of garlic's aroma, but some animals are also heavily attracted by it. A scent attractant for fishing lures or fish feeds containing garlic oil was patented recently (Ookubo et al., 1990; Rawlins, 1994). Many insects—onion flies, leek moths, and seed corn flies, for instance—are strongly attracted by the volatile odorous sulfur compounds, so that these can be properly used in insect traps (Auger et al., 1989, 1993; Ishikawa et al., 1978; Miller et al., 1984; Soni & Finch, 1979). Even leeches (*Hirudinea*) are strongly attracted by the garlic smell. In an experiment, when a subject's forearm was smeared either with garlic, ale, soured cream, or water (as control), the leeches showed the fastest onset of sucking and longest sucking time in those cases where garlic was used, while the others did not differ from the controls (Baerheim & Sandvik, 1994).

On the other hand, it must be mentioned in this context that garlic smell can be also a rather effective repellent for fleas, ticks, and other blood-feeding insects, especially for mosquitoes (*Culex* species), black flies, and fruit flies (*Drosophila melanogaster*), as well as for cockroaches (Amonkar & Reeves, 1970; Bhuyan et al., 1974; Catar, 1954; Honma, 1990a, 1990b; Iwanami et al., 1985; Reznik & Imbs, 1965; Weisler, 1989).

In higher animals, the preference for eating certain feed, especially garlic, may be acquired even before birth. This was found in baby rats whose mothers were fed with garlic during the last week of pregnancy. By the age of 12 days, the offspring showed a distinct preference for garlic over onions. In contrast, the offspring of mothers who had not eaten garlic showed no such preference. This implies that the fetus can experience certain smells in the womb if the odoriferous chemicals dissolve in the blood and cross the placenta (Hepper, 1988a, 1988b). Indeed, it has been shown in sheep that the odorants of garlic appear in the fetal environment shortly after it is fed to the ewe (Nolte et al., 1992).

Garlic's typical odor is favored by *alliophils* (garlic lovers) but rejected by *alliophobes* (people who hate it), and the latter are the majority, at least in Western society (except, of course, in some Mediterranean countries). Thus, attempts to hide or mask the typical garlic odor (in medicinal products) have been almost uncountable, especially in the patent literature (Table 8.1). On the other hand, attempts to mask the odor of another substance by adding garlic are truly exceptional: a Japanese patent claims the masking of an unwanted ethanol odor (!) in raw noodles which, of course, is only effective with garlic (Hokkaido Wako Shokuhin Kk., 1982). Garlic, added to the feed of broiler chicken, produces an improved flavor in the meat (Sanick, 1973).

A curious story from Africa should also be mentioned. In the folk medicine of the Yoruba people in Nigeria, there is a brew prepared mainly from the urine of cows ("cow's urine concoction"), which is used as a remedy for convulsions in children. Scientists at the Ife University in Nigeria had a closer look at this remedy, and were able to show the presence of nicotine and the constituents of garlic and onions in the brew—which is prepared by soaking varying quantities of tobacco leaves, the *Allium* plants, and other components in cow's urine. The effectiveness of this folk remedy could be explained in this way (Ojewole et al., 1982; Abraham et al., 1976; Adekile et al., 1983).

A peculiar nonmedical application of garlic is its use as corrosion protection for metal parts made of steel or aluminum that have been exposed to acids. For instance, a garlic extract was found to effect a 70–85% protection against the corrosion of steel by 2N-sulfuric acid or 0.5N-nitric acid (Srivastava & Srivastava, 1981, 1983, 1984). Another application in metallurgy that has long been known is the special affinity of sulfur compounds for heavy metals. Garlic juice has been used for the removal of lead from amalgam for the production of lead-free mercury. It was observed that the amalgamated lead

Table 8.1. Various Attempts to Prepare "Odorless Garlic" as Reported in the International Patent Literature.

Patent No.	Country	Type of Invention, Characteristics	Author, Year
94 209 737	Japan	GE, aq. acid + minerals	(Abe, 1994)
94 78 730	Japan	GE, aq. EtOH, phytic acid, polylysine	(Sakai, 1994b)
94 62 781	Japan	GE, aq. NaHCO$_3$, carboxylic acids	(Sakai, 1994a)
93 344 850	Japan	GE, aq. citric acid, bran extract	(Uchino et al., 1993)
93 304 924	Japan	GJ, soybean oil	(Morinaga, 1993)
93 43 454	Japan	GO, lemon oil	(Kominato & Azuma, 1993)
93 168 431	Japan	WP, high pressure	(Kuwata et al., 1993)
93 123 125	Japan	WP, high pressure	(Kunogi, 1993)
93 103 622	Japan	GE, heat and activated carbon	(Ishihara, 1993)
92 445 767	Japan	GE, NaHCO$_3$, EtOH, carboxylic acids	(Sakai, 1992)
92 440 876	Japan	DP, diglyceryl stearate etc, gelatin capsules	(Shimakawa, 1992)
92 341 154	Japan	DP, enzyme treatment, freeze-drying	(Irie et al., 1992)
92 309 358	Japan	GE, sesame oil, β-cyclodextrin	(Yamanaka & Kobayashi, 1992)
90 284 850	Japan	WP, EtOH, vinegar, sugar	(Samura, 1990b)
90 216 237	Japan	WP, vinegar, EtOH, sugar	(Samura, 1990a)
89 312 979	Japan	DP, *myo*-inositol hexaphosphate	(Sakai, 1989b)
89 281 051	Japan	DP, *myo*-inositol hexaphosphate, iodine	(Sakai, 1989c)
89 273 559	Japan	DP, *myo*-inositol hexaphosphate	(Sakai, 1989a)
89 218 568	Japan	WP, basic salts of Ca or Mg	(Shiga, 1989)
89 113 964	Japan	WP, menthol, alanine, organic acids	(Chiyoma, 1989)
89 104 142	Japan	WP, vitamin B$_1$, excipients	(Nishimura, 1989)
88 309 160	Japan	WP, enzyme treatment, vinegar	(Fujii, 1988)
88 283 550	Japan	GJ, heat, oleic acid, lactose	(Okada, 1988)
87 134 059	Japan	DP, enzyme treatment, spray-drying	(Sumi et al., 1987)
86 91 128	Japan	DP, cyclodextrin, freeze-drying	(Kawashima, 1986)
86 104 761	Japan	WP, sweet *sake*, vinegar, citric acid	(Fujiwara, 1986)
85 259 157	Japan	WP, silica, phytic acid, ZnSO$_4$	(Sakai, 1985)
84 224 664	Japan	WP, vitamin C, saponin, dried plants	(Osaka Yakuhin Kenkyu. Kk., 1984)
84 44 318	Japan	GO, β-cyclodextrin, amylase	(Riken Chem. Ind. Ltd., 1984)
83 217 596	Japan	GO, carboxyethyl-germanium-sesquioxide	(Araki, 1983)
83 179 477	Japan	WP, polyolefin, Ca or Mg oxide	(Nissan Chem. Ind. Ltd., (1983)
83 21 620	Japan	GO, β-cyclodextrin	(Riken Chem. Ind. Ltd., 1983)
82 155 982	Japan	WP, EtOH	(Hakamada, 1982)
82 29 265	Japan	WP, hot water, silicate, phytic acid	(Sakai, 1981)
82 03 341	Japan	WP, fumaric acid	(Kikkoman Shoyu Ltd., 1982)
81 164 762	Japan	WP, microwave treatment, EtOH	(Yokozawa & Onaga, 1981)
80 34 665	Japan	WP, MgO	(Yoshikawa, 1980)
79 20 159	Japan	DP, Mg(OH)$_2$	(Yoshikawa, 1979)
78 130 455	Japan	WP, acetic acid	(Takada, 1978)
78 39 504	Japan	WP, vinegar, NaCl, SiO$_2$	(Miyaji, 1978)
78 38 650	Japan	WP, carbonate solution, acetic acid	(Nagai, 1978)
78 32 147	Japan	DP, EtOH, Et$_2$O, lactose	(Horie, 1978)
77 44 400	Japan	GE, microwave treatment, flavors, sucrose	(Goku, 1977)
77 15 655	Japan	WP, heating in honey	(Mitomo, 1977)
72 00 302	Japan	WP, heating with xylose	(Uemura, 1972)
70 23 599	Japan	WP, roasting with rice bran	(Kasahara, 1970)
63 12 708	Japan	WP, heating in aq. urea	(Yamamoto, 1963)
63 6561	Japan	WP, NaClO$_3$	(Takeyama, 1963)
3 911 594	Germany	WP, vinegar, honey, aromatic plants	(Weger, 1990)
3 900 447	Germany	WP, lemon juice	(Matthess, 1990)
3 820 666	Germany	WP, aromatic plants	(Kosick, 1989)
3 743 264	Germany	WP, crème fraiche, aromatic plants	(Matthess, 1988)
3 731 675	Germany	WP, aromatic plants	(König, 1989)
3 619 570	Germany	DP, coated tablet with aromatizers	(Leusser, 1987)
3 609 116	Germany	DP, deep-freezing, β-cyclodextrin	(Kirsch, 1987)
3 525 258	Germany	WP, lemon juice, NaCl, aromatizers	(Meier, 1986)
647 067	Germany	WP, hot EtOH vapor	(Moser, 1937)
626 469	Germany	WP, EtOH, Na$_2$SO$_4$	(Sächsisches Serumwerk AG, 1936)
432 053	Germany	GJ, activated carbon	(Sandoz, 1926)
397 464	Austria	WP, aq. EtOH, microwave treatment	(Koch, 1993)
143 320	Austria	GO, cholic acid	(Riedel & De Haen, 1935)
174 460	Switzerland	GO, deoxycholic acid	(Riedel & De Haen, 1935)
599 342	France	GJ, activated carbon	(Sandoz, 1925)

Table 8.1. continued

Patent No.	Country	Type of Invention, Characteristics	Author, Year
6116	Hungary	WP, unsaturated carboxylic acids	(Lonyai et al., 1973)
4 917 881	U.S.A.	DP, ZnO	(Gray, 1990)
3 424 593	U.S.A.	GJ, citrus fruit juice, sugar	(Boxkmann et al., 1969)
3 326 698	U.S.A.	WP, dry distillation under CO_2	(Sakamoto, 1967)
2 490 424	U.S.A.	GO, bromination	(Ferguson, 1949)
1 169 697	Canada	GJ, vegetable oil	(Morinaga, 1984)
1 077 854	China	WP, aromatic oil	(Shao et al., 1993)
244 879	U.S.S.R.	WP, cooling, enzyme removal by aq. extract	(Lazarev & Ivanova, 1969)

WP = whole plant; GE = garlic extract; DP = dry powder; GJ = garlic juice; GO = garlic oil.

was especially easily transformed to lead sulfide (Banerjee, 1913). The purification of mercury with garlic juice, however, has also found practical application in Ayurvedic medicine (the traditional folk medicine of India). A modern high performance liquid chromatography (HPLC) method for this process has been developed recently (Sane et al., 1990).

An alcohol extract from garlic (allilsat) or an ammonia extract from onion/garlic waste products has been added to the medium used in the production of nickel or cobalt metal coatings. It is claimed that this increases the stability of the solution and intensifies the process (Tolok et al., 1971, 1972). No less unusual is the proposed use of garlic powder dispersed in a beeswax candle as a tobacco smoke absorbant (Fructus, 1987).

The literature on garlic, its constituents, and its use in the various fields of human activity has expanded enormously during the past decade. A recent search of the literature pertaining to garlic has revealed more than 2000 papers, the number of which published per year worldwide—in almost every language, even in Esperanto (Fujiwara, 1954)—is still increasing exponentially (Koch, 1992a).

Finally, to conclude this book, here is a true curiosity. In a computer-supported literature search with the English search keyword "garlic," one discovers a computer program called GARLIC. It deals with a system simulating energy distribution in the hot nucleus of atomic reactors, which serves as control for the processes in nuclear fission and calculations for the estimation of safety measures (Ercan et al., 1983; Hoeld & Lupas, 1983; Beraha et al., 1984, 1985). Why the creators of this program called it by this name remains a puzzle. Maybe they are fond of garlic? Thus, garlic, one of the oldest cultured plants of human civilization, has even entered nuclear technique at the end of the 20th century, though only as a name for a simulation program for the operation of nuclear reactors. If that is not an expression of appreciation!

References

Note: For western European languages, original titles are used. Other titles have been translated into English with the original language noted in parentheses. The list includes 2580 references, 2240 of them concerning garlic.

Abbruzzese, M.R., Delaha, E.C., and Garagusi, V.F. (1987) Absence of antimycobacterial synergism between garlic extract and antituberculosis drugs. Diagn. Microbiol. Infect. Dis. 8:79–85.

Abdel-Fattah, A.F. and Edrees, M. (1971) Chemical investigations on some constituents of pigmented onion skins. J. Sci. Food Agric. 22:298–300.

Abdel-Fattah, A.F. and Edrees, M. (1972) A study on the composition of garlic skins and the structural features of the isolated pectic acid. J. Sci. Food Agric. 23:871–877. Chem. Abst. 77 (1972) 138 509.

Abdel-Fattah, A.F. and Edrees, M. (1973) The pectic substances of pigmented onion skins. II. Some structural features of the pectin and pectic acid. Carbohydrate Res. 28:114–117.

Abdel-Fattah, A.F. and Khaireldin, A.A. (1970) Pectin of garlic skins. U.A.R. J. Chem. 13:27–37. Chem. Abst. 74 (1971) 63 277.

Abdo, M.S., Mansour, S.A., and El-Nahla, A.M. (1983) Effect of some feed additives on blood constituents of growing Hubbard chickens. Vet. Med. J. 31:221–231. Chem. Abst. 101 (1984) 37 558.

Abdo, M.S. and Al-Kafawi, A.A. (1969) Biological activities of *Allium sativum*. Jap. J. Pharmac. 19:1–4.

Abdou, I.A., Abou-Zeid, A.A., El-Sherbeeny, M.R., and Abou-El-Gheat, Z.H. (1972) Antimicrobial activities of *Allium sativum*, Allium cepa, Raphanus sativus, Capsicum frutescens, Eruca sativa, Allium kurrat on bacteria. Qual. Plant. Mater. Veg. 22:29–35. Chem. Abst. 78 (1973) 80 226.

Abdullaeva, A.A. (1959) Germicidal properties of volatile fractions and juices of Allium aflatunense and *Allium sativum*. Dokl. Akad. Nauk. Uzbek. (Rep. Acad. Sci. Uzbekistan), 43–45. Chem. Abst. 54 (1960) 7067 (Russian).

Abdullah, T.H., Kandil, O., Elkadi, A., and Carter, J. (1988) Garlic revisited: therapeutic for the major diseases of our times? J. Nat. Med. Assoc. 80:439–445. Chem. Abst. 109 (1988) 21 973. Int. Pharm. Abst. 26 (1989) 2286.

Abdullah, T.H., Kirkpatrick, D.V., and Carter, J. (1989) Enhancement of natural killer cell activity in AIDS with garlic. Dtsch. Ztschr. Onkologie 21:52–53.

Abe, T. (1994) Manufacture of seasoning from garlic. Japan. Patent 94 209 737. Chem. Abst. 122 (1995) 8567.

Abo El-Nil, M.M. (1977) Organogenesis and embryogenesis in callus cultures of garlic (*Allium sativum* L.). Plant Sci. Lett. 9:259–264.

Abo-Doma, M.H., Said, M.M., Shahat, N.E., Riad, R.F., and Nassar, A.R. (1991) Effect of onion and garlic oils as well as some of their sulfur constituents on lipid metabolism in albino rats. J. Drug Res. 20:1–11. Chem. Abst. 120 (1994) 95 398.

Aboul-Enein, A.M. (1986) Inhibition of tumor growth with possible immunity by Egyptian garlic extracts. Nahrung 30:161–169.

Abraham, K.O., Shankaranarayana, M.L., Raghavan, B., and Natarajan, C.P. (1976) Alliums—varieties, chemistry, and analysis. Lebensm. Wiss. Technol. 9:193–200. Chem. Abst. 85 (1976) 141 367.

Abraham, S.K. and Kesavan, P.C. (1984) Genotoxicity of garlic, turmeric, and asafoetida in mice. Mutat. Res. 136:85–88. Chem. Abst. 101 (1984) 5666.

Abramjan, A.A. and Sarkisjan, R.S. (1963) Oxydation organischer Verbindungen mit Kaliumpermanganat und Gesamtbestimmung des Schwefels. Izv. Akad. Nauk Armijan. SSR Ser. Khim. (Bull. Acad. Sci. Armenian SSR, Chem. Ser.) 16:131–135. Chem. Abst. 59 (1963) 13 343 (Russian).

Abuirmeileh, N., Yu, S.G., Qureshi, N., I-San Lin, R., and Qureshi, A.A. (1991) Suppression of cholesterogenesis by Kyolic and S-allyl cysteine. FASEB J. 5:A1756.

Adamu, I., Joseph, P.K., and Augusti, K.T. (1982) Hypolipidemic action of onion and garlic unsaturated oils in sucrose fed rats over a two-month period. Experientia 38:899–901. Chem. Abst. 97 (1982) 126 169.

Adekile, A.D., Odebiyi, O.O., Ojewole, J.A.O., and Ogunye, O. (1983) The toxic principles in cow's urine concoction (CUC). J. Trop. Pediatr. 29: 283–288.

Adetumbi, M.A. and Lau, B.H.S. (1983) *Allium sativum* (garlic)—a natural antibiotic. Med. Hypotheses 12:227–237.

Adetumbi, M.A. (1985) The antimycotic property and mode of action of *Allium sativum* (garlic). Diss. Abstr. Int. B. 46:60. Chem. Abst. 103 (1985) 119 797.

Adetumbi, M.A., Javor, G.T., and Lau, B.H.S. (1986) *Allium sativum* (garlic) inhibits lipid synthesis by Candida albicans. Antimicrob. Agents Chemoth. 30:499–501. Chem. Abst. 105 (1986) 149 641.

Adetumbi, M.A. and Lau, B.H.S. (1986) Inhibition of in vitro germination and spherulation of Coccidioides immitis by *Allium sativum*. Curr. Microbiol. 13:73–76.

Adoga, G.I. and Osuji, J. (1986) Effect of garlic oil extract on serum, liver, and kidney enzymes of rats fed on high sucrose and alcohol diets. Biochem. Int. 13:615–624. Chem. Abst. 106 (1987) 17 439.

Adoga, G.I. (1987) The mechanism of the hypolipidemic effect of garlic oil extract in rats fed on high sucrose and alcohol diets. Biochem. Biophys. Res. Commun. 142:1046–1052. Chem. Abst. 106 (1987) 155 247.

Adoga, G.I. and Ohaeri, C.O. (1991) Effect of garlic oil on prothrombin, thrombin, and partial thromboplastin times in streptozotocin-induced diabetic rats. Med. Sci. Res. *19*:407–408. Chem. Abst. 115 (1991) 198 189.

Afzal, M., Hassan, R.A.H., El-Kazimi, A.A., and Fattah, R.M.A. (1985) *Allium sativum* in the control of atherosclerosis. Agric. Biol. Chem. *49*: 1187–1188. Chem. Abst. 103 (1985) 21 431.

Agarwal, K.C. (1992) Garlic: antithrombotic actions. Drug News & Persp. *5*:592–594.

Agarwal, R.K., Dewar, H.A., Newell, D.J., and Das, B. (1977) Controlled trial of the effect of cycloalliin on the fibrinolytic activity of venous blood. Atherosclerosis *27*:347–351. Chem. Abst. 87 (1977) 111 919.

Agarwala, S.C., Murti, C.R.K., and Shirvastava, D.L. (1952) Enzyme inhibition in relation to drug action. I. Effect of certain antibiotics on urease. J. Sci. Ind. Res. *11B*:165–169. Chem. Abst. 46 (1952) 9714.

Agrawal, P. and Rai, B. (1948) Effect of bulb extracts of onion and garlic on soil bacteria. Acta Bot. Indica *12*:45–49.

Agrawal, R., Mohanty, B., Singh, N.S., and Patwardhan, M.V. (1991) Parameters affecting allicin formation as a secondary metabolite by static cultures of *Allium sativum*. J. Sci. Food Agric. *57*:155–162. Chem. Abst. 116 (1992) 55 654.

Ahmad, W., Syed, Q., Hussain, A., Ghuman, M.A., and Zaidi, S.A. (1993) Evaluation of different garlic extracts for antibacterial activity. Sci. Int. [Lahore] *5*:385–386. Chem. Abst. 122 (1995) 5069.

Ahmad, Y. (1986) Hypocholesterolemic effect of *Allium sativum* Linn. and its potential protective action against coronary heart disease. In: Natural product chemistry. A.U. Rahman, ed. Springer, Berlin, pp. 23–44.

Ahmed, K.H. and Benigno, D.A. (1984) Control of the eriophid mite Aceria-Tulipae, the cause of tangletop malady and the vector of garlic mosaic virus. Indian J. Plant Prot. *12*:153.

Ahmed, N. and Sultan, K. (1984) Fungitoxic effect of garlic (*Allium sativum*) on treatment of jute Corchorus Capsularis seed. Bangladesh J. Bot. *13*:130.

Ahujarai, P.L. and Bhatia, B. (1984) Heat tolerance of carbon tetrachloride–treated animals and its modification by some agents. Int. J. Biometeorol. *28*:85–92. Chem. Abst. 101 (1984) 145 306.

Akaranta, O. and Odozi, T.O. (1986) Antioxidant properties of red onion skin (Allium cepa) tannin extract. Agricultural Wastes *18*:299–303.

Akema, R., Okazaki, N., and Takizawa, K. (1987) Antibacterial substance in commercial Allium plants. Kanagawa-ken Eisei Kenkyusho Hokoku, 39–40. Chem. Abst. 108 (1988) 201 780.

Aki, O. (1993) A study of bath preparations containing the garlic extract-vitamin B1-complex: a study of its characteristics and clinical application for the treatment of atopic dermatitis. Fure gur ansnjanarn (Fragrance J.) *21*:47–57. Chem. Abst. 118 (1993) 197 751 (Japanese).

Al-Bekairi, A.M., Shah, A.H., and Qureshi, S. (1990) Effect of *Allium sativum* on epididymal spermatozoa, estradiol-treated mice, and general toxicity. J. Ethnopharmacol. *29*:117–125. Int. Pharm. Abst. 28 (1991) 2243.

Al-Bekairi, A.M., Qureshi, S., and Shah, A.H. (1991) Toxicity studies on Allium cepa, its effect on estradiol treated mice and on epididymal spermatozoa. Fitoterapia *62*:301–306.

Albertus Magnus. (1867) De Vegetabilibus, Libri VII, historiae naturalis pars XVIII. K. Jessen, Berlin.

Al-Delaimy, K.S. and Ali, S.H. (1970) Antibacterial action of vegetable extracts on the growth of pathogenic bacteria. J. Sci. Food Agric. *21*:110–112.

Alexander, M.M. and Sulebele, G.A. (1973) Pectic substances in onion and garlic skins. J. Sci. Food Agric. *24*:611–615. Chem. Abst. 79 (1973) 17 156.

Ali, M. and Thomson, M. (1995) Consumption of a garlic clove a day could be beneficial in preventing thrombosis. Prostaglandins Leukotrienes Essent. Fatty Acids *53*:211–212.

Alfonso, N. and Lopez, E. (1960) Bestimmungsmethoden für den Geruchswert mexikanischer Knoblaucharten. Z. Lebensm. Unters. Forsch. *111*:410–413. Chem. Abst. 54 (1960) 13 487.

Ali, A.A., El-Shanawany, M.A., Rhamadan, M.A., and El-Tewasy, O.M. (1990) Study of mucilage and pectins of Pancratium sickenbergeri Asch et. SFTH and *Allium sativum* L. Bull. Pharm. Sci. Assuit Univ. *13*:1–8. Chem. Abst. 115 (1991) 203 270.

Ali, M. and Mohammed, S.Y. (1986) Selective suppression of platelet thromboxane formation with sparing of vascular prostacyclin synthesis by aqueous extract of garlic in rabbits. Prostaglandins Leukotrienes Med. *25*:139–146.

Ali, M., Afzal, M., Hassan, R.A.H., Farid, A., and Burka, J.F. (1990a) Comparative study of the in vitro synthesis of prostaglandins and thromboxanes in plants belonging to Liliaceae family. Gen. Pharmacol. *21*:273–276. Chem. Abst. 113 (1990) 3299.

Ali, M., Thomson, M., Alnaqeeb, M.A., Al-Hassan, J.M., Khater, S.H., and Gomes, S.A. (1990b) Antithrombotic activity of garlic: its inhibition of the synthesis of thromboxane-B2 during infusion of arachidonic acid and collagen in rabbits. Prostaglandins Leukotrienes Essent. Fatty Acids *41*:95–99.

Ali, M., Afzal, M., Abul, Y.S., Saleh, J.A., Gubler, C.J., and Dhami, M.S.I. (1991) Changes in the levels of lactic dehydrogenase and trans-ketolase in liver and red cells of rats after treatment with garlic extracts. J. Environ. Sci. Health *A26*:1–11. Chem. Abst. 114 (1991) 162 995.

Alkiewicz, J. and Lutomski, J. (1992) Die Untersuchungen zur Elimination von Candida albicans durch die Chloroformfraktion des Knoblauchsaftes. Herba Pol. *38*:79–83. Int. Pharm. Abst. 30 (1993) 13 152.

Allais, C. (1987) L'ail fait-il fuir les vampires? Recherche *18*:1413–1414.

Al-Nagdy, S., Abdel Rahman, M.O., and Heiba, H.I. (1986) Extraction and identification of different prostaglandins in Allium cepa. Comp. Biochem. Physiol., C: Comp. Pharmacol. Toxicol. *85C*:163–166. Chem. Abst. 105 (1986) 222 788.

Al-Nagdy, S.A., Abdel-Rahman, M.O., and Heiba, H.I. (1988) Evidence for some prostaglandins in *Allium sativum* extracts. Phytother. Res. *2*:196–197. Chem. Abst. 110 (1989) 170 228.

Amagase, H. and Milner, J.A. (1993a) Impact of various sources of garlic and their constituents on 7,12-dimethylbenz[a]anthracene binding to mammary cell DNA. Carcinogenesis *14*:1627–1631. Chem. Abst. 119 (1993) 225 096.

Amagase, H. and Milner, J.A. (1993b) Impact of dietary lipids on the ability of garlic to inhibit 7,12-dimethylbenz(a)anthracene (DMBA) binding to mammary DNA. FASEB J. *7*:A69.

Amer, M.A. and Brisson, G.J. (1973) Selenium in human foodstuffs collected at the Ste-Foy (Quebec) food market. J. Inst. Can. Sci. Technol. Aliment. *6*:184–187.

Amer, M., Taha, M., and Tosson, Z. (1980) The effect of aqueous garlic extract on the growth of dermatophytes. Int. J. Dermatol. *19*:285–287.

Amonkar, S.V. and Banerj, A. (1971) Isolation and characterization of larvicidal principle of garlic. Science *174*:1343–1344. Chem. Abst. 76 (1972) 109 198.

Amonkar, S.V. and Reeves, E.L. (1970) Mosquito control with active principle of garlic, *Allium sativum*. J. Econ. Entomol. *63*:1172–1175.

Anantakrishnan, C.P. and Venkataraman, P.R. (1940a) The chemistry of garlic (*Allium sativum* L.). I. The nitrogen complex. Proc. Indian Acad. Sci. *12B*:268–276. Chem. Abst. 35 (1941) 3684.

Anantakrishnan, C.P. and Venkataraman, P.R. (1940b) The chemistry of garlic (*Allium sativum* L.). II. The phosphorus distribution. Proc. Indian Acad. Sci. *12B*:277–282. Chem. Abst. 35 (1941) 3684.

Anantakrishnan, C.P. and Venkataraman, P.R. (1941) The chemistry of garlic (*Allium sativum* L.). III. The reserve polysaccharides. Proc. Indian Acad. Sci. *13B*:129–133. Chem. Abst. 35 (1941) 5160.

Andreotta, R. (1967) Hair dressing. France. Patent 1 481 008. Chem. Abst. 67 (1967) 111 344.

Anonym. (1976) The herbs and the heart. Nutr. Rev. *34*:43–44.

Anonym. (1980) Hemmt Knoblauch Tumorwuchs? Spiegel *34*:265.

Anonym. (1983) Wirkungsmechanismus des Knoblauchs weitgehend aufgeklärt. Heilkunst 3.

Anonym. (1984) Knoblauch steigert die Leistung des Nervensystems. Mod. Leben Nat. Heilen *109*:44.

Anonym. (1985) Homöopathie: Allgemein gültige Erläuterungen. Pharm. Rundsch. *10*:42–44.

Anonym. (1986a) Der Fall der schwarzfleckigen Puppen. Chem. Unserer Zeit *20*:101.

Anonym. (1986b) The garlic touch. Pharm. J. *236*:383.

Anonym. (1988) Wann wirken Knoblauch-Präparate? Z. Phytother. *9*:19–20.

Anonym. (1992) Knoblauchpräparate in "Test": Zum überwiegenden Teil teure Placebos. Dtsch. Apoth. Ztg. *132*:643–644.

Anonym. (1993) Babys lieben Knoblauchgeschmack. Das freie Medikament (April).

Antsupova, T.P. and Polozhii, A.V. (1987) Presence of alline in some species of Allium L. in the Buryat Republic. Rastit. Resur. *23*:436–439. Chem. Abst. 107 (1987) 214 830 (Russian).

Anussorn-Nitisara, N., Vongratanastit, T., and Wuthi-Udomlert, M. (1979) Pharmaceutical preparation of *Allium sativum*. Varasarn Paesachasarthara *6*:31–38. Chem. Abst. 92 (1980) 47 166.

Apitz-Castro, R., Cabrera, S., Cruz, M.R., Ledezma, E., and Jain, M.K. (1983a) Effects of garlic extract and of three pure components isolated from it on human platelet aggregation, arachidonate metabolism, release reaction, and platelet ultrastructure. Thromb. Res. *32*:155–169. Chem. Abst. 100 (1984) 17 422.

Apitz-Castro, R., Cabrera, S., and Vargas, J.R. (1983b) Inhibition of the platelet reaction by essential garlic oil in vitro. Fed. Proc. *42*:1991.

Apitz-Castro, R., Escalante, J., Vargas, R., and Jain, M.K. (1986a) Ajoene, the antiplatelet principle of garlic, synergistically potentiates the antiaggregatory action of prostacyclin, forskolin, indomethacin, and dipyridamole on human platelets. Thromb. Res. *42*:303–311. Chem. Abst. 105 (1986) 401.

Apitz-Castro, R., Ledezma, E., Escalante, J., and Jain, M.K. (1986b) The molecular basis of the antiplatelet action of ajoene: direct interaction with the fibrinogen receptor. Biochem. Biophys. Res. Commun. *141*:145–150.

Apitz-Castro, R.J. and Jain, M.K. (1987) Preparation and formulation of (E,Z)-5,5,9-trithiadodeca-1,6,11-triene 9 oxides as antithrombotics. U.S.A. Patent 4 665 088. Chem. Abst. 107 (1987) 39 194.

Apitz-Castro, R., Ledezma, E., Escalante, J., Jorquera, A., Piñate, F.M., Moreno-Rea, J., Carrillo, G., and Leal, O. (1988) Reversible prevention of platelet activation by (E,Z)-4,5,9 trithiadodeca-1,6,11-triene 9-oxide (ajoene) in dogs under extracorporeal circulation. Arzneim. Forsch. *38*:901–904.

Apitz-Castro, R., Jain, M.K., Bartoli, F., Ledezma, E., Ruiz, M.C., and Salas, R. (1991) Evidence for direct coupling of primary agonist-receptor interaction to the exposure of functional IIb–IIIa complexes in human blood platelets: results from studies with the antiplatelet compound ajoene. Biochim. Biophys. Acta *1094*:269–280. Chem. Abst. 115 (1991) 229 284.

Apitz-Castro, R., Badimon, J.J., and Badimon, L. (1992) Effect of ajoene, the major antiplatelet compound from garlic, on platelet thrombus formation. Thromb. Res. *68*:145–155.

Apitz-Castro, R., Badimon, J.J., and Badimon, L. (1994) A garlic derivative, ajoene, inhibits platelet deposition on severely damaged vessel wall in an in vivo porcine experimental model. Thromb. Res. 75:243–250.

Appleton, J.A. and Tansey, M.R. (1975) Inhibition of growth of zoopathogenic fungi by garlic extract. Mycologia 67:882–885.

Aqel, M.B., Gharaibah, M.N., and Salhab, A.S. (1991) Direct relaxant effects of garlic juice on smooth and cardiac muscles. J. Ethnopharmacol. 33:13–19. Int. Pharm. Abst. 29 (1992) 1705.

Araki, M., Yokota, Y., Kuga, M., Chin, S., Fujikawa, F., Nakajima, K., Fujii, H., Tokuoka, A., and Hirota, Y. (1952) Anthelminthics. Yakugaku Zasshi (J. Pharm. Soc. Japan) 72:979–982. Chem. Abst. 46 (1952) 10 430.

Araki, Y. (1983) Odorless garlic oil. Japan. Patent 83 217 596. Chem. Abst. 100 (1983) 137 695.

Aramaki, Y., Kobayashi, T., Furuno, K., Ishikawa, I., Suzuoki, Z., and Shintani, S. (1959) Thiamine tetrahydrofurfuryl disulfide: several biological activities of thiamine tetrahydrofurfuryl disulfide. Bitamin [Kyoto] 16:240–244 (Japanese).

Archbold, H.K. (1940) Fructosans in the monocotyledons: a review. New Phytol. 39:185–219. Chem. Abst. 34 (1940) 7335.

Arguello, J.A., Bottini, R., Luna, R., De Bottini, G.A., and Racca, R.W. (1983) Dormancy in garlic (*Allium sativum* L.) cv. Rosado Paraguayo. I. Levels of growth substances in "seed cloves" under storage. Plant Cell Physiol. 24:1559–1593. Chem. Abst. 100 (1984) 48 699.

Arguello, J.A., De Bottini, G.A., Luna, R., and Bottini, R. (1986) Dormancy in garlic (*Allium sativum* L.) cv. Rosado Paraguayo. II. The onset of the process during plant ontogeny. Plant Cell Physiol. 27:553–557. Chem. Abst. 104 (1986) 222 098.

Ariga, I. and Hideka, K. (1985) Chemical synthesis of allyl methyl trisulfide, a component of garlic oil. Bull. Coll. Agr. & Vet. Med. Nihon Univ., 68–74. Chem. Abst. 103 (1985) 5867 (Japanese).

Ariga, T. and Oshiba, S. (1981a) Inhibition of platelet aggregation by garlic oil components—separation and identification of an effective substance. Igaku no Ayumi 118:859–862 (Japanese).

Ariga, T. and Oshiba, S. (1981b) Effects of the essential oil components of garlic cloves on rabbit platelet aggregation. Igaku to Seibutsugaku (Medicine & Biology) 102:169–174. Chem. Abst. 95 (1981) 126 665.

Ariga, T., Oshiba, S., and Tamada, T. (1981) Platelet aggregation inhibitor in garlic. Lancet 1:150–151.

Ariga, T. and Kase, H. (1986) Composition of essential oils of the genus Allium and their inhibitory effect on platelet aggregation. Bull. Coll. Agric. & Vet. Med. Nihon Univ. 43:170–175 (Japanese).

Arime, M. and Deki, M. (1983) Components of sugars in garlic. Kanzei Chuo Bunsekishoho 23:89–93. Chem. Abst. 99 (1983) 21 112.

Arora, R. and Palta, J.P. (1991) A loss in the plasma membrane ATPase activity and its recovery coincides with incipient freeze-thaw injury and postthaw recovery in onion bulb scale tissue. Plant Physiol. 95:846–852.

Arora, R.C., Arora, S., and Gupta, R.K. (1981) The long-term use of garlic in ischemic heart disease. Atherosclerosis 40:175–179.

Arora, R.C., Arora, S., and Nigam, P. (1985) Rationale of garlic use in ischemic heart disease? Mater. Med. Pol. 17:48–50.

Arora, R.C. and Arora, S. (1981) Comparative effect of clofibrate, garlic, and onion on alimentary hyperlipemia. Atherosclerosis 39:447–452.

Artacho, M.R., Olea Serrano, M.F., and Ruiz Lopez, M.D. (1992a) Comparative study by gas chromatography-mass spectrometry of methods for the extraction of sulfur compounds in Allium cepa L. Food Chem. 44:305–308. Chem. Abst. 117 (1992) 46 842.

Artacho, M.R., Ruiz, M.D., Olea, F., and Olea, N. (1992b) A preliminary study of the action of the genus Allium on thyroid iodine-131 uptake in rats. Rev. Esp. Fisiol. 48:59–60. Chem. Abst. 119 (1993) 40 296.

Arzamastsev, E.V. (1993) Immunomodulating properties of garlic powder. Unpublished data presented at the International Conference "New Mechanisms for Preventive and Therapeutic Actions of Garlic," Nuremberg, 11–22 January 1993.

Asada, H., Torimoto, N., and Takaoka, A. (1992) Plant pigments as teaching materials. III. Onion (Allium cepa). Kagaku to Kyoiku 40:316–319. Chem. Abst. 117 (1992) 89 552.

Aschoff, L. (1900) Das Knoblauchlied aus dem Bower Manuscript. Janus 5:493–501.

Askar, A. and Bielig, H.J. (1983) Selenium content of food consumed by Egyptians. Food Chem. 10:231–234. Chem. Abst. 98 (1983) 124 499.

Atal, C.K. and Sethi, J.K. (1961) Occurrence of amino acids and alliin in the Indian Alliums (garlics). Curr. Sci. 30:338–340. Chem. Abst. 56 (1962) 1524.

Atroshenko, Y.S. (1954) Treatment of children suffering from whooping cough with phytoncides of garlic. Sov. Med. 18:33–34 (Russian).

Attrep, K.A., Mariani, J.M., and Attrep, M. (1973) Search for prostaglandin A1 in onion. Lipids 8:484–486. Chem. Abst. 79 (1973) 113 211.

Attrep, K.A., Bellman, W.P., Sr., Attrep, M., Jr., Lee, J.B., and Braselton, W.F., Jr. (1980) Separation and identification of prostaglandin A1 in onion. Lipids 15:292–297. Chem. Abst. 93 (1980) 201 003.

Auer, W., Eiber, A., Hertkorn, E., Höhfeld, E., Köhrle, U., Lorenz, A., Mader, F., Merx, W., Otto, G., Schmid-Otto, B., and Taubenhelm, H. (1989) Hypertonie und Hyperlipidämie: In leichteren Fällen hilft auch Knoblauch. Multizentrische placebokontrollierte Doppelblind-Studie zur lipid- und blutdrucksenkenden Wirkung eines Knoblauchpräparates. Allgemeinarzt 11:205–208.

Auer, W., Eiber, A., Hartkorn, E., Höhfeld, E., Köhrle, U., Lorenz, A., Mader, F., Merx, W., Otto, G., and Schmid-Otto, B. (1990) Hypertension and hyperlipidemia: garlic helps in mild cases. Br. J. Clin. Pract. (Sympos. Suppl.) 69:3–6.

Aufreiter, U., Fuchs, M., and Urban, E. (1992) Synthese von 5,6-Dihydroxy-4,7-methano-isobenzofuranonen. Arch. Pharm. [Weinheim] *325*:579–584.

Auger, J., Lecomte, C., and Thibout, E. (1989) Leek odor analysis by gas chromatography and identification of the most active substance for the leek moth, Acrolepiopsis assectella. J. Chem. Ecol. *15*: 1847–1854.

Auger, J., Lalau-Keraly, F.X., and Belinsky, C. (1990) Thiosulfinates in vapour phase are stable and they can persist in the environment of Allium. Chemosphere *21*:837–843.

Auger, J., Boscher, J., Lages, B., Postaire, E., and Viel, C. (1992) Differences et similitudes des composés secondaires chez deux espèces d'Allium: Allium vineale L. et Allium ursinum L. Bull. Soc. bot. France *139*:61–66. Chem. Abst. 117 (1992) 208 902.

Auger, J., Lecomte, C., and Thibout, E. (1993a) Les composés soufres des Alliums: leurs activités biologiques chez les insectes et leur production. Acta bot. Gallica *140*:157–168. Chem. Abst. 120 (1994) 187 179.

Auger, J., Mellouki, F., Vannereau, A., Boscher, J., Cosson, L., and Mandon, N. (1993b) Analysis of Allium sulfur amino acids by HPLC after derivatization. Chromatographia *36*:347–350.

Augusti, K.T. and Mathew, P.T. (1973) Effect of long-term feeding of the aqueous extracts of onion (Allium cepa Linn.) and garlic (*Allium sativum* Linn.) on normal rats. Indian J. Exp. Biol. *11*:239–241. Chem. Abst. 80 (1974) 36 140.

Augusti, K.T. (1974a) Effect on alloxan diabetes of allyl propyl disulphide obtained from onion. Naturwissenschaften *61*:172–173. Chem. Abst. 81 (1974) 58 238.

Augusti, K.T. (1974b) Lipid lowering effect of allyl propyl disulfide isolated from Allium cepa on long term feeding to normal rats. Indian J. Biochem. Biophys. *11*:264–265. Chem. Abst. 82 (1975) 138 195.

Augusti, K.T. and Mathew, P.T. (1974) Lipid lowering effect of allicin (diallyl disulphide-oxide) on long-term feeding to normal rats. Experientia *30*:468–470. Chem. Abst. 81 (1974) 72 661.

Augusti, K.T., Roy, V.C.M., and Semple, M. (1974) Effect of allyl propyl disulfide isolated from onion (Allium cepa) on glucose tolerance of alloxan diabetic rabbits. Experientia *30*:1119–1120. Chem. Abst. 82 (1975) 68 299.

Augusti, K.T. (1975) Studies on the effect of allicin (diallyl disulphide-oxide) on alloxan diabetes. Experientia *31*:1263–1265. Chem. Abst. 84 (1976) 25 967.

Augusti, K.T. and Benaim, M.E. (1975) Effect of essential oil of onion (allyl propyl disulphide) on blood glucose, free fatty acid, and insulin levels of normal subjects. Clin. Chim. Acta *60*:121–123. Chem. Abst. 83 (1975) 22 593.

Augusti, K.T., Benaim, M.E., Dewar, H.A., and Virden, R. (1975) Partial identification of the fibrinolytic activators in onion. Atherosclerosis *21*:409–416. Chem. Abst. 83 (1975) 48 117.

Augusti, K.T. and Mathew, P.T. (1975) Effect of allicin on certain enzymes of liver after a short-term feeding to normal rats. Experientia *31*:148–149. Chem. Abst. 83 (1975) 484.

Augusti, K.T. (1976) Chromatographic identification of certain sulfoxides of cysteine present in onion (Allium cepa Linn.) extract. Curr. Sci. *45*:863–864. Chem. Abst. 86 (1977) 52 701.

Augusti, K.T. (1977a) Cysteine-onion oil interaction, its biological importance, and the separation of interaction products by chromatography. Indian J. Exp. Biol. *15*:1223–1224. Chem. Abst. 88 (1978) 84 820.

Augusti, K.T. (1977b) Hypocholesterolaemic effect of garlic (*Allium sativum* Linn.). Indian J. Exp. Biol. *15*:489–490.

Auler, H. (1936) Ernährung und Krebs. Ernährung [Vienna] *1*:150–167. Chem. Zbl. 1938, II, 536.

Aust, D. (1931) Der Knoblauch als sicheres Heilmittel der Ruhrfolgen. Zentralbl. Inn. Med. *52*:193–195.

Austin, S.J. and Schwimmer, S. (1971) L-Gamma-Glutamyl peptidase activity in sprouted onion. Enzymologia *40*:273–285. Chem. Abst. 75 (1971) 59 803.

Awasthi, P.B. and Leathers, C.R. (1984) Effect of aqueous garlic extract on growth of Coccidioides immitis and two Arthroconidia-producing fungi. Acta Bot. Indica *12*:22–25.

Awazu, S., Horie, T., Kodera, Y., Nagae, S., Matsuura, H., and Itakura, Y. (1992) Polysulfide compounds as lipid peroxidation inhibitor containing the polysulfide compounds as active ingredient. Eur. Patent 464 521 A2. Chem. Abst. 116 (1992) 136 231.

Aye, R.D. (1989) Garlic preparations and processing. Cardiol. Prac. *7*:7–8.

Azarkova, A.F., Stikhin, V.A., Cherkasov, O.A., and Maisuradze, N.I. (1983) Diosgenin from Allium nutans and A. cernuum. Khim. Pir. Soedin. (Chem. Nat. Compd.) *5*:653. Chem. Abst. 100 (1984) 135 771 (Russian).

Azarkova, A.F., Stikhin, V.A., Kabanov, V.S., Khotsialova, L.I., Maisuradze, N.I., Cherkasov, O.N., Rabinovich, A.M., and Ivanov, V.B. (1984) Study of plants from Dioscorea Costus and Allium genera with respect to diosgenin content. Khim. Farm. Zh. (Chem. Pharm. J.) *18*:188–191. Chem. Abst. 100 (1984) 135 886 (Russian).

Azarkova, A.F., Cherkasov, O.A., Stikhin, V.A., Kabanov, V.S., Maisuradze, N.I., and Melnikova, T.M. (1985) Dynamics of the production of vegetative mass and diosgenin content of Allium nutans. Khim. Farm. Zh. (Chem. Pharm. J.) *19*:1364–1367 (Russian).

Azarkova, A.F., Kabanov, V.S., Cherkasov, O.A., Melnikova, T.M., Maisuradze, N.I., and Zadorozhnyi, A.M. (1986) Isolation of diosgenin from Allium nutans L. Khim. Farm. Zh. (Chem. Pharm. J.) *20*:1334–1337. Chem. Abst. 106 (1987) 125 759 (Russian).

Azzouz, M.A. and Bullerman, L.B. (1982) Comparative antimycotic effects of selected herbs, spices, plant components, and commercial antifungal agents. J. Food Prot. 45:1298–1301.

Babushkin, G.A. (1952) Treatment of dysentery in children with phytoncides. Pediatriya, 69–70 (Russian).

Bachem, C. (1928) Allisatin und seine Verwendung in der Praxis. Med. Klin. 1:24–26.

Backer, H.J. and Kloosterziel, H. (1954) Esters thiosulfiniques. Rec. Trav. Chim. Pays Bas 73:129–139.

Bacon, J.S.D. (1959) The trisaccharide fraction of some monocotyledons. Biochem. J. 73:507–514.

Bae, R.N. and Lee, S.K. (1990) Factors affecting greening and its control methods in chopped garlic. Hanguk Wonye Hakhoechi 31:358–362. Chem. Abst. 115 (1991) 113 171.

Baer, A.R. and Wargovich, M.J. (1989) Role of ornithine decarboxylase in diallyl sulfide inhibition of colonic radiation injury in the mouse. Cancer Res. 49:5073–5076. Chem. Abst. 111 (1989) 170 172.

Baerheim, A. and Sandvik, H. (1994) Effect of ale, garlic, and soured cream on the appetite of leeches. Brit. Med. J. 309:1689.

Baghurst, K.I., Raj, M.J., and Truswell, A.A. (1977) Onions and platelet aggregation. Lancet 1:101.

Bai, W., Deng, B., and Cai, X. (1994) Study on the CGC/AAS and analysis of the species of trace selenium in garlic oil. Fenxi Shiyanshi 13:9–12. Chem. Abst. 121 (1994) 92 022.

Bailey, J.H. and Cavallito, C.J. (1948) The reversal of antibiotic action. J. Bacteriol. 55:175–182.

Baker, H., Thomson, A.D., Frank, O., and Leevy, C.M. (1974) Absorption and passage of fat- and water-soluble thiamin derivatives into erythrocytes and cerebrospinal fluid of man. Am. J. Clin. Nutr. 27:676–680.

Bakhsh, R. and Chughtai, M.I.D. (1984) Influence of garlic on serum cholesterol, serum triglycerides, serum total lipids, and serum glucose in human subjects. Nahrung 28:159–163.

Bakhsh, R. and Chughtai, M.I.D. (1985) Comparative study of onion and garlic on serum cholesterol, liver cholesterol, prothrombin time, and fecal sterols excretion in male albino rats. J. Chem. Soc. Pak. 7:285–288. Chem. Abst. 104 (1986) 108 382.

Baldrati, G., Cagna, D., and Giannone, L. (1970) Contributo alla conoscenza dell 'olio essenziale d'aglio. Ind. Conserve 45:125–130. Chem. Abst. 74 (1971) 11 919.

Ballard-Barbash, R., Krebs-Smith, S., and Subar, A.F. (1995) Re: vegetables, fruit, and colon cancer in the Iowa Women's Health Study. Am. J. Epidemiol. 141:84–86.

Banerjee, A.K. (1976) Effect of aqueous extract of garlic on arterial blood pressure of normotensive and hypertensive rats. Artery [Fulton, Mich.] 2:369–373.

Banerjee, C.M., Atsegbaghan, A., and Reed, L.C. (1982) Effects of allicin on blood glucose and 2:3 diphosphoglycerate level in rats. Fed. Proc. 41:1569.

Banerjee, M.N. (1913) Die Einwirkung der Sulfide der Allylgruppe auf Bleiamalgam und Quecksilber (Die völlige Reinigung von Quecksilber). Z. Anorg. Chem. 83:113–118. Chem. Zbl. 1913, II, 1653.

Banerji, A., Amonkar, S.V., and Bhabha Atomic Res. Ctr. (1978) Insecticidal principles of garlic. India. Patent 144 278. Chem. Abst. 92 (1980) 17 188.

Bark Hur, O. (1994) RFLP analyses of the chloroplast and nuclear genomes to establish cytoplasms, demonstrate interspecific DNA transfer, and estimate relationships among bulb-onion populations. Diss. Abst. Int. B 54:4498 Chem. Abst. 120 (1994) 262 777.

Barone, F.E. and Tansey, M.R. (1977) Isolation, purification, identification, synthesis, and kinetics of activity of the anticandidal component of Allium sativum, and a hypothesis for its mode of action. Mycologia 69:793–825. Chem. Abst. 87 (1977) 194 680.

Barowsky, H. and Boyd, L.J. (1944) The use of garlic (Allisatin) in gastrointestinal disturbances. Rev. Gastroenterol. 11:22–26.

Barrie, S.A., Wright, J.V., and Pizzorno, J.E. (1987) Effects of garlic oil on platelet aggregation, serum lipids, and blood pressure in humans. J. Orthomol. Med. 2:15–21.

Bartzatt, R., Blum, D., and Nagel, D. (1992) Isolation of garlic derived sulfur compounds from urine. Anal. Lett. 25:1217–1224. Chem. Abst. 117 (1992) 86 304.

Basra, A.S., Singh, B., and Malik, C.P. (1994) Priming-induced changes in polyamine levels in relation to vigor of aged onion seeds. Bot. Bull. Acad. Sin. 35:19–23. Chem. Abst. 121 (1994) 31 226.

Bastidas, G.J. (1969) Effect of ingested garlic on Necator americanus and Ancylostoma caninum. Am. J. Trop. Med. Hyg. 18:920–923.

Bauer, F., Vali, S., and Stachelberger, H. (1988) Zusammensetzung und Gehalt von Kohlenhydraten in Gewürzen. Chem. Mikrobiol. Technol. Lebensm. 11:181–187. Chem. Abst. 110 (1989) 22 451.

Bauer, R., Breu, W., Wagner, H., and Weigand, W. (1991) Enantiomeric separation of racemic thiosulphinate esters by high-performance liquid chromatography. J. Chromatogr. 541:464–468.

Bauhin, C. (1622) Catalogus Plantarum circa Basileam, etc. Joh. Jacob. Genathius, Basle.

Baumann, M. (1982) Die griechische Pflanzenwelt in Mythos, Kunst und Literatur. Hirmer, Munich.

Baycheva, O., Gabrashanska, M., and Damyanova, A. (1988) Mineral composition of the phytonematode Ditylenchus dipsaci (Kühn, 1857; Filipjev, 1936) and tissues from its host Allium sativum. Helminthologia 25:41–48.

Bayer, T., Breu, W., Seligmann, O., Wray, V., and Wagner, H. (1989a) Biologically active thiosulphinates and alpha-sulphinyl-disulphides from Allium cepa. Phytochemistry 28:2373–2377.

Bayer, T., Wagner, H., Block, E., Grisoni, S., Zhao, S.H., and Neszmelyi, A. (1989b) Zwiebelanes: novel biologically active 2,3-dimethyl-5,6-dithiabicyclo[2.1.1]hexane 5-oxides from onion. J. Am. Chem. Soc. *111*:3085–3086.

Becher, E. and Fussgänger, R. (1930) Über die Beeinflussung der Darmgiftausscheidung durch Allisatin. Münch. Med. Wochenschr. *77*: 2139–2141.

Becher, J.J. (1663) Parnassus Medicinalis illustratus. J. Görlins, Speyer.

Beck, E. and Grünwald, J. (1993) *Allium sativum* in der Stufentherapie der Hyperlipidämie. Med. Welt *44*:516–520.

Becker, A. and Schuphan, W. (1975) Ein Beitrag zur Biogenese und Biochemie antimikrobiell wirkender ätherischer Öle der Küchenzwiebel (Allium cepa L.). Qual. Plant. Pl. Fds. Hum. Nutr. *25*:107–169.

Becker, H. (1985) Knoblauch—nur Gewürz oder auch Phytopharmakon? Dtsch. Apoth. Ztg. *125*: 1677–1680. Chem. Abst. 103 (1985) 138 497.

Bedrintseva, V.V. (1954) Therapy with phytoncides in acute sore throat. Vestn. Otorinolaringol. Rep. *16*:53–55 (Russian).

Begum, R. and Bari, M.A. (1985) Effect of garlic oil on the pancreas of experimental diabetes in guinea pigs. Bangladesh Med. Res. Counc. Bull. *11*:64–68.

Behling, L. (1957) Die Pflanze in der mittelalterlichen Tafelmalerei. Böhlau, Weimar.

Behling, L. (1964) Die Pflanzenwelt der mittelalterlichen Kathedralen. Böhlau, Cologne.

Bekdairova, K.Z. (1981) Caffeic acid: natural inhibitor of garlic bulb growth. Izv. Akad. Nauk Kaz. SSR, Ser. Biol. (Bull. Acad. Sci. Kasach. SSR, Biol. Ser.), 20–24. Chem. Abst. 96 (1982) 3644 (Russian).

Bekdairova, K.Z. and Klyshev, L.K. (1982) Garlic essential oil and its quantitative analysis. Izv. Akad. Nauk Kaz. SSR, Ser. Biol. (Bull. Acad. Sci. Kasach. SSR, Biol. Ser.), 6–11. Chem. Abst. 96 (1982) 179 703 (Russian).

Belman, S. (1983a) Onion and garlic oils inhibit tumor promotion. Proc. Am. Assoc. Cancer Res. *24*:102.

Belman, S. (1983b) Onion and garlic oils inhibit tumor promotion. Carcinogenesis *4*:1063–1065.

Belman, S., Block, E., and Barany, G. (1986) Onion and garlic components inhibit soybean lipoxygenase. Proc. 77th Annual Meeting Am. Soc. Cancer Res., Los Angeles, 7–10 May 1986 *27*:140.

Belman, S., Block, E., Perchellet, J.P., Perchellet, E.M., and Fischer, S.M. (1987) Onion and garlic oils inhibit promotion whereas the oils enhance the conversion of papillomas to carcinomas. Proc. Am. Assoc. Cancer Res. *28*:166.

Belman, S., Solomon, J., and Segal, A. (1989) Inhibition of soybean lipoxygenase and mouse skin tumor promotion by onion and garlic components. J. Biochem. Toxicol. *4*:151–160. Chem. Abst. 112 (1990) 91 227.

Belman, S., Sellakumar, A., Bosland, M.C., Savarese, K., and Estensen, R.D. (1990) Papilloma and carcinoma production in DMBA-initiated, onion oil-promoted mouse skin. Nutr. Cancer *14*:141–148.

Belohvostov, S.D. (1949) Treatment of chronic dysentery with garlic. Sov. Med. (Soviet Med.), 16–17 (Russian).

Belval, H. (1939) Amaryllidacées et Liliacées. La réserve glucidique de l'ail et de la tubéreuse. Bull. Soc. Chim. Biol. *21*:294–297. Chem. Zbl. 1940, I, 71.

Belval, H. (1943) Sur les fructosanes du genre Allium. Bull. Soc. Chim. Biol. *25*:64–67. Chem. Abst. 38 (1944) 2361.

Belval, H., Grandchamp-Chaudun, A., and Merac, M.L. (1948) The constitution of the fructosans of the genus Allium. C. R. Acad. Sci. Paris *227*: 1403–1405. Chem. Abst. 44 (1950) 1913.

Benezra, C., Ducombs, G., Sell, Y., and Foussereau, J. (1985) Plant contact dermatitis. Dekker, Toronto.

Beraha, D., Lupas, O., and Ploegert, K. (1984) Modeling of reactor control and protection systems in the core simulator program GARLIC. Report BMI-1984-044, *GRS-A-924*:1–75. Chem. Abst. 102 (1985) 174 584. Energy Res. Abst. 10 (1985) 4683.

Beraha, D., Lupas, O., and Ploegert, K. (1985) Investigations to guarantee the admissible power densities in the core of pressurized power reactors. Report BMI-1984-051, *GRS-A-973*:13175. Chem. Abst. 103 (1985) 13 175 (German).

Berchenko, N.V. and Trifonova, T.K. (1951) Result of the short-term influence of the volatile phytoncide fraction from garlic on some properties of Brucella bovis. Sbornik Naukh. Trud. Minsk. Gos. Vet. -Zootekh. Inst. (Trans. Minsk. State Med. Inst.) *4*:95–96 (Russian).

Berendes, J. (1891) Die Pharmacie bei den alten Culturvölkern. Historisch-kritische Studien. Tausch & Grosse, Halle a. Saale.

Berendes, J. (1902) Des Pedanios Dioskurides aus Anazarbos Arzneimittellehre in fünf Büchern, übersetzt und mit Erklärungen versehen. Enke, Stuttgart.

Berendes, J. (1907) Das Apothekenwesen, seine Entstehung und geschichtliche Entwickelung bis zum XX. Jahrhundert. Ferd. Enke, Stuttgart.

Berger, F. (1960) Bulbus Allii sativi. In: Handbuch der Drogenkunde. W. Maudrich, Wien, pp. 29–35.

Berggren, D. and Fiskesjö, G. (1987) Aluminum toxicity and speciation in soil liquids: experiments with Allium cepa L. Envir. Toxicol. Chem *6*:771–779. Chem. Abst. 107 (1987) 192 536.

Bernhard, R.A. (1968) Comparative distribution of volatile aliphatic disulfides derived from fresh and dehydrated onions. J. Food Sci. *33*:298–304. Chem. Abst. 69 (1968) 66 261.

Bessoule, J.J., Lessire, R., and Cassagne, C. (1989) Partial purification of the acyl-CoA elongase of Allium porrum leaves. Arch. Biochem. Biophys. *268*:475–484.

Bessoule, J.J., Testet, E., and Cassagne, C. (1994) Cloning of new isoform of a DNAJ protein from Allium porrum epidermal cells. Plant Physiol. Biochem. *32*:723–727.

Betz, E., Weidler, R., and Heinle, H. (1992) Effects of garlic in an experimental model of atherosclerosis. Eur. J. Clin. Res. *3a*:10–11.

Betz, E. and Weidler, R. (1989) Die Wirkung von Knoblauchextrakt auf die Atherogenese bei Kaninchen. In: Die Anwendung aktueller Methoden in der Arteriosklerose-Forschung. Deutsche Gesellschaft für Arterioskleroseforschung, pp. 304–311.

Beuth, J., Ko, H.L., and Pulverer, G. (1994) Angewandte Lektinologie. Neue therapeutische Konzepte in Onkologie und Infektiologie. Dtsch. Apoth. Ztg. *134*:2331–2342.

Bhat, P.G. and Pattabiraman, T.N. (1979) Solubilization and purification of a particular hexokinase from garlic (*Allium sativum*) bulbs. Indian J. Biochem. Biophys. *16*:284–287. Chem. Abst. 92 (1980) 54 034.

Bhat, P.G. and Pattabiraman, T.N. (1980) Separation and purification of an invertase and an inulase from germinating garlic (*Allium sativum* L.) bulbs. Indian J. Biochem. Biophys. *17*:338–343. Chem. Abst. 94 (1981) 60 449.

Bhatia, B. and Ahujarai, P.L. (1984) Cold tolerance in CCl4-treated rats and its modification by administration of garlic oil and glucose. Int. J. Biometeorol. *28*:93–99. Chem. Abst. 101 (1984) 145 307.

Bhatia, I.S., Satyanarayana, M.N., and Srinivasan, M. (1955) Carbohydrases in garlic. J. Sci. Ind. Res. *14C*:93–96. Chem. Abst. 50 (1956) 10 194.

Bhushan, S., Sharma, S.P., Singh, S.P., Agrawal, S., Indrayan, A., and Seth, P. (1979) Effect of garlic on normal blood cholesterol level. Indian J. Physiol. Pharmacol. *23*:211–214. Chem. Abst. 92 (1980) 57 257.

Bhuyan, M., Saxena, B.N., and Rao, K.M. (1974) Repellent property of oil fraction of garlic, *Allium sativum* Linn. Indian J. Exp. Biol. *12*:575–576.

Bielzer, U. (1992) Untersuchungen zur antimikrobiellen Wirksamkeit von Knoblauch (*Allium sativum* L.). Dissertation, Inst. Med. Mikrobiol. & Hygiene, University of Cologne (Abstract).

Bilgrami, K.S., Sinha, K.K., and Sinha, A.K. (1992) Inhibition of aflatoxin production and growth of Aspergillus flavus by eugenol and onion and garlic extracts. Indian J. Med. Res. *96*:171–175.

Billau, H. (1961) Hyperlipämie als ein Pathogenetischer Faktor für die Arteriosklerose und ihre Beeinflussung durch Knoblauch-Wirkstoffe bei Kaninchen. Dissertation, University of Giessen.

Bilyk, A. and Sapers, G.M. (1985) Distribution of quercetin and kaempferol in lettuce, kale, chive, garlic chive, leek, horseradish, red radish, and red cabbage tissues. J. Agric. Food Chem. *33*:226–228. Chem. Abst. 102 (1985) 111 776.

Bimmermann, A., Weingart, K., and Schwartzkopff, W. (1988) *Allium sativum*: Studie zur Wirksamkeit bei Hyperlipoproteinämie. Therapiewoche *38*:3885–3890.

Bimmermann, A., Schwartzkopff, W., and Baeyer, H.V. (1990) Zur Wirksamkeit von *Allium sativum* (Knoblauch). Münch. Med. Wochenschr. *132*: 219–222.

Binding, G.J. (1970) About Garlic. Thorsons Publishers, Wellingborough.

Bitsch, R. and Bitsch, I. (1989) Lipidlösliche Vitaminderivate. Wirkungsweise und Anwendungsmöglichkeiten. Dtsch. Apoth. Ztg. *129*:65–68. Chem. Abst. 110 (1989) 107 410.

Bityukov, V.A., Bogolepov, G.G., and Shin, V.C. (1982) Lysozyme in Gemüsepflanzen. Tr. Kuban. Skh. Int. (Trans. Kuban. Project Int.) *197*:73–79. Chem. Abst. 97 (1982) 36 130. Ref. Zh. Rastenievod 1982, 555 317 (Russian).

Blakely, S.R., Mislo, B.L., Brown, E.D., Jenkins, M.Y., and Mitchell, G.V. (1993) Gender differences in the induction of hepatic detoxification enzymes in garlic-fed hypercholesterolemic or iron-loaded mature rats. FASEB J. *7*:A864.

Blania, G. and Spangenberg, B. (1991) Formation of allicin from dried garlic (*Allium sativum*): a simple HPTLC method for simultaneous determination of allicin and ajoene in dried garlic and garlic preparations. Planta Med. *57*:371–375. Chem. Abst. 115 (1991) 239 858.

Bleumink, E., Doeglas, H.M.G., Klokke, A.H., and Nater, J.P. (1972) Allergic contact dermatitis to garlic. Br. J. Dermatol. *87*:6–9.

Bleumink, E. and Nater, J.P. (1973) Contact dermatitis to garlic: crossreactivity between garlic, onion, and tulip. Arch. Dermatol. Forsch. *247*:117–124.

Bligh, E.G. and Dyer, W.J. (1959) A rapid method of total lipid extraction and purification. Can. J. Biochem. Physiol. *37*:911–917.

Block, E., Penn, R.E., and Revelle, L.K. (1979) Flash vacuum pyrolysis studies. 7. Structure and origin of the onion lachrymatory factor: a microwave study. J. Am. Chem. Soc. *101*:2200–2201. Chem. Abst. 91 (1979) 19 835.

Block, E., Bazzi, A.A., Lambert, J.B., Wharry, S.M., Andersen, K.K., Dittmer, D.C., Patwardhan, B.H., and Smith, D.J.H. (1980a) Carbon-13 and oxygen-17 nuclear magnetic resonance studies in organosulfur compounds: the four-member red ring sulfone effect. J. Org. Chem. *45*:4807–4810. Chem. Abst. 93 (1980) 238 386.

Block, E., Revelle, L.K., and Bazzi, A.A. (1980b) Chemistry of sulfides. 5. The lachrymatory factor of the onion: an NMR study. Tetrahedron Lett. *21*:1277–1280. Chem. Abst. 93 (1980) 91 827.

Block, E., Ahmad, S., Jain, M.K., Crecely, R.W., Apitz-Castro, R., and Cruz, M.R. (1984) (E,Z)-Ajoene: a potent antithrombotic agent from garlic. J. Am. Chem. Soc. *106*:8295–8296. Chem. Abst. 102 (1985) 21 151.

Block, E. (1985) The chemistry of garlic and onions. Sci. Am. *252*:114–119.

Block, E. (1986) Antithrombotic agent of garlic: a lesson from 5000 years of folk medicine. In: Folk medicine: the art and the science. Steiner, R.P., ed. American Chemical Society, Washington, D.C., pp. 125–137.

Block, E., Ahmad, S., Catalfamo, J.L., Jain, M.K., and Apitz-Castro, R. (1986) The chemistry of alkyl thiosulfinate esters. IX. Antithrombotic organosulfur compounds from garlic: structural, mechanistic, and synthetic studies. J. Am. Chem. Soc. *108*: 7045–7055. Chem. Abst. 106 (1987) 32 678.

Block, E., Iyer, R., Grisoni, S., Saha, C., Belman, S., and Lossing, F.P. (1988) Lipoxygenase inhibitors from the essential oil of garlic: Markovnikov addition of the allyldithio radical to olefins. J. Am. Chem. Soc. *110*:7813–7827. Chem. Abst. 109 (1988) 210 780.

Block, E. (1992a) Die Organoschwefelchemie der Gattung Allium und ihre Bedeutung für die organische Chemie des Schwefels. Angew. Chem. *104*:1158–1203. Chem. Abst. 117 (1992) 191 027.

Block, E. (1992b) The organosulfur chemistry of the genus Allium: implications for the organic chemistry of sulfur. Angew. Chem. Int. Ed. Engl. *31*:1135–1178. Chem. Abst. 117 (1992) 191 027.

Block, E., Naganathan, S., Putnam, D., and Zhao, S.H. (1992a) Allium chemistry: HPLC analysis of thiosulfinates from onion, garlic, wild garlic (ramsoms), leek, scallion, shallot, elephant (great-headed) garlic, chive, and Chinese chive: uniquely high allyl to methyl ratios in some garlic samples. J. Agric. Food Chem. *40*:2418–2430. Chem. Abst. 117 (1992) 250 231.

Block, E. and Purcell, P.F. (1992) Onion and heartburn. Am. J. Gastroenterol. *87*:679.

Block, E., Putnam, D., and Zhao, S.H. (1992b) Allium chemistry: GC-MS analysis of thiosulfinates and related compounds from onion, leek, scallion, shallot, chive, and Chinese chive. J. Agric. Food Chem. *40*:2431–2438. Chem. Abst. 117 (1992) 250 064.

Block, E. (1993) Flavor artifacts. J. Agric. Food Chem. *41*:692.

Block, E., Naganathan, S., Putnam, D., and Zhao, S.H. (1993) Organosulfur chemistry of garlic and onion: recent results. Pure Appl. Chem. *65*:625–632. Chem. Abst. 118 (1993) 232 663.

Block, E. and Calvey, E.M. (1994) Facts and artifacts in Allium chemistry. ACS Symp. Ser. *564*:63–69. Chem. Abst. 121 (1994) 299 341.

Block, E. (1995) The health benefits of organosulfur and organoselenium compounds in garlic (*Allium sativum*): recent findings. In: Hypernutritious foods. ACS Symp. Ser. J. W. Finley, D. J. Armstrong, S. F. Robinson, S. Nagy, eds. Washington, DC: American Chemical Society Books (in press).

Blum, L.E.B. and Gabardo, H. (1993) Chemical control of garlic rust in Curitibanos/SC, Brazil. Fitopatol. Bras. *18*:230–232. Chem. Abst. 119 (1993) 219 553 (Portuguese).

Bobboi, A., Augusti, K.T., and Joseph, P.K. (1984) Hypolipidemic effects of onion oil and garlic oil in ethanol-fed rats. Indian J. Biochem. Biophys. *21*:211–213. Chem. Abst. 101 (1984) 89 537.

Bock, H. (1595) Hieronymus Bock's New Kreutterbuch. Joh. Richelius, Strassburg, pp. 277–281.

Bock, H., Mohmand, S., Hirabayashi, T., and Semkov, A. (1982a) Thioacrolein: Das stabilste C3H4S-Isomere und sein PE-spektroskopischer Nachweis in der Gasphase. Chem. Ber. *115*:1339–1348.

Bock, H., Mohmand, S., Hirabayashi, T., and Semkow, A. (1982b) Thioacrolein. J. Am. Chem. Soc. *104*:312–313.

Böcker, O.E. (1938) Über die keimtötende Wirkung des Knoblauchs (*Allium sativum*). Z. Hyg. Infektionskrankh. *121*:166–180.

Bockmann, C., Nelson, R.S., and Klein, W.A. (1969) Modified Allium condiments. U.S.A. Patent 3 424 593. Chem. Abst. 70 (1969) 95 605.

Body, S.C. (1986) A taste of allicin? Anaesth. Intensive Care *14*:94.

Boegl, W. and Heide, L. (1985) Chemiluminescence measurements as an identification method for gamma-irradiated foodstuffs. Radiat. Phys. Chem. *25*:173–185. Chem. Abst. 103 (1985) 213 517.

Boelens, M., De Valois, P.J., Wobben, H.J., and Van der Gen, A. (1971) Volatile flavor compounds from onion. J. Agric. Food Chem. *19*:984–991. Chem. Abst. 75 (1971) 117 305.

Bogatskaya, L.N. (1953) Influence of the phytoncides of garlic and of mustard oil on the sensitivity of peripheral blood vessels against adrenaline. Byull. Biol. Med. (Bull. Biol. Med.) *35*:55–57 (Russian).

Bogin, E. (1973) Studies on the effect of antibacterial compounds from garlic on biological membranes. Proc. Int. Congr. Biochem., 9th, 1973, p. 271.

Bogin, E., Abrams, M., and Earon, Y. (1984) Effect of garlic extract on red blood cells. J. Food Prot. *47*:100–101.

Bogin, E. and Abrams, M. (1976) The effect of garlic extract on the activity of some enzymes. Food Cosmet. Toxicol. *14*:417–419.

Böhmova, B., Duhova, V., and Miklovicova, M. (1988) Evaluation of the effect of some chemical mutagens on garlic (*Allium sativum* L.). Acta Fac. Rer. Nat. Univ. Comenianae, Genet. [Bratislava] *19*:3–8. Chem. Abst. 110 (1989) 2622.

Boissier, E. (1881) Flora orientalis sive ennumeratio plantarum in oriente a Graecia et Aegypto ad Indiae fines. H. Georg, Geneva, Basle.

Bojko, S.P. (1962) Therapeutical activity of Alifid in the treatment of arteriosclerotics. In: Lekarstvennye Sredstva iz Rastenij. Gossudarst. Izdat. Med. Moscow, pp. 155–163 (Russian).

Bojs, G. and Svensson, A. (1988) Contact allergy to garlic used for wound healing. Contact Dermatitis *18*:179–181.

Bolton, S., Null, G., and Troetel, W.M. (1982) The medical uses of garlic: fact and fiction. Am. Pharm. NS *22*:40–43.

Bonem, P. (1928) Zur therapeutischen Verwendung des Knoblauchs. Dtsch. Med. Wochenschr. *54*: 1243–1254. Chem. Zbl. 1928, II, 1461.

Bonfante-Fasolo, P., Vian, B., Perotto, S., Faccio, A., and Knox, J.P. (1990) Cellulose and pectin localization in roots of mycorrhizal Allium porrum: labeling continuity between host cell wall and interfacial material. Planta *180*:537–547. Chem. Abst. 112 (1990) 195 311.

Bongartz, U. and Grünwald, J. (1992) Schützen Knoblauch-Präparate vor Arteriosklerose? Aerztl. Praxis 44:8–9.

Borda, J.M. and Bozzola, C. (1961) Keratosis in fingertips on contact with Liliaceae. Arch. Argent. Dermatol. 11:293–299.

Bordia, A. and Bansal, H.C. (1973) Essential oil of garlic in prevention of atherosclerosis. Lancet 2:1491–1492.

Bordia, A., Bansal, H.C., Arora, S.K., Rathore, A.S., Ranawat, R.V.S., and Singh, S.V. (1974) Effect of the essential oil (active principle) of garlic on serum cholesterol, plasma fibrinogen, whole blood coagulation time, and fibrinolytic activity in alimentary lipaemia. J. Assoc. Physicians India 22:267–270.

Bordia, A., Bansal, H.C., Arora, S.K., and Singh, S.V. (1975b) Effect of the essential oils of garlic and onion on alimentary hyperlipemia. Atherosclerosis 21:15–19. Chem. Abst. 82 (1975) 138 187.

Bordia, A., Arora, S.K., Kothari, L.K., Jain, K.C., Rathore, B.S., Rathore, A.S., Dube, M.K., and Bhu, N. (1975a) The protective action of essential oils of onion and garlic in cholesterol-fed rabbits. Atherosclerosis 22:103–109.

Bordia, A., Joshi, H.K., Sanadhya, Y.K., and Bhu, N. (1977b) Effect of essential oil of garlic on serum fibrinolytic activity in patients with coronary artery disease. Atherosclerosis 28:155–159. Chem. Abst. 88 (1978) 16 128.

Bordia, A., Verma, S.K., Vyas, A.K., Khabya, B.L., Rathore, A.S., Bhu, N., and Bedi, H.K. (1977a) Effect of essential oil of onion and garlic on experimental atherosclerosis in rabbits. Atherosclerosis 26:379–386. Chem. Abst. 86 (1977) 183 242.

Bordia, A. (1978) Effect of garlic on human platelet aggregation in vitro. Atherosclerosis 30:355–360. Chem. Abst. 89 (1978) 191 283.

Bordia, A. and Joshi, H.K. (1978) Garlic on fibrinolytic activity in cases of acute myocardial infarction part 2. J. Assoc. Physicians India 26:324–326.

Bordia, A., Sanadhya, S.K., Rathore, A.S., and Bhu, N. (1978) Essential oil of garlic on blood lipids and fibrinolytic activity in patients of coronary artery disease, part 1. J. Assoc. Physicians India 26:327–331.

Bordia, A. and Verma, S.K. (1978) Garlic on the reversibility of experimental atherosclerosis. Indian Heart J. 30:47–50.

Bordia, A. and Verma, S.K. (1980) Effect of garlic feeding on regression of experimental atherosclerosis in rabbits. Artery [Fulton, Mich.] 7:428–437.

Bordia, A. (1981) Effect of garlic on blood lipids in patients with coronary heart disease. Am. J. Clin. Nutr. 34:2100–2103.

Bordia, A., Sharma, K.D., Parmar, Y.K., and Verma, S.K. (1982) Protective effect of garlic oil on the changes produced by 3 weeks of fatty diet on serum cholesterol, serum triglycerides, fibrinolytic activity, and platelet adhesiveness in man. Indian Heart J. 34:86–88.

Bordia, A. (1986) Klinische Untersuchung zur Wirksamkeit von Knoblauch. Apoth. Magazin. 4:128–131.

Bordia, A. (1989) Knoblauch und koronare Herzkrankheit: Wirkungen einer dreijährigen Behandlung mit Knoblauchextrakt auf die Reinfarkt-und Mortalitätsrate. Dtsch. Apoth. Ztg. 129:16–17.

Borkowski, B. (1970) Bulbus Allii sativi—Cebula czonku pospolitego czosnek. In: Zarys Farmakognozji. Panstwowy Zaklad Wydawnictw Lekarskich, Warsaw, pp. 311–313 (Polish).

Born, G.V.R., Haslam, R.J., Goldman, M., and Lowe, R.D. (1965) Comparative effectiveness of adenosine analogues as inhibitors of blood-platelet aggregation and as vasodilators in man. Nature 205:678–680. Chem. Abst. 62 (1965) 12 330.

Born, G.V.R. and Cross, M.J. (1963) The aggregation of blood platelet. J. Physiol. [London] 168:178–195. Chem. Abst. 59 (1963) 15 679.

Borukh, I.F. and Demkevich, L.I. (1976) Pigment substances of some garlic varieties of the western forest-steppe of the Ukraine. Iz. Vyssh. Uchebn. Zaved. Pishch. Tekhnol., 47–49. Chem. Abst. 85 (1976) 119 571 (Russian).

Boscher, J. and Auger, J. (1991) L'Allium ampeloprasum (var. bulbilliferum Lloyd) de l'Ile d'Yeu (Vendée) est chimiquement un ail et non un poireau. Bull. Soc. Bot. France 138:315–320.

Bose, S. and Shrivastava, A.N. (1961) Soluble carbohydrates from onion (Allium cepa Linn.). Sci. Cult. [India] 27:253.

Boullin, D.J. (1981) Garlic as a platelet inhibitor. Lancet 1:776–777.

Bouthelier, V. and Cattaneo, P. (1987) Sobre contenidos y composiciones acidicas de lipidos totales de hojas de plantas comestibles. An. Soc. Quim. Argent. 75:159–172.

Bowman, A.B., Hughes, B.G., and Murray, B.K. (1995) Diallyl disulfide, a component of cooked garlic, inhibits the intracellular localization of p21 RAS to the cell membrane in the colon cancer cell line. FASEB J. 9:A1281

Bozzini, A. (1991) Discovery of an Italian fertile tetraploid line of garlic. Econ. Bot. 45:436–438.

Bradford, M.M. (1976) A rapid and sensitive method for the quantitation of microgram quantities of protein utilizing the principle of protein-dye binding. Anal. Biochem. 72:248–254.

Brady, J.F., Li, D., Ishizaki, H., and Yang, C.S. (1988) Effect of diallyl sulfide on rat liver microsomal nitrosamine metabolism and other monooxygenase activities. Cancer Res. 48:5937–5940. Chem. Abst. 109 (1988) 222 059.

Brady, J.F., Ishizaki, H., Fukuto, J.M., Lin, M.C., Fadel, A., Gapac, J.M., and Yang, C.S. (1991a) Inhibition of cytochrome P-450 2E1 by diallyl sulfide and its metabolites. Chem. Res. Toxicol. 4:642–647. Chem. Abst. 115 (1991) 225 605.

Brady, J.F., Wang, M.H., Hong, J.Y., Xiao, F., Li, Y., Yoo, J.S.H., Ning, S.M., Lee, M.J., Fukuto, J.M., Gapac, J.M., and Yang, C.S. (1991b) Modulation of rat hepatic microsomal monooxygenase enzymes and cytotoxicity by diallyl sulfide. Toxicol. Appl. Pharmacol. 108:342–354. Chem. Abst. 114 (1991) 220 920.

Brahmachari, H.D. and Augusti, K.T. (1962a) Effects of orally effective hypoglycaemic agents from plants on alloxan diabetes. J. Pharm. Pharmacol. 14:617.

Brahmachari, H.D. and Augusti, K.T. (1962b) Hypoglycaemic agents from Indian indigenous plants. J. Pharm. Pharmacol. 13:381–382. Chem. Abst. 56 (1962) 1525.

Brahmachari, H.D. and Augusti, K.T. (1962c) Orally effective hypoglycaemic agents from plants. J. Pharm. Pharmacol. 14:254–255. Chem. Abst. 57 (1962) 967.

Brahmachari, H.D. and Augusti, K.T. (1962d) Hypoglycaemic agent from onions. J. Pharm. Pharmacol. 13:128.

Braun, H. (1949) Bulbus Allii sativi. In: Pharmakologie. Vol 3. Wissenschaftl. Verlagsges, Stuttgart, p. 211.

Brechenmacher, J.K. (1929) Deutsche Sippennamen. Stuttgart.

Breinlich, J. (1950) Die Bestimmung des ätherischen Ölschwefels in schwefelölhaltigen Pflanzen unter besonderer Berücksichtigung des Knoblauchöls, Oxydation des Allyl-Senföls mit alkalischer Jodlösung. Pharm. Zentralhalle 89:217–223. Chem. Abst. 45 (1951) 1300.

Bretschneider, E. (1882) Botanicon Sinicum. Notes on Chinese botany from native and western sources. J. N. China Br. R. Asiatic Soc. 1882–1896.

Brewitt, B. and Lehmann, B. (1991) Lipidregulierung durch standardisierte Naturarzneimittel. Multizentrische Langzeitstudie an 1209 Patienten. Kassenarzt 5:47–55.

Briand, C.H. and Kapoor, B.M. (1989) The cytogenetic effects of sodium salicylate on the root meristem cells of *Allium sativum* L. Cytologia 54:203–209. Chem. Abst. 111 (1989) 169 092.

Brodnitz, M.H., Pollock, C.L., and Vallon, P.P. (1969) Flavor components of onion oil. J. Agric. Food Chem. 17:760–763. Chem. Abst. 71 (1969) 57 578.

Brodnitz, M.H., Pascale, J.V., and Van Derslice, L. (1971) Flavor components of garlic extract. J. Agric. Food Chem. 19:273–275. Chem. Abst. 74 (1971) 98 446.

Brodnitz, M.H. and Pollock, C.L. (1970) Gas chromatographic analysis of distilled onion oil. Food Technol. 24:78–80. Chem. Abst. 72 (1970) 65 446.

Brosche, T. (1989) Therapeutische Wirkungen einer Knoblauchzubereitung auf den Lipidstatus Geriatrischer Probanden. Med. Welt 40:1233–1237.

Brosche, T., Platt, D., and Dorner, H. (1990) The effect of a garlic preparation on the composition of plasma lipoproteins and erythrocyte membranes in geriatric subjects. Br. J. Clin. Pract. 44:12–19.

Brosche, T. and Platt, D. (1991) Garlic. Br. Med. J. 303:785.

Brosche, T. and Platt, D. (1993) Zur immunomodulatorischen Wirkung von Knoblauch (*Allium sativum* L.). Med. Welt 44:309–313.

Brosche, T. and Platt, N. (1994) Knoblauchtherapie und zellulaere Immunabwehr im Alter. Z. Phytother. 15:23–24.

Brown, J., Albers, J.J., Fisher, L.D., Schaefer, F.M., Lin, J.T., Caplan, C., Zhao, X.Q., Bisson, B.D., Fitzpatrick, V.F., and Dodge, H.T. (1990) Regression of coronary artery disease as a result of intensive lipid-lowering therapy in men with high levels of apolipoprotein B. New Engl. J. Med. 323:1289–1298.

Brunel-Capelle, G. and DeSerres, M. (1967) L'arginase: sa répartition chez les plantes à bulbe: Etude de quelques espèces cultivées du genre Allium. C. R. Acad. Sci. Paris 264D:54–57. Chem. Abst. 66 (1967) 102 450.

Brunfels, O. (1532) Herbarum vivae eicones ad naturae imitationem, summa cum diligentia et artificio effigiatae, etc. Ioannes Schottus, Strassburg.

Brüning, H. (1931a) Über die Wirksamkeit einiger neuer Wurmmittel. Med. Klin., 1312.

Brüning, H. (1931b) Über Wurmkuren bei Kindern. Fortschr. Med. 36:141–142.

Buchwald, H., Varco, R.L., and Matts, J.P. (1990) Effect of partial ileal bypass surgery on mortality and morbidity from coronary heart disease in patients with hypercholesterolemia. New Engl. J. Med. 323:946–955.

Buck, C., Donner, A.P., and Simpson, H. (1982) Garlic oil and ischemic heart disease. Int. J. Epidemiol. 11:294–295.

Bugaret, Y. and Marin, H. (1992) Control of white rot of garlic. Phytoma 443:57–60. Chem. Abst. 118 (1993) 75 246 (French).

Buiatti, E., Palli, D., Decarli, A., Amadori, D., Avellini, C., Bianchi, S., Biserni, R., Cipriani, F., Cocco, P., Giacosa, A., Marubini, E., Puntoni, R., Vindigni, C., Fraumeni, J., Jr., and Blot, W. (1989) A case-control study of gastric cancer and diet in Italy. Int. J. Cancer 44:611–616.

Bulatov, P.K., Zlybnikov, D.M., Fedoseev, G.B., and Khan-Fimina, W.A. (1965) The use of phytoncides for the treatment of patients with different inflammatory diseases of respiratory organs. Sov. Med. 28:86–90 (Russian).

Burden, A.D., Wilkinson, S.M., Beck, M.H., and Chalmers, R.J.G. (1994) Garlic-induced systemic contact dermatitis. Contact Dermatitis 30:299–300.

Burger, R.A., Warren, R.P., Lawson, L.D., and Hughes, B.G. (1993) Enhancement of in vitro human immune function by *Allium sativum* L. (garlic) fractions. Int. J. Pharmacogn. 31:169–174. Chem. Abst. 120 (1994) 95 106.

Burgess, J.F. (1952) Occupational dermatitis due to onion and garlic. Can. Med. Assoc. J. 66:275.

Burks, J.W. (1954) Classic aspects of onion and garlic dermatitis in housewives. Ann. Allergy 12:592–596.

Burnham, B.E. (1995) Garlic as a possible risk for postoperative bleeding. Plast. Reconstr. Surg. 95:213.

Butz, P., Tauscher, B., and Wolf, S. (1994) Ultra-high pressure processing of onions: chemical and sensory changes. Food Sci. Technol. [London] 27: 463–467. Chem. Abst. 121 (1994) 299 626.

Caesar, W. (1993) 450 Jahre "New Kreüterbuch" von Leonhart Fuchs. Dtsch. Apoth. Ztg. 133: 4800–4802.

Cai, X.J., Uden, P.C., Block, E., Zhang, X., Quimby, B.D., and Sullivan, J.J. (1994a) Allium chemistry: identification of natural abundance organoselenium volatiles from garlic, elephant garlic, onion, and Chinese chive using headspace gas chromatography with atomic emission detection. J. Agric. Food Chem. 42:2081–2084. Chem. Abst. 121 (1994) 254 286.

Cai, X.J., Uden, P.C., Sullivan, J.J., Quimby, B.D., and Block, E. (1994b) Headspace gas chromatography with atomic emission and mass selective detection for the determination of organoselenium compounds in elephant garlic. Anal. Proc. Anal. Comm. 31:325–327. Chem. Abst. 122 (1995) 51 428.

Cai, X.J., Block, E., Uden, P.C., Quimby, B.D., and Sullivan, J.J. (1995a) Allium chemistry: identification of natural abundance organoselenium compounds in human breath after ingestion of garlic using gas chromatography with atomic emission detection. J. Agric. Food Chem. 43:1751–1753.

Cai, X.J., Block, E., Uden, P.C., Zhang, X., Quimby, B.D., and Sullivan, J.J. (1995b) Allium chemistry: identification of selenamino acids in ordinary and selenium-enriched garlic, onion, and broccoli using gas chromatography with atomic emission detection. J. Agric. Food Chem. 43:1754–1757.

Cai, Y. (1991) Anticryptococcal and antiviral properties of garlic. Cardiol. Pract. 9:11.

Caldwell, D.R. and Danzer, C.J. (1988) Effects of allyl sulfides on the growth of predominant gut anaerobes. Curr. Microbiol. 16:237–241. Chem. Abst. 108 (1988) 146 953.

Caldwell, S.H., Jeffers, L.J., Narula, O.S., Lang, E.A., Reddy, K.R., and Schiff, E.R. (1992) Ancient remedies revisited: does *Allium sativum* (garlic) palliate the hepatopulmonary syndrome? J. Clin. Gastroenterol. 15:248–250.

Calvey, E.M., Matusik, J.E., White, K.D., Betz, J.M., Block, E., Littlejohn, M.H., Naganathan, S., and Putman, D. (1994a) Off-line supercritical fluid extraction of thiosulfinates from garlic and onion. J. Agric. Food Chem. 42:1335–1341. Chem. Abst. 121 (1994) 81 200.

Calvey, E.M., Roach, J.A.G., and Block, E. (1994b) Supercritical fluid chromatography of garlic (*Allium sativum*) extracts with mass spectrometric identification of allicin. J. Chromatogr. Sci. 32:93–96. Chem. Abst. 120 (1994) 186 732.

Camara, A.F.A., Pasqual, M., and Pinto, J.E.B.P. (1989) Obtencao de plantas de alho (*Allium sativum* L.) cultivares chines e chonan, a partir de meristemas in vitro. Cienc. Prat. 13:97–102. Chem. Abst. 114 (1991) 76 926 (Portuguese).

Camerarius, J. (1588) Hortus medicus et philosophicus, etc. Frankfurt-on-Main.

Campolmi, P., Lombardi, P., Lotti, R., and Sertoli, A. (1982) Immediate and delayed sensitization to garlic. Contact Dermatitis 8:352–353.

Candolle, A.D. and German translation by Goeze, E. (1884) Der Ursprung der Culturpflanzen. Brockhaus, Leipzig.

Canduela, V., Mongil, I., Carrascosa, M., Docio, S., and Cagigas, P. (1995) Garlic: always good for health? Brit. J. Dermatology 132:161–162.

Caporaso, N., Smith, S.M., and Eng, R.H.K. (1983) Antifungal activity in human urine and serum after ingestion of garlic (*Allium sativum*). Antimicrob. Agents Chemother. 23:700–702.

Cappelletti, E.M., Trevisan, R., and Caniato, R. (1982) External antirheumatic and antineuralgic herbal remedies in the traditional medicine of northeastern Italy. J. Ethnopharmacol. 6:161–190.

Carl, M., McKnight, R.S., Scott, B., and Lindegren, C.C. (1939) Physiological effects of garlic and derived substances. Am. J. Hyg. 29:32–35.

Carson, J.F. and Wong, F.F. (1959a) Separation of aliphatic disulfides and trisulfides by gas-liquid partition chromatography. J. Org. Chem. 24: 175–179. Chem. Abst. 53 (1959) 16 933.

Carson, J.F. and Wong, F.F. (1959b) A colour reaction for thiosulphinates. Nature 183:1673. Chem. Abst. 53 (1959) 19 672.

Carson, J.F. and Wong, F.F. (1961a) The volatile flavor components of onions. J. Agric. Food Chem. 9:140–143. Chem. Abst. 55 (1961) 17 944.

Carson, J.F. and Wong, F.F. (1961b) Isolation of (+)-S-methyl-L-cysteine sulfoxide and of (+)-S-propyl-L-cysteine sulfoxide from onions as their N-(2,4-dinitrophenyl) derivatives. J. Org. Chem. 26: 4997–5000. Chem. Abst. 56 (1962) 14 385.

Carson, P. (1968) In praise of onions. Br. Med. J. III:683.

Cashin-Hemphill, L., Mack, J.W., Pogoda, J.M., Sanmarco, M.E., Azen, S.P., and Blankenhorn, D.H. (1990) Beneficial effects of colestipol-niacin on coronary atherosclerosis. JAMA 264:3013–3017.

Caspari, W. (1936) Über den Einfluß der Kost auf das Wachstum von Impfgeschwülsten. Z. Krebsforsch. 43:255–263.

Castelli, W.O. (1988) Cardiovascular disease in women. Am. J. Obstet. Gynecol. 158:1553–1560.

Catar, G. (1954) Effect of plant extracts on Ixodes ricinus. Bratisl. Lek. Listy (Bratislava Medical News) 34:1004–1010. Chem. Abst. 49 (1955) 2002.

Cavallito, C.J. (1946) Relationship of thiol structures to reaction with antibiotics. J. Biol. Chem. 164:29–34. Chem. Abst. 41 (1947) 94.

Cavallito, C.J. and Bailey, J.H. (1944) Allicin, the antibacterial principle of *Allium sativum*. I. Isolation, physical properties, and antibacterial action. J. Am. Chem. Soc. 66:1950–1951. Chem. Abst. 39 (1945) 323.

Cavallito, C.J., Bailey, J.H., and Buck, J.S. (1945) The antibacterial principle of *Allium sativum*. III. Its precursor and "essential oil of garlic." J. Am. Chem. Soc. 67:1032–1033. Chem. Abst. 39 (1945) 3325.

Cavallito, C.J., Buck, J.S., and Suter, C.M. (1944) Allicin, the antibacterial principle of *Allium sativum*. II. Determination of the chemical structure. J. Am. Chem. Soc. 66:1952–1954.

Ceci, L.N., Curzio, O.A., and Pomilio, A.B. (1992a) Gamma-glutamyl transpeptidase/gamma-glutamyl peptidase in sprouted *Allium sativum*. Phytochemistry 31:441–444. Chem. Abst. 116 (1992) 231 875.

Ceci, L.N., Curzio, O.A., and Pomilio, A.B. (1992b) Effects of irradiation and storage on the gamma-glutamyl transpeptidase activity of garlic bulbs cv 'Red'. J. Sci. Food Agric. 59:504–510. Chem. Abst. 117 (1992) 232 629.

Celsus, A.C. (1906) De Medicina Libri Octo, Lipsiae 1859. Vieweg, Braunschweig.

Cessna, A.J. (1991a) Residues of triallate in garlic (*Allium sativum* L.) cloves following preplant incorporation. Can. J. Plant Sci. 71:1257–1261. Chem. Abst. 116 (1992) 150 271.

Cessna, A.J. (1991b) The HPLC determination of residues of maleic hydrazide in cloves of garlic bulbs following foliar application. Pestic. Sci. 33:169–176. Chem. Abst. 116 (1992) 82 480.

Cessna, A.J. (1991c) Residue analysis of garlic (*Allium sativum* L.) cloves following a postemergence application of linuron. Can. J. Plant Sci. 71:951–955. Chem. Abst. 115 (1991) 254 696.

Cha, C.W. (1987) A study on the effect of garlic to the heavy metal poisoning of rat. J. Korean Med. Sci. 2:213–223. Chem. Abst. 109 (1988) 18 322.

Chagovets, R.V. and Zalesskii, Y.A. (1951) Biochemische Charakterisierung der Antibiotika aus höheren Pflanzen. 3. Mitt.: Redoxwirkung der Knoblauch-Phytonzide. Byull. Biol. Med. (Bull. Biol. Med.) 31:185–188 (Russian).

Chanderbhan, R., Brockington, R., and Whittaker, P. (1993) Effect of dietary garlic extract on blood lipids in the rat. FASEB J. 7:A864.

Chandorkar, A.G. and Jain, P.K. (1972) Analysis of hypotensive actions of *Allium sativum* (garlic). Indian J. Physiol. Pharmacol. 17:132–133.

Chang, J.I. (1988) Studies on growth characteristics of "Sanghai Early" garlic. 2. Effects of storage of seed bulb and light break on seasonal change of endogenous ABA (abscisic acid)- and GA-like substances. Nunmunjip-Cheju Taehak, Chayon Kwahakpyon 27:13–20. Chem. Abst. 111 (1989) 20 945.

Chang, M.L.W. and Johnson, M.A. (1980) Effect of garlic on carbohydrate metabolism and lipid synthesis in rats. J. Nutr. 110:931–936. Chem. Abst. 93 (1980) 44 639.

Chaudhuri, B.N., Mukherjee, S.K., Mongia, S.S., and Chakravarty, S.K. (1984) Hypolipidemic effect of garlic and thyroid function. Biomed. Biochim. Acta 43:1045–1047. Chem. Abst. 101 (1984) 209 754.

Chaudhury, D.S., Sreenivasamurthy, V., Jayaraj, P., Sreekantiah, K.R., and Johar, D.S. (1962) Therapeutic usefulness of garlic in leprosy. J. Indian Med. Assoc. 39:517–520.

Chauhan, L.S., Garg, J., Bedi, H.K., Gupta, R.C., Bomb, B.S., and Agarwal, M.P. (1982) Effect of onion, garlic, and clofibrate on coagulation and fibrinolytic activity of blood in cholesterol fed rabbits. Indian Med. J. 76:126–128.

Chaurasia, L.C. (1979) Induction of mitotic and amitotic anomalies in *Allium sativum* by urea. Indian J. Exp. Biol. 17:118–119. Chem. Abst. 90 (1979) 133 704.

Chen, J. and Tang, Y. (1991) Chemical constituents and antiatherosclerotic action of garlic. Zhongguo Yaolixue Tongbao 7:88–91. Chem. Abst. 115 (1991) 269 762 (Chinese).

Chen, L., Lee, M., Hong, J.Y., Huang, W., Wang, E., and Yang, C.S. (1994) Relationship between cytochrome P450 2E1 and acetone catabolism in rats as studied with diallyl sulfide as an inhibitor. Biochem. Pharmacol. 48:2199–2205.

Chen, Q.H., Lu, W.G., and Wang, W.M. (1990) Study on preparation and character of microcapsules of garlic oil by spray technique. Zhongguo Yiyao Gongye Zazhi (Chin. J. Pharm.) 21:540–543. Chem. Abst. 114 (1991) 214 335 (Chinese).

Chen, S. and Snyder, J.K. (1987) Molluscicidal saponins from Allium vineale. Tetrahedron Lett. 28:5603–5606. Chem. Abst. 108 (1988) 147 115.

Chen, S. and Snyder, J.K. (1989) Diosgenin-bearing, molluscicidal saponins from Allium vineale: an NMR approach for the structural assignment of oligosaccharide units. J. Org. Chem. 54:3679–3689. Chem. Abst. 11 (1989) 54 186.

Cheng, H.H. and Tung, T.C. (1981) Effect of allithiamine on sarcoma-180 tumor growth in mice. Taiwan I Hsueh Hui Tsa Chih 80:385–393. Chem. Abst. 95 (1981) 197 366 (Chinese).

Cherkasov, O.A., Azarkova, A.F., Stikhin, V.A., Kabanov, V.S., and Melnikova, T.M. (1985) Diosgenin content of organs of Allium nutans introduced into the Moscow region. Rastit. Resur. (Plant Resources) 21:455–458 (Russian).

Cherkasov, O.A., Azarkova, A.F., Kabanov, V.S., and Maisuradze, N.I. (1990) Diosgenin content in inflorescences of two forms of Allium nutans L. grown in the Moscow district. Rastit. Resur. (Plant Resources) 26:76–80. Chem. Abst. 112 (1990) 155 307.

Chermsiri, C., Watanabe, H., Attajarusit, S., Tuntiwarawit, J., and Kaewroj, S. (1995) Effect of boron sources on garlic (*Allium sativum* L.) productivity. Biol. Fertil. Soils 20:125–129.

Chevastelon, M.R. (1894) Contributions à l'étude des hydrates de carbone: étude chimique et physiologique de ceux contenus dans l'ail, l'échalote et l'oignon. Dissertation, Faculty of Science, University of Paris (Abstract).

Chhabra, S.K. and Rao, A.R. (1994) Transmammary exposure of mouse pups to *Allium sativum* (garlic) and its effect on the neonatal hepatic xenobiotic metabolizing enzymes of mice. Nutr. Res. *14*:195–210.

Chi, M.S. (1982) Effects of garlic products on lipid metabolism in cholesterol-fed rats. Proc. Soc. Exp. Biol. Med. *171*:174–178. Chem. Abst. 98 (1983) 33 651.

Chi, M.S., Koh, E.T., and Stewart, T.J. (1982) Effects of garlic on lipid metabolism in rats fed cholesterol or lard. J. Nutr. *112*:241–248. Chem. Abst. 96 (1982) 161 393.

Chiou, C.H., Bennink, M.R., and Dimitrov, N.V. (1995) Effect of a garlic extract on prostaglandin E metabolite (PGE-M) excretion. FASEB J. *9*:A992.

Chiyoma, K. and Sanwa Food Kk. (1989) Menthol, organic acids, and alanine in deodorization of vegetables. Japan. Patent 89 113 964. Chem. Abst. 112 (1990) 97 294.

Cho, H.O., Kwon, J.H., Byun, M.W., and Yoon, H.S. (1984) Batch scale storage of garlic by irradiation combined with natural low temperature. Hanguk Sikpum Kwahak Hoechi (Korean J. Food Sci. Technol.) *16*:66–70. Chem. Abst. 101 (1984) 150 133 (Korean).

Choi, J.H. and Byun, D.S. (1986) Studies on anti-aging action of garlic, *Allium sativum* L. I. Comparative study of garlic and ginseng components on anti-aging action. Hanguk Saenghwa Hakhoechi (Korean Biochem. J.) *19*:140–146. Chem. Abst. 105 (1986) 91 305.

Choi, J.N., Ahn, J.H., and Lee, J.S. (1993) Molecular cloning of cDNAs for Korean garlic viruses. Hanguk Nonghwahak Hoechi (J. Korean Agric. Chem. Soc.) *36*:315–319. Chem. Abst. 120 (1994) 24 511.

Choi, J.S., Kim, J.Y., Lee, J.H., Young, H.S., and Lee, T.W. (1992) Isolation of adenosine and free amino acid composition from the leaves of Allium tuberosum. Hanguk Yongyang Siklyong Hakhoechi (J. Korean Soc. Food Nutri.) *21*:286–290. Chem. Abst. 119 (1993) 71 085.

Choi, Y., Choi, J.N., Ueom, T.S., and Choi, Y.D. (1992) Molecular cloning of cDNA for garlic mosaic virus genome. Sikmul Hakhoe Chi *35*:253–257. Chem. Abst. 119 (1993) 87 531.

Chomatova, S., Turkova, V., and Klozova, E. (1990) Protein complex and esterase isoenzyme patterns of *Allium sativum* L. cultivars and clones-regenerants. Biol. Plant. *32*:321–331.

Chopra, R.N. (1933) *Allium sativum*. In: Indigenous drugs of India: their medical and economic aspects. The Art Press, Calcutta, pp. 273–276.

Chowdhury, A.K.A., Ahsan, M., Islam, S.N., and Ahmed, Z.U. (1991) Efficacy of aqueous extract of garlic and allicin in experimental shigellosis in rabbits. Indian J. Med. Res. *A93*:33–36.

Choy, Y.M., Kwok, T.T., Fung, K.P., and Lee, C.Y. (1983) Effect of garlic, Chinese medicinal drugs, and amino acids on growth of Erlich ascites tumor cells in mice. Am. J. Chin. Med. *11*:69–73.

Christensen, R.E., Beckman, R.M., and Birdsall, J.J. (1968) Spices and other condiments: some mineral elements of commercial spices and herbs as determined by direct reading emmission spectroscopy. J. Assoc. Off. Anal. Chem. *51*:1003–1010. Chem. Abst. 69 (1968) 85 543.

Christoffel, H. (1930a) Knoblauch zur Erleichterung der kochsalzarmen Diät. Münch. Med. Wochenschr. *77*:1865.

Christoffel, H. (1930b) Knoblauch zur Erleichterung der kochsalzarmen Diät. Schweiz. Med. Wochenschr., 902.

Chu, T.C., Ogidigben, M., Han, J.C., and Potter, D.E. (1993) Allicin-induced hypotension in rabbit eyes. J. Ocul. Pharmacol. *9*:201–209.

Chung, I.S. (1985) Protective effect of garlic on ethanol-induced gastric mucosal injury in rats. J. Cathol. Med. Coll. *38*:1211–1224.

Chung, J.G., Garrett, L.R., Byers, P.E., and Cuchens, M.A. (1989) A survey of the amount of pristane in common fruits and vegetables. J. Food Compos. Anal. *2*:22–27. Chem. Abst. 111 (1989) 213 561.

Chutani, S.K. and Bordia, A. (1981) The effect of fried versus raw garlic on fibrinolytic activity in man. Atherosclerosis *38*:417–421.

Ciplea, A.G. and Richter, K.D. (1988) The protective effect of *Allium sativum* and crataegus on isoprenaline-induced tissue necroses in rats. Arzneim. Forsch. *38*:1583–1592.

Cipriani, F., Buiatti, E., and Palli, D. (1991) Gastric cancer in Italy. Ital. J. Gastroenterol. *23*:429–435.

Claeys, M., Üstünes, L., Laekeman, G., Herman, A.G., Vlietinck, A.J., and Özer, A. (1986) Characterization of prostaglandin E-like activity isolated from plant source (Allium cepa). Prog. Lipid Res. *25*:53–58.

Clarus, G.L. (1815) Handbuch der speciellen Arzneimittellehre. Leipzig.

Clayton, S., Born, G.V.R., and Cross, M.J. (1963) Inhibition of the aggregation of blood platelets by nucleosides. Nature *200*:138–139.

Clusius, C.A. (1576) Rariorum plantarum historia. Ioannes Moretus, Antwerp.

Cochran, F.D. (1950) A study of the species hybrid, Allium ascalonicum × Allium fistulosum and its backcrossed progenies. J. Am. Soc. Horticult. Sci. *55*:293–296.

Codignola, A., Verotta, L., Spanu, P., Maffei, M., Scannerini, S., and Bonfante-Fasolo, P. (1989) Cell wall bound phenols in roots of vesicular-arbuscular mycorrhizal plants. New Phytol. *112*:221–228. Chem. Abst. 111 (1989) 171 157.

Coley-Smith, J.R. (1986) A comparison of flavor and odor compounds of onion, leek, garlic, and Allium fistulosum in relation to germination of sclerotia of Sclerotium cepivorum. Plant Pathol. *35*:370–376.

Coley-Smith, J.R. and King, E.J. (1969) The production by species of Allium of alkyl sulphides and their effect on germination of sclerotia of Sclerotium cepivorum. Berk. Ann. Appl. Biol. *64*:289–301.

Collin, H.A. and Britton, G. (1993) Allium cepa L. (onion): in vitro culture and the production of flavor. Biotechnol. Agric. Forestry 5:23–40. Chem. Abst. 120 (1994) 75 707.

Collin, H.A. and Musker, D. (1988) Allium compounds [Phytochemicals in plant cell cultures]. Cell Cult. Somatic Cell Genet. Plants 5:475–493. Chem. Abst. 109 (1988) 208 229.

Collip, J.B. (1923a) An apparent synthesis in the normal animal of a hypoglycemia producing principle: animal passage of the principle. J. Biol. Chem. 58:163–208.

Collip, J.B. (1923b) Glucokinin: a new hormone present in plant tissue. J. Biol. Chem. 56:513–531.

Collip, J.B. (1923c) Glucokinin: second paper. J. Biol. Chem. 57:65–78.

Conner, D.E., Beuchat, L.R., Worthington, R.E., and Hitchcock, H.L. (1984) Effects of essential oils and oleoresins of plants on ethanol production, respiration, and sporulation of yeasts. Int. J. Food Microbiol. 1:63–74. Chem. Abst. 101 (1984) 226 691.

Conner, D.E. and Beuchat, L.R. (1984a) Inhibitory effects of plant oleoresins on yeasts. Microb. Assoc. Interact. Food, Proc. 12th Int. IUMS-ICFMH Symp. 1983, 447–451. Chem. Abst. 103 (1985) 140 509.

Conner, D.E. and Beuchat, L.R. (1984b) Sensitivity of heat-stressed yeasts to essential oils of plants. Appl. Environ. Microbiol. 47:229–233.

Conner, D.E. and Beuchat, L.R. (1984c) Effects of essential oils from plants on growth of food spoilage yeasts. J. Food Sci. 49:429–434.

Cook-Mozaffari, P.J., Azordegan, F., Day, N.E., Ressicaud, A., Sabai, C., and Aramesh, B. (1979) Oesophageal cancer studies in the Caspian littoral of Iran: results of a case-control study. Br. J. Cancer 39:293–309.

Couturier, P., Kreuter, M., and Loupi, J. (1980) Allergie à la poussière d'ail. A propos de deux observations. Rev. Fr. Allerg. 20:145–147.

Couturier, P. and Bousquet, J. (1982) Occupational allergy secondary to inhalation of garlic dust. Allergy Clin. Immunol. 70:145.

Cowan, J.W., Saghir, A.R., and Salji, J.P. (1967) Antithyroid activity of onion volatiles. Aust. J. Biol. Sci. 20:683–685. Chem. Abst. 67 (1967) 52 491.

Criss, W., Fakunle, J., Deu, B., and Knight, E. (1983) Inhibition of hepatoma growth and hepatoma guanylate cyclase activity with a garlic polypeptide. Modulation Mediation Cancer Vitam., 300–303. Chem. Abst. 100 (1983) 173 615.

Criss, W.E., Fakunle, J., Knight, E., Adkins, J., Morris, H.P., and Dhillon, G. (1982) Inhibition of tumor growth with low dietary protein and with dietary garlic extracts. Fed. Proc. 41:281.

Croci, C.A., Arguello, J.A., and Orioli, G.A. (1987) Effect of gamma rays on seed cloves of garlic (Allium sativum L.) at post harvest: reversion by exogenous growth regulators. Environ. Exp. Bot. 27:1–5. Chem. Abst. 106 (1987) 115 776.

Croci, C.A., Arguello, J.A., and Orioli, G.A. (1990) Effect of gamma rays on sprouting of seed cloves of garlic (Allium sativum L.): levels of auxin-like substances and growth inhibitors. Environ. Exp. Bot. 30:9–15. Chem. Abst. 112 (1990) 114 920.

Croci, C.A., Arguello, J.A., Curvetto, N.R., and Orioli, G.A. (1991) Changes in peroxidases associated with radiation-induced sprout inhibition in garlic (Allium sativum L.). Int. J. Radiat. Biol. 59:551–557. Chem. Abst. 114 (1991) 138 972.

Croci, C.A., Arguello, J.A., and Orioli, G.A. (1994) Biochemical changes in garlic (Allium sativum L.) during storage following gamma-irradiation. Int. J. Radiat. Biol. 65:263–266. Chem. Abst. 120 (1994) 211 556.

Cronin, E. (1987) Dermatitis of the hands in caterers. Contact Dermatitis 17:265–269.

Crooke, W.M., Knight, A.H., and MacDonald, I.R. (1960) Cation-exchange capacity and pectin gradients in leek root segments. Plant Soil 13:123–127. Chem. Abst. 55 (1961) 9579.

Cruess, W.V. (1944) Enzymes in dehydrated vegetables. Science 100:148. Chem. Abst. 38 (1944) 5320.

Cueva, J. and Duran, J.C. (1955) Dermatitis on garlic. Rev. Med. Hosp. Gen. [Mexico] 18:29–34.

Culpeper, N. (1649) A physical directory. London.

Currier, H.B. (1945) Photometric estimation of volatile sulfur in onions as a criterion of pungency. Food Res. 10:177–186. Chem. Abst. 39 (1945) 4702.

Curzio, D.A. and Ceci, L.N. (1984) Evaluation of ethereal extracts of irradiated garlic. Food Chem. 14:287–293. Chem. Abst. 101 (1984) 109 239.

Curzio, O.A., Croci, C.A., and Ceci, L.N. (1986) The effects of radiation and extended storage on the chemical quality of garlic bulbs. Food Chem. 21:153–159. Chem. Abst. 105 (1986) 189 697.

DAB 6, E.B.6. (1926) Deutsches Arzneibuch, Ergänzungsband, Neudruck 1938. R. v. Decker, Berlin, Neudruck 1938.

Dababneh, B.F.A. and Al-Delaimy, K.S. (1984) Inhibition of Staphylococcus aureus by garlic extract. Lebensm. Wiss. Technol. 17:29–31.

Daiichi. (1983) Flavonol glycosides from Allium tuberosum as antiallergy and fibrinolysis-inducing agents. Japan. Patent 83 99 498. Chem. Abst. 99 (1983) 218 578.

Daitsch, J., Brujis, R.G., Kaminsky, A., and Bucher, N. (1961) Professional dermatitis on contact with Allium sativum. Semana med. [Buenos Aires] 119:1767–1768 (Spanish).

Dale, S. (1693) Pharmacologia seu Manuductio ad Materiam medicam, etc. Smith & Walford, London.

Dalvi, R.R. (1992) Alterations in hepatic phase I and phase II biotransformation enzymes by garlic oil in rats. Toxicol Lett. 60:299–305.

Dalvi, R.R. and Saluhnke, D.K. (1993) An overview of medicinal and toxic properties of garlic. J. Maharashtra Agric. Univ. 18:378–381. Chem. Abst. 120 (1994) 330 895.

Daly, J.W. (1982) Adenosine receptors: targets for future drugs. J. Med. Chem. 25:197–207.

Damrau, F. and Ferguson, E.A. (1949) The action of carminatives. Rev. Gastroenterol. 16:411–419. Chem. Abst. 43 (1949) 5114. Ref. in Schweiz. Apoth. Ztg. 88 (1950) 274.

Damrau, F. and Ferguson, E.A. (1950) Knoblauch bei gastrointestinalen Störungen. Schweiz. Apoth. Ztg. 88:274.

Danielak, R. (1973) Glucosinolates and sulfides of garlic. Chromatogr. Cienkowarstwowa Anal. Farm., 123–126. Chem. Abst. 83 (1975) 48 249 (Polish).

Danilenko, U.A. and Epshtein, M.M. (1953) Effect of phytoncides of garlic and mustard oils on alkaline phosphatase and invertase. Ukrain. Biokhim. Zh. (Biochem. J.) 25:106–109. Chem. Abst. 47 (1953) 12 439 (Ukrainian).

Dankert, J., Tromp, T.F.J., De Vries, H., and Klasen, H.J. (1979) Antimicrobial activity of crude juices of Allium ascalonicum, Allium cepa, and *Allium sativum*. Zentralbl. Bakteriol. Parasitenkd. Infektionskrankh. Hyg. (Part I) A245:229–239.

Darbyshire, B., Henry, R.J., Melhuish, F.M., and Hewett, R.K. (1979) Diurnal variations in nonstructural carbohydrates, leaf extension, and leaf cavity carbon dioxide concentrations in Allium cepa L. J. Exp. Bot. 30:109–118.

Darbyshire, B. and Henry, R.J. (1979) The association of fructans with high percentage dry weight in onion cultivars suitable for dehydrating. J. Sci. Food Agric. 30:1035–1038.

Darbyshire, B. and Henry, R.J. (1981) Differences in fructan content and synthesis in some Allium species. New Phytol. 87:249–256.

Das, I., Khan, N.S., and Sooranna, S.R. (1995a) Nitric oxide synthase activation is a unique mechanism of garlic action. Biochem. Soc. Trans. 23:S136. Chem. Abst. 122 (1995) 186 284.

Das, I., Khan, N.S., and Sooranna, S.R. (1995b) Potent activation of nitric oxide synthetase by garlic: a basis for its therapeutic applications. Curr. Med. Res. Opinion 13:257–263.

Das, N.N., Das, A., and Mukherjee, A.K. (1977a) Characterization of the polysaccharides of garlic (*Allium sativum*) bulbs. I. Structure of the D-galactan isolated from garlic (*Allium sativum*) bulbs. Carbohydr. Res. 56:337–349.

Das, N.N., Das, A., and Mukherjee, A.K. (1977b) Structure of the D-galactan isolated from garlic (*Allium sativum*) bulbs. Carbohydrate Res. 56:337–349.

Das, N.N. and Das, A. (1978) Characterization of the polysaccharides of garlic (*Allium sativum*) bulbs. II. Structure of the D-fructan isolated from garlic (*Allium sativum*) bulbs. Carbohydr. Res. 64:155–167. Chem. Abst. 89 (1978) 126 115.

Das, S.N., Pramanik, A.K., Mitra, S.K., and Mukherjee, B.N. (1982) Effect of garlic pearls (Ranbaxyl) on blood cholesterol level in normal dogs. Indian Vet. J. 59:937–938. Chem. Abst. 98 (1982) 70 847.

Das, T., Choudhury, A.R., Sharma, A., and Talukder, G. (1993a) Modification of cytotoxic effects of inorganic arsenic by a crude extract of *Allium sativum* L. in mice. Int. J. Pharmacog. 31:316–320.

Das, T., Roychoudhury, A., Sharma, A., and Talukder, G. (1993b) Modification of clastogenicity of three known clastogens by garlic extract in mice in vivo. Environ. Mol. Mutagen. 21:383–388. Chem. Abst. 119 (1993) 88 693.

Das, V.S.R. and Rao, J.V.S. (1964) Phenolic acids of onion plant, Allium cepa. Curr. Sci. 35:471–472. Chem. Abst. 61 (1964) 11 005.

Datta, N.L., Krishnamurthi, A., and Siddiqui, S. (1948) Antibiotic principles of *Allium sativum* (lahsan). J. Sci. Ind. Res. 7B:42. Chem. Abst. 42 (1948) 7830.

Dausch, J.G. and Nixon, D.W. (1990) Garlic: a review of its relationship to malignant disease. Prev. Med. 19:346–361. Chem. Abst. 113 (1990) 184 048.

Davenport, H.W. (1982) Physiology of the digestive tract. An introductory text. Year Book Medical Publ., Chicago.

Davis, L.E., Shen, J.K., and Cai, Y. (1990) Antifungal activity in human cerebrospinal fluid and plasma after intravenous administration of *Allium sativum*. Antimicrob. Agents Chemother. 34:651–653. Int. Pharm. Abst. 28 (1991) 2126.

Davis, L.E., Shen, J.K., and Royer, R.E. (1994) In vitro synergism of concentrated *Allium sativum* extract and amphotericin B against Cryptococcus neoformans. Planta Med. 60:546–549. Chem. Abst. 122 (1995) 101 423.

De A Santos, O.S. and Grünwald, J. (1993) Effect of garlic powder tablets on blood lipids and blood pressure: a six month placebo controlled, double blind study. Br. J. Clin. Res. 4:37–44.

DeBoer, L.W.V. and Folts, J.D. (1989) Garlic extract prevents acute platelet thrombus formation in stenosed canine coronary arteries. Am. Heart J. 117:973–975.

de Carvalho, V.D., Chalfoun, S.M., and Leite, I.P. (1987) Effect of curing method on quality of some garlic varieties. Pesqui. Agropecu. Bras. 22:733–740. Chem. Abst. 110 (1989) 37 966.

de Carvalho, V.D., Chalfoun de Souza, S.M., Patto de Abreu, C.M., and de Rezende Chagas, S.J. (1991) Storage time and quality of garlic, amarante CV. Pesqui. Agropecu. Bras. 26:1679–1684 (Portuguese).

de Carvalho, V.D., Desouza, S.M.C., and Botrel, N. (1993) Effect of postharvest treatments, packing type, and storage time on conservation and quality of garlic. Pesqui. Agropecu. Bras. 28:987–992 (Portuguese).

De Halacsy, E. (1904) Allium L. In: Conspectus Florae Graecae. G. Engelmann, Leipzig, pp. 240–262.

Deineko, G.I. (1985) Lipids, fatty acids, and carbohydrates in Allium species. Rastit. Resur. 21:221–229. Chem. Abst. 103 (1985) 34 823.

Deininger, R. and Wagner, H. (1976) Extraction of therapeutic substances from garlic. Germany. Patent 2 528 075. Chem. Abst. 84 (1976) 184 882.

Deki, M. (1983) Separation of some components in imported agricultural products by high performance liquid chromatography. Kanzei Chuo Bunsekishoho 23:15–25. Chem. Abst. 99 (1983) 20 982 (Japanese).

Delaha, E.C. and Garagusi, V.F. (1985) Inhibition of mycobacteria by garlic extract (Allium sativum). Antimicrob. Agents Chemother. 27:485–486. Chem. Abst. 102 (1985) 218 177.

de Martin, R., Knasmueller, S., and Weniger, P. (1986) Antimutagene Wirkung von Knoblauch im Ames-Test. Österr. Forsch. Zent. Seibersdorf Rep. Nr.4354:1–9. Chem. Abst. 104 (1986) 223 865.

Demitsch, W. (1889) Historische Studien aus dem Pharmakologischen Institut der Universität Dorpat. University of Verlag, Halle.

Demkevich, L.I. (1981) Amino acid composition of garlic. Vopr. Khraneniya Otsenki Kachestva Plodoovoshch. Tovarov M. (Probl. Preservation & Process. Quality of Fruit & Vegetable Products), 95–97. Chem. Abst. 97 (1982) 125 937 (Russian).

Demkevich, L.I. (1985) Content of pectic compounds in garlic leaves and bulbs. Izv. Vyssh. Uchebn. Zaved. Pishch. Tekhnol. (Bull. Inst. High. Stud. Food Technol.), 120. Chem. Abst. 104 (1986) 66 010 (Russian).

Deng, D. (1993) Modulation of mutagenic drug-induced unscheduled DNA synthesis (UDS) in primary rat hepatocytes by diallyl trisulfide. Chin. J. Oncol. 15:423–426. Chem. Abst. 121 (1994) 99 192 (Chinese).

Deng, D.J., Mueller, K., Kasper, P., and Mueller, L. (1994) Effect of diallyl trisulfide on induction of UDS by mutagenic drugs in primary rat hepatocytes. Biomed. Environ. Sci. 7:85–90.

DeResende, M.L.V., Zambolim, L., and DeCruz, J. (1987) Efficacy of fungicides for the control of white rot of garlic (Allium sativum) in relation to the level of Sclerotium cepivorum sclerotia in the soil. Fitopatol. Bras. 12:71–78. Chem. Abst. 108 (1988) 33 501 (Portuguese).

Deruaz, D., Soussan Marchal, F., Joseph, I., Desage, M., Bannier, A., and Brazier, J.L. (1994) Analytical strategy by coupling headspace gas chromatography, atomic emission spectrometric detection and mass spectrometry application to sulfur compounds from garlic. J. Chromatogr. A. 677:345–364. Chem. Abst. 121 (1994) 229 147.

Deshpande, R.G., Khan, M.B., Bhat, D.A., and Navalkar, R.G. (1993) Inhibition of Mycobacterium avium complex isolates from AIDS patients by garlic (Allium sativum). J. Antimicrob. Chemother. 32:623–626.

Devaki, T., Venmadhi, S., and Govindaraju, P. (1992) Alterations in protein metabolism in ethanol-ingested rats treated with garlic. Med. Sci. Res. 20:725–727. Chem. Abst. 118 (1993) 2328.

Devasagayam, T.P.A., Pushpendran, C.K., and Eapen, J. (1982) Diallyl disulphide induced changes in microsomal enzymes of suckling rats. Indian J. Exp. Biol. 20:430–432.

DeWitt, J.C., Notermans, S., Gorin, N., and Kampelmacher, E.H. (1979) Effect of garlic oil or onion oil on toxin production by Clostridium botulinum in meat slurry. J. Food Prot. 42:222–224. Chem. Abst. 90 (1979) 185 096.

Dhar, D.C. (1951) Anti-oxidant concentrates from edible plant materials. J. Indian Chem. Soc. Ind. News Ed. 14:175–176. Chem. Abst. 46 (1952) 11 497.

Dhillon, G., Deu, B., Sahai, A., and Criss, W.E. (1981) Inhibitor of the in vivo and in vitro guanylate cyclase activity from garlic. Proc. Am. Assoc. Cancer Res. 22:17.

Didry, N., Pinkas, M., and Dubreuil, L. (1987) Activité antibactérienne d'espèces du genre Allium. Pharmazie 42:687–688. Chem. Abst. 108 (1988) 52 691.

Didry, N., Dubreuil, L., and Pinkas, M. (1992) Antimicrobial activity of naphthoquinones and Allium extracts combined with antibiotics. Pharm. Acta Helv. 67:148–151. Chem. Abst. 117 (1992) 66 404. Int. Pharm. Abst. 30 (1993) 5665.

Dieffenbach. (1902) Die Landwirtschaft der Israeliten. Feld und Wald.

Dierbach, J.H. (1824) Die Arzneimittel des Hippokrates oder Versuch einer systematischen Aufzählung der in allen hippokratischen Schriftenvorkommenden Medikamenten. Neue Akademische Buchhandlung Karl Groos, Heidelberg.

Dindonis, L.L. and Miller, J.R. (1980) Host-finding responses of onion and seedcorn flies to healthy and decomposing onions and several synthetic constituents of onion. Environ. Entomol. 9:467–472.

Dioscorides. (1610) Kräuterbuch deb uralten unnd in aller Welt berühmtesten Griechischen Scribenten PEDACII DIOSCORIDIS ANAZARBAEI, etc. Conrad Corthoys, Frankfurt-on-Main.

DiPaolo, J.A. and Carruthers, C. (1960) The effect of allicin from garlic on tumor growth. Cancer Res. 20:431–434. Chem. Abst. 55 (1961) 3844.

Dittmar, C. (1939) Über einige chemotherapeutisch bei Impftumoren wirksame Verbindungen. Z. Krebsforsch. 49:515–524.

Dixit, V.P. and Joshi, S. (1982) Effects of chronic administration of garlic (Allium sativum Linn.) on testicular function. Indian J. Exp. Biol. 20:534–536.

Dmitriev, A.P., Tverskoi, L.A., and Grodzinskii, D.M. (1986) Active mechanisms of onion phytoimmunity. Fiziol. Biokhim. Kult. Rast. (Physiol. Biochem. Plant Culture) 18:452–459. Chem. Abst. 105 (1986) 222 954 (Russian).

Dmitriev, A.P., Kovtun, A.V., and Tverskoy, L.A. (1988a) Cibulins as markers of active defense responses in onion. Fiziol. Biokhim. Kult. Rast. (Physiol. Biochem. Plant Culture) 20:353–358. Chem. Abst. 109 (1988) 126 028.

Dmitriev, A.P., Malinovskii, Y.Y., and Grodzinsky, D.M. (1988b) Induction of the synthesis of pathogen-dependent proteins in onion tissues. Dokl. Akad. Nauk Ukrain. SSR, Ser. B: Geol., Khim. Biol. Nauki (Rep. Acad. Sci. Ukrain. SSR), 64–68. Chem. Abst. 110 (1989) 132 315.

Dmitriev, A.P., Tverskoy, L.A., Kolesnikov, A.M., Kovtun, A.V., and Grodzinsky, D.M. (1989) Tsibulins: a new group of onion antibiotic substances. Prikl. Biokhim. Mikrobiol. 25:232–245 (Russian).

Dmitriev, A.P., Tverskoy, L.A., Kozlovsky, A.G., and Grodzinsky, D.M. (1990) Phytoalexins from onion and their role in disease resistance. Physiol. Mol. Plant Pathol. 37:235–244. Chem. Abst. 114 (1991) 160 912.

Döbler, H.F. (1975) Die Germanen. Bertelsmann Lexikon Verlag, Gütersloh.

Dobrzynska, K. and Podbielkowska, M. (1991) Reaction of meristematic cells to penicillin, cloxacillin, and streptomycin. Acta Pol. Pharm. 48:61–66. Chem. Abst. 118 (1993) 73 172.

Dodoens, J.R. (1557) Histoire des plantes, en laquelle est contenue la description entière des herbes, etc. Jean Loe, Antwerp.

Doetsch, H. (1989) Knoblauch: altes und neues Mittel gegen Krebs. Naturarzt 129:5–7.

Dombray and Vlaicovitch. (1924) Baktericides Vermögen des Lauchs (*Allium sativum*). C. R. Seances Soc. Biol. 90:1428–1429. Chem. Zbl. 1924, II, 683.

Domjan, G., Knasmueller, S., Szakmary, A., Wottawa, A., and Varga, L. (1988) Inhibition of lipid peroxidation in human serum by aqueous garlic extract. Österr. Forsch. Zentr. Seibersdorf. Report, 1–7.

Don, G. (1832) A monograph of the genus Allium. Memoirs Wernerian Nat. Hist. Soc. 6:1–102 (partly in Latin).

Dong, W. and Zhang, W. (1986) Ultrastructural localization of adenosine triphosphatase activity in the phloem of garlic scape. Zhiwu Xuebao (Acta Bot. Sinica) 28:441–443. Chem. Abst. 105 (1986) 187 660 (Chinese).

Dorant, E., van den Brandt, P.A., Goldbohm, R.A., Hermus, R.J.J., and Sturmans, F. (1993) Garlic and its significance for the prevention of cancer in humans: a critical view. Br. J. Cancer 67:424–429.

Dorant, E. (1994) Onion and leek consumption, garlic supplement use, and the incidence of cancer. Dissertation, Maastricht University, Netherlands (Abstract).

Dorant, E., van den Branch, P.A., Goldbohm, R.A., Hermus, R.J.J., and Sturmans, F. (1994a) Agreement between interview data and self-administered questionnaire on dietary supplement use. Eur. J. Clin. Nutr. 45:180–188.

Dorant, E., van den Brandt, P.A., and Goldbohm, R.A. (1994b) A prospective cohort study on Allium vegetable consumption, garlic supplement use, and the risk of lung carcinoma in the Netherlands. Cancer Res. 54:6148–6153.

Dorant, E., van den Brandt, P.A., and Goldbohm, R.A. (1995) Allium vegetable consumption, garlic supplement intake, and female breast carcinoma incidence. Breast Cancer Res. Treatment 33:163–170.

Dorsch, W. (1986) Chronisches Asthma: helfen Zwiebeln? Med. Trib. 24:50.

Doutremepuich, C., Gamba, G., Refauvelet, J., and Quilichini, R. (1985) Action de l'oignon, Allium cepa L., sur l'hémostase primaire chez le volontaire sain avant et après absorption d'un repas riche en lipides. Ann. Pharm. Fr. 43:273–280.

Drawert, F. and Görg, A. (1975) Über die elektrophoretische Differenzierung und Klassifizierung von Proteinen. VI. Disk-Elektrophorese und isoelektrische Fokussierung in Polyacrylamid-Gelen von Proteinen und Enzymen aus Tomaten, Gurken, Zuckermais und Zwiebeln. Z. Lebensm. Unters. Forsch. 159:23–30.

Drobnik, R. (1938a) Beiträge zur Kenntnis des Knoblauchs in seiner Verwendung in der Heilkunde. Dissertation, Institut für allgemeine und experimentelle Pathologie, University of Wien (Abstract).

Drobnik, R. (1938b) Zur Geschichte des Knoblauchs in der Heilkunde im Wandel der Zeiten. Pharm. Monatsh. [Vienna] 19:11–14.

Du, C.T., Wang, P.L., and Francis, F.J. (1974) Cyanidin-3-laminarioside in Spanish red onion (Allium cepa L.). J. Food Sci. 39:1265–1266.

Du, C.T. and Francis, F.J. (1975) Anthocyanins of garlic (*Allium sativum* L.). J. Food Sci. 40:1101–1102.

Dubrova, G.B. (1950) Action of garlic phytocides on molds. Mikrobiologiya (Microbiology) 19:229–234. Chem. Abst. 44 (1950) 8424 (Russian).

Duhova, V., Miklovicova, M., and Böhmova, B. (1986) Cytogenetic activity of some pesticides used in garlic. Acta Fac. Rer. Nat. Univ. Comenianae, Genet., 13–20. Chem. Abst. 108 (1988) 107 877.

Dumas-Gaudot, E., Grenier, J., Furlan, V., and Asselin, A. (1992) Chitinase, chitosanase, and b-1,3-glucanase activities in Allium and Pisum roots colonized by Glomus species. Plant Sci. 84:17–24. Chem. Abst. 117 (1992) 66 783.

Durbin, R.D. and Uchytil, T.F. (1971) Role of allicin in the resistance of garlic to Penicillium species. Phytopathol. Mediterr. 10:227–230. Chem. Abst. 77 (1972) 31 662.

Duris, I., Koska, M., Zaviacic, M., Mikulecky, M., and Holly, D. (1981) Precancer and cancer of the stomach. Bratisl. Lek. Listy (Bratislava Medical News) 76:529–640 (Slovakian).

Dwivedi, C., Rohlfs, S., Jarvis, D., and Engineer, F.N. (1992) Chemoprevention of chemically induced skin tumor development by diallyl sulfide and diallyl disulfide. Pharmaceut. Res. 9:1668–1670.

Dwivedi, S.K. and Sharma, M.C. (1985) Therapeutic evaluation of an indigenous drug formulation against scabies in pigs. Indian J. Vet. Med. 5:97–100.

Ebbell, B. (1937) The Papyrus Ebers. Copenhagen.

Ebers, G. (1875) Papyrus Ebers, das hermetische Buch über die Arzneimittel der alten Ägypter in hieratischer Schrift. Leipzig.

Ebert, K. (1982) Knoblauch (*Allium sativum*). In: Arznei- und Gewürzpflanzen. Vol. 2. Wiss. Verlagsges, Stuttgart, pp. 123–124.

Edelsbacher, A., Fleischhacker, W., Mostler, U., and Urban, E. (1988) Synthese potentiell antibiotischer Lactone. Sci. Pharm. *56*:11.

Edelsbacher, A., Urban, E., and Weidenauer, W. (1992) Preparation of enantiomerically pure 5,6-dihydroxy-isobenzofuranones and 5,6-dihydroxy-4,4-methano-isobenzofuranones. Montsh. Chem. *123*:741–747.

Edelstein, A.J. (1950) Dermatitis caused by garlic. Arch. Dermatol. Syph. *61*:111.

Edwards, S.J., Musker, D., Collin, H.A., and Britton, G. (1994a) The analysis of S-alk(en)yl-L-cysteine sulphoxides (flavour precursors) from species of Allium by high performance liquid chromatography. Phytochem. Anal. *5*:4–9. Chem. Abst. 120 (1994) 240 117.

Edwards, S.J., Britton, G., and Collin, H.A. (1994b) The biosynthetic pathway of the S-alk(en)yl-L-cysteine sulphoxides (flavour precursors) in species of Allium. Plant Cell Tissue Organ Culture *38*:181–188.

Effertz, B. and Weissenböck, G. (1976) Ber. Dtsch. Bot. Ges. *89*:413.

Egen-Schwind, C., Eckard, R., Jekat, F.W., and Winterhoff, H. (1992a) Pharmacokinetics of vinyldithiins, transformation products of allicin. Planta Med. *58*:8–13. Chem. Abst. 116 (1992) 227 582.

Egen-Schwind, C., Eckard, R., and Kemper, F.H. (1992b) Metabolism of garlic constituents in the isolated perfused rat liver. Planta Med. *58*:301–305. Chem. Abst. 117 (1992) 184 242.

Egger, K. (1937) Einwirkung von *Allium sativum* bei Breslauinfektion. Dissertation, Medical Faculty, University of Munich.

Eichenberger, W. and Grob, E.C. (1970) Über die quantitative Bestimmung von Sterinderivaten in Pflanzen und die intrazelluläre Verteilung der Steringlycoside in Blättern. FEBS Lett. *11*:177–180.

Eichenberger, W. and Menke, W. (1966) Sterine in Blättern und Chloroplasten. Z. Naturforsch. *B21*:859–867. Chem. Abst. 66 (1967) 17 079.

Ejzer, G.S. (1954) Treatment of dysentery with phytoncides from garlic in combination with other drugs. Voen. Med. (Military Med.) *12*:36–39 (Russian).

Elbl, G. (1991) Chemisch-biologische Untersuchungen pflanzlicher Inhibitoren des Angiotensin I-converting Enzyms (ACE), insbesondere die der Arzneipflanzen Lespedeza capitata (Michx.) und Allium ursinum (L.). Dissertation, University of Munich (Abstract).

El-Hadidy, Z.A., Moawad, F.G., Abdel-Naiem, F.M., and Attalah, R.K. (1981) Chemical constituents of garlic volatile oil. Ain Shams Univ. Fac. Agric. Res. Bull. [Cairo] *1426*:1–24.

Ellman, G.L. (1959) Tissue sulfhydryl groups. Arch. Biochem. Biophys. *82*:70–77. Chem. Abst. 53 (1959) 15 184.

Ellmore, G.S. and Feldberg, R.S. (1994) Alliin lyase localization in bundle sheaths of the garlic clove (*Allium sativum*). Am. J. Bot. *81*:89–94. Chem. Abst. 120 (1994) 265 918.

El Mofty, M.M., Sakr, S.A., Essawy, A., and Gawad, H.S.A. (1994) Preventive action of garlic. Nutr. Cancer *21*:95–100. Chem. Abst. 120 (1994) 317 618.

Elnima, E.I., Ahmed, S.A., Mekkawi, A.G., and Mossa, J.S. (1983) The antimicrobial activity of garlic and onion extracts. Pharmazie *38*:747–748.

El-Oksh, I.I., Abdel-Kader, A.S., Wally, Y.A., and El-Kholly, A.F. (1971) Comparative effects of gamma-irradiation and maleic hydrazide on storage of garlic. J. Am. Soc. Hort. Sci. *96*:555–561.

Engeland, R.L. (1991) Growing great garlic: the definitive guide for organic gardeners and small farmers. Filaree Productions, Okanogan, Washington.

Epshtein, M.M. and Frolkis, V.V. (1953) Influence of the phytoncides of garlic and of synthetic thiols on the system acetylcholine-cholinesterase. Med. Zh. Akad. Nauk Ukrain. RSR (Med. J. Acad. Sci. Ukr. RSR) *23*:73–76 (Ukrainian).

Erbach, D. (1928) Über die Behandlung von Darmerkrankungen mit Knoblauch (*Allium sativum*). Münch. Med. Wochenschr. *75*:87.

Ercan, Y., Hoeld, A., and Lupas, O. (1983) A digital code for real-time calculation of the transient behaviour of nodal and global core and plant parameters of BWR nuclear power plants. Report GRS-39, Ord. No. DE83750179, 1–46. Chem. Abst. 99 (1983) 44 988. Energy Abst. 8 (1983) 20 329.

Erdmann, H. (1931) Zur Therapie der Oxyuriasis. Dtsch. Med. Wochenschr. *33*:1413.

Erken, D. (1969) Schonende Blutdrucksenkung durch Pflanzenstoffe. Über die klinische Prüfung eines Phytotherapeutikums. Dtsch. Apotheker *21*:22–24.

Erlichman, L.N. (1949) Influence of the antibiotics of garlic (phytoncides) on the biotransformation of (−)-tyrosine in kidney tissue. Ukrain. Biokhim. Zh. (Ukrain. Biochem. J.) *21*:15–24 (Ukrainian).

Ernst, E. (1981) Therapie mit Knoblauch? Theorien über ein volkstümliches Heilprinzip. Münch. Med. Wochenschr. *123*:1537–1538.

Ernst, E. (1983) Beeinflussung der Arteriosklerose durch *Allium sativum*? Dtsch. Apoth. Ztg. *123*:625–626.

Ernst, E. (1984) Antiarteriosklerotische Prinzipien von *Allium sativum* (Knoblauch). Notabene Medici *14*:23–27.

Ernst, E., Weihmayr, I., and Matrai, A. (1985) Garlic and blood lipids. Br. Med. J. *291*:139.

Ernst, E., Weihmayr, T., and Matrai, A. (1986) Knoblauch plus Diät senkt Serumlipide. Ärztl. Prax. *38*:1748–1749.

Ernst, E. (1987) Cardiovascular effects of garlic (*Allium sativum*): a review. Pharmatherapeutica *5*:83–89. Int. Pharm. Abst. 25 (1988) 13 668.

Ernst, E. (1989) Knoblauch und Arteriosklerose. Naturwiss. Rdsch. *42*:106–197.

Esanu, V. and Prahoveanu, E. (1983) The effect of garlic extract, applied as such or in association with NaF, on experimental influenza in mice. Rev. Roum. Med. Virol. 34:11–17.

Esler, G. and Coley-Smith, J.R. (1983) Flavour and odour characteristics of species of Allium in relation to their capacity to stimulate germination of sclerotia of Sclerotium cepivorum. Plant Pathol. 32:13–22.

Espagnacq, L., Bertoni, G., and Morard, P. (1988) Evolution des sucres au cours de la maturation de l'ail. Agrochimica 32:441–450. Chem. Abst. 112 (1990) 18 928.

Essman, E.J. (1984) The medicinal uses of herbs. Fitoterapia 55:279–289. Chem. Abst. 103 (1985) 31 840.

Etoh, T. (1983) Germination of seeds obtained from a clone of garlic, Allium sativum L. Proc. Japan Acad. 59B:83–87.

Etoh, T. (1985) Studies on sterility in garlic, Allium sativum L. Mem. Fac. Agr. Kagoshima Univ. 21:77–132.

Etoh, T., Noma, Y., Nishitarumizu, Y., and Wakomoto, T. (1988) Seed productivity and germinability of various clones collected in Soviet Central Asia. Mem. Fac. Agr. Kagoshima Univ. [Japan] 24:29–139.

Etoh, T. and Ogura, H. (1977) A morphological observation on the formation of abnormal flowers in garlic (Allium sativum L.). Mem. Fac. Agr. Kagoshima Univ. 13:77–88.

Etoh, T. and Ogura, H. (1981) Peroxidase isozymes in the leaves of various clones of garlic, Allium sativum L. Mem. Fac. Agric. Kagoshima Univ. 17:71–77.

Ettala, T. and Virtanen, A.I. (1962) On the labelling of sulphur-containing amino acids and gamma-glutamylpeptides after injection of labelled sulphate into onion (Allium cepa). Acta Chem. Scand. 16:2061–2063. Chem. Abst. 58 (1963) 8239.

Euler, H.V. and Lindeman, G. (1949) Zur Biochemie der Tumorentwicklung und der Tumorhemmung. Ark. Kemi 1:87–97.

Fakunle, J.B. (1983) Evaluation of garlic extracts on cancer growth in rats fed diets with varying levels of protein. Diss. Abstr. Int. B. 44:116. Chem. Abst. 99 (1983) 121 108.

Falleroni, A.E., Zeiss, C.R., and Levitz, D. (1981) Occupational asthma secondary to inhalation of garlic dust. J. Allergy Clin. Immunol. 68:156–160.

Farag, S. (1991) Improving fresh and processed garlic by irradiation under different storage conditions. Nahrung 35:421–429.

Farber, L. (1957) The chemical evaluation of the pungency of onion and garlic by the content of volatile reducing substances. Food Technol. 11:621–624. Chem. Abst. 52 (1958) 11 310.

Farjaudon, N., Guilloteau, P., Lasselain, M.J., and Pareyre, C. (1988) Morphometric study of "floating pole" anaphase cells induced by tubulosine in a meristematic root cell population (Allium sativum L.). Cytobios 54:17–23.

Farkas, J. (1987) Decontamination, including parasite control, of dried, chilled, and frozen foods by irradiation. Acta Aliment. 16:351–384.

Farkas, J. and Andrassy, E. (1988) Comparative analysis of spices decontaminated by ethylene oxide or gamma irradiation. Acta Aliment. 17:77–94. Chem. Abst. 109 (1988) 91 423.

Farkas, M.C. and Farkas, J.N. (1974) Hemolytic anemia due to ingestion of onions in a dog. J. Am. Anim. Hosp. Assoc. 10:65.

Farva, D., Goji, I.A., Joseph, P.K., and Augusti, K.T. (1986) Effects of garlic oil on streptozotocin-diabetic rats maintained on normal and high fat diets. Indian J. Biochem. Biophys. 23:24–27. Chem. Abst. 104 (1986) 206 018.

Favaron, F., Castiglioni, C., and Di Lenna, P. (1993) Inhibition of some rot fungi polygalacturonases by Allium cepa L. and Allium porrum L. extracts. J. Phytopathol. 139:201–206.

Fehri, B., Aiache, J.M., Korbi, S., Mokni, M., Said, M.B., Memmi, A., Hizaoui, B., and Boukef, K. (1991) Effets toxiques engendrés par une administration réitérée d'Allium sativum L. J. Pharm. Belg. 46:363–374. Int. Pharm. Abst. 29 (1992) 7240.

Feldberg, R.S., Chang, S.C., Kotik, A.N., Nadler, M., Neuwirth, Z., Sundstrom, D.C., and Thompson, N.H. (1988) In vitro mechanism of inhibition of bacterial cell growth by allicin. Antimicrob. Agents Chemother. 32:1763–1768. Chem. Abst. 110 (1989) 36 674.

Feng, D., Yang, P., and Yang, Z. (1990) Determination of germanium by flow injection potentiometric stripping analysis. Jinan Daxue Xuebao 11:41–45. Chem. Abst. 114 (1991) 239 483 (Chinese).

Feng, D., Yang, P., and Yang, Z. (1991) Determination of germanium by potentiometric stripping analysis and adsorption potentiometric stripping analysis. Talanta 38:1493–1498. Chem. Abst. 116 (1992) 119 984.

Feng, Z.-H., Zhang, G.-M., Hao, T.-L., Zhou, B., Zhang, H., and Jiang, Z.-Y. (1994) Effect of diallyl trisulfide on the activation of T-cell and macrophage-mediated cytotoxicity. J. Tongji Med. Univ. 14:142–147.

Fenwick, D.E. and Oakenfull, D. (1983) Saponin content of food plants and some prepared foods. J. Sci. Food Agric. 34:186–191. Chem. Abst. 98 (1983) 196 549.

Fenwick, G.R. and Hanley, A.B. (1985a) The genus allium, part 1. CRC Crit. Rev. Food Sci. Nutr. 22:199–271.

Fenwick, G.R. and Hanley, A.B. (1985b) The genus allium, part 2. CRC Crit. Rev. Food Sci. Nutr. 22:273–377.

Ferguson, E.A. (1949) Brominated garlic carminative. U.S.A. Patent 2 490 424. Chem. Abst. 44 (1950) 2185.

Fernandez de Corres, L., Leanizbarruta, I., Munoz, D., and Corrales, J.L. (1983) Dermatitis alergica de contacto por ajo, primula, frullania y compuestas. Allergol. Immunpathol. 12:291–299.

FHM 61. (1961) Propulsora de Homeopatia. In:

Farmacopea Homeopatica Mexicana. Vol. 3. L.G. Sandoval, Mexico.

Fields, M., Lewis, C.G., and Lure, M.D. (1992) Garlic oil extract ameliorates the severity of copper deficiency. J. Am. Coll. Nutr. *11*:334–339. Chem. Abst. 117 (1992) 169 957.

Fischer, H. (1929) Mittelalterliche Pflanzenkunde. Verlag Münchner Drucke, Munich.

Fischer, W. (1936) Schlussfolgerungen über Knoblauchpräparate. Pharm. Ztg. *81*:707–708.

Fisher, A.A. (1986) Contact dermatitis. Lea & Febiger, Philadelphia.

Fiskesjö, G. (1985) The Allium test as a standard in environmental monitoring. Hereditas *102*:99–112.

Fiskesjö, G. (1988) The Allium test: an alternative in environmental studies: the relative toxicity of metal ions. Mutation Res. *197*:243–260.

Fiskesjö, G. (1993) The Allium test: a potential standard for the assessment of environmental toxicity. ASTM Spec. Tech. Publ. *2*:331–345. Chem. Abst. 121 (1994) 294 451.

Fitzpatrick, F.K. (1954) Plant substances active against mycobacterium tuberculosis. Antibiot. Chemother. [Washington, D.C.] *4*:528–536.

Flamm, S. (1935) Wurmkuren bei Schwangeren. Knoblauch-, Rainfarn-, Kürbiskuren. Hippokrates, 867–868.

Fleischhacker, W., Rachenzentner, A., and Urban, E. (1991) Synthese von 5,6-Dihydroxy-tetrahydro-isobenzofuranonen. Sci. Pharm. *59*:18.

Fletcher, R.D., Parker, B., and Hasset, M. (1974) Inhibition of coagulase activity and growth of Staphylococcus aureus by garlic extracts. Folia Microbiol. *19*:494–497. Chem. Abst. 82 (1975) 133 495.

Fleury, M.G. (1932) Action de l'ail sur les cultures de bacille coli et les cultures de bacille typhique. Bull. Soc. Pharm. Bordeaux *70*:190–194.

Fliermans, C.B. (1973) Inhibition of Histoplasma capsulatum by garlic. Mycopathol. Mycol. Appl. *50*:227–231.

Fodor, L. (1993) Klinische Prüfung eines pflanzlichen Antihypertonikums. Aerztezeitsch. Naturheilverf. *34*:991–992.

Fortunatov, M.N. (1952) Experimental use of phytoncides for therapeutic and prophylactic purposes. Vopr. Pediatr. Okhr. Materin. Det. (Probl. Pediat. Matern. Care Children) *20*:55–58. Chem. Abst. 46 (1952) 8812 (Russian).

Fortunatov, M.N. (1955) On the activity of the phytoncides from garlic in the human organism upon peroral administration. Farmakol. Toksikol. [Moscow] *18*:43–46. Chem. Abst. 50 (1956) 9626 (Russian).

Foushee, D.B., Ruffin, J., and Banerjee, U. (1982) Garlic as a natural agent for the treatment of hypertension: a preliminary report. Cytobios *34*:145–152.

Fraas, C. (1845) Synopsis plantarum florae classicae oder übersichtliche Darstellung der in den klassischen Schriften der Griechen und Römer vorkommenden Pflanzen. E.A. Fleischmann, Munich.

Francois, L.E. (1991) Yield and quality responses of garlic and onion to excess boron. Hort. Sci. *26*:547–549. Chem. Abst. 115 (1991) 28 225.

Frank von Wörd, S. (1531) Chronica. Strassburg.

Franz, H. (1988) The ricin story. Adv. Lectin Res. *1*:10–25.

Freeman, F. and Kodera, Y. (1995) Garlic chemistry: stability of S-(2-propenyl) 2-propene-1-sulfinothioate (allicin) in blood, solvents, and simulated physiological fluids. J. Agric. Food Chem. *43*:2332–2338.

Freeman, F., Ma, X.B., and Lin, R.I.S. (1993) Garlic chemistry: peroxidation of bis-(2-propenyl) disulfide, bis-(2-propenyl) trisulfide, and their monoxide derivatives. Sulfur Lett. *15*:253–262. Chem. Abst. 119 (1993) 72 421.

Freeman, G.G. (1979) Factors affecting flavour during growth, storage, and processing of vegetables. In: Progress in flavour research. D.G. Land and H.E. Nursten, eds. Applied Science Publ., London, pp. 225–243.

Freeman, G.G. and McBreen, F. (1973) A rapid spectrophotometric method of determination of thiolsulphinate in onion (Allium cepa) and its significance in flavour studies. Biochem. Soc. Trans. *1*:1150–1152. Chem. Abst. 80 (1974) 58 574.

Freeman, G.G. and Mossadeghi, N. (1971) Influence of sulphate nutrition on the flavour components of garlic (*Allium sativum*) and wild onion (A. vineale). J. Sci. Food Agric. *22*:330–334. Chem. Abst. 75 (1971) 109 341.

Freeman, G.G. and Mossadeghi, N. (1973) Relation between water regime and flavor strength in watercress (Rorippa nasturtium aquaticum), cabbage (Brassica oleracea capitata), and onion (Allium cepa). J. Hortic. Sci. *48*:365–378. Chem. Abst. 80 (1974) 57 472.

Freeman, G.G. and Whenham, R.J. (1974a) Changes in onion (Allium cepa) flavor components resulting from some post-harvest processes. J. Sci. Food Agric. *25*:499–515. Chem. Abst. 81 (1974) 62 310.

Freeman, G.G. and Whenham, R.J. (1974b) Flavor changes in dry bulb onions during overwinter storage at ambient temperature. J. Sci. Food Agric. *25*:517–520. Chem. Abst. 81 (1974) 62 309.

Freeman, G.G. and Whenham, R.J. (1975a) A survey of volatile components of some Allium species in terms of S-alk(en)yl-L-cysteine sulphoxides present as flavour precursors. J. Sci. Food Agric. *26*:1869–1886. Chem. Abst. 84 (1976) 104 001.

Freeman, G.G. and Whenham, R.J. (1975b) The use of synthetic (±)-S-1-propyl-L-cysteine sulphoxide and of alliinase preparations in studies of flavour changes resulting from processing of onion (Allium cepa L.). J. Sci. Food Agric. *26*:1333–1346. Chem. Abst. 84 (1976) 72 765.

Freeman, G.G. and Whenham, R.J. (1975c) A rapid spectrophotometric method of determination of thiopropanal S-oxide (lachrymator) in onion (Allium cepa L.) and its significance in flavour studies. J. Sci. Food Agric. *26*:1529–1543. Chem. Abst. 84

(1976) 72 699.
Frolov, A.F. and Mishenkova, Y.L. (1970) Inhibitory effect of some preparations from higher plants on reproduction of influenza virus in vitro and in vivo. Mikrobiol. Zh. (Microbiol. J.) 32:628–633. Chem. Abst. 74 (1971) 74 916 (Russian).
Fromtling, R.A. and Bulmer, G.S. (1978) In vitro effect of aqueous extract of garlic (*Allium sativum*) on the growth and viability of Cryptococcus neoformans. Mycologia 70:397–405.
Fructus, L. (1987) Anti-tobacco candle containing natural products to absorb smoke and tobacco odors. France. Patent 2 593 398. Chem. Abst. 108 (1988) 10 628.
Fu, N., Huang, L., Quan, L., and Yan, L. (1993) Antioxidant action of garlic oil and allitridi. Zhongguo Yixue Kexueyuan Xuebao 15:295–301. Chem. Abst. 120 (1994) 124 814 (Chinese).
Fuchs, L. (1543) New Kreüterbuch, etc. Michael Isingrin, Basle.
Fujii, T. and Taisho Shokuhin Co. Ltd. (1988) Modification of garlic using enzymes and acetic acid for prevention of foul breath. Japan. Patent 88 309 160. Chem. Abst. 110 (1989) 153 092.
Fujii, Y., Furukawa, M., Hayakawa, Y., Sugahara, K., and Shibuya, T. (1991) Survey of Japanese medicinal plants for the detection of allelopathic properties. Weed Res. [Japan] 36:36–42 (Japanese).
Fujime, Y., Kudou, R., and Okuda, N. (1993) Breeding and propagation of good seedlings of garlic by tissue culture. 1. In vitro formation of shoot and bulblet in garlic. Shokubutsu Soshiki Baiyo 10:9–16. Chem. Abst. 119 (1993) 133 383.
Fujita, M., Endo, M., and Sano, M. (1990) Purification and characterization of alliin lyase from Welsh onion, Allium fistulosum L. Agric. Biol. Chem. 54:1077–1079. Chem. Abst. 113 (1991) 19 913.
Fujiwara, K. (1986) Deodorization of garlic. Japan. Patent 86 104 761. Chem. Abst. 105 (1986) 77 825.
Fujiwara, K. and Murakami, F. (1960) Studies on the nutritional value of Allium plants: formation of allicin-protein complex. Bitamin [Kyoto] 20:136–139. Chem. Abst. 61 (1964) 15 093 (Japanese).
Fujiwara, M. (1954) Allithiamine, Alinamin, and other derivatives. Medicina Rev. [Chiba, Japan] 2:28–31. Chem. Abst. 49 (1955) 3341 (Esperanto).
Fujiwara, M., Nanjo, H., Arai, T., and Suzuoki-Ziro. (1954a) "Allithiamine," a newly found derivative of vitamin B1: the effect of allithiamine on living organism. J. Biochem. [Japan] 41:273–285. Chem. Abst. 49 (1955) 11 795.
Fujiwara, M., Watanabe, H., and Matsui, K. (1954b) "Allithiamine," a newly found derivative of vitamin B1. I. Discovery of allithiamine. J. Biochem. [Japan] 41:29–39. Chem. Abst. 48 (1954) 7129.
Fujiwara, M., Yoshimura, M., and Tsuno, S. (1955) "Allithiamine," a newly found derivative of vitamin B1. III. On the allicin homologues in the plants of the Allium species. J. Biochem. [Japan] 42:591–601. Chem. Abst. 50 (1956) 1132.
Fujiwara, M., Yoshimura, M., Tsuno, S., and Murakami, F. (1958) "Allithiamine," a newly found derivative of vitamin B1. IV. On the alliin homologues in the vegetables. J. Biochem. [Japan] 45:141–149. Chem. Abst. 52 (1958) 13 011.
Fujiwara, M., Itokawa, Y., Uchino, H., and Inoue, K. (1972) Antihypercholesterolemic effect of a sulfur-containing amino acid, S-methyl-L-cysteine sulfoxide, isolated from cabbage. Experientia 28:254–255. Chem. Abst. 77 (1972) 83 501.
Fujiwara, M. (1976) Allithiamine and its properties. J. Nutr. Sci. Vitaminol. 22:57–62.
Fujiwara, M. and Natata, T. (1967) Induction of tumour immunity with tumour cells treated with extract of garlic (*Allium sativum*). Nature 216:83–84.
Fujiwara, M. and Watanabe, H. (1952) Allithiamine, a newly found compound of vitamin B1. Proc. Imp. Acad. [Tokyo] 28:156–158. Chem. Abst. 46 (1952) 7629.
Fujiwara, W. (1976) Allithiamine and its properties. J. Nutr. Sci. Vitaminol. 22:57–62.
Fulder, S. and Blackwood, J. (1991) Garlic: nature's original remedy. Healing Arts Press, Rochester, Vermont.
Fuleki, T. (1971) Anthocyanins in red onion, Allium cepa. J. Food Sci. 36:101–104.
Funke, H. (1977) Senföl-Glykosid-Drogen. Naturheilpraxis 30:92–104.
Gadkari, J.V. and Joshi, V.D. (1991) Effect of ingestion of raw garlic on serum cholesterol level, clotting time, and fibrinolytic activity in normal subjects. J. Postgrad. Med. 37:128–131.
Gaffen, J.D., Tavares, I.A., and Bennett, A. (1984) The effect of garlic extracts on contractions of rat gastric fundus and human platelet aggregation. J. Pharm. Pharmacol. 36:272–274.
Gaind, K.N., Dar, R.N., Chopra, B.M., and Kaul, R.N. (1965) Spectrophotometric estimation of alliin in Allium species. Indian J. Pharm. 27:199–200. Chem. Abst. 63 (1965) 11 255.
Galenus. (1948) Werke des Galenos. Marquardt, Stuttgart.
Galetto, W.G. and Hoffman, P.G. (1976a) Synthesis and flavor evaluation of several alkylfuranones found in Allium species (onion, shallots, leeks). J. Agric. Food Chem. 24:854–856.
Galetto, W.G. and Hoffman, P.G. (1976b) Synthesis and flavor evaluation of some alkylthiophenes: volatile components of onion. J. Agric. Food Chem. 24:852–854.
Gareis, K. (1895) Die Landgüterordnung Kaiser Karls des Groben (Capitulare de villis vel curtis imperii). J. Guttentag, Berlin.
Gargouri, Y., Moreau, H., Jain, M.K., de Haas, G.H., and Verger, R. (1989) Ajoene prevents fat digestion by human gastric lipase in vitro. Biochim. Biophys. Acta 1006:137–139. Chem. Abst. 112 (1990) 498.
Gargouri, Y., Salah, A.B., and Verger, R. (1992) Importance de la lipase gastrique humaine dans la

digestion des lipides alimentaires. Rev. Fr. Corps. Gras *39*:207–213.
Garty, B.Z. (1993) Experience and reason—briefly recorded: garlic burns. Pediatrics *91*:658–659.
Gassmann, B. (1992) Knoblauch: Lebensmittel und Modedroge? Ernähr. Umsch. *39*:444–449.
Gebhardt, R. (1991) Hemmung der Cholesterin-Biosynthese in Primärkulturen von Rattenhepatozyten durch einen wäbrigen Knoblauchextrakt. Arzneim. Forsch. *41*:800–804. Chem. Abst. 115 (1991) 150 132. Congress abstract in: Eur. J. Clin. Res. 3A (1992) 11.
Gebhardt, R. (1993) Multiple inhibitory effects of garlic extracts on cholesterol biosynthesis in hepatocytes. Lipids *28*:613–619. Chem. Abst. 119 (1993) 131 302.
Gebhardt, R., Beck, H., and Wagner, K.G. (1994) Inhibition of cholesterol biosynthesis by allicin and ajoene in rat hepatocytes and HepG2 cells. Biochim. Biophys. Acta *1213*:57–62. Chem. Abst. 121 (1994) 73 516.
Gebhardt, R. (1995) Amplification of palmitate-induced inhibition of cholesterol biosynthesis in cultured rat hepatocytes by garlic-derived organosulfur compounds. Phytomedicine *2*:29–34.
Geier, H. and Woitowitz, H.J. (1988) Milch+Knoblauch: Schützt das vor Bleivergiftung? Med. Trib. *20*:24.
Geithner, R. (1931) Beitrag zur Behandlung gastrointestinaler Störungen. Ther. Ggw., 524–525.
Geoffroy, S.F. (1761) Abhandlung von der Materia medica oder von der Kenntnib, der Kraft, der Wahl, und dem Gebrauch einfacher Arzneimittel, etc. Carl Ludw. Jacobi, Leipzig.
George, G., Fellous, R., Schippa, C., and Cozzolino, F. (1991) Apport complémentaire de l'ICN à l'étude de certaines huiles essentielles. Riv. ital. EPPOS, 139–149. Chem. Abst. 115 (1991) 189 461.
George, K.C., Amonkar, S.V., and Eapen, J. (1973) Effect of garlic oil on incorporation of amino acids into proteins of Culex pipiens quinquefasciatus Say larvae. Chem. Biol. Interac. *6*:169–175. Chem. Abst. 79 (1973) 770.
George, K.C. and Eapen, J. (1973) In vivo studies on the effect of garlic oil in mice. Toxicology *1*:337–344. Chem. Abst. 81 (1974) 46 137.
George, K.C. and Eapen, J. (1974) Mode of action of garlic oil: effect on oxidative phosphorylation in hepatic mitochondria of mice. Biochem. Pharmacol. *23*:931–936. Chem. Abst. 81 (1974) 58 960.
Gerhardt, U. and Blat, P. (1984) Dynamische Mebmethode zur Ermittlung der Fettstabilität. Einflub von Gewürzen und Zusatzstoffen. Fleischwirtschaft *64*:484–486. Chem. Abst. 101 (1984) 37 312.
Gesner, C. (1555) De rariis et admirandis herbis. Zurich.
Gessner, O. (1953) Die Gift- und Arzneipflanzen von Mitteleuropa (Pharmakologie, Toxikologie, Therapie). Carl Winter, Heidelberg.

Ghannoum, M.A. (1988) Studies on the anticandidal mode of action of *Allium sativum* (garlic). J. Gen. Microbiol. *134*:2917–2924. Chem. Abst. 110 (1989) 36 660.
Ghannoum, M.A. (1990) Inhibition of candida adhesion to buccal epithelial cells by an aqueous extract of *Allium sativum* (garlic). J. Appl. Bacteriol. *68*:163–169.
Ghosh, A., Sharma, A., and Talukder, G. (1992) Toxic effects of cesium chloride on plant chromosomes. J. Indian Bot. Soc. *71*:229–232. Chem. Abst. 120 (1994) 317 591.
Gilroy. (1987) The complete garlic lovers' cookbook. Celestial Arts, Berkeley, California.
Glaser, E. (1937) Die Beurteilung der Aphrodisiaca und Antiaphrodisiaca bei biologischer Prüfung durch den Glaser-Haempelschen Fischtest. Sci. Pharm. *8*:1–8.
Glaser, E. (1940) Über Inhaltsstoffe des Knoblauchs, welche für seine Verwendung als Heilmittel von Bedeutung sind. Hippokrates *11*:169–171.
Glaser, E. and Drobnik, R. (1939) Beiträge zur Kenntnis der Wirkstoffe des Knoblauchs. Arch. Exp. Pathol. Pharmakol. *193*:1–9. Chem. Abst. 34 (1940) 3283.
Goda, Y., Shibuya, M., and Sankawa, U. (1987) Inhibitors of the arachidonate cascade from Allium chinense and their effect on in vitro platelet aggregation. Chem. Pharm. Bull. *35*:2668–2674. Chem. Abst. 107 (1987) 190 622.
Goku, K. (1977) Odorless garlic wine. Japan. Patent 77 44 400. Chem. Abst. 88 (1978) 61 136.
Goldberg, M.T. and Josephy, P.D. (1987) Studies on the mechanism of action of diallyl sulfide, an inhibitor of the genotoxic effects of cyclophosphamide. Can. J. Physiol. Pharmacol. *65*:467–471.
Goldschmiedt, M. and Feldman, M. (1993) Gastric secretion in health and disease. In: Gastrointestinal disease. M.H. Sleisenger and J.S. Fordtran, eds. Vol. 5. WB Saunders, Philadelphia, p. 531.
Gonzalez Fandos, E., Garcia Lopez, M.L., Sierra, M.L., and Otero, A. (1994) Staphylococcal growth and enterotoxins (A–D) and thermonuclease synthesis in the presence of dehydrated garlic. J. Appl. Biol. *77*:549–552. Chem. Abst. 122 (1995) 51 109.
Gori, O. and Ferri, S. (1982) Ultrastructural study of the microspore development in *Allium sativum* clone piemonte. J. Ultrastruct. Res. *79*:341.
Gorovits, M.B., Khristulas, F.S., and Abubakirov, N.K. (1971) Alliogenin and alliogenin *b*-D-glucopyranoside from Allium giganteum. Khim. Prir. Soedin. (Chem. Nat. Compd.) *7*:434–442. Chem. Abst. 75 (1971) 141 102 (Russian).
Gorovits, M.B., Kelginbaev, A.N., Khristulas, F.S., and Abubakirov, N.K. (1973a) Oxidation of alliogenin and *b*-chlorogenin by N-bromosuccinimide. Khim. Prir. Soedin. (Chem. Nat. Compd.) *9*:562–563. Chem. Abst. 80 (1974) 37 385 (Russian; English cover-to-cover translation, p. 534).

Gorovits, M.B., Khristulas, F.S., and Abubakirov, N.K. (1973b) Steroid saponins and sapogenins of Allium. IV. Karatavigenin, a new sapogenin from Allium karataviense. Khim. Prir. Soedin. (Chem. Nat. Compd.) 9:747–749. Chem. Abst. 82 (1975) 108 809 (Russian).

Goryachenkova, E.V. (1952) Enzyme in garlic which forms allycine (allyinase), a protein with phosphopyridoxal. Dokl. Akad. Nauk SSSR (Rep. Acad. Sci. USSR) 87:457–460. Chem. Abst. 47 (1953) 4928 (Russian).

Goto, F., Konishi, T., Ido, Y., Ootsuki, H., and Katsuyama, K. (1994) Methods for preparation of high-purity DNA from onions. Japan. Patent 94 197 762. Chem. Abst. 121 (1994) 223 650.

Goto, T. and Fujino, M. (1967) On the selenium content in foods and its analysis. Eiyo to Shokuryo 20:311–313. Chem. Abst. 69 (1968) 1897 (Japanese).

Graham, S., Marshall, J., Haughey, B., Mittelman, A., Swanson, M., Zielezny, M., Byers, T., Wilkinson, G., and West, D. (1994) Dietary epidemiology of cancer of the colon in western New York. Am. J. Epidemiol. 128:490–503.

Grandmaison, J., Olah, G.M., Van Calsteren, M.R., and Furlan, V. (1993) Characterization and localization of plant phenolics likely involved in the pathogen resistance expressed by endomycorrhizal roots. Mycorrhiza 3:155–164. Chem. Abst. 120 (1994) 266 026.

Granroth, B. (1968) Separation of Allium sulfur amino acids and peptides by thin-layer electrophoresis and thin-layer chromatography. Acta Chem. Scand. 22:3333–3335. Chem. Abst. 70 (1969) 84 813.

Granroth, B. (1970) Biosynthesis and decomposition of cysteine derivatives in onion and other Allium species. Ann. Acad. Sci. Fenn. (Ser. A, II) 154:1–71. Chem. Abst. 73 (1970) 84 658.

Granroth, B. (1974) Partial purification of cysteine synthase (O-acetylserine sulfhydrylase) from onion (Allium cepa). Acta Chem. Scand. B28:813–814. Chem. Abst. 82 (1975) 27 719.

Granroth, B. and Sarnesto, A. (1974) Synthesis of S-substituted cysteine derivatives by the cysteine synthase (O-acetylserine sulfhydrylase) of onion (Allium cepa). Acta Chem. Scand. 28:814–815. Chem. Abst. 82 (1975) 27 822.

Granroth, B. and Virtanen, A.I. (1967a) S-(2-Carboxypropyl)-cysteine and its sulfoxide as precursors in the biosynthesis of cycloalliin. Acta Chem. Scand. 21:1654–1656. Chem. Abst. 67 (1967) 106 001.

Granroth, B. and Virtanen, A.I. (1967b) On the biosynthesis of cycloalliin in the onion bulb. Suom. Kemistil. B. 40:103–104.

Gray, D.S. (1990) Dehydrated garlic preparations and buffer compounds to alleviate digestively induced aftereffects. U.S.A. Patent 4 917 881.

Grechkin, A.N., Fazliev, F.N., and Mukhtarova, L.S. (1995) The lipoxygenase pathway in garlic (Allium sativum) bulbs: detection of the novel divinyl ether oxylipins. FEBS Lett. 371:159–162.

Greenstock, D. (1982) Garlic in the blood. New Sci. 95:22.

Grill, E., Winnacker, E.L., and Zenk, M.H. (1987) Phytochelatins, a class of heavy metal-binding peptides from plants, are functionally analogous to metallothioneins. Proc. Natl. Acad. Sci. USA, 84:439–443.

Grimm, J. (1876) Deutsche Mythologie. Ferd. Dümmlers Verlagsbuchhandlung, Berlin.

Grünwald, J. (1989) Alliin: der Qualitätsstandard des Knoblauchs. Dtsch. Apoth. Ztg. 129:1198B.

Grünwald, J., Hübner, W.D., and Schulz, V. (1991) Prophylaxis of arteriosclerosis and therapy of hyperlipidemia: two indications for the use of garlic powder tablets. Mol. Biol. Atheroscler., 645–646. Chem. Abst. 118 (1993) 246 681.

Grünwald, J. (1992) Knoblauch: Cholesterinsenkende Wirkung doppelblind nachgewiesen. Dtsch. Apoth. Ztg. 132:1356.

Grünwald, J., Heede, J., Koch, H., Albrecht, R., Knudsen, O., Rasmussen, N., Sottrup, P.C., Thusgaard, G., and Vejlgaard, T.F. (1992) Effects of garlic powder tablets on blood lipids and blood pressure. The Danish Multicenter Kwai Study. Eur. J. Clin. Res. 3:179–186.

Grzybowski, R., Siuchninska, U., and Trojanowska-Bielak, H. (1988) The effect of selected chemical preservatives on the antibiotic activity of garlic and horseradish extracts. Przem. Spozyw. 42:145–147. Chem. Abst. 109 (1988) 189 004 (Polish).

Grzybowski, R. and Lewicka, B. (1987) Effect of the phytoncides of garlic, onion, and horseradish on the temperature-dependent activity of viruses and bacteria. Przem. Spozyw. 41:165–166 (Polish).

Gstirner, F. (1955) Prüfung und Verarbeitung von Arzneidrogen. Springer, Heidelberg.

Gu, Y.Q., Liu, Y.Y., Yang, X.H., Chen, D., and Fu, F.H. (1988) Effect of the essential oils of Allium cepa L. var. agregatum Don and Allium macrostemon Bunge on arachidonic acid metabolism. Yaoxue Xuebao 23:8–11. Chem. Abst. 108 (1988) 216 054 (Chinese).

Gudi, V.A. and Singh, S.V. (1991) Effect of diallyl sulfide, a naturally occurring anti-carcinogen, on glutathione-dependent detoxification enzymes of female CD-1 mouse tissues. Biochem. Pharmacol. 42:1261–1265. Chem. Abst. 115 (1991) 197 894.

Guenther, E. (1952) Oil of Garlic. In: The essential oils. Vol. 6. Van Nostrand, Princeton, New Jersey, pp. 67–69.

Guevara, B.Q., Solevilla, R.C., Mantaring, N.M., Remulla, R.A., and Santos, P.S. (1983) The antifungal properties of Allium sativum Linn. grown in the Philippines. Acta Manilana Ser. A:1–14.

Guillaume, A. and Wadie, J.A. (1950) Sur les préparations galéniques obtenues avec les gousses d'ail. Prod. Pharm. 5:421–425.

Gulati, A., Prakash, S., and Gupta, S.P. (1994) Genotoxicity testing for relative efficacy of selected pesticides on Allium cepa. J. Environ. Biol. 15:89–95. Chem. Abst. 122 (1995) 25 568.

Güntzel-Lingner, H. (1941) Der Knoblauch. Kleine Heilpflanzen-Monographien No. 1. Zinnser & Co., Leipzig.

Guo, H. and Cui, L. (1990) Mass spectrometric study on components in the volatile oil of garlic. Fenxi Ceshi Tongbao 9:11–16. Chem. Abst. 114 (1991) 22 600 (Chinese).

Guo, N.L., Lu, D.P., Woods, G.L., Reed, E., Zhou, G.Z., Zhang, L.B., and Waldman, R.H. (1993) Demonstration of the anti-viral activity of garlic extract against human cytomegalovirus in vitro. Chin. Med. J. 106:93–96.

Gupta, J.B. and Godhwani, J.L. (1984) Modification of immunological response by garlic, guggal, and tumeric in albino rats. Indian J. Pharmacol. 16:62–63.

Gupta, K.C. and Viswanathan, R. (1955) Combined action of streptomycin and chloramphenicol with plant antibiotics against tubercle bacilli. I. Streptomycin and chloramphenicol with cepharanthine. II. Streptomycin and allicin. Antibiot. Chemother. [Washington, D.C.] 5:24–27. Chem. Abst. 49 (1955) 7732.

Gupta, M.K., Mittal, S.R., Mathur, A.K., and Bhan, A.K. (1993) Garlic: the other side of the coin. Int. J. Cardiol. 38:333.

Gupta, N.K. (1988) Hypolipidemic action of garlic unsaturated oils in irradiated mice. Natl. Acad. Sci. Lett. [India] 11:401–403. Chem. Abst. 111 (1989) 190 407.

Gupta, N.N., Mehrotra, R.M.L., and Sircar, A.R. (1966) Effect of onion on serum cholesterol, blood coagulation factors, and fibrinolytic activity in alimentary lipemia. Indian J. Med. Res. 54:48–53.

Gupta, P.P., Khetrapal, P., and Ghai, C.L. (1987) Effect of garlic on serum cholesterol and electrocardiogram of rabbit consuming normal diet. Indian J. Med. Sci. 41:6–11.

Gupta, R. and Sharma, N.K. (1993) A study of the nematocidal activity of allicin, an active principle in garlic, Allium sativum L., against root-knot nematode, Meloidogyne incognita (Kofoid & White, 1919) Chitwood, 1949. Int. J. Pest Management 39:390–392.

Gusev, S.P. and Grishina, L.V. (1963) The chemical composition of different varieties of garlic. Tr. Moskov. Inst. Narod. Khoz. (Trans. Moscow. Inst. Natl. Econ.) 24:30–34. Chem. Abst. 62 (1965) 1012 (Russian).

Güven, K.C., Aktulga, A., and Cetin, E.T. (1972) Thin-layer chromatography and antibacterial activity of garlic oil. Eczacilik Bull. 14:51–61. Chem. Abst. 78 (1973) 80 159 (Turkish).

Gwilt, P.R., Lear, C.L., Tempero, M.A., Birt, D.D., Greandjean, A.C., Ruddon, R.W., and Nagel, D.L. (1994) The effect of garlic extract on human metabolism of acetaminophen. Cancer Epidemiol., Biomarkers Prev. 3:155–160. Chem. Abst. 121 (1994) 26 172.

HAB 34. (1934) Allium sativum, Knoblauch. In: Homöopathisches Arzneibuch. Vol. 2. W. Schwabe, ed. Leipzig, p. 54.

HAB 53. (1953) Allium sativum, Knoblauch. In: Homöopathisches Arzneibuch. Vol. 3. W. Schwabe, ed. Berlin, pp. 55–56.

HAB 78. (1978) Allium sativum. In: Homöopathisches Arzneibuch. Deutscher Apotheker Verlag, Stuttgart, pp. 159–161.

Haber, D., Siess, M.H., De Waziers, I., Beaune, P., and Suschetet, M. (1994) Modification of hepatic drug metabolizing enzymes in rat fed naturally occurring allyl sulfides. Xenobiotica 24:169–182.

Haber, D., Siess, M.-H., Canivenc-Lavier, M.-C., Le Bon, A.-M., and Suschetet, M. (1995) Differential effects of dietary diallyl sulfide and diallyl disulfide on rat intestinal and hepatic drug-metabolizing enzymes. J. Toxicol. Environ. Health 44:423–434.

Hadacova, V., Klozova, E., Hadac, E., Turkova, V., and Pitterova, K. (1981) Comparison of esterase isoenzyme patterns in seeds of some Allium species and in cultivars of Allium cepa L. Biol. Plantarum [Prague] 23:174–181.

Hadacova, V., Vackova, K., Klozova, E., Kutacek, M., and Pitterova, K. (1983) Cholinesterase activity in some species of the Allium genus. Biol. Plantarum [Prague] 25:209–215.

Hadacova, V., Turkova, V., Klozova, E., and Pitterova, K. (1985) Comparison of esterase isoenzymes and protein patterns in Allium montanum F. W. Schmidt from various localities. Biol. Plantarum [Prague] 27:28–33. Chem. Abst. 102 (1985) 201 179.

Hadjiolov, D., Fernando, R.C., Schmeiser, H.H., Wiebler, M., Hadjiolov, N., and Pirajnov, G. (1993) Effect of diallyl sulfide on aristolochic acid-induced forestomach carcinogenesis in rats. Carcinogenesis 14:407–410.

Haenel, I.H., Gerriets, E., and Rieche, A. (1962) Nutritional effects in chicks of synthetic compounds which liberate mustard oil. Arch. Geflügelkd. 26:33–50. Chem. Abst. 57 (1962) 5080.

Haenen, J.M. (1984) Histoire d'aulx. Presse Med. 13:745.

Haenszel, W., Kurihara, M., Segi, M., and Lee, R.K.C. (1972) Stomach cancer among Japanese in Hawaii. J. Natl. Cancer Inst. 49:969–988.

Hagen, T.M., Wierzbicka, G.T., Sillau, A.H., Bowman, B.B., and Jones, D.P. (1990) Bioavailability of dietary glutathione: effect on plasma concentration. Am. J. Physiol. 259:G524–G529.

Hager. (1969) Allium. In: Hagers Handbuch der phramazeutischen Praxis. Vol. 2. Springer, Berlin, pp. 1210–1217.

Hager, H. (1861) Medicamenta homoeopathica et isopathica omnia. E. Günther, Lesnae.

Hahn, S. and Koh, K. (1949) Studies on the resistance of intestinal parasites' eggs and larvae. Korean J. Intern. Med. 1:6–10.

Hahnemann, S. (1833) Organon der Heilkunst. Haug Verlag, Heidelberg.

Hakamada, K. (1982) Odorless garlic wine. Japan. Patent 82 155 982. Chem. Abst. 98 (1983) 15 745.

Halevy, A., Levi, Y., Shnaker, A., and Orda, R. (1990) Auto-vampirism: an unusual case of anaemia. J. Roy. Soc. 82:630–631; Med. Trib. 22 (1990) No. 15, 6.

Hall, J. (1986) Biochemical explanations for folk tales: vampires and werewolves. Trends Biochem. Sci. 11:31.

Haller, A. von (1745) De Allii genere naturali libellus. Göttingen.

Haltrich, J. and Wolff, J. (1885) Zur Volkskunde der Siebenbürger Sachsen. Carl Graeser, Vienna.

Hamamoto, A. and Mazelis, M. (1986) The C-S-lyases of higher plants: isolation and properties of homogeneous cystine lyase from broccoli (Brassica oleracea var. botrytis) buds. Plant Physiol. 80:702–706.

Han, J. (1993) Highlights of the cancer chemoprevention studies in China. Prev. Med. 22:712–722.

Han, J., Lawson, L., Chu, T.C., Potter, D., Han, G., and Han, P. (1993) Modification of catalytic properties of chicken liver fructose 1,6-bisphosphatase by allicin. Biochem. Mol. Biol. Int. 31:1007–1015. Chem. Abst. 120 (1994) 318 157.

Han, J., Lawson, L., Han, G., and Han, P. (1995) A spectrophotometric method for quantitative determination of allicin and total garlic thiosulfinates. Anal. Biochem. 225:157–160. Chem. Abst. 122 (1995) 182 424.

Han, N., Liu, B., and Wang, M. (1992) Effect of allicin on antioxidases in mice. Yingyang Xuebao (Acta Nutr. Sinica) 14:107–108. Chem. Abst. 117 (1992) 205 145 (Chinese).

Hanafy, M.S.M., Shalaby, S.M., El Fouly, M.A.A., El Aziz, M.I.A., and Soliman, F.A. (1994) Effect of garlic on lead contents in chicken tissues. Dtsch. Tieraerztl. Wochenschr. 101:157–158.

Hanley, A.B. and Fenwick, G.R. (1985) Cultivated Alliums. J. Plant Foods 6:211–238. Chem. Abst. 105 (1986) 132 207.

Hansen, S.H., Björnsdottir, I., Bager, S., and Poulsen, M.N. (1993) Nyt stoff til hvidlögsdebatten. Farmaci 100:6–8.

Harborne, J.B. (1986) The natural distribution in angiosperms of anthocyanins acylated with aliphatic dicarboxylic acids. Phytochemistry 25: 1887–1894. Chem. Abst. 105 (1986) 149 755.

Hardman, R. (1991) From ginseng to garlic, physalis to frogs. Pharm. J. 247:542.

Harenberg, J., Giese, C., and Zimmermann, R. (1988) Effect of dried garlic on blood coagulation, fibrinolysis, platelet aggregation, and serum cholesterol levels in patients with hyperlipoproteinemia. Atherosclerosis 74:247–249.

Harkness, B. and Fox, H.M. (1944) Ornamental Alliums for North American gardens. Herbertia 11:313–319.

Harmatha, J., Mauchamp, B., Arnault, C., and Slama, K. (1987) Identification of a spirostane-type saponin in the flowers of leek with inhibitory effects on growth of leek-moth larvae. Biochem. Systematics Ecology 15:113–116.

Harris, C.M., Mitchell, S.C., Waring, R.H., and Hendry, G.L. (1986) The case of the black-speckled dolls: an occupational hazard of unusual sulphur metabolism. Lancet 1:492–493.

Harris, L.J. (1979) The book of garlic. Pajandrum/Aris, Los Angeles.

Hartwell, J.L. (1960) Plant remedies for cancer. Cancer Chemother. Rep. 7:19–24.

Hasegawa, Y., Kikuchi, N., Kawashima, Y., Ono, Y., Shimizu, K., and Nishiyama, M. (1983) Clinical effects of Kyolepin against various complaints in the field of internal medicine. J. New Remedies [Japan] 32:365.

Hatanaka, H. and Kaneda, Y. (1980) Enzymic assay of scordinin as the main tonic principle active in garlic used in health foods. Nippon Eiseigaku Zasshi (Jap. J. Hyg.) 35:746–751. Chem. Abst. 94 (1981) 137 918.

Hausen, B.M. (1988) Allergiepflanzen – Pflanzenallergene. Handbuch und Atlas der allergieinduzierenden Wild- und Kulturpflanzen. Kontaktallergene. Ecomed, Landsberg, Germany.

Havey, M.J. (1992a) Restriction enzyme analysis of the nuclear 45s ribosomal DNA of six cultivated Alliums (Alliaceae). Plant Syst. Evol. 181:45–55. Chem. Abst. 117 (1992) 248 696.

Havey, M.J. (1992b) Restriction enzyme analysis of the chloroplast and nuclear 45s ribosomal DNA of Allium sections Cepa and Phyllodolon (Alliaceae). Plant Syst. Evol. 183:17–31.

Havranek, P. (1974) The effect of virus diseases on the yield of common garlic. Sbor. UVTI ochr. rostl. 10:251–256 (Czech).

Havranek, P. and Novak, F.J. (1973) The bud formation in the callus cultures of Allium sativum L. Z. Pflanzenphysiol. 68:308–318.

Hayashi, T., Sano, K., and Ohsumi, C. (1993) Gas chromatographic analysis of alliin in the callus tissues of Allium sativum. Biosci. Biotech. Biochem. 57:162–163. Chem. Abst. 118 (1993) 120 144.

Hayes, M.A., Rushmore, T.H., and Goldberg, M.T. (1987) Inhibition of hepatocarcinogenic responses to 1,2-dimethylhydrazine by diallyl sulfide, a component of garlic oil. Carcinogenesis 8:1155–1157.

Hecker, G. (1815) Practische Arzneimittellehre oder die Kunst, die Krankheiten des Menschen zu heilen. Hennings, Erfurt.

Hegi, G. (1939) Allium L. – Lauch. In: Illustrierte Flora von Mittel-Europa. Pichler, Vienna, pp. 213–234.

Hegnauer, R. (1963) Chemotaxonomie. Birkhäuser, Basel.

Hehn, V. (1894) Kulturpflanzen und Hausthiere in ihrem Übergang aus Asien nach Griechenland und Italien sowie in das übrige Europa. Historisch-linguistische Skizzen. Borntraeger, Berlin.

Heide, L. and Boegl, W. (1985) Chemiluminescence measurements on 20 spices. A method of detecting treatment with ionizing radiation. Z. Lebensm. Unters. Forsch. 181:283–288. Chem. Abst. 103 (1985) 213 526 (German).

Heikal, H.A., El Dashlouty, M.S., and Saied, S.Z. (1972) Biochemical, histological, and technological changes occurring during the production of sausage from camel meat and beans. Agric. Res. Rev. 50:243–252.

Heinle, H. and Betz, E. (1994) Effects of dietary garlic supplementation in a rat model of atherosclerosis. Arzneim. Forsch. 44:614–617. Chem. Abst. 121 (1994) 170 243.

Heinze, R. (1939) Das Kräuterkäsgericht (Moretum). Antike 15:76–88.

Helm, J. (1956) Die zu Würz- und Speisezwecken kultivierten Arten der Gattung Allium. Kulturpflanze 4:130.

Henry, R.J. and Darbyshire, B. (1979) The distribution of fructan metabolizing enzymes in the onion plant. Plant Sci. Lett. 14:155–158.

Henry, R.J. and Darbyshire, B. (1980) Sucrose:sucrose fructosyltransferase and fructan:fructan fructosyltransferase from Allium cepa. Phytochemistry 19:1017–1020.

Hepper, P.G. (1988a) Garlic-bred rats. New Sci. 119:48.

Hepper, P.G. (1988b) Adaptive fetal learning: prenatal exposure to garlic affects postnatal preferences. Animal Behaviour 36:935–936.

Hermann, F. (1945) Sections and subsections of the genus Allium in Europe. Herbertia 12:71–72.

Herodotus. (1963) Herodot Historien. Ernst Heimeran, Munich.

Herrmann, K. (1956a) Über das Vorkommen von Kaffeesäure und Chlorogensäure im Obst und Gemüse. Naturwiss. 43:109.

Herrmann, K. (1956b) Über Kaffeesäure und Chlorogensäure. Pharmazie 11:433–449.

Herrmann, K. (1956c) Über die Quercetinglykosid der Zwiebel (Allium cepa L.). Naturwiss. 43:158–159.

Herrmann, K. (1977) Übersicht über nichtessentielle Inhaltsstoffe der Gemüsearten. II. Cruciferen (Kohlarten, Radieschen, Rettiche, Speiserüben, Kohlrüben, Meerrettich) sowie Gramineen (Zwiebeln, Porree, Schnittlauch, Knoblauch, Spargel). Z. Lebensm. Unters. Forsch. 165:151–164. Chem. Abst. 88 (1978) 34 478.

Hess, D. (1994) Meister um das "mittelalterliche Hausbuch." Studien zur Hausbuchmeisterfrage. Philipp von Zabern, Mainz.

Hess, H., Mehn, S., Schoenmann, H., and R.P. Scherer Gmbh. (1987) Oral preparations of garlic and method of production. Germany. Patent 3 541 304 A1. Chem. Abst. 107 (1988) 76 450.

Hess, J.W. (1860) Kaspar Bauhins Leben und Charakter. Beitr. vaterländ. Gesch. 7:105–176.

Heyn, K. (1978) Plant-compatible oils for use with herbicides. Germany. Patent 2 701 129. Chem. Abst. 89 (1978) 124 586.

Hikino, H., Tohkin, M., Kiso, Y., Namiki, T., Nishimura, S., and Takeyama, K. (1986) Antihepatotoxic actions of Allium sativum bulbs. Planta Med. 53:163–168. Chem. Abst. 105 (1986) 146 180.

Hildegard von Bingen. (1955) Ursachen und Behandlung der Krankheiten (causae et curae). K.F. Haug, Ulm.

Hiller, K. (1964) Antimikrobielle Stoffe in Blütenpflanzen. Pharmazie 19:167–188. Chem. Abst. 60 (1964) 14 331.

Hills, L.D. (1972) Will garlic replace DDT? Org. Garden. Farm. (Sept.)

Hinoki, M., Nakanishi, K., Kishimoto, S., Ushio, N., Kitamura, H., Higashitsuji, H., Hayashi, M., Tamaki, S., Uehara, N., Matsuura, K., and Saijo, H. (1981) Olfactory vertigo: a neurological approach. Aggressologie 22B:45–57.

Hintzelmann, U. (1935a) Ein Beitrag zur Wertbestimmung von Allium-Präparaten. Arch. Exp. Pathol. Pharmakol. 178:480–485. Chem. Zbl. 1936, I, 2772. Ref. in Münch. Med. Wschr. 82 (1935) 1419.

Hintzelmann, U. (1935b) Zur pharmakologischen Wertbestimmung von Alliumpräparaten. Fortsch. Therap. 11:359–361. Chem. Zbl. 1935, II, 883.

Hippokrates. (1897) Hippokrates, Sämtliche Werke, ins Deutsche übersetzt und ausführlich commentiert. Verlag Dr. H. Lüneburg, Munich.

Hirano, K. and Hirano, S. (1984) Production of garlic powder. Japan. Patent 84 203 465. Chem. Abst. 102 (1985) 165 551.

Hirao, S., Yamamoto, R., and Nishino, T. (1994) Collagenase solution containing Allium sativum extract for hepatocyte separation. Japan. Patent 94 237 763. Chem. Abst. 122 (1995) 27 269.

Hirao, Y., Sumioka, I., Nakagami, S., Yamamoto, M., Hatono, S., Yoshida, S., Fuwa, T., and Nakagawa, S. (1987) Activation of immunoresponder cells by the protein fraction from aged garlic extract. Phytother. Res. 1:161–164. Chem. Abst. 109 (1988) 423.

Hirsch, A.F., Piantadosi, C., and Irvin, J.L. (1965) Potential anticancer agents. II. The synthesis of some nitrogen mustard containing sulfones and thiosulfinates. J. Med. Chem. 8:10–14. Chem. Abst. 62 (1965) 5215.

Hitokoto, H., Morozumi, S., Wauke, T., Sakai, S., and Ueno, I. (1978) Inhibitory effects of condiments and herbal drugs on the growth and toxin production of toxigenic fungi. Mycopathologia 66:161–167.

Hitokoto, H., Morozumi, S., Wauke, T., Sakai, S., and Kurata, H. (1980) Inhibitory effects of spices on growth and toxin production of toxigenic fungi. Appl. Environ. Microbiol. 39:818–822. Chem. Abst. 93 (1980) 44 284.

Hizume, M. (1994) Allodiploid nature of Allium wakegi Araki revealed by genomic in situ hybridization and localization of 5S and 18S rDNAs. Jap. J. Genet. 69:407–415. Chem. Abst. 122 (1995) 76 902.

Hjorth, N. and Roed-Petersen, J. (1976) Occupational protein contact dermatitis in food handlers. Contact Dermatitis 2:28–42.

Ho, M.F. and Mazelis, M. (1993) The C-S lyases of higher plants, determination of homology by immunological procedures. Phytochemistry 34: 625–629.

Hobbs, C. (1993) Garlic: the pungent panacea. Pharm. Hist. *34*:152–157. Int. Pharm. Abst. 30 (1993) 3123.

Hoeld, A. and Lupas, O. (1983) Real-time simulation of the transient behaviour of local and global pressurized water reactor core and plant parameters. Nucl. Sci. Eng. *85*:396–417. Chem. Abst. 100 (1984) 27 205.

Hoernle, R. (1897) The Bower Manuscript Facsimile Lewes, Nagari Transcript; romanised transliteration and English translation with notes, 1893–1897. Archaeological Survey of India, Calcutta.

Hoffmann-Ostenhof, O. and Keck, K. (1951) Pflanzliche Stoffwechselprodukte als Mitosegifte. 2. Diallyldisulfid, ein Bestandteil der Lauchöle, als mitosestörende Substanz. Monatsh. Chem. *82*:562–564. Chem. Abst. 45 (1951) 9615.

Höfler, M. (1908) Volksmedizinische Botanik der Germanen. R. Ludwig, Vienna.

Hofmann, H. and Held, U. (1953) Die Nebenwirkungen der Anthelmintica. Pharmazie *8*:24–28.

Hokkaido Wako Shokuhin Kk. (1982) Ethanol odor masking in raw noodles. Japan. Patent 82 202 259. Chem. Abst. 98 (1983) 124 519.

Holtmeier, H.J. (1993) Cholesterin: Harmlos oder gefährlich? Gazette Medicale *14*:372–374.

Holuby, J.L. (1884) Knoblauch (*Allium sativum* L.) als Volksheilmittel bei den Slovaken Nordungarns. Dtsch. bot. Monatsschr. *2*:7–9.

Holzgartner, H., Schmidt, U., and Kuhn, U. (1992) Wirksamkeit und Verträglichkeit eines Knoblauchpulver-Präparates im Vergleich mit Bezafibrat (Comparison of the efficacy and tolerance of a garlic preparation vs. bezafibrate). Arzneim. Forsch. *42*:1473–1477. Congress abstract in Eur. J. Clin. Res. 3A (1992) 8.

Holzhey, M., Roth, H.H., and Höpfner, V. (1992) Use of allicin-urotropin as an internal drug for the treatment of infection and cancer in people and animals. Germany. Patent 4 024 155. Chem. Abst. 116 (1992) 201 109.

Homer. (1971) Werke in zwei Bänden, Bd.1: Ilias, Bd.2: Odissee. Aufbau Verlag, Berlin.

Hong, J.Y., Wang, Z.Y., Smith, T.J., Zhou, S., Shi, S., Pan, J., and Yang, C.S. (1992) Inhibitory effects of diallyl sulfide on the metabolism and tumorigenicity of the tobacco-specific carcinogen 4-(methylnitrosamino)-1-(3-pyridyl)-1-butanone (NNK) in A/J mouse lung. Carcinogenesis *13*:901–904.

Hong, S.K., Koh, S.D., Shin, H.K., and Kim, K.S. (1992) Effects of garlic oil, garlic juice, and ally sulfide on the responsiveness of dorsal horn cell in the cat. Hanyang Uidae Haksulchi *12*:621–633. Chem. Abst. 119 (1993) 40 844.

Honma, T. (1990a) Boric acid-containing repellents for cockroach. Japan. Patent 90 233 604. Chem. Abst. 114 (1991) 57 565.

Honma, T. (1990b) Control of cockroaches by compositions containing attractants and boric acid. Japan. Patent 90 169 505. Chem. Abst. 114 (1991) 37 810.

Hooper, D. (1938) On chinese medicine: drugs of Chinese pharmacies in Malaya Gardens Bulletin. Straits Settlements. Weinheim.

Hoops, G. (1905) Waldbäume und Kulurpflanzen im germanischen Altertum. K.J. Trübner, Strassburg.

Hopff, W. and Prokop, O. (1992) Erklärung zur Homöopathie. Dtsch. Apoth. Ztg. *132*:1630–1631.

Hoppen, V.R., Auricchio, M.T., and Batistic, M.A. (1989) Thin-layer chromatographic characterization of preparation from *Allium sativum* L. Rev. Inst. Adolpho Lutz *49*:5–10. Chem. Abst. 112 (1990) 204 802 (Portuguese).

Hörhammer, L., Wagner, H., Seitz, M., and Vejdelek, Z.J. (1968) Zur Wertbestimmung von Knoblauchpräparaten. I. Mitteilung: Chromatographische Untersuchungen über die genuinen Inhaltsstoffe von *Allium sativum* L. Pharmazie *23*:462–467. Chem. Abst. 70 (1969) 26 377.

Hörhammer, L. (1970) Teeanalyse. Springer, Berlin.

Horie, I. (1978) Odorless garlic product. Japan. Patent 78 32 147. Chem. Abst. 89 (1978) 74 464.

Horie, T., Murayama, T., Mishima, T., Itoh, F., Minamide, Y., Fuwa, T., and Awazu, S. (1989) Protection of liver microsomal membranes from lipid peroxidation by garlic extract. Planta Med. *55*:506–508. Chem. Abst. 112 (1990) 117 539.

Horie, T., Awazu, S., Itakura, Y., and Fuwa, T. (1992) Identified diallyl polysulfides from an aged garlic extract which protects the membranes from lipid peroxidation. Planta Med. *58*:468–469. Chem. Abst. 118 (1993) 52 390.

Horton, G.M.J., Fennell, M.J., and Prasad, B.M. (1991) Effect of dietary garlic (*Allium sativum*) on performance, carcass composition, and blood chemistry changes in broiler chickens. Can. J. Anim. Sci. *71*:939–942.

Hotzel, A. (1936) Über Knoblauchpräparate. Pharm. Ztg. *81*:440.

HPI 71. (1971) *Allium sativum*. In: Homeopathic Pharmacopoeia of India. Vol. 1. Government of India, Ministry of Health, pp. 45–46.

HPUS 79. (1979) *Allium sativum*. In: The homeopathic pharmacopoeia of the United States. Vol. 8. American Institute of Homeopathy, Falls Church, Virginia, pp. 60–61.

Hsu, J.P., Jeng, J.G., and Chen, C.C. (1993) Identification of a novel sulfur compound from the interaction of garlic and heated edible oil. Proc. Int. Conf. Prog. Flavour Precursor Stud., 391–394. Chem. Abst. 121 (1994) 81 405.

Hu, P., Anderson, M.D., and Wargovich, M.J. (1990) Protective effect of diallyl sulfide, a natural extract of garlic, on MNNG-induced damage of rat glandular stomach mucosa. Zhonghua Zhongliu Zazhi (Chin. J. Cancer) *12*:429–431. Chem. Abst. 114 (1991) 156 749 (Chinese).

Hu, P.J. and Wargovich, M.J. (1989) Effect of diallyl sulfide on MNNG-induced nuclear aberrations and ornithine decarboxylase activity in the glandular stomach mucosa of the Wistar rat. Cancer Lett.

47:153–158. Chem. Abst. 113 (1990) 17 601.
Huang, J.K., Wang, C.S., and Chang, W.H. (1981) Studies on the antioxidative activities of spices grown in Taiwan. Chung-kuo Nung Yeh Hua Hsueh Hui ChiH (J. Chin. Agric. Chem. Soc.) 19:200–207. Chem. Abst. 97 (1982) 143 289 (Chinese).
Hughes, B.G., Murray, B.K., North, J.A., and Lawson, L.D. (1989) Antiviral constituents from *Allium sativum*. Planta Med. 55:114.
Hughes, B.G. and Lawson, L.D. (1991) Antimicrobial effects of *Allium sativum* L. (garlic), Allium ampeloprasum (elephant garlic), and Allium cepa L. (onion), garlic compounds and commercial garlic supplement products. Phytother. Res. 5:154–158. Chem. Abst. 115 (1991) 247 549.
Huh, K., Choi, C.W., Cho, S.Y., and Kim, S.H. (1983) Effect of *Allium sativum* L. on the hepatic xanthine oxidase activity in rat. J. Resource Dev., Yeungnam Univ. 2:111.
Huh, K., Lee, S.I., and Park, J.M. (1985a) Effect of garlic on the hepatic xanthine oxidase activity in rats. Hanguk Saenghwa Hakhoechi (Korean Biochem. J.) 18:209–214. Chem. Abst. 104 (1986) 17 005.
Huh, K., Nam, S.H., Park, J.M., and Chang, U.K. (1985b) Effect of garlic on liver microsomal aniline hydroxylase activity in mouse. J. Resource Dev., Yeungnam Univ. 4:71.
Huh, K., Park, J.M., and Lee, S.I. (1985c) Effect of garlic on the hepatic glutathione-S-transferase and glutathione peroxidase activity in rat. Arch. Pharmacol. Res. [Seoul] 8:197–203.
Huh, K., Lee, S.I., Park, J.M., and Kim, S.H. (1986a) Effect of diallyl disulfide on the hepatic glutathione-S-transferase activity in rat: diallyl disulfide effect on the glutathione-S-transferase. Arch. Pharmacol. Res. [Seoul] 9:205–209. Chem. Abst. 106 (1987) 131 525.
Huh, K., Lee, S.I., Park, J.M., and Kim, S.H. (1986b) Effect of garlic on the purine metabolic pathway. Yakhak Hoechi (J. Pharm. Soc. Korea) 30:62–67. Chem. Abst. 105 (1986) 132 751.
Hunan Medical College. (1980) Garlic in cryptococcal meningitis: a preliminary report of 21 cases. Chin. Med. J. 93:123–126.
Hunt, J. (1980) Determination of total sulfur in small amounts of plant material. Analyst [London] 105:83–85.
Huq, F., Saha, G.C., Begum, F., and Adhikary, S. (1991) Studies on *Allium sativum* Linn. (garlic). II. Chemical investigation on garlic oil. Bangladesh J. Sci. Ind. Res. 26:41–51. Chem. Abst. 117 (1992) 169 765.
Huss, R. (1938) Ein Versuch zur Prophylaxe der Kinderlähmung mittels eines Knoblauchpräparates. Wien. Med. Wochenschr. 88:697–698.
Hussain, S.P., Jannu, L.N., and Rao, A.R. (1990) Chemopreventive action of garlic on methylcholanthrene-induced carcinogenesis in the uterine cervix of mice. Cancer Lett. 49:175–180.
Hwang, J.I., Bae, E.S., and Cha, C.W. (1986) A study on the protective effect of Korean garlic on the albino rat, chronically exposed to methylmercury. Koryo Taehakkyo Uikwa Taehak Nonmunjip (Korea Univ. Med. J.) 23:121–130. Chem. Abst. 105 (1986) 110 102.
Hyams, E. (1971) Plants in the service of man: 10,000 years of domestication. J.M. Dent & Sons, London.
Iberl, B., Winkler, G., Müller, B., Jansen, H., and Knobloch, K. (1989) On the quantitative determination of allicin and alliin from garlic by HPLC. Planta Med. 55:640–641.
Iberl, B., Winkler, G., and Knobloch, K. (1990a) Products of allicin transformation: ajoenes and dithiins, characterization and their determination by HPLC. Planta Med. 56:202–211. Chem. Abst. 113 (1990) 130 796.
Iberl, B., Winkler, G., Müller, B., and Knobloch, K. (1990b) Quantitative determination of allicin and alliin from garlic by HPLC. Planta Med. 56:320–326.
Ibn al Baithar. (1840) Grosse Zusammenstellung über die Kräfte der bekannten einfachen Heil- und Nahrungsmittel, etc. Hallberger, Stuttgart.
Igue, T., Pavezi, R.T., and Paulo, E.M. (1982) Sampling in herbicide experiments. Planta Daninha 5:14–19. Chem. Abst. 100 (1984) 187 151 (Portuguese).
Ikeda, K. (1969a) Studies on the nutritional value of Allium plants. LII. Nutritional significance of the ingestion of garlic (a supplement). Bitamin [Kyoto] 40:263–267. Chem. Abst. 72 (1970) 1204.
Ikeda, K. (1969b) Nutritional value of Allium plants. L. Hemocyte affinity of thiamine propyl disulfide. Bitamin [Kyoto] 40:251–259. Chem. Abst. 80 (1974) 1202 (Japanese).
Ikpeazu, O.V., Agusti, K.T., and Joseph, P.K. (1987) Hypolipidemic effect of garlic extracts mixed with 3% ethanol in rats fed sucrose-high fat diet. Indian J. Biochem. Biophys. 24:252–253. Chem. Abst. 107 (1987) 235 325.
Ikram, M. (1972) A review on chemical and medicinal aspects of *Allium sativum*. Pak. J. Sci. Ind. Res. 15:81–86. Chem. Abst. 78 (1973) 47 683.
Ilyushenko, V.P. and Shchegolkov, V.N. (1990) Sensitivity of the Allium test to the presence of heavy metals in an aqueous medium. Khim. Tekhnol. Vody 12:275–278. Chem. Abst. 113 (1990) 72 675 (Russian).
Imai, J., Ide, N., Nagae, S., Moriguchi, T., Matsuura, H., and Itakura, Y. (1994) Antioxidant and radical scavenging effects of aged garlic extract and its constituents. Planta Med. 60:417–420. Chem. Abst. 122 (1995) 71 940.
Inagaki, N., Matsunaga, H., Kawano, T., Maekawa, S., and Terabun, M. (1992) In vitro micropropagation of Allium giganteum R.I. callus and shoot formation and regeneration of plantlet through in vitro culture of emerged young leaves. Kobe Daigaku Nogakubu Kenkyu Hokoku 20:47–53. Chem. Abst. 118 (1993) 75 300.
Inouye, S., Goi, H., Miyauchi, K., Muraki, S., Ogihara,

M., and Iwanami, Y. (1983) Inhibitory effect of volatile constituents of plants on the proliferation of bacteria. Bokin Bobai *11*:609–615. Chem. Abst. 100 (1983) 48 439.

Ionescu, C.N., Ichim, A., and Zingher, S. (1954) Allicin, a bactericide derived from *Allium sativum*. Acad. Repub. Pop. Rom. Fil. Cluj. Stud. Cercet. Chem. *2*:213–221. Chem. Abst. 50 (1956) 9994 (Romanian).

Ip, C., Lisk, D.J., and Stoewsand, G.S. (1992) Mammary cancer prevention by regular garlic and selenium-enriched garlic. Nutr. Cancer *17*: 279–286.

Ip, C., Lisk, D.J., and Scimeca, J.A. (1994) Potential of food modification in cancer prevention. Cancer Res. *54*:S1957–S1959.

Ip, C. and Lisk, D.J. (1993) Bioavailability of selenium from selenium-enriched garlic. Nutr. Cancer *20*:129–137. Chem. Abst. 120 (1994) 162 296.

Ip, C. and Lisk, D.J. (1994a) Characterization of tissue selenium profiles and anticarcinogenic responses in rats fed natural sources of selenium-rich products. Carcinogenesis *15*:573–576.

Ip, C. and Lisk, D.J. (1994b) Enrichment of selenium in allium vegetables for cancer prevention. Carcinogenesis *15*:1881–1885. Chem. Abst. 121 (1994) 254 456.

Irie, T., Sugiura, M., Kutsuna, T., Konishi, T., Tagoyama, Y., and Shin Nippon Kagaku Kogyo Co. Ltd. (1992) Manufacture of odorless garlic powders with enzymes and vitamin C. Japan. Patent 92 341 154. Chem. Abst. 118 (1993) 168 022.

Isenberg, N. and Grdinic, M. (1973) Thiosulfinates. Int. J. Sulfur Chem. *3*:307–320. Chem. Abst. 79 (1973) 125 319.

Isensee, H., Rietz, B., and Jacob, R. (1993) Cardioprotective actions of garlic (*Allium sativum*). Arzneim. Forsch. *43*:94–98.

Ishiguro, K. (1963) Studies on pantothenic acid intake. I. Pantothenic acid content in Japanese foods. Tohoku J. Exp. Med. *78*:375–380. Chem. Abst. 59 (1963) 2095.

Ishihara, T. and Bizen Kasei Kk. (1993) Manufacture of odorless garlic extracts with activated carbon and bath preparations containing the extracts. Japan. Patent 93 5103 622. Chem. Abst. 119 (1993) 71 273.

Ishikawa, Y., Ikeshoji, T., and Matsumoto, Y. (1978) A propylthio moiety essential to the oviposition attractant and stimulant of the onion fly, Hylemya antiqua Meigen. Appl. Ent. Zool. *13*:115–122.

Ismailov, A.I., Tagiev, S.A., and Rasulov, E.M. (1976) Steroidal saponins and sapogenins from Allium rubellum and A. albanum. Khim. Prir. Soedin. (Chem. Nat. Compd.) *12*:550–551. Chem. Abst. 85 (1976) 189 199 (Russian; English cover-to-cover translation, p. 495).

Ismailov, A.I. and Aliev, A.M. (1974) Study of sapogenins from the white onion growing in Azerbaijan. Uch. Zap., Azerb. Gos. Med. Inst. *37*:60–64. Chem. Abst. 87 (1977) 180 731 (Russian).

Ismailov, A.I. and Aliev, A.M. (1976) Determination of the steroidal saponins in the white-flowered onion growing in Azerbaijan. Farmatsiya [Moscow] (Pharmacy) *25*:17–20. Chem. Abst. 84 (1976) 176 138 (Russian).

Isovich, J.M., L'Abbe, K.A., and Castelleto, R.E. (1992) Colon cancer in Argentina. Int. J. Cancer *51*:851–857.

Ito, N. and Pola Chem. Ind. Inc. (1990) Nail lacquers containing gamma-oryzanol, scordinin, and/or dibutylhydroxytoluene. Japan. Patent 2 290 806. Chem. Abst. 114 (1991) 108 721.

Itoh, T., Tamura, T., Mitsuhashi, T., and Matsumoto, T. (1977) Sterols of Liliaceae. Phytochemistry *16*:140–141.

Itokawa, Y. (1963a) Distribution of thiamine propyl disulfide-35S in red cells and plasma. Bitamin [Kyoto] *28*:568–573. Chem. Abst. 61 (1964) 16 505.

Itokawa, Y. (1963b) The mechanism of intestinal absorption of thiamine propyl disulfide-35S. Bitamin [Kyoto] *28*:574–577. Chem. Abst. 61 (1964) 16 505.

Itokawa, Y. (1963c) Determination of thiamine propyl disulfide-35S and several fundamental experiments. Bitamin [Kyoto] *28*:554–557. Chem. Abst. 61 (1964) 16 505.

Itokawa, Y. (1963d) Urinary excretion of 35S after administration of thiamine propyl disulfide-35S (inner) and thiamine-35S. Bitamin [Kyoto] *28*: 564–567. Chem. Abst. 61 (1964) 16 505.

Itokawa, Y., Uchino, H., and Nishino, N. (1971) Antihypercholesterolemic effect of S-methylcysteine sulfoxide. II. Effect of S-methylcysteine sulfoxide on fat metabolism in experimental hypercholesterolemia of rats. Eiyo to Shokuryo *24*:481–484. Chem. Abst. 77 (1972) 14 029 (Japanese).

Itokawa, Y., Inoue, K., Sasagawa, S., and Fujiwara, M. (1973) Effect of S-methylcysteine sulfoxide, S-allylcysteine sulfoxide and related sulfur-containing amino acids on lipid metabolism of experimental hypercholesterolemic rats. J. Nutr. *103*:88–92. Chem. Abst. 78 (1973) 56 857.

Iwanami, Y. (1981) Inhibiting effects of volatile constituents of plants on pollen growth. Experientia *37*:1280–1281.

Iwanami, Y., Asashima, M., Takemura, M., and Ichikura, M. (1982) Volatile plant constituents cause behavioral changes ultimately leading to death in the newt Cynops pyrrhogaster. Annot. Zool. Jpn. *55*:134–142.

Iwanami, Y., Cho, A., Togashi, S., and Asashima, M. (1985) Effects of exposure to volatile substances from plants on the behavior of Drosophila. Yokohama-shiritsu Daigaku Ronso, Shizen Kagaku Keiretsu *36*:15–33. Chem. Abst. 106 (1987) 45 697 (Japanese).

Jacob, R., Ehrsam, M., Ohkubo, T., and Rupp, H. (1991) Antihypertensive und Kardioprotektive Effekte von Knoblauchpulver (*Allium sativum*). Med. Welt *42*:39–41.

Jacob, R., Isensee, H., Rietz, B., Makdessi, S., and Sweidan, H. (1993a) Cardioprotection by dietary interventions in animal experiments: effect of garlic and various dietary oils under the conditions of experimental infarction. Pharm. Pharmacol. Lett. *3*:124–127. Chem. Abst. 123 (1994) 162 425.

Jacob, R., Isensee, H., Rietz, B., Makdessi, S., and Sweidan, H. (1993b) Cardioprotection by dietary interventions in animal experiments: effect of garlic and various dietary oils under the conditions of experimental infarction. Pharm. Pharmacol. Lett. *3*:131–134. Chem. Abst. 120 (1994) 268 947

Jacobsen, J.V., Bernhard, R.A., Mann, L.K., and Saghir, A.R. (1964) Infrared spectra of some asymmetric disulfides produced by Allium. Arch. Biochem. Biophys. *104*:473–477. Chem. Abst. 61 (1964) 14 826.

Jacobsen, J.V., Yamaguchi, Y., Mann, L.K., Howard, F.D., and Bernhard, R.A. (1968) An alkyl-cysteine sulfoxide lyase in Tulbaghia violacea and its relation to other alliinase-like enzymes. Phytochemistry *7*:1099–1108.

Jäger, H. (1955) Quantitative Bestimmung von Allicin in Frischem Knoblauch. Arch. Pharm. [Weinheim] *288*:145–148. Chem. Abst. 49 (1955) 12 772.

Jäger, W., Koch, H.P., and Pfaff, K. (1992) Standard allicin adsorbates on silicates. Phytother. Res. *6*:149–151. Chem. Abst. 117 (1992) 258 309.

Jäger, W., Hofeneder, M., Urban, E., and Koch, H.P. (1993) Zytotoxische Aktivität neuer cyclischer Lactone im "Hefe-Test". Pharmazie *48*:553–554. Chem. Abst. 119 (1993) 221 450.

Jain, A.K., Vargas, R., Grünwald, J., Kirby, G.S., and McMahon, F.G. (1992) Lipid effects of Kwai (garlic tablets): a controlled study. Clin. Pharmacol. Ther. *51*:164.

Jain, A.K., Vargas, R., Gotzkowsky, S., and McMahon, F.G. (1993) Can garlic reduce levels of serum lipids? A controlled clinical study. Am. J. Med. *94*:632–635.

Jain, M.K., Scanzello, C., and Apitz-Castro, R. (1988) Wirkung des Knoblauchs: Wahrheit und Dichtung. Molekulare Grundlagen des überlieferten Brauchtums. Chemie Uns Zeit *22*:193–200. Chem. Abst. 110 (1989) 82 324.

Jain, M.K. and Apitz-Castro, R. (1987) Garlic: molecular basis of the putative "vampire-repellant" action and on matters related to heart and blood. Trends Biochem. Sci. *12*:252–254. Chem. Abst. 107 (1987) 108 682.

Jain, M.K. and Apitz-Castro, R. (1993) Garlic: a product of spilled ambrosia. Curr. Sci. *65*:148–156. Chem. Abst. 119 (1993) 194 937.

Jain, P.K., Chandorkar, A.G., Bulakh, P.M., Reddy, B.V., Ranade, S.M., and Mathur, V.P. (1977) Observations on effect of *Allium sativum* on some haematological values in rabbits and human volunteers. J. Shivaji Univ. (Science) *17*:121–123.

Jain, R.C. and Andleigh, H.S. (1969) Onion and blood fibrinolytic activity. Br. Med. J. *1*:514.

Jain, R.C. (1971a) Effect of onion on serum cholesterol, lipoproteins, and fibrinolytic activity of blood in alimentary lipaemia. J. Assoc. Physicians India *19*:305–310.

Jain, R.C. (1971b) Study of effects of onion feeding on fibrinolytic activity of blood. J. Assoc. Physicians India *19*:301–303.

Jain, R.C. (1971c) Effect of butter fat and onion on coagulability of blood. Indian J. Med. Sci. *25*:598–600.

Jain, R.C. and Sachdev, K.N. (1971) A note on hypoglycemic action of onion in diabetics. Curr. Med. Pract. *15*:901–902.

Jain, R.C., Vyas, C.R., and Mahatma, O.P. (1973) Hypoglycemic action of onion and garlic. Lancet *2*:1491.

Jain, R.C. and Vyas, C.R. (1974) Hypoglycaemia action of onion on rabbits. Br. Med. J. *2*:730. Chem. Abst. 82 (1975) 197.

Jain, R.C. (1975a) Onion and garlic in experimental atherosclerosis. Lancet *1*:1240. Chem. Abst. 83 (1975) 56 469.

Jain, R.C. (1975b) Onion and garlic in experimental cholesterol atherosclerosis in rabbits: I. Effect on serum lipids and development of atherosclerosis. Artery [Fulton, Mich.] *1*:115–125.

Jain, R.C. and Vyas, C.R. (1975a) Garlic in alloxan-induced diabetic rabbits. Am. J. Clin. Nutr. *28*:684–685. Chem. Abst. 83 (1975) 130 435.

Jain, R.C. and Vyas, C.R. (1975b) Effect of onion (*Allium cepa*) and garlic (*Allium sativum*) on experimentally induced hypercholesterolemia in rabbits. Artery [Fulton, Mich.] *1*:363.

Jain, R.C. (1976) Onion and garlic in experimental cholesterol-induced atherosclerosis. Indian J. Med. Res. *74*:1509–1515.

Jain, R.C. and Konar, D.B. (1976a) Garlic oil in experimental atherosclerosis. Lancet *1*:918. Chem. Abst. 83 (1975) 56 469.

Jain, R.C. and Konar, D.B. (1976b) Onion and garlic in experimental cholesterol atherosclerosis in rabbits: effect on serum proteins and development of atherosclerosis. Artery [Fulton, Mich.] *2*:531–539.

Jain, R.C. (1977) Effect of garlic on serum lipids, coagulability, and fibrinolytic activity of blood. Am. J. Clin. Nutr. *30*:1380–1381. Chem. Abst. 87 (1977) 199 738.

Jain, R.C. and Konar, D.B. (1977) Blood sugar lowering activity of garlic (*Allium sativum* Linn.). Medikon *6(3)*:15–18. Chem. Abst. 87 (1977) 116 795.

Jain, R.C. and Vyas, C.R. (1977) Onion and garlic in atherosclerotic heart disease. Medikon *6(5)*:12–18. Chem. Abst. 87 (1977) 116 795.

Jain, R.C. (1978) Effect of alcoholic extract of garlic in atherosclerosis. Am. J. Clin. Nutr. *31*:1982–1983. Chem. Abst. 90 (1979) 21 259.

Jain, R.C. and Konar, D.B. (1978) Effect of garlic oil in experimental cholesterol atherosclerosis. Atherosclerosis 29:125–129. Chem. Abst. 89 (1978) 70 826.

Jain, R.C. (1993) Antitubercular activity of garlic oil. Indian Drugs 30:73–75. Chem. Abst. 119 (1993) 329.

Jamaluddin, M.P., Krishnan, L.K., and Thomas, A. (1988) Ajoene inhibition of platelet aggregation: possible mediation by a hemoprotein. Biochem. Biophys. Res. Commun. 153:479–486.

Jang, J.J., Cho, K.J., Lee, Y.S., and Bae, J.H. (1991) Modifying responses of allyl sulfide, indole-3-carbinol, and germanium in a rat multi-organ carcinogenesis model. Carcinogenesis 12:691–695.

Jankov, S.J. (1961) Freie Aminosäuren in Gemüsesäften. Confructa 1:63–64. Chem. Abst. 55 (1961) 27 692.

Janot, M.M. and Laurin, J. (1930) Action hypoglycémiante des bulbes de Allium cepa L. C. R. Hebd. Séances Acad. Sci. 191:1098–1100.

Jansen, H., Müller, B., and Knobloch, K. (1987) Allicin characterization and its determination by HPLC. Planta Med. 53:559–562.

Jansen, H., Müller, B., and Knobloch, K. (1989) Characterization of an alliin lyase preparation from garlic (Allium sativum). Planta Med. 55:434–439. Chem. Abst. 112 (1990) 18 059.

Jastrow, M. (1914) The medicine of the Babylonians and Assyrians. Proc. Roy. Soc. Med., Sect. Hist. Med., 109–176.

Jezowa, L., Rafinski, T., and Wrocinski, T. (1966) Investigations on the antibiotic activity of Allium sativum L. Herba Pol. 12:3–13 (Polish).

Jirousek, L. and Jirsak, J. (1956) Antifungal properties of organic polysulphides from cabbage. Naturwissenschaften 43:375–376.

Jirovetz, L., Jäger, W., Koch, H.P., and Remberg, G. (1992a) Investigations of volatile constituents of the essential oil of Egyptian garlic (Allium sativum L.) by means of GC-MS and GC-FTIR. Z. Lebensm. Unters. Forsch. 194:363–365. Chem. Abst. 117 (1992) 89 146.

Jirovetz, L., Koch, H.P., Jäger, W., and Remberg, G. (1992b) Investigations of German onion oil by GC-FID, GC-MS, and GC-FTIR. Pharmazie 47:455–456. Chem. Abst. 117 (1992) 157 495.

Jirovetz, L., Ecker, G., Jäger, W., and Heiss, T. (1993) Untersuchungen des Gehalts verschiedener Zwiebelchargen an Blei, Cadmium, Quecksilber und Selen mittels Atomabsorptions-Spektroskopie. Ernährung/Nutrition 17:265–266.

Joachim, H. (1890) Papyros Ebers. Das älteste Buch über Heilkunde. Aus dem Aegyptischen zum erstenmal vollständig übersetzt. Georg Reimer, Berlin.

Joardar, M. (1988) Cytotoxicity of certain environmental agents to plant systems. J. Indian Bot. Soc. 67:183–185. Chem. Abst. 113 (1990) 54 069.

Jocelyn, P.C. (1987) Chemical reduction of disulfides. Methods in Enzymology 143:246–256.

Johnson, M.G. and Vaughn, R.H. (1969) Death of Salmonella typhimurium and Escherichia coli in the presence of freshly reconstituted dehydrated garlic and onion. Appl. Microbiol. 17:903–905.

Jood, S., Kapoor, A.C., and Singh, R. (1993) Evaluation of some plant products against Trogoderma granarium Everts in stored maize and their effects on nutritional composition and organoleptic characteristics of kernels. J. Agric. Food Chem. 41:1644–1648. Chem. Abst. 119 (1993) 202 180.

Joseph, P.K., Rao, K.R., and Sundaresh, C.S. (1989) Toxic effects of garlic extract and garlic oil in rats. Indian J. Exp. Biol. 27:977–979.

Joseph, P.K. and Sundaresh, C.S. (1989) Serum urea and enzymes in rat on intragastric administration of garlic oil and garlic extracts. Curr. Sci. 58:1409–1410. Chem. Abst. 113 (1990) 96 298.

Joshi, D.J., Dikshit, R.K., and Mansuri, S.M. (1987) Gastrointestinal actions of garlic oil. Phytother. Res. 1:140–141.

Joslyn, M.A. and Peterson, R.G. (1958) Reddening of white onion bulb purees. J. Agric. Food Chem. 6:754–765. Chem. Abst. 53 (1959) 13443.

Joslyn, M.A. and Peterson, R.G. (1960) Reddening of white onion tissue. J. Agric. Food Chem 15:72–76. Chem. Abst. 55 (1961) 3868.

Joslyn, M.A. and Sano, T. (1956) The formation and decomposition of green pigment in crushed garlic tissue. Food Res. 21:170–183. Chem. Abst. 50 (1956) 7344.

Jummel, F. (1940) Über die Verwendung von Allium sativum in der Paradentosebehandlung. Zahnärztl. Rundsch., 1682–1685.

Jung, E.M., Jung, F., Mrowietz, C., Kiesewetter, H., Pindur, G., and Wenzel, E. (1991) Influence of garlic powder on cutaneous microcirculation: a randomized, placebo-controlled, double-blind, crossover study in apparently healthy subjects. Arzneim. Forsch. 41:626–630.

Jung, F., Kiesewetter, H., Mrowietz, C., Pindur, G., Heiden, M., Miyashita, C., and Wenzel, E. (1989a) Akutwirkung Eines Zusammengesetzten Knoblauchpräparates auf die Fliebfähigkeit des Blutes. Z. Phytother. 10:87–91.

Jung, F., Wolf, S., Kiesewetter, H., Mrowietz, C., Pindur, G., Heiden, M., Miyashita, C., Wenzel, E., and Reim, M. (1989b) Wirkung von Knoblauch auf die Fliebfähigkeit des Blutes. Ergebnisse Placebokontrollierter Pilotstudien an gesunden Probanden. Natur-& Ganzheitsmed. 190–196.

Jung, F., Jung, E.M., Mrowietz, C., Kiesewetter, H., and Wenzel, E. (1990a) Influence of garlic powder on cutaneous microcirculation: a randomised, placebo-controlled, double-blind, crossover study in apparently healthy subjects. Br. J. Clin. Pract. 44:30–35.

Jung, F., Jung, E.M., Pindur, G., and Kiesewetter, H. (1990b) Effect of different garlic preparations on the fluidity of blood, fibrinolytic activity, and peripheral microcirculation in comparison with placebo. Planta Med. 56:668.

Jung, F. and Kiesewetter, H. (1991) Einfluß einer Fettbelastung auf Plasmalipide und Kapilläre Hautdurchblutung unter Knoblauch. Med. Welt 42:14–17.

Jung, K.Y., Do, J.C., and Son, K.H. (1993) The structures of two diosgenin glycosides isolated from the subterranean parts of Allium fistulosum. Hanguk Yongyang Siklyong Hakhoechi (J. Korean Soc. Food Nutr.) 22:313–316. Chem. Abst. 121 (1994) 31 075.

Jungmayr, P. (1994) Knoblauch bei peripherer arterieller Verschlußkrankheit? Med. Monatsschr. Pharm. 17:317–318.

Juvenal. (1858) Die Satiren des D. Iunius Iuvenalis. Lateinischer Text mit metrischer Übersetzung und Erläuterungen. Wilhelm Engelmann, Leipzig.

Kabelik, J. (1970) Antimikrobielle Eigenschaften des Knoblauchs. Pharmazie 25:266–270. Chem. Abst. 73 (1970) 117 195.

Kabelik, J. and Hejtmankova-Uhrova, N. (1968) The antifungal and antibacterial effects of certain drugs and other substances. Vet. Med. [Prague] 13: 295–303. Chem. Abst. 70 (1969) 36 264 (Czech).

Kade, F. and Miller, W. (1993) Standardised garlic-ginkgo combination product improves well-being: a placebo controlled double blind study. Eur. J. Clin. Res. 4:49–55.

Kagawa, K., Matsutaka, H., Yamaguchi, Y., and Fukuhama, C. (1986) Garlic extract inhibits the enhanced peroxidation and production of lipids in carbon tetrachloride–induced liver injury. Jpn. J. Pharmacol. 42:19–26. Chem. Abst. 105 (1986) 147 768.

Kajita, H. (1957) Nachweis von Suchtmitteln im Harn. III. Über Kuppelung der Bestandteile von Knoblauch (bsd. des Allicins) mit Phenyl-methyl-propylamin. Seikagaku [Tokyo] (Biochemistry) 29:236–242. Chem. Abst. 55 (1961) 5631.

Kajiyama, G. (1982) Clinical studies of Kyolepin. Jpn. J. Clin. Rep. 16:1515.

Kaku, H., Goldstein, I.J., Van Damme, E.J.M., and Peumans, W.J. (1992) New mannose-specific lectins from garlic (Allium sativum) and ramsons (Allium ursinum) bulbs. Carbohydr. Res. 229:347–353. Chem. Abst. 117 (1992) 108 144.

Kallio, H., Tuomola, M., Pessala, R., and Vilkki, J. (1990) Headspace GC analysis of volatile sulfur and carbonyl compounds in chive and onion. 6th Weurman Symposium on Flavour Science & Technology, 57–60. Chem. Abst. 114 (1991) 205 700.

Kalra, C.L. (1987) Harvesting, handling, storage, chemistry, pharmacological properties, and technology of garlic (Allium sativum L.): a review. Indian Food Packer 41:56–80. Chem. Abst. 107 (1987) 174 452.

Kamanna, V.S. and Chandrasekhara, N. (1980) Fatty acid composition of garlic (Allium sativum Linnaeus) lipids. J. Am. Oil Chem. Soc. 57:175–176. Chem. Abst. 93 (1980) 68 996.

Kamanna, V.S. and Chandrasekhara, N. (1982) Effect of garlic (Allium sativum Linn.) on serum lipoproteins and lipoprotein cholesterol levels in albino rats rendered hypercholesteremic by feeding cholesterol. Lipids 17:483–488.

Kamanna, V.S. and Chandrasekhara, N. (1984) Hypocholesteremic activity of different fractions of garlic. Indian J. Med. Res. 79:580–583. Chem. Abst. 101 (1984) 37 549.

Kamanna, V.S. and Chandrasekhara, N. (1986) Lipid composition of garlic. Fette Seifen Anstrichm. 88:136–139. Chem. Abst. 104 (1986) 223 869.

Kamenetskaya, R.P. (1952) On the epidemiological efficacy of garlic's phytoncides in bacterial dysentery. Voen. Med. (Military Med.) 10:11–15 (Russian).

Kameoka, H., Iida, H., Hashimoto, S., and Miyazawa, M. (1984) Sulphides and furanones from steam volatile oils of Allium fistulosum and Allium chinense. Phytochemistry 23:155–158. Chem. Abst. 100 (1984) 206 475.

Kametani, T., Fukumoto, K., and Umezawa, O. (1959) Studies on anticancer agents. I. Synthesis of various alkyl thiosulfinates and their tumor-inhibiting effect. Yakugaku Kenkyu (Jap. J. Pharm. Chem.) 31:60–74. Chem. Abst. 54 (1960) 11 018 (Japanese).

Kandil, O., Abdullah, T., Tabuni, A.M., and Elkadi, A. (1988) Potential role of Allium sativum in natural cytotoxicity. Arch. AIDS Res. 1:230–231.

Kandil, O.M., Abdullah, T.H., and Elkadi, A. (1987) Garlic and the immune system in humans: its effect on natural killer cells. Fed. Proc. 46:441.

Kandziora, J. (1988a) Antihypertensive Wirksamkeit und Verträglichkeit eines Knoblauch-Präparates. Ärztl. Forsch. 35:1–8.

Kandziora, J. (1988b) Blutdruck- und lipidsenkende Wirkung eines Knoblauch-Präparates in Kombination mit einem Diuretikum. Ärztl. Forsch. 35:3–8.

Kaneta, M., Hikichi, H., Endo, S., and Sugiyama, N. (1980) Identification of flavones in thirteen Liliaceae species. Agric. Biol. Chem. 44: 1405–1406. Chem. Abst. 93 (1980) 110 584.

Kanezawa, A., Nakagawa, S., Sumiyoshi, H., Masamoto, K., Harada, H., Nakagami, S., Date, S., Yokota, A., Nishikawa, M., and Fuwa, T. (1984) General toxicity test of garlic extract preparation contained vitamins (Kyoleopin). Oyo Yakuri (Pharmacometrics) 27:909–929. Chem. Abst. 101 (1984) 163 493 (Japanese).

Kanngiesser, G. (1911) Arch. Gesch. Naturw. Techn. 111:81.

Kapferer, R. and Sticker, G. (1934) Die Werke des Hippokrates. Die hippokratische Schriftensammlung in neuer deutscher Übersetzung. Hippokrates Verlag, Stuttgart.

Kappenberg, F.J. and Glasl, H. (1990) Quantitative DC zur Standardisierung von Allium sativum L. Pharm. Z. Wiss. 3:189–193. Chem. Abst. 114 (1991) 205 619.

Karagiannis, C.S. and Pappelis, A.J. (1993) Ethylene is a selective ribosomal cistron regulator in Allium cepa epidermal cells. Mech. Ageing Dev. 72: 199–211. Chem. Abst. 121 (1994) 31 190.

Karagiannis, C.S. and Pappelis, A.J. (1994) Effect of ethylene on selective ribosomal cistron regulation in quiescent and senescent onion leaf base tissue. Mech. Ageing Dev. 75:141–149. Chem. Abst. 122 (1995) 51 524.

Karaioannoglou, P.G., Mantis, A.J., and Panetsos, A.G. (1977) The effect of garlic extract on lactic acid bacteria (Lactobacillus plantarum) in culture media. Lebensm. Wiss. Technol. 10:148–150.

Kararah, M.A., Barakat, F.M., Mikhail, M.S., and Fouly, H.M. (1985) Pathophysiology in garlic cloves inoculated with Bacillus subtilis, Bacillus pumilus, and Erwinia carotovora. Egypt. J. Phytopathol. 17:131–140. Chem. Abst. 107 (1987) 57 673.

Karawya, M.S., Abdel Wahab, S.M., El-Olemy, M.M., and Farrag, N.M. (1984) Diphenylamine, an antihyperglycemic agent from onion and tea. J. Nat. Prod. 47:775–780. Chem. Abst. 101 (1984) 226 916.

Kardashev, M.V., Berkovich, M.S., and Shvartsberg, A.M. (1952) Treatment of angina with phytoncides of garlic. Voen. Med. (Military Med.), 74–75 (Russian).

Kasahara, I. (1970) Sake from garlic. Japan. Patent 70 23 599. Chem. Abst. 74 (1971) 75 212.

Kasai, T., Nishitoba, T., Shiroshita, Y., and Sakamura, S. (1984) Several 4-substituted glutamic acid derivatives and small peptides in some Liliaceae plants. Agric. Biol. Chem. 48:2271–2278. Chem. Abst. 101 (1984) 226 885.

Kasai, T. and Kiriyama, S. (1987) Acidic amino acid fraction of Allium bakeri (Rakkyo) bulbs. Nippon Nogei Kagaku Kaishi 61:1289–1291. Chem. Abst. 108 (1988) 52 848.

Kasakowa, A.N. (1953a) Experimentelle Untersuchung der Wirksamkeit von trockenem Knoblauch bei der Behandlung eiternder Wunden. Zh. Mikrobiol. Epidemiol. Immunobiol., 18–19. Chem. Zbl. 1954, II, 10 520 (Russian).

Kasakowa, A.N. (1953b) Weitere Untersuchungen der bakteriellen Eigenschaften von trockenem Knoblauch. Zh. Mikrobiol. Epidemiol. Immunobiol., 17–18. Chem. Zbl. 1954, II, 10 505 (Russian).

Kasuga, S., Kanesawa, A., and Nakagawa, S. (1988) Extraction of ajoene from garlic for treatment of liver diseases. Japan. Patent 88 08 328. Chem. Abst. 108 (1988) 210 184.

Kathe, J. (1905) Das ätherische Öl im Knoblauch, ein neues, angeblich antituberkulöses Specifikum. Dissertation, University of Halle-Wittenberg (Abstract).

Katoch, K. and Sogani, R.K. (1979) Risk factors in coronary artery disease. Rajasthan Med. J. 18:195–198.

Kaul, P.L. and Prasad, M.C. (1990) Hypocholesterolemic and antiatherosclerotic effects of garlic (*Allium sativum* L.) in goats: an experimental study. Indian Vet. J. 67:1112–1115.

Kaur, J., Srivastava, V.K., Srivastava, R.K., Mehrotra, M.L., and Prasad, D.N. (1982) Garlic oil in experimental atherosclerosis. Indian J. Pharmacol. 14:125.

Kawakishi, S. (1991) Inhibitors of platelet aggregation in food plants. Gendai Igaku 39:337–342. Chem. Abst. 116 (1992) 207 202 (Japanese).

Kawakishi, S. and Morimitsu, Y. (1994) Sulfur chemistry of onions and inhibitory factors of the arachidonic acid cascade. ACS Symp. Ser. I., Food Phytochemicals for Cancer Prevention 546: 120–127. Chem. Abst. 120 (1994) 75 736.

Kawasaki, C. (1963) Modified thiamine compounds. Vitamins & Hormones 21:69–111.

Kawashima, H., Ochiai, Y., and Shuzenji, H. (1986) Anti-fatigue effect of aged garlic extract in athletic club students. Clin. Rep. [Japan] 20:111–127.

Kawashima, K., Mimaki, Y., and Sashida, Y. (1991a) Steroidal saponins from Allium giganteum and A. aflatunense. Phytochemistry 30:3063–3067. Chem. Abst. 115 (1991) 252 122.

Kawashima, K., Mimaki, Y., and Sashida, Y. (1991b) Schubertosides A–D, new (22S)-hydroxycholestane glycosides from Allium schubertii. Chem. Pharm. Bull. 39:2761–2763. Chem. Abst. 116 (1992) 170 124.

Kawashima, K., Mimaki, Y., and Sashida, Y. (1993) Steroidal saponins from the bulbs of Allium schubertii. Phytochemistry 32:1267–1272. Chem. Abst. 118 (1993) 251 485.

Kawashima, Z. (1986) Manufacture of odorless garlic powder. Japan. Patent 86 91 128. Chem. Abst. 105 (1986) 113 943.

Kawecki, Z. and Krynska, W. (1968) Growth dynamics of lowland garlic during the growing season and the respective changes in the levels of carbohydrates. Biul. Warczywniczy, 365–375. Chem. Abst. 72 (1970) 75 729.

Kazaryan, R.A., Kocherginskaya, S.A., and Goryachenkova, E.V. (1979) Alliinase interaction with inhibitors. Bioorg. Khim. (Bioorgan. Chem.) 5:1691–1699. Chem. Abst. 92 (1980) 54 070 (Russian). English translation: Sov. J. Bioorg. Chem. 1980 5:1257–1264.

Keck, K., Frisch-Niggemeyer, W., Ascher, D., and Hoffmann-Ostenhof, O. (1951) Inhaltsstoffe des Knoblauchs und ihre Wirkungen. I. Über Substanzen, welche den Austritt des Chromatins aus Zellkernen im Wurzelmeristem von Allium-Arten bewirken (Kurze Mitteilung). Monatsh. Chem. 82:755–758. Chem. Abst. 46 (1952) 3612.

Keck, K. and Hoffmann-Ostenhof, O. (1951) Pflanzliche Stoffwechselprodukte als Mitosegifte. 1. Mitosehemmende und -störende Substanzen in wäbrigen Auszügen aus Allium cepa (Speisezwiebel). Monatsh. Chem. 82:559–562. Chem. Abst. 45 (1951) 9615.

Kedzia, W., Lutomski, J., Muszynski, Z., and Kedzia, B. (1969) A microbial determination of the antibacterial activity of dragées with lyophilized garlic. Herba Pol. 15:222–226. Chem. Abst. 73 (1970) 28 968 (Polish).

Kehr, A.E. and Schaeffer, G.W. (1976) Tissue culture and differentiation of garlic. Hort. Sci. *11*:422–423.

Kelginbaev, A.N., Gorovits, M.B., Khamidkhodzhaev, S.A., and Abubakirov, N.K. (1973) Steroid saponins and sapogenins of Allium. V. Neoapigenin from Allium giganteum. Khim. Prir. Soedin. (Chem. Nat. Compd.) *9*:438 (Russian; English cover-to-cover translation, p. 416).

Kelginbaev, A.N., Gorovits, M.B., and Abubakirov, N.K. (1975) Steroid saponins and sapogenins of Allium. VIII. Structure of gantogenin. Khim. Prir. Soedin. (Chem. Nat. Compd.) *11*:521–522. Chem. Abst. 84 (1976) 74 513 (Russian; English cover-to-cover translation, pp. 546–547).

Kelginbaev, A.N., Gorovits, M.B., Gorovits, T.T., and Abubakirov, N.K. (1976) Allium steroidal saponins and sapogenins. IX. Structure of aginoside. Khim. Prir. Soedin. (Chem. Nat. Compd.) *12*:480–486. Chem. Abst. 86 (1977) 43 972 (Russian).

Keller, A. (1988) Die Abortiva in der römischen Kaiserzeit. Deutscher Apotheker Verlag, Stuttgart.

Kempski, H.W. (1967) Zur kausalen Therapie chronischer Helminthen-Bronchitis. Med. Klin. *62*:259–260.

Kendler, B.S. (1987) Garlic (*Allium sativum*) and onion (Allium cepa): A review of their relationship to cardiovascular disease. Prev. Med. *16*:670–685.

Kenzelmann, R. and Kade, F. (1993) Limitation of the deterioration of lipid parameters by a standardized garlic-ginkgo combination product. Arzneim. Forsch. *43*:978–981.

Kerekes, M., Nicoara, D., and Csögör, S. (1974) Effect of *Allium sativum* L. on serum cholesterol of healthy individuals. In: Atherosclerosis III: Proceedings of the Third International Symposium. G. Schettler and A. Weizel, eds. Springer-Verlag, Berlin, pp. 912.

Kerekes, M.F. and Feszt, T. (1975) Effect of *Allium sativum* L. on serum cholesterol. Artery [Fulton, Mich.] *1*:325–326.

Kereselidze, E.V., Pkheidze, T.A., and Kemertelidze, E.P. (1970) Diosgenin from Allium albidum. Khim. Prir. Soedin. (Chem. Nat. Compd.) *6*:378. Chem. Abst. 73 (1970) 117 181 (Russian).

Keys, A. (1980) Wine, garlic, and CHD in seven countries. Lancet *1*:145–146.

Khaletskii, A.M. and Reznik, M.B. (1957) Investigation of the volatile substances from garlic. Zh. Obshch. Khim. (J. Gen. Chem.) *27*:1727–1730. Chem. Abst. 52 (1958) 1317 (Russian).

Khan, H.H., Bibi, N., and Zia, G.M. (1985) Trace metal contents of common spices. Pak. J. Sci. Ind. Res. *28*:234–237.

Khan, K.I., Khan, F.Z., and Nazar, S. (1985) The antimicrobial activity of *Allium sativum* (garlic), Allium cepa (onion), and Raphanus sativus (radish). J. Pharm. Punjab Univ. Lahore *6*:59–71.

Khan, M.R. and Mahmood, N. (1984) A study of lipase activity of garlic. J. Nat. Sci. Math. *24*:75–82. Chem. Abst. 102 (1985) 218 411.

Khanna, P., Sharma, A., Kaushik, P., and Chaturvedi, P. (1989) Trigonellin from three plant sps. in vivo and in vitro cell cultures. Indian Drugs *26*:334–336. Chem. Abst. 111 (1989) 171 164.

Kharkina, G.A. (1951) Relative activity of the phytoncides from garlic on phytopathogenic bacteria. Mikrobiologiya (Microbiology) *20*:434–437. Chem. Abst. 46 (1952) 10 299 (Russian).

Khodzhaeva, M.A., Ismailov, Z.F., Kondratenko, E.S., and Shashkov, A.S. (1982) Allium carbohydrates. II. New type of glucofructans from *Allium sativum*. Khim. Prir. Soedin. (Chem. Nat. Compd.), 23–28. Chem. Abst. 96 (1982) 196 527 (Russian).

Khodzhaeva, M.A. and Ismailov, Z.F. (1979) Allium carbohydrates. I. Isolation and characterization of the polysaccharides. Khim. Prir. Soedin. (Chem. Nat. Compd.), 137–142. Chem. Abst. 91 (1979) 171 673 (Russian).

Khodzhaeva, M.A. and Kondratenko, E.S. (1983) Allium carbohydrates. III. Characteristics of the carbohydrates of the genus Allium. Khim. Prir. Soedin. (Chem. Nat. Compd.), 228–229 (Russian).

Khodzhaeva, M.A. and Kondratenko, E.S. (1984a) Carbohydrates of Allium. V. Glucofructans from Allium cepa. Khim. Prir. Soedin. (Chem. Nat. Compd.), 105–106. Chem. Abst. 100 (1984) 171 604 (Russian).

Khodzhaeva, M.A. and Kondratenko, E.S. (1984b) Carbohydrates of Allium. VI. Glucofructanes from Allium cepa. Khim. Prir. Soedin. (Chem. Nat. Compd.), 383–384. Chem. Abst. 102 (1985) 21 249 (Russian).

Khodzhaeva, M.A. and Kondratenko, E.S. (1985) Carbohydrates of Allium. VIII. Polysaccharides of Allium coeruleum. Khim. Prir. Soedin. (Chem. Nat. Compd.), 17–21. Chem. Abst. 102 (1985) 201 171 (Russian).

Khristulas, F.S., Gorovits, M.B., Luchanskaya, V.N., and Abubakirov, N.K. (1970) New steroidal sapogenin from Allium giganteum. Khim. Prir. Soedin. (Chem. Nat. Compd.) *6*:489. Chem. Abst. 74 (1972) 10 356 (Russian).

Khristulas, F.S., Gorovits, M.B., and Abubakirov, N.K. (1974) Steroid saponins and sapogenins of Allium. VI. 3-O-b-D-glucopyranoside of karatavigenin B. Khim. Prir. Soedin. (Chem. Nat. Compd.), 530–531. Chem. Abst. 82 (1975) 82 955 (Russian).

Kice, J.L. and Rogers, T.E. (1974) Mechanisms of the alkaline hydrolysis of aryl thiolsulfinates and thiolsulfonates. J. Am. Chem. Soc. *96*:8009–8015. Chem. Abst. 82 (1975) 15 893.

Kiesewetter, H. (1990) Knoblauch: Naturheilmittel von Tradition und Aktualität. Kassenarzt *12*:46–51.

Kiesewetter, H., Jung, F., Mrowietz, C., Pindur, G., Heiden, M., and Wenzel, E. (1990) Effects of garlic on blood fluidity and fibrinolytic activity: a randomised, placebo-controlled, double-blind study. Br. J. Clin. Pract. *44*:24–29.

Kiesewetter, H., Jung, E.M., Jung, F., and Wenzel, E. (1991a) Steigerung der peripheren Durchblutung mit Knoblauch. Magazin Forschung Univ. Saarland., 40–44.

Kiesewetter, H., Jung, F., Pindur, G., Jung, E.M., Mrowietz, C., and Wenzel, E. (1991b) Effect of garlic on thrombocyte aggregation, microcirculation, and other risk factors. Int. J. Clin. Pharmacol. Ther. Toxicol. 29:151–155.

Kiesewetter, H., Jung, F., Jung, E.M., Blume, J., Mrowietz, C., Birk, A., Koscielny, J., and Wenzel, E. (1993a) Effects of garlic coated tablets in peripheral arterial occlusive disease. Clin. Investigator 71:383–386.

Kiesewetter, H., Jung, F., Jung, E.M., Mrowietz, C., Koscielny, J., and Wenzel, E. (1993b) Effect of garlic on platelet aggregation in patients with increased risk of juvenile ischaemic attack. Eur. J. Clin. Pharmacol. 45:333–336.

Kiesewetter, H., Jung, F., Mrowietz, C., and Wenzel, E. (1993c) Wirkung von Knoblauch (Allium sativum L.), insbesondere rheologische und hämostaseologische Effekte. Hämostaseologie 13:43–52.

Kiesewetter, H. and Jung, F. (1991) Beeinflubt Knoblauch die Atherosklerose? Med. Welt 42:21–23.

Kiesewetter, H. and Jung, F. (1992) Kann Knoblauch die Atherosklerose verhindern? Argumente & Fakten, 6–10.

Kihara, Y. (1929) Kohlenhydrate in der Zwiebel von Allium scorodoprasum L. I. Proc. Imperial Acad. Tokyo 5:348–350. Chem. Zbl. 1930, I, 696 (Japanese).

Kihara, Y. (1936) Carbohydrates in the bulbs of Allium. VII. Derivatives of scorodose. J. Agric. Chem. Soc. Jpn. 12:1044–1048. Chem. Abst. 31 (1937) 3013 (Japanese).

Kihara, Y. (1939) Scorodose. J. Agric. Chem. Soc. Jpn. 15:348–352.

Kikkoman Shoyu Co. Ltd. (1982) Deodorization and bleaching of garlic. Japan. Patent 82 03 341. Chem. Abst. 96 (1982) 179 806.

Kim, B.D., Song, D.B., and Cha, C.W. (1987a) A study on the effect of garlic, 2,3-dimercaptosuccinic acid and N-acetyl-DL-penicillamine on excretion of cadmium in rat. Koryo Taehakkyo Uikwa Taehak Nonmunjip (Korea Univ. Med. J.) 24:223–236. Chem. Abst. 107 (1987) 192 526 (Korean).

Kim, D.S., Ahn, B.W., Yeum, D.M., Lee, D.H., Kim, S.B., and Park, Y.H. (1987b) Degradation of carcinogenic nitrosamine formation factor by natural food components: nitrite-scavenging effects of vegetable extracts. Hanguk Susan Hakhoechi (Bull. Korean Fish. Soc.) 20:463–468. Chem. Abst. 120 (1994) 5473 (Korean).

Kim, D.Y., Rhee, C.O., and Kim, Y.B. (1981) Characteristics of polyphenol oxidase from garlic (Allium sativum L.). Hanguk Nonghwahak Hoechi (J. Korean Agric. Chem. Soc.) 24:167–173. Chem. Abst. 97 (1982) 2671.

Kim, H.K., Jo, K.S., Kwon, D.Y., and Park, M.H. (1992) Effects of drying temperature and sulfiting on the qualities of dried garlic slices. Hanguk Nonghwahak Hoechi (J. Korean Agric. Chem. Soc.) 35:6–9. Chem. Abst. 117 (1992) 89 077.

Kim, M.R., Yun, J.H., and Sok, D.E. (1994) Correlation between pungency and allicin content of pickled garlic during aging. Hanguk Yongyang Siklyong Hakhoechi (J. Korean Soc. Food Nutrit.) 23:805–810. Chem. Abst. 122 (1995) 131 529.

Kim, S.H., Kim, J.O., Lee, S.H., Park, K.Y., Park, H.J., and Chung, H.Y. (1991) Antimutagenic compounds identified from the chloroform fraction of garlic (Allium sativum). Hanguk Yongyang Siklyong Hakhoechi (J. Korean Soc. Food Nutr.) 20:253–259. Chem. Abst. 116 (1992) 67 012.

Kim, S.K., Bae, E.S., and Cha, C.W. (1984) A study on the effect of garlic on the toxicity of cadmium in rats. Koryo Taehakkyo Uikwa Taehak Nonmunjip (Korea Univ. Med. J.) 21:65–76. Chem. Abst. 101 (1984) 185 460 (Korean).

Kim, S.M., Wu, C.M., Kubota, K., and Kobayashi, A. (1995) Effect of soybean oil on garlic volatile compounds isolated by distillation. J. Agric. Food Chem. 43:449–452.

Kim, Y.S., Lee, M.Y., and Park, Y.M. (1992) Purification and characterization of chitinase from green onion. Hanguk Saenghwa Hakhoechi (Korean Biochem. J.) 25:171–177. Chem. Abst. 117 (1992) 126 987.

Kim, Y.S., Park, K.S., and Kim, J.G. (1993) Purification and characterization of b-galactosidase from green onion. Korean Biochem. J. 26:602–608. Chem. Abst. 120 (1994) 100 119.

Kimm, S.W. and Park, S.C. (1982) Evidences for the existence of antimutagenic factors in edible plants. Korean J. Biochem. 14:47–59. Chem. Abst. 98 (1983) 159 324.

Kimura, Y. and Yamamoto, K. (1964) Cytological effect of chemicals on tumors: influence of crude extracts from garlic and some related species on MTK-sarcoma III. Gann 55:325–329. Chem. Abst. 63 (1965) 1089.

Kintia, P.K., Degtiaryova, L.P., Balashova, N.N., and Shvets, S.A. (1986) Sterols and steroidal glycosides of bulb onion seeds. Proc. FECS 3rd. Int. Conf. Chem. & Biotechnol. of Biol. Act. Natl. Prods., 16–21 September 1985, Sofia, Bulgaria 5:166–170.

Kintya, P.K. and Degtyareva, L.P. (1989) Steroidal glycosides of onion seeds. Structure of ceposide D. Khim. Prir. Soedin. (Chem. Nat. Compd.) 25:139–140. Chem. Abst. 11 (1989) 93 884.

Kirsch, H. (1987) Stable preparations from herbs or drugs with cyclodextrin complexes. Germany. Patent 3 609 116. Chem. Abst. 108 (1988) 173 544.

Kirsten, D. and Meister, W. (1985) Berufsbedingte Knoblauchallergie. Allergologie 8:511–512. Ref. in Ärzte-Ztg. 5 (1986) No. 15, 11.

Kisu, Y. (1985) Comparison of selenium contents in organic agricultural products with those in agricultural products on the market. Seikatsu Kagaku Kenkyusho Kenkyu Hokoku 18:17–21. Chem. Abst. 104 (1986) 128 755 (Japanese).

Kit, S.M. and Orlik, G.G. (1960) Über die blutzuckersenkende Wirkung einiger Arzneipflanzen. Vrach. Delo (Therapeutics), 617–622 (Russian).

Kitada, Y., Tamase, K., Sasaki, M., and Yamazoe, Y. (1988) Determination of alliin in garlic by high performance liquid chromatography with a fluorescence detector. Naraken Eisei Kenkyusho Nenpo *1989*:37–39. Chem. Abst. 112 (1990) 234 061 (Japanese).

Kitagawa, M. and Amano, A. (1935) Antiseptic action of garlic. Bul. Sci. Fak. Terkultura, Kjusu Imp. Univ. Fukuoka, Japan *6*:299–304. Chem. Abst. 30 (1936) 3019.

Kitagawa, M. and Hirano, S. (1937) Untersuchungen über die antiseptische Wirkung von Knoblauch. Bul. Sci. Fak. Terkultura, Kjusu Imp. Univ. Fukuoka, Japan *7*:296. Chem. Zbl. 1938, I, 1023 (Japanese).

Kitagawa, M. and Noda, I. (1939) Untersuchungen über die antiseptische Wirkung von Knoblauch. Bul. Sci. Fak. Terkultura, Kjusu Imp. Univ. Fukuoka, Japan, *8*:349. Chem. Zbl. 1941, II, 232 (Japanese).

Kitahara, S. and Fujinaga Pharm. Co. Ltd. (1977) Garlic preparations for metal detoxication. Japan. Patent 77 72 810. Chem. Abst. 87 (1977) 206 498.

Kitahara, S. (1974) A reagent for urinary glucose detection. Japan. Patent 74 120 693. Chem. Abst. 82 (1975) 121 267.

Kiviranta, J., Huovinen, K., and Hiltunen, R. (1986) Variation of flavonoids in Allium cepa. 34th Ann. Cong. Med. Plant Res. Hamburg, 22–27 September 1986.

Kiviranta, J., Huovinen, K., and Hiltunen, R. (1988) Variation of phenolic substances in onion. Acta Pharm. Fenn. *97*:67–72. Chem. Abst. 110 (1989) 56 280.

Kiviranta, J., Huovinen, K., Seppänen-Laakso, T., Hiltunen, R., Karppanen, H., and Kilpeläinen, M. (1989) Effects of onion and garlic extracts on spontaneously hypertensive rats. Phytother. Res. *3*:132–135.

Kleijnen, J., Knipschild, P., and Ter Riet, G. (1989) Garlic, onions, and cardiovascular risk factors: a review of the evidence from human experiments with emphasis on commercially available preparations. Br. J. Clin. Pharmacol. *28*:535–544.

Klein, P. and Souverein, C. (1954) Über die mikrobiologische Auswertung des Enzym-Substratsystems Alliin-Alliinase. Biochem. Z. *326*:123–131. Chem. Abst. 50 (1956) 4259.

Kloppenburg-Versteegh, J. (1937) Wenken en Raadgevingen betreffende het gebruik van Indische planten, vruchten enz. (Tips and advice concerning the use of Indian plants, fruit, etc.). s'Gravenhage, Netherlands.

Klosa, J. (1949) Über einige die Keimung von Samen und das Wachstum von Bakterien hemmende Substanzen aus Vegetabilien. II. Pharmazie *4*:574–577. Chem. Abst. 45 (1951) 1650.

Klosa, J. (1951) Hormonartige Körper in der Küchenzwiebel (Allium cepa). Seifen Fette Öle Wachse *77*:166–167. Chem. Zbl. 1952, I, 3510.

Knasmüller, S., de Martin, R., Wottawa, A., and Szakmary, A. (1986) Besitzt Knoblauch Inhaltsstoffe, die vor Krebserkrankungen schützen? Österr. Apoth. Ztg. *40*:830–832.

Knasmüller, S., de Martin, R., Domjan, G., and Szakmary, A. (1989) Studies on the antimutagenic activities of garlic extract. Environ. Mol. Mutagen. *13*:357–365. Chem. Abst. 111 (1989) 92 184.

Knobloch, K., Winkler, G., Lohmüller, E., Landshuter, J., Wiegand, D., Haupt, W., and Reith, J. (1993) Flavor precursors and their enzymic turnover in Allium species. Proc. Int. Conf. Progress Flavour Precursor Studies *1992*:175–183. Chem. Abst. 121 (1994) 78 428.

Knudsen, M.V. and Mavala, S.A. (1968) Composition for strengthening nails. U.S.A. Patent 3 382 151. Chem. Abst. 69 (1968) 12 888.

Knypl, J.S. and Janas, K.M. (1990) The stimulatory effect of 1-amino-2-phenylethyl phosphonic acid on growth and phenylalanine ammonia lyase activity in Allium cepa L. Acta Physiol. Plant. *12*:127–130. Chem. Abst. 115 (1991) 46 177.

Ko, M. (1991) Insoles containing allicin for trichophytosis control in feet. Japan. Patent 91 215 419. Chem. Abst. 115 (1991) 263 503.

Kobayashi, A., Itagaki, R., Tokitomo, Y., and Kubota, K. (1994) Changes of aroma character of irradiated onion during storage. Nippon Shokuhin Kogyo Gakkaishi *41*:682–686. Chem. Abst. 122 (1995) 30 053.

Koch, H.P. (1985a) Der Knoblauch, ein wirksames und sicheres Arzneimittel für den Handverkauf und zur Selbstmedikation. Österr. Apoth. Ztg. *39*:781–787. Chem. Abst. 103 (1985) 189 044.

Koch, H.P. (1985b) Der Knoblauch, ein wirksames und sicheres Arzneimittel für den Handverkauf und zur Selbstmedikation. Apoth. J. 16–29.

Koch, H.P. and Ritschel, W.A. (1986) Synopsis der Biopharmazie und Pharmakokinetik. Ecomed, Landsberg/Lech.

Koch, H.P. (1987) "Knoblauchpräparate": Eine dringende Klarstellung. Dtsch. Apoth. Ztg. *127*: 367–369. Chem. Abst. 106 (1987) 182 479.

Koch, H.P. (1988) "Knoblauch": Wie dosiert man richtig? Dtsch. Apoth. Ztg. *128*:408–412.

Koch, H.P. and Hahn, G. (1988) Knoblauch: Grundlagen der therapeutischen Anwendung von Allium sativum L. Urban & Schwarzenberg, Munich.

Koch, H.P., Jäger, W., Brauner, A., and Roth, S. (1988) Selen im Knoblauch und in Knoblauchpräparaten: Knoblauch als Lieferant des biologisch wichtigen Spurenelementes. Dtsch. Apoth. Ztg. *128*:993–995.

Koch, H.P. (1989) Kann man Knoblauch "geruchlos" machen? Dtsch. Apoth. Ztg. *129*:1991–1996.

Koch, H.P. and Jäger, W. (1989) Knoblauch: Allicin-Freisetzung aus frischem und getrocknetem Knoblauch und einigen daraus hergestellten Fertigarzneimitteln. Dtsch. Apoth. Ztg. *129*: 273–276. Chem. Abst. 110 (1989) 219 160.

Koch, H.P. and Jäger, W. (1990) Knoblauch-Ölmazerate: Analytische Bewertung von Knoblauchzubereitungen in öliger Lösung. Dtsch. Apoth. Ztg. *130*:2469–2474. Chem. Abst. 114 (1991) 49 678.

Koch, H.P. (1991) Analytische Bewertung von Knoblauch-Ölmazeraten. Dtsch. Apoth. Ztg. *131*:15–16.

Koch, H.P. (1992a) Schon Säuglinge lieben Knoblauch. Österr. Apoth. Ztg. *46*:821.

Koch, H.P. (1992b) "Hormonwirkungen" bei Allium-Arten: Historische Berichte und moderne wissenschaftliche Erkenntnisse. Z. Phytother. *13*:177–188.

Koch, H.P. (1992c) Epidemiologie der Knoblauchforschung: Das Auf und Ab im Spiegel der wissenschaftlichen Publikationen. Dtsch. Apoth. Ztg. *132*:2103–2106. Int. Pharm. Abst. 30 (1993) 10 095.

Koch, H.P. (1992d) Wie "sicher" ist Knoblauch? Toxische, allergische und andere unerwünschte Nebenwirkungen. Dtsch. Apoth. Ztg. *132*: 1419–1428. Chem. Abst. 118 (1993) 32 397.

Koch, H.P. (1992e) Metabolismus und Pharmakokinetik der Inhaltsstoffe des Knoblauchs: Was Wissen wir darüber? Z. Phytother. *13*:83–90.

Koch, H.P., Jäger, W., Groh, U., and Plank, G. (1992a) In vitro inhibition of adenosine deaminase by flavonoids and related compounds: new insight into the mechanism of action of flavonoids. Meth. Find. Exp. Clin. Pharmacol. *14*:413–417. Chem. Abst. 118 (1993) 160 770.

Koch, H.P., Jäger, W., Hysek, J., and Körpert, B. (1992b) Garlic and onion extracts: in vitro inhibition of adenosine deaminase. Phytother. Res. *6*:50–52.

Koch, H.P. (1993a) Garlicin: fact or fiction? The antibiotic substance from garlic (*Allium sativum* L.). Phytother. Res. *7*:278–280. Chem. Abst. 120 (1994) 73 329.

Koch, H.P. (1993b) Zehn Fragen zur Homöopathie. Dtsch. Apoth. Ztg. *133*:4700–4701.

Koch, H.P. (1993c) Method for obtaining therapeutic plant extract. Austria. Patent 397 464 B.

Koch, H.P. (1993d) Knoblauchforschung in einer Sackgasse? Z. Phytother. *14*:274.

Koch, H.P. (1993e) Saponine in Knoblauch und Küchenzwiebel. Dtsch. Apoth. Ztg. *133*:3733–3743. Chem. Abst. 121 (1994) 244 554.

Koch, H.P., Jäger, W., Groh, U., Hovie, J.E., Plank, G., Sedlak, U., and Praznik, W. (1993) Carbohydrates from garlic bulbs (*Allium sativum* L.) as inhibitors of adenosine deaminase enzyme activity. Phytother. Res. *7*:387–389. Chem. Abst. 120 (1994) 292 686.

Koch, H.P. (1994a) Method for obtaining therapeutic plant extract. European Patent 592 382 A1.

Koch, H.P. (1994b) Die Küchenzwiebel—eine zu Unrecht vernachlässigte Arzneipflanze. Pharm. Uns. Zeit *23*:333–339.

Koch, H.P., Aichinger, A., Bohne, B., and Plank, G. (1994) In vitro inhibition of adenosine deaminase by a group of steroid and triterpenoid compounds. Phytother. Res. *8*:109–111. Chem. Abst. 118 (1993) 160 770; 121 (1994) 295 743.

Koch, H.P. (1995) "Moly"—der Zauberlauch der griechischen Mythologie. Geschichte der Pharmazie *47*:34–44.

Koch, J., Berger, L., and Vieregge-Reiter, C. (1989) Allicin in Knoblauch—*Allium sativum* L.—und Knoblauchpräparaten: Gehaltsbestimmung mittels Headspace-Gaschromatographie. Planta Med. *55*:327–331. Chem. Abst. 111 (1989) 140 603.

Koczwara, M. (1949) New Polish saponin-containing plants. Publ. Pharm. Comm., Pol. Acad. Sci. *1*:65–102. Chem. Abst. 46 (1952) 10 548 (Polish).

Kodera, Y., Matsuura, H., Yoshida, S., Sumida, T., Itakura, Y., Fuwa, T., and Nishino, H. (1989) Allixin, a stress compound from garlic. Chem. Pharm. Bull. *37*:1656–1658. Chem. Abst. 112 (1990) 73 738.

Kodera, Y. (1991) Method for preparing an S-allylcysteine-containing composition. Europe. Patent 429 080 A1. Chem. Abst. 115 (1991) 120 023.

Koenigsfeld, H. and Prausnitz, C. (1913) Über Wachstumshemmung der Mäusekarzinome durch Allylderivate. Dtsch. Med. Wochenschr., 2466–2468.

Kohmura, H., Ikeda, Y., and Sakai, A. (1994) Cryopreservation of apical meristems of Japanese shallot (*Allium wakegi* A.) by vitrification and subsequent high plant regeneration. Cryo-Lett. *15*:289–298.

Kojima, R., Ohnishi, S.T., and Toyama, Y. (1994) Protection of doxorubicin-induced cardiotoxicity by aged garlic extracts. FASEB J. *8*:A605.

Kokas, E.V. and Ludany, G.V. (1933) Die Wirkung der Gewürzmittel auf die Bewegung der Darmzotten und die Glykoseresorption. Arch. Exp. Pathol. Pharmakol. *169*:140–145.

Kominato, H. (1971) Scordine from garlic. Japan. Patent 71 14 918. Chem. Abst. 75 (1971) 52 793.

Kominato, J., Nishimura, S., and Takeyama, Y. (1986) Gastrointestinal absorption acceleration of pharmaceutical amino acids by garlic extract for the treatment of hepatopathy. Japan. Patent 86 155 322. Chem. Abst. 105 (1986) 158 869.

Kominato, J. and Azuma, Y. (1993) Masking of unpleasant odor of alkyl sulfide drugs. Japan. Patent 93 43 454. Chem. Abst. 118 (1993) 240 969.

Kominato, J. and Ohira, H. (1987) Oral preparations containing useful microorganisms and scordinins, oxoamidins, or nicotinic acid derivatives as additives. Japan. Patent 62 212 324. Chem. Abst. 108 (1988) 210 208.

Kominato, K. (1953) Separation of a crystalline sulfur-containing glucoside and its phosphoric acid ester from allium plants. Japan. Patent 53 1696. Chem. Abst. 48 (1954) 3644.

Kominato, K. (1959) Prosthetic groups and coenzymes with vitamin-like activity from plants and animal tissue. Germany. Patent 1 063 334. Chem. Abst. 55 (1959) 18 025.

Kominato, K. (1960) Effective component from Allium schoenoprasum. Japan. Patent 60 2398. Chem. Abst. 54 (1960) 20 099.

Kominato, K. (1969a) Studies on biological active component in garlic (Allium scorodoprasm L. or *Allium sativum*) I. Thioglycoside. Chem. Pharm. Bull. *17*:2193–2197. Chem. Abst. 72 (1970) 24 543.

Kominato, K. (1969b) Studies on biological active component in garlic (Allium scorodoprasm L. or *Allium sativum*) II. Chemical structure of scordinin A1. Chem. Pharm. Bull. *17*:2198–2200. Chem. Abst. 72 (1970) 35 728.

Kominato, K. (1970a) Scordinines A and B from garlic. Japan. Patent 70 12 876. Chem. Abst. 73 (1970) 69 836.

Kominato, K. (1970b) Separation and purification of metabolism regulators from natural products. Japan. Patent 70 18 679. Chem. Abst. 73 (1970) 73 134.

Kominato, K. (1972a) The wonder of garlic—history of the investigations on garlic: discovery of scordinin. Sohban-sha, Tokyo.

Kominato, K. (1972b) Scordine extraction. Japan. Patent 72 29 966. Chem. Abst. 77 (1972) 130 583.

Kominato, K. (1977) Scordinin flavoring agent. Japan. Patent 77 30 587. Chem. Abst. 87 (1977) 199 483.

Kominato, K. and Kominato, M. (1965) Mercury compound of scordinin. Japan. Patent 65 26 691. Chem. Abst. 64 (1966) 9733.

Kominato, K. and Kominato, M. (1972a) Separation of scordinine A1, A2, and B from garlic. Japan. Patent 72 15 115. Chem. Abst. 77 (1972) 45 755.

Kominato, K. and Kominato, M. (1972b) 3-((3-(N-(4-Carboxy-3-hydroxyphenyl)glutaminyl-cysteinyl-S-(3-(4-S-allyl-4-thiafructofuranosyl)guanidino)-homocysteinyl)-3-(hydroxy(2-sulfoethyl)amino)phosphinyl)-1-(carboxymethyl)guanidino)methyl)-5-(2-hydroxyethyl)-4-methyl-thiazolium. Japan. Patent 72 22 229. Chem. Abst. 77 (1972) 102 254.

Kominato, K. and Kominato, Y. (1970) Thiocornin und Thiamamidincysteinphosphat sowie Verfahren zu ihrer Herstellung. Ger. Offen., 1–23. Chem. Abst. 72 (1970) 55 899.

Kominato, K. and Kominato, Y. (1973) Silkworm attractant, scordinin A, from garlic. Japan. Patent 73 87 009. Chem. Abst. 81 (1974) 562.

Kominato, Y., Takeuchi, Y., and Kominato, K. (1971) Thiamacornin, thiamamidine, and free thiamamidine from scordinin. Japan. Patent 71 29 507. Chem. Abst. 75 (1971) 143 988.

Kominato, Y., Nishimura, S., and Takeyama, K. (1976a) Studies on the biologically active component of garlic (Allium scorodoprasum L. or *Allium sativum*). I. Isolation and bacteriostatic effect of scordinin A1 and its decomposition product. Oyo Yakuri (Pharmacometrics) *11*:941–944. Chem. Abst. 88 (1978) 99 210.

Kominato, Y., Nishimura, S., and Takeyama, K. (1976b) Studies on biological active component in garlic (Allium scorodoprasm L. or *Allium sativum*): thiamine retaining effects of scordinin A, B mixture and antibacterial activities of scordinin A1's decomposition product. Oyo Yakuri (Pharmacometrics) *12*:571–577. Chem. Abst. 88 (1978) 131 391.

König, F.K. (1989) Oral garlic preparation with at least one odor-reducing component. Germany. Patent 3 731 675.

König, F.K. and Schneider, B. (1986) Knoblauch bessert Durchblutungsstörungen. Ärztl. Prax. *38*:344–345.

König, J. (1879) Chemie der menschlichen Nahrungs- und Genubmittel. Springer, Berlin.

Kononko, L.N., Yemchenko, N.L., and Vashchenko, N.V. (1984) Use of onion and garlic preserved in refrigerators with air drying by lithium chloride. Vrach. Delo (Therapeutics), 111–113 (Russian).

Kononkov, P.F. (1953) On the problem of generating seeds from garlic. Sad Ogorod [Moscow] (Garden & Truck Garden), 38–40 (Russian).

Konvicka, O. (1973) Die Ursachen der Sterilität von *Allium sativum* L. Biol. Plantarum [Prague] *15*:144–149.

Konvicka, O., Niehaus, F., and Fischbeck, G. (1978) Untersuchungen über die Ursachen der Pollensterilität bei *Allium sativum* L. Z. Pflanzenzücht. *80*:265–276.

Konvicka, O. (1983) Knoblauch: eine Gewürz- und Heilpflanze. Naturwiss. Rundsch. *36*:209–215. Chem. Abst. 99 (1983) 67 470.

Konvicka, O. (1984) Zum mitotischen Hemmungseffekt von Allium-sativum-Extrakt. Cytologia *49*:761–769.

Konvicka, O. (1989) Selenium enrichment of garlic for pharmaceutical purposes. Germany. Patent 3 737 566. Chem. Abst. 112 (1990) 84 154.

Korotkov, V.M. (1953) On the use of phytoncides in the surgery of putride abscesses. Khirurgiya [Moscow] (Surgery) *29*:88–90 (Russian).

Korotkov, W.M. (1966) Effect of garlic juice on blood pressure. Vrach. Delo (Therapeutics), 123 (Russian).

Koscielny, J., Jung, E.M., Jung, F., Mrowietz, C., Pindur, G., Kiesewetter, H., and Wenzel, E. (1991a) Wirkung verschiedener Knoblauchpräparate auf die Fließfähigkeit des Blutes: Bei Knoblauchtrockenpulver signifikante Wirkung festgestellt. Med. Welt *42*:29–31.

Koscielny, J., Jung, F., Jung, E.M., Pindur, G., Kiesewetter, H., and Wenzel, E. (1991b) Garlic and capillary blood flow: beneficial effects of Kwai on blood fluidity, fibrinolytic activity, and capillary flow. Cardiol. in Prac. (June suppl.):9–10.

Kosick, F.W. (1989) Garlic preparation with a strong odor-reducing effect. Germany. Patent 3 820 666.

Kosyan, A.M. (1985) Spectrophotometric determination of garlic organic sulfides. Khim. Farm. Zh. (Chem. Pharm. J.) *19*:1463–1465. Chem. Abst. 104 (1986) 65 203 (Russian).

Kotlinska, T., Havranek, P., Navratil, M., Gerasimova, L., Pimakhov, A., and Neikov, S. (1990) Collecting onion, garlic, and wild species of Allium in central Asia, USSR. Plant Genet. Resources Newsletter *83/84*:31–32.

Koul, A.K. and Gohil, R.N. (1970) Causes averting sexual reproduction in *Allium sativum* Linn. Cytologia 35:197–202.

Kourennoff, P.M. (1970) Arteriosclerosis: garlic treatment. In: Russian folk medicine. Allen, London, pp. 16–17.

Kourounakis, P.N. and Rekka, E.A. (1991) Effect on active oxygen species of alliin and *Allium sativum* (garlic) powder. Res. Comm. Chem. Pathol. Pharmacol. 74:249–252. Chem. Abst. 116 (1992) 40 280. Congress abstract in: Eur. J. Clin. Res. 3A (1992) 3.

Kramarenko, L.E. (1951) Some notes to the therapy of lambiasis with the phytoncides of onion and garlic. Voen. Med. (Military Med.), 44–46 (Russian).

Kramer, H. (1984) Leser-Anfrage aus der Praxis: Gibt es Knoblauch-Allergien? Allergologie 5:204–205.

Krauss, F.S. (1885) Sitte und Brauch der Südslaven. Alfred Hölder, Vienna.

Kravets, S.D., Vollerner, Y.S., Gorovits, M.B., Shashkov, A.S., and Abubakirov, N.K. (1986a) Spirostan and furostan type steroids from plants of the genus Allium. XXI. Structure of alliospiroside A and alliofuroside A from Allium cepa. Krim. Prir. Soedin. (Chem. Nat. Compd.) 22:188–196. Chem. Abst. 106 (1987) 15 707 (Russian).

Kravets, S.D., Vollerner, Y.S., Gorovits, M.B., Shashkov, A.S., and Abubakirov, N.K. (1986b) Steroids of the spirostane and furostane series from plants of the genus Allium. XXII. The structure of alliospiroside B from Allium cepa. Khim. Prir. Soedin. (Chem. Nat. Compd.) 22:589–592. Chem. Abst. 106 (1987) 135 236 (Russian).

Kravets, S.D., Vollerner, Y.S., Gorovits, M.B., and Abubakirov, N.K. (1990) Steroids of spirostane and furostane series in herbs of the genus Allium. Khim. Prir. Soedin. (Chem. Nat. Compd.) 26:429–443. Chem. Abst. 114 (1991) 98 080 (Russian).

Kreitmair, H. (1936) Pharmakologische Versuche mit einigen einheimischen Pflanzen. In: Jahresbericht über die Neuerungen auf den Gebieten der Pharmacotherapie und Pharmazie. E. Merck, ed. Darmstadt, pp. 102–110.

Kretschmer, W. (1935a) Zur Knoblauchfrage. Münch. Med. Wochenschr., 1613–1614.

Kretschmer, W. (1935b) Als Armeepathologe in Serbien und Makedonien 1916–1917. Dtsch. Med. Wochenschr. 61:1893–1896.

Kretschmer, W. (1935c) Knoblauchfrage. Dtsch. Med. Wochenschr. 61:1825.

Krishnaiah, K. and Mohan, N.J. (1983) Control of cabbage pests by new insecticides. Indian J. Entomol. 45:222–228. Chem. Abst. 103 (1985) 137 079.

Krishnamurthy, K. and Sreenivasamurthy, V. (1956) Garlic. Bull. Cent. Food Technol. Res. Inst. Mysore 5:264–267. Chem. Abst. 51 (1957) 3057.

Kritchevsky, D. (1975) Effect of garlic oil on experimental atherosclerosis in rabbits. Artery [Fulton, Mich.] 1:319–323. Chem. Abst. 84 (1976) 41 631.

Kritchevsky, D., Tepper, S.A., Morrisey, R., and Klurfeld, D. (1980) Influence of garlic oil on cholesterol metabolism in rats. Nutr. Rep. Int. 22:641–645. Chem. Abst. 94 (1981) 46 030.

Kritchevsky, D. (1991) The effect of dietary garlic on the development of cardiovascular disease. Trends Food Sci. Technol. 2:141–144. Chem. Abst. 116 (1992) 104 885.

Kröning, F. (1964) Garlic as an inhibitor for spontaneous tumors in mice. Acta Unio Int. Cancrum 20:855–856. Chem. Abst. 61 (1964) 15 206.

Kruczkowska, M. (1958) Garlic extract in the treatment of tooth canals and chronic periodontitis. Czas. Stomatol. (Stomatolog. J.) 11:567–573 (Polish).

Krynska, W. and Kawecki, Z. (1970) Varying level of saccharides during seasonal growth of mountain garlic (one-year study). Biul. Warczywniczy 11:239–248. Chem. Abst. 75 (1971) 60 022 (Polish).

Kubo, S. and Kishida, K. (1956) Microbiological use of the microrespirometer technique with Cartesian sinker. 2. Influence of allyl compounds on Ehrlich carcinoma cells. J. Kyoto Prefect. Med. Univ. 60:409–410. Chem. Zbl. 1958, II, 14 342 (Japanese).

Kubota, K. and Ken, S. (1927) A method of manufacturing from garlic an injection for tuberculosis. Great Britain. Patent 264 960. Chem. Zbl. 1928, I, 1070.

Kuhn, A. (1932) Wertbestimmung knoblauchhaltiger Produkte. In: Jahrbuch Madaus, pp. 41–43.

Kumar, A., Kumar, S., and Yadav, M.P. (1981a) Antirickettsial property of *Allium sativum* Linn.: in vivo and in vitro studies. Indian J. Anim. Res. 15:93–97.

Kumar, C.A., Saxena, K.K., Gupta, C., Gopal, R., Singh, R.C., Juneja, S., Srivastava, R.K., and Prasad, D.N. (1981b) *Allium sativum*: effect of three weeks feeding in rats. Indian J. Pharmacol. 13:91.

Kumar, A. and Sharma, V.D. (1982) Inhibitory effect of garlic (*Allium sativum* Linn.) on enterotoxigenic Escherichia coli. Indian J. Med. Res. 76:66–70.

Kumar, R.V.O., Banerji, A., Kurup, C.K.R., and Ramasarma, T. (1991) The nature of inhibition of 3-hydroxy-3-methylglutaryl CoA reductase by garlic-derived diallyl disulfide. Biochim. Biophys. Acta 1078:219–225. Chem. Abst. 115 (1991) 88 143.

Kunogi, M. (1993) Deodorization of plants of Allium with high pressure. Japan. Patent 5 123 125. Chem. Abst. 119 (1993) 71 271.

Kupiecki, F.P. and Virtanen, A.I. (1960) Cleavage of alkyl cysteine sulphoxides by an enzyme in onion (Allium cepa). Acta Chem. Scand. 14:1913–1918. Chem. Abst. 56 (1962) 9102.

Kupinic, M. (1979) Microbiological determination of antibiotics. Arh. Farm. [Zagreb] 29:411–418. Chem. Abst. 93 (1980) 179 158 (Croatian).

Kupinic, M., Petricic, J., and Lulic, B. (1980) Bulbus Allii sativi (bulb of garlic), semiquantitative microbial evaluation. Acta Pharm. Jugoslav. 30:205–209.

Kurnakov, B.A. (1952) Action of the phytoncides of garlic on parasitical protozoa. Dokl. 11go Naukh. Konf. Gossudarst. Med. Inst. Tomsk (Rep. 11th Sci. Conf. State Inst. Tomsk), 50–51 (Russian).

Kuwata, K., Kono, A., Kanda, T., Yamauchi, T., and Kobe Steel Ltd. (1993) Deodorization of Allium by high pressure treatment. Japan. Patent 5 168 431. Chem. Abst. 119 (1993) 158 844.

Kwon, J.H., Byun, M.W., and Cho, H.O. (1984) Sterilization of garlic powder by irradiation. Hanguk Sikpum Kwahak Hoechi (Korean J. Food Sci. Technol.) 16:139–142. Chem. Abst. 101 (1984) 129 129 (Korean).

Kwon, J.H., Yoon, H.S., Byun, M.W., and Cho, H.O. (1988) Chemical changes in garlic bulbs resulting from ionizing energy treatment at sprout-inhibition dose. Hanguk Nonghwahak Hoechi (J. Korean Agric. Chem. Soc.) 31:147–153. Chem. Abst. 109 (1988) 229 089 (Korean).

Kwon, J.H., Choi, J.U., and Yoon, H.S. (1989) Sulfur-containing components of gamma-irradiated garlic bulbs. Radiat. Phys. Chem. 34:669–672. Chem. Abst. 112 (1990) 97 219.

Laakso, I., Seppänen-Laakso, T., Hiltunen, R., Müller, B., Jansen, H., and Knobloch, K. (1989) Volatile garlic odor components: gas phases and adsorbed exhaled air analysed by headspace gas chromatography-mass spectrometry. Planta Med. 55:257–261. Chem. Abst. 111 (1989) 95 853.

Lachmann, G., Lorenz, D., Radeck, W., and Steiper, M. (1994) Untersuchungen zur Pharmakokinetik der mit 35S markierten Knoblauchinhaltsstoffe Alliin, Allicin und Vinyldithiine. Arzneim. Forsch. 44:734–743. Chem. Abst. 121 (1994) 244 982.

Lackhovsky, G. (1932) Geheimnisse des Lebens. Munich.

Laland, P. and Havrevold, O.W. (1933) Zur Kenntnis des per os wirksamen blutzuckersenkenden Prinzips der Zwiebeln (*Allium sativum*). Z. Physiol. Chem. 221:180–196.

Lammerink, J. (1990) Effects of fumigation, cold storage, and fungicide treatment of planting cloves on yield and quality of garlic. N. Z. J. Crop Hortic. Sci. 18:55–59. Chem. Abst. 113 (1990) 206 660.

Lancaster, J.E., McCallion, B.J., and Shaw, M.L. (1986) The dynamics of the flavour precursors, the S-alk(en)yl-L-cysteine sulphoxides, during leaf blade and scale development in the onion (Allium cepa). Physiol. Plant. 66:293–297.

Lancaster, J.E., Dommisse, E.M., and Shaw, M.L. (1988) Production of flavour precursors [S-alk(en)yl-L-cysteine sulphoxides] in photomixotrophic callus of garlic. Phytochemistry 27:2123–2124. Chem. Abst. 109 (1988) 125 944.

Lancaster, J.E. and Boland, M.J. (1990) Flavor biochemistry. In: Onions and allied crops. Vol. 3. J.L. Brewster and H.D. Rabinowitch, eds. CRC Press, Boca Raton, Florida, pp. 33–72.

Lancaster, J.E. and Collin, H.A. (1981) Presence of alliinase in isolated vacuoles and of alkyl cysteine sulphoxides in the cytoplasm of bulbs of onion (Allium cepa). Plant Sci. Lett. 22:169–176.

Lancaster, J.E. and Kelly, K.E. (1983) Quantitative analysis of the S-alk(en)yl-L-cysteine sulphoxides in onion (Allium cepa L.). J. Sci. Food Agric. 34:1229–1235. Chem. Abst. 100 (1984) 21 605.

Lancaster, J.E. and Shaw, M.L. (1989) Gamma-glutamyl peptides in the biosynthesis of S-alk(en)yl-L-cysteine sulphoxides (flavour precursors) in Allium. Phytochemistry 28:455–460. Chem. Abst. 110 (1989) 189 471.

Lancaster, J.E. and Shaw, M.L. (1991) Metabolism of gamma-glutamyl peptides during development, storage, and sprouting of onion bulbs. Phytochemistry 30:2857–2859. Chem. Abst. 115 (1991).

Lancaster, J.E. and Shaw, M.L. (1994) Characterization of purified gamma-glutamyl transpeptidase in onions: evidence for in vivo role as a peptidase. Phytochemistry 36:1351–1358. Chem. Abst. 121 (1994) 251 329.

Landshuter, J., Lohmüller, E., Winkler, G., and Knobloch, K. (1992) Comparative biochemical studies on a purified C-S-lyase preparation from wild garlic, Allium ursinum. Planta Med. 58:A666.

Lang, Y., Wang, C., Wang, D., and Shanghai No. 2 Pharm. Factory. (1989) Preparation of garlicin. People's Republic of China. Patent 1 034 201. Chem. Abst. 113 (1990) 114 644.

Lang, Y.J. and Chang, K.J. (1981) Studies on the active principle of garlic, *Allium sativum* L. Chung Tsao Yao (Chin. Trad. Herbal Drugs) 12:4–6. Chem. Abst. 95 (1981) 86 205 (Chinese).

Lansbury, P.T. and La Clair, J.J. (1993) A stereoselective total synthesis of alliacane lactones. Tetrahedron Lett. 34:4431–4434. Chem. Abst. 120 (1994) 107 362.

Larner, A.J. (1995) How does garlic exert its hypocholesterolemic action? The tellurium hypothesis. Medical Hypothesis 44:295–297.

Lata, S., Saxena, K.K., Bhasin, V., Saxena, R.S., Kumar, A., and Srivastava, V.K. (1991) Beneficial effects of *Allium sativum*, Allium cepa, and Commiphora mukul on experimental hyperlipidemia and atherosclerosis: a comparative evaluation. J. Postgrad. Med. 37:132–135.

Latchinian-Sadek, L. and Ibrahim, R.K. (1991) Partial purification and some properties of a ring B-O-glucosyltransferase from onion bulbs. Phytochemistry 30:1767–1771.

Latsinik, E.Y. and Kuperman, E.O. (1955) Combination therapy in acute dysentery with phytoncides of garlic and sulfanilamides. Vrach. Delo (Therapeutics), 539–540 (Russian).

Lau, B.H.S., Adetumbi, M.A., and Sanchez, A. (1983) *Allium sativum* (garlic) and atherosclerosis: a review. Nutr. Res. 3:119–128.

Lau, B.H.S., Marsh, C.L., Barker, G.R., Woolley, J., and Torrey, R. (1985) Effects of biological response modifiers on murine bladder tumor. Nat. Immun. Cell Growth Reg. 4:260–261.

Lau, B.H.S., Woolley, J.L., Marsh, C.L., Barker, G.R., Koobs, D.H., and Torrey, R.R. (1986a) Superiority of intralesional immunotherapy with Corynebacterium parvum and *Allium sativum* in control of murine transitional cell carcinoma. J. Urol. *136*:701–705.

Lau, B.H.S., Woolley, J.L., Marsh, C.L., and Koobs, D.H. (1986b) Superiority of intralesional immunotherapy in control of murine transitional cell carcinoma. Abst. Microbiol. (U.S.) *86*:103.

Lau, B.H.S., Lam, F., and Wang-Cheng, R. (1987) Effect of an odor-modified garlic preparation on blood lipids. Nutr. Res. *7*:139–149. Chem. Abst. 107 (1987) 6292.

Lau, B.H.S. (1989) Detoxifying, radioprotective, and phagocyte-enhancing effects of garlic. Int. Clin. Nutr. Rev. *9*:27–31.

Lau, B.H.S., Tadi, P.P., and Tosk, J.M. (1990) *Allium sativum* (garlic) and cancer prevention. Nutr. Res. *10*:937–948.

Lau, B.H.S., Tadi, P.P., and Teel, R.W. (1991a) Organosulfur compounds of garlic as dietary anticarcinogens. J. Am. Coll. Nutr. *10*:546.

Lau, B.H.S., Yamasaki, T., and Gridley, D.S. (1991b) Garlic compounds modulate macrophage and T-lymphocyte functions. Mol. Biother. *3*:103–107.

Lautier, R. and Wendt, V. (1985) Kontaktallergie auf Alliaceae. Fallbeschreibung und Literaturübersicht. Dermatosen Beruf Umwelt *33*:213–215. Ref. in: Dtsch. Apoth. Ztg. 126 (1986) No. 17, 845.

Lawson, L.D. and Hughes, B.G. (1989) Analysis of aqueous garlic extract and garlic products by HPLC. Planta Med. *55*:639.

Lawson, L.D. and Hughes, B.G. (1990) trans-1-Propenyl thiosulfinates: new compounds in garlic homogenates. Planta Med. *56*:589.

Lawson, L.D., Wang, Z.J., and Hughes, B.G. (1991a) Identification and HPLC quantitation of the sulfides and dialk(en)yl thiosulfinates in commercial garlic products. Planta Med. *57*:363–370. Chem. Abst. 115 (1991) 263 558.

Lawson, L.D., Wang, Z.Y.J., and Hughes, B.G. (1991b) Gamma-glutamyl-S-alkylcysteines in garlic and other Allium species: precursors of age-dependent trans-1-propenyl thiosulfinates. J. Nat. Prod. *54*:436–444. Chem. Abst. 115 (1991) 48 001.

Lawson, L.D., Wood, S.G., and Hughes, B.G. (1991c) HPLC analysis of allicin and other thiosulfinates in garlic clove homogenates. Planta Med. *57*:263–270. Chem. Abst. 115 (1991) 179 355.

Lawson, L.D. and Hughes, B.G. (1992) Characterization of the formation of allicin and other thiosulfinates from garlic. Planta Med. *58*:345–350.

Lawson, L.D., Ransom, D.K., and Hughes, B.G. (1992) Inhibition of whole blood platelet aggregation by compounds in garlic clove extracts and commercial garlic products. Thromb. Res. *65*:141–156. Chem. Abst. 116 (1992) 166 029.

Lawson, L.D. (1993) Bioactive organosulfur compounds of garlic and garlic products: role in reducing blood lipids. In: Human medicinal agents from plants. Vol. ACS Symp. Ser. 534. A.D. Kinghorn and M.F. Balandrin, eds. Am. Chem. Soc. Books, Washington, D.C., pp. 306–330. Chem. Abst. 119 (1993) 216 610.

Lawson, L.D. and Wang, Z.J. (1993) Pre-hepatic fate of the organosulfur compounds derived from garlic (*Allium sativum*). Planta Med. *59*:A688.

Lawson, L.D. and Wang, Z.J. (1994) (previously unpublished).

Lawson, L.D. and Wang, Z.Y.J. (1995) Changes in the organosulfur compounds released from garlic during aging in water, dilute ethanol, or dilute acetic acid. J. Toxicology *14*:214.

Lazarev, I.Z. and Ivanova, O.I. (1969) Two-component garlic extract. U.S.S.R. Patent 244 879. Chem. Abst. 72 (1970) 30 446.

Lebinski, G. (1927) Beitrag zur Therapie akuter und chronischer Darmkatarrhe. Klin. Wochenschr. *6*:2119–2120.

Leclerc, H. (1918) Histoire de l'ail. Janus *23*:167–191.

Ledebour, C.F. (1853) Allium L. In: Flora Rossica sive enumeratio plantarum in totius Imperii Rossici provinciis Europaeis, Asiaticis et Americanis hucusque observatarum. E. Schweizerbart, Stuttgart, pp. 161–190.

Lee, D.G., Min, J.G., and Cha, C.W. (1985) A study on the effect of garlic in the inhibitory action of cadmium on ALAD activities in human blood in vitro. Koryo Taehakkyo Uikwa Taehak Nonmunjip (Korea Univ. Med. J.) *22*:135–141. Chem. Abst. 103 (1985) 191 016 (Korean).

Lee, E.S., Steiner, M., and Lin, R. (1994) Thioallyl compounds: potent inhibitors of cell-proliferation. Biochim. Biophys. Acta *1221*:73–77.

Lee, H.S., Bae, E.S., and Cha, C.W. (1985) The effect of garlic on pathological damage of testis due to cadmium poisoning. Koryo Taehakkyo Uikwa Taehak Nonmunjip (Korea Univ. Med. J.) *21*:39–47. Chem. Abst. 102 (1985) 144 325 (Korean).

Lee, K.R., Kim, J.H., and Park, D.C. (1994) Studies on the callus culture of garlic and the formation of alliin. Hangug Yongyang Siklyong Hakhoechi *20*:1–4. Chem. Abst. 117 (1992) 66 837 (Korean).

Lee, S.H. and Kim, S.D. (1988) Effect of various ingredients of kimchi on kimchi fermentation. Hanguk Yongyang Siklyong Hakhoechi (J. Korean Soc. Food Nutr.) *17*:249–254. Chem. Abst. 111 (1989) 113 987 (Korean).

Lee, W.M. and Lee, K.T. (1980) Lack of effect of phenformin, pyridinolcarbamate combined with a decalcifying agent, and an onion-garlic mixture on growth and necrosis of aortic atherosclerosis in balooned swine. Exp. Mol. Pathol. *33*:345–348. Chem. Abst. 94 (1981) 76 789.

Lee, Y.O., Cha, C.W., and Rhim, K.H. (1986) A study on the effect of garlic, D-penicillamine, and N-acetyl-DL-penicillamine against cadmium poisoning in rat. Koryo Taehakkyo Uikwa Taehak Nonmunjip (Korea Univ. Med. J.) *23*:43–53. Chem. Abst. 107 (1987) 72 244 (Korean).

Lee, Y.S. (1967) Comparative study of the effect of allicin and arsenite on albino rats, with special regard to the effect on body weight, hemoglobin, and hepatic histology. New Med. J. [Seoul] 9:99–101. Chem. Abst. 66 (1967) 84 264.

Lee, Y.S. and Jang, J.J. (1991) Modifying effect of garlic and red pepper extracts on diethylnitrosamine-induced hepatocarcinogenesis. Envir. Mutag. Carcinog. 11:21–28. Chem. Abst. 117 (1992) 62 469.

Leeser, O. (1984) Pflanzliche Arzneistoffe. In: Lehrbuch der Homöopathie. Haug, Heidelberg, pp. 1030–1037.

Legendre, M.G., Dupuy, H.P., Rayner, E.T., and Schuller, W.H. (1980) Rapid instrumental technique for the analysis of volatiles in salad dressing. J. Am. Oil Chem. Soc. 57:361–362. Chem. Abst. 93 (1980) 237 185.

Legnani, C., Frascaro, M., Guazzaloca, G., Ludovici, S., Cesarano, G., and Coccheri, S. (1993) Effects of a dried garlic preparation on fibrinolysis and platelet aggregation in healthy subjects. Arzneim. Forsch. 43:119–122.

Lehmann, A. (1930) Untersuchungen über *Allium sativum* (Knoblauch). Arch. Exp. Pathol. Pharmakol. 147:245–264. Chem. Zbl. 1930, II, 88.

Lehmann, B. and Brewitt, B. (1991) Lowering of blood lipid values with standardised garlic powder drug: long range, multicentre study. Cardiol. in Pract. (June suppl.):18

Leighton, T., Ginther, C., Fluss, L., Harter, W.K., Cansado, J., and Notario, V. (1992) Molecular characterization of quercetin and quercetin glycosides in Allium vegetables: their effects on malignant cell transformation. ACS Symp. Ser. 507:220–238. Chem. Abst. 118 (1993) 79 723.

Lembo, G., Balato, N., Patruno, C., Auricchio, L., and Ayala, F. (1991) Allergic contact dermatitis due to garlic (*Allium sativum*). Contact Dermatitis 25:330–331.

Lenz, H.O. (1859) Botanik der alten Griechen und Römer, deutsch in Auszügen aus deren Schriften nebst Anmerkungen. E.F. Thienemann, Gotha.

Lercari, B. (1982) The effect of far-red light on the photoperiodic regulation of carbohydrate accumulation in Allium cepa L. Physiol. Plant. 54:475–479.

Lerda, D. (1992) The effect of lead on Allium cepa L. Muttation Res. 281:89–92. Chem. Abst. 116 (1992) 123 112.

Lesnikov, E.P. (1947) Data on the fungicidal and fungistatic action in vitro of phytoncides of onion and garlic on geotrichoid. Byull. Eksptl. Biol. Med. (Bull. Exp. Biol. Med.) 24:70–72. Chem. Abst. 42 (1948) 4701 (Russian).

Letham, D.S., Stevenson, K.R., and Tao, G.Q. (1993) Transgenic plants expressing an heterologous isopentenyl transferase gene. Int. Appl. Patent 93 07 292. Chem. Abst. 119 (1993) 5124.

Leusser, L. (1987) Knoblauch-Dragee. Germany. Patent 3 619 570. Chem. Abst. 108 (1988) 173 560.

Levan, A. (1936) Die Zytologie von Allium cepa × fistulosum. Hereditas 21:195–214.

Levi, F., Franceschi, S., Negri, E., and LaVecchia, C. (1993a) Dietary factors and the risk of endometrial cancer. Cancer 71:3575–3581.

Levi, F., LaVecchia, C., Gulie, C., and Negri, E. (1993b) Dietary factors and breast cancer risk in Vaud, Switzerland. Nutr. Cancer 19:327–335.

Levitin, I.A. (1954) Alcoholic extract from garlic as deodorizing agent in purulent processes of the lungs. Sov. Zdravookhr. Kirg. [Frunze] (Health Care in Soviet Kirgizia), 57–58 (Russian).

Lewin, G. and Popov, I. (1994) Antioxidant effects of aqueous garlic extract. 2. Inhibition of the Cu^{2+}–initiated oxidation of low density lipoproteins. Arzneim. Forsch. 44:604–607. Chem. Abst. 121 (1994) 170 498.

Li, H.L. (1969) The vegetables of ancient China. Econ. Bot. 23:253–260.

Li, K.H., Bundus, R.H., and Noznick, P.P. (1967) Prevention of pink color in white onions. U.S.A. Patent 3 352 691. Chem. Abst. 68 (1968) 21 062.

Li, X., Yin, W., Shao, L., and Lou, C. (1986) Contrast in enzyme activities between the two kinds of epidermal cells in garlic protective sheath during senescing. Zhiwu Xuebao (Acta Bot. Sinica) 28:175–178. Chem. Abst. 105 (1986) 3663 (Chinese).

Liakopoulou-Kyriakides, M. (1985) Relation between the structure of alliin analogues and their inhibitory effect on platelet aggregation. Phytochemistry 24:1593–1594. Chem. Abst. 103 (1985) 115 690.

Liakopoulou-Kyriakides, M., Sinakos, Z., and Kyriakidis, D.A. (1985) Identification of alliin, a constituent of Allium cepa with an inhibitory effect on platelet aggregation. Phytochemistry 24:600–601. Chem. Abst. 103 (1985) 395.

Lichtenstein, L.M., Pickett, W.C., and Johns Hopkins University (1985) Treatment of allergies and inflammatory conditions. Europe. Patent 153 881. Chem. Abst. 104 (1986) 45 756.

Lichtner, M. (1984) Knoblauch, Kräuter und Oliven. In: Spezialitäten der provencalischen Küche. Vol. 2. Weingarten, Württemberg.

Liebe, S. and Quader, H. (1994) Myosin in onion (Allium cepa) bulb scale epidermal cells: involvement in dynamics of organelles and endoplasmatic reticulum. Physiol. Plant. 90:114–124. Chem. Abst. 120 (1994) 187 239.

Lin, R.I. (1991) Cardiovascular protective properties of garlic and some Chinese herbs. J. Am. Coll. Nutr. 10:564.

Lin, X., Liu, J., and Wu, K. (1986) The blocking effect of garlic extract on moulds-mediated nitrosation. Acta Nutr. Sin. 8:262.

Lin, X., Li, L., Zhang, Q., and Mei, X. (1991) The preventive effect of garlic against toxicity of dimethylnitrosamine in rats fed aminopyrine and nitrite. Yingyang Xuebao (Acta Nutr. Sinica) 13:126–132. Chem. Abst. 116 (1992) 150 591 (Chinese).

Lin, X.Y., Liu, J., and Milner, J.A. (1992) Dietary garlic powder suppresses the in vivo formation of DNA adducts induced by N-nitroso compounds in liver and mammary tissues. FASEB J. 6:A1392.

Lin, X.Y., Liu, J.Z., and Milner, J.A. (1994) Dietary garlic suppresses DNA-adducts caused by N-nitroso compounds. Carcinogenesis 15:349–352. Chem. Abst. 120 (1994) 190 358.

Linde, E.I. and Rodina, W.J. (1956) Anwendung der Knoblauch-Phytonzide bei verschiedenen HNO-Erkrankungen. Vestn. Otorinolaringol. (Otorhinolaryngol. Messenger) 18:97–98 (Russian).

Lindner, K., Szente, L., and Szejtli, J. (1981) Food flavouring with b-cyclodextrin-complexed flavour substances. Acta Aliment. 10:175–186. Chem. Abst. 96 (1981) 5067.

Lindner, K. (1982) Using cyclodextrin aroma complexes in the catering. Nahrung 26:675–680. Chem. Abst. 97 (1982) 214 428.

Link, K.P., Angell, H.R., and Walker, J.C. (1929) The isolation of protocatechuic acid from pigmented onion scales and its significance in relation to disease resistance in onions. J. Biol. Chem. 81:369–375. Chem. Zbl. 1929, II, 313.

Link, K.P. and Walker, J.C. (1933) The isolation of catechol from pigmented onion scales and its significance in relation to disease resistance in onions. J. Biol. Chem. 379–383.

Linnaeus, C. (1753) Allium. In: Caroli Linnaei Species Plantarum, etc. Holmiae, 1907 volume, reprinted, pp. 294–302.

Linne, C. (1770) Systema naturae per regna tria naturae, secundum classes, ordines, genera, species cum characteribus et differentiis. I. Thomae de Trattnern, Vienna.

Lio, G. and Agnoli, R. (1927) Action of Allium sativum on smooth muscle. Arch. Int. Pharmacodyn. Ther. 33:400–408. Chem. Abst. 23 (1929) 2487.

Liu, D., Jiang, W., and Li, M. (1992) Effects of trivalent and hexavalent chromium on root growth and cell division of Allium cepa. Hereditas 117:23–29. Chem. Abst. 117 (1992) 246 775.

Liu, D., Jiang, W., and Li, D. (1993) Effects of aluminum ion on root growth, cell division, and nucleoli of garlic (Allium sativum L.). Environ. Pollut. 82:295–299. Chem. Abst. 119 (1993) 243 449.

Liu, D., Jiang, W., Guo, L., Hao, Y., Lu, C., and Zhao, F. (1994) Effects of nickel sulfate on root growth and nucleoli in root tip cells of Allium cepa. Israel J. Plant Sci. 42:143–148.

Liu, J., Lin, X., Li, C., and Peng, S. (1985) The blocking effect of garlic on dimethylnitrosamine formation mediated by Fusarium moniliforme. Shandong Yixueyuan Xuebao 23:31–34. Chem. Abst. 103 (1985) 173 751 (Chinese).

Liu, J., Lin, X., Peng, S., Song, P., and Hu, J. (1986a) The blocking effect of garlic extract on the in vitro chemical formation of nitrosamines. Acta Nutr. Sin. 8:9.

Liu, J., Lin, X., Wu, K., and Peng, S. (1986b) The mechanism of blocking effect of garlic on the formation of nitrosamines. Acta Nutr. Sin. 8:327–334 (Chinese).

Liu, J., Mei, X., Lin, X., Gao, H., Li, X., Ma, Z., Xia, H., and Liu, S. (1989) The quantitative measurement of 26 elements in garlic with DCP-AES method. Yingyang Xuebao (Acta Nutr. Sinica) 11:159–162. Chem. Abst. 112 (1990) 75 410 (Chinese).

Liu, J., Lin, R.I., and Milner, J.A. (1992a) Inhibition of 7,12-dimethylbenz[a]anthracene-induced mammary tumors and DNA adducts by garlic powder. Carcinogenesis 13:1847–1851. Chem. Abst. 118 (1993) 5964.

Liu, J., Lin, X.Y., and Milner, J.A. (1992b) Dietary garlic powder increases glutathione content and glutathione S-transferase activity in rat liver and mammary tissues. FASEB J. 6:A1493.

Liu, J. and Lin, X. (1985) The blocking effect of garlic extract on bacteria-mediated nitrosation. Acta Acad. Med. Shandong Yixueyuan Xuebao 23:56–59.

Liu, J. and Milner, J. (1990) Influence of dietary garlic powder with and without selenium supplementation on mammary carcinogen adducts. FASEB J. 4:A1175.

Loeper, M., Debray, M., and Chailley-Bert. (1921) Recherches expérimentales sur l'hypotension par les produits alliaces. C. R. Séances Soc. Biol. Ses Fil. 85:160–161. Chem. Zbl. 1921, III, 1509.

Loeper, M. and Debray, M. (1921) L'action hypotensive de la teinture d'ail. Ann. Med. Interne 37:1032–1037.

Logacheva, L.I. (1953) Investigation on the mechanism of action upon the microflora of the skin of the phytoncides from garlic. Zh. Mikrobiol. Epidemiol. Immunobiol. 24:16–17. Chem. Zbl. 1954, II, 10 520 (Russian).

Lohmüller, E.M., Landshuter, J., and Knobloch, K. (1994a) Purification and characterization of a C-S-lyase from ramson, the wild garlic, Allium ursinum. Planta Med. 60:343–347. Chem. Abst. 122 (1995) 100 228.

Lohmüller, E.M., Landshuter, J., and Knobloch, K. (1994b) On the isolation and characterization of a C-S-lyase preparation from leek, Allium porrum. Planta Med. 60:337–342. Chem. Abst. 122 (1995) 100 227.

Lonitzer, A. (1557) Adami Loniceri Kreuterbuch/New zugericht, etc. Frankfurt-on-Main.

Lonyai, P., Darvas, F., and Timar, G. (1973) Destruction of the odors of garlic and onion. Hungarian Patent. Teljes 6116. Chem. Abst. 80 (1974) 69 402.

Lorand, A. (1934) Diätetische Krebsverhütung. Med. Klin., 1030–1031.

Lorenzato, D. (1984) Control tests on the noxious mites in stored garlic (Allium sativum L.). Agron. Sulriograndense 20:153–165. Chem. Abst. 105 (1986) 170 882 (Portuguese).

Loret, V. (1904) L'ail chez les anciens Egyptiens. Sphinx 8:135–147.

Loskutov, A.M. (1950a) Influence of garlic preparations upon diuresis and blood pressure. Farmakol. Toksikol. [Moscow] 13:11–12 (Russian).

Loskutov, A.M. (1950b) Use of garlic in veterinary practice. Veterinariya [Moscow] 27:53 (Russian).

Louria, D.B., Lavenhar, M., Kaminski, T., and Eng, R.H.K. (1989) Garlic (Allium sativum) in the treatment of experimental cryptococcosis. J. Med. Vet. Mycol. 27:253–256.

Lu, D.P., Guo, N.L., and Jin, N.R. (1988) Efficacy of garlic extract together with placental gammaglobulin against interstitial pneumonia after bone marrow transplantation. Experim. Hematol. 16:484.

Lu, D.P. (1994) Bone marrow transplantation in the People's Republic of China. Bone Marrow Transpl. 13:703–704.

Lu, D.P., Guo, N.L., Jin, N.R., Zheng, H., Lu, X.J., Shi, Q., Shan, F.X., Jiang, B., Tang, H., and Liu, M.Y. (1990) Allogeneic bone marrow transplantation for the treatment of leukemia. Chin. Med. J. 103:125–130.

Lu, J., Gao, M., and Xang, M. (1992) ESR study of scavenging effects on active oxygen radicals of garlic and its extracts. Zhongguo Yaoxue Zazhi 27:339–342. Chem. Abst. 117 (1992) 143 405 (Chinese).

Lu, L., Qian, Y., and Wu, L. (1992) Colorimetric determination of germanium in garlic by acid digestion under reflux and extraction. Yingyang Xuebao (Acta Nutr. Sinica) 14:426–429. Chem. Abst. 119 (1993) 26 904.

Ludeke, B.I., Dominé, F., Ohgaki, H., and Kleihues, P. (1992) Modulation of N-nitrosomethylbenzylamine bioactivation by diallyl sulfide in vivo. Carcinogenesis 13:2467–2470.

Ludlam, H. (1962) A biography of Dracula: the life story of Bram Stoker. London.

Lukes, T.M. (1971) Thin-layer chromatography of cysteine derivatives of onion flavor compounds and the lacrimatory factor. J. Food Sci. 36:662–664. Chem. Abst. 75 (1971) 18 703.

Lukes, T.M. (1986) Factors governing the greening of garlic puree. J. Food Sci. 51:1577,1582. Chem. Abst. 106 (1987) 83 215.

Luley, C., Lehmann-Leo, W., Möller, B., Martin, T., and Schwartzkopff, W. (1986) Lack of efficacy of dried garlic in patients with hyperlipoproteinemia. Arzneim. Forsch. 36:766–768.

Luley, C.H. and Schwartzkopff, W. (1985) Knoblauchpillen als Antiatherosklerotikum? Klin. Wochenschr. 63:144.

Lun, Z.R., Burri, C., Menzinger, M., and Kaminsky, R. (1994) Antiparasitic activity of diallyl trisulfide (Dasuansu) on human and animal pathogenic protozoa (Trypanosoma sp. Entamoeba histolytica, Giardia lamblia) in vitro. Ann. Soc. Belg. Med. Tropic. 74:51–59.

Lutomski, J., Adamczewski, B., and Ostrowska, B. (1968) Modification of the Jaeger method for determining alliin in garlic and its preparations. Farm. Pol. 24:575–580. Chem. Abst. 70 (1969) 109 191 (Polish).

Lutomski, J., Kedzia, W., and Adamczewski, B. (1969a) The antibacterial activity of garlic bulbs from different domestic cultures. Herba Pol. 15:363–368 (Polish).

Lutomski, J., Kedzia, W., Adamczewski, B., and Muszynski, Z. (1969b) Adaptability of chemical and microbiological methods for the determination of pharmacological activity of raw material and preparations of garlic (Allium sativum). Farm. Pol. 25:995–998. Chem. Abst. 73 (1970) 7291 (Polish).

Lutomski, J. (1980) Die Bedeutung von Knoblauch und Knoblauch-Präparaten in der Phytotherapie. Pharm. Unserer Zeit 9:45–50.

Lutomski, J. (1983) Das Wichtigste über Knoblauch und Knoblauch-Präparate. Dtsch. Apoth. Ztg. 123:623–624.

Lutomski, J. (1984) Klinische Untersuchungen zur therapeutischen Wirksamkeit von Ilja Rogoff Knoblauchpillen mit Rutin. Z. Phytother. 5:938–942.

Lutomski, J. (1987) Components and biological properties of some Allium species. Institute of Medicinal Plants 27:1–58.

Lutomski, J., Kedzia, B., and Holderna, E. (1988) Knoblauchpräparat Wang 1000—Beurteilung der antimikrobiellen Aktivität. Dtsch. Apoth. Ztg. 128:XXXVII–XXXIX.

Lutomski, J. and Mrozikiewicz, A. (1989) Klinische Beurteilung der antisklerotischen Wirkung des Präparats Graf's Knoblauch-Hausmittel. Herba Pol. 35:193–200.

Lutomski, J. and Mrozikiewicz, A. (1992) Die optimalen Wirkungseffekte einiger Knoblauchpräparate auf der Basis polnischer therapeutischer Untersuchungen. Herba Pol. 38:93–103. Chem. Abst. 118 (1993) 160 403. Int. Pharm. Abst. 30 (1993) 13 149.

Lybarger, J.A., Gallagher, J.S., Pulver, D.W., Litwin, A., Brooks, S., and Bernstein, I.L. (1982) Occupational asthma induced by inhalation and ingestion of garlic. J. Allergy Clin. Immunol. 69:448–454. Ref. in Dermatosen 30 (1982) No. 6, 176.

Ma, X., Gu, Y., and Fu, J. (1990) Biosynthesis of LTB4 and selection of its inhibitors. Baiqiuen Yike Daxue Xuebao 16:222–225. Chem. Abst. 114 (1991) 240 172 (Chinese).

Maass, H.I. and Klaas, M. (1995) Infraspecific differentiation of garlic (Allium sativum L.) by isozyme and RAPD markers. Theor. Appl. Genet. 91:89–97.

Mabrouk, S.S. and El Shayeb, N.M.A. (1980) Aflatoxin production on some Egyptian agricultural food commodities. Chem. Mikrobiol. Technol. Lebensm. 6:167–170. Chem. Abst. 94 (1981) 43 808.

MacDonald, K.L., Spengler, R.F., Hatheway, C.L., Hargrett, N.T., and Cohen, M.L. (1985) Type A botulism from sauteed onions: clinical and epidemiologic observations. JAMA 253:1275–1278.

Machado, P.A., Duran, M.G., Cross, J.D., and Santos, D.C. (1948) Garlicina—Um novo antibiotico. An. Paul. Med. Cir. [Sao Paulo] *55*:93–115. Chem. Abst. 46 (1952) 4180 (Portuguese).

Madamba, P.S., Driscoll, R.H., and Buckle, K.A. (1993) Bulk density, porosity, and resistance to airflow of garlic slices. Drying Technol. *11*:1837–1854. Chem. Abst. 120 (1994) 302 495.

Madaus, G. (1938) Lehrbuch der Biologischen Heilmittel. G. Thieme, Leipzig.

Madaus, G. (1976) Lehrbuch der biologischen Heilmittel. G. Olms, Hildesheim.

Mader, F.H. (1990a) Treatment of hyperlipidaemia with garlic-powder tablets. Arzneim. Forsch. *40*:1111–1116. Int. Pharm. Abst. 28 (1991) 9602.

Mader, F.H. (1990b) Hyperlipidämie Behandlung mit Knoblauch Dragees: Doppelblind Studie mit 261 Patienten in 30 Fachpraxen für Allgemeinmedizin. Der Allgemeinsarzt, *12*:435–440.

Madhavi, D.L., Prabha, T.N., Singh, N.S., and Patwardhan, M.V. (1991) Biochemical studies with garlic (*Allium sativum*) cell cultures showing different flavour levels. J. Sci. Food Agric. *56*:15–24. Chem. Abst. 115 (1991) 254 564.

Madsen, B.C. and Murphy, J.R. (1981) Flow injection and photometric determination of sulfate in rainwater with methylthymol blue. Anal. Chem. *53*:1924–1926.

Mahajan, V.M. (1983) Antimycotic activity of different chemicals, Chaksine iodide and garlic. Mykosen *26*:94–99. Chem. Abst. 98 (1983) 157 754.

Mahato, S.B., Ganguly, A.N., and Sahu, N.P. (1982) Steroid saponins. Phytochemistry *21*:959–978.

Mahler, P. and Pasterny, K. (1924) Klinische Beobachtung über Insulinwirkung beim Diabetes mellitus. Med. Klin. *11*:335–338.

Makheja, A.N., Vanderhoek, J.Y., and Bailey, J.M. (1979) Inhibition of platelet aggregation and thromboxane synthesis by onion and garlic. Lancet *1*:781. Chem. Abst. 91 (1979) 117 504.

Makheja, A.N., Vanderhoek, J.Y., Bryant, R.W., and Bailey, J.M. (1980) Altered arachidonic acid metabolism in platelets inhibited by onion or garlic extracts. Adv. Prostaglandin Thromboxane Res. *6*:309–312. Chem. Abst. 93 (1980) 63 206.

Makheja, A.N., Bailey, J.M., and Low, C.E. (1981) Biological nature of platelet inhibitors from *Allium cepa*, *Allium sativum*, and *Auricularia polytricha*. Thromb. Haemostasis *46*:148.

Makheja, A.N. and Bailey, J.M. (1990) Antiplatelet constituents of garlic and onion. Agents Actions *29*:360–363. Chem. Abst. 112 (1990) 95 574.

Makowsky, L. (1936) Zur Frage der Beeinflussung von Impfgeschwülsten der Maus durch Gewürzstoffe und Colibakterien. Z. Immunitätsforsch. Exp. Ther. *89*:423–430.

Maleszka, R., Lutomski, J., Swiatlowska-Gorna, B., and Rzepeczka, B. (1991) Study on extending of the activity spectrum of a garlic preparation against Candidiasis. Herba Pol. *37*:85–88. Int. Pharm. Abst. 30 (1993) 1741.

Malet, J.C. (1993) Control of garlic rust. Phytoma *451*:34–36. Chem. Abst. 119 (1993) 153 907.

Malik, Z.A. and Siddiqui, S. (1981) Hypotensive effect of freeze-dried garlic (*Allium sativum*) sap in dog. J. Pak. Med. Assoc. *31*:12–13.

Malova, A.V. (1950) On the pharmacology of garlic. Farmakol. Toksikol. [Moscow] *13*:9–10 (Russian).

Malova, A.V. (1951) On the pharmacology of garlic: on the influence of a garlic extract on blood pressure. Farmakol. Toksikol. [Moscow] *14*:49 (Russian).

Malpathak, N.P. and David, S.B. (1986) Flavor formation in tissue cultures of garlic (*Allium sativum* L.). Plant Cell Rep. *5*:446–447. Chem. Abst. 106 (1987) 116 571.

Mand, J.K., Gupta, P.P., Soni, G.L., and Singh, R. (1985) Effect of garlic on experimental atherosclerosis in rabbits. Indian Heart J. *37*:183–188.

Mandal, S. and Choudhuri, D.K. (1982) Cholesterol metabolism in Lohita grandis Gray (Hemiptera: Pyrrhocoridae: Insecta): effect of corpora allatectomy and garlic extract. Curr. Sci. *51*:367–369. Chem. Abst. 97 (1982) 36 423.

Mangum, P.D. and Peffley, E.B. (1994) Inheritance of ADH, 6-PGDH, PGM, and SKDH in Allium fistulosum L. J. Am. Soc. Hort. Sci. *119*:335–338. Chem. Abst. 121 (1994) 153 509.

Mankarios, A.T., Jones, C.F.G., Jarvis, M.C., Threlfall, D.R., and Friend, J. (1979) Hydrolysis of plant polysaccharides and GLC analysis of their constituent neutral sugars. Phytochemistry *18*:419–422. Chem. Abst. 91 (1979) 105 155.

Mano, D. (1962a) The inhibitory action of some plant extracts on bacterial growth. III. The inhibitory action of licorice root fractions on bacterial growth and the increased potential of resistance in bacteria. Nippon Saikingaku Zasshi *17*:938–940. Chem. Abst. 64 (1966) 10 122.

Mano, D. (1962b) Studies on the inhibitory action of some plant extracts on bacterial growth. I. Inhibitory effect on the growth of Staphylococcus aureus and enteric bacteria, bacteria that acquired resistance to penicillin and streptomycin. Nippon Saikingaku Zasshi *17*:417–421.

Mano, D. (1962c) The inhibitory action of some plant extracts on bacterial growth. II. Changes in the susceptibility to antibiotics of the strains of bacteria adapted to culturing with a fraction from *Allium sativum*. Nippon Saikingaku Zasshi *17*:807–809. Chem. Abst. 64 (1966) 10 122.

Mansell, P. and Reckless, J.P.D. (1991) Garlic: effects on serum lipids, blood pressure, coagulation, platelet aggregation, and vasodilatation. Br. Med. J. *303*:379–380. Int. Pharm. Abst. 28 (1991) 12 997.

Mantis, A.J., Karaioannoglou, P.G., Spanos, G.P., and Panetsos, A.G. (1978) The effect of garlic extract on food poisoning bacteria in culture media. I. Staphylococcus aureus. Lebensm. Wiss. Technol. *11*:26–28.

Marburger Erklärung. (1992) Homöopathie als Irrlehre und Täuschung des Patienten. Apoth. Ztg. [Stuttgart] *9*:1–8.

Marcovici, E. (1915) *Allium sativum* als Therapeutikum bei chronischem und akut infektiösem Darmkatarrh. Wien. Klin. Wochenschr. *33*:894–896.

Marcovici, E. and Pribram, E. (1915) Klinische und experimentelle Untersuchungen über die Wirkung von *Allium sativum* und daraus hergestellten Präparaten (Allphen) bei infektiösen Darmkrankheiten. Wien. Klin. Wochenschr. *28*:995–999. Ref. in Münch. Med. Wschr. 33 (1915) 1395.

Marcovici, E. and Schmitt, M. (1915) Zur Therapie der Cholera asiatica. Wien. Klin. Wochenschr. *28*:789–790.

Margolina, M.I., Karut, T.A., and Kandyba, S.G. (1951) Action of garlic phytoncides on tuberculosis bacilli. Trans. Ukrain. Inst. Epidemiol. *18*:123–131 (Russian).

Markov, S. (1964) Micromethod for the quantitative determination of sulfur in glucosides of garlic. Farmatsiya [Sofia] (Pharmacy) *14*:36–38. Chem. Abst. 62 (1965) 12 140 (Bulgarian).

Marks, H.S., Anderson, J.L., and Stoewsand, G.S. (1992) Inhibition of benzo[a]pyrene-induced bone marrow micronuclei formation by diallyl thioethers in mice. J. Toxicol. Environ. Health *37*:1–9.

Marks, H.S., Anderson, J.A., and Stoewsand, G.S. (1993) Effect of S-methyl cysteine sulphoxide and its metabolite methyl methane thiosulphinate, both occurring naturally in brassica vegetables, on mouse genotoxicity. Food Chem. Toxicol. *31*:491–495.

Marsh, C.L., Torrey, R.R., Woolley, J.L., Barker, G.R., and Lau, B.H.S. (1987) Superiority of intravesical immunotherapy with Corynebacterium parvum and *Allium sativum* in control of murine bladder cancer. J. Urol. *137*:359–362.

Martin, D.J. and Pearce, R.H. (1966) Analysis of bis-dialkyl polysulfides by proton nuclear magnetic resonance. Anal. Chem. *38*:1604–1605. Chem. Abst. 65 (1966) 19 492.

Martin, N., Bardisa, L., Pantoja, C., Vargas, M., Quezada, P., and Valenzuela, J. (1994) Anti-arrhythmic profile of a garlic dialysate assayed in dogs and isolated atrial preparations. J. Ethnopharmacol. *43*:1–8. Chem. Abst. 121 (1994) 221 500.

Martinescu, E. (1981) Contact dermitis to *Allium sativum*. Rev. Med. Chir. Soc. Med. Nat. Iasi *85*:541–542 (Romanian).

Martius, A. (1951) Knoblauch. Dtsch. Apoth. Ztg. *91*:722–723. Chem. Abst. 46 (1952) 2753.

Martín, N., Bardisa, L., Pantoja, C., Román, R., and Vargas, M. (1992) Experimental cardiovascular depressant effects of garlic (*Allium sativum*) dialysate. J. Ethnopharmacol. *37*:145–149.

Maruniak, J.A., Mason, J.R., and Kostelc, J.G. (1983) Conditioned aversions to an intravascular odorant. Physiol. Behav. *30*:617–620.

Maruzella, J.C., Scrandis, D.A., Scrandis, J.B., and Grabon, G. (1960) Action of odoriferous organic chemicals and essential oils on wood-destroying fungi. Plant Dis. Reptr. *44*:789–792. Chem. Abst. 55 (1961) 12 743.

Maruzella, J.C. and Sicurella, N.A. (1960) Antibacterial activity of essential oil vapors. J. Am. Pharm. Assoc. *49*:692–694. Chem. Abst. 55 (1961) 3726.

Marzell, H. (1967) Geschichte und Volkskunde der deutschen Heilpflanzen. Wissenschaftliche Buchgesellschaft, Darmstadt.

Mascolo, N., Autore, G., Capasso, F., Menghini, A., and Fasulo, M.P. (1987) Biological screening of Italian medicinal plants for anti-inflammatory activity. Phytother. Res. *1*:28–31.

Masters, A. (1972) The natural history of the vampire. Rupert Hart-Davis, London.

Mathew, P.T. and Augusti, K.T. (1973a) Studies on the effect of allicin (diallyl disulphide-oxide) on alloxan diabetes. Part I. Hypoglycaemic action and enhancement of serum insulin effect and glycogen synthesis. Indian J. Biochem. Biophys. *10*:209–212. Chem. Abst. 81 (1974) 45 370.

Mathew, P.T. and Augusti, K.T. (1973b) Hypoglycaemic effects of onion, Allium cepa Linn. on diabetes mellitus: a preliminary report. Indian J. Physiol. Pharmacol. *19*:213–217.

Mathew, P.T. and Augusti, K.T. (1973c) Isolation of hypo- and hyperglycaemic agents from Allium cepa Linn. Indian J. Exp. Biol. *11*:573–575.

Matikkala, E.J. and Virtanen, A.I. (1958) A new sulphur-containing amino acid in onion. Suom. Kemistil. *B31*:219. Chem. Abst. 52 (1958) 12 101.

Matikkala, E.J. and Virtanen, A.I. (1962) A new gamma-glutamylpeptide, gamma-L-glutamyl-S-(prop-1-enyl)-L-cysteine, in the seeds of chives (Allium schoenoprasum). Acta Chem. Scand. *16*:2461–2462. Chem. Abst. 58 (1963) 14 339.

Matikkala, E.J. and Virtanen, A.I. (1963) New gamma-glutamylpeptides isolated from the seeds of chives (Allium schoenoprasum). Acta Chem. Scand. *17*:1799–1801. Chem. Abst. 60 (1964) 4362.

Matikkala, E.J. and Virtanen, A.I. (1964) Synthesis of 3,3'-(2-methylethylene-1,2-dithio)-dialanine, an amino acid found as gamma-glutamylpeptide in the seeds of chive (Allium schoenoprasum). Acta Chem. Scand. *18*:2009–2010.

Matikkala, E.J. and Virtanen, A.I. (1965a) Gamma-glutamylpeptidase (glutaminase) in germinating seeds of chive (Allium schoenoprasum). Acta Chem. Scand. *19*:1258–1261.

Matikkala, E.J. and Virtanen, A.I. (1965b) Gamma-glutamylpeptidase in sprouting onion bulbs. Acta Chem. Scand. *19*:1261–1262.

Matikkala, E.J. and Virtanen, A.I. (1967) On the quantitative determination of the amino acids and γ-glutamylpeptides of onion. Acta Chem. Scand. *21*:2891–2893. Chem. Abst. 68 (1968) 57 355.

Matikkala, E.J. and Virtanen, A.I. (1970) Isolation of gamma-L-glutamyl-L-arginine and gamma-L-glutamyl-S-(2-carboxy-N-propyl)-L-cysteine from onion (Allium cepa). Suom. Kemistil. B. 43:435–438. Chem. Abst. 74 (1971) 61 610.

Matkovics, B., Varga, S.I., and Matkovics, I. (1981) Properties of enzymes. VI/A. Some data on the superoxide dismutase and peroxidase contents of fruits, seeds, and different parts of plants. Acta Biol. Szeged [Hungary] 27:25–31. Chem. Abst. 97 (1982) 20 751.

Matsuda, K. (1982) Pharmaceuticals containing eritadenine from Basidiomycetes or allicin and alliin from Allium. Japan. Patent 82 142 917. Chem. Abst. 97 (1982) 203 216.

Matsuda, K. (1990) Scordinine as an index compound in the assay of processed garlic and quantitative analysis of alliin in processed garlic. Toyama-ken Yakuji Kenkyusho Nenpo 18:117–127. Chem. Abst. 117 (1992) 46 835.

Matsuda, T., Takada, N., Yano, Y., Wanibuchi, H., Otani, S., and Fukushima, S. (1994) Dose-dependent inhibition of glutathione S-transferase placental form-positive hepatocellular foci induction in the rat by methyl propyl disulfide and propylene sulfide from garlic and onions. Cancer Lett. 86:229–234. Chem. Abst. 122 (1995) 545.

Matsukawa, T., Yurugi, S., and Matsuoka, T. (1953) Products of the reaction between thiamine and ingredients of the plants of Allium genus: detection of allithiamine and its homologs. Science 118:325–327. Chem. Abst. 48 (1954) 3488.

Matsukawa, T. and Yurugi, S. (1952a) On the structure of allithiamine. Proc. Imp. Acad. [Tokyo] 28:146–149.

Matsukawa, T. and Yurugi, S. (1952b) Studies on vitamin B1 and related compounds. XXXVIII. Isolation of allithiamine. Yakugaku Zasshi (J. Pharm. Soc. Japan) 72:1602–1604. Chem. Abst. 47 (1953) 9332 (Japanese).

Matsukawa, T. and Yurugi, S. (1952c) Studies on vitamin B1 and related compounds. XXXIX. Structure of allithiamine. Yakugaku Zasshi (J. Pharm. Soc. Japan) 72:1616–1619. Chem. Abst. 47 (1953) 9330.

Matsuura, H., Ushiroguchi, T., Tsuyoshi, I., Hayashi, N., and Fuwa, T. (1988) A furostanol glycoside from garlic, bulbs of Allium sativum L. Chem. Pharm. Bull. 36:3659–3663. Chem. Abst. 110 (1989) 54 478.

Matsuura, H., Morita, T., Gokuchi, T., Itakura, Y., and Hayashi, N. (1989a) Antifungal steroid saponins from Allium. Japan. Patent 89 224 396. Chem. Abst. 112 (1990) 125 183.

Matsuura, H., Ushiroguchi, T., Itakura, Y., and Fuwa, T. (1989b) Further studies on steroidal glycosides from bulbs, roots, and leaves of Allium sativum L. Chem. Pharm. Bull. 37:2741–2743. Chem. Abst. 112 (1990) 232 505.

Matsuura, H., Ushiroguchi, T., Itakura, Y., and Fuwa, T. (1989c) A furostanol glycoside from Allium chinense G. Don. Chem. Pharm. Bull. 37:1390–1391. Chem. Abst. 111 (1989) 228 948.

Matsuzawa, M., Kawai, H., and Hosogai, Y. (1984) Selenium content of foods of plant origin. Joshi Eiyo Daigaku Kiyo 13:141–143. Chem. Abst. 98 (1983) 214 381 (Japanese).

Matthess, K. (1988) Garlic preparation with an odor-reducing effect. Germany. Patent 3 743 264 A1.

Matthess, K. (1990) Garlic preparation with a strong odor-reducing effect. Germany. Patent 3 900 447 A1.

Mattioli, P.A. (1526) Kreutterbuch deb... D. Petri Andreae Matthioli, etc. Jacob Fischer, Frankfurt-on-Main.

Matysek, V. and Behounkova, L. (1968) Antibiotische Eigenschaften des Knoblauchs. Acta Univ. Palacki. Olomuc. 40:53–58. Chem. Zbl. 1968, II, 1264.

Maurizio, A. (1928) Die Geschichte unserer Pflanzennahrung von den Urzeiten bis zur Gegenwart. Paul Parey, Berlin.

Maurya, A.K. and Singh, S.V. (1991) Differential induction of glutathione transferase isoenzymes of mice stomach by diallyl sulfide, a naturally occurring anticarcinogen. Cancer Lett. 57:121–129.

May, S. (1926) Zur Behandlung arteriosklerotischer Beschwerden. Fortschr. Ther. 2:762–764.

Mayer, R.A. (1951) Allium sativum L., Knoblauch, eine sehr wertvolle Arznei- und Gewürzpflanze. Pharmazie 6:680–686.

Mayerhofer, E. (1934) Nutzen und Schaden der Knoblauchanwendung im Kindesalter mit besonderer Berücksichtigung der Coeliakie. Arch. Kinderheilkd. 102:106–116.

Mayeux, P.R., Agrawal, K.C., Tou, J.S.H., King, B.T., Lippton, H.L., Hyman, A.L., Kadowitz, P.J., and McNamara, D.B. (1988) The pharmacological effects of allicin, a constituent of garlic oil. Agents Actions 25:182–190. Chem. Abst. 109 (1988) 104 186.

Mazelis, M. (1963) Demonstration and characterization of cysteine sulfoxide lyase in the Cruciferae. Phytochemistry 2:15–22. Chem. Abst. 58 (1963) 11 625.

Mazelis, M. and Creveling, R.K. (1975) Purification and properties of S-alkyl-L-cysteine lyase from seedlings of Acacia farnesiana Willd. Biochem. J. 147:485–491.

Mazelis, M. and Crews, L. (1968) Purification of the alliin lyase of garlic, Allium sativum L. Biochem. J. 108:725–730. Chem. Abst. 69 (1968) 64 7125.

Mazza, G., Ciaravolo, S., Chiricosta, G., and Celli, S. (1992) Volatile flavour components from ripening and mature garlic bulbs. Flavour Fragrance J. 7:111–116. Chem. Abst. 117 (1992) 149 676.

McCully, R.S. (1964) Vampirism: historical perspective and underlying process in relation to a case of auto-vampirism. J. Nerv. Ment. Dis. 139:140–152.

McFadden, J.P., White, I.R., and Rycroft, R.J.G. (1992) Allergic contact dermatitis from garlic. Contact Dermatitis 27:333–334.

McKnight, R.S. and Lindegren, C.C. (1936) Bactericidal effects of vapors from crushed garlic on Mycobacterium leprae. Proc. Soc. Exp. Biol. Med. 35:477–479.

McMahon, F.G. and Vargas, R. (1993) Can garlic lower blood pressure? A pilot study. Pharmacother. 13:406–407. Int. Pharm. Abst. 31 (1994) 170. Congress abstract in: Eur. J. Clin. Res. 3A (1992), 8–9.

McRary, W.L. and Slattery, M.C. (1945) The colorimetric determination of fructosan in plant material. J. Biol. Chem. 157:161–167.

Mei, X., Wang, M.L., Xu, H.X., Pan, X.Y., Gao, C.Y., Han, N., and Fu, M.Y. (1982) Garlic and gastric cancer: the influence of garlic on the level of nitrate and nitrite in gastric juice. Acta Nutr. Sin. 4:53–56.

Mei, X., Wang, M., Li, T., Gao, C., Han, N., Fu, M., Lin, B., and Nie, H. (1985) Garlic and gastric cancer. II. The inhibitory effect of garlic on nitrate-reducing bacteria and their production of nitrite in gastric juice. Yingyang Xuebao (Acta Nutr. Sinica) 7:173–177. Chem. Abst. 104 (1986) 33 429 (Chinese).

Mei, X., Lin, X., Liu, J., Song, P., Hu, J., and Liang, X. (1989) Garlic inhibition of the formation of N-nitrosoproline in the human body. Acta Nutr. Sin. 11:141–145. Chem. Abst. 112 (1990) 53 927.

Meier, B., Hasler, A., and Keller, B. (1989) Analytical strategies for the standardization of ginkgo and garlic preparations. Planta Med. 55:639–640.

Meier, G. (1986) Garlic preparation with an odor-reducing effect. Germany. Patent 3 525 258.

Melkozerowa, T.W. (1978) Study on the action of natural garlic preparations for the treatment of parodontosis. Zdravookhr. Beloruss. (Health Care), 81–82. (Russian).

Mellouki, F., Vannereau, A., Auger, J., Marcotte, J.L., and Cosson, L. (1994) Flavor production in tissue cultures of chive (A. schoenoprasum L.): callus structure and flavor production. Plant Sci. 95:165–173.

Melzig, M.F., Krause, E., and Franke, S. (1995) Inhibition of adenosine deaminase activity of aortic endothelial cells by extracts of garlic (Allium sativum L.). Pharmazie 50:359–361. Chem. Abst. 123 (1995) 47 625.

Meng, C.L. and Shyu, K.W. (1990) Inhibition of experimental carcinogenesis by painting with garlic extract. Nutr. Cancer 14:207–217. Chem. Abst. 115 (1991) 43 950.

Meng, Y., Lu, D., Guo, N., Zhang, L., and Zhou, G. (1993) Anti-HCMV effect of garlic components. Bingduxue Zazhi (Virologica Sinica) 8:147–150. Chem. Abst. 120 (1994) 86 207 (Chinese).

Menke, G. (1994) Wird Vitamin B1 nach oraler Gabe resorbiert? Dtsch. Apoth. Ztg. 130:491–494.

Mennella, J.A. and Beauchamp, G.K. (1991) Maternal diet alters the sensory qualities of human milk and the nursling's behavior. Pediatrics 88:737–744. Ref. in: Österr. Apoth. Ztg. 46 (40) 821 (1992).

Mennella, J.A. and Beauchamp, G.K. (1993) The effects of repeated exposure to garlic-flavored milk on the nursling's behavior. Pediatr. Res. 34:805–808.

Menon, I.S., Kendal, R.Y., Dewar, H.A., and Newell, D.J. (1968) Effect of onions on blood fibrinolytic activity. Br. Med. J. 3:351–352.

Menon, I.S. (1969) Fresh onions and blood fibrinolysis. Br. Med. J. 1:845.

Menon, I.S. (1970) Onions and blood fibrinolysis. Br. Med. J. II:421.

Mezger, J. (1984) Gesichtete Homöopathische Arzneimittellehre. Haug, Heidelberg.

Micera, G. and Dessi, A. (1988) Chromium adsorption by plant roots and formation of long-lived Cr(V) species: an ecological hazard? J. Inorg. Biochem. 34:157–166. Chem. Abst. 110 (1989) 2462.

Micera, G. and Dessi, A. (1989) Oxovanadium(IV) adsorption by plant roots: ESR identification of mobile and immobilized species. J. Inorg. Biochem. 35:71–78. Chem. Abst. 110 (1989) 72 724.

Michahelles, E. (1974) Über neue Wirkstoffe aus Knoblauch (Allium sativum L.) und Küchenzwiebel (Allium cepa L.). Dissertation, University of Munich.

Mie, Y. (1983) Preliminary study on the quality of sugar-preserved fermented garlic preparations. Shipin Kexue [Beijing], 5–12. Chem. Abst. 100 (1983) 173 285 (Chinese).

Miething, H. (1985) Allicin und Öl in Knoblauchzwiebeln: HPLC-Gehaltsbestimmung. Dtsch. Apoth. Ztg. 125:2049–2050. Chem. Abst. 103 (1985) 210 187.

Miething, H. (1988) HPLC analysis of the volatile oil of garlic bulbs. Phytother. Res. 2:149–151. Chem. Abst. 110 (1989) 6560.

Miething, H. and Thober, H. (1985) Knoblauch: eine sehr alte Arzneipflanze. Apoth. J. 10:42–48.

Miller, J.R., Harris, M.O., and Breznak, J.A. (1984) Search for potent attractants of onion flies. J. Chem. Ecol. 10:1477–1488. Chem. Abst. 102 (1985) 1999.

Mimaki, Y., Kawashima, K., Kanmoto, T., and Sashida, Y. (1993) Steroidal glycosides from Allium albopilosum and A. ostrowskianum. Phytochemistry 34: 799–805. Chem. Abst. 120 (1994) 4701.

Mimaki, Y., Nikaido, T., Matsumoto, K., Sashida, Y., and Ohmoto, T. (1994) New steroidal saponins from the bulbs of Allium giganteum exhibiting potent inhibition of cAMP phosphodiesterase activity. Chem. Pharm. Bull. 42:710–714. Chem. Abst. 121 (1994) 31 162.

Minami, T., Boku, T., Inada, K., Morita, M., and Okazaki, Y. (1989) Odor components of human breath after the ingestion of grated raw garlic. J. Food Sci. 54:763–765. Chem. Abst. 111 (1989) 54 556.

Mirelman, D. (1987) Inhibition of growth of Entamoeba histolytica by allicin, the active principle of garlic extract (Allium sativum). J. Infect. Dis. 156:243–244. Ref.in Aerztezeitschr. Naturheilverf. 29 (1988) 88–89.

Mirelman, D. and Varon, S. (1987) Knoblauch für die Amöbiasis-Therapie (Referat). Neue Aerztliche, No. 59, p. 10.

Mirghis, E. and Mirghis, R. (1991) Phytohormonal composition for garlic reproduction by micropropagation. Romania. Patent 102 059. Chem. Abst. 120 (1994) 99 453.

Mirhadi, S.A., Ahuja, S.P., and Gupta, P.P. (1983) Effect of garlic (*Allium sativum* L.) on the lipid composition and lipid biosynthesis in tissues of rabbits during experimental atherosclerosis. J. Nuc. Agric. Biol. *12*:33–37. Chem. Abst. 99 (1983) 138 614.

Mirhadi, S.A., Singh, S., and Gupta, P.P. (1986) Effect of garlic supplementation to atherogenic diet on collagen biosynthesis in various tissues of rabbits. Indian Heart J. *38*:373–377.

Mirhadi, S.A., Singh, S., and Gupta, P.P. (1991) Effect of garlic supplementation to cholesterol-rich diet on development of atherosclerosis in rabbits. Indian J. Exp. Biol. *29*:162–168. Chem. Abst. 115 (1991) 157 797.

Mirhadi, S.A. and Singh, S. (1987) Effect of garlic extract on in vitro uptake of Ca and HPO4 by matrix of sheep aorta. Indian J. Exp. Biol. *25*:22–23.

Mirhadi, S.A. and Singh, S. (1988) Effect of atherogenic diet on synthesis and excretion of cholesterol by intestinal wall of rabbits in vitro. Indian J. Exp. Biol. *26*:74–75.

Misekow, R.W. and Fabian, F.W. (1953) Proteolytic enzymes of garlic. Food Res. *18*:1–5. Chem. Abst. 47 (1953) 10 762.

Mitchell, J.C. (1980) Contact sensitivity to garlic (Allium). Contact Dermatitis *6*:356–357.

Mitomo, K. (1977) Garlic deodorization. Japan. Patent 77 15 655. Chem. Abst. 88 (1978) 73 287.

Mitsui, Y. (1980) Enhancement of enzyme action. Japan. Patent 80 131 388. Chem. Abst. 94 (1981) 43 329.

Mittal, M.M., Mittal, S., Sarin, J.C., and Sharma, M.L. (1974) Effects of feeding onion on fibrinolysis, serum cholesterol, platelet aggregation and adhesion. Indian J. Med. Sci. *28*:144–148.

Miyaji, A. (1978) Odorless garlic powder. Japan. Patent 78 39 504. Chem. Abst. 90 (1979) 53 412.

Miyoshi, A., Hasegawa, Y., Yamamoto, T., Omata, S., Harada, H., Nakazawa, S., Kita, S., Moriga, M., and Kajiyama, G. (1984) Effect of Kyoleopin on various unidentified complaints followed on internal disease. Shinryou to Shin-yaku (Treatment & New Medicine) *21*:1806–1820 (Japanese).

Mizoi, M., Sawayama, S., Kawabata, A., and Homma, S. (1992) Browning reaction of onion by heating. Nippon Eiyo Shokuryo Gakkaishi *45*:441–447. Chem. Abst. 118 (1993) 21 344 (Japanese).

Mizuno, T., Kinpyo, T., and Harada, K. (1957) Studies on the carbohydrates of Allium species. III. The carbohydrate components in garlic, *Allium sativum* Linn. Nippon Nogei Kagaku Kaishi (Trans. Jpn. Agric. Soc.) *31*:572–574. Chem. Abst. 53 (1959) 22 275 (Japanese).

Mizutani, J., Tahara, S., and Nishimura, H. (1979) Allium plant flavor. Kagaku to Seibutsu *17*:814–820. Chem. Abst. 92 (1980) 126 997 (Japanese).

Mizutani, M., Umezawa, H., and Kuramasu, S. (1981) Studies on the agglutinability and hemolytic activity of various phytohemagglutinins to the red blood cells of mice. Jpn. J. Zootech. Sci. *52*:88–96.

Mizutani, M., Umezawa, H., and Kuramasu, S. (1985) Studies on agglutinability and hemolytic activity of various lectins on red blood cells in chicken. Jpn. J. Zootech. Sci. *54*:525–534.

Mo, H., Van Damme, E.J.M., Peumans, W.J., and Goldstein, I.J. (1993) Purification and characterization of a mannose-specific lectin from shallot (Allium ascalonicum) bulbs. Arch. Biochem. Biophys. *306*:431–438. Chem. Abst. 120 (1994) 3075.

Mochizuki, E., Nakayama, A., Kitada, Y., Saito, K., Nakazawa, H., Suzuki, S., and Fujita, M. (1988) Liquid chromatographic determination of alliin in garlic and garlic products. J. Chromatogr. *455*:271–277.

Mohamed-Yasseen, Y. and Splittstoesser, W.E. (1992) Regeneration of onion (Allium cepa) bulbs in vitro. PGRSA Quarterly *20*:76–82. Chem. Abst. 117 (1992) 85 168.

Mohammad, S.F., Brown, S.J., Chuang, H.Y.K., and Mason, R.G. (1980) Isolation, characterization, identification, and synthesis of an inhibitor of platelet function from *Allium sativum* (garlic). Fed. Proc. *39*:543A.

Mohammad, S.F. and Woodward, S.C. (1986) Characterization of a potent inhibitor of platelet aggregation and release reaction isolated from *Allium sativum* (garlic). Thromb. Res. *44*:793–806. Chem. Abst. 106 (1987) 61 014.

Molina, C., Lachaussée, R., Jeanneret, A., Petit, R., and Grouffal, C. (1983) Histoire d'aulx. Presse Med. *NF.12*:1941. Ref. in Dermatosen 32 (1984) No. 2, 40.

Möllering, H. and Bergmeyer, H.U. (1974) Adenosin. In: Methoden der enzymatischen Analyse. Vol. 3. H.U. Bergmeyer, ed. Verlag Chemie, Weinheim, pp. 1967–1970.

Molnar, B., Botz, L., and Szabo, L.G. (1991) Phytochemical assessment of garlic and garlic products by thin-layer chromatography (TLC) with densitometric determination of alliin. Acta Pharm. Hung. *61*:146–152. Chem. Abst. 119 (1993) 103 442 (Hungarian).

Moore, G.S. and Atkins, R.D. (1977) The fungicidal and fungistatic effects of an aqueous garlic extract on medically important yeast-like fungi. Mycologia *69*:341–348.

Morck, H. (1988) Knoblauch bei Hyperlipoproteinämie einsetzbar. Pharm. Ztg. *133*:21.

Morck, H. and Schulz, V. (1992) Knoblauch: Neuer Wirkungsmechanismus. Pharm. Ztg. *137*:4096–4097.

Moreno, M.L., Ferrero, M.L., and De la Torre, C. (1994) The pattern of polypeptides in proliferating cells of Allium cepa L. roots during cold and heat conditions. J. Plant. Physiol. *143*:759–762. Chem. Abst. 121 (1994) 78 613.

Moriconi, D.N., Conci, V.C., and Nome, S.F. (1989) In vitro platelet regeneration from callus in garlic (*Allium sativum* L.). Phyton (Buenos Aires) *49*:97–103. Chem. Abst. 114 (1991) 242 703.

Moriguchi, T., Takashina, K., Chu, P.J., Saito, H., and Nishiyama, N. (1994) Prolongation of life span and improved learning in the senescence accelerated mouse produced by aged garlic extract. Biol. Pharm. Bull. *17*:1589–1594.

Morimitsu, Y., Morioka, Y., and Kawakishi, S. (1992) Inhibitors of platelet aggregation generated from mixtures of Allium species and/or S-alk(en)nyl-L-cysteine sulfoxides. J. Agric. Food Chem. *40*:368–372. Chem. Abst. 116 (1992) 127 290.

Morimitsu, Y. and Kawakishi, S. (1991) Optical resolution of 1-(methyl-sulfinyl)-propyl alk(en)yl disulfides, inhibitors of platelet aggregation isolated from onion. Agric. Biol. Chem. *55*:889–890.

Morinaga, J. (1993) Soybean oil for removal of garlic odor. Japan. Patent 93 304 924. Chem. Abst. 120 (1994) 132 821.

Morinaga, M. and Tomoda, M. (1984) Deodorized liquid garlic extract. Canada. Patent 1 169 697. Chem. Abst. 101 (1984) 150 215.

Morioka, N., Sze, L.L., Morton, D.L., and Irie, R.F. (1993) A protein fraction from aged garlic enhances cytotoxicity and proliferation of human lymphocytes mediated by interleukin-2 and concanavalin A. Cancer Immunol. Imunother. *37*:316–322. Chem. Abst. 120 (1994) 153 276.

Morita, T., Ushiroguchi, T., Hayashi, N., Matsuura, H., Itakura, Y., and Fuwa, T. (1988) Steroidal saponins from elephant garlic, bulbs of Allium ampeloprasum L. Chem. Pharm. Bull. *36*:3480–3486. Chem. Abst. 110 (1989) 54 474.

Morozov, A.S. (1950) Effect of gramicidin and the phytoncides of onion and garlic on the synthetic and hydrolytic activity of red clover invertase. Dokl. Akad. Nauk SSSR (Rep. Acad. Sci. USSR) *70*:269–270. Chem. Abst. 45 (1951) 4789 (Russian).

Morris, J., Burke, V., Mori, T.A., Vandongen, R., and Beilin, L.J. (1995) Effects of garlic extract on platelet aggregation: a randomized placebo-controlled double-blind study. Clin. Experim. Pharmacol. Physiol. *22*:414–417.

Morse, D.L., Pickard, L.K., Guzewich, J.J., Devine, B.D., and Shayegani, M. (1990) Garlic-in-oil associated botulism: episode leads to product modification. Am. J. Public Health *80*:1372–1373.

Moser, H. (1937) Geruchsarme Präparate aus Lauch- oder Senföl enthaltenden Drogen. Germany. Patent 647 067, 1–2. Chem. Zbl. 1937, II, 2033.

Moser, H. (1948) Fermentprobleme in der galenischen Pharmazie. Pharmazie *3*:433–439. Chem. Abst. 43 (1949) 5901.

Mossa, J.S. (1985) A study on the crude antidiabetic drugs used in Arabian folk medicine. Int. J. Crude Drug Res. *23*:137–145.

Most, G.F. (1843) Encyklopädie der gesamten Volksmedicin oder Lexikon der vorzüglichsten und wirksamsten Haus- und Volksarzneimittel aller Länder. Brockhaus, Leipzig.

Mostler, U. and Urban, E. (1989) Synthese und Konfigurationszuordnung eines potentiell antimikrobiellen 5,6-Dihydroxyisobenzofuranons. Monatsh. Chem. *120*:349–355.

Mubarak, A.M. and Kulatilleke, C.P. (1990) Sulfur constituents of neem seed volatiles: a revision. Phytochemistry *29*:3351–3352. Chem. Abst. 114 (1991) 78 677.

Müller, A.L. and Virtanen, A.I. (1965) Über die Biosynthese von Cycloalliin. Acta Chem. Scand. *19*:2257–2258. Chem. Abst. 64 (1966) 11 558.

Müller, B. (1989) Analytische Bewertung von Knoblauchpräparaten. Dtsch. Apoth. Ztg. *129*:2500–2504. Chem. Abst. 112 (1990) 42 708.

Müller, B. (1990) Garlic (*Allium sativum*): quantitative analysis of the tracer substances alliin and allicin. Planta Med. *56*:589–590.

Müller, B. and Aye, R.D. (1991) Analytische Methoden zur Standardisierung von Frischknoblauch und Knoblauchpulverpräparaten. Dtsch. Apoth. Ztg. *131*:8–10.

Müller, B. and Ruhnke, A. (1993) Qualitätsprüfung von Knoblauchpulver. Vorschläge für eine Arzneibuchmonographie. Dtsch. Apoth. Ztg. *133*:2177–2187. Chem. Abst. 121 (1994) 163 739.

Müller, B.H. (1991) Beitrag zur Charakterisierung und Analytik von Alliin und *Allium sativum* und seinen Zubereitungen. Dissertation, University of Erlangen (Abstract).

Müller, J. (1982) Die pflanzlichen Heilmittel bei Hildegard von Bingen. O. Müller, Salzburg.

Müller-Dietz, H., Kraus, E.M., and Rintelen, K. (1969) Allium. In: Arzneipflanzen in der Sowjetunion. Vol. 5, pp. 118–119.

Müller-Dietz, H. and Rintelen, K. (1960) Allium. In: Arzneipflanzen in der Sowjetunion. Vol. 1, pp. 34–45.

Münch, W. (1966) Allium cepa (Küchenzwiebel), *Allium sativum* (Knoblauch), Allium ursinum (Bärenlauch) und ihre Bedeutung für die Therapie. Allg. Homöopath. Ztg. *211*:8–14.

Munday, R. and Manns, E. (1994) Comparative toxicity of prop(en)yl disulfides derived from Alliaceae: possible involvement of 1-propenyl disulfides in onion-induced hemolytic anemia. J. Agric. Food Chem. *42*:959–962. Chem. Abst. 120 (1994) 237 743.

Munoz, C. and Escaff, M. (1987) Multiplicacion del ajo (*Allium sativum* L.) a partir de bulbillos obtenidos in vitro. Simiente *57*:106.

Murakami, F. (1960a) Studies on the nutritional value of Allium plants. XXXVI. Decomposition of alliin homologues by microorganisms and formation of substance with thiamine masking activity. Bitamin [Kyoto] *20*:126–131. Chem. Abst. 61 (1964) 15 093 (Japanese).

Murakami, F. (1960b) Studies on the nutritional value of Allium plants. XXXVII. Decomposition of alliin homologues by acetone-powdered enzyme preparation of Bacillus subtilis. Bitamin [Kyoto] 20:131–135. Chem. Abst. 61 (1964) 15 093 (Japanese).

Murakami, F. and Tazoe, K. (1960) The nutritional value of Allium plants. XXXV. Influences of allithiamine and other vitamins upon the liver tissue respiration in acute hydrocyanic acid poisoning. Bitamin [Kyoto] 20:123–125. Chem. Abst. 61 (1964) 15 093 (Japanese).

Murthy, N.B.K. and Amonkar, S.V. (1974) Effect of a natural insecticide from garlic (Allium sativum L.) and its synthetic form (diallyl-disulphide) on plant pathogenic fungi. Indian J. Exp. Biol. 12:208–209. Chem. Abst. 81 (1974) 164 537.

Mütsch-Eckner, M. (1991) Isolierung, Analytik und biologische Aktivität von Aminosäuren und Dipeptiden aus Allium sativum L. Dissertation, Eidgenössische Technische Hockschule, Zürich.

Mütsch-Eckner, M., Meier, B., Wright, A.D., and Sticher, O. (1992a) Gamma-glutamyl peptides from Allium sativum bulbs. Phytochemistry 31: 2389–2391. Chem. Abst. 117 (1992) 207 371.

Mütsch-Eckner, M., Sticher, O., and Meier, B. (1992b) Reversed-phase high-performance liquid chromatography of S-alk(en)yl-L-cysteine derivatives in Allium sativum including the determination of (+)-S-allyl-L-cysteine sulphoxide, gamma-L-glutamyl-S-allyl-L-cysteine and gamma-L-glutamyl-S-(trans-1-propenyl)-L-cysteine. J. Chromatogr. 625:183–190. Chem. Abst. 118 (1993) 79 575.

Mütsch-Eckner, M., Erdelmeier, C.A.J., and Sticher, O. (1993) A novel amino acid glycoside and three amino acids from Allium sativum. J. Nat. Prod. 56:864–869. Chem. Abst. 119 (1993) 113 409.

Nadkarni, K.M. (1954) Indian Materia Medica. Vol. 1. Popular Book Depot, Bombay.

Nagae, S., Ushijima, M., Hatano, S., Imai, J., Kasuga, S., Matsuura, H., Itakura, Y., and Higashi, Y. (1994) Pharmacokinetics of the garlic compound S-allyl cysteine. Planta Med. 60:214–217. Chem. Abst. 121 (1994) 99 040.

Nagai, H. (1978) Removal of odor and pungent taste from garlic. Japan. Patent 78 38 650. Chem. Abst. 89 (1978) 89 132.

Nagai, K. (1973a) Experimental studies on the preventive effect of garlic extract against infection with influenza virus. J. Jpn. Assoc. Infect. Dis. 47:321–325 (Japanese).

Nagai, K. (1973b) Experimental studies on preventive effect of garlic extract against infections with influenza and Japanese encephalitis viruses in mice. J. Jpn. Assoc. Infect. Dis. 47:111–115 (Japanese).

Nagai, K., Nakagawa, S., Nojima, S., and Mimori, H. (1975) Effect of aged garlic extract on glucose tolerance test in rats. Basic Pharmacol. Ther. 3:45–53 (Japanese).

Nagai, K. and Osawa, S. (1974) Cholesterol lowering effect of aged garlic extract in rats. Basic Pharmacol. Ther. 2:41–50.

Nagakubo, T., Nagasawa, A., and Ohkawa, H. (1993) Micropropagation of garlic through in vitro bulblet formation. Plant Cell Tissue Organ Cult. 32:175–183. Chem. Abst. 119 (1993) 2995.

Nagasawa, A. and Finer, J.J. (1988a) Induction of morphogenic callus cultures from leaf tissue of garlic. Hort. Sci. 23:1068–1070.

Nagasawa, A. and Finer, J.J. (1988b) Development of morphogenic suspension cultures of garlic (Allium sativum L.). Plant Cell Tissue Organ Culture 15:183–187.

Nagda, K.K., Ganeriwal, S.K., Nagda, K.C., and Diwan, A.M. (1983) Effect of onion and garlic on blood coagulation and fibrinolysis in vitro. Indian J. Physiol. Pharmacol. 27:141–145.

Naito, S., Yamaguchi, N., and Yokoo, Y. (1981a) Antioxidative activities of vegetables of Allium species: studies on natural antioxidant. J. Jpn. Soc. Food Sci. Technol. 28:291–296. Chem. Abst. 96 (1982) 50 860 (Japanese).

Naito, S., Yamaguchi, N., and Yokoo, Y. (1981b) Studies on natural antioxidant. III. Fractionation of antioxidant activity from garlic extract. Nippon Shokuhin Kogyo Gakkaishi (J. Jpn. Soc. Food Sci. Technol.) 28:465–470. Chem. Abst. 96 (1982) 50 860 (Japanese).

Nakagawa, S., Sumiyoshi, H., Masamoto, K., Kanezawa, A., Harada, H., Nakagami, S., Date, S., Yokota, A., Nishikawa, M., and Fuwa, T. (1984) Acute and subacute toxicity tests of a ginseng and garlic preparation containing vitamin B1 (Leopin-Five). Oyo Yakuri (Pharmacometrics) 27:1133–1150 (Japanese).

Nakagawa, S., Yoshida, S., Hirao, Y., Kasuga, S., and Fuwa, T. (1985) Cytoprotective activity of components of garlic, ginseng, and ciuwjia on hepatocyte injury induced by carbon tetrachloride in vitro. Hiroshima J. Med. Sci. 34:303–309. Chem. Abst. 104 (1986) 124 397.

Nakagawa, S., Kasuga, S., and Matsuura, H. (1989) Prevention of liver damage by aged garlic extract and its components in mice. Phytother. Res. 3:50–53. Chem. Abst. 111 (1989) 146 760.

Nakamura, K. and Tamura, G. (1990) Isolation of serine acetyltransferase complexed with cysteine synthase from Allium tuberosum. Agric. Biol. Chem. 54:649–656. Chem. Abst. 113 (1990) 2378.

Nakamura, S. and Tahara, M. (1977) Determination of species and cultivars on the basis of electrophoretic patterns of seed protein and seed enzymes. II. Allium species, cucumber and melon, carrot, pea, and garden bean. Engei Gakkai Zasshi 46:233–244. Chem. Abst. 91 (1979) 87 347.

Nakata, C., Nakata, T., and Hishikawa, A. (1970) An improved colorimetric determination of thiolsulfinate. Anal. Biochem. 37:92–97. Chem. Abst. 73 (1970) 106 145.

Nakata, T. (1973) Effect of fresh garlic extract on tumor growth. Nippon Eiseigaku Zasshi (Jap. J. Hyg.) 27:538–543. Chem. Abst. 79 (1973) 111 680 (Japanese).

Nakata, T. and Fujiwara, M. (1975) Adjuvant action of garlic sugar solution in animals immunized with Ehrlich ascites tumor cells attenuated with allicin. Gann 66:417–419.

Narvaiz, P. (1995) Chemiluminescence measurements on irradiated garlic powder by the single photon counting technique. Radiat. Phys. Chem. 45:203–206. Chem. Abst. 121 (1994) 299 501.

Nassauer, G. (1990) Ein 60jähriger schlägt die Asse der Tour-de-France: Knoblauch als Doping: Mit dem Rad durch ganz Europa! Kronen-Zeitung (Vienna), 10 August 1990, p. 14.

Nasseh, M.O. (1982) Zur Wirkung von Knoblauch-Extrakt auf Syrphus corollae F., Chrysopa carnea Steph. und Coccinella septempunctata L. Z. Angew. Entomol. 94:123–126.

Nasseh, M.O. (1983) Wirkung von Rohextrakten aus *Allium sativum* L. auf Getreideblattläuse Sitobion avebae F. und Rhopalosiphum padi L. sowie die Grüne Pfirsichblattlaus Mycus persicae Sulz. Z. Angew. Entomol. 95:228–230.

Nath, A., Sharma, N.K., Bhardwaj, S., and Thapa, C.D. (1982) Nematicidal properties of garlic. Nematologica 28:253–255.

Neil, A. and Silagy, C. (1994) Garlic: its cardio-protective properties. Curr. Opin. Lipidol. 5:6–10. Chem. Abst. 120 (1994) 215 784.

Neves, R.G. (1964) Contact dermatitis to bulbs of Liliaceae. Anais Bras. Dermatol. 39:23–31 (Portuguese).

Nielson, K.K., Mahoney, A.W., Williams, L.S., and Rogers, V.C. (1991) x-Ray fluorescence measurements of Mg, P, S, Cl, K, Ca, Mn, Fe, Cu, and Zn in fruits, vegetables, and grain products. J. Food Comp. Analysis 4:39–51.

Nikolaeva, V.G. (1979) Pflanzen, die bei den Völkern der UdSSR zur Behandlung von infizierten Wunden verwendet werden. Farmatsiya [Moscow] (Pharmacy) 28:46–49 (Russian).

Nishii, K., Kominato, K., and Kominato, Y. (1970) 3-Creatininyl-4-methyl-5-(b-hydroxyethyl)thiazolium bromide. Japan. Patent 70 24 779. Chem. Abst. 74 (1971) 31 750.

Nishimura. H. and Seibutsu Shigen Kenkyusho Kk. (1989) Preparation of deodorizing food containing garlic and vitamin B1. Japan. Patent 89 104 142. Chem. Abst. 111 (1989) 152 492.

Nishimura, H., Wijaya, C.H., and Mizutani, J. (1988) Volatile flavor components and antithrombotic agents: vinyldithiins from Allium victorialis L. J. Agric. Food Chem. 36:563–566. Chem. Abst. 108 (1988) 203 507.

Nishimura, S. and Kominato, K. (1971) Scordinin halides. Japan. Patent 71 15 619. Chem. Abst. 75 (1971) 49 069.

Nishino, H., Iwashima, A., Itakura, Y., Matsuura, H., and Fuwa, T. (1989) Antitumor-promoting activity of garlic extracts. Oncology 46:277–280. Chem. Abst. 111 (1989) 89 940.

Nishino, H., Nishino, A., Satomi, Y., Takayasu, J., Hasegawa, T., Tokuda, H., Fukuda, T., Tanaka, H., and Shibata, S. (1990a) Antitumor promoter principles in Allium spp. Kyto furitsu Ika Daigaku Zasshi (J. Kyoto Pref. Univ. Med.) 99:1159–1164. Chem. Abst. 114 (1990) 135 660.

Nishino, H., Nishino, A., Takayasu, J., Iwashima, A., Itakura, Y., Kodera, Y., Matsuura, H., and Fuwa, T. (1990b) Antitumor-promoting activity of allixin, a stress compound produced by garlic. Cancer J. 3:20–21. Chem. Abst. 112 (1990) 191 529.

Nishino, H. (1993) Cancer preventive agents in processed garlic. Spec. Publ. R. Soc. Chem. 123:290–294. Chem. Abst. 119 (1993) 224 735.

Nissan Chem. Ind. Ltd. (1983) Food-deodorizing packaging materials. Japan. Patent 83 179 477. Chem. Abst. 100 (1983) 66 959.

Nitschke, D. and Smoczkiewiczowa, A. (1976) Saponins in garlic and onion. Zesz. Nauk. Akad. Ekon. Poznan, Ser. I (Sci. Notes Econ. Acad. Poznan) 69:104–108. Chem. Abst. 87 (1977) 164 252 (Polish).

Nock, L.P. and Mazelis, M. (1986) The carbon-sulfur lyases of higher plants: preparation and properties of homogeneous alliin lyase from garlic (*Allium sativum*). Arch. Biochem. Biophys. 249:27–33. Chem. Abst. 105 (1986) 148 620.

Nock, L.P. and Mazelis, M. (1987) The carbon-sulfur lyases of higher plants: direct comparison of the physical properties of homogeneous alliin lyase of garlic (*Allium sativum*) and onion (Allium cepa). Plant Physiol. 85:1079–1083. Chem. Abst. 108 (1988) 127 279.

Nock, L.P. and Mazelis, M. (1989) The carbon-sulfur lyases from higher plants. V. Lack of homology between the alliin lyases of garlic and onion. Phytochemistry 28:729–731. Chem. Abst. 110 (1989) 151 364.

Noda, K., Taniguchi, H., Suzuki, S., and Hirai, S. (1983) Comparison of the selenium contents of vegetables of the genus Allium measured by fluorometry and neutron activation analysis. Agric. Biol. Chem. 47:613–615. Chem. Abst. 98 (1983) 177 589.

Noda, K., Isozaki, S., and Taniguchi, H. (1985) Growth promoting and inhibiting effects of spices on Escherichia coli. Nippon Shokuhin Kogyo Gakkaishi (J. Jpn. Soc. Food Sci. Technol.) 32:791–796. Chem. Abst. 104 (1986) 108 088 (Japanese).

Noda, M., Tanaka, M., Seto, Y., Aiba, T., and Oku, C. (1988) Occurrence of cholesterol as a major sterol component in leaf surface lipids. Lipids 23:439–444. Chem. Abst. 109 (1988) 35 293.

Noether, P. (1925) Pharmakologische Untersuchung des Knoblauchs (*Allium sativum*). Münch. Med. Wochenschr. 72:1641–1642. Chem. Zbl. 1926, I, 437.

Nojiri, H., Toyomasu, T., Yamane, H., Shibaoka, H., and Murofushi, N. (1993) Qualitative and quantitative analysis of endogenous gibberellins in onion plants and their effects on bulb development. Biosci. Biotech. Biochem. 57:2031–2035. Chem. Abst. 120 (1994) 158 817.

Nolte, D.L., Provenza, F.D., Callan, R., and Panter, K.E. (1992) Garlic in the ovine fetal environment. Physiol. Behav. 52:1091–1093.

Nomura, J., Nishizuka, Y., and Hayaishi, O. (1963) S-alkylcysteinase: enzymatic cleavage of S-methyl-L-cysteine and its sulfoxide. J. Biol. Chem. 238:1441–1446.

Nomura, Y., Maeda, M., Tsuchiya, T., and Makara, K. (1994) Efficient production of interspecific hybrids between Allium chinense and edible Allium spp. through ovary culture and pollen storage. Breeding Sci. 44:151–155.

Nortje, P.F. and Henrico, P.J. (1985) Evaluation of herbicides for long-term weed control in garlic (Allium sativum L.). S. Afr. J. Plant Soil 2:85–88. Chem. Abst. 107 (1987) 129 062.

Nour, R.A., Schneider, K., and Urban, E. (1992) Synthesis of a 4,7-dihydroxyperhydroisobenzofuran-1-one and comparison with Zwergal's structure of "Garlicin". Liebigs Ann. Chem., 383–386. Chem. Abst. 116 (1992) 214 211.

Novak, F.J. and Havranek, P. (1975) Attempts to overcome the sterility of common garlic (Allium sativum L.). Biol. Plantarum [Prague] 17:376–379.

Novak, F.J. (1978) In vitro techniques in garlic (Allium sativum L.) breeding. In: Use of Tissue Cultures in Plant Breeding. Czech. Acad. Sci. Inst. Exp. Bot., Prague, pp. 165–181.

Novak, F.J. (1980) Phenotype and cytological status of plants regenerated from callus cultures of Allium sativum. Z. Pflanzenzücht. 84:250.

Novak, F.J. (1981) Chromosomal characteristics of long-term callus cultures of Allium sativum. L. Cytologia [Tokyo] 46:371.

Novak, F.J., Havel, L., and Dolezel, J. (1982) In vitro breeding system of Allium. In: Plant tissue culture. A. Fujiwara, ed. Proc. 5th Intern. Congress of Plant Tissue Culture, Tokyo.

Novak, F.J. (1983) Production of garlic (Allium sativum L.) tetraploids in shoot-tip in vitro culture. Z. Pflanzenzüchtg. 91:329–333.

Novak, F.J. (1984) Somaclonal variation in garlic tissue culture as a new breeding system. Proc. 3rd Eucarpia Allium Symp., 39–43.

Novak, F.J., Havel, L., and Dolezel, J. (1985) Allium. In: Handbook of plant cell culture. Techniques and applications. D.A. Evans, W.R. Sharp, and P.V. Ammirato, eds. Macmillan, New York, pp. 419–456.

Novikov, I.E. (1954) Therapy with the volatile fractions of the phytoncides from garlic of anginas, influenza, and serious catarrhs of the upper respiratory tract. Voen. Med. (Military Med.) 12:56–59 (Russian).

Novikova, M.A., Levi, I.S., and Khokhlov, A.S. (1957) On the antitumoral action of alliin. Antibiotiki [Moscow] 2:41–46. Chem. Zbl. 1958, I, 4811 (Russian).

Nunez, A.J., Iglesias, I., and Magraner, J. (1985) Headspace constituents of Allium species preserved by cobalt-10 gamma irradiation. Rev. Cienc. Quim. 16:285–288. Chem. Abst. 107 (1987) 95 465 (Spanish).

Oakenfull, D. (1981) Saponins in food: a review. Food Chem. 6:19–40.

Oaks, D.M., Hartmann, H., and Dimick, K.P. (1964) Analysis of sulfur compounds with electron capture/hydrogen flame dual channel gas chromatography. Anal. Chem. 36:1560–1565. Chem. Abst. 61 (1964) 6404.

Oefele, F.V. (1902) Keilschriftmedizin in Parallelen. Vorderasiat. Ges., Leipzig.

Oelkers, B., Diehl, H., and Liebig, H. (1992) In vitro inhibition of cytochrome P-450 reductases from pig liver microsomes by garlic extracts. Arzneim. Forsch. 42:136–139.

Ogawa, H., Suezawa, K., Meguro, T., and Sasagawa, S. (1993) Effect of garlic powder on lipid metabolism in stroke-prone spontaneously hypertensive rats. Nippon Eiyo, Shokuryo Gakkaishi 46:417–423. Chem. Abst. 120 (1994) 29 982.

Ogston, D. (1983) Effect of onion and garlic on hemostatic function. In: The phasiology of hemostasis. Croom Helm, London, pp. 311–312.

Ogston, D. (1985) Nutritional influence on the fibrinolytic system. Proc. Nutr. Soc. 44:379–384.

Ohashi, K. (1965) Garlic extract. Japan. Patent 68 84. Chem. Abst. 68 (1968) 94 752.

Ohga, S. (1988) Biotechnological studies of edible mushroom cultivation. VII. Growth stimulating activity for Lentinus edodes by hot-water extracts of Allium fistulosum and nucleic acid–related compounds therein as the principle active components. Mokuzai Gakkaishi 34:745–752. Chem. Abst. 110 (1989) 54 641.

Ohm, H.T., Song, D.B., and Cha, C.W. (1986) A study on the protective effect of garlic, DMSA (2,3-dimercaptosuccinic acid) and BAL to cadmium poisoning in rat. Koryo Taehakkyo Uikwa Taehak Nonmunjip (Korea Univ. Med. J.) 23:109–119. Chem. Abst. 105 (1986) 92 667 (Korean).

Ohnishi, S.T. and Kojima, R. (1994) In vitro anti-oxidant activity of an aged garlic extract. FASEB J. 8:A596.

Ohsumi, C., Hayashi, T., Kubota, K., and Kobayashi, A. (1993a) Volatile flavor compounds formed in an interspecific hybrid between onion and garlic. J. Agric. Food Chem. 41:1808–1810. Chem. Abst. 119 (1993) 202 140.

Ohsumi, C., Hayashi, T., and Sano, K. (1993b) Formation of alliin in the culture tissues of Allium sativum oxidation of S-allyl-L-cysteine. Phytochemistry 33:107–111. Chem. Abst. 119 (1993) 45 317.

Ohsumi, C., Kojima, A., Hinata, K., Etoh, T., and Hayashi, T. (1993c) Interspecific hybrid between

Allium cepa and *Allium sativum*. Theor. Appl. Genet. 85:969–975.

Ohsumi, C. and Hayashi, T. (1994a) The oligosaccharide units of the xyloglucans in the cell walls of bulbs of onion, garlic, and their hybrid. Plant Cell Physiol. 35:963–967. Chem. Abst. 121 (1994) 200 941.

Ohsumi, C. and Hayashi, T. (1994b) Carbohydrate analysis of an interspecific hybrid between onion and garlic. Biosci. Biotechnol. Biochem. 58:959–960. Chem. Abst. 121 (1994) 31 169.

Ojewole, J.A.O., Adekile, A.D., and Odebiyi, O.O. (1982) Pharmacological studies on the Nigerian herbal preparation. I. Cardiovascular actions of cow's urine concoction and its individual components. Int. J. Crude Drug Res. 20:71–85.

Oka, Y., Kiriyama, S., and Yoshida, A. (1973) Sterol composition of vegetables. J. Jpn. Soc. Food Nutr. 26:121–128 (Japanese).

Oka, Y., Kiriyama, S., and Yoshida, A. (1974) Sterol composition of spices and cholesterol in vegetable foodstuffs. J. Jpn. Soc. Food Nutr. 27:347–355. Chem. Abst. 82 (1975) 153 998 (Japanese).

Okada, A. and Fuji Sangyo Co. Ltd. (1988a) Bath preparations containing allithiamine or allithiamine-containing garlic extracts. Japan. Patent 88 307 812. Chem. Abst. 110 (1988) 160 249.

Okada, A. and Fuji Sangyo Co. Ltd. (1988b) Preparation of odorless garlic compositions containing oily substances. Japan. Patent 88 283 550. Chem. Abst. 111 (1989) 76 772.

Okada, A. and Fujiwara K. (1987) Stable Vitamin B1 additive for fish feeds. Japan. Patent 87 36 153. Chem. Abst. 107 (1987) 6095.

Okada, A. and Fujiwara, K. (1989) Allithiamine-containing ointments. Japan. Patent 89 61 419. Chem. Abst. 111 (1989) 120 934.

Okada, K. and Miyagaki, H. (1983) Effect of Kyolepin on fatigue and non-specific complaints: clinical study. Clin. Rep. 17:2173–2183.

Okuyama, T., Shibata, S., Hoson, M., Kawada, T., Osada, H., and Noguchi, T. (1986) Effect of oriental plant drugs on platelet aggregation. III. Effect of Chinese drug "Xiebai" on human platelet aggregation. Planta Med. 52:171–175. Chem. Abst. 105 (1986) 145 674.

Okuyama, T., Fujita, K., Shibata, S., Hoson, M., Kawada, T., Masaki, M., and Yamate, N. (1989) Effects of Chinese drugs "Xiebai" and "Dasuan" on human platelet aggregation (Allium bakeri, A. sativum). Planta Med. 55:242–244. Chem. Abst. 111 (1989) 108 764.

Okwute, S.K., Ndukwe, G.I., Watanabe, K., and Ohno, N. (1986) Isolation of griffonilide from the stem bark of Bauhinia thonningii. J. Nat. Prod. 49:716–717. Chem. Abst. 105 (1986) 232 294.

Olschok, H.W. (1981) Zur fibrinolytischen Wirksamkeit eines Knoblauchpräparates. Dissertation, University of Frankfurt-on-Main (Abstract).

Omidiji, O. (1993) Flavonol glycosides in the wet scale of the deep purple onion (Allium cepa L. cv. Red Creole). Discovery Innovation 5:139–141. Chem. Abst. 120 (1994) 162 112.

Omidiji, O. and Ehimidu, J. (1990) Changes in the content of antibacterial isorhamnetin 3-glucoside and quercetin 3'-glucoside following inoculation of onion (Allium cepa L. cv. Red Creole) with Pseudomonas cepacia. Physiol. Mol. Plant Pathol. 37:281–292. Chem. Abst. 114 (1991) 141 801.

Ontyd, J. and Schrader, J. (1984) Measurement of adenosine, inosine, and hypoxanthine in human plasma. J. Chromatogr. 307:404–409.

Ookubo, T., Matsumoto, M., Kanetake, M., and Yamazaki, N. (1990) Fish feeds containing essential oils. Japan. Patent 90 207 758. Chem. Abst. 114 (1991) 22 846.

Opdyke, D.L.J. (1978) Monographs on fragrance raw materials. Food Cosmet. Toxicol. 16:723–727. Chem. Abst. 79 (1973) 111 680.

Oppikofer, F. (1947) Influence d'un adsorbant (charbon) sur l'activité biologique de l'ail. Une incompatibilité à eviter. Schweiz. Apoth. Ztg. 85:849–850. Chem. Abst. 42 (1948) 2398.

Orekhov, A.N. (1992) Direct anti-atherosclerotic and anti-atherogenic effects of garlic. Eur. J. Clin. Res. 3A:6–7.

Orekhov, A.N., Tertov, V.V., Sobenin, I.A., and Pivovarova, E.M. (1995) Direct anti-atherosclerosis–related effects of garlic. Ann. Medicine 27:63–65.

Ornish, D., Brown, S.E., Scherwitz, L.W., Billings, J.H., Armstrong, W.T., Ports, T.A., McLanahan, S.M., Kirkeeide, R.L., Brand, R.J., and Gould, K.L. (1990) Can lifestyle change reverse coronary heart disease? Lancet 336:129–133.

Örsi, F. (1976) Determination of the content of active ingredients in some garlic-containing foods. Elelmiszervizsgalati Közl. (Food Anal. Comm.) 22:23–29. Chem. Abst. 86 (1977) 41 891 (Hungarian).

Orth-Wagner, S. (1986) Moderne Phytotherapie. Knoblauch (*Allium sativum*). Dtsch. Apotheker 38:42–47.

Ortner, L. (1929) Über die Verwendung des Knoblauchs bei Arteriosklerose. Ärztl. Rundsch., 249.

Orzechowski, G. (1933) Knoblauchöl bei experimenteller Arterienverkalkung. Klin. Wochenschr. 12:509.

Orzechowski, G. and Schreiber, E. (1934) Knoblauchöl bei subakuter Vergiftung durch Vitamin D. Arch. Exp. Pathol. Pharmakol. 175:265–283. Chem. Zbl. 1935, I, 107.

Osaka Yakuhin Kenkyusho Kk. (1984) Deodorization of garlic. Japan. Patent 84 224 664. Chem. Abst. 102 (1985) 202 932.

Osborn, E.M. (1943) On the occurence of antibacterial substances in green plants. Br. J. Exp. Pathol. 24:227–231.

Oshiba, S., Ariga, T., Sawai, H., Imai, H., and Endoh, E. (1981) The inhibitory effect of garlic oil on platelet aggregation. J. Physiol. Soc. Jpn. *43*:407.

Osiander, J.F. (1829) Volksarzneymittel. Meyer, Tübingen.

Osman, S.A. (1984) Chemical and biological studies of onion and garlic in an attempt to isolate a hypoglycaemic extract. Proc. Asian Symp. Med. Plants Spices, 4th, 1980, 117.

Ostrowska, B. (1987) Method for quantitative determination of alliin in garlic (*Allium sativum* L.). Herba Pol. *33*:1–15. Chem. Abst. 109 (1988) 188 886 (Polish).

Osumi, C., Hayashi, T., Sano, T., and Ajinomoto Co. Inc. (1989) Manufacture of alliin and methylcysteine sulfoxide by cell culture of *Allium sativum*. Japan. Patent 89 257 487. Chem. Abst. 112 (1990) 156 758.

Ozek, O. (1945) On the effect of garlic on bacteria. Istanbul Seririyati *27*:156–161. Chem. Abst. 40 (1946) 6547 (Turkish).

Ozturk, Y., Aydin, S., Kosar, M., and Baser, K.H.C. (1994) Endothelium-dependent and independent effects of garlic on rat aorta. J. Ethnopharmacol. *44*:109–116.

Pai, S.T. and Platt, M.W. (1995) Antifungal effects of *Allium sativum* (garlic) extract against the Aspergillus species involved in otomycosis. Lett. Appl. Microbiol. *20*:14–18.

Pan, J., Hong, J.Y., Ma, B.L., Ning, S.M., Paranawithana, S.R., and Yang, C.S. (1993) Transcriptional activation of cytochrome P450 2B1/2 genes in rat liver by diallyl sulfide, a compound derived from garlic. Arch. Biochem. Biophys. *302*:337–342. Chem. Abst. 118 (1993) 224 965.

Pan, X. (1985) Comparison of cytotoxic effect of fresh garlic, diallyl trisulfide, 5-fluorouracil (5-FU), mitomycin C (MMC) and cis-DDP on two lines of gastric cancer cells. Zhonghua Zhongliu Zazhi (Chin. J. Cancer) *7*:103–105. Chem. Abst. 103 (1985) 153 510 (Chinese).

Pan, X., Li, F., Yu, R., Xie, G., Wang, H., Zhao, L., and Zhang, Q. (1988) Experimental chemotherapy of human gastric cancer cell lines in vitro and in nude mice. Chin. J. Oncol. *10*:15–18. Chem. Abst. 109 (1988) 31 680 (Chinese).

Pandey, N.D., Singh, S.R., and Tewari, G.C. (1976) Use of some plant powders, oils, and extracts as protectants against pulse beetle, Callosobruchus chinensis Linn. Indian J. Entomol. *38*:110–113.

Pandey, R., Chandel, K.P.S., and Rao, S.R. (1992) In vitro propagation of Allium tuberosum Rottl. ex. Spreng. by shoot proliferation. Plant Cell Rep. *11*:211–214. Chem. Abst. 117 (1992) 64 749.

Panosyan, A.G. (1981) The search for prostaglandins and prostaglandin-like compounds. Khim. Prir. Soedin. (Chem. Nat. Compd.) *17*:102–108.

Pant, R., Agrawal, H.C., and Kapur, A.S. (1962) The water-soluble sugar and total carbohydrate content of onion (Allium cepa), garlic (*Allium sativum*) and turnip (Brassica rapa). Flora *152*:530–533. Chem. Abst. 59 (1962) 8051.

Pantoja, C.V., Chiang, L.C., Norris, B.C., and Concha, J.B. (1991) Diuretic, natriuretic,, and hypotensive effects produced by *Allium sativum* (garlic) in anaesthetized dogs. J. Ethnopharmacol. *31*:325–331. Int. Pharm. Abst. 28 (1991) 13 738.

Papageorgiou, C., Corbet, J.P., Menezes Brandao, F., Pecegueiro, M., and Benezra, C. (1983) Allergic contact dermatitis to garlic (*Allium sativum* L.) identification of the allergens: the role of mono-, di-, and trisulfides present in garlic, a comparative study in man and animal (guinea pig). Arch. Dermatol. Res. *275*:229–234. Chem. Abst. 99 (1983) 138 366.

Papayannopoulos, G. (1969) Garlic. Lancet *2*:962.

Paracelsus von Hohenheim. (1928) Theophrast von Hohenheim, gen. Paracelsus, Sämtl. Werke., Jena.

Parish, R.A., McIntire, S., and Heimbach, D.M. (1987) Garlic burns: a naturopathic remedy gone awry. Pediatr. Emerg. Care *3*:258–260.

Park, J.S. and Cha, C.W. (1984) A study on the effect of garlic on the toxicity of phenyl mercuric acetate in rats. Koryo Taehakkyo Uikwa Taehak Nonmunjip (Korea Univ. Med. J.) *21*:49–58. Chem. Abst. 102 (1985) 144 326 (Korean).

Park, K.Y., Kim, S.H., Suh, M.J., and Chung, H.Y. (1991) Inhibitory effects of garlic on the mutagenicity in Salmonella system and on the growth of HT-29 human colon carcinoma cells. Hanguk Sikpum Kwahak Hoechi (Korean J. Food Sci. Technol.) *23*:370–374. Chem. Abst. 119 (1993) 47 893.

Park, M.H., Kim, J.P., and Kwon, D.J. (1988) Physicochemical characteristics of components and their effects on freezing point depression of garlic bulbs. Hanguk Sikpum Kwahak Hoechi (Korean J. Food Sci. Technol.) *20*:205–212. Chem. Abst. 111 (1989) 213 555 (Korean).

Park, S.C., Park, J.B., Juhnn, Y.S., Kimm, S.W., and Lee, K.Y. (1981) Evidence for the existence of antimutagenic factor(s) in edible plants. Korean J. Biochem. *13*:180–181.

Park, S.Y., Yun, Y.H., and Kim, H.U. (1980) Studies on the effects of several spices on the growth of Lactobacillus casei YIT9018. Korean J. Anim. Sci. *22*:301–308 (Korean).

Park, Y.B. (1987) Effect of day length on gibberellic acid and nucleic acid content in northern and southern type garlics. Nunmunjip-Cheju Taehak, Chayon Kwahappyon *24*:23–32. Chem. Abst. 111 (1989) 4398 (Korean).

Park, Y.B. (1988) Effect of growth regulators on growth, bulbing, and secondary growth in garlic. Nunmunjip-Cheju Taehak, Chayon Kwahakpyon *27*:29–38. Chem. Abst. 111 (1989) 2633 (Korean).

Park, Y.B. and Lee, B.Y. (1992) Effect of storage temperature on changes in carbohydrate and endogenous hormones in garlic bulbs. Hanguk Wonye Hakhoechi *33*:442–451. Chem. Abst. 119 (1993) 71 132.

Parry, R.J. and Sood, G.R. (1989) Investigations of the biosynthesis of trans-(+)-S-1-propenyl-L-cysteine sulfoxide in onions (Allium cepa). J. Am. Chem. Soc. *111*:4514–4515.

Parry, R.J. and Lii, F.-L. (1991) Investigations of the biosynthesis of trans-(+)-S-1-propenyl-L-cysteine sulfoxide. Elucidation of the stereochemistry of the oxidative decarboxylation process. J. Am. Chem. Soc. *113*:4704–4706.

Parthasarathi, K. and Sastry, C.A. (1959) Amino acid composition of common varieties of onion and garlic. Indian J. Pharm. *21*:283–285.

Pasricha, J.S. and Guru, B. (1979) Preparation of an appropriate antigen extract for patch tests with garlic. Arch. Dermatol. *115*:230.

Pasteur, L. (1858) Mémoire sur la fermentation appelée lactique. Mem. Soc. Imp. Sci. Agric. Lille 5:13–26. Also in: Ann. Chim. Phys., Ser. 3, 52:404–418 (1858).

Paszewski, A. and Jarosz, J. (1978) Antimicrobial action of garlic (*Allium sativum* L.) and garlic preparations produced in Poland. Ann. Univ. Mariae Curie-Sklodowska *33D*:415–422. (Polish).

Patel, R.B., Gandhi, T.P., Chakravarthy, B.K., Patel, R.J., Pundarikakshudu, K., and Dhyani, H.K. (1986) Antibacterial activity of phenolic and nonphenolic fractions on some Indian medicinal plants. Indian Drugs *23*:595–597. Chem. Abst. 105 (1986) 222 770.

Pena, N., Auro, A., and Sumano, H. (1988) A comparative trial of garlic, its extract, and ammonium-potassium tartrate as anthelminthics in carp. J. Ethnopharmacol. *24*:199–203. Int. Pharm. Abst. 27 (1990) 947.

Peng, J., Wu, Y., Yao, X., Okuyama, T., and Narui, T. (1992) Two new steroidal saponins from the bulbs of Allium macrostemon Bunge. Chin. Chem. Lett. *3*:285–286. Chem. Abst. 117 (1992) 108 116.

Peng, J., Wang, X., and Yao, X. (1993) Studies of two new furostanol glycosides from Allium macrostemon Bunge. Chin. Chem. Lett. *4*:141–144. Chem. Abst. 120 (1994) 158 785.

Peng, J.P., Yao, X.S., Okada, Y., and Okuyama, T. (1994) Further studies on new furostanol saponins from the bulbs of Allium macrostemon. Chem. Pharm. Bull. *42*:2180–2182.

Penn, R.E., Block, E., and Revelle, L.K. (1978) Methanesulfenic acid. J. Am. Chem. Soc. *100*:3622–3623.

Pentz, R., Guo, Z., Kress, G., Müller, D., Müller, B., and Siegers, C.P. (1990) Standardisation of garlic powder preparations by the estimation of free and hydrolysable SH groups. Planta Med. *56*:691.

Pentz, R., Guo, Z., and Siegers, C.P. (1991) Bioverfügbarkeit und Stoffwechsel von Thiokomponenten aus Verschiedenen Knoblauchpräparaten. Med. Welt *42*:46–47.

Pentz, R., Guo, Z., Müller, B., Aye, R.D., and Siegers, C.P. (1992) Standardisierung von Knoblauchpräparaten. Untersuchung von 22 Fertigarzneimitteln. Dtsch. Apoth. Ztg. *132*:1779–1782. Chem. Abst. 118 (1993) 175 572.

Perchellet, J.P., Perchellet, E.M., Abney, N.L., Zirnstein, J.A., and Belman, S. (1986) Effects of garlic and onion oils on glutathione peroxidase activity, the ratio of reduced/oxidized glutathione and ornithine decarboxylase induction in isolated mouse epidermal cells treated with tumor promoters. Cancer Biochem. Biophys. *8*:299–312. Chem. Abst. 106 (1987) 43 559.

Perchellet, J.P., Perchellet, E.M., and Belman, S. (1990) Inhibition of DMBA-induced mouse skin tumorigenesis by garlic oil and inhibition of two tumor-promotion stages by garlic and onion oils. Nutr. Cancer *14*:183–193. Chem. Abst. 115 (1991) 43 949.

Peretto, R., Favaron, F., Bettini, V., De Lorenzo, G., Marini, S., Alghisi, P., Cervone, F., and Bonfante, P. (1992) Expression and localization of polygalacturonase during outgrowth of lateral roots in Allium porrum L. Planta *188*:164–172.

Perkovskaya, G.Y., Beider, A.M., and Dmitriev, A.P. (1991) Induction of disease resistance in onion by biotic elicitors. Biopolim. Kletka (Bioploymers of the Cell) *7*:91–94. Chem. Abst. 115 (1991) 252 253 (Russian).

Perrin, C., Ky, T.M., and Lem, N.V. (1983) Traitement des ruptures traumatiques du tympan. J. Fr. Oto-Rhino-Laryngol. *32*:565–570.

Perseca, T. and Parvu, M. (1986) Free and protein amino acids in some species of plants used for food. Contrib. Bot., 255–261. Chem. Abst. 106 (1987) 195 005.

Peterssone, E.Y. (1958) The isolation of the phytoncides of garlic with sunflower oil. Med. Promyshl. SSSR (Med. Ind. USSR) *12*:28–33. Chem. Zbl. 1959, II, 8247 (Russian).

Petkov, V. (1962) Über die Pharmakodynamik einiger in Bulgarien wildwachsender bzw. angebauter Arzneipflanzen. Z. Ärztl. Fortbild., 430–440.

Petkov, V., Stoev, V., Bakalov, D., and Petev, L. (1965) "Satal," a Bulgarian drug to be used as a medicamentous agent in industrial lead poisoning. Gigiena Truda Prof. Zabolev. (Hygien. Rep.) *9*:42–49. Chem. Abst. 63 (1965) 8948 (Bulgarian).

Petkov, V. (1966) Pharmakologische und klinische Untersuchungen des Knoblauchs. Dtsch. Apoth. Ztg. *106*:1861–1867. Chem. Abst. 66 (1967) 74 823.

Petkov, V. (1979) Plants with hypotensive antiatheromatous and coronarodilatating action. Am. J. Chin. Med. *7*:197–236. Chem. Abst. 92 (1980) 140 250.

Petkov, V. (1986) Bulgarian traditional medicine: a source of ideas for phytopharmacological investigations. J. Ethnopharmacol. *15*:121–132.

Petkov, V. and Kushev, V. (1966) Influence of garlic on the excretion and the accumulation of 198Au in some organs. Rentgenol. Radiol. [Sofia] *5*:89–93. Chem. Abst. 65 (1966) 11 202 (Bulgarian).

Petricic, J. and Lulic, B. (1977a) Antimicrobial efficiencies and stabilities of active components of garlic (*Allium sativum* L.). Acta Pharm. Jugosl. *27*:35–41. Chem. Abst. 86 (1977) 165 841 (Croatian).

Petricic, J. and Lulic, B. (1977b) Garlic (*Allium sativum* L.): chemistry, action, and use. Farm. Glas. (Pharm. Rep.) *33*:415–423. Chem. Abst. 88 (1978) 158 305 (Croatian).

Petroff, C. (1963) Zur Frage der Einwirkung von Phytonciden auf Oxytrichiden. Zentralbl. Bakteriol. Parasitenkd. Infektionskrankh. Hyg., Abt. II, Part II *117*:70–73. Chem. Abst. 62 (1965) 15 131.

Petry, J.J. (1995) Garlic and postoperative bleeding. Plastic Recon. Surg. *96*:483–484.

Peyer, W. (1927) Über den Knoblauch. Süddtsch. Apoth. Ztg. *67*:573–575.

Peyer, W. and Remund, K. (1928) Medizinisches aus Martial. Veröffentl. Schweiz. Ges. Gesch. Med. *6*:27–33.

Pfaff, K. (1991) Allicin-Freisetzung und Lagerungsstabilität. Bestimmung anhand von frischem Knoblauch und Knoblauchpulver mit einer substanzspezifischen HPLC-Methode. Dtsch. Apoth. Ztg. *131*:12–15.

Pharmacopoea Helvetica VI. (1971) Bulbus Allii sativi recens, Frische Knoblauchknolle, Bulbe d'ail frais, Bulbo di aglio fresco. Eidgenössisches Departement des Innern, Bern, Switzerland.

Phelps, S. and Harris, W.S. (1993) Garlic supplementation and lipoprotein oxidation susceptibility. Lipids *28*:475–477. Chem. Abst. 119 (1993) 48 346. Congress abstract in: Eur. J. Clin. Res. 3A (1992) 4–5.

Phillips, C. and Poyser, N.L. (1972) Inhibition of platelet aggregation by onion extracts. Lancet, 1051–1052.

PHP 72. (1872) *Allium sativum*. In: Pharmacopoea homoepathica polyglottica. W. Schwabe, ed. Dr. Willmar Schwabe, Leipzig, pp. 135.

Pictet, A. (1877) L'oignon et l'ail. In: Les origines Indo-Européennes ou les Aryas primitifs. Essai de paléontologie linguistique. Vol. 2. Sandoz & Fischbacher, Paris, pp. 368–374.

Pino, J., Rosado, A., and Gonzalez, A. (1991) Volatile flavour components of garlic essential oil. Acta Aliment. *20*:163–171. Chem. Abst. 118 (1993) 58 348.

Pino, J.A. (1992) Headspace sampling methods for the volatile components of garlic (*Allium sativum*). J. Sci. Food Agric. *59*:131–133. Chem. Abst. 117 (1992) 110 292.

Piotrowski, G. (1948) L'ail en thérapeutique. Praxis *37*:488–492.

Pisko, J.E. (1896) Gebräuche bei der Geburt und Behandlung der Neugeborenen bei den Albanesen. Mitt. Anthropol. Ges. Wien *26*:141–146.

Pizzocaro, F., Caffa, F., Gasparoli, A., and Fedeli, E. (1985) Antioxidative properties of some aromatic herbs on sardine muscle and oil. Riv. Ital. Sostanze Grasse *62*:351–356. Chem. Abst. 104 (1985) 50 028 (Italian).

Pkheidze, T.A., Kereselidze, E.V., Kachukhashvili, T.N., and Kemertelidze, E.P. (1967) Steroid sapogenins of some Georgian plants. Tr. 1-go Vses. Sezda Farm., 215–221. Chem. Abst. 75 (1971) 115 868 (Russian).

Platenius, H. (1935) A method for estimating the volatile sulphur content and pungency of onions. J. Agric. Res. *51*:847–853. Chem. Zbl. 1936, I, 3597.

Platt, D., Brosche, T., Jacob, B.G., and Schwandt, P. (1992) Cholesterin-senkende Wirkung von Knoblauch? Dtsch. Med. Wochenschr. *117*:962–963.

Plengvidhya, C., Chinayon, S., Sitprija, S., Pasatrat, S., and Tankeyoon, M. (1988) Effects of spray dried garlic preparation on primary hyperlipoproteinemia. J. Med. Assoc. Thailand *71*:248–252.

Plinius. (1881) Die Naturgeschichte des Cajus Plinius Secundus. Gressner & Schramm, Leipzig.

Plinius Secundus der Ältere. (1977) Naturalis Historiae. Heimeran, Kempten.

Pliny. (1961) Natural History. English translation, in 10 volumes. William Heinemann Ltd., London.

Pobozsny, K., Tetenyi, P., Hethelyi, I., Kocsar, L., and Mann, V. (1979) Biologically active substances: investigations into the prostaglandin content of Allium species. Herba Hung. *18*:71–81. Chem. Abst. 92 (1980) 211 834.

Pooler, M.R. and Simon, P.W. (1993) Characterization and classification of isozyme and morphological variation in a diverse collection of garlic clones. Euphytica *68*:121–130. Chem. Abst. 120 (1994) 159 038.

Pooler, M.R. and Simon, P.W. (1994) True seed production in garlic. Sexual Plant Reproduction *7*:282–286.

Popov, I., Blumstein, A., and Lewin, G. (1994) Antioxidant effects of aqueous garlic extract. 1. Direct detection using photochemoluminescence. Arzneim. Forsch. *44*:602–604. Chem. Abst. 121 (1994) 170 497.

Prasad, D.N., Bhattacharya, S.K., and Das, P.K. (1966) A study of antiinflammatory activity of some indigenous drugs in albino rats. Indian J. Med. Res. *54*:582–590. Chem. Abst. 65 (1966) 7847.

Prasad, G., Sharma, V.D., and Kumar, A. (1982) Efficacy of garlic (*Allium sativum*) therapy against experimental dermatophytosis in rabbits. Indian J. Med. Res. *75*:465–467.

Prasad, G. and Sharma, V.D. (1980) Efficacy of garlic (*Allium sativum*) treatment against experimental candidiasis in chicks. Br. Vet. J. *136*:448–451.

Prasad, G. and Sharma, V.D. (1981) Antifungal property of garlic (*Allium sativum* Linn.) in poultry feed substrate. Poult. Sci. *60*:541–545.

Prasad, K., Laxdal, V.A., Yu, M., and Raney, B.L. (1995) Antioxidant activity of allicin, an active principle in garlic. Mol. Cell. Biochem. *148*:183–189.

Prins, H. (1984) Vampirism: legendary or clinical phenomenon? Med. Sci. Law *24*:283–293.

Pros, J.S. (1979) Virtudes curativas del ajo. Editorial Sintes S.A., Barcelona.

Pruthi, J.S., Girdhari, L., and Subrahmanyan, V. (1959a) Chemistry and technology of garlic powder. Food Sci. (Mysore) 8:429–431. Chem. Abst. 54 (1960) 15 754.

Pruthi, J.S., Singh, L.J., and Girdhari, L. (1959b) Some technological aspects of dehydration of garlic: a study of some factors affecting the quality of garlic powder during dehydration. Food Sci. (Mysore) 8:441–444. Chem. Abst. 54 (1960) 15 745.

Pruthi, J.S., Singh, L.J., and Girdhari, L. (1959c) Determination of the critical temperature of dehydration of garlic. Food Sci. (Mysore) 8:436–440. Chem. Abst. 54 (1960) 15 754.

Pruthi, J.S., Singh, L.J., Indiramma, K., Sankaran, A.N., and Girdhari, L. (1959d) Effect of nitrogen packing on storage temperature on the quality of garlic powder. Food Sci. (Mysore) 8:461–464. Chem. Abst. 54 (1960) 15 755.

Pruthi, J.S., Singh, L.J., Kalbag, S.S., and Girdhari, L. (1959e) Effect of different methods of dehydration on the quality of garlic powder. Food Sci. (Mysore) 8:444–448. Chem. Abst. 54 (1960) 15 754.

Pruthi, J.S., Singh, L.J., and Lal, G. (1959f) Thermal stability of alliinase and enzymatic regeneration of flavor in odorless garlic powder. Curr. Sci. 28:403–404. Chem. Abst. 54 (1960) 10 180.

Pruthi, J.S., Singh, L.J., Ramu, S.D.V., and Girdhari, L. (1959g) Pilot-plant studies on the manufacture of garlic powder. Food Sci. (Mysore) 8:448–452. Chem. Abst. 54 (1960) 15 755.

Pruthi, J.S., Singh, L.J., and Lal, G. (1960) Non-enzymatic browning in garlic powder during storage. Food Sci. (Mysore) 9:243–247. Chem. Abst. 55 (1961) 5802.

Pruthi, J.S. (1980) Spices and condiments: chemistry, microbiology, and technology. In: Advanced food research, Supplement 4. Academic Press, New York.

Purseglove, J.W. (1988) Alliaceae. In: Tropical crops. Longman, New York, pp. 37–57.

Pushpendran, C.K., Devasagayam, T.P.A., Banerji, A., and Eapen, J. (1980a) Cholesterol-lowering effect of Allitin in suckling rats. Indian J. Exp. Biol. 18:858–861. Chem. Abst. 93 (1980) 179 732.

Pushpendran, C.K., Devasagayam, T.P.A., Chintalwar, G.J., Banerji, A., and Eapen, J. (1980b) The metabolic fate of [35S]-diallyl disulphide in mice. Experientia 36:1000–1001. Chem. Abst. 93 (1980) 125 399.

Pushpendran, C.K., Devasagayam, T.P.A., and Eapen, J. (1982) Age-related hyperglycaemic effect of diallyl disulphide in rats. Indian J. Exp. Biol. 20:428–429.

Qian, Y.X., Shen, P.J., Xu, R.Y., Liu, G.M., Yang, H.Q., Lu, Y.S., Sun, P., Zhang, R.W., Qi, L.M., and Lu, Q.H. (1986a) Spermicidal effect in vitro by the active principle of garlic. Contraception 34:295–302.

Qian, Y.X., Shen, P.J., Xu, R.Y., Lu, Y.S., and Liu, G.M. (1986b) Spermicidal effect in vitro of garlicin, active principle of garlic. Yaoxue Tongbao (Bull. Pharmacol.) 21:708–710. Int. Pharm. Abst. 25 (1988) 11 162 (Chinese).

Qureshi, A.A., Abuirmeileh, N., Burger, W.C., Din, Z.Z., and Elson, C.E. (1983a) Effect of AMO 1618 on cholesterol and fatty acid metabolism in chickens and rats. Atherosclerosis 46:203–216.

Qureshi, A.A., Abuirmeileh, N., Din, Z.Z., Elson, C.E., and Burger, W.C. (1983b) Inhibition of cholesterol and fatty acid biosynthesis in liver enzymes and chicken hepatocytes by polar fractions of garlic. Lipids 18:343–348. Chem. Abst. 99 (1983) 21 339.

Qureshi, A.A., Din, Z.Z., Abuirmeileh, N., Burger, W.C., Ahmad, Y., and Elson, C.E. (1983c) Suppression of avian hepatic lipid metabolism by solvent extracts of garlic: impact on serum lipids. J. Nutr. 113:1746–1755.

Qureshi, A.A., Crenshaw, T.D., Abuirmeileh, N., Peterson, D.M., and Elson, C.E. (1987) Influence of minor plant constituents on porcine hepatic lipid metabolism. Atherosclerosis 64:109–115.

Rabinkov, A., Zhu, X.Z., Grafi, G., Galili, G., and Mirelman, D. (1994) Alliin lyase (alliinase) from garlic (Allium sativum). Biochemical characterization and cDNA cloning. Appl. Biochem. Biotechnol. 48:149–171. Chem. Abst. 122 (1995) 26 419.

Rachenzentner, A. and Urban, E. (1992) Synthese von 5,6-Dihydroxy-tetrahydro-isobenzofuranonen. Arch. Pharm. [Weinheim] 325:101–105.

Ragab, M.M., Kararah, M.A., Osman, M.E., and Mostafa, M.A. (1984) Effect of infection with Fusarium solani on garlic cloves with special reference to the biochemical changes. Egypt. J. Phytopathol. 16:23–33. Chem. Abst. 106 (1987) 83 221.

Raghavan, B., Abraham, K.O., and Shankaranarayana, M.L. (1983) Chemistry of garlic and garlic products. J. Sci. Ind. Res. 42:401–409. Chem. Abst. 99 (1983) 174 247.

Raghavan, B., Abraham, K.O., and Shankaranarayana, M.L. (1986) Flavor losses during dehydration of garlic and onion. PAFAI J. 8:11–15. Chem. Abst. 105 (1986) 96 208.

Rainov, N.G. and Burkert, W. (1993) Spontaneous shrinking of a macroprolactinoma. Neurochirurgia 36:17–19.

Raj, K.P.S., Agrawal, Y.K., and Patel, M.R. (1980) Analysis of garlic for its metal contents. J. Indian Chem. Soc. 57:1121–1122. Chem. Abst. 94 (1981) 71 635.

Rakhimbaev, I.R. and Olshanskaya, R.V. (1981) First identification of natural gibberellins in garlic. Izv. Akad. Nauk Kaz. SSR [Alma Ata] (Bull. Acad. Sci. Kasach. SSR), 17–22 (Russian).

Rakhimov, D.A. and Khodzhaeva, M.A. (1990) Allium carbohydrates. X. Glucofructans of Allium karataviense. Khim. Prir. Soedin. (Chem. Nat. Compd.) 26:110–111. Chem. Abst. 112 (1990) 213 911 (Russian).

Ramos, R.S., Sinigaglia, C., Issa, E., and Chiba, S. (1987) Chemical control of rust (Puccinia allii (D.C.) Rud.) and purple blotch (Alternaria porri (Ell.) Cif.) of garlic (*Allium sativum* L.). Summa Phytopathol. *13*:197–209. Chem. Abst. 108 (1988) 182 113 (Portuguese).

Rao, A.R., Sadhana, A.S., and Goel, H.C. (1990) Inhibition of skin tumors in DMBA-induced complete carcinogenesis system in mice by garlic (*Allium sativum*). Indian J. Exp. Biol. *28*:405–408.

Rao, E.V. and Rao, M.A. (1978) Chemical and biological properties of the odoriferous principles of garlic and onion. Indian Drugs Pharm. Ind. *13*:10–12. Chem. Abst. 89 (1978) 208 736.

Rao, H.T.R., Humar, U., Jayrajan, P., Devi, G., Shakunthala, V.T., Srinath, U., Rai, A.Y., Basavaraju, M., and Prasad, V.S. (1981) Effect of garlic extracts on the cardiac, smooth, and skeletal muscle of the frog. Indian J. Physiol. Pharmacol. *25*:303.

Rao, M.A. and Rao, E.V. (1982) Estimation of alliin in Adenocalymma alliaceum and comparison with that of garlic. Indian Drugs *20*:6–7. Chem. Abst. 98 (1982) 104 310.

Rao, P.L.N. and Verma, S.C.L. (1952) Antibiotic principle of *Allium sativum*; structure of allicin; preparation and properties of diphenyl disulfide oxide. J. Indian Inst. Sci. *34*:315–321. Chem. Abst. 47 (1953) 9290.

Rao, R.R., Rao, S.S., and Venkataraman, P.R. (1946) Investigations on plant antibiotics. I. Studies on allicin, the antibacterial principle of *Allium sativum* (garlic). J. Sci. Ind. Res. *1B*:31–35. Chem. Abst. 41 (1947) 2461.

Rao, V.K. and Jones, H.M. (1984) Selenium content of condiments and nuts. Fed. Proc. *43*:868.

Räsänen, L., Kuusisto, P., Penttilä, M., Nieminen, M., Savolainen, J., and Lehto, M. (1994) Comparison of immunological tests in the diagnosis of occupational asthma and rhinitis. Allergy Eur. J. Allergy Clin. Immunol. *49*:342–347.

Rashid, A., Hussain, M., and Khan, H.H. (1986) Bioassay for prostaglandin-like activity of garlic extract using isolated rat fundus strip and rat colon preparation. J. Pak. Med. Assoc. *36*:138–141.

Rashid, A. and Khan, H.H. (1985) The mechanism of hypotensive effect of garlic extract. J. Pak. Med. Assoc. *35*:357–362.

Ratner, D. (1916) Ein hygienisch-talmudisches Hausmittel wieder zu Ehren gebracht. Hygien. Rdsch. *26*:165–166.

Ravnikar, M., Zel, J., Plaper, I., and Spacapan, A. (1993) Jasmonic acid stimulates shoot and bulb formation of garlic in vitro. J. Plant Growth Regul. *12*:73–77. Chem. Abst. 120 (1994) 158 953.

Rawlins, D.W. (1994) Fish attractant for fishing lures. U.S.A. Patent 5 277 918. Chem. Abst. 120 (1994) 99 477.

Razumovich, M.B. (1979) Physiological activity of volatile phytoncides. Fitontsidy (Phytoncides) *8*:185–189. Chem. Abst. 97 (1982) 88 710 (Russian).

Reddy, B.S., Rao, C.V., Rivenson, A., and Kelloff, G. (1993) Chemoprevention of colon carcinogenesis by organosulfur compounds. Cancer Res. *53*:3493–3498.

Rees, L.P., Minney, S.F., Plummer, N.T., Slater, J.H., and Skyrme, D.A. (1993) A quantitative assessment of the antimicrobial activity of garlic (*Allium sativum*). World J. Microbiol. Biotechnol. *9*:303–307.

Reeve, V.E., Bosnic, M., Rozinova, E., and Boehm-Wilcox, C. (1993) A garlic extract protects from ultraviolet B (280–320 nm) radiation–induced suppression of contact hypersensitivity. Photochem. Photobiol. *58*:813–817. Chem. Abst. 120 (1994) 100 678.

Regel, E. (1875) Alliorum adhuc cognitorum Monographia—St. Petersburg 1875. Acta Horti Petropol. *3*:1–266.

Regel, E. (1887) Allii species Asiae centralis in Asia media a Turcomana desertisque Aralensibus et Caspicis usque ad Mongoliam crescentes. St. Petersburg.

Reichart, A., Novak, F.J., and Tanasch, L. (1985) In vitro plant regeneration from leaf explants in garlic (*Allium sativum* L.). Proc. 4th Eucarpia Allium Symp., 1–9.

Reicks, M., Johanning, J., and Tatini, S. (1993) Dietary fat modulation of the inhibition of rat hepatic CYP2E1 activity by diallyl sulfide. FASEB J. *7*:A864.

Reicks, M., Yin, J., and Crankshaw, D. (1994) Inhibition of rat liver microsomal CYP2E1 catalytic activity by garlic organosulfur compounds. FASEB J. *8*:A718.

Reimers, F., Smolka, S.E., Werres, S., Plank-Schumacher, K., and Wagner, G. (1993) Effect of ajoene, a compound derived from *Allium sativum*, on phytopathogenic and epiphytic micro-organisms. Z. Pflanzenkrankh. Pflanzenschutz *100*:622–633. Chem. Abst. 121 (1994) 198 409.

Reinhard, K.H. (1984) Eine "anrüchige" Angelegenheit mit Tradition und Zukunft. Wissenswertes über den Knoblauch. Dtsch. Apotheker *36*:435–440.

Reinhard, K.H. (1986) Botanik und Geschichte des Knoblauchs. Naturheilpraxis *39*:384–388.

Rekka, E.A. and Kourounakis, P.N. (1994) Investigation of the molecular mechanism of the antioxidant activity of some *Allium sativum* ingredients. Pharmazie *49*:539–540. Chem. Abst. 121 (1994) 73 827.

Renapurkar, D.M. and Deshmukh, P.B. (1984) Pulicidal activity of some indigenous plants. Insect Sci. Its Appl. *5*:101–102.

Rendu, F., Daveloose, D., Debouzy, J.C., Bourdeau, N., Levy-Toledano, S., Jain, M.K., and Apitz-Castro, R. (1989) Ajoene, the antiplatelet compound derived from garlic, specifically inhibits platelet release reaction by affecting the plasma membrane internal microviscosity. Biochem. Pharmacol. *38*:1321–1328. Chem. Abst. 111 (1989) 464.

Rengwalska, M.M. and Simon, P.W. (1986) Laboratory evaluation of pink root and Fusarium basal rot resistance in garlic. Plant Disease *70*:670–672.

Rennenberg, H. (1982) Glutathione metabolism and possible biological roles in higher plants. Phytochemistry *21*:2771–2781.

Rescke, A. and Herrmann, K. (1982) Vorkommen von 1-O-Hydroxicinnamyl-b-D-glucosen im Gemüse. I. Phenolcarbonsäure-Verbindungen des Gemüses. Z. Lebensm. Unters. Forsch. *174*:5–8. Chem. Abst. 96 (1982) 161 085.

Ressin, W. (1985) Der Einfluß von Alligoa plus auf psychische und physische Parameter bei geriatrischen Patienten. Dtsch. Apotheker *37*:596–599.

Reuter, H.D. (1980) Knoblauch als Arteriosklerosehemmer. Ärztl. Prax. *32*:847–850.

Reuter, H.D. (1983a) Antiarteriosklerotische Wirkung von Knoblauchinhaltsstoffen. Therapiewoche *33*:2474–2487. Chem. Abst. 99 (1983) 33.

Reuter, H.D. (1983b) Im Knoblauch steckt ein Plättchen-Aggregationshemmer. Ärztl. Prax. *35*:1913.

Reuter, H.D. (1986) Knoblauch (*Allium sativum*): Neue pharmakologische Ergebnisse einer "uralten" Arzneipflanze. Z. Phytother. *7*:99–106.

Reuter, H.D. (1987) The effect of herbal drugs and of isolated compounds from medicinal plants on the functional activity of human platelets. Herba Pol. *33*:275–280.

Reuter, H.D. (1988) Spektrum *Allium sativum* L. Aesopus Verlag, Stuttgart.

Reuter, H.D. (1990) Knoblauch: Lassen sich die Risikofaktoren der Arteriosklerose beeinflussn? PTA heute *4*:416–424.

Reuter, H.D. (1991) Spektrum *Allium sativum* L. Aesopus GmbH, Basle.

Reuter, H.D. (1993a) Garlic (*Allium sativum* L.) in the prevention and treatment of atherosclerosis. Br. J. Phytother. *3*:3–9.

Reuter, H.D. (1993b) Neues über Fischöl, Knoblauch, Weibdorn u.a. Z. Phytother. *14*:22–23.

Reuter, H.D. and Sendl, A. (1994) *Allium sativum* and Allium ursinum: chemistry, pharmacology, and medicinal applications. In: Economic and medicinal plant research. Academic Press, New York, pp. 54–113.

Reznik, P.A. and Imbs, Y.G. (1965) Ixodid ticks and phytoncides. Zool. Zh. (Zool. J.) *44*:1861–1864 (Russian).

Rhee, M.G., Cha, C.W., and Bae, E.S. (1985) A study on the chronological changes of rat tissues and the effect of garlic in acute methyl mercury poisoning. Koryo Taehakkyo Uikwa Taehak Nonmunjip (Korea Univ. Med. J.) *22*:153–163. Chem. Abst. 103 (1985) 136 665 (Korean).

Ribeiro, R., Melo, M.M.R., Barros, F.D., Gomes, C., and Trolin, G. (1986) Acute antihypertensive effect in conscious rats produced by some medicinal plants used in the state of Sao Paulo Brazil. J. Ethnopharmacol. *15*:261–270.

Rich, G.E. (1982) Garlic an antibiotic? Med. J. Aust. *1*:60.

Rico, J.T. (1926) Sur les propriétés anti-helminthiques de l'*Allium sativum*. C. R. Séances Soc. Biol. Ses Fil. *95*:1597–1599.

Ridley, C. (1900) The botany and materia medica of the Bible. Pharm. J. *64*:527.

Riedel, I.D. and De Haen, E. (1935) Therapeutically effective components from sulfur-containing volatile oils. Austria. Patent 143 320. Chem. Zbl. 1936, I, 1917.

Riedel, J.D. and De Haen, E. (1935) Method of an odorless crystalline product that is one of the effective components of garlic oil. Switzerland. Patent 174 460.

Rietz, B., Isensee, H., Strobach, H., Makdessi, S., and Jacob, R. (1993) Cardioprotective actions of wild garlic (Allium ursinum) in ischemia and reperfusion. Mol. Cell. Biochem. *119*:143–150.

Riken Chem. Ind. Co. Ltd. (1982) Pharmaceuticals for control of Filaria in dogs. Japan. Patent 82 192 318. Chem. Abst. 98 (1983) 59 946.

Riken Chem. Ind. Co. Ltd. (1983) Preparation of odorless inclusion compounds containing medicinal garlic extract. Japan. Patent 83 21 620. Chem. Abst. 98 (1983) 185 572.

Riken Chem. Ind. Co. Ltd. (1984) Drug-cyclodextrin inclusion compound formulations containing amylase. Japan. Patent 84 44 318. Chem. Abst. 100 (1984) 215 562.

Rinne, F. (1892) Knoblauch Anwendungen. Dissertation, University of Dorpat (Abstract).

Rinneberg, A.L. and Lehmann, B. (1989) Therapie der Hyperlipoproteinämie mit einem standardisierten Knoblauchpulver-Arzneimittel. Kassenarzt *45*:40–47.

Riva, M., Di Cesare, L.F., and Schiraldi, A. (1993) Microwave and traditional technology to prepare garlic aromatized olive oil. Dev. Food Sci. *32*:327–338. Chem. Abst. 119 (1993) 179 859.

Rocchietta, S. (1958) Derivato dell'allicina, principo attivo dell'aglio ad azione inibitrice della crescita dei tumori. Boll. Chim. Farm. *97*:162.

Rockwell, P. and Raw, I. (1979) A mutagenic screening of various herbs, spices, and food additives. Nutr. Cancer *1*:10–15. Chem. Abst. 92 (1980) 92 862.

Rode, H., De Wet, P.M., and Cywes, S. (1989) The antimicrobial effect of *Allium sativum* L. (garlic). S. Afr. J. Sci. *85*:462–463.

Romanyuk, N.M. (1954) The influence of the antibiotics of garlic on the activity of proteolytic enzymes of malignant tumors of humans and experimental animals. Ukrain. Biokhim. Zh. (Biochem. J.) *24*:53–60. Chem. Abst. 48 (1954) 6007 (Ukrainian).

Rook, A. (1960) Plant dermatitis. Br. Med. J. *2*:1771–1774.

Rook, A. (1962) Plant dermatitis in general practice. Practitioner 188:627–638.

Roos, E. (1925) Über die Verwendbarkeit des Knoblauchs (Allium sativum) als Darmheilmittel. Münch. Med. Wochenschr. 72:1637–1641. Chem. Zbl. 1926, I, 437.

Rose, K.D., Croissant, P.D., Parliament, C.F., and Levin, M.B. (1990) Spontaneous spinal epidural hematoma with associated platelet dysfunction from excessive garlic ingestion. Neurosurgery 26: 880–882.

Roser, D. (1990) Garlic. Lancet 335:114–115.

Rotzsch, W., Richter, V., Rassoul, F., and Walper, A. (1992) Postprandiale Lipämie unter Medikation von Allium sativum. Arzneim. Forsch. 42:1223–1227. Int. Pharm. Abst. 30 (1993) 11 294. Congress abstract in: Eur. J. Clin. Res. 3A (1992), 9.

Roychoudhury, A., Das, T., Sharma, A., and Talukder, G. (1993) Use of crude extract of garlic (Allium sativum L.) in reducing cytotoxic effects of arsenic in mouse bone marrow. Phytother. Res. 7:163–166.

Rudat, K.D. (1957) The antibiotic substances of higher plants active against Bacterium pyocyaneum (Pseudomonas aeruginosa). J. Hyg. Epidemiol. Microbiol. Immunol. [Prague] 1:213–224. Chem. Abst. 54 (1960) 19 868.

Rudat, K.D. (1969) Vergleichende Untersuchungen über die antibakterielle Wirksamkeit verschiedener Lauchgewächse und Cruciferen-Arten. Qual. Plant. Mater. Veg. 18:29–43.

Rüdiger, H. and Gabius, H.J. (1993) Lectinologie. Geschichte, Konzepte und pharmazeutische Bedeutung. Dtsch. Apoth. Ztg. 133:2371–2381.

Ruffin, J. and Hunter, S.A. (1983) An evaluation of the side effects of garlic as an antihypertensive agent. Cytobios 37:85–89.

Ruiz, R., Hartman, T.G., Karmas, K., Lech, J., and Rosen, R.T. (1994) Breath analysis of garlic-borne phytochemicals in human subjects: combined adsorbent trapping and short-path thermal desorption gas chromatography–mass spectrometry. ACS Symp. Ser. 546:102–119. Chem. Abst. 120 (1994) 128 965.

Rundquist, C. (1910) Pharmakochemische Untersuchung von Bulbus Allii. Apoth. Ztg. 25:105.

Runkova, L.V. and Talieva, M.N. (1970) Chlorogenic acid in species of the genus Allium. Fiziol. Biokhim. Kult. Rast. (Physiol. Biochem. Plant Culture) 2:544–547. Chem. Abst. 74 (1971) 136 629 (Russian).

Saari, J.C. and Schultze, M.O. (1965) Cleavage of S-(1,2-dichlorovinyl)-L-cysteine by Escherichia coli B. Arch. Biochem. Biophys. 109:595–602.

Sächsisches Serumwerk AG Dresden. (1936) Method of production of high quality, odorless, iodine-containing Allium preparations. Germany. Patent 626 469. Chem. Zbl. 1936, I, 3719.

Sadek, I.A. and Abdul-Salam, F. (1994) Effect of diallyl sulfide on toad liver tumor induced by 7,12-dimethylbenz(a)anthracene. Nutr. Res. 14:1513–1521.

Sadhana, A.S., Rao, A.R., Kucheria, K., and Bijani, V. (1988) Inhibitory action of garlic oil on the initiation of benzo[a]pyrene-induced skin carcinogenesis in mice. Cancer Lett. 40:193–197. Chem. Abst. 109 (1988) 87 943.

Safarli, S.R. (1955) Experimental treatment of corneal burns with phytoncidonaphthalene emulsion. Vestn. Oftal'mol. (Opthalmol. Rep.) 34:17–19. Chem. Abst. 50 (1956) 7310 (Russian).

Safronov, A.P. (1963) A new method for detecting organic sulphoxides. Zh. Anal. Khim. (J. Anal. Chem.) 18:548–550. Chem. Abst. 59 (1963) 12 189 (Russian).

Saghir, A.R., Mann, L.K., Bernhard, R.A., and Jacobsen, J.V. (1964) Determination of aliphatic mono- and disulfides in Allium by gas chromatography and their distribution in the common food species. Proc. Am. Soc. Hortic. Sci. 84:386–398. Chem. Abst. 62 (1965) 4592.

Saghir, A.R., Cowan, J.W., and Salji, J.P. (1966) Goitrogenic activity of onion volatiles. Nature 211:87.

Saghir, A.R., Cowan, J.W., and Salji, J.P. (1967) Antithyroid activity of volatile components of Allium. 3rd Symp. Hum. Nutr. Health Near East 1967, 154–161. Chem. Abst. 73 (1970) 54 116.

Saghir, A.R., Cowan, J.W., and Salji, J.P. (1968) The molecular structure of sulfides in relation to antithyroid activity. Eur. J. Pharmacol. 2:399–402. Chem. Abst. 68 (1968) 103 527.

Sagmeister, T. (1987) Knoblauch gegen Atom-Tod (Reportage). Neue Kronen-Zeitung [Vienna], 30–31.

Sainani, G.S., Desai, D.B., and More, K.N. (1976) Onion, garlic, and atherosclerosis. Lancet 2:575–576.

Sainani, G.S., Desai, D.B., Gorhe, N.H., Natu, S.M., Pise, D.V., and Sainani, P.G. (1979a) Effect of dietary garlic and onion on serum lipid profile in Jain community. Indian J. Med. Res. 69:776–780. Chem. Abst. 91 (1979) 55 212.

Sainani, G.S., Desai, D.B., Gorhe, N.H., Natu, S.M., Pise, D.V., and Sainani, P.G. (1979b) Dietary garlic, onion, and some coagulation parameters in Jain community. J. Assoc. Physicians India 27:707–712.

Sainani, G.S., Desai, D.B., Gorhe, N.H., Pise, D.V., and Sainani, P.G. (1979c) Effect of garlic and onion on important lipid and coagulation parameters in alimentary hyperlipaemia. J. Assoc. Physicians India 27:57–64.

Sainani, G.S., Desai, D.B., Natu, M.N., Katrodia, K.M., Valame, V.P., and Sainani, P.G. (1979d) Onion, garlic, and experimental atherosclerosis. Jap. Heart J. 20:351–357.

Saito, F., Urushibata, O., and Murao, T. (1982) Contact dermatitis from plants for the last 6 years. Skin Res. 24:238–249 (Japanese).

Saito, K., Horie, M., Yoji, H., Nose, N., Mochizuki, E., and Nakazawa, H. (1988) Gas chromatographic determination of alliin in garlic products. Eisei Kagaku 34:536–541. Chem. Abst. 110 (1989) 133 806 (Japanese).

Saito, K., Horie, M., Hoshino, Y., Nose, N., Mochizuki, E., Nakazawa, H., and Fujita, M. (1989) Determination of allicin in garlic and commercial garlic products by gas chromatography with flame photometric detection. J. Assoc. Off. Anal. Chem. 72:917–920. Chem. Abst. 112 (1990) 34 492.

Sakai, I. (1981) Deodorization of garlic. Japan. Patent 82 29 265. Chem. Abst. 96 (1982) 198 185.

Sakai, I. (1985) Deodorization of garlic by an aqueous solution containing silica, phytic acid, and zinc compounds. Japan. Patent 85 259 157. Chem. Abst. 104 (1986) 167 222.

Sakai, I. (1989a) Manufacture of odorless garlic powder using meso-inositol hexaphosphate. Japan. Patent 89 273 559. Chem. Abst. 112 (1990) 97 274.

Sakai, I. (1989b) Preparation of odorless garlic powder for health food. Japan. Patent 89 312 979. Chem. Abst. 113 (1990) 39 177.

Sakai, I. (1989c) Iodized garlic and its preparation. Japan. Patent 89 281 051. Chem. Abst. 113 (1990) 57 787.

Sakai, I. (1992a) Odor-free garlic with good preservability and its manufacture. Japan. Patent 92 445 767. Chem. Abst. 116 (1992) 254 443.

Sakai, I. (1992b) Allicin-containing pesticide for golf course greens. Japan. Patent 92 05 211. Chem. Abst. 116 (1992) 168 344.

Sakai, I. (1994a) Food preservatives containing epsilon-polylysine and plant alkaloids and manufacture of the food preservatives. Japan. Patent 94 78 730. Chem. Abst. 121 (1994) 33 789.

Sakai, I. (1994b) Removal of odor from vegetables like garlic and onions. Japan. Patent 94 62 781. Chem. Abst. 121 (1994) 7855.

Sakamoto, Y. (1967) Deodorizing garlic. U.S.A. Patent 3 326 698. Chem. Abst. 67 (1967) 63 124.

Saleem, Z.M. and Al-Delaimy, K.S. (1982) Inhibition of Bacillus cereus by garlic extracts. J. Food Prot. 45:1007–1009.

Salji, J.P., Cowan, J.W., and Saghir, A.R. (1971) The antithyroid activity of Allium volatiles in the rat: in vitro studies. Eur. J. Pharmacol. 16:251–253. Chem. Abst. 76 (1972) 82 006.

Salton, M.R.J. (1964) The bacterial cell wall. Elsevier, Amsterdam.

Samikov, K., Shakirov, R., Antsupova, T.P., and Yunusov, S.Y. (1986) Allium alkaloids. Khim. Prir. Soedin. (Chem. Nat. Compd.) 22:383. Chem. Abst. 105 (1986) 102 094 (Russian).

Samson, R.R. (1982) Effects of dietary garlic and temporal drift on platelet aggregation. Atherosclerosis 44:119–120.

Samura, M. and Samusan Kk. (1990a) Alcohol-containing odorless garlic and odorless garlic wine. Japan. Patent 90 284 850.

Samura, M. and Samusan Kk. (1990b) Odorless garlic and liquid thereof. Japan. Patent 90 216 237.

San-Blas, G., San-Blas, F., Gil, F., Mariño, L., and Apitz-Castro, R. (1989) Inhibition of growth of the dimorphic fungus Paracoccidioides brasiliensis by ajoene. Antimicrob. Agents Chemother. 33:1641–1644. Chem. Abst. 111 (1989) 191 344.

San-Blas, G., Mariño, L., San-Blas, F., and Apitz-Castro, R. (1993) Effect of ajoene on dimorphism of Paracoccidioides brasiliensis. J. Med. Vet. Mycol. 31:133–141.

Sanchez, M.A., Bertoni, M.H., and Cattaneo, P. (1988) Sobre contenidos y composiciones acidicas de lipidos totales de bulbos y raices de plantas comestibles. Anal. Asoc. Quim. Argent. 76:227–235. Chem. Abst. 110 (1989) 37 964.

Sandhu, D.K., Warraich, M.K., and Singh, S. (1980) Sensitivity of yeasts isolated from cases of vaginitis to aqueous extracts of garlic. Mykosen 23:691–698.

Sandoz. (1925) Procédé pour la production de préparations d'allium. France. Patent 599 342.

Sandoz. (1926) Method of production of Allium preparations. Germany. Patent 432 053. Chem. Zbl. 1926, II, 1550.

Sane, R.T., Chakraborty, M., and Ramachandran, J. (1990) HPLC study of the Ayurvedic process of purification of mercury using garlic juice. Indian Drugs 28:81–85. Chem. Abst. 114 (1991) 30 098.

Sanfilippo, G. (1946) Blood calcium and chemical composition of bones during prolonged administration of garlic juice. Boll. Soc. Ital. Biol. Sper. 22:282–283. Chem. Abst. 41 (1947) 528 (Italian).

Sanick, I.H. (1973) Supplementary feed for improved chicken flavor in broiler meat. Britain. Patent 1 330 209. Chem. Abst. 79 (1973) 145 146.

Sanick, I.H. (1974) Food preservative containing cinnamaldehyde and Allium extracts. Germany. Patent 2 423 076. Chem. Abst. 82 (1975) 154 047.

Sanick, I.H. (1975) Preserving foods. Australia. Patent 499 390. Chem. Abst. 91 (1974) 18 617.

Saratikov, A.S. and Khomjakov, A.F. (1952) Influence of garlic's phytoncides on metal-contenting enzymes. Biul. Eksper. Biol. Med. (Bull. Exp. Biol. Med.) 33:54–57 (Russian).

Saratikov, A.S. and Plakhova, N.B. (1950) On the mechanism of action of garlic phytoncides. Farmakol. Toksikol. [Moscow] 13:3–6 (Russian).

Sarkar, A.R. and De, M.K. (1981) Histopathological studies on the effects of garlic on experimental atherosclerosis. Indian J. Pathol. Microbiol. 24:261–266.

Sarter, H., Deumig, H.J., and Frey, G. (1991) Knoblauch in Ölmazeraten. Charakterisierung und pharmazeutische Qualität. Dtsch. Apoth. Ztg. 131:716–719. Chem. Abst. 115 (1991) 99 426.

Sas, G., Scheffer, K., Heiss, G., and Bräuer, H. (1988) Knoblauchpulver senkt den Cholesterin-Spiegel. Aerztl. Praxis 40:2894–2896.

Sasaki, H. (1957) Certain drugs (thio and thiol derivatives in particular) with action against Candida albicans. Igaku Kenkyu 27:2679–2692. Chem. Abst. 52 (1958) 11 266 (Japanese).

Sashida, Y., Kawashima, K., and Mimaki, Y. (1991) Novel polyhydroxylated steroidal saponins from Allium giganteum. Chem. Pharm. Bull. 39:698–703. Chem. Abst. 115 (1991) 46 059.

Sato, H., Konoma, K., and Sakamura, S. (1979) Phytotoxins produced by onion pink root fungus. Agric. Biol. Chem. 43:2409–2311. Chem. Abst. 92 (1980) 72 340.

Satoh, Z. (1952a) Effect of Allium diets on bacterial synthesis of thiamine in the intestinal tract. I. Test on dietary intake of Allium plants. Vitamins [Japan] 5:184–190. Chem. Abst. 48 (1954) 237.

Satoh, Z. (1952b) Effect of Allium diets on bacterial synthesis of thiamine in the intestinal tract. II. Influence of ethereal oil of Allium on thiamine metabolism. Vitamins [Japan] 5:296–305. Chem. Abst. 48 (1954) 237.

Satoh, Z. (1952c) Effect of Allium diets on bacterial synthesis of thiamine in the intestinal tract. III. Effects of thiamine-deficient diet containing the ethereal oil of Allium on thiamine metabolism in the human body. Vitamins [Japan] 5:306–312. Chem. Abst. 48 (1954) 237.

Satoh, Z. (1952d) Effect of Allium diets on bacterial synthesis of thiamine in the intestinal tract. IV. Influence of cooking of the Allium plants on the bacterial synthesis of thiamine in the intestinal tract. Vitamins [Japan] 5:495–499. Chem. Abst. 48 (1954) 237.

Satoh, Z. (1952e) Effect of Allium diets on bacterial synthesis of thiamine in the intestinal tract. V. Influence of oral administration of organic sulfur compounds on the bacterial synthesis of thiamine in the intestinal tract. Vitamins [Japan] 5:500–502. Chem. Abst. 48 (1954) 237.

Satoh, Z. (1952f) Effect of Allium diets on bacterial synthesis of thiamine in the intestinal tract. VI. Influence of administration of garlic oil on erythrocytes. Vitamins [Japan] 5:503–505. Chem. Abst. 48 (1954) 237.

Satoh, Z. (1952g) Effect of Allium diets on bacterial synthesis of thiamine in the intestinal tract. VII. The in vitro synthesis of thiamine from garlic oil by Escherichia coli. Vitamins [Japan] 5:585–589. Chem. Abst. 48 (1954) 237.

Satoh, Z. (1952h) Effect of Allium diets on bacterial synthesis of thiamine in the intestinal tract. VIII. Influence of various sulfur compounds on the bacterial synthesis of thiamine. Vitamins [Japan] 5:589–593. Chem. Abst. 48 (1954) 237.

Savitri, A., Bhavanishankar, T.N., and Desikachar, H.S. (1986) Effect of spices on in vitro gas production by Clostridium perfringens. Food Microbiol. 3:195–199.

Saxena, K.K., Gupta, B., Kulshrestha, V.K., Srivastava, R.K., and Prasad, D.N. (1979) Garlic in stress induced myocardial damage. Indian Heart J. 31:187–188.

Saxena, K.K., Gupta, B., Kulshrestha, V.K., Srivastava, R.K., and Prasad, D.N. (1980) Effect of garlic pretreatment on isoprenaline-induced myocardial necrosis in albino rats. Indian J. Physiol. Pharmacol. 24:233–236.

Schardt, D. and Liebman, B. (1995) Powder wise ... pill foolish. Nutrition Action Health Lett. 22:4–55.

Scharfenberg, K., Wagner, R., and Wagner, K.G. (1990) The cytotoxic effect of ajoene, a natural product from garlic, investigated with different cell lines. Cancer Lett. 53:103–108. Chem. Abst. 114 (1991) 283.

Scheibe, W. (1958) Über die antibiotische Wirkung des Knoblauchs. Medizinische 2:1633–1634. Chem. Abst. 53 (1959) 1538.

Schiefer, H. (1953) Preparation for treatment of animal diseases. Austria. Patent 176 065. Chem. Abst. 47 (1953) 10 814.

Schimmer, O., Krüger, A., Paulini, H., and Haefele, F. (1994) An evaluation of 55 commercial plant extracts in the Ames mutagenicity test. Pharmazie 49:448–451.

Schindel, L. (1934) Über pflanzliche Choleretica. Arch. Exp. Pathol. Pharmakol. 175:313–321.

Schinzel, W. and Graf, T. (1979) Höhenkrankheit und Knoblauch. Dtsch. Ärztebl. 76:1974–1975.

Schlechtendal, D.F.L.V., Langethal, L.E., and Schenk, E. (1880) Flora von Deutschland. Fr. E. Köhler, Gera-Untermhaus.

Schleich, C. and Henze, G. (1990) Trace analysis of germanium: polarographic behaviour and determination by adsorptive stripping voltammetry. Fresenius Z. Anal. Chem. 338:145–148. Chem. Abst. 114 (1991) 35 041.

Schlesinger, K. (1926) Knoblauch (Allium sativum) als Heilmittel bei Arteriosklerose. Wien. Med. Wochenschr. 76:1076–1077.

Schmidt, P.W. and Marquardt, U. (1936) Über den antimykotischen Effekt ätherischer Öle von Lauchgewächsen und Kreuzblütlern auf pathogene Hautpilze. Zentralbl. Bakteriol. Parasitenkd. Infektionskrankh. Hyg., Abt. 1, Part I 138:104–128. Chem. Zbl. 1937, I, 1708.

Schmidtlein, H. and Herrmann, K. (1975) Über die Phenolsäuren des Gemüses. IV. Hydroxyzimtsäuren und Hydroxybenzoesäuren weiterer Gemüsearten und der Kartoffeln. Z. Lebensm. Unters. Forsch. 159:255–263. Chem. Abst. 84 (1976) 57 534.

Schmiedeberg, O. (1879) Über ein neues Kohlehydrat. Hoppe-Seyler's Z. Physiol. Chem. 3:112–133.

Schneider, W. (1974) Lexikon zur Arzneimittelgeschichte pflanzlicher Drogen. Govi, Frankfurt.

Schöffler, P. (1485) Hortus Sanitatis/Garten der Gesundheit. Mainz.

Schormüller, J. (1968) Obst, Gemüse, Kartoffeln, Pilze. In: Handbuch der Lebensmittelchemie. L. Acker, ed. Springer, Berlin.

Schroeder, H.A., Buckman, J., and Balassa, J.J. (1967) Abnormal trace elements in man: tellurium. J. Chron. Dis. 20:147–161.

Schubert, M. (1931) Zur Behandlung von Darmerkrankungen mit Knoblauch. Med. Welt 1:18.

Schultz, O.E. and Mohrmann, H.L. (1965a) Beitrag zur Analyse der Inhaltsstoffe von Knoblauch: Allium sativum L. 1. Mitteilung: Dünnschichtchromatographie des Knoblauchöls. Pharmazie 20:379–381. Chem. Abst. 63 (1965) 8116.

Schultz, O.E. and Mohrmann, H.L. (1965b) Beitrag zur Analyse der Inhaltsstoffe von Knoblauch: *Allium sativum* L. 2. Mitteilung: Gaschromatographie des Knoblauchöls. Pharmazie *20*:441–447. Chem. Abst. 63 (1965) 9744.

Schultze-Heubach, D. (1928) Erfahrungen mit Allisatin bei Darmerkrankungen. Münch. Med. Wochenschr. *75*:1379.

Schulz, K.H. and Hausen, B.M. (1975) Kontaktekzeme durch Pflanzen und Hölzer. Hautarzt *26*:92–96.

Schuphan, W. and Schwerdtfeger, E. (1971) Arginine as a nitrogen reserve in onions. Ernaehr. Umsch. *18*:228. Chem. Abst. 75 (1971) 117 302.

Schwahn, H. (1928) Hypertonie und Knoblauch. Schweiz. Med. Wochenschr. *58*:104–106.

Schweinfurth, G. (1887) Die letzten botanischen Entdeckungen in den Gräbern Ägyptens. In: Botanische Jahrbücher der Systematik, Pflanzengeschichte und Pflanzengeographie. A. Engler, ed. Wilhelm Engelmann, Leipzig, pp. 1–16.

Schwimmer, S., Carson, J.F., Makower, R.U., Mazelis, M., and Wong, F.F. (1960) Demonstration of alliinase in a protein preparation from onion. Experientia *16*:449–450. Chem. Abst. 55 (1961) 10 592.

Schwimmer, S., Ryan, C.A., and Wong, F.F. (1964) Specificity of L-cysteine sulfoxide lyase and partially competitive inhibition by S-alkyl-L-cysteines. J. Biol. Chem. *239*:777–782. Chem. Abst. 60 (1964) 7082.

Schwimmer, S. (1969) Characterization of S-propenyl-L-cysteine-sulfoxide as the principal endogenous substrate of L-cysteine sulfoxide lyase of onion. Arch. Biochem. Biophys. *130*:312–320.

Schwimmer, S. (1971) Enzymic conversion of gamma-L-glutamyl cysteine peptides to pyruvic acid, a coupled reaction for enhancement of onion flavor. J. Agric. Food Chem. *19*:980–983. Chem. Abst. 75 (1971) 117 306.

Schwimmer, S. and Austin, S.J. (1971a) Gamma glutamyl transpeptidase of sprouted onion. J. Food Sci. *36*:807–811. Chem. Abst. 75 (1971) 115 647.

Schwimmer, S. and Austin, S.J. (1971b) Enhancement of pyruvic acid release and flavor in dehydrated Allium powders by gamma glutamyl transpeptidases. J. Food Sci. *36*:1081–1085. Chem. Abst. 76 (1972) 71 174.

Schwimmer, S. and Friedman, M. (1973) Genesis of volatile sulphur-containing food flavours. Flavour Industry *3*:137–145. Chem. Abst. 76 (1972) 152 074.

Schwimmer, S. and Guadagni, D.G. (1962) Relation between olfactory threshold concentration and pyruvic acid content of onion juice. J. Food Sci. *27*:94–97. Chem. Abst. 60 (1964) 4700.

Schwimmer, S. and Guadagni, D.G. (1967) Cysteine induced odor intensification in onions and other foods. J. Food Sci. *32*:405–408. Chem. Abst. 61 (1967) 99 045.

Schwimmer, S. and Kjaer, A. (1960) Purification and specifity of the C-S-lyase of Albizzia lophanta. Biochim. Biophys. Acta *42*:316–324.

Schwimmer, S. and Mazelis, M. (1963) Characterization of alliinase of Allium cepa (onion). Arch. Biochem. Biophys. *100*:66–73. Chem. Abst. 58 (1963) 8185.

Schwimmer, S. and Weston, W.J. (1961) Enzymatic development of pyruvic acid in onion as a measure of pungency. J. Agric. Food Chem. *9*:301–304. Chem. Abst. 56 (1962) 9174.

Sebastian, K.L., Zacharias, N.T., and Philip, B. (1979) The hypolipidemic effect of onion (Allium cepa Linn.) in sucrose fed rabbits. Indian J. Physiol. Pharmacol. *23*:27–30.

Sediyama, M.A.N., Francisco da Silva, J., Cardoso, A.A., and Casali, V.W.D. (1992) Tolerance of garlic (*Allium sativum* L.) BGH 492 to the herbicides prometryn and oxadiazon. Rev. Ceres *39*:21–30. Chem. Abst. 117 (1992) 207 042 (Portuguese).

Seel, H. (1952) Der physiologische und pharmakologische Wert einiger einheimischer Bittermittel und Gewürzpflanzen. Pharmazie *7*:837–854.

Seelert, K. (1989) Ernährung und Krebs—besteht ein Zusammenhang? Dtsch. Apoth. Ztg. *129*: 2578–2579.

Sekine, T., Ando, K., Machida, M., and Kanaoka, Y. (1972) Fluorescent thiol reagents. V. Microfluorometry of thiol compounds with a fluorescent-labeled maleimide. Anal. Biochem. *48*:557–567. Chem. Abst. 77 (1972) 45 974.

Selby, C., Galpin, I.J., and Collin, H.A. (1979) Comparison of the onion plant (Allium cepa) and onion tissue culture. I. Alliinase activity and flavour precursor compounds. New Phytol. *83*:351–359.

Selby, C., Turnbull, A., and Collin, H.A. (1980) Comparison of the onion plant (Allium cepa) and onion tissue culture. II. Stimulation of flavour precursor synthesis in onion tissue cultures. New Phytol. *84*:307–312.

Seligmann, S. (1910) Der böse Blick und Verwandtes. Ein Beitrag zur Geschichte des Aberglaubens aller Zeiten und Völker. Hermann Barsdorf, Berlin.

Semm, H. (1987) Therapeutische Wirkung von Ilja-Rogoff-Knoblauchpillen mit Rutin bei Patienten mit Fettstoffwechselstörungen. Pharm. Rundsch. *3*:28–30.

Semmler, F.W. (1892a) Über das Ätherische Öl des Knoblauchs (*Allium sativum*). Arch. Pharm. [Weinheim] *230*:434–443.

Semmler, F.W. (1892b) Das ätherische Öl der Küchenzwiebel. Arch. Pharm. [Weinheim] *230*: 443–448.

Sen, S.K. and Rao, C.V.N. (1966) Studies on pectic substances in onion (Allium cepa Linn.). Indian J. Appl. Chem. *29*:127–129.

Sendl, A. (1992) Chemisch-analytische und pharmakologische Untersuchungen von Allium ursinum L. und Sedum telephium L. Dissertation, University of Munich.

Sendl, A., Elbl, G., Steinke, B., Redl, K., Breu, W., and Wagner, H. (1992) Comparative pharmacological investigations of Allium ursinum and *Allium sativum*. Planta Med. *58*:1–7. Chem. Abst. 116 (1992) 248 053.

Sendl, A. and Wagner, H. (1990) Comparative chemical and pharmacological investigations of extracts of Allium ursinum (wild garlic) and *Allium sativum* (garlic). Planta Med. 56:588–589.

Sendl, A. and Wagner, H. (1991) Isolation and identification of homologues of ajoene and alliin from bulb-extracts of Allium ursinum. Planta Med. 57:361–362. Chem. Abst. 115 (1991) 252 109.

Sepulveda, D. (1938a) Un caso de dermatitis provocado por el contacto con ajo. Arch. Hosp. Clin. Nin. Roberto Rio 8:64.

Sepulveda, D. (1938b) Un caso de dermatosis por contacto con ajos. Rev. Chilena Pediat. [Santiago de Chile] 9:339–343.

Serrano, A., Cordoba, F., Gonzalez-Reyes, J.A., Navas, P., and Villalba, J.M. (1994) Purification and characterization of two distinct NAD(P)H dehydrogenases from onion (Allium cepa L.) root plasma membrane. Plant Physiol. 106:87–96.

Sethi, S.C. and Aggarwal, J.S. (1957) Stabilization of edible fats by spices. III. J. Sci. Ind. Res. [New Delhi] 16:181–182.

Seuri, M., Taivanen, A., Ruoppi, P., and Tukiainen, H. (1993) Three cases of occupational asthma and rhinitis caused by garlic. Clin. Exp. Allergy 23:1011–1014.

Shah, S.A. and Vohora, S.B. (1990) Boron enhances anti-arthritic effects of garlic oil. Fitoterapia 61:121–126. Chem. Abst. 113 (1990) 204 565.

Shalinsky, D.R., McNamara, D.B., and Agrawal, K.C. (1989) Inhibition of GSH-dependent PGH2 isomerase in mammary adenocarcinoma cells by allicin. Prostaglandins 37:135–148.

Shankaranarayana, M.L., Abraham, K.O., Raghavan, B., and Natarajan, C.P. (1981) Determination of flavor strength in Alliums (onion and garlic). Indian Food Packer 35:3–8. Chem. Abst. 95 (1981) 95 597.

Shannon, S., Yamaguchi, M., and Howard, F.D. (1967a) Precursors involved in the formation of pink pigments in onion purees. J. Agric. Food Chem. 15:423–426. Chem. Abst. 67 (1967) 31 728.

Shannon, S., Yamaguchi, M., and Howard, F.D. (1967b) Reactions involved in formation of a pink pigment in onion purees. J. Agric. Food Chem. 15:417–422. Chem. Abst. 67 (1967) 31 727.

Shao, J., Kang, J., Zhang, Q., and Kunming General Hospital. (1993) Deodorization process for garlic. People's Republic of China. Patent 1 077 854. Chem. Abst. 120 (1994) 268 831.

Sharafatullah, T., Khan, M.I., and Ahmad, S.I. (1986) Diuretic action of garlic extract in anesthetized normotensive dogs. J. Pak. Med. Assoc. 36:280–282.

Sharma, A., Pawdal-Desai, S.R., Tewari, G.M., and Bandyopadhyay, C. (1981) Factors affecting antifungal activity of onion extractives against aflatoxin-producing fungi. J. Food Sci. 46:741–744. Chem. Abst. 94 (1981) 203 184.

Sharma, C.P. and Nirmala, N.V. (1985) Effects of garlic extract and of three pure components isolated from it on human platelet aggregation, arachidonate metabolism, release reaction and platelet ultrastructure—comments. Thromb. Res. 37:489–490. Chem. Abst. 102 (1985) 178 877.

Sharma, C.P. and Sunny, M.C. (1988) Effects of garlic extracts and of three pure components isolated from it on human platelet aggregation, arachidonate metabolism, release reaction and platelet ultrastructure—comments. Thromb. Res. 52:493–494. Chem. Abst. 110 (1989) 69 166.

Sharma, K.K., Chowdhury, N.K., and Sharma, A.L. (1975) Long term effect of onion on experimentally-induced hypercholesteremia and consequently decreased fibrinolytic activity in rabbits. Indian J. Med. Res. 63:1629–1634.

Sharma, K.K., Sharma, A.L., Dwived, K.K., and Sharma, P.K. (1976) Effect of raw and boiled garlic on blood cholesterol in butter fat lipaemia. Indian J. Nutr. Diet. 13:7–10. Chem. Abst. 85 (1976) 41 021.

Sharma, K.K., Gupta, R.K., Gupta, S., and Samuel, K.C. (1977) Antihyperglycemic effect of onion: effect on fasting blood sugar and induced hyperglycemia in man. Indian J. Med. Res. 65:422–429.

Sharma, K.K., Sharma, S.P., and Arora, R.C. (1978) Some observations on the mechanism of fibrinolytic enhancing effect of garlic during alimentary lipemia in man. J. Postgrad. Med. 24:98–102.

Sharma, K.K. and Sharma, S.P. (1979) Effect of onion and garlic on serum cholesterol of normal subjects. Mediscope 22:134–136. Chem. Abst. 92 (1980) 127 498.

Sharma, V.D., Sethi, M.S., Kumar, A., and Rarotra, J.R. (1977) Antibacterial property of *Allium sativum* Linn.: in vivo and in vitro studies. Indian J. Exp. Biol. 15:466–468.

Shashikanth, K.N., Basappa, S.C., and Sreenivasamurthy, V. (1981) Stimulatory factors of garlic. Indian J. Biochem. Biophys. 18:79–80.

Shashikanth, K.N., Basappa, S.C., and Sreenivasamurthy, V. (1984) A comparative study of raw garlic extract and tetracycline on caecal microflora and serum proteins of albino rats. Folia Microbiol. 29:348–352.

Shashikanth, K.N., Basappa, S.C., and Sreenivasamurthy, V. (1985) Allicin concentration in the gut of rats and its influence on the microflora. J. Food Sci. Technol. 22:440–442. Chem. Abst. 104 (1986) 223 782.

Shashikanth, K.N., Basappa, S.C., and Sreenivasamurthy, V. (1986) Effect of feeding raw and boiled garlic (*Allium sativum* L.) extracts on the growth, caecal microflora, and serum proteins of albino rats. Nutr. Rep. Int. 33:313–319. Chem. Abst. 104 (1986) 128 842.

Sheela, C.G. and Augusti, K.T. (1992) Antidiabetic effects of S-allyl cysteine sulfoxide isolated from garlic *Allium sativum* Linn. Indian J. Exp. Biol. 30:523–526. Chem. Abst. 117 (1992) 163 661.

Sheela, C.G. and Augusti, K.T. (1995) Antiperoxide effects of S-allylcysteine sulphoxide isolated from *Allium sativum* Linn. and gugulipid in cholesterol diet fed rats. Ind. J. Exp. Biol. 33:337–341.

Sheen, L.Y., Lin, S.Y., and Tsai, S.J. (1992) Odor assessments for volatile compounds of garlic and ginger essential oils by sniffing gas chromatography. Zhongguo Nongye Huaxue Huizhi (J. Chin. Agric. Chem. Soc.) 30:14–24. Chem. Abst. 117 (1992) 130 135.

Shen, L., Wang, Y., and Feng, L. (1983) Study and preparation of allicin microcapsule. Zhongcaoyao (Chin. Trad. Herbal Drugs) 14:161–164. Chem. Abst. 98 (1983) 221 761 (Chinese).

Sheo, H.J., Lim, H.J., and Jung, D.L. (1993) Effects of onion juice on toxicity of lead in rat. Hanguk Yongyang Siklyong Hakhoechi (J. Korean Soc. Food Nutr.) 22:138–143. Chem. Abst. 120 (1994) 76 284 (Korean).

Shiga, Y. (1989) Manufacture of odorless garlic using basic salts. Japan. Patent 89 218 568. Chem. Abst. 112 (1990) 97 301.

Shimakawa, T. and Japan Health Summit Kk. (1992) Manufacture of soft capsules using Poem S100 as emulsifier. Japan. Patent 92 440 876. Chem. Abst. 117 (1992) 25 093.

Shin, M. and Sakihama, N. (1986) All angiosperms have two molecular species of ferredoxin. Plant Physiol. 80:143.

Shoetan, A., Augusti, K.T., and Joseph, P.K. (1984) Hypolipidemic effects of garlic oil in rats fed ethanol and a high lipid diet. Experientia 40:261–263. Chem. Abst. 100 (1984) 152 352.

Shoji, S., Furuishi, K., Yanase, R., Miyazaka, T., and Kino, M. (1993) Allyl compounds selectively killed human immunodeficiency virus (type 1)–infected cells. Biochem. Biophys. Res. Commun. 194:610–621.

Shrivastava, A.K. and Singh, K.V. (1982) Letter to the Editor. Indian Drugs 19:245.

Shu, X.O., Zheng, W., Potischman, N., Brinton, L.A., Hatch, M.C., Gao, Y.T., and Fraumeni, J.F. (1993) A population-based case-control study of dietary factors and endometrial cancer in Shanghai, People's Republic of China. Am. J. Epidemiol. 137:155–165.

Shyu, K.W. and Meng, C.L. (1987) The inhibitory effect of oral administration of garlic on experimental carcinogenesis in hamster buccal pouches by DMBA painting. Proc. Natl. Sci. Counc. B. ROC 11:137–147.

Sial, A.Y. and Ahmad, S.I. (1982) Study of the hypotensive action of garlic extract on experimental animals. J. Pak. Med. Assoc. 32:237–239.

Sibthorp, J. (1823) Allium. In: Flora Graeca. London, pp. 221–227.

Siddiqui, A.M., Hashmi, R.S., and Pawar, S.S. (1988) Effect of garlic oil administration on hepatic microsomal mixed function oxidase system in adult male and female rats. Med. Sci. Res. 16:777–779. Chem. Abst. 109 (1988) 229 348.

Sieg, H. (1953) Gottessegen der Kräuter. E. Staneck, Berlin.

Siegel, G., Emden, J., Schnalke, F., Walter, A., Rueckborn, K., and Wagner, K.G. (1991a) Wirkungen von Knoblauch auf die Gefaessregulation. Medizin. Welt 42 (Suppl. 7a):32–34.

Siegel, G., Walter, A., Schnalke, F., Schmidt, A., Buddecke, E., Loirand, G., and Stock, G. (1991b) Potassium channel activation, hyperpolarization, and vascular relaxation. Z. Kardiol. 80:9–24.

Siegel, G., Emden, J., Wenzel, K., Mironneau, J., and Stock, G. (1992) Potassium channel activation in vascular smooth muscle. In: Excitation-contraction coupling in skeletal, cardiac, and smooth muscle. G.B. Frank et al., eds. Plenum Press, New York, pp. 53–72.

Siegers, C.P. (1987) Toxikologie der Phytopharmaka. Z. Phytother. 8:110–113.

Siegers, C.P. (1989) Toxikologische Bewertung von Knoblauch und Knoblauchinhaltsstoffen. Dtsch. Apoth. Ztg. Suppl.15:11–13.

Siegers, C.P. (1992) Free radical reactions and atherosclerosis. Eur. J. Clin. Res. 3A:2.

Siegers, C.P. (1993) Neues zur antiarteriosklerotischen Wirkung des Knoblauchs. Z. Phytother. 14:21–22.

Silagy, C. and Neil, A. (1994) Garlic as a lipid lowering agent: a meta-analysis. J. R. Coll. Physicians London 28:39–45.

Silber, W. (1933) Über die Wirkung des Knoblauchs auf Experimentelle Arterienverkalkung. Klin. Wochenschr. 12:509.

Simmonds, N.W. (1976) Evolution of crop plants. Longman, London.

Simon, D. (1932) Behandlung der Ruhrfolgen mit Allisatin. Med. Klin. 3:86–87.

Simonis, W.C. (1965) Die einkeimblättrige Pflanze. Haug, Ulm.

Simons, L.A., Balasubramaniam, S., and Konigsmark, M. (1995) On the effect of garlic on plasma lipids and lipoproteins in mild hypercholesterolaemia. Atherosclerosis 113:219–225.

Singh, K.V. and Deshmukh, S.K. (1984) Volatile constituents from members of Liliaceae and spore germination of Microsporum gypseum complexes. Fitoterapia 55:297–299. Chem. Abst. 103 (1985) 3558.

Singh, K.V. and Shukla, N.P. (1984) Activity on multiple resistant bacteria of garlic (Allium sativum) extract. Fitoterapia 55:313–315.

Singh, L.J., Pruthi, J.S., Sankaran, A.N., Indiramma, K., and Girdhari, L. (1959a) Effect of type of packaging and storage temperature on flavour and colour of garlic powder. Food Sci. (Mysore) 8:457–461. Chem. Abst. 54 (1960) 15 755.

Singh, L.J., Pruthi, J.S., Sreenivasamurthi, V., and Girdhari, L. (1959b) Effect of regional variability in garlic on the quality of garlic powder. Food Sci. (Mysore) 8:431–436. Chem. Abst. 54 (1960) 15 754.

Singh, L.J., Pruthi, J.S., Sreenivasamurthi, V., Swaminathan, M., and Subrahmanyan, V. (1959c) Effect of type of packaging and storage temperature on allyl sulphide, total sulphur, antibacterial activity, and volatile reducing substances in garlic powder. Food Sci. (Mysore) 8:453–457. Chem. Abst. 54 (1960) 15 755.

Singh, M. and Singh, S.P. (1994) Some environmental studies on toxicity of manganese metal. Proc. Acad. Environ. Biol. 3:53–57. Chem. Abst. 122 (1995) 3241.

Singh, S.B. and Abrol, I.P. (1985) Effect of exchangeable sodium percentage on growth yield and chemical composition of onion (Allium cepa) and garlic (*Allium sativum*). J. Indian Soc. Soil Sci. 33:358–361.

Singh, S.C. (1984) A note on some home remedies available from kitchen stock in Eastern Uttar Pradesh. J. Econ. Tax. Bot. 5:149–150.

Singh, V., Kumar, A., and Singh, S.P. (1984) Effect of normal saline, potassium permanganate, and garlic extract on healing of contaminated wounds in buffalo-calves. Indian J. Anim. Sci. 54:41–45.

Sinha, A. (1959) Chemical examination of Allium cepa Linn., Part I. Study of glycosidic and sugar fractions. Indian J. Appl. Chem. 22:89–91.

Sinha, A. and Sanyal, A.K. (1959) Separation and estimation of sugar components of Allium cepa Linn. (N. O. Liliaceae) by paper chromatography. Curr. Sci. 25:281–282.

Sinha, N.K., Guyer, D.E., Gage, D.A., and Lira, C.T. (1992) Supercritical carbon dioxide extraction of onion flavors and their analysis by gas chromatography–mass spectrometry. J. Agric. Food Chem. 40:842–845. Chem. Abst. 116 (1992) 234 139.

Sinha, S.M., Pasricha, J.S., Sharma, R.C., and Kandhari, K.C. (1977) Vegetables responsible for contact dermatitis of the hands. Arch. Dermatol. 113:776–779.

Siqueira, W.J., Medina Filho, H.P., Lisbao, R.S., and Fornasier, J.B. (1985) Morphological and electrophoretic characterization of garlic clones. Bragantia 44:357–374. Chem. Abst. 106 (1987) 2992 (Portuguese).

Sitprija, S., Plengvidhya, C., Kangkaya, V., Bhuvapanich, S., and Tunkayoon, M. (1987) Garlic and diabetes mellitus phase II clinical trial. J. Med. Assoc. Thailand 70:223–227.

Skapska, S. and Karwowska, K. (1990) Method for garlic evaluation with respect to usefulness in flavor extracts production. I. Preliminary results. Pr. Inst. Lab. Badaw. Przem. Spozyw. 41:91–104. Chem. Abst. 114 (1991) 22 602 (Polish).

Sklan, D., Berner, Y.N., and Rabinowitch, H.D. (1992) The effect of dietary onion and garlic on hepatic lipid concentrations and activity of antioxidative enzymes in chicks. J. Nutr. Biochem. 3:322–325. Chem. Abst. 117 (1992) 89 410.

Small, L.D., Bailey, J.H., and Cavallito, C.J. (1947) Alkyl thiolsulfinates. J. Am. Chem. Soc. 69:1710–1713. Chem. Abst. 41 (1947) 6196.

Small, L.D., Bailey, J.H., and Cavallito, C.J. (1949) Comparison of some properties of thiolsulfonates and thiolsulfinates. J. Am. Chem. Soc. 71:3565–3566. Chem. Abst. 44 (1950) 1011.

Smith, P.K., Krohn, R.I., Hermanson, G.T., Mallia, A.K., Gartner, F.H., Provenzano, M.D., Fujimoto, E.K., Goeke, N.M., Olson, B.J., and Klenk, D.C. (1985) Measurement of protein using bicinchoninic acid. Anal. Biochem. 150:76–85.

Smoczkiewicz, M.A., Nitschke, D., and Stawinski, T.M. (1977) Mikrobestimmung von Spuren von Saponinglykosiden in "Nicht-Saponinführendem" Pflanzenmaterial. Mikrochim. Acta 2:597–605. Chem. Abst. 88 (1978) 34 014.

Smoczkiewicz, M.A., Lutomski, J., Nitschke, D., and Wieladek, H. (1978) Determination of traces of saponin components in plants of the species Allium. 11th IUPAC Int. Symp. Chem. Nat. Prod. 2:488–489. Chem. Abst. 91 (1979) 207 426.

Smoczkiewicz, M.A., Nitschke, D., and Wieladek, H. (1982) Microdetermination of steroid and triterpene saponin glycosides in various plant materials. I. Allium species. Mikrochim. Acta [Vienna] II:43–53. Chem. Abst. 97 (1982) 35 652.

Smoczkiewiczowa, M.A., Lutomski, J., and Nitschke, D. (1981) Chemical and pharmacological characterization of the onion (Allium cepa L.). Herba Pol. 27:169–188. Chem. Abst. 97 (1982) 159 467 (Polish).

Smoczkiewiczowa, A., Ostrowska, B., Jasiczak, J., Olszak, M., and Rychlinska, H. (1990) Method of estimation of diphenylamine as hypoglycemic agent in onion. Herba Pol. 36:97–109. Chem. Abst. 116 (1992) 79 662 (Polish).

Smoczkiewiczowa, M.A. and Nitschke, D. (1975) Studies on saponins in garlic (*Allium sativum*). Zesz. Nauk Akad. Ekon. Poznan, Ser. I (Sci. Notes Econ. Acad. Poznan) 62:43–48. Chem. Abst. 85 (1976) 59 614 (Polish).

Smoczkiewiczowa, A. and Nitschke, D. (1978a) Study of saponins and sapogenins in onions. Zesz. Nauk. Akad. Ekon. Poznan., Ser. I (Sci. Notes Acad. Econ. Poznan) 73:40–43. Chem. Abst. 90 (1979) 200 256 (Polish).

Smoczkiewiczowa, A. and Nitschke, D. (1978b) Flavonoids in onions. Zesz. Nauk. Akad. Ekon. Poznan., Ser. I (Sci. Notes Acad. Econ. Poznan) 73:35–39. Chem. Abst. 91 (1979) 16 662 (Polish).

Smoczkiewiczowa, A. and Wieladek, H. (1978) Experiments for establishing the presence of saponins in Allium porrum. Zesz. Nauk. Akad. Ekon. Poznan., Ser. I (Sci. Notes Acad. Econ. Poznan) 73:44–47 (Polish).

Sodimu, O., Joseph, P.K., and Augusti, K.T. (1984) Certain biochemical effects of garlic oil on rats maintained on high fat-high cholesterol diet. Experientia 40:78–80. Chem. Abst. 100 (1984) 119 812.

Sogani, R.K. and Katoch, K. (1981) Correlation of serum cholesterol levels and incidence of myocardial infarction with dietary onion and garlic eating habits. J. Assoc. Physicians India 29:443–446.

Soh, C.T. (1960) The effects of natural food-preservative substances on the development and survival of intestinal helminth eggs and larvae. II. Action an Ancylostoma duodenale larvae. Am. J. Trop. Med. Hyg. 9:8–10.

Sokolov, N.I. (1954) Use of the phytoncides of onion and garlic with gynecological diseases. Veterinariya [Moscow] 31:51–52 (Russian).

Song, C.S., Kim, J.H., Kim, E.S., and Lee, P.H. (1963a) A blood anticoagulant substance from garlic (Allium sativum). I. Its preparation and studies on its anticoagulant effect. Yonsei Med. J. 4:17–20.

Song, C.S., Kim, J.H., and Rhee, D.J. (1963b) Blood anticoagulant substance from garlic extraction, and physical and chemical properties. Yonsei Med. J. 3:114.

Song, C.S., Kim, Y.S., Lee, D.J., and Nam, C.C. (1963c) A blood anticoagulant substance from garlic (Allium sativum). II. Chemical analysis and studies on the biochemical and pharmacological effects. Yonsei Med. J. 4:21–26. Chem. Abst. 61 (1964) 8800.

Song, P. and Peffley, E.B. (1994) Plant regeneration from suspension cultures of Allium fistulosum and an A. fistulosum × A. cepa interspecific hybrid. Plant Sci. 98:63–68. Chem. Abst. 121 (1994) 5317.

Song, T.B., Bae, E.S., and Yum, Y.T. (1987) A study on the effect of garlic and 2,3-dimercaptosuccinic acid on the cadmium poisoning of pregnant rat. Koryo Taehakkyo Uikwa Taehak Nonmunjip (Korea Univ. Med. J.) 24:237–245. Chem. Abst. 107 (1987) 192 527.

Soni, K.B., Rajan, A., and Kuttan, R. (1992) Reversal of aflatoxin-induced liver damage by turmeric and curcumin. Cancer Lett. 66:115–121.

Soni, S.K. and Finch, S. (1979) Laboratory evaluation of sulfur-bearing chemicals as attractants for larvae of the onion fly, Delia antiqua (Meigen) (Diptera: Anthomyiidae). Bull. Entomol. Res. 69:291–298. Chem. Abst. 91 (1979) 118 597.

Sopova, M., Sekovski, Z., and Jovanovska, M. (1985) Cytological effects of nicotine on root-tip cells of Allium sativum L. Acta Biol. Med. Exp. 10:43–49. Chem. Abst. 105 (1986) 185 658.

Souci, S.W., Fachmann, W., and Kraut, H. (1986) Food composition and nutrition tables 1986/87. Wissenschaftl. Verlagsges, Stuttgart.

Souza, R.J. and Casali, D.V.W. (1991) Influencia do nitrogenio e cycocel na cultura do alho (Allium sativum L.). Cienc. Prat. 15:69–78. Chem. Abst. 118 (1993) 54 233 (Portuguese).

Souza, R.J. and Casali, V.W.D. (1992a) Effects of PIX (mepiquat chloride) growth regulator upon productivity and over-sprouting of garlic (Allium sativum L.). Cienc. Prat. 16:219–223. Chem. Abst. 122 (1995) 3467.

Souza, R.J. and Casali, V.W.D. (1992b) Influence of paclobutrazol on commercial characteristics of garlic (Allium sativum L.). Cienc. Prat. 16:246–257. Chem. Abst. 122 (1995) 3468.

Spanu, P. and Bonfante-Fasolo, P. (1988) Cell wall–bound peroxidase activity in roots of mycorrhizal Allium porrum. New Phytol. 109:119–124. Chem. Abst. 109 (1988) 146 372.

Spare, C.G. and Virtanen, A.I. (1964) On the occurrence of free selenium-containing amino acids in onion (Allium cepa). Acta Chem. Scand. 18:280–282.

Sparnins, V.L., Mott, A.W., Barany, G., and Wattenberg, L.W. (1986) Effects of allyl methyl trisulfide on glutathione S-transferase activity and BP-induced neoplasia in the mouse. Nutr. Cancer 8:211–215. Chem. Abst. 105 (1986) 90 912.

Sparnins, V.L., Barany, G., and Wattenberg, L.W. (1988) Effects of organosulfur compounds from garlic and onions on benzo[a]pyrene-induced neoplasia and gluthathione S-transferase activity in the mouse. Carcinogenesis 9:131–134. Chem. Abst. 109 (1988) 31 640.

Spice, R.N. (1976) Hemolytic anemia associated with ingestion of onions in a dog. Can. Vet. J. 17:181–183.

Spilkova, J. and Hubik, J. (1988) Biologische Wirkungen von Flavonoiden. I. Pharm. Uns. Zeit 17:1–9.

Spilkova, J. and Hubik, J. (1992) Biologische Wirkungen von Flavonoiden. II. Pharm. Uns. Zeit 21:174–182.

Spinka, J. and Stampfer, J. (1956) Therapeutic garlic preparations. Austria. Patent 187 239. Chem. Abst. 51 (1957) 3095.

Spivak, M.Y. (1962) On the use of phytoncides of garlic and onion for the treatment of tumorous patients. Vopr. Onkol. (Problems in Oncology) 8:93–96 (Russian).

Sprecher, E. (1986a) Allium sativum L.: Wundermittel oder Arzneipflanze? Pharm. Ztg. 131:3161–3168.

Sprecher, E. (1986b) Knoblauchzehen auf dem Prüfstand. Dtsch. Apoth. Ztg. 126:1265–1267.

Sprecher, E. (1986c) Allium sativum L.: Wundermittel oder Arzneipflanze? In: Schriftenreihe. Bundesapothekerkammer, Bonn, pp. 209–225.

Sprecher, E. (1987) Allium sativum L.: Wundermittel oder Arzneipflanze? Schweiz. Apoth. Ztg. 125:605–609.

Sprecher, E. (1989) Wirksamkeit von Knoblauchpräparaten erwiesen? Med. Monatsschr. Pharm. 12:125–126.

Sprengel, K. (1822) Theophrast's Naturgeschichte der Gewächse. Joh. Friedr. Hammerich, Altona.

Sreenivasamurthy, V., Sreekantiah, K.R., and Johar, D.S. (1961) Studies on the stability of allicin and alliin present in garlic. J. Sci. Ind. Res. 20C:292–295. Chem. Abst. 56 (1962) 9177.

Sreenivasamurthy, V., Sreekantiah, K.R., Jayaraj, A.P., and Johar, D.S. (1962) A preliminary report on the treatment of acute lepromatous neuritis with garlic. Leprosy India 34:171–174.

Sreenivasamurthy, V. and Krishnamurthy, K. (1959) Place of spices and aromatics in Indian dietary. Food Sci. (Mysore) 8:284–288. Chem. Abst. 54 (1960) 1764.

Srinivas, N. and Rao, P.V.V.P. (1993) Toxicology of arsenic(III) and arsenic(V) on Allium cepa. Recent Trends Biotechnol., 144–149. Chem. Abst. 121 (1994) 75 598.

Srinivasan, M., Bhatia, I.S., and Satyanarayana, M.N. (1953) Carbohydrates of garlic (*Allium sativum* L.) and onion (Allium cepa L.). Curr. Sci. 22:208–209. Chem. Abst. 48 (1954) 2945.

Srinivasan, M. and Bhatia, I.S. (1954) Glucofructosan from Polianthes tuberosa Linn. and garlic (*Allium sativum* Linn.). Curr. Sci. 23:192–193. Chem. Abst. 49 (1955) 1150.

Srinivasan, M.R. and Srinivasan, K. (1995) Hypocholesterolemic efficacy of garlic-smelling flower Adenocalymma alliaceum Miers in experimental rats. Ind. J. Exp. Biol. 33:64–66.

Srinivasan, V. (1969) A new anti-hypertensive agent? Lancet 2:800.

Srivastava, H.C. and Mathur, P.B. (1956) Studies on the smoke-curing of garlic. Bull. Cent. Food Technol. Res. Inst. Mysore 5:286–287.

Srivastava, K. and Srivastava, P. (1981) Studies on plant materials as corrosion inhibitors. Br. Corros. J. 16:221–223. Chem. Abst. 96 (1982) 167 040.

Srivastava, K.C. (1984a) Aqueous extracts of onion, garlic, and ginger inhibit platelet aggregation and alter arachidonic acid metabolism. Biomed. Biochim. Acta 43:S335–S346.

Srivastava, K.C. (1984b) Effects of aqueous extracts of onion, garlic, and ginger on platelet aggregation and metabolism of arachidonic acid in the blood vascular system: in vitro study. Prostaglandins Leukotrienes Med. 13:227–235.

Srivastava, K.C. (1986a) Onion exerts antiaggregatory effects by altering arachidonic acid metabolism in platelets. Prostaglandins Leukotrienes Med. 24:43–50.

Srivastava, K.C. (1986b) Evidence for the mechanism by which garlic inhibits platelet aggregation. Prostaglandins Leukotrienes Med. 22:313–321.

Srivastava, K.C. and Justesen, U. (1989) Isolation and effects of some garlic components on platelet aggregation and metabolism of arachidonic acid in human blood platelets. Wien. Klin. Wochenschr. 101:293–299.

Srivastava, K.C. and Mustafa, T. (1989) Spices: antiplatelet activity and prostanoid metabolism. Prostaglandins Leukotrienes Essent. Fatty Acids 38:255–266.

Srivastava, K.C. and Mustafa, T. (1993) Pharmacological effects of spices: eicosanoid modulating activities and their significance in human health. Biomed. Rev. 2:15–29.

Srivastava, K.C. and Tyagi, O.D. (1993) Effects of a garlic-derived principle (ajoene) on aggregation and arachidonic acid metabolism in human blood platelets. Prostaglandins Leukotrienes Essent. Fatty Acids 49:587–595.

Srivastava, K.S., Perera, A.D., and Saridakis, H.O. (1982) Bacteriostatic effects of garlic sap on gram negative pathogenic bacteria: an in vitro study. Lebensm. Wiss. Technol. 15:74–76.

Srivastava, M., Nityanand, S., and Kapoor, N.K. (1984) Effect of hypocholesterolemic agents of plant origin on catecholamine biosynthesis in normal and cholesterol fed rabbits. J. Biosci. 6:277–282. Chem. Abst. 102 (1985) 72 692.

Srivastava, M.M., Juneja, A., Dass, S., Srivastava, R., Srivastava, S., Mishra, S., Srivastav, S., Singh, V., and Prakash, S. (1994) Studies on the uptake of trivalent and hexavalent chromium by onion (Allium cepa). Chem. Speciation Bioavaliability 6:27–30. Chem. Abst. 121 (1994) 276 800.

Srivastava, P. and Srivastava, K. (1983) Inhibition of corrosion of mild steel in nitric acid by garlic. Corros. Maint. 6:149–152. Chem. Abst. 100 (1983) 107 390.

Srivastava, P. and Srivastava, K. (1984) Inhibition of corrosion of mild steel in nitric acid by garlic. Proc. Natl. Acad. Sci. India, Sect. A, 54:141–147. Chem. Abst. 103 (1985) 199 325.

Srivastava, U.S., Jaiswal, A.K., and Abidi, R. (1985) Juvenoid activity in extracts of certain plants. Curr. Sci. 54:576–578.

Stacchini, A. and Mangegazzini, C. (1987) Piante aromatiche in alimentazione: criteri per valutare la loro sicurezza d'uso. Boll. Chim. Farm. 126:88–92. Int. Pharm. Abst. 26 (1989) 2091.

Stan, H.J. and Christall, B. (1991) Residue analysis of onions and other foodstuffs with a complex matrix using two-dimensional capillary-GC with three selective detectors. Fresenius J. Anal. Chem. 339:395–398. Chem. Abst. 114 (1991) 183 924.

Standen, O.D. (1953) Experimental chemotherapy of oxyuriasis. Br. Med. J. 2:757–758. Chem. Abst. 48 (1954) 1569.

Starke, H. and Herrmann, K. (1976) Flavonole und Flavone der Gemüsearten. VII. Flavonole des Porrees, Schnittlauchs und Knoblauchs. Z. Lebensm. Unters. Forsch. 161:25–30. Chem. Abst. 85 (1976) 59 568.

Stearn, W. (1944) Notes on the genus Allium in the Old World. Its distribution, names, literature, classification, and gardenworthy species. Herbertia 11:11–34.

Stearn, W. (1978) European species of Allium and allied genera of Alliaceae: a synonymic enumeration. Ann. Musei Goulandris 4:83–198.

Stearn, W. (1980) Allium L. Flora Europaea 5:49–69.

Steele, V.E., Kelloff, G.J., Wilkinson, B.P., and Arnold, J.T. (1990) Inhibition of transformation in cultured rat tracheal epithelial cells by potential chemopreventive agents. Cancer Res. 50:2068–2074.

Steer, B.T. (1982) The effect of growth temperature on dry weight and carbohydrate content of onion (Allium cepa L. cv. Creamgold) bulbs. Aust. J. Agric. Res. 33:559–563.

Steiner, M. and Lin, R.I. (1994) Cardiovascular and lipid changes in response to aged garlic extract ingestion. J. Am. Coll. Nutr. 13:524.

Steinmetz, K.A., Kushi, L.H., Bostick, R.M., Folsom, A.R., and Potter, J.D. (1994) Vegetables, fruit, and colon cancer in the Iowa Women's Health Study. Am. J. Epidemiol. 139:1–15.

Steinmetz, K.A. and Potter, J.D. (1991) Vegetables, fruit, and cancer: Epidemiology. Cancer Causes Control 2:325–357.

Steinmetz, K.A. and Potter, J.D. (1993) Food group consumption and colon cancer in the Adelaide case-control study. I. Vegetables and fruits. Int. J. Cancer 53:711–719.

Steinmetz, K.A. and Potter, J.D. (1995) Two of the authors reply. Am. J. Epidemiol. 141:85–86.

Stejskal, R., Urban, E., and Völlenkle, H. (1991) Synthese und Konfigurationszuordnung von Potentiell Antimikrobiellen 5,6-Dihydroxyisobenzofuranonen. Monatsh. Chem. 122:145–156.

Stephenson, R.M. (1949) Aspects of modern onion and garlic dehydration. Food Technol. 3:364–366. Chem. Abst. 44 (1955) 10 955.

Sticher, O. (1991) Beurteilung von Knoblauchpräparaten. Dtsch. Apoth. Ztg. 131:403–413. Chem. Abst. 114 (1991) 192 341.

Sticher, O. and Mütsch-Eckner, M. (1991) Aminosäuren und Gamma-Glutamylpeptide aus Knoblauch. Dtsch. Apoth. Ztg. 131:3–5.

St. Louis, M.E., Peck, S.H.S., Bowering, D., Morgan, G.B., Blatherwick, J., Banerjee, S., Kettyls, G.D.M., Black, W.A., Milling, M.E., Hauschild, A.H.W., Tauxe, R.V., and Blake, P.A. (1988) Botulism from chopped garlic: delayed recognition of a major outbreak. Ann. Med. Interne 108:363–368.

Stöger, R. (1967) Gezielte interne Krebsbehandlung. Kombinierte Hyperergisierungstherapie. Med. Welt 18:3183–3188.

Stöger, R. (1968) Wie stehen derzeit die Aussichten für ein ideales Krebsheilmittel? Prophyl. 7:2–11.

Stöger, R. (1970) Gezielte interne postoperative Tumortherapie in der Praxis. Prakt. Arzt 24:358–378.

Stöger, R. (1976) Gezielte interne Nachbehandlung nach Brustkrebsoperationen. Kombinierte Hyperergisierungstherapie. Wien. Med. Wochenschr. 126:121–126.

Stoianova-Ivanova, B., Tzutzulova, A., and Caputto, R. (1980) On the hydrocarbon and sterol composition in the scales and the fleshy part of Allium sativum Linnaeus bulbs. Riv. ital. EPPOS 62:373–376. Chem. Abst. 94 (1982) 61 791.

Stoianova-Ivanova, B. and Tzutzulova, A.M. (1974) On the composition of higher fatty acids in the scales and the interior of Allium sativum L. bulbs. Dokl. Bolg. Akad. Nauk. (Rep. Bulgarian Acad. Sci.) 27:503–506.

Stoker, B. (1966) Dracula. Jarrolds, London.

Stoll, A. and Seebeck, E. (1947) Über Alliin, die genuine Muttersubstanz des Knoblauchöls. Experientia 3:114–115. Chem. Abst. 41 (1947) 4893.

Stoll, A. and Seebeck, E. (1948) Über Alliin, die genuine Muttersubstanz des Knoblauchöls. 1. Mitteilung über Allium-Substanzen. Helv. Chim. Acta 31:189–210. Chem. Abst. 42 (1948) 4136.

Stoll, A. and Seebeck, E. (1949a) Über den enzymatischen Abbau des Alliins und die Eigenschaften der Alliinase. 2. Mitteilung über Allium-Substanzen. Helv. Chim. Acta 32:197–205. Chem. Abst. 43 (1949) 3482.

Stoll, A. and Seebeck, E. (1949b) Über die Spezifität der Alliinase und die Synthese mehrerer dem Alliin verwandter Verbindungen. 3. Mitteilung über Alliin-Substanzen. Helv. Chim. Acta 32:866–876. Chem. Abst. 43 (1949) 6576.

Stoll, A. and Seebeck, E. (1950a) Über spezifische Inhaltsstoffe des Knoblauchs: Alliin, Alliinase und Allicin. Sci. Pharm. 18:61–79. Chem. Abst. 44 (1950) 11 033.

Stoll, A. and Seebeck, E. (1950b) Die Synthese des natürlichen Alliins. Experientia 6:330. Chem. Abst. 45 (1950) 2864.

Stoll, A. and Seebeck, E. (1951a) Die Synthese des natürlichen Alliins und seiner drei optisch aktiven Isomeren. 5. Mitteilung über Allium Substanzen. Helv. Chim. Acta 34:481–487. Chem. Abst. 45 (1951) 7523.

Stoll, A. and Seebeck, E. (1951b) Chemical investigations on alliin, the specific principle of garlic. Adv. Enzymol. 11:377–400. Chem. Abst. 46 (1952) 585.

Stoll, A. and Seebeck, E. (1951c) Sur la specifité de l'alliinase provenant de l'*Allium sativum*. C. R. Acad. Sci. Paris 232:1441–1442. Chem. Abst. 45 (1951) 7619.

Strecker, J. (1930) Zur Behandlung der Darmbeschwerden im Klimakterium. Zentralb. Gynäkol., 1690–1692.

Strobel, M., N'Diaye, B., Padonou, F., and Marchand, J.P. (1978) Les dermites de contact d'origine végétale (à propos de 10 cas observés à Dakar). Bull. Soc. Med. Afr. Noire Lgue Frse 23:124–127.

Strübing, E. (1967) Knoblauch in alten Zeiten. Zur Diätetik und Ernährung der Menschen. Ernährungsforsch. 12:585–623.

Strunk, E. (1951) Behandlung von chron. Dysenterie mit Knoblauch. Österr. Apoth. Ztg. 5:271–272.

Sturm, D. (1968) Von denen Vampiren oder Menschensaugern. Dichtungen und Dokumente. Hanser, Munich.

Sucur, M. (1980) Effect of garlic on serum lipids and lipoproteins in patients suffering from hyperlipoproteinemia. Diabetol. Croat. 9:323–338 (Croatian).

Sugihara, J. and Cruess, W.V. (1945) Observations on the oxidase of garlic. Fruit Prod. J. Am. Food Manuf. 24:297–298. Chem. Abst. 40 (1946) 6508.

Sugii, M., Nagasawa, S., and Suzuki, T. (1963a) Biosynthesis of S-methyl-L-cysteine and S-methyl-L-cysteine sulfoxide from methionine in garlic. Chem. Pharm. Bull. 11:135–136. Chem. Abst. 58 (1963) 14 447.

Sugii, M., Suzuki, T., and Nagasawa, S. (1963b) Isolation of (–)S-propenyl-L-cysteine from garlic. Chem. Pharm. Bull. 11:548–549. Chem. Abst. 59 (1963) 6509.

Sugii, M., Suzuki, T., Kakimoto, T., and Kato, J. (1964a) Studies on the sulfur-containing amino acids and the related compounds in garlic: I.

Assimilation of sulfate (S35) in garlic. Bull. Inst. Chem. Res. Kyoto Univ. 42:246–251. Chem. Abst. 62 (1965) 5582.

Sugii, M., Suzuki, T., Nagasawa, S., and Kawashima, K. (1964b) Isolation of gamma-L-glutamyl-S-allylmercapto-L-cysteine and S-allylmercapto-L-cysteine from garlic. Chem. Pharm. Bull. 12:1114–1115. Chem. Abst. 61 (1964) 16 437.

Suh, S.K. and Park, H.G. (1993) Rapid multiplication through immature bulbil cultures of garlic. Hanguk Wonye Hakhoechi 34:173–178. Chem. Abst. 120 (1994) 25 522 (Korean).

Suh, S.K. and Park, H.G. (1994) Effects of temperature pretreatment, growth regulators, and antibiotic treatments on anther cultures of various cultivars of garlic. Hanguk Wonye Hakhoechi 35:337–344. Chem. Abst. 121 (1994) 294 993 (Korean).

Sumi, N., Yamashita, H., and Meiji Seika Kaisha Ltd. (1987) Garlic powder manufacture. Japan. Patent 87 134 059. Chem. Abst. 107 (1987) 133 031.

Sumi, S., Tsuneyoshi, T., and Furutani, H. (1993) Novel rod-shaped viruses isolated from garlic, *Allium sativum*, possessing a unique genome organization. J. Gen. Virol. 74:1879–1885. Chem. Abst. 120 (1994) 4132.

Sumiyoshi, H., Kanezawa, A., Masamoto, K., Harada, H., Nakagami, S., Yokota, A., Nishikawa, M., and Nakagawa, S. (1984) Chronic toxicity test of garlic extract in rats. J. Toxicol. Sci. 9:61–75 (Japanese).

Sumiyoshi, H. and Wargovich, M.J. (1989) Garlic (*Allium sativum*): A review of its relationship to cancer. Asia Pac. J. Pharmacol. 4:133–140.

Sumiyoshi, H. and Wargovich, M.J. (1990) Chemoprevention of 1,2-dimethylhydrazine-induced colon cancer in mice by naturally occurring organosulfur compounds. Cancer Res. 50:5084–5087. Chem. Abst. 113 (1990) 126 061.

Sun, C., Mo, H., Yu, L., Xu, J., and Shen, Z. (1987) Purification and properties of *Allium sativum* lectin. Shengwu Huaxue Yu Shengwu Wuli Xuebao (Acta Biochim. Biophys. Sinica) 19:188–194. Chem. Abst. 107 (1987) 233 171 (Chinese).

Sun, C. and Yu, L. (1986) Lectins from Allium plants. Shengwu Huaxue Yu Shengwu Wuli Xuebao (Acta Biochim. Biophys. Sinica) 18:213–215. Chem. Abst. 105 (1986) 131 940 (Chinese).

Sun, Q., Zhang, X., and Gu, Y. (1988) Studies on prostaglandins in plants. 2. Separation and determination of prostglandin A1 in the longstamen onion (Allium macrostemon) and the common onion (Allium cepa) by HPLC. Zhongcaoyao (Chin. Trad. Herbal Drugs) 19:249–250. Chem. Abst. 109 (1988) 145 565 (Chinese).

Sun, Q.L., Gu, Y.Q., Yang, X.H., Chen, G.R., and Zhou, R. (1988) Studies on prostaglandins in plants. 1. Isolation and identification of prostaglandin A1 from common onion (Allium cepa). Zhongcaoyao (Chin. Trad. Herbal Drugs) 19:146–147. Int. Pharm. Abst. 27 (1990) 5258 (Chinese).

Sun, Q.L. (1991) Studies on prostaglandins in plants. 3. Isolation and identification of prostaglandin A1 and B1 from longstamen onion (Allium macrostemon). Zhongcaoyao (Chin. Trad. Herbal Drugs) 22:150–152. Chem. Abst. 115 (1991) 179 329; Int. Pharm. Abst. 29 (1992) 3393 (Chinese).

Sundaram, S.G. and Milner, J.A. (1992) Antitumor effects of organosulfur compounds present in garlic against canine mammary tumor cells. FASEB J. 6:A1391.

Sundaram, S.G. and Milner, J.A. (1993) Impact of organosulfur compounds in garlic on canine mammary tumor cells in culture. Cancer Lett. 74:85–90.

Sundaram, S.G. and Milner, J.A. (1994) Organosulfur compounds in processed garlic alter the in vitro growth of human tumor cell lines. FASEB J. 8:A426.

Sundaram, S.G. and Milner, J.A. (1995) Diallyl disulfide in garlic oil inhibits both in vitro and in vivo growth of human colon tumor cells. FASEB J. 9:A869.

Sunter, W.H. (1991) Warfarin and garlic. Pharm J. 246:722.

Sutaria, P.B. and San Diego, L. (1982) Essential amino acid analysis of selected Philippine vegetables and fruits. Philipp. J. Sci. 111:45–55. Chem. Abst. 98 (1983) 3717.

Suzuki, M. and Motoyoshi, K. (1966) Pharmacological studies of scordinin, a principle extracted from Allium plants. I. Nippon Yakurigaku Zasshi (Folia Pharmacol. Jap.) 62:105–114. Chem. Abst. 67 (1967) 72 328 (Japanese).

Suzuki, S., Hirai, S., and Noda, K. (1982) Determination of selenium in herb plants by neutron activation analysis using a coincidence counting method. Bunseki Kagaku 31:67–71. Chem. Abst. 96 (1982) 120 991 (Japanese).

Suzuki, T., Sugii, M., and Kakimoto, T. (1961a) New gamma-glutamyl peptides in garlic. Chem. Pharm. Bull. 9:77–78. Chem. Abst. 55 (1961) 21 258.

Suzuki, T., Sugii, M., Kakimoto, T., and Tsuboi, N. (1961b) Isolation of (–) S-allyl-L-cysteine from garlic. Chem. Pharm. Bull. 9:251–252. Chem. Abst. 55 (1961) 23 702.

Suzuki, T., Sugii, M., and Kakimoto, T. (1962a) Gamma-L-glutamyl-S-allyl-L-cysteine, a new gamma-glutamyl peptide in garlic. Chem. Pharm. Bull. 10:345–346. Chem. Abst. 58 (1963) 4646.

Suzuki, T., Sugii, M., and Kakimoto, T. (1962b) Metabolic incorporation of L-valine-[C14] into S-(2-carboxypropyl)glutathione and S-(2-carboxypropyl) cysteine in garlic. Chem. Pharm. Bull. 10:328–331. Chem. Abst. 58 (1962) 11 561.

Suzuki, Y. and Uchida, K. (1984) Three forms of alpha-glucosidase from Welsh onion (Allium fistulosum L.). Agric. Biol. Chem. 48:1343–1345.

Svendsen, L., Rattan, S.I.S., and Clark, B.F.C. (1994) Testing garlic for possible anti-aging effects on long term growth characteristics, morphology, and macromolecular synthesis of human fibroblasts in culture. J. Ethnopharmacol. 43:125–134.

Swain, A.R., Dutton, S.P., and Truswell, A.S. (1985) Salicylates in foods. J. Am. Diet. Assoc. 85:950–960. Chem. Abst. 103 (1985) 159 239.

Sweet, W.J. and Mazelis, M. (1987) Homogeneous alkylcysteine lyase of Acacia farnesiana: fresh seedlings vs. acetone powder. Phytochemistry 26:945–948.

Swetschnikow, W.A. and Bechterewa, S.W. (1931) Über die direkte Wirkung des Knoblauchs (Allium sativum) auf das Herz und das Blutgefäbsystem. Z. Gesamte Exp. Med. 76:596–619.

Sykow, M.P. (1954) The activity of phytoncides on pathogenic incapsulated bacteria. Zh. Mikrobiol. Epidemiol. Immunobiol., 82. Chem. Zbl. 1955, I, 613 (Russian).

Szilagyi, A. (1952) Preparation and properties of allicin. Yearbook Inst. Agr. Chem. Technol., Univ. Tech. Sci. Budapest 3:178–180. Chem. Abst. 49 (1955) 14 275.

Szybejko, J., Zukowski, A., and Herbec, R. (1982) A unusual cause of small intestine obturation. Wiad. Lek. (Med. Sci.) 35:163–164 (Polish).

Szymona, M. (1952) Effect of phytoncides of Allium sativum on the growth and respiration of some pathogenic fungi. Acta Microbiol. Pol. 1:5–23. Chem. Abst. 47 (1953) 2412.

Tabernaemontanus, J.T. (1613) D. Iacobi Theodori Tabernaemontani New und vollkommen Kräuterbuch, etc. Matthias Becker, Frankfurt-on-Main.

Tada, M., Hiroe, Y., Kiyohara, S., and Suzuki, S. (1988) Nematicidal and antimicrobial constituents from Allium grayi Regel and Allium fistulosum L. var. caespitosum. Agric. Biol. Chem. 52:2383–2385.

Tadi, P.P., Teel, R.W., and Lau, B.H.S. (1990) Anticandidal and anticarcinogenic potentials of garlic. Int. Clin. Nutr. Rev. 10:423–429.

Tadi, P.P., Teel, R.W., and Lau, B.H.S. (1991) Organosulfur compounds of garlic modulate mutagenesis, metabolism, and DNA binding of aflatoxin B1. Nutr. Cancer 15:87–95. Chem. Abst. 115 (1991) 108 271.

Tagiev, G.A. (1967) Some histochemical studies during the medical treatment of infected wounds by plant antibiotics. Azerb. Med. Zh. (Azerbaijan. Med. J.) 44:82–85. Chem. Abst. 67 (1967) 62 810 (Azerbaijani).

Tahara, S., Miura, Y., and Mizutani, J. (1979) Hydantoin derivatives of S-alk(en)yl-L-cysteine-S-oxides. III. Antimicrobial L-5-alk(en)ylthiomethyl-hydantoin-(±)-S-oxides: mode of action. J. Agric. Biol. Chem. 43:919–924. Chem. Abst. 91 (1979) 117 871.

Tahara, S. and Mizutani, J. (1979) Hydantoin derivatives of S-alk(en)yl-L-cysteine-S-oxides. V. L-5-Alk(en)ylthiomethylhydantoin-(±)-S-oxides: non-enzymatical precursors of fresh flavors of Allium plants. Agric. Biol. Chem. 43:2021–2028. Chem. Abst. 92 (1980) 56 998.

Tajima, K. and Tominaga, S. (1985) Dietary habits and gastro-intestinal cancers: a comparative case-control study of stomach and large intestinal cancers in Nagoya, Japan. Jpn. J. Cancer Res. 76:705–716.

Takada, M. (1978) Deodorization of garlic. Japan. Patent 78 130 455. Chem. Abst. 90 (1979) 102 266.

Takada, N., Matsuda, T., Otoshi, T., Yano, Y., Otani, S., Hasegawa, T., Nakae, D., Konishi, Y., and Fukushima, S. (1994) Enhancement by organosulfur compounds from garlic and onions of diethylnitrosamine-induced glutathione S-transferase positive foci in the rat liver. Cancer Res. 54:2895–2899. Chem. Abst. 121 (1994) 33 571.

Takagi, H. (1990) Effect of nutrient, growth regulator, and temperature for in vitro growth and organogenesis of excised buds from garlic cloves. Yamagat Daigaku Kiyo Nogaku 11:187–200. Chem. Abst. 113 (1990) 187 520.

Takasugi, N., Kotoo, K., Fuwa, T., and Saito, H. (1984) Effect of garlic on mice exposed to various stresses. Oyo Yakuri (Pharmacometrics) 28:991–1002.

Takasugi, N., Kira, K., and Fuwa, T. (1986) Effects of garlic extract preparation containing vitamins (Kyoleopin) and a ginseng-garlic preparation containing vitamin B1 (Leopin-five) on mice exposed to stresses. Oyo Yakuri (Pharmacometrics) 31: 967–976. Chem. Abst. 105 (1986) 114 052.

Takeyama, F. (1963) Deodorization of garlic. Japan. Patent 63 6561. Chem. Abst. 60 (1964) 6144.

Takeyama, H., Hoon, D.S.B., Saxton, R.E., Morton, D.L., and Irie, R.F. (1993) Growth inhibition and modulation of cell markers of melanoma by S-allyl cysteine. Oncology 50:63–69.

Takizawa, H. (1989) Insecticidal baits containing boric acid for cockroaches. Japan. Patent 89 216 907. Chem. Abst. 112 (1990) 193 804.

Tamura, M. (1994) Extraction of antimicrobial compounds from Allium sativum. Japan. Patent 94 192 116. Chem. Abst. 121 (1994) 263 666.

Tan, Z., Xu, J., Chai, B., Jing, W., Zhu, S., and Xue, P. (1994) Significance of flavonoids in the classification of Allium. Sichuan Daxue Xuebao, Ziran Kexueban 31:119–122. Chem. Abst. 121 (1994) 129 905.

Tanaka, M. (1982) Clinical studies of Kyolepin on complaints following treatment of gynecological malignancies. J. New Remedies [Japan] 31:1349 (Japanese).

Tansey, M.R. and Appleton, J.A. (1975) Inhibition of fungal growth by garlic extract. Mycologia 67:409–413.

Tao, G. and Fang, Z. (1993) Determination of trace and ultra-trace amounts of germanium in environmental samples by preconcentration in a graphite furnace using a flow injection hydride generation technique. J. Anal. At. Spectrom. 8:577–584.

Taskhodzhaev, B., Samikov, K., Yagudaev, M.R., Antsupova, T.P., Shakirov, R., and Yunusov, S.Y. (1985) Structure of alline. Khim. Prir. Soedin. (Chem. Nat. Compd.) 21:687–691. Chem. Abst. 104 (1986) 126 519 (Russian).

Tatami, R., Inone, N., Itoh, H., Kishino, B., Koga, N., Nakashima, Y., Nishide, T., Okamara, K., Saito, Y., Teramoto, T., Yasugi, T., Yamamoto, A., and Goto, Y. (1992) Regression of coronary atherosclerosis by combined LDL-apheresis and lipid-lowering drug therapy in patients with familial hypercholesterolemia: a multicenter study. Atherosclerosis 95:1–13.

Tatarintsev, A.V., Vrzheshch, P.V., Yershuv, D.E., Schegolev, A., Turgiyev, A.S., Karamov, E.V., Kornilayeva, G.V., Makarova, T.V., Fedorov, N.A., and Varfolomeyev, S.D. (1992) Ajoene blockade of integrin-dependent processes in the HIV-infected cell system. Vestn. Rossiiskoi Akad. Med. Nauk (J. Russ. Acad. Med. Sci.) 11:6–10.

Taubmann, G. (1934) Therapie mit Drogen. Med. Klin. 32:1067–1069.

Tawaraya, K. and Saito, M. (1994) Effect of vesicular-arbuscular mycorrhizal infection on amino acid composition in roots of onion and white clover. Soil Sci. Plant Nutr. [Tokyo] 40:339–343. Chem. Abst. 121 (1994) 276 935.

Tazoe, K. (1960a) The nutritional value of Allium plants. XXI. Effect of allithiamine against acute hydrocyanic acid poisoning. Bitamin [Kyoto] 20:97–101. Chem. Abst. 61 (1964) 15 093 (Japanese).

Tazoe, K. (1960b) The nutritional value of Allium plants. XXII. The method of measurement for the activity of mouse liver rhodanese. Bitamin [Kyoto] 20:102–104. Chem. Abst. 61 (1964) 15 093 (Japanese).

Tazoe, K. (1960c) The nutritional value of Allium plants. XXIII. Influences of allithiamine and other vitamins upon mouse liver rhodanese. Bitamin [Kyoto] 20:105–111. Chem. Abst. 61 (1964) 15 093 (Japanese).

Tazoe, K. (1960d) The nutritional value of Allium plants. XXIV. Influences of garlic and related sulfur compounds upon mouse liver rhodanese. Bitamin [Kyoto] 20:112–122. Chem. Abst. 61 (1964) 15 093 (Japanese).

Tchernychev, B., Rabinkov, A., Mirelman, D., and Wilchek, M. (1995) Natural antibodies to alliinase (Alliin lyase) and mannose-specific lectin from garlic (*Allium sativum*) in human serum. Immunol. Lett. 47:53–57.

Tchilov, K., Stantchev, L., and Vladimirov, V. (1951) Die Knoblauchbehandlung der Hochdruckkrankheit. Suvrem. Med. 9:46–60 (Bulgarian).

Tchilov, K., Stantchev, L., and Vladimirov, V. (1956) Die Behandlung der Hochdruckkrankheit mit Folallin. Abh. Med. Inst. Bulg. Akad. Wiss. 13:71–81.

Teel, R.W. (1993) Effect of phytochemicals on the mutagenicity of the tobacco-specific nitrosamine 4-(methylnitrosamino)-1-(3-pyridyl)-1-butanone (NNK) in Salmonella typhimurium strain TA1535. Phytother. Res. 7:248–251.

Teleky-Vamossy, G. and Petro-Turza, M. (1986) Evaluation of odor intensity versus concentrations of natural garlic oil and some of its individual aroma compounds. Nahrung 30:775–782. Chem. Abst. 106 (1987) 3984.

Tempel, K.H. (1962) Über den Einfluß von Knoblauch auf die Experimentelle Cholesterin-Atherosklerose. Med. Ernähr. 3:197–199.

Tergit, G. (1963) Kaiserkron und Päonien rot. Droemer, Munich.

Tewari, G.M. and Bandyopadhyay, C. (1975) Quantitative evaluation of lachrymatory factor in onion by thin-layer chromatography. J. Agric. Food Chem. 23:645–647. Chem. Abst. 83 (1975) 112 462.

Tewari, G.M. and Bandyopadhyay, C. (1977) Pungency and lachrymatory factor as a measure of flavor strength of onions. Lebensm. Wiss. Technol. 10:94–96.

Thomas, D.J. and Parkin, K.L. (1991) Immobilization and characterization of C-S-lyase from onion (Allium cepa) bulbs. Food Biotechnol. 5:139–159. Chem. Abst. 116 (1992) 17 781.

Thomas, D.J. and Parkin, K.L. (1994) Quantification of Alk(en)yl-L-cysteine sulfoxides and related amino acids in Alliums by high-performance liquid chromatography. J. Agric. Food Chem. 42:1632–1638. Chem. Abst. 121 (1994) 103 313.

Thompson, R. (1923) Assyrian medical texts from the originals in the British Museum. London.

Thompson, R. (1924) The Assyrian herbal. London.

Thoms, H. (1926) Handbuch der praktischen und wissenschaftlichen Pharmazie. Urban & Schwarzenberg, Berlin.

Thomson, A.D., Frank, O., Baker, H., and Leevy, C.M. (1971) Thiamine propyl disulfide: absorption and utilization. Ann. Int. Med. 74:529–534.

Tilger, A. (1929) Über die Verabreichung von Knoblauch zu Heilzwecken. Münch. Med. Wochenschr. 76:18.

Timonin, M.I. and Thexton, R.H. (1950) The rhizosphere effect of onion and garlic on soil microflora. Soil Sci. Soc. Am. Proc. 15:186–189.

Tinao, M.M. and Terren, R.C. (1955) Acciones del *Allium sativum* y associaciones sobre motilidad uterina. Arch. Inst. Farmacol. Exp. Madrid 8:127–147. Chem. Abst. 50 (1956) 17 161.

Titta, A. (1904) L'alliotherapie de Boschetti. Rec. Med. Vet. 81:252–253.

Tobkin, H.E. and Mazelis, M. (1979) Alliin lyase: preparation and characterization of the homogeneous enzyme from onion bulbs. Arch. Biochem. Biophys. 193:150–157. Chem. Abst. 90 (1979) 147 538.

Tokarska, B. and Karwowska, K. (1983) The role of sulphur compounds in evaluation of flavouring value of some plant raw materials. Nahrung 27:443–447. Chem. Abst. 99 (1983) 69 118.

Tokin, B. (1943) Phytoncides or plant bactericides: effect of phytoncides upon protozoa. Am. Rev. Sov. Med. 1:239–241.

Tokitomo, Y. and Kobayashi, A. (1992) Isolation of the volatile components of fresh onion by thermal desorption cold trap capillary gas chromatography. Biosci. Biotechnol. Biochem. *56*:1865–1866. Chem. Abst. 118 (1993) 79 579.

Tolok, V.K., Zemskov, G.V., and Smekh, E.V. (1971) Solution for the chemical deposition of nickel coatings. U.S.S.R. Patent 290 963. Chem. Abst. 74 (1971) 145 685.

Tolok, V.K., Zemskov, G.V., and Smekh, E.V. (1972) Stabilizer for the chemical deposition of nickel and cobalt coatings. U.S.S.R. Patent 348 589. Chem. Abst. 78 (1973) 19 605.

Tongia, S.K. (1984) Effects of intravenous garlic juice *Allium sativum* on rat electrocardiogram. Indian J. Physiol. Pharmacol. *28*:250–252.

Toohey, J.I. (1986) Persulfide sulfur is a growth factor for cells defective in sulfur metabolism. Biochem. Cell Biol. *64*:758–765.

Toohey, J.I. (1989) Sulphane sulphur in biological systems: a possible regulatory role. Biochem. J. *264*:625–632.

Török, B., Belagyi, J., Rietz, B., and Jacob, R. (1994) Effectiveness of garlic on the radical activity in radical generating systems. Arzneim. Forsch. *44*:608–611. Chem. Abst. 121 (1994) 170 499.

Toropzev, I. and Kamnev, I. (1946) Certain data as to the nature of phytoncides. C. R. Acad. Sci. URSS [Moscow] *51*:373–375. Chem. Abst. 40 (1946) 6676 (Russian).

Torrescasana, E.U.d. (1946) Estudio experimental sobre la farmacologia de los principios activos del "*Allium sativum*". Rev. Esp. Fisiol. [Barcelona] *2*:6–31. Chem. Abst. 41 (1947) 2172.

Traub, H.L. (1968) The subgenera, sections, and subsections of Allium L. Plant Life *24*:147–163.

Traub, H.P. (1972) Genus Allium L.: subgenera, sections and subsections. Plant Life *28*:132–138.

Tsai, Y., Cole, L.L., Davis, L.E., Lockwood, S.J., Simmons, V., and Wild, G.C. (1985) Antiviral properties of garlic: in vitro effects on influenza B, herpes simplex, and coxsackie viruses. Planta Med. *51*:460–461. Chem. Abst. 104 (1986) 61 572.

Tschirch, A.V. and Lippmann, E.O. (1933) Handbuch der Pharmakognosie. Tauchnitz, Leipzig.

Tsuboi, S., Kishimoto, S., and Ohmori, S. (1989) S-(2-carboxypropyl)glutathione in vegetables of Liliflorae. J. Agric. Food Chem. *37*:611–615. Chem. Abst. 110 (1989) 211 042.

Tsuji, M., Fujimori, K., Nakano, T., and Okuno, T. (1991) Identification of volatile sulfur compounds in plants. Hyogo-kenritsu Kogai Kenkyusho Kenkyu Hokoku, 106–108. Chem. Abst. 117 (1992) 250 230 (Japanese).

Tsuno, S. (1958a) The nutritional value of Allium plants. XVI. Alliinase in Allium plants. Bitamin [Kyoto] *14*:659–664. Chem. Abst. 55 (1961) 12 569 (Japanese).

Tsuno, S. (1958b) The nutritional value of Allium plants. XIX. Thiamine content of Allium plants. Bitamin [Kyoto] *14*:676–677. Chem. Abst. 55 (1961) 12 569 (Japanese).

Tsuno, S., Murakami, F., Tazoe, K., and Kikumoto, S. (1960) The nutritional value of Allium plants. XXX. Isolation of methiin. Bitamin [Kyoto] *20*:93–96. Chem. Abst. 61 (1964) 15 093 (Japanese).

Turnheim, K. (1985) Adenosin irritiert den Darm. Österr. Apoth. Ztg. *39*:724.

Tutakne, M.A., Satyanarayanan, G., Bhardwaj, J.R., and Sethi, I.C. (1983) Sporotrichosis treated with garlic juice: a case report. Indian J. Dermatol. Venereol. *28*:41–45.

Tuyns, A.J., Kaaks, R., Haelterman, M., and Riboli, E. (1992) Diet and gastric cancer: a case-control study in Belgium. Int. J. Cancer *51*:1–6.

Tverskoi, L., Dmitriev, A., Kozlovskii, A., and Grodzinskii, D. (1991) Two phytoalexins from Allium cepa bulbs. Phytochemistry *30*:799–800. Chem. Abst. 115 (1991) 25 962.

Tynecka, Z., Z.Szcesniak, Z., and Glowniak, K. (1993) The effect of various environmental conditions on the antimicrobial activity of Allium ursinum. Planta Med. *59*:701.

Tynecka, Z. and Gos, Z. (1973) The inhibitory action of garlic (*Allium sativum* L.) on growth and respiration of some microorganisms. Acta Microbiol. Pol. *5*:51–62. Chem. Abst. 79 (1973) 63 491.

Tynecka, Z. and Gos, Z. (1975) The fungistatic activity of garlic (*Allium sativum* L.) in vitro. Ann. Univ. Mariae Curie-Sklodowska *30D*:5–13.

Tynecka, Z. and Skwarek, T. (1974) The influence of garlic (*Allium sativum* L.) on tissue cultures infected with Candida albicans. Farm. Pol. *30*:531–538 (Polish).

Tynecka, Z. and Szymona, O. (1972) The effect of certain SH-group inhibitors on the growth and respiration of Staphylococcus strains. Ann. Univ. Mariae Curie-Sklodowska *27D*:57–70. Chem. Abst. 80 (1974) 78 744.

Ubaid, R.H. (1989) Plant tissue culture and prostaglandin production from range of Allium species. Diss. Abstr. Int. B. *49*:2951 Chem. Abst. 111 (1989) 4292.

Uchino, K., Mizuno, T., Sugyama, J., and Kawaguchi, K. (1993) Deodorants containing bran extracts for foods. Japan. Patent 93 344 850. Chem. Abst. 120 (1994) 190 209.

Ueda, Y., Sakaguchi, M., Hirayama, K., Miyajima, R., and Kimizuka, A. (1990) Characteristic flavor constituents in water extract of garlic. Agric. Biol. Chem. *54*:163–169. Chem. Abst. 112 (1990) 196 844.

Ueda, Y., Kawajiri, H., Miyamura, N., and Miyajima, R. (1991) Content of some sulfur-containing components and free amino acids in various strains of garlic. Nippon Shokuhin Kogyo Gakkaishi (J. Jpn. Soc. Food Sci. Technol.) *38*:429–434. Chem. Abst. 115 (1991) 155 089.

Ueda, Y., Tsubuku, T., and Miyajima, R. (1994) Composition of sulfur-containing components in onion and their flavor characters. Biosci. Biotech. Biochem. 58:108–110.

Uemori, T. (1929) Pharmacological investigation of Allium sativum. Folia Pharmacol. Jpn. 9:21–26. Chem. Abst. 24 (1930) 2191.

Uemura, C. (1972) Deodorization of garlic and thiamine. Japan. Patent 72 00 302. Chem. Abst. 76 (1972) 139 272.

Ulloa, M., Corgan, N., and Dunford, M. (1994) Chromosome characteristics and behaviour differences in Allium fistulosum L., A. cepa L., their F-1 hybrid, and selected backcross progeny. Theor. Appl. Genet. 85:567–571.

Ulloa-Godinez, M. (1993) A cytogenetic, isoenzyme, and morphological study on interspecific progenies from Allium fistulosum × Allium cepa crosses. Diss. Abst. Int. B. 54:1760. Chem. Abst. 120 (1994).

Umalakshimi, K. and Devaki, T. (1992) Effect of garlic oil on mitochondrial lipid peroxidation induced by ethanol. Med. Sci. Res. 20:435–437. Chem. Abst. 117 (1992) 106 145.

Unnikrishnan, M.C., Soudamini, K.K., and Kuttan, R. (1990) Chemoprotection of garlic extract toward cyclophosphamide toxicity in mice. Nutr. Cancer 13:201–207. Chem. Abst. 113 (1990) 126 123.

Unnikrishnan, M.C. and Kuttan, R. (1990) Tumour reducing and anticarcinogenic activity of selected spices. Cancer Lett. 51:85–89.

Upadhyay, M.P., Manandhar, K.L., and Shrestha, R.B. (1980) Antifungal activity of garlic against fungi isolated from human eyes. J. Gen. Appl. Microbiol. 26:421–424.

Urban, E., Jäger, W., and Koch, H.P. (1989) Modellverbindungen für "Garlicin" und ihre antimikrobielle Wirksamkeit im Hefe-Wuchstest. Sci. Pharm. 57:451–455. Chem. Abst. 114 (1991) 3279.

Urbina, J.A., Marchan, E., Lazardi, K., Visbal, G., Apitz-Castro, R., Gil, F., Aguirre, T., Piras, M.M., and Piras, R. (1993) Inhibition of phosphatidylcholine biosynthesis and cell proliferation in Trypanosoma cruzi by ajoene, an antiplatelet compound isolated from garlic. Biochem. Pharmacol. 45:2381–2387.

Urushibara, S., Okuno, T., and Matsumoto, T. (1991) New flavonol glycosides, major determinants inducing the green fluorescence in the guard cells of Allium cepa. Tennen Yuki Kagobutsu Toronkai Koen Yoshishu 33:541–548. Chem. Abst. 116 (1992) 170 202 (Japanese).

Urushibara, S., Kitayama, Y., Watanabe, T., Okuno, T., Watarai, A., and Matsumoto, T. (1992) New flavonol glycosides, major determinants inducing the green fluorescence in the guard cells of Allium cepa. Tetrahedron Lett. 33:1213–1216. Chem. Abst. 116 (1992) 231 906.

USP XXII. (1990) United States Pharmacopeia XXII. The United States Pharmacopeial Convention Inc. Rockville, Maryland, pp. 1577–1578, 1788–1789.

Üstünes, L., Claeys, M., Laekeman, G., Herman, A.G., Vlietinck, A.J., and Özer, A. (1985) Isolation and identification of two isomeric trihydroxy octadecenoic acids with prostaglandin E–like activity from onion bulbs (Allium cepa). Prostaglandins 29:847–864.

Utsumi, I., Harada, K., Kobayashi, H., Kohno, K., Yasuda, K., Kondo, Y., and Hirao, H. (1962a) Studies on thiamine disulfide. II. Biological properties of O-benzoylthiamine disulfide. J. Vitaminol. 8:213–219.

Utsumi, I., Harada, K., and Kono, K. (1962b) Interaction of thiamine and its related compounds with protein. VIII. Reaction between thiamine and the reaction products of allicin-related compounds with protein. Bitamin [Kyoto] 26:299–302. Chem. Abst. 60 (1964) 10 984 (Japanese).

Vainberg, Z.T. (1952) Einflub der Phytonzide des Knoblauchs und der synthetischen Senföle auf die Pyruvatdehydrogenase des Nervengewebes. Ukrain. Biokhim. Zh. (Biochem. J.) 24:65–68 (Ukrainian).

Valadaud-Barrieu, D. (1983) A micronucleus induction test of Allium sativum. Differentiation of clastogenic and mitoclastic substances. Mutat. Res. 119:55–58. Chem. Abst. 98 (1983) 102 375.

Van Damme, E.J.M., Goldstein, I.J., and Peumans, W.J. (1991) A comparative study of mannose-binding lectins from the Amaryllidaceae and Alliaceae. Phytochemistry 30:509–514. Chem. Abst. 114 (1991) 118 563.

Van Damme, E.J.M., Smeets, K., Torrekens, S., Van Leuven, F., Goldstein, I.J., and Peumans, W.J. (1992a) The closely related homomeric and heterodimeric mannose-binding lectins from garlic are encoded by one-domain lectin genes, respectively. Eur. J. Biochem. 206:413–420. Chem. Abst. 118 (1993) 206 462.

Van Damme, E.J.M., Smeets, K., Torrekens, S., Van Leuven, F., and Peumans, W.J. (1992b) Isolation and characterization of alliinase cDNA clones from garlic (Allium sativum L.) and related species. Eur. J. Biochem. 209:751–757. Chem. Abst. 119 (1993) 42 302.

Van Damme, E.J.M., Smeets, K., Engelborghs, I., Aelbers, H., Balzarini, J., Pusztai, A., and Van Leuven, F. (1993a) Cloning and characterization of the lectin cDNA clones from onion, shallot, and leek. Plant Mol. Biol. 23:365–376. Chem. Abst. 120 (1994) 70 560.

Van Damme, E.J.M., Smeets, K., Torrekens, S., Van Leuven, F., and Peumans, W.J. (1993b) The mannose-specific lectins from ramsons (Allium ursinum L.) are encoded by three sets of genes. Eur. J. Biochem. 217:123–129. Chem. Abst. 119 (1993) 197 030.

Van Damme, E.J.M., Willems, P., Torrekens, S., Van Leuven, F., and Peumans, W.J. (1993c) Garlic (Allium sativum) chitinases: characterization and molecular cloning. Physiol. Plant. 87:177–186. Chem. Abst. 119 (1993) 3906.

Van den Bergh, R.L. and Kelly, J.F. (1964) Vampirism: a review with new observations. Arch. Gen. Psychiat. *11*:543–547.

Van Ketel, W.G. and De Haan, P. (1978) Occupational eczema from garlic and onion. Contact Dermatitis *4*:53–64.

Van Wyk, D. (1967) Die Kohlenhydrate im Strukturproteid der Chloroplasten von Allium porrum. Z. Naturforsch. *22*:690. Chem. Abst. 67 (1967) 87 744.

Vanderhoek, J.Y., Makheja, A.N., and Bailey, J.M. (1980) Inhibition of fatty acid oxygenases by onion and garlic oils: evidence for the mechanism by which these oils inhibit platelet aggregation. Biochem. Pharmacol. *29*:3169–3173. Chem. Abst. 94 (1981) 115 124.

Vardosanidze, M.G., Gurielidze, K.G., Pruidze, G.N., and Paseshnichenko, V.A. (1991) The substrate specifity of Allium erubescens b-glucosidase. Biokhimiya [Moscow] *56*:2025–2031. Chem. Abst. 116 (1992) 36 804.

Vardosanidze, M.G., Gurielidze, K.G., Pruidze, G.N., and Paseshnichenko, V.A. (1992) Some properties of Allium erubescens b-glucosidase catalyzing decomposition of oligofurostanosides. Prikl. Biokhim. Mikrobiol. *28*:698–702. Chem. Abst. 118 (1993) 228 792.

Varga, L.v. (1938) Die Wirkung verschiedener Reize auf den Magenchemismus. Arch. Verdau. Krankh. *62*:14–23.

Vatsala, T.M., Singh, M., and Murugesan, R.G. (1980) Effects of onion in induced atherosclerosis in rabbits. I. Reduction of arterial lesions and lipid levels. Artery *7*:519–530.

Veien, N.K., Hattel, T., Justesen, O., and Norholm, A. (1983) Causes of eczema in the food industry. Dermatosen Beruf Umwelt *31*:84–86.

Velazquez, B.L., Sanchez, B., Murias, F., and Mijan, C.D. (1958) Garlic extract as an oxytocic substance. Arch. Inst. Farmacol. Exp. Madrid *10*:10–14. Chem. Abst. 53 (1959) 20 519 (Spanish).

Velazquez, B.L. and Rodriguez, J.M.O. (1955a) Action of garlic, corticotropin, and cortisone on vaginal estrus. Arch. Inst. Farmacol. Exp. Madrid *8*:5–9. Chem. Abst. 50 (1956) 17 160.

Velazquez, B.L. and Rodriguez, J.M.O. (1955b) Elimination of 17-keto steroids following the action of garlic extract. Arch. Inst. Farmacol. Exp. Madrid *8*:10–22. Chem. Abst. 50 (1956) 17 160.

Velisek, J., De Vos, R.H., and Schouten, A. (1993) HPLC determination of alliin and its transformation products in garlic and garlic-containing phytomedical preparations. Potravin. Vedy *11*:445–453. Chem. Abst. 121 (1994) 18 165.

Venmadhi, S. and Devaki, T. (1992) Studies on some liver enzymes in rats ingesting ethanol and treated with garlic oil. Med. Sci. Res. *20*:729–731. Chem. Abst. 118 (1993) 2329.

Venugopal, M.S. and Narayanan, V. (1981) Effects of Allitin on the green peach aphid (Myzus persicae Sulzer). Int. Pest Control *23*:130–131.

Vera Lopez, P., Ruiz Rejon, C., Lozano, R., and Ruiz Rejon, M. (1990) Effects of thiram on the mitotic division rhythm in roots of *Allium sativum* L. Cytobios *62*:135–139.

Vernin, G., Metzger, J., Fraisse, D., and Scharff, C. (1986) GC-MS (EI, PCI, NCI) computer analysis of volatile sulfur compounds in garlic essential oils: application of the mass fragmentometry SIM technique. Planta Med. *52*:96–101. Chem. Abst. 105 (1986) 48 808.

Vernin, G. and Metzger, J. (1991) GC-MS (EI, PCI, NCI, SIM) SPECMA bank analysis of volatile sulfur compounds in garlic essential oils. In: Essential oils and waxes. H.F. Linskens and J.F. Jackson, eds. Springer, New York, pp. 99–130. Chem. Abst. 117 (1992) 232 498.

Verschuren, W.M.M., Jacobs, D.R., Bloemberg, B.P.M., Kromhout, D., Menotti, A., Aravanis, C., Blackburn, H., Buzina, R., Dontas, A.S., Fidanza, F., Karvonen, M.J., Nedeljkovic, S., Nissinen, A., and Toshima, H. (1995) Serum total cholesterol and long-term coronary heart disease mortality in different cultures: twenty-five-year follow-up of the seven countries study. J. Am. Med. Assoc. *274*:131–136.

Vimala, M. and Sharma, R. (1988) Dietary protein modifies garlic induced changes in phospholipid and cholesterol contents in rat brain. Res. Bull. Panjab Univ. Sci. *39*:135–137. Chem. Abst. 111 (1989) 213 713.

Vinokurov, S.I., Bronz, L.M., and Korsak, S.E. (1947) Biochemical characteristics of antibiotics of higher plants. I. Inhibition of some oxidation processes, which are catalyzed by heavy metals, by phytoncides of onion and garlic. Byull. Eksp. Biol. Med. (Bull. Exp. Biol. Med.) *23*:296–300. Chem. Abst. 42 (1948) 6864 (Russian).

Virgil. (1880) Publius Virgilius Maro. Deutsch in der Versweise der Urschrift. Langenscheidt, Berlin.

Virtanen, A.I. (1958a) Antimikrobiell wirksame Substanzen in Kulturpflanzen. Angew. Chem. *70*:544–552. Chem. Abst. 53 (1959) 18 385.

Virtanen, A.I. (1958b) Antimicrobial substances in cultured plants and their significance for the plants and for the nutrition of man. Schweiz. Z. Allg. Pathol. Bakteriol. *12*:970–993. Chem. Abst. 53 (1959) 2369.

Virtanen, A.I. (1962a) Organische Schwefelverbindungen in Gemüse- und Futterpflanzen. Angew. Chem. *74*:374–382.

Virtanen, A.I. (1962b) Some organic sulfur compounds in vegetables and fodder plants and their significance in human nutrition. Angew. Chem. Int. Ed. Engl. *1*:299–306.

Virtanen, A.I., Hatanaka, M., and Berlin, M. (1962) Gamma-L-Glutamyl-S-n-propylcystein in Knoblauch (*Allium sativum*). Suom. Kemistil. B. *35*:52. Chem. Abst. 57 (1962) 6323.

Virtanen, A.I. (1965) Studies on organic sulphur compounds and other labile substances in plants. Phytochemistry *4*:207–228. Chem. Abst. 63 (1965) 917.

Virtanen, A.I. (1969) Antimikrobielle und antithyreoide Stoffe in einigen Nahrungspflanzen. Qual. Plant. Mater. Veg. *18*:8–28. Chem. Abst. 72 (1970) 99 173.

Virtanen, A.I. and Matikkala, E.J. (1958) A new sulphur-containing amino acid in onion. Suom. Kemistil. B. *31*:191.

Virtanen, A.I. and Matikkala, E.J. (1959a) The structure and synthesis of cycloalliin isolated from Allium cepa. Acta Chem. Scand. *13*:623–626. Chem. Abst. 55 (1961) 3592.

Virtanen, A.I. and Matikkala, E.J. (1959b) The isolation of S-methylcysteine-sulphoxide and S-n-propylcysteine sulphoxide from onion (Allium cepa) and the antibiotic activity of crushed onion. Acta Chem. Scand. *13*:1898–1900.

Virtanen, A.I. and Matikkala, E.J. (1960a) New gamma-glutamyl peptides in onion (Allium cepa). I. Gamma-Glutamylphenylalanine and gamma-glutamyl-S-(b-carboxy-b-methylethyl)-cysteinylglycine. Suom. Kemistil. B. *33*:83–84. Chem. Abst. 55 (1961) 1809.

Virtanen, A.I. and Matikkala, E.J. (1960b) Neue Gamma-Glutamylpeptide in der Zwiebel (Allium cepa). Z. Physiol. Chem. *322*:8–20. Chem. Abst. 55 (1961) 12 547.

Virtanen, A.I. and Matikkala, E.J. (1961a) The structure of the gamma-glutamyl peptide 4 isolated from onion (Allium cepa) gamma-L-glutamyl-S-(1-propenyl)-cystein-sulphoxide. Suom. Kemistil. B. *34*:84.

Virtanen, A.I. and Matikkala, E.J. (1961b) Evidence for the presence of gamma-glutamyl-S-(1-propenyl)-sulphoxide and cycloalliin as original compounds in onion (Allium cepa). Suom. Kemistil. B. *34*:114. Chem. Abst. 56 (1962) 10 585.

Virtanen, A.I. and Matikkala, E.J. (1961c) New gamma-L-glutamyl peptides in onion (Allium cepa). Suom. Kemistil. B. *34*:53–54.

Virtanen, A.I. and Matikkala, E.J. (1962) Gamma-L-glutamyl-S-(prop-1-enyl)-L-cysteine in the seeds of chives (Allium schoenoprasum). Suom. Kemistil. B. *35*:245. Chem. Abst. 58 (1963) 7135.

Virtanen, A.I. and Mattila, I. (1961a) Gamma-L-glutamyl-S-allyl-L-cysteine in garlic (*Allium sativum*). Suom. Kemistil. B. *34*:44. Chem. Abst. 56 (1962) 10 266.

Virtanen, A.I. and Mattila, I. (1961b) On the gamma-glutamylpeptides in garlic (*Allium sativum*). Suom. Kemistil. B. *34*:73. Chem. Abst. 56 (1962) 10 585.

Virtanen, A.I. and Spåre, C.G. (1961) Isolation of the precursor of the lachrymatory factor in onion (Allium cepa). Suom. Kemistil. B. *34*:72. Chem. Abst. 56 (1962) 10 585.

Vitalyeva, G.M., Brodskii, R.I., Zelenskaya, W.L., and Peshkova, E.G. (1968) Application of the phytoncides of garlic for the treatment of pulpitis and periodontitis. Stomatologia [Moscow] *47*:83–84 (Russian).

Viterbo, A., Rabinowitch, H.D., and Altman, A. (1992) Plant regeneration from callus cultures of Allium trifoliatum subsp. hirsutum and assessment of genetic stability by isozyme polymorphism. Plant Breed. *108*:265–273. Chem. Abst. 117 (1992) 167 785.

Vlaykovitch, M. (1924) L'ail (*Allium sativum*) en thérapeutique. Ses effets dans la tuberculose, recherche de son pouvoir bactéricide, détermination de son équivalent toxique. Dissertation, Faculty of Medicine, University of Nancy (Abstract).

Voigt, M. and Wolf, E. (1986) Knoblauch: HPLC-Bestimmung von Knoblauchwirkstoffen in Extrakten, Pulver und Fertigarzneimitteln. Dtsch. Apoth. Ztg. *126*:591–593. Chem. Abst. 104 (1986) 193 276.

Voisin, H. (1984) Materia medica des homöopathischen Praktikers. Haug, Heidelberg.

Voldrich, M., Vopatova, H., and Hrdlicka, J. (1989) Yellow flavonoids in onions. Prum. Potravin *40*:125–126. Chem. Abst. 111 (1989) 56 098 (Czech).

Vollerner, Y.S., Abdullaev, N.D., Gorovits, M.B., and Abubakirov, N.K. (1983a) Steroidal saponins and sapogenins of Allium. XVIII. Structure of karatavioside B. Khim. Prir. Soedin. (Chem. Nat. Compd.) *19*:197–201 (Russian).

Vollerner, Y.S., Abdullaev, N.D., Gorovits, M.B., and Abubakirov, N.K. (1983b) Steroidal saponins and sapogenins of Allium. XIX. Structure of karatavigenin C. Khim. Prir. Soedin. (Chem. Nat. Compd.) *19*:736–740. Chem. Abst. 100 (1984) 171 548 (Russian; English cover-to-cover translation, pp. 699–703).

Vollerner, Y.S., Abdullaev, N.D., Gorovits, M.B., and Abubakirov, N.K. (1984) Steroidal saponins and sapogenins of Allium. XX. Structure of karatavioside E and F. Khim. Prir. Soedin. (Chem. Nat. Compd.) *20*:69–73 (Russian).

Vollerner, Y.S., Kravets, S.D., Gorovits, M.B., and Abubakirov, N.K. (1988) Spirostane and furostane steroids from plants of the genus Allium. XXIV. Structure of anzurogenin A from Allium suvorovii and A. stipitatum. Khim. Prir. Soedin. (Chem. Nat. Compd.) *24*:68–73. Chem. Abst. 109 (1988) 51 596 (Russian).

Vollerner, Y.S., Kravets, S.D., Shashkov, A.S., Gorovits, M.B., and Abubakirov, N.K. (1989) Steroids of the spirostane and furostane series from the genus Allium. XXVI. Structure of anzurogenin C and anzuroside from fruits of Allium suvorovii and A. stipitatum. Khim. Prir. Soedin. (Chem. Nat. Compd.) *25*:505–510. Chem. Abst. 112 (1990) 95 499 (Russian).

Vollerner, Y.S., Kravets, S.D., Shashkov, A.S., Tashkhodzhaev, B., Gorovits, M.B., Yagudaev, M.R., and Abubakirov, N.K. (1991) Steroids of the spirostane and furostane series from plants of the genus Allium. XXVII. Alliosterol and allosides A and B from Allium suvorovii and Allium stipitatum, structural analogs of furostanols. Khim. Prir. Soedin. (Chem. Nat. Compd.) *27*:231–241. Chem. Abst. 117 (1992) 86 649 (Russian).

Vollrath, R.E., Walton, L., and Lindegren, C.C. (1936) Baktericide Eigenschaften des Acroleins. Proc. Soc. Exp. Biol. Med. 36:55–58. Chem. Zbl. 1937, I, 4822.

Vonderbank, H. (1950) Die Antibiotika auber Penicillin. Pharmazie 5:210–217. Chem. Abst. 44 (1950) 8063.

Vorberg, C. and Schneider, B. (1990) Therapy with garlic: results of a placebo-controlled, double-blind study. Br. J. Clin. Pract. 44:7–11.

Vorberg, G. and Schneider, B. (1990) Therapie mit Knoblauch. Ergbenisse einer placebokontrollierten Doppelblindstudie. Natur-& Gazheitsmed. 3:62–66.

Vorwahl, H. (1923) Knoblauch als Aphrodisiakum. Arch. Gesch. Med. 14:127–128.

Vvedensky, A. (1946) Flora of USSR. Herbertia 11:1944.

Wäfler, U., Shaw, M.L., and Lancaster, J.E. (1994) Effect of freezing upon alliinase activity in onion extracts and pure enzyme preparations. J. Sci. Food Agric. 64:315–318. Chem. Abst. 120 (1994) 268 670.

Wagner, H., Bladt, S., and Zgainski, E.M. (1983) Drogenanalyse. Dünnschichtchromatographische Analyse von Arzneidrogen. Springer, Berlin.

Wagner, H., Bladt, S., and Zgainski, E.M. (1984) Plant drug analysis. Springler-Verlag, Berlin.

Wagner, H., Wierer, M., and Fessler, B. (1987) Effects of garlic constituents on arachidonate metabolism. Planta Med. 53:305–306. Chem. Abst. 107 (1987) 89 878.

Wagner, H. and Sendl, A. (1990) Bärlauch und Knoblauch: Vergleichende chemische und pharmakologische Untersuchungen von Bärlauch- und Knoblauchextrakten. Dtsch. Apoth. Ztg. 130:1809–1815. Int. Pharm. Abst. 28 (1991) 9272.

Wagner, H., Breu, W., Redl, K., Sendl, A., and Steinke, B. (1991) Wirkung von Allium-Arten auf den ArachidonsΣurestoffwechsel und die Thrombozytenaggregation. Med. Welt 42:37–38.

Wagner, J.A. (1793) Ammian Marcellin, aus dem Lateinischen übersetzt und mit erläuternden Anmerkungen begleitet. Joh. Chr. Hermann, Frankfurt-on-Main.

Wahlroos, O. and Virtanen, A.I. (1965) Volatiles from chives (Allium schoenoprasum). Acta Chem. Scand. 19:1327–1332. Chem. Abst. 64 (1966) 3954.

Waliszewski, S.M. and Waliszewski, K.N. (1986) Gas chromatographic determination of triazophos (hostathion) in garlic. Z. Anal. Chem. 325:393. Chem. Abst. 106 (1987) 17 117.

Walker, J.C. (1925) Further studies on the toxicity of juice extracted from succulent onion scales. J. Agric. Res. 30:175–187.

Walker, J.C. and Stahmann, M.A. (1955) Chemical nature of disease resistance in plants. Annu. Rev. Plant Physiol. 6:351–366. Chem. Abst. 49 (1955) 1417.

Walper, A. (1992a) Knoblauch: eine gut geprüfte Naturarznei zur Vorbeugung der Arerisklerose. Oesterr. Apoth. Ztg. 46:269–271.

Walper, A. (1992b) Pflanzliche Lipidsenker: was können sie? Favorit: Knoblauchpräparate. pta in der Apotheke 21:117–120.

Walper, A., Rassoul, F., Purschwitz, K., and Schulz, V. (1994) Effizienz einer Diaetempfehlung und einer zusaetzlichen Phytotherapie mit Allium sativum bei leichter bis maessiger Hypercholesterinaemie. Med. Welt 45:327–332.

Walton, L., Herbold, M., and Lindegren, C.C. (1936) Bactericidal effects of vapors from crushed garlic. Food Res. 1:163–169. Chem. Zbl. 1938, I, 1023.

Wang, S., Lo, S., and Xiang, D. (1989) Studies on long-lived mRNA. IV. Long-lived mRNA in bulbs of Allium sativum L. Hunan Shifan Daxue Ziran Kexue Xuebao 12:147–150. Chem. Abst. 112 (1990) 4722.

Wang, T.G., You, W.C., Henderson, B.E., and Blot, W.J. (1985) A case-control study of stomach cancer in Shandong province. Natl. Cancer Inst. Monogr. 69:9–10.

Wang, W., Tan, Z., and Ji, S. (1988) Abscisic acid in senescence of garlic (Allium sativum) scape. Zhiwu Xuebao (Acta Bot. Sinica) 30:169–175. Chem. Abst. 109 (1988) 89 881 (Chinese).

Wang, W., Tang, J., and Peng, A. (1989) The isolation, identification, and the bioactivities of selenoproteins in selenium-rich garlic. Shengwu Huaxue Zazhi 5:229–234. Chem. Abst. 111 (1989) 95 847 (Chinese).

Wargovich, M.J. (1987) Diallyl sulfide, a flavor component of garlic (Allium sativum), inhibits dimethylhydrazine-induced colon cancer. Carcinogenesis 8:487–489. Chem. Abst. 107 (1987) 168 374.

Wargovich, M.J. (1988) New dietary anticarcinogens and prevention of gastrointestinal cancer. Dis. Colon. Rectum 31:72–75.

Wargovich, M.J., Stephens, L.C., and Gray, K. (1988a) Complete inhibition of nitrosomethylbenzylamine-induced esophageal cancer by diallyl sulfide. Proc. Am. Soc. Cancer Res. 29:136.

Wargovich, M.J., Woods, C., Eng, V.W.S., Stephens, L.C., and Gray, K. (1988b) Chemoprevention of N-nitrosomethylbenzylamine-induced esophageal cancer in rats by the naturally occurring thioether, diallyl sulfide. Cancer Res. 48:6872–6875. Chem. Abst. 110 (1989) 19 694.

Wargovich, M.J. (1992) Inhibition of gastrointestinal cancer by organosulfur compounds in garlic. Cancer Chemoprevention, 195–203. Chem. Abst. 118 (1993) 246 682.

Wargovich, M.J. and Eng, V.W.S. (1989) Rapid screening of organosulfur agents for potential chemopreventive activity using the murine NA assay. Nutr. Cancer 12:189–193.

Wargovich, M.J. and Goldberg, M.T. (1985) Diallyl sulfide: a naturally occurring thioether that inhibits carcinogen-induced nuclear damage to colon epithelial cells in vivo. Mutat. Res. 143:127–129. Chem. Abst. 103 (1985) 83 294.

Wargovich, M.J. and Imada, O. (1993) Esophageal carcinogenesis in the rat: a model for aerodigestive tract cancer. J. Cell. Biochem. *Supplement 17F*:91–94.

Warner, J. (1994) Garlic wards off undead bacteria. New Sci., No. 5, p. 17.

Warshafsky, S., Kamer, R.S., and Sivak, S.L. (1993) Effect of garlic on total serum cholesterol: a meta-analysis. Ann. Intern. Med. *119*:599–605.

Watanabe, H. (1953a) Nutritional value of garlic. I. The masking effect of garlic upon thiamine and the biochemical properties of the masked thiamine. Vitamins [Japan] *6*:121–125. Chem. Abst. 47 (1953) 10 077.

Watanabe, H. (1953b) Nutritional value of garlic. IV. Extraction of the thiamine masked by garlic. Vitamins [Japan] *6*:132–133. Chem. Abst. 47 (1953) 10 077.

Watanabe, T., Iwata, S., and Otani, Y. (1963) Studies on active principles in garlic. I. Distributions of enzymes in garlic and the occurrence of unknown material having reducing power. Osaka Shiritsu Daigaku Kaseigakubo Kiyo *11*:1–8.

Watanabe, T. and Komada, K. (1966) Measurement of allicin with N-ethylmaleimide. Agric. Biol. Chem. *29*:418–419. Chem. Abst. 65 (1966) 3703.

Watt, B.K. and Merrill, A.L. (1963) Composition of foods: raw, processed, prepared, agricultural. U.S. Department of Agriculture, Washington, D.C.

Wattenberg, L.W. (1985) Chemoprevention of cancer. Cancer Res. *45*:1–8.

Wattenberg, L.W., Hanley, A.B., Barany, G., Sparnins, V.L., Lam, L.K.T., and Fenwick, G.R. (1986) Inhibition of carcinogenesis by some minor dietary constituents. In: Diet, nutrition, and cancer. Y. Hayashi, ed. Japan Sci. Soc. Press, Tokyo/VNC Sci. Press, Tokyo, pp. 193–203.

Wattenberg, L.W., Sparnins, V.L., and Barany, G. (1989) Inhibition of N-nitrosodiethylamine carcinogenesis in mice by naturally occurring organosulfur compounds and monoterpenes. Cancer Res. *49*:2689–2692. Chem. Abst. 111 (1989) 17 246.

Wattenberg, L.W. (1992) Inhibition of carcinogenesis by minor dietary constituents. Cancer Res. *52*:2085s–2091s.

Watts, G.F., Lewis, B., Brunt, J.N.H., Lewis, E.S., Coltart, D.J., Smith, L.D.R., Mann, J.I., and Swan, A.V. (1992) Effects on coronary artery disease of lipid lowering diet, or diet plus colestyramine in the St. Thomas atherosclerosis study (STARS). Lancet *339*:563–569.

Weber, H. (1962) Botanik, eine Einführung für Pharmazeuten und Mediziner. Wiss. Verlagsges, Stuttgart.

Weber, N.D., Andersen, D.O., North, J.A., Murray, B.K., Lawson, L.D., and Hughes, B.G. (1992) In vitro virucidal effects of *Allium sativum* (garlic) extract and compounds. Planta Med. *58*:417–423. Chem. Abst. 118 (1993) 73 174.

Weger, N.P. (1990) Garlic preparation. Germany. Patent 3 911 594 A1.

Wehr, A. (1993) Knoblauch und Fettstoffwechsel: Neue Aspekte? Z. Phytother. *14*:298.

Weinberg, D.S., Manier, M.L., Richardson, M.D., Haibach, F.G., and Rogers, T.S. (1992) Identification and quantification of anticarcinogens in garlic extract and licorice root extract powder. J. High Resol. Chromatogr. *15*:641–654. Chem. Abst. 118 (1993) 190 195.

Weinberg, D.S., Manier, M.L., Richardson, M.D., and Hailbach, F.G. (1993) Identification and quantitation of organosulfur compliance markers in a garlic extract. J. Agric. Food Chem. *41*:37–41. Chem. Abst. 118 (1993) 45 870.

Weisberger, A.S. and Pensky, J. (1957) Tumor-inhibiting effects derived from an active principle of garlic (*Allium sativum*). Science *126*:1112–1114.

Weisberger, A.S. and Pensky, J. (1958) Tumor inhibition by a sulfhydryl-blocking agent related to an active principle of garlic (*Allium sativum*). Cancer Res. *18*:1301–1308. Chem. Abst. 53 (1959) 4576.

Weisenberger, H., Grube, H., Koenig, E., and Pelzer, H. (1972) Isolation and identification of the platelet aggregation inhibitor present in the onion, Allium cepa. FEBS Lett. *26*:105–108. Chem. Abst. 77 (1972) 162 867.

Weisler, R. (1989) Systemic insect repellent composition comprising vitamin B1 and allyl sulfide. U.S.A. Patent 4 876 090. Chem. Abst. 112 (1990) 212 499.

Weiss, E. (1941) A clinical study of the effect of desiccated garlic on intestinal flora. Med. Rec. *153*:404–408.

Weiss, R.F. (1957) Phytotherapie der Arteriosklerose. Hippokrates [Stuttgart], 247–251.

Weiss, R.F. (1974) Lehrbuch der Phytotherapie. Hippokrates, Stuttgart.

Weiss, R.F. (1982) Phytotherapie bei Altersleiden. Ärztez. Naturheilverf. *23*:341–345.

Weiss, R.F. (1983) Arjuveda-Medizin und Phytotherapie. Z. Phytother. *4*:615–620.

Weiss, R.F. (1984a) Wirkung des Knoblauchs auf arteriosklerotische Netzhautveränderungen des Auges. Apoth. J. *6*:71.

Weiss, R.F. (1984b) Netzhautveränderungen des Auges durch Knoblauch günstig zu beeinflussen. Volksgesundheit, 11–12.

Weiss, R.F. (1986) Neues vom Knoblauch. Ärztez. Naturheilverf. *27*:206–209.

Weissenböck, G., Schnabl, H., Scharf, H., and Sachs, G. (1987) On the properties of fluorescing compounds in guard and epidermal cells of Allium cepa L. Planta *171*:88–95.

Welch, C., Wuarin, L., and Sidell, N. (1992) Antiproliferative effect of the garlic compound S-allylcysteine on human neuroblastoma cells in vitro. Cancer Lett. [Shannon, Ireland] *63*:211–219. Chem. Abst. 117 (1992) 39 962.

Wen, G.Y., Mato, A., Malik, M.N., Jenkins, E.C., Sheikh, A.M., and Kim, K.S. (1995) Light and electron microscopic immunocytochemical localization of two major proteins in garlic bulb. J. Cell. Biochem. 58:481–489.

Wendelbo, P. (1971) Alliaceae. In: Flora des iranischen Hochlandes und der umrahmenden Gebirge. Persien, Afghanistan, Teile von West-Pakistan, Nord-Iraq, Azerbaidjan, Turkmenistan. K.H. Rechinger, ed. Akad. Druck-& Verlagsanstalt, Graz, Austria, pp. 1–130.

Wertheim, E. (1929) Derivatives for the identification of mercaptans. J. Am. Chem. Soc. 51:3661–3664. Chem. Zbl. 1930, I, 812.

Wertheim, T. (1844) Untersuchung des Knoblauchöls. Ann. Chem. Pharm. 51:289–315. Pharm. Centralbl. 15 (1844), 833–848.

Wertheim, T. (1845) Über den Zusammenhang zwischen Senföl und Knoblauchöl. Ann. Chem. Pharm. 55:297–304.

Weslowski, E. (1910) Die Vampirsage im rumänischen Volksglauben. Zeitschr. österr. Volkskunde 16:209–216.

Wheeler, D.M. and Follett, J.M. (1991) Effect of aluminum on onions, asparagus, and squash. J. Plant. Nutr. 14:897–912. Chem. Abst. 115 (1991) 225 713.

Whitaker, J.R. (1976) Development of flavor, odor, and pungency in onion and garlic. Adv. Food Res. 22:73–133. Chem. Abst. 85 (1976) 107 509.

Whitaker, J.R. and Mazelis, M. (1991) Enzymes important in flavor development in the Alliums. In: Food enzymology. P.F. Fox, ed. Elsevier, London, pp. 479–497. Chem. Abst. 116 (1992) 233 993.

Wierzbicka, M. (1984) Ultrastructural location of lead in the cell walls of Allium cepa L. roots. Postepy Biol. Komorki 11:509–511.

Wierzbicka, M. (1986) The effect of lead on the ultrastructure changes in the root tip of onion: Allium cepa L. Folia Histochem. Cytobiol. 24:340–341.

Wierzbicka, M. (1987a) Lead accumulation and its translocation barriers in roots of Allium cepa L.: autoradiographic and ultrastructural studies. Plant Cell Environ. 10:17–26.

Wierzbicka, M. (1987b) Lead translocation and localization in Allium cepa roots. Can. J. Bot. 65:1851–1860.

Wierzbicka, M. (1988) Mitotic disturbances by low doses of inorganic lead. Caryologia 41:143–160. Chem. Abst. 110 (1989) 90 390.

Wierzbicka, M. (1994) Resumption of mitotic activity in Allium cepa L. root tips during treatment with lead salts. Environ. Exp. Bot. 34:173–180. Chem. Abst. 121 (1994) 51 864.

Wiesenfeld, P., Chanderbhan, R., and Whittaker, P. (1993) Blood parameters and Fe status in rats fed dietary garlic extract. FASEB J. 7:A743.

Wilcox, B.F., Joseph, P.K., and Augusti, K.T. (1984) Effects of allylpropyl disulphide isolated from Allium cepa Linn. on high-fat fed rats. Indian J. Biochem. Biophys. 21:214–216.

Wilkie, S.E., Isaac, P.G., and Slater, R.J. (1993) Random amplified polymorphic DNA (RAPD) markers for genetic analysis in Allium. Theor. Appl. Genet. 86:497–504.

Williams, H.H., Erickson, B.N., Beach, E.F., Macy, I.G., Avrin, I., Shepherd, M., Souders, H., Teague, D.M., and Hoffman, O. (1941) Biochemical studies of the blood of dogs with N-propyl disulfide anemia. J. Lab. Clin. Med. 26:996–1008. Chem. Abst. 35 (1941) 3317.

Williams, K.J. and Leung, D.W.M. (1993) Chitinase induction in onion tissue cultures. Plant Cell, Tissue Organ Cult. 32:193–198. Chem. Abst. 118 (1993) 251 535.

Wills, E.D. (1956) Enzyme inhibition by allicin, the active principle of garlic. Biochem. J. 63:514–520. Chem. Abst. 50 (1956) 15 612.

Winkler, G., Iberl, B., and Knobloch, K. (1992a) Reactivity of allicin and its transformation products with sulfhydryl groups, disulfide groups, and human blood. Planta Med. 58:A665.

Winkler, G., Lohmüller, E.M., Landshuter, J., Weber, W., and Knobloch, K. (1992b) Schwefelhaltige Leitsubstanzen in Knoblauchpräparaten: Quantitative HPLC-Bestimmung. Dtsch. Apoth. Ztg. 132:2312–2317. Chem. Abst. 119 (1993) 56 281.

Winter, M., Brandl, W., and Herrmann, K. (1987) Bestimmung von Hydroxyzimtsäure-Derivaten in Gemüse. Z. Lebensm. Unters. Forsch. 184:11–16. Chem. Abst. 106 (1987) 154 883.

Winterhoff, H. (1987) Endokrinologisch wirksame Phytopharmaka. Z. Phytother. 8:169–171.

Winterhoff, H. and Egen-Schwind, C. (1991) Die Wirkung von Knoblauch auf das Rattenendokrinium. Med. Welt 42:44.

Woenig, F. (1886) Die Pflanzen im alten Aegypten, ihre Heimat, Geschichte, Kultur und ihre mannigfache Verwendung im sozialen Leben, in Kultus, Sitten, Gebräuchen, Medizin und Kunst. Wilh. Friedrich, Leipzig.

Wolf, J.E. (1927) Über die Behandlung von Darmstörungen Tuberkulöser mit Allisatin. Schweiz. Med. Wochenschr. 57:488–490.

Wolf, S. and Reim, M. (1990) Effect of garlic on conjunctival vessels: a randomised, placebo-controlled, double-blind trial. Br. J. Clin. Pract. 44:36–39.

Won, T. and Mazelis, M. (1989) The C-S-lyases of higher plants: purification and characterization of homogeneous alliin lyase of leek (Allium porrum). Physiol. Plant. 77:87–92. Chem. Abst. 112 (1990) 32 449.

Wu, D. (1992) Studies of browning prevention and long-term preservation of condiment garlic. Zhongguo Tiaoweipin, 13–16. Chem. Abst. 119 (1993) 70 914 (Chinese).

Wu, J.L.P. and Wu, C.M. (1981) High-performance liquid chromatographic separation of shallot volatile oil. J. Chromatogr. 214:234–236. Chem. Abst. 95 (1981) 202 144.

Wu, Y., Peng, J., Yao, X., Okuyama, T., and Narui, T. (1992) Macrostemonoside A: a new spirostane saponin from Allium macrostemon Bge. Shenyang

Yaoxueyuan Xuebao 9:69–70. Chem. Abst. 117 (1992) 230 139.

Xiguang, L. (1986) Ultrastructural study of the effects of Bulbus Allii and some other drugs on Staphylococcus aureus. Chung hsih i chieh ho tsa chih [China] 6:710.

Xu, W., Wong, M., Huang, X.S., and Yhu, Y. (1991) Effects of garlicin on MMS-CY–induced sex-linked recessive lethal mutations. Zhongguo Yaolixue Yu Dulixue Zazhi (Chin. J. Pharmacol. Toxicol.) 5:76–78. Chem. Abst. 115 (1991) 43 967 (Chinese).

Xu, Z. (1981) Quantitation of allicin by oxygen flask combustion analysis. Yaoxue Tongbao (Bull. Pharmacol.) 16:38–39. Chem. Abst. 96 (1982) 57 851 (Chinese).

Xue, H.M., Araki, H., Shi, L., and Yakuwa, T. (1991a) Somatic embryogenesis and plant regeneration in basal plate and receptacle derived-callus cultures of garlic (Allium sativum L.). Engei Gakkai Zasshi (J. Jap. Soc. Hort. Sci.) 60:627–634. Chem. Abst. 116 (1992) 102 781.

Xue, H.M., Araki, H., and Yakuwa, T. (1991b) Varietal difference of embryonic callus induction and plant regeneration in garlic (Allium sativum L.). Unknown [Japan] 8:166–170 (Japanese).

Yacobson, L.M. (1936) The isolation of bacteriophage from vegetables and fruits. Zh. Mikrobiol. Epidemiol. Immunobiol. 17:584–585. Chem. Abst. 31 (1937) 6689 (Russian).

Yakovlev, A.I. and Zvyagin, S.G. (1950) Influence of phytoncides on virus influenza A. I. Action of the volatile components from garlic and onion on virus influenza A. Byull. Biol. Med. (Bull. Biol. Med.) 29:384–387 (Russian).

Yamada, Y., Azuma, K., and Hirozawa, A. (1976) In vitro effects of allicin on phagocytic activity of rabbit polymorphonuclear leucocytes. Igaku to Seibutsugaku (Medicine & Biology) 92:515–518. Chem. Abst. 88 (1978) 182 629 (Japanese).

Yamada, Y. and Azuma, K. (1975a) The in vitro action of allicin against dermatophytes. Igaku to Seibutsugaku (Medicine & Biology) 91:237–241. Chem. Abst. 84 (1976) 130 883 (Japanese).

Yamada, Y. and Azuma, K. (1975b) Morphological observations on the action of allicin against Trichophyton mentagrophytes. Igaku to Seibutsugaku (Medicine & Biology) 91:242–247. Chem. Abst. 84 (1976) 130 884 (Japanese).

Yamada, Y. and Azuma, K. (1975c) The in vitro antimycotic action of allicin against Candida, Cryptococcus, and Aspergillus. Igaku to Seibutsugaku (Medicine & Biology) 91:199–203. Chem. Abst. 84 (1976) 130 885 (Japanese).

Yamada, Y. and Azuma, K. (1977) Evaluation of the in vitro antifungal activity of allicin. Antimicrob. Agents Chemother. 11:743–749. Chem. Abst. 87 (1977) 48 492.

Yamamoto. M. (1963) Deodorization of garlic. Japan. Patent 63 12 708. Chem. Abst. 60 (1964) 8565.

Yamanaka, S. and Kobayashi, T. (1992) Inclusion compounds for garlic deodorization. Japan. Patent 92 309 358. Chem. Abst. 118 (1993) 108 848.

Yamasaki, S. (1973) Effects of adenosine and unsaturated fatty acid on the metabolism of cholesterol. Rinsho Byori 21:129–135. Chem. Abst. 79 (1973) 16 232.

Yamasaki, T., Teel, R.W., and Lau, B.H.S. (1991) Effect of allixin, a phytoalexin produced by garlic, on mutagenesis, DNA-binding, and metabolism of aflatoxin B1. Cancer Lett. 59:89–94.

Yamasaki, T., Li, L., and Lau, B.H.S. (1994) Garlic compounds protect vascular endothelial cells from hydrogen peroxide–induced oxidant injury. Phytother. Res. 8:408–412. Chem. Abst. 122 (1995) 71 956.

Yamato, O. (1994) Hemolytic anemia induced by Allium cepa in dogs. Kagaku to Seibutsu 32:628–629. Chem. Abst. 121 (1994) 273 878.

Yan, X., Wang, Z., and Barlow, P. (1992) Quantitative estimation of garlic oil content in garlic oil based health products. Food Chem. 45:135–139. Chem. Abst. 117 (1992) 110 283.

Yan, X., Wang, Z., and Barlow, P. (1993) Quantitative determination and profiling of total sulphur compounds in garlic health products using a simple GC procedure. Food Chem. 47:289–294. Chem. Abst. 119 (1993) 115 654.

Yang, C.S., Hong, J.Y., and Wang, Z.Y. (1993) Inhibition of nitrosamine-induced tumorigenesis by diallyl sulfide and tea. In: Food and cancer prevention. K. Waldron, ed. Royal Society of Chemistry, United Kingdom.

Yang, C.S., Smith, T.J., and Hong, J.Y. (1994a) Cytochrome P-450 enzymes as targets for chemoprevention against chemical carcinogenesis and toxicity: opportunities and limitations. Cancer Res. (Suppl.) 54:1982s–1986s.

Yang, C.S., Wang, Z.Y., and Hong, J.Y. (1994b) Inhibition of tumorigenesis by chemicals from garlic and tea. In: Advances in experimental medicine and biology: diet and cancer; markers, prevention, and treatment. M.M. Jacobs, ed. Plenum, New York, pp. 113–122. Chem. Abst. 122 (1995) 22 889.

Yang, G.C., Yasaei, P.M., and Page, S.W. (1993) Garlic as anti-oxidants and free radical scavengers. Yaowu Shipin Fenxi 1:357–364. Chem. Abst. 121 (1994) 26 825.

Yang, K.Y. and Shin, H.S. (1982) Lipids and fatty acid composition of garlic (Allium sativum Linnaeus). Hanguk Sikpum Kwahak Hoechi (Korean J. Food. Sci. Technol.) 14:388–393. Chem. Abst. 98 (1983) 124 455 (Korean).

Yang, M., Wang, K., Gao, L., Han, Y., Lu, J., and Zou, T. (1992) Exploration for a natural selenium supplement: characterization and bioactivities of Se-containing polysaccharide from garlic. J. Chin. Pharm. Sci. 1:28–32. Chem. Abst. 118 (1993) 77 092.

Yang, W., Li, C., Liu, H., Hu, Z., Zhu, J., Chen, J., and Jin, Q. (1987) Study on inclusion complexes of garlicin with beta-cyclodextrin. Zhongguo Yaoke Daxue

Xuebao (J. Chin. Pharm. Univ.) *18*:293–296. Chem. Abst. 108 (1988) 118 832 (Chinese).

Yano, M. (1958a) Nutritional value of Allium plants. XX. Thiamine excretion in urine and stool following oral ingestion of allithiamine. Vitamins [Japan] *15*:606–612. Chem. Abst. 56 (1962) 14 694.

Yano, M. (1958b) Nutritional value of Allium plants. XXI. Excretion of thiamine in urine and stool following parenteral injection of allithiamine. Vitamins [Japan] *15*:613–616. Chem. Abst. 56 (1962) 14 694.

Yano, M. (1958c) Nutritional value of Allium plants. XXII. Variations in the thiamine blood level following oral and parenteral administration of allithiamine. Vitamins [Japan] *15*:617–621. Chem. Abst. 56 (1962) 14 694.

Yano, M. (1958d) Nutritional value of Allium plants. XXIII. The effect of allithiamine doses on the amount of pyruvic and lactic acids in the blood. Vitamins [Japan] *15*:622–624. Chem. Abst. 56 (1962) 14 694.

Yano, M. (1958e) Nutritional value of Allium plants. XXIV. The effect of allithiamine doses on the amount of pyruvic and lactic acids in the urine. Vitamins [Japan] *15*:625–627. Chem. Abst. 56 (1962) 14 695.

Yao, X. and Peng, J. (1994) New steroidal saponins from Allium macrostemon and their inhibitory effects on human platelet aggregation. Proc. 3rd. Int. Cong. Enthnopharmacology, Beijing, China, 6–10 Sepetmpber 1994.

Yasseen, Y.M., Barringer, S.A., and Splittstoesser, W.E. (1995) In vitro bulb production from Allium spp. In vitro Cell. & Dev. Biol. *31*:51–52.

Yeh, Y.Y. and Yeh, S.M. (1994) Garlic reduces plasma lipids by inhibiting hepatic cholesterol and triacylglycerol synthesis. Lipids *29*:189–193. Chem. Abst. 120 (1994) 241 010.

Yellin, S.A., Davidson, B.J., Pinto, J.T., Sacks, P.G., Qiao, C., and Schantz, S.P. (1994) Relationship of glutathione and glutathione-S-transferase to cisplatin sensitivity in human head and neck squamous carcinoma cell lines. Cancer Lett. *85*:223–232.

Yg, G., Liu, Y.Y., Yang, X.H., Chen, D., and Fu, F.H. (1988) Effect of Allium cepa L. var. agregatum Don. and Allium macrostemon Bunge on arachidonic acid metabolism. Yao Hsueh Pao (Acta Pharm. Sinica) *23*:8–11 (Chinese).

Yokoyama, K., Uda, N., Takasugi, N., and Fuwa, T. (1986) Anti-stress effects of garlic extract preparation containing vitamins (Kyoleopin) and ginseng-garlic preparation containing vitamin B1 (Leopin-Five) in mice. Oyo Yakuri (Pharmacometrics) *31*:977–984. Chem. Abst. 105 (1986) 114 053.

Yokoyama, K., Yoshii, M., Takasugi, N., and Fuwa, T. (1988) Effects of garlic extract preparation containing vitamins (Kyoleopin) and ginseng-garlic preparation containing vitamin B1 (Leopin-Five) on peripheral blood circulation of animals. Oyo Yakuri (Pharmacometrics) *36*:301–308. Chem. Abst. 110 (1989) 74 171 (Japanese).

Yokozawa, H. and Onaga, K. (1981) Odorless garlic powder. Japan. Patent 81 164 762. Chem. Abst. 96 (1982) 121 210.

Yoshida, S., Hirao, Y., and Nakagawa, S. (1984) Mutagenicity and cytotoxicity tests of garlic. J. Toxicol. Sci. *9*:77–86 (Japanese).

Yoshida, S., Kasuga, S., Hayashi, N., Ushiroguchi, T., Matsuura, H., and Nakagawa, S. (1987) Antifungal activity of ajoene derived from garlic. Appl. Environ. Microbiol. *53*:615–617. Chem. Abst. 106 (1987) 152 894.

Yoshida, T., Saito, T., and Kadoya, S. (1987) New acylated flavonol glucosides in Allium tuberosum Rottler. Chem. Pharm. Bull. *35*:97–107. Chem. Abst. 107 (1987) 36 573.

Yoshikawa, K., Hadame, K., Saitoh, K., and Hijikata, T. (1979) Patch tests with common vegetables in hand dermatitis patients. Contact Dermatitis *5*:274–275.

Yoshikawa, Y. (1979) Deodorization of garlic with magnesium hydroxide. Japan. Patent 79 20 159. Chem. Abst. 90 (1979) 185 223.

Yoshikawa, Y. (1980) Removal of garlic odor. Japan. Patent 80 34 665. Chem. Abst. 94 (1981) 29 075.

Yoshimura, M. (1958a) The nutritional value of Allium plants. X. Studies to prove existence of allicin and its homologs. Bitamin [Kyoto] *14*:627–632. Chem. Abst. 55 (1961) 12 568 (Japanese).

Yoshimura, M. (1958b) The nutritional value of Allium plants. XIII. Allithiamine formation in the presence of S-allyl-L-cysteine. Bitamin [Kyoto] *14*:647–649. Chem. Abst. 55 (1961) 12 568 (Japanese).

Yoshimura, M. (1958c) The formation of allithiamine. Bitamin [Kyoto] *14*:640–646. Chem. Abst. 55 (1961) 12 568.

Yoshimura, M., Tsuno, S., and Murakami, F. (1958) Studies on the nutritional value of Allium plants. XV. Alliin homologs in Allium plants. Bitamin [Kyoto] *14*:654–658. Chem. Abst. 55 (1961) 12 568 (Japanese).

Yoshimura, M. and Arai, T. (1958) Studies on the nutritional value of Allium plants. XIV. Some doubt on the theory of thiamine synthesis in the intestinal canal by the intake of garlic. Bitamin [Kyoto] *14*:650–653.

You, W.C., Blot, W.J., Chang, Y.S., Ershow, A.G., Yang, Z.T., An, Q., Henderson, B., Xu, G.W., Fraumeni, J.F., Jr., and Wang, T.G. (1988) Diet and high risk of stomach cancer in Shandong, China. Cancer Res. *48*:3518–3523.

You, W.C., Blot, W.J., Chang, Y.S., Ershow, A., Yang, Z.T., An, Q., Henderson, B.E., Fraumeni, J.F., Jr., and Wang, T.G. (1989) Allium vegetables and reduced risk of stomach cancer. J. Natl. Cancer Inst. *81*:162–164.

Yu, T.H., Wu, C.M., and Chen, S.Y. (1989a) Effects of pH adjustment and heat treatment on the stability and the formation of volatile compounds of garlic. J. Agric. Food Chem. *37*:730–734. Chem. Abst. 110 (1989) 211 221.

Yu, T.H., Wu, C.M., and Liou, Y.C. (1989b) Volatile compounds from garlic. J. Agric. Food Chem. 37:725–730. Chem. Abst. 110 (1989) 211 222.

Yu, T.H., Wu, C.M., and Ho, C.T. (1993) Volatile compounds of deep-oil fried, microwave-heated, and oven-baked garlic slices. J. Agric. Food Chem. 41:800–805. Chem. Abst. 119 (1993) 253 714.

Yu, T.H., Lee, M.H., Wu, C.M., and Ho, C.T. (1994a) Volatile compounds generated from thermal interaction of 2,4-decadienal and the flavor precursors of garlic. ACS Symp. Ser. 558:61–76. Chem. Abst. 121 (1994) 299 461.

Yu, T.H., Lin, L.Y., and Ho, C.T. (1994b) Volatile compounds of blanched, fried-blanched, and baked-blanched garlic slices. J. Agric. Food Chem. 42:1342–1347. Chem. Abst. 121 (1994) 7658.

Yu, T.H., Wu, C.M., and Ho, C.T. (1994c) Meat-like flavor generated from thermal interactions of glucose and alliin or deoxyalliin. J. Agric. Food Chem. 42:1005–1009.

Yu, T.H., Wu, C.M., and Ho, C.T. (1994d) Volatile compounds generated from thermal interactions of inosine-5'-monophosphate and alliin or deoxyalliin. ACS Symp. Ser. 564:188–198. Chem. Abst. 121 (1994) 299 457.

Yu, T.H., Wu, C.M., Rosen, R.T., Hartman, T.G., and Ho, C.T. (1994e) Volatile compounds generated from thermal degradation of alliin and deoxyalliin in an aqueous solution. J. Agric. Food Chem. 42:146–153. Chem. Abst. 120 (1994) 53 066.

Yu, T.H. and Ho, C.T. (1993) Chemistry and stability of sulfur-containing compounds in the genus Allium. Dev. Food Sci. 33:501–546. Chem. Abst. 121 (1994) 7502.

Yu, T.H. and Wu, C.M. (1989a) Stability of allicin in garlic juice. J. Food Sci. 54:977–981.

Yu, T.H. and Wu, C.M. (1989b) Effects of pH on the formation of flavour compounds of disrupted garlic. J. Chromatogr. 462:137–145. Chem. Abst. 110 (1989) 152 961.

Yurugi, S. (1954a) Studies on vitamin B1 and related compounds. LVII. Reaction between thiamine and ingredients of Allium genus plants. 3. Reaction of L-cysteine and allithiamine. Yakugaku Zasshi (J. Pharm. Soc. Japan) 74:511–514. Chem. Abst. 49 (1955) 8300 (Japanese).

Yurugi, S. (1954b) Studies on vitamin B1 and related compounds. LVIII. Reaction between thiamine and ingredients of Allium genus plants. 4. Detection of allithiamine and its homologs. Yakugaku Zasshi (J. Pharm. Soc. Japan) 74:514–519. Chem. Abst. 49 (1955) 8301.

Yurugi, S. (1954c) Studies on vitamin B1 and related compounds. LV. Reaction between thiamine and ingredients of Allium genus plants. 1. Detection and isolation of allithiamine-like compounds. Yakugaku Zasshi (J. Pharm. Soc. Japan) 74:502–506. Chem. Abst. 49 (1955) 8300 (Japanese).

Yurugi, S., Kawasaki, H., and Noguchi, S. (1956) Studies on vitamin B1 and related compounds. LXIX. Syntheses of allithiamine homologs. Yakugaku Zasshi (J. Pharm. Soc. Japan) 75:498–501. Chem. Abst. 50 (1956) 5679.

Zabel, J., Pawelczyk, K., and Liburska-Lugowska, D. (1979) Treatment of crural ulcers with ointment with lyophilized garlic. Przegl. Dermatol. 66:567–570 (Polish).

Zacharias, N.T., Augusti, K.T., Sebastian, K.L., and Philip, B. (1980) Hypoglycemic and hypolipidaemic effects of garlic in sucrose fed rabbits. Indian J. Physiol. Pharmacol. 24:151–154.

Zaghloul, S., El-Sabban, F., Radwan, G., Fahim, M., and Singh, S. (1995) Garlic retards hyperthermia-induced thrombosis in PIAL microcirculation of the mouse. FASEB J. 9.

Zalewski, S. (1960) Investigation of the antioxidant action of condiments added to lard. Gospod. Miesna, 151–154. Chem. Abst. 57 (1962) 15 661.

Zander, R. (1993) Handwörterbuch der Pflanzennamen. Eugen Ulmer, Stuttgart.

Zelikoff, J.T., Atkins, N.M., Jr., and Belman, S. (1986) Stimulation of cell growth and proliferation in NIH-3T3 cells by onion and garlic oils. Cell Biol. Toxicol. 2:369–378. Chem. Abst. 108 (1988) 106 054.

Zelikoff, J.T. and Belman, S. (1985) Effect of onion and garlic oil on 3T3 cell transformation. In Vitro 21:41A.

Zeller, A.P. (1985) Das grobe Buch vom Knoblauch. Kulturgeschichte, Anbau, Heilmittel, Rezepte. Delphin, Munich.

Zhang, M., Yuan, L., and Tian, J. (1994) Determination of selenium in rice, the bulb of garlic, and tea by flame AAS with slotted quartz tube. Guangpuxue Yu Guangpu Fenxi 14:101–104, 120. Chem. Abst. 122 (1995) 8266.

Zhang, W. (1985) Determination of allicin content of garlic and its preparations. Zhongcaoyao (Chin. Trad. Herbal Drugs) 16:17–19. Chem. Abst. 104 (1986) 18 714.

Zhang, Y., Chen, X., and Yu, Y. (1989a) Antimutagenic effect of garlic, tannic acid, and cinnamaldehyde. Zhejiang Yike Daxue Xuebao (J. Zhejiang Md. Univ.) 18:201–204. Chem. Abst. 112 (1990) 134 196 (Chinese).

Zhang, Y.S., Chen, X.R., and Yu, Y.N. (1989b) Antimutagenic effect of garlic (Allium sativum L.) on 4NQO-induced mutagenesis in Escherichia coli WP2. Mutat. Res. 227:215–219. Chem. Abst. 112 (1990) 72 030.

Zhang, Y., Yuan, Z., Chen, X., and Yu, Y. (1990) Inhibition of MNNG-induced mutations by garlic and diallyl trisulfide. Yingyang Xuebao (Acta Nutr. Sinica) 12:243–247. Chem. Abst. 115 (1991) 2935 (Chinese).

Zhang, Y., Li, H., Guo, Y., and Chen, L. (1993) Studies on cell culture and superoxide dismutase production of Allium sativum L. Huanan Ligong Daxue Xuebao, Ziran Kexueban 21:91–94. Chem. Abst. 120 (1994) 268 253 (Chinese).

Zhang, Y., Li, H., Guo, Y., and Yao, R. (1994) Studies on growth and SOD accumulation of *Allium sativum* L. cells in suspension culture. Huanan Ligong Daxue Xuebao, Ziran Kexueban 22:67–73. Chem. Abst. 121 (1994) 253 760 (Chinese).

Zhao, F., Chen, H., Shen, Y., Liu, Z., Chen, Y., Sun, X., Cheng, G., and Lang, L. (1982) Study of synthetic allicin on the prevention and treatment of atherosclerosis. Yingyang Xuebao (Acta Nutr. Sinica) 4:109–116. Chem. Abst. 98 (1983) 209 844 (Chinese).

Zhao, S. and Wang, L. (1983) Transformation of the main components of garlic during low-temperature controlled-atmosphere storage. Xibei Shifan Xueyuan Xuebao, Ziran Kexueban, 70–79. Chem. Abst. 100 (1983) 33 449 (Chinese).

Zhao, Z., Shen, Z., Wang, G., Tian, C., and Zhang, H. (1988) Dietary fiber and its components in foods. Yingyang Xuebao (Acta Nutr. Sinica) 10:56–61. Chem. Abst. 109 (1988) 148 212 (Chinese).

Zhaosheng, Y., Yingli, S., Xianfang, L., Jiazhen, L., Suzhen, Y., Yinglin, L., Xinyi, C., and Laigang, F. (1984) Effect of allitridi on platelet aggregation, a preliminary study. J. Trad. Chin. Med. 4:29–32.

Zhejiang. (1986) The effect of essential oil of garlic on hyperlipemia and platelet aggregation: an analysis of 308 cases. J. Trad. Chin. Med. 6:117–120.

Zheng, W., Blot, W.J., Shu, X.O., Gao, Y.T., Ji, B.T., Ziegler, R.G., and Fraumeni, J.F., Jr. (1992) Diet and other risk factors for laryngeal cancer in Shanghai, China. Am. J. Epidemiol. 136:178–191.

Zhou, J., Qi, R., and Zhang, M. (1988) Growth suppression of human leukemic cells in vitro by garlicin [ethyl ethanethiosulfinate]. Shandong Yike Daxue Xuebao 26:43–47. Chem. Abst. 110 (1989) 50 897 (Chinese).

Ziegler, S.J., Meier, B., and Sticher, O. (1987) Determination of alliin in garlic by RP-HPLC/electrochemistry and pre-column derivatization. Pharm. Weekl. Sci. Ed. 9:248.

Ziegler, S.J., Meier, B., and Sticher, O. (1991) Knoblauchanalytik: Neue Möglichkeiten für die qualitative und quantitative Bestimmung genuiner Inhaltsstoffe. Schweiz. Apoth. Ztg. 129:186–190. Chem. Abst. 110 (1989) 219 154.

Ziegler, S.J. and Sticher, O. (1988) Electrochemical, fluorescence, and UV detection for HPLC analysis of various cysteine derivatives. J. High Resolut. Chromatogr. 11:639–646.

Ziegler, S.J. and Sticher, O. (1989) HPLC of S-alk(en)yl-L-cysteine derivatives in garlic including quantitative determination of (+)-S-allyl-L-cysteine sulfoxide (alliin). Planta Med. 55:372–378. Chem. Abst. 112 (1990) 11 976.

Zimina, T.A., Kaznelson, I.A., and Zhilin, S.I. (1963) Phytonzide Eigenschaften der Küchenzwiebel des Knoblauchs und einiger anderer Pflanzen Sachalins. Izv. Sib. Otd. Akad. Nauk SSSR (Bull. Siberian Div. Acad. Sci. USSR), 47–52 (Russian).

Zimmermann, W. (1985) Der obere Dünndarm. Eine Phytotherapiestudie. Therapiewoche 35:1592–1602.

Zimmermann, W. and Zimmermann, B. (1990a) Reduction in elevated blood lipids in hospitalised patients by a standardised garlic preparation. Br. J. Clin. Pract. 44:20–23.

Zimmermann, W. and Zimmermann, B. (1990b) Senkung erhöhter Blutfette durch ein Knoblauch-Präparat. Offene Studie an stationären Patienten. Bayer. Internist 10:40–43.

Zoghbi, M.G.B., Ramos, L.S., Maia, J.G.S., da Silva, M.L., and Luz, A.I.R. (1984) Volatile sulfides of the Amazonian garlic bush. J. Agric. Food Chem. 32:1009–1010. Chem. Abst. 101 (1984) 126 818.

Zoumas, C., Amagase, H., and Milner, J.A. (1992) Impact of dietary lipid, protein, and methionine on the ability of garlic to decrease binding of 7,12-dimethylbenz(a)anthracene to mammary DNA. FASEB J. 6:A1392.

Zwergal, A. (1952) Beitrag zur Kenntnis der Inhaltsstoffe des Knoblauchs, *Allium sativum* L. Pharmazie 7:245–256. Chem. Abst. 47 (1952) 3224.

Index

A

Abscisic acid, 91
Absorption, 213
   of adenosine, 153
   garlic as enhancer of, 220
   of Machado's garlicin, 220
   of sulfur constituents of garlic, 213, 214, 216
Acetic acid, garlic aged in dilute, 106–107, 106f, 110
Active agents, in garlic, 37, 211–212. *see also* specific ingredients e.g. Alliin
Acyl-CoA elongase, 78
Adenosine, in garlic, 90, 90f
   absorption of, 153, 220
   analytical evaluation of, 132
   antithrombotic effects of, 157–158, 161
   blood pressure, effect on, 153
   degradation by ADA, 211–212
   in garlic juice, 112
   side effects of, 226
Adenosine deaminase, 153, 211–212
Adenosine triphosphatase, 77
Africa, 231
Aged alcoholic garlic extract, 37, 93t, 103–106, 104t
   antibiotic effects of, 171
   anticancer effects of, 185–186, 187
   antihepatotoxic effects of, 202
   antiinflammatory effects of, 193
   antioxidant effects of, 189
   antithrombotic effects, 158
   cholesterol, effect on, 148
   constituents of, influence of, 218
   ingredients of, 113
   odorless, 114, 209–210
   toxicity of, 223
   uses of, 110t
Ajoene, 62–64, 63f, 134, 212, 214
   analytical evaluation of, 127–128
   antibiotic effects of, 163, 163t, 172
   antioxidant effects of, 188
   antithrombotic effects of, 159–160
   blood pressure, effect on, 153–154
   cysteine, reaction with, 65
   in ether-extracted oil, 102
   in oil-macerated garlic products, 101–102, 102t, 113
   stability of, 64
Alcohol, extract of garlic aged in dilute. *see* Aged alcoholic garlic extract
Alcohol dehydrogenase, 78
Alcohols, as alliinase inhibitors, 53
Alkaloids, 81
Alkenyl disulfides, toxicity of, 222
Alkyl furanones, 68
Alkylcysteine sulfoxides, 117, 117t
Alkylcysteines, 117, 117t
S-Alkylcysteines, 69–70, 69f, 74
S-Alkylmercaptocysteines, 117, 117t

Allergies
   to garlic, 226–227, 227t
   garlic as treatment for, 204
Allicin, 38, 53–69, 114, 134, 211–212
   as absorption enhancer, 220
   ajoenes, transformation to, 62–64
   allergies to, 226
   alliin transformed to, 40, 53–55, 55t, 94, 113
   allyl sulfides, transformation to, 59–62, 61f
   analytical evaluation of, 118–122, 119f, 120t, 121t
   antibiotic effects of, 163, 163t, 165–167, 168, 172
   anticancer effects of, 16–187
   antiinflammatory effects of, 192
   antioxidant effects of, 188–189, 190–191
   antiparasitic effects of, 173–174
   antithrombotic effects of, 157, 158, 162
   blood pressure, effect on, 153
   cholesterol, effect on, 147–148
   cooking, effects of, 68–69
   cysteine and other thiols, reactions with, 65, 66f
   discovery of, 39–40
   drying of garlic, effect of, 94
   enzyme activities, effect on, 199–201
   formation from alliin, 53–56, 55f
   garlic bulbs, fresh, yield of, 56
   in garlic preparations, 95–100, 96t, 103, 106, 112, 132–133
   hepatotoxicity of, 222
   hypoglycemic effects of, 195
   immunomodulatory effects of, 192
   metabolism of, 215, 219
   names for, 40, 40f
   as neuralgia treatment, 207
   odor of garlic and, 56, 66–68, 67t, 115, 211
   organic solvents, transformation in, 62–64, 63f
   pharmacokinetics of, 213–216
   physical and spectral properties of, 56–57
   stability of, 57–59, 58t, 61f, 112, 113
   standardization of, 56–57
   stomach irritation caused by, 225
   sulfhydryl enzymes, reactions with, 65
   synthesis of, 57, 75
   toxicity of, 222, 223
   undiluted, transformation of, 62, 62t
   vinyldithiins, transformation to, 62–64
Alliin, 45, 46–48, 81, 134, 211–212
   allicin formed from, 40, 53–55, 55f, 94, 113
   alliinase as lysis of, 48–53
   analytical evaluation of, 116–118
   antibiotic effects of, 163
   antihepatotoxic effects of, 201
   antioxidant effects of, 191
   antithrombotic effects of, 160
   blood pressure, effect on, 153
   content in garlic, 42t
   cutting of garlic clove, effect of, 94
   discovery of, 40

Alliin—*continued*
  enzyme activities, effect on, 200
  in garlic preparations, 103–105, 107, 109, 110, 132–133
  hypoglycemic effects of, 195
  metabolism of, 215, 216
  natural sources of, 47
  as neuralgia treatment, 207
  parent compounds of, 41
  pharmacokinetics of, 215–216
  structure, 43f
  synthesis of, 72, 74, 75f
  thiosulfinates, transformation to, 48–53, 50f, 94
  transformation by cooking, 69
  variation in plants' content of, 48, 49f
  whole plant distribution of, 44t, 48
Alliinase, 47, 48–53, 53, 75, 77, 134, 211–212
  discovery of, 40
  drying of garlic, effect of, 94
  in garlic preparations, 100, 103, 106, 109, 110–111
  inactivated by cooking, 68
  inactivation by gastric juice, 95, 96, 111–112
  inhibition of, 52–53, 53f, 54t
  location in garlic clove, 51f
  methyl-specific, 52
  odor of, 211
Alliin-derived sulfur-containing compounds, tertiary, 109
Allithiamine, 70–71
  as cyanide poisoning antidote, 205
  pharmacokinetics of, 219
Allium genus
  botanical classification of, 28, 28t
  overview of, 25–28
Allium sativum L.. *see* Garlic
Allium species
  characteristics of species, 25–27
  number of, 25, 25t
  overview of, 26t
Allium ursinum L.. *see* Wild garlic
Allyl mercaptan, 67, 213–214
  antioxidant effects, 191
  odor of, 115
Allyl methanethiosulfinate
  organic solvents, transformation in, 64
Allyl methyl di-, tri-, and tetrasulfides, 59, 65, 66, 67–68, 112
Allyl 1-propenyl disulfide, 65, 66
Allyl propyl disulfide
  cysteine, reaction with, 65
Allyl sulfides, 38–41, 59–62, 99
  anticancer effects of, 187
  antioxidant effects of, 190–191
  antithrombotic effects of, 157
  odor of, 115
Allyl thiosulfinates, stability of, 59
S-Allylcysteine, 128
  analytical evaluation of, 131
  anticancer effects of, 185, 187
  antihepatotoxic effects, 201, 202
  antioxidant effects of, 191
  in extract of garlic, 104–105
  pharmacokinetics of, 216
S-Allylcysteine sulfoxide. *see* Alliin
S-Allylmercaptocysteine, antihepatotoxic effects of, 201, 202
Allyls, synthesis of, 72–74, 75f
Aluminum, corrosion protection of, 231
Amino acid glycoside, 77
Amino acid synthases, 78–79
Amino acids, in garlic, 80, 81t, 103, 104, 104t
Angiotensin-converting enzyme, 154
Anglo-Saxons, 11
Animals, effect of garlic odor on, 174–175, 231
Antiarteriosclerotic effects of garlic, 38, 187. *see also* Blood pressure, effects of garlic on; Heart and circulatory system, garlic's effect on
Antibacterial effects of garlic, 38, 39, 75, 91–92, 163t, 164–168
Antibiotic effects of garlic, 39, 40, 162–176, 175t
  antibacterial effects. *see* Antibacterial effects of garlic
  antifungal effects, 38, 75, 168–172
  antiparasitic effects, 173–174
  antiprotozoal effects, 172
  antiviral effects, 172–173
  insecticidal and repellent effects, 174–175
  intestinal or dermal, 213–214
  of Machado's garlicin, 91, 175–176
Anticancer effects of garlic. *see* Cancers, effects of garlic on
Anticoagulant, interaction with, 227
Antifungal effects of garlic, 38, 75, 168–172
Antihepatotoxic effects of garlic, 187, 201–202
Antiinflammatory effects of garlic, 192–193, 193t
Antimutagenic factors, in garlic, 181–182
Antioxidant effects of garlic, 187–191
  active compounds, 190–191
  in vitro studies, 187–190
  in vivo studies, 190
Antiparasitic effects of garlic, 173–174
Antiprotozoal effects of garlic, 172
Antithrombotic effects of garlic, 38, 156–162, 161t
Antiviral effects of garlic, 172–173
Aphids
  garlic as repellent, 174
  infesting garlic plants, 36
Aqueous extract of garlic
  antibiotic effects of, 163, 165, 169, 170
  anticancer effects of, 179–180
  antioxidant effects of, 188
  antithrombotic effects of, 157
  blood pressure, effect on, 153–154
  enzyme activities, effect on, 199
  fibrinolytic activity, effect on, 154
  hormone-like effects of, 196
  hypoglycemic effects of, 194
Arabs, 7–8
Argentina, 34
Arginase, 79

Arginine, 80
Arteriosclerosis. *See* Antiarteriosclerotic effects of garlic
Asia, 2–5, 34
Aspartic, 80
Assyrians, 6
Asthma, caused by garlic, 226–227

B

Babylonians, 6
Bacillus cepivorus Delacr., 36
Bacteria
   alliinase activity, 52
   antibacterial effects of garlic. *see* Antibacterial effects of garlic
Balkan countries, 20, 23, 34, 36
Bioavailability of drugs, 213
Biopharmaceutics, effective constituents of garlic, 213–220
   absorption enhancer, garlic as, 220
   allithiamine, 219
   Machado's garlicin, 220
   sulfur components of garlic, 213–218
   vinyldithiins, 218–219
Biotransformation, of drug effective ingredient, 213
Black-spotted dolls, case of, 230
Blood coagulation, effect of garlic on, 154–156
Blood flow, effect of garlic on, 155–156
Blood pressure, effects of garlic on, 148–154, 152t
   toxic effects of treatment, 223
Boron, in garlic, 88
Botanical characterization of garlic, 25–34
Botrytis, garlic infected with, 35
Botulism, caused by garlic, 226
Breast milk, odor of garlic in, 230–231
Brood bulbils, cultivation from, 34
Bulb smut, 35
Bulbs, garlic, 27f, 30, 31f, 33–34, 37
   alliciin yield of fresh, 56
   alliin synthesis in, 74
   drying of, 110, 111
   post-harvest treatment of, 93
Burns, garlic as treatment for, 206

C

Cancers, effects of garlic on, 176–187
   active compounds, 181, 186–187
   animal and in vitro studies, 178–186
   antioxidant effect, 187
   epidemiological studies, 176–178, 177t
Capsules, garlic, 37. *see also* Garlic powder tablets
   microcapsules, 116
   stomach acid-resistant coating for, 52, 115–116
Carbohydrases, 78
Carbohydrates, in garlic, 76–77
   analytical evaluation of, 132
   metabolism enzymes, 78
Cardiovascular system. *see* Heart and circulatory system, garlic's effect on
Catabolism, of active constituents of garlic, 214
Catalase, 77, 190

Celts, 11
China, 34, 36, 106
   ancient period, use of garlic in, 3, 4–5
Cholesterol, garlic consumption to control, 136–148, 210
Cholinesterase, 78
Chopped garlic, uses of, 110
Circulatory system. *see* Heart and circulatory system, garlic's effect on
Coagulation inhibitor from garlic, toxicology of, 224–225
Cobalt coatings, 233
Contaminants, ecological hazards of, 228
Cooking, effects on garlic sulfur compounds of, 68–69, 163
Corrosion protection, garlic used for, 231
Crushed garlic
   adenosine in, 90
   oil-soluble compounds in, 84t
   sulfur compounds in, 42t, 46t
   thiosulfinate transformations in water/steam, 59–62
   uses of, 110
   vapor of, composition of, 65–68, 67t
Cultivation of garlic, 34–36
Cut herb, 34
Cycloalliin, 47, 156
Cysteine(s), 41, 80. *see also* γ-Glutamylcysteines
   S-alkylcysteines, 69–70, 69f
   in garlic preparations, 109
   odor perceptions and, 114
   thiosulfinate reactions with, 65
Cysteine sulfoxides, 46–48, 130
   alliinase as lysis of, 48–53, 53
   analytical evaluation of, 116–118, 117t
   thiosulfinates, transformation to, 48–53, 50f, 94
   variation in plants' content of, 48, 49f
   whole plant distribution of, 44t
Czechoslovakia, 34

D

Danish name derivation, 2
Deodorized garlic extracts. *see* Odorless garlic preparations
Deoxyribonuclease, 79
Dermatitis, caused by garlic, 225
Dialkyl thiosulfinates, 53, 179
Diallyl sulfides (mono-, di-, tri-, and tetra-), 39, 59, 62, 64, 65, 66–68, 99, 121, 212, 214–215, 215f, 217–218, 218t
   antibiotic effects of, 163, 163t, 167, 170, 172, 174
   anticancer effects of, 182–185, 186–187
   antiinflammatory effects of, 192
   antioxidant effects of, 188, 189
   antiparasitic effects of, 173–174
   antithrombotic effects of, 159
   cysteine, reaction with, 65
   in extracts of garlic, 102, 103
   pharmacokinetics of, 216–217, 218t
   in steam-distilled oil, 112, 133
   toxicity of, 222

324 Index

Diallyl sulfides—*continued*
  as toxin antidote, 206
Diarrhea, caused by garlic, 226
Dipeptides, 80–81
Diphhenylamine, 223
Diseases of garlic plants, 35
Distribution
  of drug effective ingredient, 213
  of Machado's garlicin, 220
  of sulfur constituents of garlic, 216–218
Disulfides, 65, 68, 99. *see also* Allyl methyl
  di-, tri-, and tetrasulfides; Diallyl sulfides
  in extract of garlic, 103
  in steam-distilled garlic oil, 112, 113
  thiosulfinates transformed to, 64
Dosing of garlic, 208–209
Dracula, illness of, 229–230
Drug effect, 213
Drug interactions, 227
Drugs
  bioavailability of, 213
  degraded, 213
Dry garlic powder. *see* Garlic powder
Dyspepsia, effects of garlic on, 202–204

E

Eastern Europe, 20, 22, 34
Ecological hazards of garlic, 227–228
Eczema, caused by garlic, 226
Effective constituents of garlic,
  biopharmaceutics of, 213–220
Egypt, 5–6, 34
Elimination
  of drug effective ingredient, 213
  of Machado's garlicin, 220
  of odor carrier, 214
  of sulfur constituents of garlic, 213–215, 230
Enzyme activities, effects of garlic on, 199–201
Enzymes, in garlic, 77–79
  inhibitor, 79
  pickled garlic, 106
Esophagitis, 225
Essential oil. *see* Steam-distilled garlic
Ethanol
  in aged alcoholic garlic extract, 114
  dry extracts of garlic using, 111–112
Ether extract of garlic, 102
  antithrombotic effects of, 157, 161
  cholesterol, effect on, 146, 147
  hypoglycemic effects of, 193
S-Ethylcysteine, 201
Ethylene, 91
Europe, history of garlic use in, 11–19, 20, 22–24
Extracts of garlic
  aged alcoholic. *see* Aged alcoholic garlic extract
  antibiotic effects of, 163, 165–166, 169, 170, 171
  anticancer effects of, 179–180
  antiinflammatory effects of, 192–193
  antioxidant effects of, 188
  antiparasitic effects of, 174

  antithrombotic effects of, 157, 158
  blood pressure, effect on, 153–154
  cholesterol, effect on, 146, 147
  dry, with conservation of alliin, 110t, 111–112
  enzyme activities, effect on, 199
  fibrinolytic activity, effect on, 154–155
  garlic extract factor (GEF), 156
  hormone-like effects of, 196
  hypoglycemic effects of, 193–194, 195
  odorless, 209–211
  toxicity of, 222, 223

F

Fatty acids, in garlic, 82–83, 132
Ferredoxin, 78
Fibrinolysis, effect of garlic on, 154–156, 155t
Flavonoids, in garlic, 89, 89f, 211, 212
Folk medicine, garlic in, 19–24, 198, 229–230, 231
France, 23, 34
Fructans, in garlic, 76, 211
Fungi, effects of garlic on, 7 5, 168–172

G

α-Galactosidase, 78
Garlic
  analytical determination of constituents, 116–131
  biopharmaceutics of effective constituents, 213–220
  botanical description of, 28–34
  commercially processed. *see* Processing of garlic
  composition, generally, 37–38, 38t
  cooking, effects of, 68–69, 163
  cultivation of, 34–36
  derivation of name, 1–2
  dosing of, 208–209
  flowering plant, 27f, 31, 32f
  forms of, 37
  genetic propagation of, 33
  growth changes in, 31, 31f
  local irritation by, 225–226
  method of consumption of, 52
  micro-flora of soil and, 36
  regions native to, 34
  subspecies of, 28–29, 29–30f
  wild plants. *see* Wild garlic
Garlic breath, 65–68, 67t, 214
Garlic bulbs. *see* Bulbs, garlic
Garlic cloves
  alliin synthesis in, 74
  alliinase location in, 51f
  cutting, effect of, 94
  drying, effect of, 94, 95t
  processing of. *see* Processing of garlic
  storage of, 153
  total known sulfur compounds in, 41
  γ-glutalmylcysteines in, 41
GARLIC computer program, 233
Garlic extract factor (GEF), 156
Garlic juice
  antibiotic effects of, 167–168, 170, 171
  anticancer effects of, 176, 179

fibrinolytic activity, effect on, 154–155
hypoglycemic effects of, 194
ingredients of, 112
uses of, 110t
Garlic oils. *see* Oils, garlic
Garlic powder, 34, 48, 93–99, 93t, 111t, 134
  S-alkylcysteines in, 70
  alliin/alliinase system, 109
  antibacterial effects of, 164
  antioxidant effects of, 189
  blood flow, effect on,1 55
  cysteine-containing-glutamylpeptide systems, 109
  extracts with conservation of alliin, 111–112
  heating, effect of, 163
  immunomodulatory effects of, 191
  ingredients of, 110–111
  side effects of, 226
  as spice, 95
  stability of sulfur compounds in, 94–95
  standardization of, 132–133
  storage of, 111
  tablets. *see* Garlic powder tablets
  as tobacco smoke absorbent, 233
  uses of, 110t
Garlic powder tablets, 37, 95–99, 96t
  cholesterol, effect on, 141, 143–145, 147
  effective allicin yield, 95–99, 96t
  method of consumption of, 52
  odor in, 115
  stomach acid-resistant, 52, 114–115
Garlic preparations, 37, 109–134, 110t. *see also* Processing of garlic; specific preparations, e.g. Garlic powder
  alliin/alliinase system, 109
  analytical determination of constituents, 116–131
  assessment of ingredients, 109
  comparison of, 114
  cysteine-containing-glutamylpeptide systems, 109
  discoloration of, 230
  dosing of, 208–209
  garlic-gingko combination product, 145–146
  main ingredients of, 110–116
  odorless, 114–116
  standardization of, 132–134, 208
  value of, 114
Garlic salt, 37
Garlic touch, 230
Garlic-ginko combination product, 145–146
Garlicin, Machado's. *see* Machado's garlicin
Gas chromatography (GC)
  analysis of garlic constituents using, 118, 120–121, 124–125, 126–127, 128
  standardization of garlic products using, 132, 133
Gastric juice,111–112, 199, 202, 213
Gastric lipase, 65
Gastric pain, garlic as cause of, 52
Gastrointestinal ailments
  cancer, 176–178, 186, 221
  garlic as treatment for, 22, 202–204
  infections, 199

intestinal occlusion, caused by garlic, 225
stomach irritated by garlic, 225
Genus Allium. *see* Allium genus
German name derivation, 1–2
Germanium, 87, 87t, 182
Germany, 11, 13–16, 19, 34
Gibberellins, 91
Gilroy, California, 34
Glucose, in garlic, 76–77
β-Glucosidase, 78
β-O-Glucosyltransferase, 78
Glutamic acids, 80
γ-Glutamylcysteines, 41–46, 70, 74, 130
  abundance of, 41, 42t, 44
  analytical evaluation of, 128–129, 129f
  blood pressure, effect on, 153, 154
  content of, 44–45, 45f, 46t
  discovery of, 41
  in extract of garlic, 104–105
  in garlic juice, 112
  in garlic powder tablets, 95, 99
  in pickled garlic, 106–107
  polymers of, as phytochelatins, 81
  as storage compounds, 45–46
  structure, 41, 43f
  synthesis of, 72–74
  whole plant distribution of, 44t
γ-Glutamylpeptidase, 78–79, 104
γ-Glutamylpeptides, 41, 78–79, 211
  analytical evaluation of, 128–129, 129f
  in garlic powder, 133
γ-Glutamyl-S-alkylcysteines, 81, 132
γ-L-Glutamyl-transpeptidase, 78–79
Glutathione, 65, 190
Glycoside, amino acid, 77
Gray mold, garlic infection with, 35
Greece, 8–9
Guanylate cyclase, 79
Guanosine, in garlic, 90

H

Hangover, garlic as remedy for,22
Heart and circulatory system, garlic's effect on, 135–162
  antithrombotic effects, 156–162, 161t
  blood coagulation, fibrinolysis, and blood flow, 154–156, 155t
  blood pressure, effects on, 148–154, 152t, 223
  cholesterol and lipid-lowering effects, 136–148
  odorless garlic preparations, 210
Heartburn, 225
Heavy metals
  ecological hazards of, 228
  as poison, garlic as antidote for, 205–206
  treated with garlic, 231, 233
Hexasulfides, in steam-distilled garlic oil, 112
Hexokinase, 78
High performance liquid chromatography (HPLC)
  analysis of garlic constituents using,117–118, 119, 121–122, 125–127, 128, 132

High performance liquid chromatography—*continued*
  standardization of garlic products using, 132–133
History of garlic use, 2–19
  ancient period, 2–10
  medieval to modern periods, 11–19
Homeopathy, garlic in, 207–208
Homocysteine, and odor perceptions, 114
Hormone-like effects of garlic, 196–197
Hungary, 34
Hydrocarbons, in garlic, 84
Hydrolysis, to thiols, 130–131
Hypertension. *see* Blood pressure, effects of garlic on
Hypoglycemic effects of garlic, 193–196, 194f

I

Immunomodulatory effects of garlic, 191–192
India, 3–4, 22, 34, 36
Indigestion, effects of garlic on, 202–204
Infected wounds, garlic as treatment for, 206–207
Inflammation, effects of garlic on, 192–193, 193t
Insecticidal effects of garlic, 174–175, 231
Intestinal occlusion, caused by garlic, 225
Inulase, 78
Invertase, 78
Iran, 7
Irritation, by garlic, 225–226
Isoalliin, 40, 45, 46–48
  alliinase and, 51
  analytical evaluation of, 131
  content in garlic, 42t
  source of, 41, 72–73
  structure, 43f
  synthesis of, 75f
  whole plant distribution of, 44t, 48
Israel, 7
Italy, 2, 34

J

Japan, 5
Jasmonic acid, 91

K

Knoblauch, derivation of, 1–2
Knolau hybrid, 33
Korea, 4
Kwai tablets. *see* Garlic powder tablets

L

Lead poisoning, garlic as antidote for, 205–206
Lead-free mercury, production of, 231, 233
Lectins, in garlic, 80, 91
Legendary plant, garlic as, 19–24
Liberation, of drug effective ingredient, 213
Lignin, 89
Lipase, 79
Lipids
  fatty acid content, 82–83
  in garlic, 81–84, 83t, 84t
  garlic consumption to reduce level of, 38, 136–148, 208, 210, 221, 221t

Lipophilic compounds, in garlic, 84
Liver and liver enzymes
  antihepatotoxic effects of garlic, 187, 201–202
  toxic effects of garlic on, 222, 223
Lysozyme, 79

M

Machado's garlicin, 91–92
  as antibiotic, 175–176
  pharmacokinetics of, 220
  toxicology of, 224
Magical plant, garlic as, 19–24
Mass spectrometry (MS), analysis of garlic constituents using, 120
Mediterranean region, 5–10, 23, 34
Mercaptans, 171. *see also* Allyl mercaptan
Mercury, production of lead-free, 231, 233
Metabolic disturbances, effects of garlic on, 198–207
Metabolism
  of allithiamine, 219
  of carbohydrates, 78
  of drug effective ingredient, 213
  of sulfur constituents of garlic, 214–216, 219
Metallurgy, garlic applications in, 231, 233
Methiin, 40, 46–48
  alliinase and, 51, 52
  content in garlic, 42t
  structure, 43f
  synthesis of, 72–74
  transformation into dialykyl thiosulfinates of, 48
  whole plant distribution of, 44t, 48
Methionine, 80
Methyl thiosulfinates, formation of, 54–56
S-Methylcysteine, 128, 201
Methylcysteine sulfoxide. *see* Methiin
S-Methylcysteine sulfoxides. *see* Methiin
Methyl-Se-S-Allyl, 66
Microcapsules, 116
Micro-flora of soil, and garlic, 36
Microorganisms, effects of garlic on, 172
Minerals, in garlic, 86–88, 87t
Monosulfides, 65, 112, 123. *see also* Diallyl sulfides
Mystical connotations of garlic, 19

N

Nematodes, 36, 75, 174
Neuralgia, garlic as treatment for, 207
Neurotropic effects of garlic, 207
Nickel coatings, 233
Nitric oxide, effect of garlic on production of, 154
Nonsulfur components in garlic, 131–132. *see also* specific components, e.g. Adenosine
Nonsulfur compounds in garlic, 76–92
Nucleic acids, in garlic, 90–91

O

Odor, 39
  of alliciin, 56
  animals attracted/repelled by, 174–175, 231

in breast milk, 230–231
composition of, 65–68, 67t
humans attracted/repelled by, 231
of steam-distilled garlic, 99
sulfur as cause of, 38
thiopental, effect of, 230
of thiosulfinates, 56
of whole garlic, 50
Odorless garlic preparations, 114–116, 114t, 209–211, 231, 232–233t
elimination of odor carrier, 214
scordinin, 71
Oil-macerated garlic products, 93t, 99–102, 100t, 101t, 102t, 134. see also Vinyldithiins
ingredients of, 113
polysulfides in, 122
stability of compounds in, 101, 102t
standardization of, 133
uses of, 110t
Oils, garlic, 81. see also Ether extract of garlic; Oil-macerated garlic products; Steam-distilled garlic oil
antibiotic effects of, 163–164, 163t
antioxidant effects of, 189
antiparasitic effects of, 174
antithrombotic effects of, 161–162
cholesterol, effect on, 146, 147–148
dyspepsia and indigestion, effects on, 202–203
fibrinolytic activity, effect on, 154
mixtures of, 134
side effects of, 225
standardization of, 133–134
thiosulfinate reactions used in production of, 59–65, 60f
toxicity of, 222
uses of, 110t
Oligosulfides, 171
antibiotic effects of, 172
enzyme activities, effect on, 201
Organic disturbances, effects of garlic on, 198–207
Organoselenium compounds, in garlic, 88–89
Otology, use of garlic in, 206
Oxalate crystals, 33, 33f
Oxidase action, 74

P

Paper chromatography (PC), analysis of garlic constituents using, 117, 119
Parasites, effects of garlic on, 173–174
Pectinesterase, 78
Pectins, in garlic, 77
Pentasulfides, 112
γ-Peptides, 80–81. see also γ-Glutamylpeptides
Peroxidase, 77–78
Pest control, ecological hazards of, 228
Phagocytosis, effects of garlic on, 191–192
Pharmacokinetics, 213
of allicin, 213–216
of alliin, 215–216
of allithiamine, 219

of S-Allylcysteine, 216
of diallyl sulfides, 216–217, 218t
of Machado's garlicin, 220
of sulfur compounds in garlic, 213–218, 218t
of vinyldithiins, 218–219, 219t
Pharmacopeias, garlic indexed in, 134, 134t, 207–208
Phenolic acids, 89
Phenolic compounds, antithrombotic effects, 161
Phenols, 89, 211
Phenylalanine-ammonia-lyase, 77
Phosphatase, 77
Phosphoglucose isomerase, 78
Phospholipids
analytical evaluation of, 132
Phystochelatins, 81
Phythormones, 91
Phytic acids, 91, 224
Phytin, 91
Phytoalexin function, of flavonoids, 8 9
Phytosterols, 84
Pickled garlic, 106–107, 106f, 110
Pills, garlic. see Capsules, garlic; Garlic powder tablets
Plant antibiotics, 166
Plant hormones, in garlic, 91
Plants, garlic as absorption enhancer for, 220
Platelet antiaggregatory effects of garlic, 156–162, 161t
Polyfructosidase, 78
Polygalacturonase, 78
Polyphenol oxidase, 77
Polysulfides. see also Allyl methyl di-, tri-, and tetrasulfides
analytical evaluation of, 122–126, 124t, 125t, 126f
antithrombotic effects of, 159
1-propenyl compounds, 99
in steam-distilled garlic oil, 112, 113
Porphyria, 229–230
Postoperative bleeding, effect of garlic intake on, 227
Preparations, garlic. see Garlic preparations
Preservatives, ecological hazards of, 228
Processing of garlic, 92–107, 93t, 109. see also Garlic preparations
aging of garlic in dilute acetic acid, 106–107, 106f
annual production, by country, 109t
ether-extracted oil, 102
extract of garlic aged in dilute alcohol, 103–106, 104t
garlic powder products, 93–99
oil-macerated garlic, oil of, 99–102, 100t, 102t
steam-distilled garlic, oil of, 99, 100t
1-Propenyl compounds
mono- and polysulfides, 99
in steam-distilled garlic oil, 113
synthesis of, 72–74, 75f
S-1-Propenylcysteine, 104, 128
S-trans-1-Propenylcysteine sulfoxide. see Isoalliin
Propyl sulfides, 113

Propyl thiosulfinates, 113
S-Propylcysteine, 201
Propylcysteine sulfoxide, 113
Prostaglandins, in garlic, 83
Protein content of garlic, 79–80
 immunomodulatory effects of, 192
Proto-desgalactotigonin, 85
Proto-eroboside B, 85
Protozoa, effects of garlic on, 172
Pseudomedicine. see Folk medicine
Pyruvate, 114
Pyruvic acid, 116, 219

R

Religious connotations of garlic, 19
Respiratory diseases, garlic as treatment
 for, 204–205
Respiratory organs, irritation by garlic of, 226
Rome, 9–11
Russia, 22, 34, 36, 106
Rust, garlic infection with, 35

S

Salicylic acids, 89
Sapogenins, in garlic, 85–86, 85t, 86f, 132
Saponins, in garlic, 85–86, 85t, 132
Sativoside B1, 85
Scordinins, 71–72, 211
 as absorption enhancer, 220
 analytical evaluation of, 132
 antithrombotic effects of, 160–161
 toxicology of, 224
Selenium, 66, 67, 87, 87t
 anticancer effects of, 182
 increasing level of, 88–89
SH enzymes, 163, 167
Side effects of garlic, 221t. see also Toxicity of garlic
 allergies to garlic, 226–227, 227t
 of antihypertensive therapy, 223
 drug interactions, 227
 local irritation, 225–226
Skin burns, caused by garlic, 225
Skin diseases, garlic as treatment for, 206–207
South Korea, 34
Spain, 12
Spice, garlic used as, 36, 37, 95, 229
Steam, thiosulfinate transformations in, 59–62
Steam-distilled garlic oil, 3 7, 38–39, 93t, 99,
  101t, 134
 antibiotic effects of, 163–164, 163t, 171–172, 174
 antioxidant effects of, 189
 antithrombotic effects of, 162
 cholesterol, effect on, 146, 147–148
 immunomodulatory effects of, 191–192
 ingredients of, 112–113
 polysulfides in, 122–126, 125t, 126f
 production of, 81–82, 99, 100t
 stability of, 99
 standardization of, 133
 sulfides in, 59, 61

Steel, corrosion protection of, 231
Steroids, in garlic, 85–86, 85t, 175, 211
Sterols, in garlic, 4, 84t, 132
Stomach acid-resistant coating, capsules, 52, 115–116
Stomach ailments. see Gastrointestinal
 ailments
Stress relief, garlic used for, 209–210, 229
Sugars, in garlic, 56–77
Sulfenic acids, 53–56, 55f
Sulfhydryl enzymes, 65, 181, 201
Sulfides, 68, 99. see also Allyl methyl di-,
  tri-, and tetrasulfides; Allyl sulfides;
  Diallyl sulfides; Disulfides;
  Monosulfides; Polysulfides; Trisulfides
 in garlic breath, 67–68
 in oil-macerated garlic products, 113
Sulfur compounds, in garlic, 38–77, 93, 93t
 as absorption enhancers, 220
 S-alkylcysteines, 69–70, 69f
 allicin. see Allicin
 alliinase. see Alliinase
 allithiamine, 70–71, 205, 219
 allyl sulfides, 38–41
 analytical evaluation of elemental, 124t,
  129–130, 130f
 biogenesis of, 69f, 72–74, 75f
 cloves, total known sulfur compounds in, 41
 cooking, effects of, 68–69
 cysteine sulfoxides. see Cysteine sulfoxides
 discovery of, 38–41
 fibrinolytic activity, effect on, 156
 function of, 74–76
 garlic odor and, 114
 health effects of, 130–131
 insecticidal and repellent effects, 175
 pharmacokinetics of, 213–218, 218t
 porphyria, effect on, 229–230
 scordinins. see Scordinins
 stability of, 216
 thiosulfinates. see Thiosulfinates
 total sulfur content of garlic, 130
 toxicity of, 222
 in whole and crushed garlic, 42t, 46t
Sulfur compounds, in garlic
 preparations, 1 03–107, 104t, 106f, 109, 114
 stability of, 94–95
Sumerians, 3, 6
Super-critical fluid chromatography (SEC),
 analysis of garlic constituents using, 122
Superoxide dismutase, 77–78

T

Tablets, garlic powder. see Garlic powder tablets
Tellurium, in garlic, 87–88
Terpenes, 68
Tetrasulfides, 68, 99, 103. see also Allyl methyl di-,
 tri-, and tetrasulfides; Diallyl sulfides
Therapeutic effects and applications of
 garlic, 135–212
 active compounds, 211–212

antibiotic effects. *see* Antibiotic effects
    of garlic
anticancer effects, 176–187
antiinflammatory effects, 192–193
antioxidant effects, 187–191
deodorized garlic extracts, 209–211
dosing of garlic and preparations, 208–209
heart and circulatory system, effects on, 135–162
homeopathy, 207–208
hormone-like effects, 196–197
hypoglycemic effects, 193–196
immunomodulatory effects, 191–192
organic and metabolic disturbances, effect
    on, 198–207
thiamine absorption, enhancement of, 197–198
Thiamine, 70–71, 87t
    absorption enhancement, 197–198, 199f
    conversion of allithiamine to, 219
Thin-layer chromatography (TLC)
    analysis of garlic constituents
        using, 116–117, 119, 132
    pharmacokinetic study of allicin and alliin, 216
    standardization of garlic products using, 132
Thioallyl compounds, total, 107
Thiols
    hydrolysis to, 130–131
    thiosulfinate reactions with, 65
Thiopental, 230
Thiophenes, 68
Thiosulfinates, 38, 40, 40f, 53–69, 75, 84. *see
    also* Allicin
    alliin transformed to, 94
    analytical evaluation of, 118–122, 120t, 121t, 123f
    as antibacterial agents, 75
    antibiotic effects of, 163, 165, 167
    anticancer effects of, 179, 181
    as antifungal agents, 75
    antithrombotic effects of, 158
    content in garlic, 42t
    cooking, effects of, 68–69
    cysteine and other thiols, reactions with, 65
    cysteine sulfides transformed to, 48–53, 50f
    disulfides, transformation to in base, 64
    formation of, 53–56
    in garlic preparations, 99–100, 103, 113, 133
    γ-gluamylcysteines transformed to, 43f
    immunomodulatory effects of, 192
    odor of garlic and, 66–68
    organic solvents, transformation in, 62–64, 63f
    physical and spectral properties of, 56–57
    precursors of, 46–48
    reactions, 59–65
    stability of, 57–59, 113
    stomach irritation caused by, 225
    sulfides, transformation to, 68
Threonine, 80
Thrombosis, effects of garlic on, 156–162, 161t
Tibet, 4

Tissue culture, 35
Tissue necrosis, caused by garlic, 225
Tobacco smoke absorbent, garlic powder as, 233
Toxicity of garlic, 221–225
    acute, subacute, and chronic, 222–224
    coagulation inhibitor, toxicology of, 224–225
    ecological hazards, 227–228
    genotoxicity of orally fed garlic, 223–224
    Machado's garlicin, toxicology of, 224
    scordinin, toxicology of, 224
    treatment of acute toxicity, 224
Toxins, garlic as antidote for, 205–206
Transfructosidase, 78
Transpeptidases, 79
Triglycerides, effect of garlic on, 141–145
Trigonellin, 81
Trisulfides, 65, 99
    antithrombotic effects, 157
    in extract of garlic, 103
    in steam-distilled garlic oil, 112
Triterpenoids, 85–86, 85t, 211, 212

U

United States, 34, 35t, 106

V

Vampires, effect of garlic on, 230
Vegetable oil-macerate extract, 37
Veterinary medicine, garlic used in, 204
Vinyldithiins, 62–64, 63f, 134
    analytical evaluation of, 126–127
    antibiotic effects of, 163
    in garlic preparations, 101–102, 102t, 113, 133
    pharmacokinetics of, 218–219, 219t
Viruses, effects of garlic on, 172–173
Vitamins, 86, 87t, 223
    B, 219
    B_1, 70, 198
    C, 197

W

Warfarin, interaction with garlic of, 227
Warts, garlic as treatment for, 21
Water
    garlic as protection for untreated, 21
    thiosulfinate transformations in, 59–62
Well-being, preservation of, 136, 151, 229
Wild garlic, 28, 64
    antithrombotic effects, 157
    blood pressure, effect on, 154
Worms, effect of garlic on, 75, 173–174